Dictionary of Critical R

Dictionary of Critical Realism fills a vital gap in the literature. Critical realism is sometimes criticised for deploying unfamiliar specialist concepts which render it inaccessible to a wider audience. However, as Mervyn Hartwig puts it, 'Just as the tools of the various skilled trades need to be precision-engineered for specific, interrelated functions, so meta-theory requires concepts honed for specific interrelated tasks: it is impossible to think creatively at that level without them.'

The dictionary seeks to redress the problem of accessibility by thoroughly explaining all the main concepts and key developments. It has more than 500 entries on these themes, with contributions from many leading critical realists, and is thoroughly cross-referenced. However, this text does not stop at the elucidation of concepts. It incorporates surveys of critical realist work and prospects in more than fifty areas of study across the humanities and social sciences, thereby demonstrating the appropriate use and possibilities of the concepts in action.

This book will be an indispensable reference tool for anyone aspiring to understand, deploy or engage with critical realist approaches to comprehending the world. Scholars in philosophy, social theory, the social sciences and the humanities alike will relish its systematic exegesis of the dynamic and vibrant discourse of critical realism.

Mervyn Hartwig is a foundation member of the International Association for Critical Realism and founding editor of *Journal of Critical Realism*. He taught history and philosophy of the social sciences for many years in Sydney. He now lives in London.

Critical Realism: Interventions
Edited by Margaret Archer, Roy Bhaskar, Andrew Collier, Kathryn Dean, Nick Hostettler, Jonathan Joseph, Tony Lawson, Alan Norrie and Sean Vertigan

Critical realism is one of the most influential new developments in the philosophy of science and in the social sciences, providing a powerful alternative to positivism and post modernism. This series will explore the critical realist position in philosophy and across the social sciences.

Critical Realism
Essential readings
Edited by Margaret Archer, Roy Bhaskar, Andrew Collier, Tony Lawson and Alan Norrie

The Possibility of Naturalism 3rd edition
A philosophical critique of the contemporary human sciences
Roy Bhaskar

Being and Worth
Andrew Collier

Quantum Theory and the Flight from Realism
Philosophical responses to quantum mechanics
Christopher Norris

From East to West
Odyssey of a soul
Roy Bhaskar

Realism and Racism
Concepts of race in sociological research
Bob Carter

Rational Choice Theory
Resisting colonisation
Edited by Margaret Archer and Jonathan Q. Tritter

Explaining Society
Critical realism in the social sciences
Beth Danermark, Mats Ekström, Jan Ch. Karlsson and Liselotte Jakobsen

Critical Realism and Marxism
Edited by Andrew Brown, Steve Fleetwood and John Michael Roberts

Critical Realism in Economics
Edited by Steve Fleetwood

Realist Perspectives on Management and Organisations
Edited by Stephen Ackroyd and Steve Fleetwood

After International Relations
Critical realism and the (re)construction of world politics
Heikki Patomäki

Capitalism and Citizenship
The impossible partnership
Kathryn Dean

Philosophy of Language and the Challenge to Scientific Realism
Christopher Norris

Transcendence
Critical realism and God
Margaret S. Archer, Andrew Collier and Douglas V. Porpora

Critical Realist Applications in Organisation and Management Studies
Edited by Steve Fleetwood and Stephen Ackroyd

Making Realism Work
Realist social theory and empirical research
Edited by Bob Carter and Caroline New

Towards Realist Dialectical Materialism
From Kant and Hegel to Marx and Engels
Jolyon Agar

Also published by Routledge

Routledge Studies in Critical Realism
Edited by Margaret Archer, Roy Bhaskar, Andrew Collier, Kathryn Dean, Nick Hostettler, Jonathan Jospeh, Tony Lawson, Alan Norrie and Sean Vertigan

1 **Marxism and Realism**
A materialistic application of realism in the social sciences
Sean Creaven

2 **Beyond Relativism**
Raymond Boudon, cognitive rationality and critical realism
Cynthia Lins Hamlin

3 **Education Policy and Realist Social Theory**
Primary teachers, child-centred philosophy and the new managerialism
Robert Wilmott

4 **Hegemony**
A realist analysis
Jonathan Jospeh

5 **Realism and Sociology**
Anti-foundationalism, ontology and social research
Justin Cruickshank

6 **Critical Realism**
The difference it makes
Edited by Justin Cruickshank

7 **Critical Realism and Composition Theory**
Donald Judd

8 **On Christian Belief**
A defence of a cognitive conception of religious belief in a Christian context
Andrew Collier

9 **In Defence of Objectivity and Other Essays**
Andrew Collier

10 **Realism Discourse and Deconstruction**
Edited by Jonathan Joseph and John Michael Roberts

11 **Critical Realism, Post-positivism and the Possibility of Knowledge**
Ruth Groff

12 **Defending Objectivity**
Essays in honour of Andrew Collier
Edited by Margaret S. Archer and William Outhwaite

13 **Ontology of Sex**
Carrie Hull

14 **Explaining Global Poverty**
A critical realist approach
Branwen Gruffyd-Jones

Dictionary of Critical Realism

Mervyn Hartwig

LONDON AND NEW YORK

First published 2007 by Routledge
2 Park Square, Milton Park, Abingdon, Oxon, OX14 4RN

Simultaneously published in the USA and Canada
by Routledge
270 Madison Ave, New York, NY 10016

Routledge is an imprint of the Taylor & Francis Group

© 2007 Edited by Mervyn Hartwig

Typeset in Baskerville by
Florence Production Ltd, Stoodleigh, Devon
Printed and bound in Great Britain by
Antony Rowe Ltd, Chippenham, Wiltshire

All rights reserved. No part of this book may be reprinted or reproduced or utilised in any form or by any electronic, mechanical, or other means, now known or hereafter invented, including photocopying and recording, or in any information storage or retrieval system, without permission in writing from the publishers.

British Library Cataloguing in Publication Data
A catalogue record for this book is available from the British Library

Library of Congress Cataloging in Publication Data
A catalog record for this book has been requested

ISBN10: 0–415–26161–9 (hbk)
ISBN10: 0–415–26099–x (pbk)
ISBN10: 0–203–36684–0 (ebk)

ISBN13: 978–0–415–26161–6 (hbk)
ISBN13: 978–0–415–26099–2 (pbk)
ISBN13: 978–0–203–36684–4 (ebk)

Freedom is irrespective of the arrow's flight, no regarder of the river's flow.

It is not that there are the starry heavens above and the moral law within, as Kant would have it; rather, the true basis of your virtuous existence is the fact that the starry heavens are within you, and you are within them.

It may be necessary for morality to correct bad science, but it corrects it in the name of a higher norm, true freedom. And that is guided by a highest norm of all – fundamental truth.

Roy Bhaskar

Things should be made as simple as possible, but not any simpler.

Albert Einstein

Contents

List of Tables — xi
List of Figures — xiii
Contributors and Contributions — xiv
Preface — xvi

HOW TO USE THIS BOOK — 1
Cross References, References and Quotations — 1
Citation of Works by Bhaskar — 1
Citation of Works by Other Authors — 2
General Abbreviations — 2
Symbols — 3
Abbreviations Used by Critical Realist Authors — 3
   Critical realism — 3
   The causal–axiological chain (MELD[ARA] schema) — 4
   Domains — 4
   The hermeneutical circles — 4
   Dialectic — 4
   Epistemological dialectic — 5
   General — 6

DICTIONARY OF CRITICAL REALISM — 9
Works by Roy Bhaskar — 504
Other Works Referred to in the Text — 507

List of Tables

1	Polysemy and modes of absence	10
2	Transformative agency	20–1
3	Moments of truth and untruth	27
4	Fivefold alienation	33
5	Forms of alienation	35
6	Alienation mapped to the domains of reality	37
7	Irrealism and master–slave relations	51
8	*'To cause is to change is to absent is to transform and so redetermine'*	59
9	Aristotle's causes mapped to the concrete universal and the ontological–axiological chain	62
10	Centrism → triumphalism → endism and the realist/materialist critique	63
11	The concrete universal ↔ singular	74
12	Elements of the creative process in PMR > CR	88
13	Dichotomies in mainstream social thought and their resolution by critical naturalism	92
14	Limits to naturalism	93
15	Phases of development of CR mapped to its stadia	101
16	Modes of untruth in the demi-real	114
17	Key moments and figures of PMR mapped to the CR domains of reality	115
18	Key dialectics of critical realism	142–3
19	Components of the judgement form	161
20	The moments of eudaimonia	188
21	Modes of freedom and unfreedom	212
22	The fine structure of the Bhaskarian critique of Hegel	226
23	Hegel–Marx critique	228
24	The holy trinity	239
25	Four-planar social being, or human nature	244
26	Forms of CR realism/materialism	249
27	Some characteristics of the IA/EA duality	267
28	Modes of master–slave or power$_2$ relations	288
29	Key MELD concepts	297–302
30	Key TDCR and PMR concepts	307–8

31	The naturalistic *aka* anti-naturalistic fallacy	323
32	The production of perceptual meaning and meaning in general	343
33	The philosophical discourse of modernity and the CR critique	351
34	Types of emancipatory politics	366
35	Unification of the problems of philosophy by absence of the concept of absence	382–3
36	Levels of rationality	397
37	Domains of reality	401
38	The social cube, four-planar social being and alienation	421
39	Subjectivity/the self	445
40	Types of tendency, the concrete universal and the ontological–axiological chain	458
41	Modes of presence and absence of the past, future and present	461
42	Implications of identity-thinking	470
43	The relation between the cognitive and conative components of reasons for action	501

List of Figures

1	Concepts of negation	10
2	The stratification of agency	25
3	Dialectics of co-presence, departing from generative separation	85
4	A topology of critiques	106
5	The CR epistemological dialetic	176
6	Functional explanation	214
7	The structure of irrealism	269
8	Morphogenesis	319
9	The Ontogenesis of subjectivity	447
10	Transformational Model of Social Activity	469

Contributors and Contributions

Alexander, David (University of Technology, NSW), *brain*
Archer, Margaret S. (University of Warwick), *morphogenesis/morphostasis*
Blom, Björn (Umeå University), *social work* (with Stefan Morén)
Bowring, Bill (London Metropolitan University), *legal studies*
Brannan, Matthew (Keele University), *ethnography*
Brereton, Derek P. (University of Michigan), *anthropology; evolution of society; evolutionary psychology*
Brown, Andrew (University of Leeds), *economics*
Byrne, David (Durham University), *chaos/complexity theory* (with David L. Harvey)
Calder, Gideon (University of Wales, Newport), *ethics; philosophical anthropology; sociology of the body*
Carter, Bob (University of Warwick), *race, racism, ethnicity*
Clark, Alexander M. (University of Alberta), *nursing*
Clarke, Graham (University of Essex), *psychoanalysis*
Cruickshank, Justin (University of Birmingham), *essentialism*
Dean, Kathryn (University of London), *medium theory; needs; reason*
Downward, Paul (Loughborough University) *method, quantitative*
D'Souza, Radha (University of Waikato), *colonialism; neo-colonialism; post-colonialism; emancipatory movements*
Ekström, Mats (Örebro University), *media studies*
Engholm, Pär (Örebro University), *irrealism; pragmatism; TMSA; totality*
Fairclough, Norman (Lancaster University), *discourse analysis*
Faulkner, Philip (University of Cambridge), *closure; demi-reg; PRVS model*
Ferber, Michael P. (West Virginia University), *geography of religion*
Fleetwood, Steve (Lancaster University), *industrial relations; labour markets*
Groff, Ruth (Williams College), *truth*
Harvey, David L. (University of Illinois), *chaos/complexity theory* (with David Byrne)
Hostettler, Nick (University of London), *sociology of knowledge*
Høyer, Karl Georg (Oslo University College), *ecology* (with Petter Næss)
Jessop, Bob (Lancaster University), *political science; social form; space–time; strategic-relational approach*
Jones, Peter (Sheffield Hallam University), *linguistics*
Joseph, Jonathan (University of Kent), *hegemony; political theory; structuralism; post-structuralism*

Contributors and contributions xv

Kowalczyk, Ruth (Lancaster University), *management science*
Lacey, Hugh (Swarthmore College/Universidade de São Paulo), *explanatory critique; philosophy of religion* (with Douglas V. Porpora)
Lawson, Clive (University of Cambridge), *technology*
Lawson, Julie (University of Amsterdam), *urban studies*
Maton, Karl (University of Wollongong), *studies of education* (with Brad Shipway)
Mingers, John (University of Kent), *autopoiesis; system*
Minnerup, Günter (University of New South Wales), *historiography*
Morén, Stefan (Umeå University), *social work* (with Björn Blom)
Morgan, Jamie (University of Helsinki), *analytical problematic; emergence; empiricism; globalisation; idealism; identity theory; materialism; mind; political economy* (with Heikki Patomäki); *power*
Morrow, Ross (University of Technology, NSW), *social theory; sociology; sociology of sexuality*
Næss, Petter (Aalborg University), *ecology* (with Karl Georg Høyer); *urban and regional planning*
Nellhaus, Tobin (Yale University), *aesthetics; cognitive science; culture; cultural analysis; signs; semiology, semiotic*
New, Caroline (University of Bath Spa), *feminist theory; gender*
Nielsen, Peter (Roskilde University), *capitalism; critical theory; fetishism; Marxism; rational choice theory*
Norrie, Alan (University of London), *dialectical critical realism*
Norris, Christopher (University of Wales, Cardiff), *ontology; transcendental realism*
Olsen, Wendy (University of Manchester), *poverty*
Parker, Ian (Manchester Metropolitan University), *psychology*
Parker, Jenneth (London South Bank University), *sustainability*
Patomäki, Heikki (University of Helsinki), *futures studies; international relations; neoliberalism; political economy* (with Jamie Morgan)
Pinkstone, Brian (University of Western Sydney), *economic history; tendency* (with Mervyn Hartwig)
Porpora, Douglas V. (Drexel University), *philosophy of religion* (with Hugh Lacey); *social structure*
Potter, Garry (Wilfrid Laurier University), *knowledge theory of; philosophy of social science*
Pratten, Stephen B. (University of London), *contrast explanation; explanation*
Psillos, Stathis (University of Athens), *casual law; inference; Nicod's criterion; philosophy of science; realism*
Ratcliffe, Peter (University of Warwick), *migration studies*
Roberts, John M. (Brunel University), *postmodernism*
Scambler, Graham (University of London), *disability; sociology of health and medicine*
Shield, Richard (University of Leeds), *management and organisation studies; method; methodology*
Shipway, Brad (Southern Cross University), *studies of education* (with Karl Maton)
Tew, Philip (Brunel University), *literary theory*
Westerhuis, Diane (James Cook University), *social constructionism*
Williams, Malcolm (University of Plymouth), *probability*

Preface

This book intends to underlabour for critical realism (CR) and the research programmes it informs as an indispensable work of reference. It seeks above all to present and elucidate the CR constellation of concepts and categories, figures and critiques, in their complex interinanimation. The constellation is articulated along four dimensions, mimicking the 4-D reality it seeks to express: that is, both synchronically (laterally and vertically) and diachronically (horizontally), the whole constituting an evolving conceptual cubic stretch-flow or rhythmic. Lateral articulation is amply catered for by the dictionary-form itself, which necessarily presents a 'snapshot' of a developing system of thought at a moment in time, detotalising it into alphabetically arranged entries and retotalising them via a system of cross-references. Vertical and horizontal articulation present more difficulties. The former – the footprint of conceptual emergence – is brought before the reader's mind by, among other means, an array of figures displaying correspondences with the stadia of the ontological-axiological chain and with the domains of reality, the fundamental parameters of depth-stratification registered by the system. The latter – diachronic articulation – which sets synchrony in motion and must be conveyed against the grain of the fixism of the dictionary-form – is attempted by three main means. First, by devoting five of the longest entries to the developmental moments of CR – TRANSCENDENTAL REALISM, CRITICAL NATURALISM, EXPLANATORY CRITIQUE, EMANCIPATORY AXIOLOGY and DIALECTICAL CRITICAL REALISM (DCR) – along with an interlinking essay on CRITICAL REALISM itself. The reader new to CR is advised to start with these entries, together with PHILOSOPHICAL DISCOURSE OF MODERNITY (PDM). ('Critical Realism' has both an extended and a restricted sense, as embracing DCR or not; it is used in its extended sense here and throughout, except where the context indicates otherwise.) Although it is not part of my brief to treat it in detail, there is also a longish entry on the philosophy of META-REALITY (PMR), which both largely embraces and goes beyond CR, and which (or something like it) is in my view dialectically necessary for the 'completion' and integrity of the philosophical system. Second, by frequent reminders, together with illustrations, that the later moments do not annul but largely presuppose – are essentially preservative sublations of – the earlier ones. Finally, by registering dialectical development within particular concepts and noting how this points the way (where it does) to PMR. An important – and, I understand, novel – subsidiary aim is to display the concepts at work. This is essayed in some fifty 'area surveys' by specialists in the relevant field or cluster of fields (e.g., cognitive science,

Preface xvii

discourse analysis, sociology), which indicate what CR work has been done in an area, canvass possibilities and recommend further reading. Like the other entries, from which nothing but their content distinguishes them, they appear in alphabetical order.

A Dictionary of Critical Realism (*DCR*) comprises some 500 entries totalling a quarter of a million words, of which I have written two-thirds. While many distinguished scholars have contributed significantly to the development of CR across a wide range of fields (and to the writing of *DCR*), I have concentrated in my own entries on presenting the thought of Roy Bhaskar, in full confidence that the diversity and vitality of CR is registered by other contributions (as well as, to some extent, in my own entries). There is strong disagreement with Bhaskar within CR on a range of issues, registered perhaps most powerfully here in EXPLANATORY CRITIQUE and TRUTH (for a supporting view on the same issues, see CRITICAL NATURALISM and ALETHIA). This is entirely as things should be; it would be a logocentric error of the first magnitude to assume that there is a single correct interpretation of any particular move or phase of a system of thought or of the whole, let alone that it is correct in every detail concerning what it is about. My own entries concentrate on Bhaskar's work because, operating as it does at the level at which the tectonic plates of philosophical systems and traditions collide, and itself highly systematic (for truth is the whole), it above all calls for a work of reference to facilitate its critical reception. This is in my view CR's most pressing current need. Bhaskar has a reputation, especially in his dialectical work, for impossible difficulty, if not downright obscurity. However, if, as Goethe has said, reading Kant is 'like stepping into a brightly lighted room', my own experience of reading Bhaskar has been like stepping into broad sunlight next morning – after 'a hard day's night', admittedly. His writing *is* difficult and complex, for his subject-matter is; one has to work hard and perseveringly at appropriating it. But it is not obscure. Bhaskar is one of the most precise and rigorous of thinkers. Every concept is carefully defined in relation to the others and consistently deployed; and, apart from 'printer's errors' – which often turn out not to be such on closer inspection – every word in a Bhaskarian paragraph is usually just right, doing the precise task asked of it. I can relate very well to those who feel swept to sea by 'a Niagara of neologisms', as Andrew Sayer (2000: 170, 210) has put it, when they open *Plato Etc.* and/or *Dialectic: The Pulse of Freedom* (destined, in my view, to assume a place in the ranks of the truly great works of philosophy). That is the very real experience of many. But the extraordinary thing is that – except in the important sense in which 'a Niagara of neologisms' is itself a neologism – there are actually very few neologisms in Bhaskar's work. That is to say, very few new words as such; rather, old words deployed in new and precisely specified ways that none the less do not do violence to old meanings, and often deftly exploit them. (For this reason, and to furnish mnemonic aids, I have often included a brief etymology of concepts.) The attention to detail is remarkable. In what I have dubbed the Bhaskarian stretch-adjective, for example, every conjoined concept is in deliberate order: thus, the sequencing in 'anthro-ethno-ego-present-centric' reflects the fact that conceptually anthropocentrism > ethnocentrism > ego-present-centrism, but because the causal order is the reverse (egocentrism → ethnocentrism → anthropocentrism) the stretch-adjective sometimes also takes that form, in the appropriate context, i.e., 'ego-ethno-anthro-present-centric'; or a perspectival switch is effected from one form to the other. Such is the subtle precision of a good dialectician; and, if *DCR* assists readers to thrill to it rather than to experience it as obscurantism or show, it will have accomplished one of its main tasks.

There are, of course, many who hold that philosophy should be written largely in 'plain English', deploying as few specialist concepts (often decried as 'jargon') as possible, especially if – as in the present case – it is emancipatory in intent, wishing to reach a large audience. Bhaskar is clearly of the contrary school of thought. Just as the tools of the various skilled trades need to be precision-engineered for specific, interrelated functions, so meta-theory requires concepts honed for specific interrelated tasks: it is impossible to think creatively at that level without them. That there are important differences – in particular, tools are for coming to grips with the same sort of stuff as themselves, whereas concepts must often express that which is very different – hardly defeats the analogy, and there are others that might also instruct: one does not expect, say, a physicist to theorise in plain English (which is, in any case, the language of the inverse elitist, not of the person in the street), though her work has important implications for human flourishing. If philosophy is to convince as meta-theory, it must first operate at that level; the oft-heard complaint that the system of concepts is abstract is on a par with complaining that the theoretical sciences are abstract, or that water is wet, and overlooks that a general theory of being and human being is precisely necessary for the concrete sciences and that CR philosophy retotalises its conceptions to satisfy all four moments of the concrete universal and effects transitions from form to content, theory to practice. 'Jargon' speaks of lack of precision and of pretension – of the invention of concepts for effect when more homely words will do. There is none of it in Bhaskar. You can no more translate one of his turbo-charged paragraphs into plain English without grave loss than you can a lyric poem; and, if you try, you end up with a far longer paragraph, just as you take far longer to build a house if you substitute hand- for power-tools. Therefore, I have not tried. My goal is to facilitate the learning of a new language of concepts, rather than to translate it holus-bolus into actually existing language (which CR seeks in part to transform as an integral aspect of transition to a post-slave order), and to assist in advancing its reception and creative deployment to a higher level. Of course, each entry tries to explain the meaning and development of a concept as clearly as possible, and in that sense to mediate between actually existing language and abstract theory; but the concept is usually then deployed in other entries on the assumption that the reader is familiar with its CR meaning. If she is not, and wants to understand, there is no alternative to familiarising herself with it. She can take some comfort from the fact that – since each concept derives its meaning (as distinct from its reference) from its relation to and resonance with others, and its full meaning only from intermingling with the whole – understanding is cumulative. It is impossible to abolish the hermeneutical circle, or to escape it – least of all in, or by means of, a dictionary. In my own appropriation of Bhaskar, I try to follow his advice to read 'out of the text', but non-logocentrically and dialogically, as the circles of inquiry and communication demand. Where in doubt, I choose the reading that seems most adequate in terms of the developmental logic of the system. This is usually, but not necessarily, the most charitable; it sometimes involves implicit criticism and correction. In my entries, I have occasionally absented the absence of a concept or formulation where its general sense seems to be present (e.g., 'methodological unity-in-diversity', espoused by critical naturalism).

The issue of accessibility is very important, though not the only one. The work of a dictionary is only half done when it has elucidated the meaning of concepts; it must explain what their tasks are and facilitate their widespread use. The latter is attempted in particular in the 'area surveys'. One sometimes hears jibes to the effect that CR is a closed circle of concepts fetishised by an esoteric cult of acolytes. Nothing could be further from the truth. It is open and dynamically developing, deeply engaged with the major discourses and social movements of post/modernity[1] and inviting and attracting engagement. As a student of geo-history, I have, moreover, great confidence in the capacity of 'ordinary' people to appropriate and develop creatively new ways of thinking when their hour arrives, and in particular a system like CR, which only gives an abstract account of what is implicit at the level of the real in their own practices, and actually manifest on a daily basis. The passage of time in any case makes learning any system of concepts far easier; while the general theory of relativity was incomprehensible to all but a few high specialists when it was first published, today in its essentials it is grasped by many. Meanwhile, the work of popularising and substantively elaborating and developing can and should be carried forward at multiple levels. The more adequate the meta-theory that informs it the greater will be its value, not least in terms of critical feedback into the theory, which is itself complexly articulated as but one element – indispensable, though in no wise the most important – among many in the overall process of freedom.

DCR is by no means just my own project, in more senses than one. It was long dreamed of by Bhaskar and other senior CR figures, independently of me (to what extent dreams have come true remains to be seen). My heartfelt thanks to all the other contributors to the actual writing, and in particular to Jamie Morgan, then to Pär Engholm, Tobin Nellhaus and Stathis Psillos, who did far more than most, and, in the next rank, Derek Brereton, Gideon Calder, Kathryn Dean, Radha D'Souza, Ruth Groff, Bob Jessop, Jonathan Joseph, Hugh Lacey, John Mingers, Ross Morrow, Caroline New, Peter Nielsen, Alan Norrie, Christopher Norris, Heikki Patomäki, Brian Pinkstone, Garry Potter and Stephen Pratten. A number of contributors felt uneasy about mentioning their own work; they do so on my express recommendation. I also want to thank the International Association for Critical Realism, and all the *JCR* contributors, for the opportunity to found and edit *A Journal of Critical Realism*. This, along with the many productive discussions of CR I have enjoyed on the Bhaskar List (now the Critical Realism List), was undoubtedly an important jump point in the rhythmic that led to an invitation from the editorial board of the Routledge Critical Realism: Interventions series (to whose members also, many thanks) to edit/write *DCR*. A big thanks also to all the members of 'the Friday group' in Sydney, who contributed to another leap in the same rhythmic at an earlier date. At Routledge, I very much appreciate the work of Emma Hart and of Alan Jarvis, who arranged the contract and has been very patient through unforeseen delays; the 'area surveys' were included at his wise insistence. I am conscious of borrowing particular pieces of useful information or advice, otherwise unacknowledged, from David Bailey, Jake Kuiken, Richard Moody,

---

1 Except where I wish to refer to 'modernity and 'postmodernity' separately, I write 'post/modernity' throughout to highlight that postmodernity is constellationally contained within modernity as a distinct phase.

Brad Rose, Dafydd Roberts and Jan Straathof. There are lots of others – including many of those who have produced entries – whose contributions are too numerous to individuate. However, the errors my work is bound to contain (see PARADOX OF THE PREFACE) are my responsibility alone.

At a more personal level, I thank three people in particular (the 'holy trinity' of this book, who absented constraints): Sean Vertigan, for exuberant solidarity, and excited and exciting use of an earlier version of *DCR* in his research; Roy Bhaskar, who has been steadfast in support and terrific for self-esteem and courage to think; and Rachel Sharp, my wife, who rejects a concept of unconditional love but practises it better than most. Above all I thank depth-strugglers everywhere, who show us all the way.

How to use this book

CROSS-REFERENCES, REFERENCES AND QUOTATIONS

Because we are dealing with a constellation (of constellations) of concepts, I have supplied many (strategic, rather than comprehensive) cross-references (in small capitals), mainly internally to entries but sometimes also at the end, and more than a thousand cross-referencing entries. Within any one entry, a concept is normally cross-referenced only once. To economise on words, I sometimes (1) cross-reference a different form of a concept from the one its entry appears under (e.g., DIACHRONIC and PERSPECTIVAL SWITCH invite the reader to consult the entries for **diachrony/synchrony** and **perspective**, respectively); (2) supply indirect cross-references, i.e., to a cross-referencing entry (e.g., EMANCIPATORY CRITIQUE directs the reader to a cross-referencing entry for that concept, which says 'See CRITIQUE; EMANCIPATORY AXIOLOGY'); and (3) shorten a long title of an entry by use of an 'etc.' (e.g., **solipsistic/transdictive complex[es]** becomes SOLIPSISTIC ETC.), or by use of the first word only.

References to sources within an entry apply to all quotes that follow the previous reference. An 'etc.' in an entry title is sometimes used to indicate that an array of related concepts is discussed there (e.g., **freedom etc**.).

Within quotes, square brackets are those of the author of the quote, curly brackets those of the author of the entry.

For purposes of the alphabetical ordering of entries, Greek letters are treated as their Roman equivalents; thus 'dφ' is treated as 'df'.

CITATION OF WORKS BY BHASKAR

Books by Bhaskar are cited according to a system of abbreviation. To find the full citation for a book, the reader should first consult the list below, then refer to the bibliography of 'Works by Roy Bhaskar' (p. 504). Other works by Bhaskar are cited by date of publication; their full title can be found in the same bibliography.

How to use this book

Many of Bhaskar's writings are already widely referred to by acronym, some three or four letters long. In the present context, economy has dictated shorter abbreviations:

D	1993a	*Dialectic: The Pulse of Freedom*
DG	1993a	*Dialectic: The Pulse of Freedom*, glossary, pp. 393–406
EW	2000a	*From East to West*
MR	2002c	*The Philosophy of Meta-Reality*, Vol.1
P	1994	*Plato Etc.*
PF	1991a	*Philosophy and the Idea of Freedom*
PG	1994	*Plato etc.*, glossary, pp. 249–60
PN		*The Possibility of Naturalism.* Any edition
PN1	1979a	*The Possibility of Naturalism.* First edition
PN2	1989a	*The Possibility of Naturalism.* Second edition
PN3	1998e	*The Possibility of Naturalism.* Third edition
R	1989b	*Reclaiming Reality*
RM	2002c	*Reflections on Meta-Reality*
RS		*A Realist Theory of Science.* Any edition
RS1	1975c	*A Realist Theory of Science.* First edition
RS2	1978	*A Realist Theory of Science.* Second edition
RS3	1997b	*A Realist Theory of Science.* Third edition
SE	2002a	*From Science to Emancipation*
SR	1986	*Scientific Realism and Human Emancipation*

CITATION OF WORKS BY OTHER AUTHORS

CDP	2001	*The Cambridge Dictionary of Philosophy*, ed. Robert Audi
DCR		*A Dictionary of Critical Realism*
ER	1998	*Critical Realism: Essential Reading*s, ed. Margaret Archer et al.

Otherwise, works by other authors are referred to by author or editor and year of publication. Full details can be found in the bibliography of 'Works Referred to in the Text' (p. 507).

GENERAL ABBREVIATIONS

adj.	adjective
cf.	compare
CP, c.p.	*ceteris paribus* (other things being equal)
CT	critical theory
e.a.	emphasis altered (added or removed; may include punctuation)
Eng.	English
f.	formed on (etymologically); and the following (pages)
Fr.	French
Ger.	German

How to use this book 3

Gr.	Greek
iff	if and only if, i.e. necessary and sufficient
incl.	including
L.	Latin
n.	noun
n	note
neut.	neuter
pl.	plural
s.	singular
vb.	verb

SYMBOLS

=	is equal or equivalent to
≠	is not equal or equivalent to
>	greater than (constellationally embraces or contains)
≥	greater than or equal or equivalent to
⊃	materially implies (material conditional: $p \supset q$, either p is false or q is true)
⊂	is materially implied by
→	entails (strict conditional: $p \supset q$ is necessarily true), more loosely to or towards
↔	mutually entail, mutually implicate
✓	acceptance
✗	rejection
/	indecision
– or ~	not, the negation of
( )	being
(–)	non-being

ABBREVIATIONS USED BY CRITICAL REALIST AUTHORS

Note. Most, but not all, of the following are used in *DCR*. The list is provided also as an aid to reading other CR texts, particularly those by Bhaskar.

Critical realism

CR	critical realism, critical realist in either the extended or restricted sense, depending on the context, as inclusive of DCR or not
DCR	dialectical critical realism, dialectical critical realist
PMR	the philosophy of meta-Reality, and its adjectival form
TDCR	transcendental dialectical critical realism, and its adjectival form
SR	transcendental or scientific realism, and its adjectival form
TR	transcendental or scientific realism, and its adjectival form

4 *How to use this book*

The ontological–axiological or causal–axiological chain
(in CR MELD, in PMR MELDARZ or MELDARA, or the MELD or MELDARZ/MELDARA schema)

1M	first *m*oment – non-identity (structure)
2E	second *e*dge – negativity (process)
3L	third *l*evel – totality (holistic causality)
4D	fourth *d*imension – transformative agency (intentional causality)
5A	fifth *a*spect – reflexivity
6R	sixth *r*ealm – (re-)enchantment
7Z/A	seventh *z*one – *a*wakening (non-duality)

Domains

d_r	domain of the real
d_a	domain of the actual
d_+	domain of the positive
d_s	domain of the subjective (experience, concepts, signs)
d_e	domain of human experience
d_f	domain of humanly construed facts

The hermeneutical circles

C1. C of I	circle of inquiry
C2. C of C	circle of communication
C3. C of I (C)	circle of inquiry into other cultures, traditions, etc. (the second C stands for 'communication')
C4. C of I (T)	circle of inquiry into texts and text-analogues

Dialectic

d	dialectic, dialectical
d+	positive dialectic(s); domain of the positive
d–	negative dialectic(s)
da'	dialectical explanatory argument or explanation
dc'	dialectical comment
dd'	dialectical distinction
dg'	dialectical ground(s) or condition(s) of existence
dl'	dialectical limit
δ[*delta*] node	in the epistemological dialectic, the arrival of a hint of restoration of consistency by expansion or transformation of the pre-existing conceptual field
dp'	dialectical process(es)
dr°	dialectical outcome or result of a dp'
dr†	dialectical result which resolves the contradiction in dp'

How to use this book

dr′	(i) dialectical result which provides a rational resolution of the contradiction in dp′; (ii) dialectical reason; (iii) ontologically, the real reason for or alethic truth of a thing
dr″	(i), above, and which conforms to Hegelian radical preservative determinate negation
dr‴	– and affords us reconciliation to life
dr⁗	– and encourages mutual recognition in a free society
dr_a' (in dφ)	absolute reason, or the unity of theory and practice in practice (dialectical rationality)
dr_k', dr_j', dr_i'	the conditions of impossibility (dr_k'), of the conditions of possibility (dr_j'), of some significant result or phenomena (dr_i'), respectively (in Hegel)
drk′	dialectical remark
drk†	generalised dialectical remark
dr_p'	practical reason
dr_t'	theoretical reason
ds′	dialectical sublation
dt′	dialectical totality
dφ [*phi*]	dialectical praxis or practically oriented transformative negation of contradictions
dφ′	transformed transformative totalising transformist praxis or emancipatory axiology (TTTTφ); more fully, transformed transformative trustworthy totalising transformist transitional praxis (TTTTTTφ)

Epistemological dialectic

D	dialectical reason – negatively rational or dialectical thought; the epoch of scientific revolution
R	speculative reason – positively rational or literally speculative thought, aspiring to rational totality
PRT	pre-reflective thought
PRW	post-reflective wisdom
U	the understanding or analytical reason; 'normal science'
σ (*sigma*) transform	emergence and identification of an inconsistency
τ (*tau*) transform	repair of an inconsistency
ρ (*rhō*) transform	the extra-scientific inputs (the relational dialectics in science) – scientific training and the whole gamut of extra-scientific inputs
υ (*upsilon*) transform	the extra-scientific outputs (the relational dialectics out of science) – applied science and technology, (re-)appropriation by the lay community of the skills and knowledge formed in the intra-scientific domain
φ (*phi*) transform	the resolution of contradictions in practice (dφ)

6 How to use this book

General

A	action, appearances, axiological operator
AH	Achilles heel (critique)
AR	anthroporealism
B	belief
C, c	comment, complex (generative or transformational complex), condition(s), correction
CAJ	concrete axiological judgement
CEP	cultural emergent power or property
cf	cognitive function
CSM	central state materialism
CU	concrete universal (see UPMS)
D	description, diagnosis, differentiating mediation
DEA	diagnosis, explanation, action (practical problem resolution)
DET	description, explanation, transformation (normative change)
DI	depth-investigation, depth-inquiry
DIP	dialectical or dynamic integrative pluralism
dm	domain of morality
dm_a	domain of actual morality
dm_r	domain of real morality
DREI(C)	description, retroduction, elimination, identification, correction (theoretical explanation)
E, e	effect, elimination, essence, event, state of affairs
(e)	a positive state of affairs
(−e)	a negative state of affairs
EA	emancipatory axiology, extrinsic aspect (of TD)
EC	expanatory critque
EC[†]	explantory critique[†] or explanatory critical theory[†] complex (ECs + exercises in concrete utopias + theories of transition)
EF	epistemic fallacy
ef	epistemic function
ER	empirical realism
erf	empirical realist epistemic function
exp	explains
F	the factual
fef	fundamentalist epistemic function
G	generative mechanism, ground(s), operator of a generative complex or grid (C)
GCS	general conceptual scheme
I	identification, institution, the fictional, the ideological (ideology)
IA	intrinsic aspect (of TD)
ID	intransitive dimension
ideology[†]	ideology in the restricted sense, involving category mistakes
IE	the irrealist ensemble

IP	integrative pluralism
L, l	law, level, zone, stratum
M, m	meaning, mechanism, mutation
MC_1	meta-critique$_1$ (isolates an absence in a theory etc.)
MC_2	meta-critique$_2$ (explains the absence identified in MC_2)
MD	meta-critical dimension
MI	methodological individualism
mo	material object (proposition)
MOE	material object empiricism
N	neurophysiological states
O	object, overt behaviour, outcome
OF	ontic fallacy
OI	objective idealism
OM	ontological monovalence
P	causal or dispositional power, phenomenon, positivism, practice, principle, proto-scientific theory, psychological states, particular (see UPS)
PDM	philosophical discourse of modernity
PEP	personal emergent power or property
Ph	phenomenalism
Q	qualifying clause
R, r	rationalisation (justification), reason, redescription, reference, reproduction, resolution, retroduction, rhythmic
R_r/R_p	real reason/possible reason
rf	realist function
RREI(C)	resolution, redescription, retrodiction, elimination, identification, correction (applied explanation)
S	selection, situation, the social or social rules or social states, statement, structure, subject, system, singular (see UPS)
sd	sense-datum
SE	subjective empiricism
SEP	societal or structural emergent power or property
SEPM	synchronic emergent powers materialism
sf	social function
T, t	theory, thought (cognitive activity), time, totality or Totality (the cosmos as a whole)
t_1, t_2, etc.	time one (earlier than), time two (later than), etc.
TCF, TF	Tina compromise formation, Tina formation
TD	transitive dimension
T^dR	transcendent realism
TGK	theory of general knowledge
TMSA	transformational model of social activity
T/P	theory/practice
T/P inc. *or* $(-T[P])$	theory/practice inconsistency
$-(T_1[P_1])$	ideology (in the narrow sense of a false or inadequate theory)

(–(–T₁[P₁])	ideology-critique, which demonstrates the falsity or inadequacy both of a theory and of the corresponding theory–practice ensemble
TPF[SS]	theory problem-field solution set
TPK	theory of particular knowledge
TTTTφ	transformed, transformative, totalising, transformist praxis or emancipatory axiology (dφ′)
TTTTTTφ	transformed transformative trustworthy totalising transformist transitional praxis or emancipatory axiology (dφ′)
U	universal
UPS	*universality*, *particularity* (mediation), *singularity* (the schematic form of Hegel's concrete universal)
UPMS	*universality*, *processuality*, (particular) *mediation*(s), (concrete) *singularity* (the schematic form of CR's concrete universal, to which the moments of MELD correspond)

A

abduction. See INFERENCE.
aboutness. See REFERENCE; REFERENTIAL DETACHMENT; TRUTH.
absence or **real negation**. At once the major blind spot in the ANALYTICAL tradition, screened by the doctrine of ONTOLOGICAL MONOVALENCE, and the pivotal category of dialectic – whether construed as argument, the onto-logic of change or the process of freedom – absence or real negation structures and unifies the ontological–axiological chain (see MELD). *Negation* and *negativity* are central to all dialectics, but it is Bhaskar's claim that CR 'uniquely sustains an adequate account' thereof, hence of dialectic itself (*D*: 300). (The two terms are sometimes used interchangeably, but negativity is the more general concept, capturing better than negation simpliciter the dual senses of [evaluatively neutral] absence and [pejorative] ill-being. When absence – 'the simplest and most elemental concept of all' (*D*: 239) – is given an ethical inflection as ill-being, it is synonymous with negativity.) If Bhaskar's earlier work revindicates ontology, *Dialectic* revindicates negativity, thereby foregrounding questions concerning the contingency of being, not least of human social being, which the monovalent tradition sidelines, and 'emancipat{ing. . .} dialectic for (the dialectic of) emancipation' (*D*: 40).

(Real) negation or absence has a product/PROCESS (being/becoming) BIPOLARITY or HOMONYMY, issuing in a fourfold meaning or POLYSEMY corresponding to 1M–4D and the moments of the CONCRETE UNIVERSAL, as illustrated in Table 1: (1) *product* (simple absence); (2) *process* (simple absenting); (3) *process-in-product*; and (4) *product-in-process*. (1) constitutes its primary meaning as 'real DETERMINATE absence or non-being (i.e., including non-existence)' – the outcome of a process. It may refer either to the absence of any entity or feature from consciousness (e.g., the unconscious) or from a space–time region (resulting from 'DISTANCIATION or MEDIATION, death or demise', e.g., the gaps between these marks) or to simple non-existence anywhere anywhen (e.g., a yellow logarithm or Hamlet, Prince of Denmark) (*D*: 5). But it also connotes (2), the process of absenting, distanciating or mediating. If at (1) absence is 'the heart of existence', at (2) absenting is 'the hub of SPACE, TIME and CAUSALITY' (*P*: 56), with positive bipolars in EMERGENCE and CREATIVITY. Combining these two basic connotations, we have absence as (3) process-in-product (e.g., an absence of fertile land, or, desert existentially constituted by its geo-history and context) and (4) product-in-process (e.g., the desertified region in process, i.e. exercising its causal powers).

It is important to distinguish the following forms of negation, which Hegelian dialectic conflates as 'determinate negation': real negation ≥ transformative negation

10 **absence** or **real negation**

≥ radical negation ≥ linear negation. The bases for these distinctions are displayed in Figure 1.

Besides process/product polysemy, all forms of negation display real/actual, determinate/indeterminate and ontological/epistemological AMBIVALENCE, and may themselves be present in a negative (the memory of your dead mother) or a positive (her absence simpliciter) mode. *Real determinate negation* in its simplest definition is determinate

Table 1 Polysemy and modes of absence

Causal–Axiological Chain	1M Non-Identity	2E Negativity	3L Totality	4D Transformative Agency
Concrete universal ↔ singular	universality	processuality (rhythmicity)	(particular) mediations	(concrete) singularity (not just human)
Polysemy of absence	product	process	process-in-product	product-in-process
Causal modes of absence	transfactual causality	rhythmic causality	holistic causality	intentional causality
Concepts of negation	real negating process (substantial & non-substantial)	transformative negating process (substantial)	radical self-negating process	linear (dialectical reason) self-conscicously negating process
Modes of radical negation	auto-subversion	self-transformation	self-realisation	self-overcoming

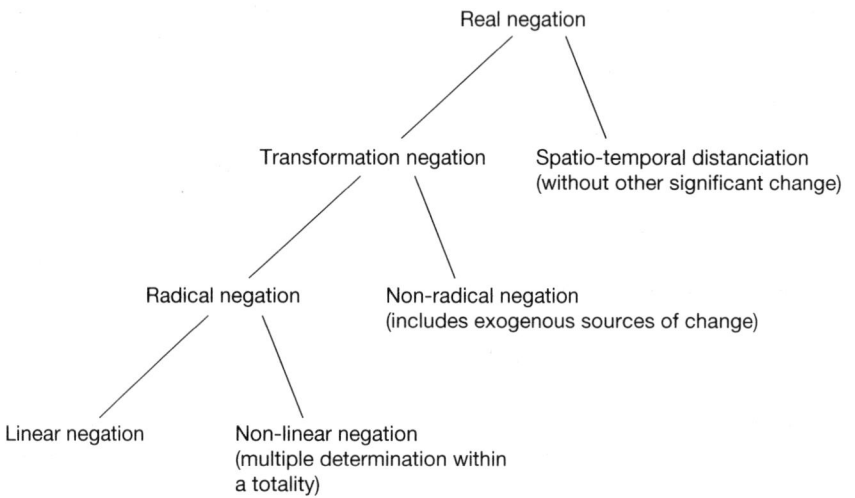

Figure 1 Concepts of negation
Source: *P*. 56.

non-being (the *presence* of an *absence* of an entity) in some determinate locale (e.g., the hole in the ozone layer), which is existentially intransitive 'relative to any possible indexicalised observer on any possible WORLD-LINE', whether or not identified or even identifiable (*D*: 38). Real determinate negation or non-being is thus not equivalent to Hegelian or Sartrean (indeterminate) nothingness (it is structured by a specific process and context), nor to logical negation (see CONTRADICTION), and 'contra Sartre {...} is no more {...} anthropic than the physical concept of force' (*D*: 239). As already indicated, however, real negation may be indeterminate to a varying degree, e.g., 'fuzzy'; and it always possesses a *moment* of indeterminacy (*indeterminate negation*) prior to the determinate result, as for example in the TRANSITION from the absence to the presence of rain, where 'not raining' is the indeterminate negation of 'raining': it leaves open whether there is still full sunshine, an imminent gale, etc. It is thus 'a moment of genuine contingency, openness, multi-possibility (and doubt)' (*D*: 31). Real negation embraces, in addition to the other forms of negation, spatio-temporal DISTANCIATION without significant change, and action-at-a-distance (including intra-action) and across voids, which is in effect non-substantial process or change. It is these features that make it more basic than transformative negation. It is thus 'the most all-encompassing concept, extending from non-existence to metacritique', and includes the main kind of absence CR is concerned with: determinate lacks and needs (e.g., lack of food in a belly or of truth in a politician, or an aporia in a theory). *Transformative negation* refers to the transformation or demise of a pre-existing entity or state of affairs, i.e., substantial process. It is consistent with exogenous sources of alteration. It is the key (but is not confined) to socio-historical dialectics, and its schema is, indeed, given by the TMSA. *Radical negation* refers to self-transformation resulting from multiple determination within an entity (*subject-* or *developmental negation*, see CONSISTENCY; FIXISM/FLUXISM). Not only is it 'obviously the pivotal concept in self-emancipation', which 'connects with "radical" in a more familiar sense', but also, in an increasingly interdependent world, all change (transformative negation) must tend towards radical (totalising) negation (*D*: 6). In addition to participating in the fourfold process/product tetrapolity, it has a corresponding 'fourfold polysemy of its own': *auto-subversion*, *self-transformation* (see also SELF-REFERENTIALITY), *self-realisation* and *self-overcoming*. Its negative forms include split and split-off or detotalisation (see ALIENATION; FISSION/FUSION). *Linear negation* is self-transformation in a unilinear sequence or line of transition. This, however, does not make it AUTOGENETIC, as in Hegel. In an open world most results are multiply and contingently determined.

The MELD meshwork again underpins and is implicated in these distinctions. Most obviously, transformative negation corresponds to 2E, and radical negation to 3L, where the concept of a healthy functioning whole links it to negativity qua ill. The differentia specifica of real negation, synchronic difference mediated by spatial distance without real change, corresponds to 1M non-identity (see ALTERITY), where absence informing desire also powers REFERENTIAL DETACHMENT. Finally, the achievement of DIALECTICAL REASON at 4D would tendentially approximate linear negation. The importance of thus distinguishing 'negating processes from self-negating processes {...} from self-consciously negating processes' (*D*: 6) should be apparent in the human world.

12 absence or real negation

There are many finer meanings and figures of negation and negativity, which are conveniently listed at *D*: 238. It should now be apparent why CR dialectics focus mainly on the major forms indicated. All four forms in their polysemy also exemplify the DUALITY *of absence* (and of presence): 'what is absent or void at or from one level, region or PERSPECTIVE may be present at another' (*D*: 5), e.g., use-value in market society is absent at the level of exchange and present at the level of consumption.

Contrary to what is sometimes implied, the argument in *Dialectic* for the category of real negation/absence and its centrality is complex and multi-pronged. In bare outline it goes as follows. (1) We concede the reality of negative existences or real non-beings (*de-onts* by contrast with ONTS or real beings) every time we insist that 'Pierre really is absent from the café', etc., for this is to make a negative *ontological* claim. Not to admit de-onts to our ontology is to commit performative contradiction. (2) We can refer to, or REFERENTIALLY DETACH, de-onts as well as onts. This establishes (though it does not constitute) their existential intransitivity. We can, indeed, refer to anything imaginable, e.g., animals that speak a human language, and, when we do, the imaginary or fictional may be inscribed within the real (the agent's or society's register of the imagined) as a distinct class of non-being capable of causally affecting us. (3) Some de-onts (e.g., Pierre's absence from the café when I was expecting to meet him) straightforwardly satisfy both CR criteria for ascribing reality: perceptual and causal. It is not the case, contrary to Kant, that one can always analyse negative into purely positive predicates: 'Pierre's absence from the café doesn't *mean* the same as his presence at home (although the latter entails the former – which is equally entailed by his death) any more than it means the same as Jean's occupying his customary place' (*D*: 7). (4) Any world which is changing – as is ours – must incorporate absence, for change is a mode of absence/absenting. (5) Likewise, any world incorporating intentional causality must incorporate absence. (6) Argument and critique, and more broadly any LEARNING PROCESS, themselves depend upon the identification and elimination of mistakes. Mistakes depend upon absences (e.g., lacunae in a theory), which their correction absents. The possibility of remedying inconsistency and incompleteness in an ongoing dialectic is a transcendentally necessary condition for science, as is the use of metaphors and analogies, drawn from the past and outside, in retroductive modelling. Specifically transcendental arguments also presuppose the category of absence in that they turn on human agency. 'Even more simply, a sentence without absences, pauses or spaces, would be unintelligible. Thus absence is a condition of any intelligibility at all' (*D*: 240). (7) Indeed, 'both conceptually and causally, all the *decisive* moments in social life are negative' (*D*: 160). Social structure is the acervative result of past praxis, and living praxis is absentive agency which may issue in the transformative negation of social structures. The axiology of freedom (see EMANCIPATORY AXIOLOGY) entails absence and absenting. Desire is propelled by absence or lack, and informed desire drives praxis on to absent constraints and ills, which, as blocks on well-being, may themselves be seen as absences. (8) Metacritically, the absenting of the absence of a concept of absence is shown to resolve most of the traditional PROBLEMS of philosophy, thus creating a prima facie case for the reality and central importance of negativity. In ways already indicated, all the basic categories of the 1M–4D chain of being–becoming can be derived from absence, which also plays a crucial role in each stadion quite independently of their generation (*D*: 248, 304–5). (9) Those who

absence or **real negation**

concede that de-onts exist often mount a rearguard argument to the effect that they are entirely dependent on positivity, which must be seen as ontologically prior to negativity or at least on a par with it: the identification of absences depends on factual discourse and a network of positive material things, and determinate transformative negation is *of* material resources. While there is a sense in which positivity and negativity do mutually presuppose each other, conceptually and causally, within any continuing process, the priority of the negative none the less asserts itself: (9.1) The identification of an ont involves 'the absenting of a pre-existent state of affairs, be it only a state of existential doubt' (*D*: 44). (9.2) 'The material world operates as a referential grid for the identification of {. . .} onts and de-onts only in virtue of their mutual exclusion relations, that is to say, their differences in space and changes in time', and both difference and change presuppose the category of absence (see ALTERITY/CHANGE). (9.3) In any continuing process, the initial and resultant product is radically constituted existentially by the negative (see CAUSALITY), and this process, when situated as a process, 'appears as the *NEGATION OF THE NEGATION*, i.e., the geo-historical transformation of geo-historical products' (*D*: 241, e.a.). In the human world this is exemplified in the TMSA. (9.4) On any depth-ontology, what is absent from actuality is definitionally far vaster and greater in possibility than the actual. (9.5) A purely positive material object world without voids is transcendentally impossible, for in such a world nothing could move. 'Transmission of energy, like information in inter-personal communication, is possible only by (substantial or non-substantial) travel across, at the very least, level-specific gaps' (*D*: 46). Therefore, non-being is constitutively essential to being; no being could exist without absence. The converse, however, does not hold logically: there could have been just nothing, a total void. This is counterfactual – beings exist – but, dialectically detaching ourselves from that premise, we can see not only that totally nothing is logically possible, but also that, *if* there was a *unique* beginning, or were to be a unique ending, of everything, it could only be from or to absolutely nothing. 'In sum, complete positivity is impossible, but sheer indeterminate negativity is not' (*D*: 47). (9.6) Therefore, far from the positive having ontological priority: 'Within the world as we know it {i.e., a world containing positive existence}, non-being is at least on a par with being. Outwith it the negative has ontological PRIMACY.' The effect of 'outwith' here (see CHIASMUS) is to suggest that absence has ontological priority over (constellationally overreaches) presence within being as a whole, that the positive is indeed only 'a tiny, but important, ripple on the surface of a sea of negativity' (*D*: 5). Elsewhere, the negative is accordingly said to have both *relative* and *absolute* physical priority over the positive. The argument for priority does not depend exclusively on the notion of an absolute beginning ex nihilo, but is constituted independently by the overall thematics of *Dialectic* demonstrating that absence is the key category of/for thinking being–becoming.

The cardinal points of the Bhaskarian account of absence are: it exists (1M); changes are absentings (2E); ills can be seen as absences, which act as constraints (3L); and empowered praxis is absenting agency which can remove remediable ills (4D) (*D*: 203). While the essence of dialectic for Hegel was to see the positive in the negative, 'it is more correct to say that it is to see *the negative in the positive*, the absent in the present, the ground in the figure, the periphery in the centre, the content obscured by the form, the living masked by the dead' (*D*: 241). Its real DEFINITION in relation to geo-history

14   absence and presence, dialect of

is 'the absenting of absences on absenting absences' (*P*: 135), where absences are construed, first, as constraints and, second, as ills. See also DIALECTICAL ARGUMENT; Collier 2001b.

absence and presence, dialectic of. See CO-PRESENCE.

absolute. (1) Unrestricted, complete or utter, as in 'absolute ecological constraint' or 'absolute or DIALECTICAL REASON'. (2) Not ontologically dependent on anything else, as in ULTIMATA or *absolute reality* as distinct from relative reality in the philosophy of META-REALITY, i.e., *the absolute*; but indispensable for the existence of everything else (relative reality). (2) may be INFINITE or unbounded. A *false absolute* (e.g., the abstract universals of money and power) is so because it is not ontologically independent and occludes that which sustains it. See also DEMI-REALITY; FINITE/INFINITE; IDEALISM; IDEOLOGY; IMMANENCE/TRANSCENDENCE; MEDIATE; TINA SYNDROME.

absolute reason. See DIALECTICAL REASON.

absolutism. An epistemological concept, *absolutism* or *epistemological ENDISM* arbitrarily halts the process of knowledge production with the assumption that we know all there is to know. It is the inverse of *epistemological* FOUNDATIONALISM or fundamentalism, which assumes we can start from scratch. Our THROWNNESS entails that neither is true.

abstract, abstraction. See CONCRETE/ABSTRACT; TOTALITY.

abstract universality or **universalism.** See CONCRETE UNIVERSAL; PDM; UNIVERSALISABILITY.

accident. See MODALITY.

Achilles heel critique. See CRITIQUE.

action. See AGENCY.

action-at-a-distance. See ABSENCE; GEO-HISTORICITY; LAGGED/LAPSED.

action theory. See EXPLANATORY CRITIQUE.

actionability. *Actionable*, capable of being accomplished by the relevant (AGENTIVE) AGENTS. The *actionability* or *feasibility principle* implies in addition a DIALECTIC OF SOLIDARITY with the addressees of a fiduciary remark, and is a vital consideration of practical reason in EXPLANATORY CRITIQUE and EMANCIPATORY AXIOLOGY. It is entailed by the requirement for theory–practice CONSISTENCY. It is non-actualist, implying that 'a thing can be done, not that it will' (*D*: 293 n). It is related to, but distinct from, the principle of PREFIGURATIONALITY, which is likewise entailed. Synonymous with AGENT-SPECIFICITY.

activity-dependence. See CRITICAL NATURALISM; SOCIAL STRUCTURE.

acts, basic. See MEDIATE; PERCEPTION.

actual, domain of. See ACTUALISM; REALITY.

actualism. *Actual* has three everyday meanings: (1) active; (2) existing in act or fact, real; (3) existing or acting at present, currently. In modal LOGIC, actualism is the view that what is actual exists, but not the possible or possible worlds, the concept of possibility being of heuristic value only.

CR usage (*the actual, actualism*) likewise takes its departure from (2) and (3); it has no connection with actualism in the sense (*c.* 1860) that all existence is active, not inert or dead. Rather, *ontological actualism* is (a) the view that causal laws are constant conjunctions of events (which overlooks that invariances are normally experimentally

actualism 15

produced results under artificially closed conditions and that closures are the rare exception outside the laboratory). This is the *generalised form of subject–object (epistemic–ontic) IDENTITY theory*, seen at work in, e.g., Descartes' 'conception of a perfect closed actualist {. . .} deductivist science of body' (*P*: 187) and the D–N model of EXPLANATION, which presuppose the de-differentiating and de-processualising ubiquity of closed systems, regularity DETERMINISM and BLOCKISM. (b) More generally, it is the view that the domain of the real (see REALITY, where $d_r > d_a$) is exhausted by the domain of the actual, i.e., possibility and necessity are reduced to an actuality comprised either of events and states of affairs (the dominant empiricist form of actualism), or of concepts and signs, or of will to power, etc., $d_r = d_a$ (the *actualist fallacy*). (A range of finer meanings is listed at *D*: 234.) At the heart of actualism is thus a determinate ABSENCE – of the concepts of natural necessity and possibility – which is the main source of PRIMAL SQUEEZE. It also loses (in theory, but not in practice) a concept of non-identity or ALTERITY, since there can be no true identity as long as the object is not literally empirical (or conceptual, etc. – see below); or it regards itself as an *achieved* identity theory. The main irrealist error at 1M (resting on illicit FUSION, together with the epistemic–ontic and anthropic fallacies, of which it is the direct consequence – hence the concept *anthropo-ACTUALISM* [*D*: 208]) – it may thus be said to give a purely actual account of reality, which complements and is complemented by the purely positive account afforded by ontological monovalence (2E) and the purely externalist account provided by ontological extensionalism (3L), leading to de-agentification (reductionist reification and dualist disembodiment) at 4D.

Ontological actualism goes hand in hand with *epistemological actualism*, which denies the independent existence and transfactual efficacy of causal powers. Actualism often simply refers to both, and sometimes additionally to *moral* or *normative actualism*, which holds that morality reduces to moral rules or laws that are universally applicable regardless of the specific context, i.e. non-dialectical, e.g., Kantian prescriptivism, utilitarianism, natural law theory. It involves both a vertical collapse (*vertical actualism*) of ontological stratification into flatness and a horizontal collapse (*horizontal actualism*) of ontological differentiation (transfactuality and openness) into closure, constituting 'an actualist blanket suffocating hope' (*P*: 107), ultimately expressing the remorselessly REIFICATORY logic of the capitalist mode of production with its refrain of more of the same (*RM*: 46; see also PROBLEM).

Since it is impossible to carry through the reduction of being to knowledge consistently, actualism is inherently DILEMMATIC. When confronted with open systems, the actualist must say that universal empirical generalisations or general principles are either (1) not laws (the transcendental realist position), or (2) not universal, or (3) not empirical or conceptual. Since they cannot be both universal and empirical, actualism, 'presupposed as self-identical with itself' (*P*: 232), must divide against itself – i.e., the illicit FUSION it rests on must give way to FISSION – necessitating, as the REALITY PRINCIPLE is encountered, a TINA COMPROMISE. *Strong actualism* (3) is so called because it adopts the heroic posture that, while there are no empirical invariances known to science, they are there to be discovered (in the scientific *JENSEITS*) and by proposing a programme of REDUCTION for science whereby behaviour in a higher order level can be predicted from knowledge of lower order levels, thus flying in the face of EMERGENCE. *Weak actualism* (2) restricts laws to closed systems, thereby

16 adequacy

providing no answer to what governs phenomena in open systems and leaving both experimental and applied science without a rationale. Weak actualism tries to 'save the empirical at the expense of regularity (science at the cost of philosophy)' (*D*: 181) (cf. empiricism); strong actualism tries to save regularity at the expense of the empirical (philosophy at the cost of science) (cf. rationalism). What Bhaskar calls Hegel's DEMI-*actual*(*ism*), whereby non-actual phenomena are accorded 'a lower ontological status as irrational existents', is a form of weak actualism (*DG*: 393).

There are many varieties of actualism besides the dominant empiricist form. Plato's EIDETIC *actualism* holds that the actual is a direct manifestation of transcendent Forms. Aristotle's *kinetic* (or kinetic [quasi-]eidetic) *actualism* is a dynamic immanentist version of eidetic actualism on which changing things are 'nothing other than the actualisation of their potentialities', i.e., there is an 'identity of pure form and sheer actuality' (HYLOMORPHISM) (*P*: 182 n, 183). *Conceptual actualism* reduces the real as such (e.g., Hegel) or the social (the hermeneutical tradition) to concepts. Hegel's conceptual (expressivist-kinetic-eidetic [*D*: 192]) actualism is additionally *transfigurative* in that it re-describes the power$_2$-stricken world as the embodiment of absolute reason. It immanentises transcendent reality, substituting spirit for structure, thereby collapsing the ID to actuality; and transcendentises actuality, substituting the ETERNISED 'logical present for a tensed geo-history' (*P*: 122).

In so far as all actualisms implicitly or explicitly equate the actual with the real, they are also (empirical, conceptual, etc.) *realisms* (see EMPIRICAL REALISM). For CR the concept of the actual, divorced from actualism, is of course unobjectionable (see REALITY).

adequacy. See HERMENEUTICS.

aesthetics. The term *aesthetics* arose in the nineteenth century as the study of beauty, but it soon came to refer to the study of art. The word possesses several corollary meanings, including the study of style and taste. The last is closest to the etymological meaning of the term, which is 'sense perception' (Gr. *aisthēsis*). Aesthetics has sometimes been treated as a synonym for the study of CULTURE, in the sense of upper-class culture. In part because of these elitist associations, during the late twentieth century aesthetics fell out of favour in scholarly circles. Reluctance to take up aesthetics was exacerbated by the increasing socio-cultural segmentation of the twentieth century, and by the powerful impact of relativist thinking. In recent decades, however, both culture and aesthetics have acquired more socially encompassing interpretations, and aesthetic discussions are returning to the scholarly realm.

Like culture, aesthetics has seldom been a subject of CR analysis. Among the many reasons for this, aesthetics' roots in idealism are hard to overcome. There are, however, several possible avenues for CR aesthetics analysis. One is the sociology of taste, a field that Bourdieu (1984) pioneered in a manner highly amenable to CR. Through questionnaires, interviews, market research and statistics, he discussed how preferences in art, entertainment, music, sports and other cultural domains follow patterns. For example, people who drive a Subaru might also tend to like Thai cuisine, publicly funded radio, and foreign movies. His analysis of these patterns demonstrated how they connect to two major dimensions of society, financial position (economic capital) and educational/intellectual attainment (cultural capital), and that these two dimensions bear a structural relationship, so that bourgeois intellectuals are the dominated fraction of the dominant class and consequently may find themselves in sympathy with the dominated class. Moreover, conflicts over taste are essentially struggles of class distinction and in part hinge on the relationship to embodiment and physicality, including by analogy the preference for or rejection of (say) abstract art.

The sociology of art and taste fits readily into the CR analyses most indebted to Marxism, such as ideological critique. It does not, however, address art perception or the aesthetic experience. It is important to understand such experience, for it is a rare person indeed who has no interest in having something appealing or emotionally compelling in their life. All peoples, societies and classes have developed art forms; and even when someone selects or only knows kitsch the need for art is there.

One of the difficulties for a CR aesthetic theory is aesthetics' long history of purely formal or stylistic analysis. Form and style are crucial features of aesthetic objects, but they must be put into the context of CR's central concern: What are the object's causal powers? Or, as John Ciardi put the question, 'How does a poem mean?'. Placing causal powers front and centre avoids prejudicing aesthetic values (the coarse and unpleasant may make for more effective art than the delicate and beautiful), recognises the import of an object's material aspects (a paraphrase simply is not the same thing as the original) and resists reducing art to personal reactions. But 'causal powers' must be taken in a rigorously CR sense: they are real structures that depend on socio-cultural contexts within which they interact with audiences, and consequently may have varied effects. For that matter, the use to which objects are put (such as Duchamp's urinal in the art gallery, or safety pins as earrings) can create or transform an aesthetic impact.

Among the contexts of interpretation are concepts of genre, which can seemingly override other concerns. There is little doubt that the worst-written novel is more likely to be identified as an artwork (albeit a bad one) than the best-written political speech. But if the speech were an extract from a novel the aesthetic evaluation would be otherwise. A key difference is fictionality, and arguably fictionality is essential to all art, including music, photography, and dance. Fictionality as either an ontological or an epistemological status is not well understood or even widely recognised within CR, since it is neither knowledge nor not-knowledge (but see ONT/DE-ONT re the ontological category). It partly involves 'framing' devices, or even literal frames, which establish the artwork as an open TOTALITY that, unlike most communicative acts, does not necessarily refer in any direct manner to states of the world. The frame is often explicit, but it can also be part of the work's structure, implied by context, or even established by the viewer's 'frame of mind'. Among other things, framing often foregrounds an artwork's metaphoric possibilities, emotional force and sensuous qualities, which return us to the etymological meaning of 'aesthetic'.

Concepts of genre are connected to one of the messiest areas of aesthetics: its intersections with philosophies of science. Art has sometimes drawn upon scientific (and 'scientific') developments: for example, perspective painting developed from the study of optics, and pointillism had a similar background. Moreover, art has always borne complicated relationships with the idea of truth. But when aesthetics emerged as a branch of philosophy it did so primarily in relation to philosophy of science. One of the most explicit connections was in aesthetic naturalism, in which positivism was applied to artistic principles. The principal idea was that art should display the effects of heredity and environment (both conceived in highly mechanistic ways), but the main

18 agency

effect was to elaborate verisimilitude down to bric-à-brac and kitchen sinks. The term 'naturalism' gave way to 'realism', a conflation that desperately confused the meaning of both. Notably, the leftwing playwright Bertolt Brecht (1964) rejected the conflation of realism into positivism, and argued for a realism that specifically illuminated social relations (that is, a realism in the CR vein) and could readily dispose of verisimilitude in order to do so. Brecht's ideas have guided and inspired many artists and cultural theorists on the left. For the most part, however, mainstream theatre has continued in the tradition of naturalism (or its closely related successor, psychological realism), and non-realist approaches have dominated painting and sculpture since the early twentieth century, such as in cubism and abstract expressionism. See also SIGNS.

TOBIN NELLHAUS

agency (f. L. *agere*, to act, do, set in motion). *Agency* is synonymous with CAUSALITY, whether in its 1M, 2E, 3L or 4D mode (cf. *PN3*: 113). However, it is usually reserved for the geo-historically most recently EMERGENT form of causality: human *intentional causality* or *agency* or *transformative praxis* or *agency*, the bearers of which are *agents*; while the home of intentional agency is at 4D, it also has 1M–3L modes, as displayed in Table 2. Since to cause is to absent, agency is also synonymous with the fourfold polysemy of ABSENCE, with its 4D form corresponding to absence qua geo-historically constituted (processes-in-)product-in-process. *Absenting absentive agency* or *transformative negating praxis* (other synonyms for the 4D concept) highlight this – agency always gets rid of (absents or transformatively negates) an absence or presence (the given) (*P*: 58; *D*: 305), and *agentive agency* (yet another synonym) underlines that intentional causality, if it is not DE-AGENTIFIED by the machinery of power$_2$ relations, is capable of adjusting the world to its will to promote flourishing via the exercise of the transformative POWER$_1$ that is analytic to it, effecting a coherence of theory and practice in practice (see DIALECTICAL REASON). Human agency is thus (more fully) *embodied intentional causal absenting* (*D*: 198, e.a.). Analytically, it has two key aspects: *causal intervention* in the world that brings about a state of affairs that would not otherwise have obtained, entailing material embodiment of the agent and the openness of the world; and *reflexive monitoring* of that intervention (*intentionality*), entailing a psychology, possession of language, etc. It is important to note that the first is 'both logically and temporally (both phylogenetically and ontogenetically) prior to the second' (*PN3*: 81) – a fact that has important extra-discoursal implications, elaborated by Archer, for the ontogeny of human agency and SUBJECTIVITY. Archer demonstrates that the properties and powers of agency are neither pre-given nor socially bestowed, but realised through (emergent from) our practical transactions and relations with our natural, practical and social environment. As such they have 'relative autonomy from biology and society alike, and causal powers to modify both of them' (2000: 87; see also 2003b).

Intentionality is the differentia specifica of praxis – the specifically human synchronic emergent power of matter, 'not just, like the other higher order animals, to initiate changes in a purposeful way, to monitor and control their performances, but to monitor the monitoring of these performances and to be capable of a commentary upon them', including retrospective and anticipatory commentary (*PN3*: 35), hence to increase such power. The specificity to humans of this capacity for *second-order monitoring* or *REFLEXIVITY* (see also CONSCIOUSNESS) is not defeated by the demonstration that other higher animals make tools, deploy sign systems, etc.; human animals make tools to make tools, sign systems to monitor sign systems, etc., recursively. Here, as elsewhere in an emergent pluriverse, there is distinction as well as connection. The reflexive deliberations or *internal conversations* of agents have been theorised by Archer (2003b: ch. 4) as the process whereby agency MEDIATES SOCIAL STRUCTURE. Intentionality (and *intentional DISPOSITIONS* – belief, affect, hope, DESIRE, WANT, etc., including preconscious [implicit, tacit] and unconscious ones) is necessarily both *about* something and oriented to the achievement of a *goal* (*TELEOLOGICAL*), i.e., it possesses a dual epistemo-practical character, presupposing at once objects and objectives (*object/ives*, including appraisal of beliefs for their suitability for acting upon – the standpoint of the AXIOLOGICAL IMPERATIVE – entailing morality as an emergent property of the world), hence REFERENTIAL DETACHMENT and the transitive dimension, and within that the INTRINSIC ASPECT (IA), which just is the intentional aspect of human agency, constellationally contained within (as part of) its extrinsic or causal aspect (EA); for it also entails that reasons that are acted upon are causes (see MIND) and the wider category of transformative agency itself – 'a species-specific ineliminable *fact*' (*D*: 210; cf. Archer 2000: 1). An *intention* is the manifestation of intentionality at the level of the concretely singular person. *Disintentionality* results from DE-AGENTIFICATION. *Split* or divided or ALIENATED *intentionality* is intentionality that is unfulfilled as a result of focusing, not on the action itself (the doing), but on its result (having and getting). *Fulfilled intentionality* (PMR) is the positive side of FREEDOM – the fulfilment of needs and potentials in aesthetic enjoyment, normally referred to as self-realisation in CR.

Acts (or *praxes*) are the specific outcomes of the exercise of the power of intentional agency, the human counterpart of EVENTS in the non-human world: things that we do, as distinct from things that happen to us or just happen (events) (cf. *PN3*: 82). They are distinguished from events by being caused by reasons; or, just as agency is a specific kind of causality, so acts are a specific kind of event. *Actions* are types (e.g., making a cup of coffee) of which acts are TOKENS (making this cup of coffee). *Basic acts* are a logical precondition for other forms of action, unmediated by any prior thought, or immediate – things that we just do, spontaneously – which PMR links to the non-dual meta-Real in the concept of SPONTANEOUS RIGHT-ACTION (e.g., *RM*: 98, 100; see also MEDIATION, PERCEPTION). *Action* (intentional action) is the PROCESS of acting generically considered, the particular form of which is *activity*. Its characteristic structure is 'present absence → orientation to future → grounding in the presence of the past → praxis' (*D*: 255; see also TENSE). It has five 'componential springs' or bases, which are constellationally contained in the concept of *reasons for acting*: the *cognitive domain* (2E), providing discursive knowledge *that* (beliefs), driven partly by the basic instinct of curiosity but also by desire to meet needs and absent constraints on flourishing; the *affective domain* (mood) (3L), the seat of the moral imagination and optation (e.g., hope), issuing in values and sentiment; the elemental *conative domain* (1M), the home of absence in the form of desire, including WILL, and of motivations and drives, producing wants; the *expressive domain* (4D), the locus of speech action, style, etc.; and the *performative domain* (4D), the arena of practical action, ERGONIC efficiency and PHRONESIS. For any successful action to occur, these must be supplemented by *competencies* (practical knowledge how, including tacit knowledge) and *facilities* (access to resources). More fundamentally, undergirding all five bases of action is an 'ego-syntonic or

Table 2 Transformative agency, with some correspondences

Causal–Axiological Chain	1M Non-Identity	2E Negativity	3L Totality	4D Transformative Agency (human concrete singularity)
Concrete universal ↔ singular (UPMS)	universality	processuality	mediation	(concrete) singularity (not just human)
Polysemy of absence	product	process (incl. subject- or developmental negation)	process-in-product	product-in-process
Causal modes of absence	transfactual causality	rhythmic causality	holistic causality	intentional causality
Transformative agency	stratification of mind (consciousness, self-consciousness, pre-conscious, unconscious)	reasons for acting flow of action/praxis	inter-/intra-action within and among persons	embodied intentional casualty (exercised action/praxis–poiesis
Components of action	will	thought/unthought	emotion	making (action and objectification)
Elements of the creative process	*raw materials*, incl. relational, semiosic (material causality)	*transcendence* (epistemological, causal efficient causality)	*formation*, shaping (formal causality)	*objectification* (final causality)
Domains of action ('componential springs')	conative	cognitive	affective, existential security underpins 1M–4D	expressive, performative system (based in trust)
Transcendental morphogenesis of action†	ground (implicit potential)	emergence (creativity)	identification, unification (love, solidarity)	agency (capacity for right-action)

Non-dual components of action (modes of transcendence)[†]	*transcendental consciousness of real self* (loss of ego)	*transcendental identification in consciousness* (away from subjectivity)	*transcendental teamwork or holism*	*transcendental agency* (absorption in activity)	
Basic human capacities	freedom	creativity	love	right-action[†]	
Modes of radical negation	auto-subversion	self-transformation	self-realisation	self-overcoming	
Components of human nature	(1) core universal	(4) changing	(2) historically specific	(3) unique individuality	
Four-planar social being or human nature (social tetrapolity)	d. (intra-subjectivity (the stratified person)	c. social relations (concept-dependent, site of social oppositionality)	b. inter-/intra-subjective (personal) relations – transactions with ourselves and others	a. material transactions with nature (making)	
Social cube		(1) power$_2$, (2) discursive, (3) normative relations intersecting in (4) ideology	(1) power$_2$, (2) communicative, (3) moral relations illusion of ego (ideological)		
Subjectivity/self	transcendentally real self or alethic self	ongoing change (geo-historical constitution)	self-consciousness or reflexive monitoring, incl. internal conversation; internal relations with others	embodied self, site of intentional causality	
An embodied self (person)	core universal human *nature*	*rhythmics* of her world-line	particular *mediations*	unique *concretely singular person*	
Qualities of the self	psychic	mental	emotional	physical	

[†]PMR concepts, implicit in CR

agency

ego-emancipatory *existential security system*' (3L), founded on relations of TRUST (*D*: 164–6; see also PRIMARY POLYADISATION). When we distinguish real reasons (which may be conscious or unconscious) for acting from reasons that are mere rationalisations or dissimulations (entailing self-deception), and include the conditions (which may be unacknowledged or unknown) and consequences (unintended as well as intended) of action, it can be seen that action is *stratified* (*D*: 164–5). While it is analytical to the concept of action that we could have acted otherwise and that it is accountable, not all actions are free in the sense of contributing to the realisation our real interests, 'which means knowing, acting on and bringing about a state of affairs satisfying them' (*PN3*: 114) – the possibility of which is, however, implicit in intentional causality, which promises a double freedom: '*from* instinctual determination and *to* produce in a planned, premeditated way' (*D*: 94, citing Marxian practical materialism).

As indicated above, and depicted in Figure 2, transformative agency or intentional causality (which is also stratified) is synonymous with *praxis*, understood broadly as characteristically involving making, doing and saying, whether as a capacity or a process. In Plato and Aristotle *praxis* is action undertaken for its own sake (doing). It is contrasted with *poiesis*, adj. *poietical* (Gr. *poiēsis*), production or creation (making), an activity which results in a product, characteristic of crafts; *poiesology* is the study of poiesis. Excellence in poiesis requires skill (*technē*), in praxis virtue (*aretē*). In Bhaskar, as in Marx, *praxis* (pl. *praxes*) is the *wider* concept, constellationally embracing all the different kinds of poiesis (aligned with creative work, labour, production) in addition to other forms of distinctively human activity or *practice* (which stands to praxis as activity to action); or praxis–poiesis constitute a constellational identity (a DUALITY). In its broadest sense, transformative agency or praxis is the specifically human capacity for world-change or -building and self-change, centrally involving imagination and reason (practical as well as theoretical), first constituted by absence in the form of desire and the NEED to produce the necessities of life (cf. Dean 2003), presupposing a grasp of intransitivity and transfactual efficacy. In recognising ourselves in the products of our praxis (OBJECTIVATIONS), we arrive at a practical solution to the PROBLEM of OBJECTIVITY/SUBJECTIVITY, prefiguring the possibility of thoroughgoing coherence of theory and practice in practice; ontologically, agency/subjectivity is constellationally contained within an overarching objectivity (Figure 2, top left), which the agent's own ontogenetic stratification, permitting a META-REFLEXIVELY TOTALISING SELF-SITUATION, enables her to see. Transitively considered, praxis should be thought of as the always already 'flow of intentional agency' (*D*: 165) entailed by the AXIOLOGICAL imperative, i.e. as the PROCESS of action, as distinct from specific acts/actions and the *social practices* within which acts/actions occur. Since it always requires a material cause, no form of praxis, including poiesis, exclusively 'makes' (creates); rather, it produces something new in reproducing and/or transforming the given (see CREATIVITY, where the elements of the creative process are elaborated, TMSA). Poiesis is, however, sometimes employed to refer to this moment of creativity in the reproduction/transformation of socio-psychological forms within praxis rather than to production or work in general. Since such forms are a precondition for praxis, we can speak of '*the poiesis of praxis*, the *making of doing*, including the recursive remaking or undoing of the structured transcendental conditions of any intentional doing at all' (*D*: 97). From a transcendental point of view, then, poiesis is the *deeper* concept.

agency 23

```
Cosmos [biosphere [nature [society [inter-subjectivity [intra-subjectivity
                          |
                    Process (action)
                          |                      {Illusory ego
            Agency    Subjectivity (self)        {Embodied self (person)
                          |                      {transcendental self
          Meta-reflexively totalising (self-)situation
                          |                          Self-narratival capacity
              Reflexive monitoring of life situation ↗
                          |                                    Accountability
        Reflexive monitoring of routinised activities along space–time paths ↗
                          |
              Conscious engagement in a space practice
                          |              ↗ Acts performed in or by it
                   Praxis–Poiesis ↗
                          =
                 Intentional causal agency
                          ↑
                Conscious/self-consciousness
                          |
                     Preconscious
                     _____
                     Unconscious
                     _____
                  Biological sub-stratum
```

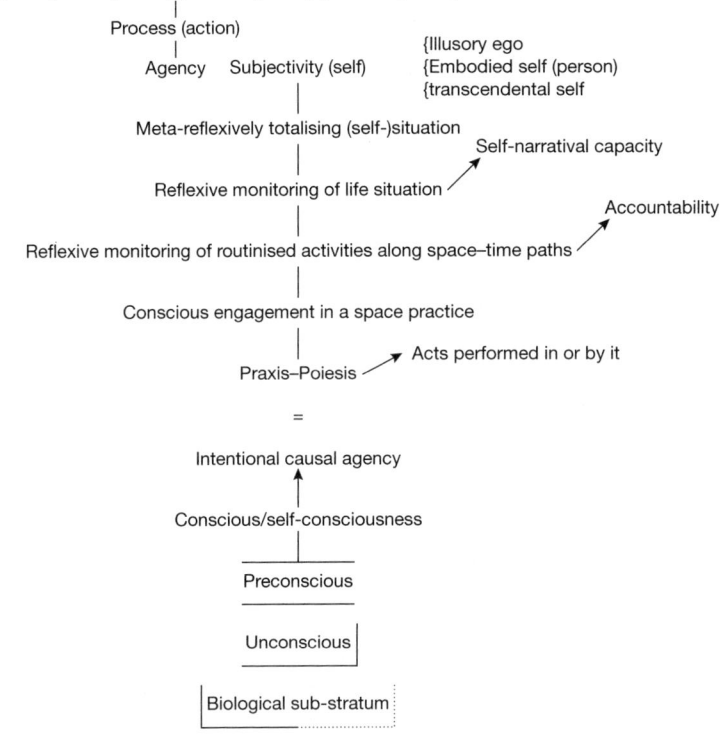

Figure 2 The stratification of agency: a moment in a person's life

Source *D*: 167, fig. 2.30 (slightly modified), which is substantially similar to fig. 2.20, 'A Stratified Model of the Self' (*D*: 149); it is reproduced at *P*: 98, where it is given the title 'Stratification of the Subject'. None the less, the figure implicitly recognises a distinction between agency and the subject/self, which PMR explicitly draws out in distinguishing between *outer action* (agency – doing things in the world) and *inner action* (work on the self) – to argue the primacy of self-change (SELF-REFERENTIALITY) in social change (*RM*: 101).

While 4D intentional agency comes at the 'end' of the ontological–axiological chain, such that 4D > 3L > 2E > 1M, there is a sense in which it is 'already there at the outset in the phenomenologicality of science' and as the supplier of the major premise from which the whole system flows, so that 'the end is implicit in the beginning'. However, 'if we go along with this rather Hegelian way of speaking, we must see the agency as a *radically transformed transformative praxis*, oriented to rationally groundable projects – ultimately flourishing in freedom' (*D*: 9). This radical development of the concept of agency, bearer of the pulse of freedom, is caught in the concepts of *dialectical praxis* (dφ) (practically oriented transformative negation of contradictions) and *emancipatory praxis*, which in its deepest sense is synonymous with *transformed, transformative, trustworthy, totalising, transformist, transitional praxis* or EMANCIPATORY AXIOLOGY (TTTTφ or dφ'). *Transcendental agency*, in which we are totally absorbed in the deployment of our agency in some task, is one of the five main forms of TRANSCENDENCE thematised in the philosophy of META-REALITY.

24 agency, dialectic of

A *social practice* is an ongoing (discursive or non-discursive) way of doing things incorporating *positions* occupied by 'functionaries' (Archerian 'actors'), which is irreducible to the activities of the people who engage in it; it is 'so to speak, the structure at work in practice' (*SR*: 129). The *position–practice system* mediates the DUALITY of praxis (conscious work or production, unconscious reproduction/transformation of social forms): 'engagement in a social activity is itself a conscious human action which may, in general, be described either in terms of the agent's reason for engaging in it or in terms of its social function or role. When praxis is seen under the aspect of process, human choice becomes functional necessity' (*PN3*: 35). *Primacy of practice*: see PRIMACY; SUBJECTIVITY. See also AGENTIVE REALISM; CONSCIOUSNESS; DETERMINISM (psychic ubiquity determinism); PROBLEM OF AGENCY.

agency, dialectic of. See EMANCIPATORY AXIOLOGY.

agency, problem of. See PROBLEM OF AGENCY.

agency and structure. See SOCIAL CONSTRUCTIONISM; SOCIAL STRUCTURE; TMSA.

agentification is the process whereby human agency is empowered to fulfil its potential, more fundamentally whereby human transformative praxis (4D) appears on the geo-historical scene as an emergent reality. *De-agentification* is the reverse process, normally taking the form in post/modernity of dualistic disembodiment and/or reductionist REIFICATION of human agency. Disembodiment/reification *de-agentifies* (detotalises) *reality* by 'dissolving' the emergent level of 4D such that there is no specifically intentional embodied causality, only the mechanistic causality of empirical realism, on the one hand, and the disembodied intentional non-causality of reasons and meanings on the other, constituting a dichotomy or DUALISM of causes vs. reasons (see CRITICAL NATURALISM; DIALECTICAL ANTAGONISTS). De-agentification manifests itself socio-substantively 'in the enervation or fragmentation of agents or groups, impotent empty selves, fissiparousness and alienating retrojective, introjective, projective, etc. modes of identification (e.g., in a fantasy world)' (*DG*: 396). Fissiparousness produces new individuals, in this case multiple selves, by FISSION. De-agentification is the human form of desingularisation. *Re-agentification* reverses the REVERSAL accomplished by de-agentification. See also CONCRETE/ABSTRACT; CONSTRAINT/ENABLEMENT; EMANCIPATORY AXIOLOGY; PROBLEM OF AGENCY; VOLUNTARISM.

agentive agency. See AGENCY.

agentive realism upholds the sui generis reality of INTENTIONAL CAUSALITY. See also AGENCY; MIND; REALISM.

agent-specificity. Another name for *ACTIONABILITY*: the goal of any praxis must be capable of being accomplished by *these* agents. It highlights that emancipatory praxis is *self*-emancipation (*D*: 171).

agonistic (f. Gr. *agōnistikos*, fit for or pertaining to contending, especially in public games, striving, struggling or fighting). Used by Bhaskar to denote an irreducible competitive componental spring of human action, important for flourishing but potentially baneful in its consequences (see ESTEEM; EUDAIMONIA; POLYAD; RECOGNITION; WANTS).

alethia (the Anglicised form of Gr. *alētheia*, truth, f. *a-*, not, + *lēthē*, (1) forgetting or forgetfulness, (2) *Lēthē*, the river of oblivion in the lower world; hence literally the undoing of oblivion). The TRUTH of things as distinct from propositions. For Bhaskar

alethia 25

'{a} *proposition* is true if and only if the state of affairs that it expresses (describes) is real' (*RS2*: 249, e.a.). This early formula for the meaning of a truth claim is never resiled from, though it is elaborated. It is equivalent in form to standard definitions of the correspondence theory of truth, *except that* 'expresses' takes the place of 'corresponds to'; it is therefore dubbed the *expressive theory of truth* (*SR*: 100), a label which, we can see in retrospect, picks out the relevant moment (the *expressive-referential*) of the more comprehensive later theory of the TRUTH TETRAPOLITY, incorporating a notion of truth as ontological and *alethic*.

Bhaskar rejects that the truth of a claim means 'correspondence' in the sense of likeness, i.e. except 'as a metaphor' (*SR*: 100), on two grounds. First, sentences (theories, etc.) and the world (except in so far as the latter consists of sentences) are radically different, such that there is no way they could resemble each other, or, if they did, of knowing that they did – 'what concept or ideational object could correspond to, say, a laser'?; to suppose otherwise is a form of anthropomorphism or IDENTITY-THINKING, closely bound up with EMPIRICISM (*PN3*: 142, 165 n65; *SR*: 254, cf. Sayer 2000a: 42). To distinguish the meaning of truth from its justification does not defeat the force of this objection, which is that any correspondence theory of truth carrying a connotation of likeness between concepts and things ('truthlikeness') must be otiose. If truth *means* correspondence in this sense, no proposition asserting correspondence can ever be *justified*. Second, EPISTEMIC RELATIVITY entails that correspondence as likeness is not just nugatory but false. Since the epistemic frameworks within which we view the world necessarily change, the same object may be (and often has been) known under different descriptions, and descriptions and objects may change independently of each other. The often radically different descriptions of a sublated theory cannot have corresponded if those of the successor theory do. The same then applies to successor theories, generating infinite regress. Or correspondence as likeness (cf. its defence in terms of 'established' meanings of the concept of truth) has false FIXIST implications. The expressive theory encounters no such problems, because in theory-change the entities of the predecessor theory readily find new expression within the successor theoretical framework. Some correspondence theorists nowadays maintain that 'correspondence' carries no connotation of likeness (cf. Collier 1994: 239–40); in that case, they can have no real objection to the notion that truth claims are expressive-referential.

The view has been put, from a correspondence perspective, that the early Bhaskar is a consensus theorist (Groff 2004: 19). However, as the above suggests, the expressive theory is arguably no more a consensus theory than the correspondence theory itself, and it is certainly not promoted as such. That scientists on the Bhaskarian account achieve their immediate objective when they referentially detach a deeper level of structure which is explanatory of a higher level does not entail that 'for a claim to be true just *is* for it to be judged to be true', as consensus theorists hold (see TRUTH). This, for Bhaskar, is its *justification*, which on epistemic relativity *must* always be intrinsic to the science concerned, but may include a level of necessary (analytic a posteriori) knowledge when a thing's dispositional properties are deduced from a statement of its nature (*RS2*: 19, et passim). (This does not preclude philosophers from reflecting on criteria of justification, including arriving at the conclusion that standards are intrinsic, nor from integrating an account of justification into an overall theory of truth.) The *meaning* of a truth claim – that it expresses/refers to a real state of affairs – is unaffected by this. This is in no way inconsistent with the theory of the truth tetrapolity – which embraces it – nor of course with what is at the heart of the notion of correspondence: that it is real states of affairs that make *propositions* true (one does not have to be a correspondence theorist, except metaphorically, to hold this). Both correspondence and the expressive-referential theory, whilst capturing the meaning of truth for any depth realism, seem deficient in that they hardly begin to provide a comprehensive theory of truth (which the truth tetrapolity aspires to provide), viewing it as a mere regulative ideal and saying very little about what it is for a proposition to correspond to or express the way the world is.

Turning to theory of the truth tetrapolity, its four components are by no means 'simply a list' or 'aggregation' of different meanings of the concept of truth, but dialectically interrelated moments in the overall DCR theory of being–becoming, as Table 3 exemplifies.

Therefore, normative-fiduciary > adequating > expressive-referential > ontological and alethic truth (*alethia*); or, inter alia, alethia (the focus of this entry) is 'a *condition* of fiduciary, adequating {and/}or expressive truth' (*D*: 385, e.a.). It is argued not to be an option: *alethic realism* – which 'opens up the possibility of *constitutive falsity*' as well as truth (*D*: 174) – like ontology as such, is inexorable. If we deny the reality of alethic truth and falsity, we inevitably engender and get ensnared in TINA FORMATIONS, i.e., epistemological and ontological *untruth* (see also IDEOLOGY; DEMI-REALITY) and disemancipation. Alethia exemplifies FREEDOM or 'the absence of heterology; it is *true* to, for, in and of itself'; and provides the key for resolving 1M PROBLEMS of philosophy and 'the non-arbitrary principle of stratification that powers the dialectic of scientific discovery' (*D*: 219, e.a.). Bhaskar rates it 'the second great discovery' made in *Dialectic* after an adequate concept of ABSENCE (*D*: 200). The basic idea is very simple. Truth is not primarily propositional (cf. Hegel). Humans do not bring truth to the world (cf. the Kantian CATEGORIES), rather the world imposes its truth on us (cf. the CR categories). Contingently true knowledge of the world is not only possible, but also an AXIOLOGICAL necessity (cf. *RM*: 49; Dean et al. 2005: 8).

To claim – as its critics have sometimes implied – that alethia *is* (i.e., is identical to) Being or the ID or the generative mechanisms (GMs) and causal powers that produce the phenomena of the world seems a FUSIONIST distortion. Truth, unlike being, GMs, etc., per se, '*is always tied to the possibility of language, theory and human practice*'; in its ontological use it stands to GM in roughly the way that 'object of knowledge' does (*SR*: 99–100). (1.1) It *does* embrace GMs and indeed all the things of the world, but *in so far as* they are epistemologically mediated or REFERENTIALLY DETACHED. This is *truth as ontological*. The ontological truth of the proposition 'there is a cat on the mat' is the referentially detached cat-being on the mat (see REFERENCE); likewise, the truth of 'x has a certain structure' is the detached structure. In the terminology of Alston (1996) and others, the proposition is a *truth-bearer*, the cat and the structure are *truth-makers* in relation to the proposition (and not, as Bhaskar's critics would have it, truth-bearers). Note that this is in relation to propositional truth only; at the level of the truth of things it makes no sense to speak of truth-bearers and truth-makers, only of truths. (1.2) *Alethic truth* then pertains to that which causally explains referentially detached things, GMs, etc. – in the case of GMs, the referentially detached GMs that explain *them*; it thus presupposes ontological truth, but not vice versa (we can referentially detach objects,

Table 3 Moments of truth and untruth

Ontological–Axiological Chain	1M Non-Identity	2E Negativity	3L Totality	4D Transformative Agency
Concrete universal ↔ singular	universality	processuality	mediation	concrete singularity
Judgement form, incl. truth	evidential	descriptive	imperatival-fiduciary	expressively veracious
Theory of alethic truth (truth tetrapolity) – truth as	(4) ontological, alethic (ID)	(2) adequating (warrantedly assertible) or epistemic (2E and 1M) (TD)	(1) normative-fiduciary (IA)	(3) expressive-referential (4D and 1M) (TD/ID)
Modes of alethic truth	ontological stratification, natural necessity, reality principle	praxis-dependent	totalising (oriented to maximising explanatory power)	contextualised by the dialectic of the singular science concerned
Qualities of truth	grounded	dynamic	totalising	context-sensitive
Theory of moral alethia – truth as	human nature, our transcendentally real selves	dialectical reason	universalisability	freedom: free flourishing of each and all
Ethical tetrapolity, logic of	(4) freedom qua universal human emancipation, the realisation of moral truth (alethia)	(2) explanatory critical theory complex	(1) fiduciariness	(3) totalising depth-praxis
Modes of falsity	(3) *of* an object or being *to* its essential nature (ID)	(1) *about* an object or being (at any one level of reality) (TD)	untrustworthiness	(2) *in* an object or being (at that level of reality) (TD/ID)

28 alethia

including structures, without knowing why they exist) (see *P*: 26 n). Alethia is thus identical to the REALITY PRINCIPLE or epistemologically mediated axiological (natural) necessity; it is this that generates Tina formations when contravened. Axiological necessity = the reality principle = alethic truth, which is also synonymous with 'real DEFINITION', which captures the essential structure of a thing. While truths are not things or properties of things, then, epistemologically mediated things are (ontological) truths, and the epistemologically mediated GMs *of* things (including other GMs) are alethic truths (cf. *SR*: 100). As such, alethic truths are the real or DIALECTICAL REASONS, grounds or causes of things – that which explains them, their *objective truth* in so far as human beings have encountered or apprehended it; they are by no means restricted to science, but are encountered in our daily lives at every level, since the dawn of geo-history. The process of their discovery constitutes the *dialectic of truth*. This has the important consequence that *alethia is dynamic and developing* – not only do structures (especially social structures) change, so also does our capacity to referentially detach them. (1.3) In virtue of multi-tiered ontological stratification, alethic truth is also the truth of truth, the alethia of alethia, *recursively*. This takes any EPISTEMOLOGICAL DIALECTIC to new explanatory levels; alethic truth may thus be said to be the *stratified form of referential detachment*. For example, that water tends to boil when heated can be explained by its electronic configuration. This is alethic truth. When we can in turn explain the configuration of the electrons, Bhaskar reasons, there can be no further practical doubt – no warrant for inductive scepticism – that water does indeed tend to boil in virtue of its electronic configuration. This is the resolution of the PROBLEM OF INDUCTION, the alethic truth of alethic truth in its scientific manifestation. The recursivity of the process of endlessly capturing deeper truths of truths is captured in the figure of the *alethic ellipse* (*P*: 26 n).

Generalising (1.2) and (1.3), alethia is the epistemologically mediated 'real reason or dialectical grounds for the phenomena of the world' (*P*: 163). That science and other practices discover objective truths is in no wise incompatible with epistemic relativity: in theory-change, the theoretical framework is transformed – negated – but genuine discoveries are preserved. It is therefore not SCIENTISTIC in the standard sense; nor is it science-centred, for it recognises other ways of knowing besides science. Nor is it incompatible with FALLIBILISM in its CR (non-Popperian) meaning, nor conducive to ABSOLUTISM: we do not thereby jump out of geo-history; we know that our theories will be revised and that we know 'something but not how much' of the potentially infinite totality of the world. On a more lateral as well as vertical scale, alethic truth is ontologically *deep truth*, grounded in DEEP STRUCTURE, and the deep truths of those truths, often highly abstract. Thus the alethia of global warming is arguably ultimately 'the degradation of being into having' (Debord), and more immediately the causal dynamics of the capitalist SYSTEM. The alethia of dialectic is the 'absenting of constraints on absenting absences', i.e., dialectic *is* because such a process is really operative in geo-history.

(1.4) At an even more abstract level, PHILOSOPHY can establish contingently necessary constitutive truths of the world, i.e., the general contours of ontology. This is CATEGORIAL and DISPOSITIONAL REALISM and, generically, alethic truth as *verification-* or *recognition-transcendent*, i.e., truth that exists independently of its verification by science but is necessarily presupposed by it (see MODALITY; TRANSCENDENTAL

REALISM; Norris 2005). It is still epistemologically mediated, but in the transitive process of philosophy, rather than of science as such.

(2) In addition to epistemologically mediated causal relations, alethic truth embraces relations between concepts and concept-dependent things, including at the limit 'transworld necessary truths' such as those established by logic or mathematics which apply to any conceivable world (see TRANSCENDENTAL REALISM). In the world as we know it, conceptual or concept-dependent alethic truths may be the causally efficacious truths of false things (e.g., of a refuted theory), and the *alethic truth of falsity* may itself be false (e.g., IRREALISM is the alethic truth of the false dichotomy between [subjective] idealism and [reductive] materialism); this is *constitutive* or *alethic falsity*, entailing a concept of *emergent falsity* (e.g., *P.* 133, 243; see also CRITIQUE; DEMI-REALITY; IDEOLOGY). Alethia also embraces ethical (practical, action-guiding) relations. Morality as ontological is actually existing morality, referentially detached. *Alethic* or *objective morality* is established by the transition from facts to values, underpinned by the unity of theory and practice (see EMANCIPATORY AXIOLOGY; EXPLANATORY CRITIQUE). The ultimate *moral* or *ethical alethia* of the human species is universal freedom. This is not a causal claim immediately, but a claim that such a goal (*alethic morality*) logically is implicit at the level of the real in human discourse and practice (see EMANCIPATORY AXIOLOGY). However, our NEED for freedom is a form of axiological (natural) necessity, and tendential DIRECTIONALITY towards such a goal is a causal process (since reasons, norms, etc., are causes) whereby directionality is inscribed in the struggle for freedom by the REALITY PRINCIPLE, i.e., alethic truth. *PMR* thematises alethic morality as also subject-specific and concretely singular (an aspect of the *ground-state* or *dharma*, the ultimate categorial structure of the human being).

It is not clear in the early work of Bhaskar how it is possible for science to come to express (aspects of) the truth of things given the somewhat stark subject–object non-identity there argued for (and reified by some commentators). *Dialectic* arguably makes good this deficit via the EPISTEMOLOGICAL DIALECTIC (which elaborates and finesses the earlier 'logic of scientific discovery') and the concept of the constellational identity of subject and object within subjectivity within an overarching objectivity, such that ideas themselves are emergent parts of the natural world (see OBJECTIVITY/SUBJECTIVITY). The philosophy of META-REALITY elaborates the underlying identity thesis and explicitly addresses the subjective condition for arriving at truth via the (conjectural) theories of generalised CO-PRESENCE and *Platonic anamnesis*: we can come to know truths of the world ultimately because the world was formed by the same powers that constituted us, such that in making fundamentally new discoveries we 'recall' what is already within us, thereby achieving a point of identity or union with what we have discovered. This echoes the etymology of alethia as the undoing of oblivion. For PMR, truth is 'a more basic conception than reality' because, unlike reality and realism, it is 'not necessarily implicated in a dualistic mode or structure of thought' and speaks of a level of truth which 'cannot be conceptualised in normal realistic, that is dualistic, terms – in this sense beyond reality' (*MR*: 50–1). This connects with the conception of truth as freedom or non-heterology already thematised in *Dialectic* and with the idea that there may be truths in principle unknowable to us (see TRANSCENDENTAL REALISM).

30 alethic ellipse

There is no evidence that the theory of alethic truth explicitly borrows from Heidegger, whose theory of truth as *alētheia* is very different, lacking as it does a realist, stratified ontology. Nor is there any necessity that it lead 'all the way to God' (Potter 2006). That is an issue it leaves open. It is in fact the absence of concepts of ontological stratification and alethic truth, as Bhaskar has consistently maintained, that leads to philosophical FIDEISM, FUNDAMENTALISM and other-worldliness. See also Norris (2005).

alethic ellipse. See ALETHIA.
alethic falsity. See ALETHIA.
alethic modalities. See MODALITY.
alethic morality. See ALETHIA; ETHICS.
alethic realism. See ALETHIA; REALISM.
alethic reason. See ALETHIA; DIALECTICAL REASON.
alethic self. The alethic self is the *transcendentally real self*, which is the deep content of praxis, linked in PMR to the ground-state. See ALIENATION; CONCRETE UTOPIANISM; EMANCIPATORY AXIOLOGY; IMMANENCE/TRANSCENDENCE; META-REALITY; SELF-REFERENTIALITY; SUBJECTIVITY.
alethic truth. See ALETHIA; ETHICS; TRUTH.
alienation (f. L. *alienare*, to take away, transfer property, sell, banish from the mind, estrange, f. *alius*, another, other) in ordinary language has all the connotations of its Latin progeniture. Although its philosophical use has roots in Judaeo-Christian and Greek thought, its explicit elaboration as a philosophical concept comes to us from Hegel via Marx (see BEAUTIFUL SOUL, ETC.; GENERATIVE SEPARATION; HEGEL–MARX CRITIQUE). In Hegel, it pertains to absolute spirit – the 'logical process {. . .} which actualises itself {in nature} by *alienating*, or becoming other than, itself and which restores its self-unity by recognising this alienation as nothing other than its own free expression or manifestation' (*D*: 17, e.a.). Marx criticised this conception for its inability to sustain the objectivity of nature and being generally, conceived as independently real and irreducible to thought (thereby anticipating the TD/ID distinction), substituted a transformed concept of 'labour' or 'praxis' (anticipative of the TMSA) for 'spirit', thematised 'alienation' (which separates us from possibilities for the flourishing of our essential human powers) as a distinct form of the 'objectification' of praxis (in which, if it is non-alienated, we can recognise our essential humanity), and demonstrated the reality of the fivefold alienation (see below) and exploitation of labour in class societies in general, and specifically in the Hegelian civil society and state. Both Marx and Hegel were profoundly influenced by 'a deep rooted theme in Judaic/Christian/neo-Platonic thought' which 'postulates an original undifferentiated unity {of human being and being, theory and practice, praxis and product, fact and value}, geo-historical diremption or diaspora and an eventual return to a non-alienated but differentiated self or unity-in-diversity'. In Marx such a conception remains as 'the counterfactual limits or poles implied by the SYSTEMATIC DIALECTICS of the commodity form' (the human need for non-alienation may set *limits to capital*), and it informs his theory of an original generative separation and the possibility of an unalienated society of free flourishing (*D*: 101–2; see also PROCESS).

Bhaskar finesses and extends the Marxian theory, such that alienation and the overcoming of alienation (*de-alienation*) are for him the alpha and omega of CR as an

Table 4 Fivefold alienation, with some MELD correspondences

Causal Chain	1M Non-Identity	2E Negativity	3L Totality	4D Transformative Agency
Components of human nature	(1) core universal	(4) changing	(2) historically specific	(3) unique individuality
Four-planar social being (human nature)	d. (intra-)subjectivity (stratified persons)	c. social relations (concept-dependent, site of social oppositionality)	b. inter-/intra-subjective (personal) relations – transactions with ourselves and others	a. material transactions with nature (making)
Alienation (fivefold) from	d. ourselves (5)	c. the nexus of social relations (4) the social (i) power$_2$, (ii) discursive (iii) normative relations intersecting in (iv)	b. each other (3) cube (i) power$_2$, (ii) communicative (iii) moral relations ideology	a. the labour process: labour and its product (-in-process) (1), material and means of production (2)
Forms of alienated (split) consciousness	extra-jection (splits the world from thought about it)	retro-jection (splits geo-historical change from the present)	intro-jection (splits the psyche)	pro-jection (splits minds from bodies)
Irrealist attitudes to master–slave relations	Stoicism (indifference)	Scepticism (denegation)	Beautiful Soul (alienation)	Unhappy Consciousness (introjection/projection)
Modes of untruth	*of* an object or being to its essential nature	*about* an object or being (at any one level of reality)	untrustworthiness	*in* an object or being (at that level of reality)
Truth tetrapolity – truth as	ontological, alethic (ID)	adequating (warrantedly assertible) or epistemic (2E and 1M) (TD)	normative-fiduciary (IA)	expressive-referential (4D and 1M) (TD/ID)

alienation

emancipatory movement in philosophy, social theory and science (see also Figure 3; DIALECTIC OF DE-ALIENATION). Alienation is the condition of 'being something other than {...} oneself or {than} what is essential and intrinsic to one's nature or identity', of having been – and being – 'separated, split, torn or *estranged* from oneself'. It is thus, in the first place, a 2E figure of ABSENCE (*alienating absence* – something essential has been absented in the course of our quest to remedy absence and incompleteness), of CONSTRAINT on well-being or ill (the normatively negative form of absence) and of split or FISSION, hence of OPPOSITIONALITY and CONTRADICTION, and is closely related conceptually to heterology in two of its meanings: not the same as oneself and not true to, for or in oneself, i.e. *heteronomy* or the *contrary of FREEDOM*. What is intrinsic to oneself is not necessarily physically internal (e.g., solidarity with others), and 'what is still essentially one's own at one level (e.g., one's humanity) may be alienated at another (e.g., by being subjugated to gross indignity)', i.e. alienation may be present at the level of the actual but absent at the level of the real (see CO-PRESENCE) (*D*: 114, 188, e.a.). Alienation is thus, inter alia, a profoundly sociohistorical concept, signifying a disjunction (the possibility of which is pinpointed by the TMSA and geo-historically was implicit in the first act of REFERENTIAL DETACHMENT), not between a fixed inner real self and one's actual self, but between what one has become (essentially is and is tending to become) and what one socially is obliged to be or thwarted from becoming (cf. *PN3*: 112). It is separation from anything that is intrinsic to our well-being, a rift or gash in four-planar social being. While alienation is thus *self-alienation* or *estrangement from SELF* ('*estrangement of part or whole of one's essence from one's self*', *D*: 360, e.a.) (cf. Marx), at the level of the concretely singular person it is *not self-inflicted* except in the truistic sense that, if we all acted in a non-alienated way, it would soon cease to exist; it is, rather, a social condition – the objectified result of human social activity – into which the concretely singular person is THROWN at birth, such that agents are dominated by social products which they do not experience as their own. As already implied, it is closely bound up with TRUTH (hence freedom), or rather the absence of truth, or untruth, hence also with EVIL (emergent error) and IDEOLOGY. Thus, as illustrated in Table 4, the different kinds of alienation correspond to the moments of the truth tetrapolity, as well as to those of the ontological–axiological chain and four-planar social being, such that the denizens of master–slave-type societies (the DEMI-REAL) are alienated in all major dimensions of their existence. We humans have a fundamental DESIRE and NEED (hence right) to experience at-oneness with the natural world and our labour process (4D), each other (3L), our social relations (2E), and ultimately our essential selves, inner and outer (embracing the structural dimension of 4D–2E phenomena) (1M, see Table 5) – i.e., a need for non-alienation (or, if we are already alienated, de-alienation); yet in all these respects we are split. While it does its absenting, thwarting and diremptive work at 2E – where it is manifest in the irrealist doctrine of ONTOLOGICAL MONOVALENCE, which absents even the concept of absence itself – the home of alienation is 3L because that which is objectively whole is sundered (detotalised) by each of its forms; its short definition can thus also be expressed as absence of totality (which *is* estrangement from self). Of course, this indulges an oxymoron: alienation does not have a home: it is a real emergent conceptual, and/or practice-dependent and/or concretely singular *false social being* which human depth-praxis may serve to disemerge. The

Table 5 Forms of alienation, with some correspondences

Domains of reality	Forms of CR realism/materialism	Features of social reality	Modes of freedom/unfreedom	Modes of truth	Modes of untruth (ID: objectively false being, i.e. falsely constituted or categorised)	Forms of alienation
Empirical/Conceptual	transcendental realism/epistemological materialism	concept-dependence conceptual formations {2E}	alienation	adequating (epistemic)	(1) *about* something – false as an account *theoretical ideology* ↑↑ ↓	(1) *conceptual alienation* – from reality in thought ↑↑ ↓
Actual	judgmental rationalism practical materialism	activity-dependence social practices {4D}	unity	expressive-referential (epistemic–ontic)	(2) *in* something – false in itself *practical ideology* ↑↑ ↓	(2) *practical alienation* – from reality in social practices ↑↑ ↓
Real	ontological or metaphysical realism/materialism	stratified nature {1M, 3L} fundamental categorial structure (may be misdescribed, misapprehended)	autonomy (true for, to and of itself)	ontological/alethic	(3) *of* something – to its essential nature *underlying generative structure*	(3) *self-alienation* – from reality in ourselves: our true intrinsic nature, inner and outer: (i) our transcendentally real selves and (ii) the social-natural totality of which they are a part

Key source: *EW*: 33–9, which, however, designates (2) '*real alienation*', inviting confusion with (3).

Note. 'False' ('untrue') is shorthand for 'false, or inadequate, misleading, incomplete, etc.'; it is consistent with being true in some respects.

34 alienation

task of emancipatory philosophy and social science is to assist in this process of disemergence (de-alienation) and retotalisation, such that our social relations do not constrain and occlude but promote the free flourishing of our intrinsic potentials (see EMANCIPATORY AXIOLOGY). Just as the possibility of alienation is situated by the TMSA, so is that of its reversal, for the model 'implies a radical transformation in our idea of a non-alienating society {...} as the immaculate product of unconditioned {...} human decisions' into a conception in which 'people self-consciously transform their social conditions of existence (the social structure) so as to maximise the possibilities for the development and spontaneous exercise of their natural (species) powers' (*PN3*: 37).

Alienation from our labour and its product (1) is caught in such notions as 'working for the boss or the firm, not for yourself', but includes anything that disguises the human agency in an outcome, e.g., the REIFICATION of the constant conjunction form; it may thus involve a double alienation: from our product and the thought that it is our product (*D*: 209). The means and materials of production (2) are anything that we work on or with to produce an outcome and, more broadly, the natural world as such. Alienation from each other (3) and our social relations (4) is just what it says, and alienation from ourselves (5) is self-alienation in the sense explained above. These forms of alienation correspond to four forms of *alienated* or split *consciousness*. *Extra-jection* splits the world from thought about it, objectivity from subjectivity, moral alethia from actually existing moralities – nature from history, singular person from social whole, agency from structure, causes from reasons, theory from practice, facts from values (*de-evaluating* and disenchanting reality), essential selves from egos (e.g., the babyisation produced by the relentless huckstering of the commodity). *Retro-jection* splits geo-historical change from the present, exemplified in ETERNISATION and REIFICATION. *Intro-jection* splits the psyche, as when slaves identify with masters. *Pro-jection* splits mind from body, subjectivity from objectivity, as in the daydreaming of subjects$_2$ in a fantasy world made for slaves, or the FETISHISM of facts that are the alienated products of the fetishiser's own mind (*D*: 89).

Bhaskar, again borrowing from Marx, also distinguishes three forms of alienation (dimensions of each of the moments of fivefold alienation) corresponding to the domains of reality, forms of realism, key features of social reality and modes of freedom/ unfreedom and of truth/untruth, as displayed in Table 5.

Each form of alienation is an intransitive social product implicated in all our transitive activities, whereby we are separated from something essential to the flourishing of human being. In Marx, (1)–(3) correspond to inverted or false consciousness or ideology, inverted or false appearances or phenomenal forms, and inverted or false 'inner relations' or generative structures, respectively (see, e.g., Larrain 1979: 122–9). Any irrealist theory, premised on categorial error, exemplifies (1), e.g., the absenting or splitting-off of being from thought in DUALISM, obscuring the totality of which it is a part. Examples of (2) include the FACT-FORM and the wage-form. They are objectively misleading: their falsity is independent of the absence at (1) of a true conceptualisation of them (*EW*: 36). While the fact-form, involving spontaneously occurring objective mystification produced by 'the conceptual and cognitive structures in terms of which our apprehension of things is organised', is probably (unlike the wage-form) necessary for 'any conceivable mode of discursive

Table 6 Alienation mapped to the domains of reality, with key correspondences

Domains of Reality	Real *experiences, concepts and signs events mechanisms*	Actual *experiences, concepts and signs events*	Empirical *(the semiosic) experiences, concepts and signs*
Realms of Reality	demi-reality relative reality *absolute reality*	demi-reality *relative reality* [absolute reality]	demi-reality [relative reality] [absolute reality]
Meta-Philosophical Principle	dualism duality *identity* (non-duality, non-alienation)	dualism *duality* (without alienation) [identity]	dualism (alienation) [duality] [identity]
Ontological Principle	irrealism realism *truth*	irrealism *realism* [truth]	irrealism [realism] [truth]
Orientation to Being	evading being thinking being *being being*	evading being *thinking being* [being being]	evading being [thinking being] [being being]
Subjectivity–Objectivity relation	diremption (alienation) expressive unity *unity-in-diversity*	diremption (alienation) *expressive unity* [unity-in-diversity]	diremption (alienation) [expressive unity] [unity-in-diversity]
Modes of alienation	conceptual practical *self-alienation*	conceptual *practical* [self-alienation]	conceptual [practical] [self-alienation]
Modes of Freedom/Unfreedom	alienation unity *autonomy* (identity) (true for, to and in itself)	alienation *unity* [autonomy]	alienation [unity] [autonomy]
Modes of Truth	adequating (epistemic) expressive-referential *alethic*	adequating (epistemic) *expressive-referential* [alethic]	adequating (epistemic) [expressive-referential] [alethic]
Modes of Untruth	about something in something *of something to its essential nature*	about something *in something* [of something to its essential nature]	about something [in something] [of something to its essential nature]

Note. Upper levels presuppose and are either limited or false in relation to lower levels, which they function to occlude. Square-bracketed levels are not given in the concept of unbracketed levels but are presupposed by it.

alienation

social life', we are not fated to remain dominated by false consciousness *of* this form (*SR*: 283–5). While each form of alienation depends on and is immediately explained by a constitutive absence – of reality in thought, in social practices and in ourselves, respectively – (2) is typically the deeper explanation of (1), as well as reinforced by it in a quasi-functional manner, and is itself in turn explained by, and reinforces, (3). Thus (3) ultimately explains (2) and (1), which are dependent for their existence and power on it, and the whole constitutes a self-reinforcing interlocking chain of avidya-TINA FORMATION – e.g., by splitting us off from reality, irrealism also reinforces the fivefold alienation underpinning power$_2$ relations, which its own dichotomous nature reflects. Examples of (3) include any persons who are theory/practice inconsistent such that they are internally riven and untrue to themselves (their inner nature) (cf. *D*: 66) or who have been alienated from their potentialities (*D*: 282); and social relations (an aspect of our outer nature) that are untrue to the essential nature of human being, as any master–slave-type relations are. However, (3) is itself in turn explained by fundamental conceptual (category) mistakes sedimented in discursive ideologies premised on ignorance of our true nature (knowledge of which has been alienated in the sense of *banished* from our minds): by *alienation (absence) of self-consciousness* (of the self in consciousness), and therefore *alienation from a key distinctive feature of human being*, REFLEXIVE *self-CONSCIOUSNESS* (*EW*: 38; cf. *D*: 44). Metacritically, this takes the form of a dialectical chain:

> absence (incompleteness) → categorial error (ignorance, illusion – alienation from the self in consciousness) → contradiction (theory/practice inconsistencies) and split (generative separation, division) → dualism and alienation from what is axiologically necessary → Tina syndrome (objectified or emergent error and illusion) → multiply compounded in the demi-real. (Adapted from *EW*: 5 [key source.])

All forms of alienation thus depend upon the reality of FREE WILL (in an open, non-regularity-determined world we are free to make mistakes, and must learn by trial and error) and the objectified consequences of committing category mistakes and being untrue to ourselves. This does not, however, entail geo-historical idealism. Category mistakes and the logic of the TINA SYNDROME that they entrain are necessary but not sufficient conditions for alienation to occur geo-historically and to be maintained synchronically. While the alienator certainly commits categorial error, especially in the form of ontological monovalence, he must do the work of generative separation and have the power$_2$ to carry it through in a definite historical context and to reinforce its synchronic articulation. Alienation (or, if you like, the fall) thus should not be thought of as pertaining to some remote past. It is now. Nor should it be thought of as alienation from an ideal, rather from (part of) what we really are essentially and are tending to – and contingently may, through emancipatory praxis – become; it is the 'wicked spell cast over the world' that prevents people from realising their real utopian longings (Adorno 1964: 4).

Table 6 highlights the causal efficacy and emergent reality of alienation, and displays some additional key correspondences.

See also EXPRESSIVISM; NEGATION OF THE NEGATION; PROBLEM OF ALIENATION.

alloplastic/autoplastic (Gr. *allos/autos* + *plassein*) literally, other-forming or -moulding and self-forming, respectively. Used by Bhaskar to distinguish *transformative* praxis, which is directed at changing things other than one's self, from *transformed* praxis, which involves self-change. See also EMANCIPATORY AXIOLOGY.

alter-globalisation. See DIRECTIONALITY.

alterity/change. Key 1M and 2E categories, respectively, at once irreducible to and presupposing each other, interconnected dialectically as modes of ABSENCE or real negation and (within that) of DISTANCIATION. *Change* is absence qua CAUSALITY and PROCESS (paradigmatically substantial, but also non-substantial) – structures and things considered under the aspect of their story in time. *Alterity* (*non-identity* or *difference*) is absence qua product – structures and things considered under the aspect of their differentiation in space. It is conceptually tied closely to HETEROLOGY (non-homology); avoidance of homology is the key to the solution to the 1M PROBLEMS of philosophy. The two concepts come together in the concept of SPACE–TIME and causality as a 'tensed triunity' (*D*: 114), within which the identities of things are constituted relationally by their position within a system of 'changing differentiations and differentiating changes' (*D*: 240).

Contra the analytical and monovalent tradition, '{c}hange cannot be analysed in terms of difference because it presupposes the idea of a continuing thing in a tensed process', and difference itself typically presupposes change (see EMERGENCE). None the less, conversely, contra the irrealist dialectical tradition, '{d}ifference cannot be analysed in terms of change because it includes the idea of two or more non-identical tokens' which is transcendentally necessary for existential INTRANSITIVITY (hence REFERENTIAL DETACHMENT and PRIMARY POLYADISATON) to occur (*D*: 45). Thus contradiction, which is essential for change, presupposes constraint by at least one other element; if there was a unique beginning to everything, difference must have been irreducibly present at the outset. TOTALITY (3L) presupposes the internal relationality of distinct elements, and human praxis (4D) presupposes both change and difference (an objectified result in the ID, entraining the possibility of ALIENATION). On the irreducibility of difference – 'sheer alterity or otherness (that is, real determinate other-being), not analysable in terms of change', e.g., the non-identity of words and things – DCR is thus in full agreement with Derrida (*D*: 6).

This is not of course to deny that changes that produce differences and differences that change occur. Nor is it incompatible with the notion of an underlying identity at the level of possibilia thematised in *Dialectic* and the philosophy of META-REALITY, which is precisely *identity-in-difference*. There is a *constellational unity of identity and change* in the spatio-temporal continuity of any entity (e.g., the dispositional identity of a person with her changing causal powers), which by perspectival switch is also a *constellational unity of identity and difference* in entity relationism (e.g., the constitution of a person by her relations with others). At the point of TRANSITION and in entity RELATIONISM we must say that a thing both is and is not itself – the constellational unities 'encompass identities – of identity and change and identity and difference: blows at the heart of analytic reason where, in Bishop Butler's famous pronouncement, "everything is what it is and not another thing"' (*D*: 183).

alterology. See FREEDOM, ETC.

altruism. See ESTEEM; INTEREST; POLYAD; SOLIDARITY.

38 ambiguity

ambiguity (f. L. *amb-*, around, around about + *agere*, to act). A 2E figure of OPPOSITIONALITY signifying 'vague indeterminacy of positions' (*P*: 242). Sometimes used synonymously with BIVALENCE and HOMONYMY.

ambivalence. See DUPLICITY; VALENCE.

American critical realism. See CRITICAL REALISM.

amour de soi, amour-propre. See ESTEEM.

anaclitic rationality. See REASON.

analogical grammar. See GRAMMAR; THEORY, ETC.

analogue, analogy. See THEORY, ETC.

analytic/synthetic (Gr. *ana* + *luein*, to break up, and *syn* + *tithenai*, to put together, respectively). A statement is analytic if the predicate is contained in the concept of the subject, otherwise it is synthetic. 'All bachelors are unmarried' is analytic; 'All bachelors are happy' is synthetic. The former is logically necessary, the latter logically contingent (it adds something not contained in the concept of the subject). The analytic/synthetic distinction is a logico-semantic one which must itself be distinguished from the epistemological A PRIORI/A POSTERIORI and the modal or metaphysical *NECESSITY/ contingency* distinctions.

analytical dualism. See CRITICAL NATURALISM; DUALITY; MORPHOGENESIS.

analytical problematic (*P*: 132–40; *DG*: 394; *D*: 188–95). The perspective in philosophy informed by, rather than simply reducible to expressions of, a model of reasoning based on the formation and analysis of varieties of propositional forms according to the principles of *non-CONTRADICTION, IDENTITY* and *excluded middle*. Non-contradiction is the basic axiom of Aristotle's first philosophy (*Metaphysics* bk 1) and states that an attribute cannot be both a definite thing P and not that thing (not-P). Aristotle qualifies this with the conditions 'at the same time' and 'in the same respect'. He argues that non-contradiction is not a provable hypothesis but the very condition of reasoned argument, since for any statement of assertion to have determinate meaning (including the denial of non-contradiction) it must at least intend to be itself unambiguous, and thus non-contradictory, and thus accept the principle. Aristotle uses non-contradiction to develop a *syllogistic mode of argument*. A syllogism is a structure of sentences that makes a claim or assertion. It is a form of deduction in which one analyses the validity of the construction of the argument – does it contain a subject sentence and a predicate sentence where the structure seeks to confirm or deny the predicate of the subject? If it follows, the form of argument is termed valid. The aim is to ensure that the form of argument itself is not in error and that no false conclusion can be drawn from true premises (though this does not imply that either conclusions or premises are real or actual). The concept of identity follows from the structure of argument in the sense that for P to be true or false it must be P or not-P only. Different derivations of identity are permitted, but this basic principle applies. Since non-contradiction is taken to imply that for any proposition P it is true that P or not-P (otherwise there is a contradiction and something about the argument may be invalid), the form of argument cannot allow intermediates – hence an *excluded middle*.

Following Aristotle, analytical reasoning has undergone a great deal of development. For example, Aristotle focuses on single-assertion predicate–subject forms rather than on more complex groupings, MODALITIES and syntactical variations. Moreover, the narrow field of modern formal LOGIC has a variety of forms (languages, approaches such as fuzzy logic, etc.) and has had particular cycles of critique of its relations to Aristotle's work. The significance of the analytical problematic, however, is in its impact when generalised as a way of viewing the world. It is important to note that this in no way implies the rejection of reasoned argument or that unintended contradiction within an argument can be a source of error. On this level the principle of non-contradiction, rather than necessarily Aristotle's focus on attributes and his justification, is an unobjectionable regulatory commitment of most forms of argument (not including satire). The significant point is that the peculiarities of analytical reasoning frame how many philosophers think about how the world is to be interrogated and thus limit how the world can be theorised to be – hence the relevance of Althusser and Balibar's concept of the PROBLEMATIC as 'what forbids and represses the reflection of the field on its object' (1970: 26).

To think about the world primarily in terms of statements amenable to analysis through forms of analytical reasoning is to think that reality conforms to a method based in the significance of isolated decontextualised statements or assertions – this is a form of ACTUALISM. It can be a linguistic reduction if it is a focus on syntax and semantics in terms of truth and falsity and a well-formed argument, or an empiricist reduction if it develops statements in an object language of correspondence to sense experience/observation. Since the problem of identity is one of identifying that P is P in an actualistic way, self-identity is effectively disconnected through method from time, contingency and variation, despite the acknowledgement of them by Aristotle and thereafter. Acknowledgement is not in itself a systematic conceptual adjustment that accommodates the acknowledgement. The result of reduction is a form either of EPISTEMIC or of ONTIC FALLACY, unable to deal with aspects of complex PROCESS, RELATIONALITY and change. One might then argue that the way non-contradiction is deployed is itself contradicted by the failure of method to capture recognisable aspects of reality. Method, rather than non-contradiction per se, has damaging ontological implications. Furthermore, as a focus on self-identity statements, actualism is also a form of ONTOLOGICAL MONOVALENCE. The focus is on is/are and derivation statements. This can lead to a false conception not only of what is, but also of the EMERGENCE of what is from what has been, what might have been and what may be, and its significance – since these cannot adequately be accounted for without an understanding of context, relations and process. In this sense, actualism is also a form of REIFICATION of concepts and FIXISM or homeostasis of reality. The problematic deriving from analytical reason developing out of Aristotle's method thus contains characteristics of theory–practice contradiction. Those characteristics are important above and beyond any particular exposition of formal method because that method provides a tacit frame of thought for thinking less formally about proof and philosophical argument. In one sense it is intrinsic to how we think of philosophy. At the same time it is also part of the fault-lines that define alternatives resulting in PRIMAL SQUEEZE problems. More specifically, those characteristics can be explored through immanent CRITIQUE in order to develop the case for a different form of ontology. For example, the analytical problematic is one source in Bhaskar's development of an alternative concept of identity argued in terms of concepts of the significance of non-identity (1M), negativity (2E) and open totality (3L) within dialectics. See also TABLE OF OPPOSITION.

JAMIE MORGAN

40   analytical reinstatement (Hegel's)

analytical reinstatement (Hegel's). See EPISTEMOLOGICAL DIALECTIC; HEGEL–MARX CRITIQUE; NEGATION OF THE NEGATION.

Anamnēsis. See ALETHIA; MENO'S PARADOX.

anankastic (f. Gr. *Anankē*, necessity). Von Wright (1963) distinguishes between social rules and *anankastic* rules which depend solely on the operation of natural laws. Human practices such as mathematics, language-learning, puzzles and problems are *para-anankastic* in that, while they go beyond considerations of natural necessity, they are governed by something analogous to it, hence we can expect 'right' and 'wrong' solutions and moves (see *PN3*: 40, 144). This usage exploits both senses of the Greek preposition *para*, 'beside' or 'beyond'. *Anankastic propositions* state that something is a necessary condition of something else. In psychiatry, anankastic personality disorder is compulsive.

anomaly (f. Gr. *anōmalos*, irregular, uneven, f. *a* -, not + *homalos*, even, equal). A 2E figure of OPPOSITIONALITY signifying incompatibility with a pattern established within some theoretical framework. Close to, but distinct from, PARADOX, which violates a canonical principle. See also EPISTEMOLOGICAL DIALECTIC.

antagonism. See CONTRADICTION; DIALECTICAL ANTAGONISTS; OPPOSITION.

anthropic fallacy, anthropomorphic fallacy. See ANTHROPISM; IRREALISM.

anthropism (f. Gr. *anthrōpos*, human being, pl. *anthrōpoi*, human kind) 'incorporates, subjectively, *anthropocentrism* – literally, taking man as the centre or goal of the cosmos – and, objectively, projective *anthropomorphism* – painting or interpreting the cosmos in the image of human being' (*DG*: 394). Collier (2003b) uses *ontocentrism* as the antonym of anthropocentrism. The *anthropic* or *anthropomorphic fallacy* – 'a deep-seated anthropocentric/anthropomorphic bias in irrealist philosophies and Western thought generally' – is 'the exegesis of being in terms of human being' (*D*: 205). Anthropism 'underpins *anthroporealism* in its various guises – empirical, conceptual, Nietzschean (will-to-power) realism. Anthroporealism is implicit in subject–object IDENTITY or equivalent theory, where it involves a tacit exchange of epistemic and ontic fallacies' (*DG*: 394). In rejecting anthropism, CR carries through the Copernican revolution – which demonstrates that the universe does not revolve around humans – in philosophy.

Anthroporealism (of which EMPIRICAL REALISM is the dominant form) is a prime example of illicit FUSION, involving the exchange of non-equivalents (thoughts and things) or *anthroporealist exchanges*, in which 'a naturalised (and so incorrigibly ETERNISED) science is purchased at the price of a humanised nature' (*SR*: 23); hence it symbiotically parallels the logic of commodification and commodity FETISHISM. Any regional geo-historical variant of anthropocentrism (e.g., Eurocentrism) is an *ethnocentrism* (which accounts for the sequencing of the first two terms in the stretch-adjective Bhaskar deploys to refer to the subjective and empirical realist elements in Hegel's philosophical position: 'anthro-ethno-ego-present-centric' [*D*: 91]) (the causal sequencing is the reverse: 'ego → ethno → anthropism' [*D*: 358]). While anthropocentrism and ethnocentrism appear to be transhistorically applicable categories, they take on an extreme form (the latter in the guise of *nationalism*) in post/modernity.

The anthropic fallacy is even more fundamental to IRREALISM than the EPISTEMIC FALLACY, which it underpins even as it is concealed by it. It is 'the common unifying bias' in the failure to distinguish the transitive from the intransitive dimension (i.e., in the denegation of ontology) and the domain of the actual from the domain of the real (i.e., in the denegation of ontological stratification and transfactuality) – hence the concept *anthropo-ACTUALISM* or the denial of necessity in being (*D*: 208). In some Bhaskarian formulations the epistemic fallacy accordingly *is* (ultimately) the anthropic fallacy; or, the epistemic fallacy is itself anthropic (*D*: 378, 382). Even Heidegger commits it. 'For in *Being and Time* being is always mediated by *Dasein* or human being; and, in his later works, he re-thematises ontology in terms of its human traces from the pre-Socratics to the contemporary age of nihilism and technology' (*D*: 205). Anthroporealism and *transcendent realism* (which defines [a zone of] being in terms of its *inaccessibility* to human being, and is in part generated by anthroporealism) constitute the *irrealist ensemble*. Both anthroporealism and the epistemic fallacy it incorporates are, however, entailed by the even more cardinal fallacy of ONTOLOGICAL MONOVALENCE.

A non-anthropic world of structures, mechanisms and laws is presupposed by scientific experiment, and, indeed, by human agency as such – i.e., a 'world, which has occurred and will come again to pass, without [humans]' (*D*: 230). For human agency, discourse and thought presuppose the possibility of REFERENTIAL DETACHMENT. The 'first act of referential detachment' (*EW*: 24 n4) constituted the dawn of specifically human (self-)consciousness and praxis, hence of GEO-HISTORY. It entails the capacity to be non-anthropic: to *absent* 'oneself or one's discursive act or one's pre-linguistic intuitive acts from what they are about' (*D*: 230). Thus to be human is among other things precisely to possess the potential to be non-anthropic (and non-monovalent), to recognise the cosmic contingency and precariousness of human being. In what are only apparently paradoxical reversals, irrealist anthropism at 1M thus has as its 'condition and result' *anti-humanism* at 4D (the 'de-agentifying reification of facts and their conjunctions', which both reflects and supports the reification of human beings as labour-power), whereas realist non-anthropism at 1M entrains a (re-agentifying) dialectic to *HUMANISM* at 4D (*D*: 206, 301). Emancipation is among other things liberation from the labyrinthine coils of the TINA SYNDROME of anthropic irrealism, which banishes thought of what we essentially are and could be. See also AGENCY; SCANDAL OF PHILOSOPHY.

anthropocentrism. See ANTHROPISM.

anthropology. Investigates humanness via four interrelated sub-disciplines. *Physical anthropology* studies primatology, human palaeontology, environmentally conditioned biological plasticity, and population genetics. *Linguistics* focuses on communication's emergence and development, historical and existing language taxonomy, speech in relation to ethnicity, and usage micropolitics. *Archaeology* uncovers and analyses concrete evidence of historic and prehistoric settlements, and the technological and social adaptations they represent. And *ethnology* describes and theorises existing socio-cultural forms and interrelationships. Ethnographers have described some 6,000 cultures since methodical ethnography began in the mid-nineteenth century. These range from small, mobile, technologically simple foraging bands, through hierarchically organised, segmented, territorially and culturally distinguishable communities, to modern states and global polities facilitated by mass transportation and communication.

The concept linking these sub-disciplines is adaptation. What distinguishes the generally human adaptive mode? How are particular socio-cultural adaptations manifested? Because we know that (1) the basis of all socio-cultural forms derive from

42 **anthropology**

a single, small, Pleistocene, East African population, (2) those forms now display wide but limited variation, and (3) evolution is a change in form over time, the theory of evolution has informed investigations of cognition, language, the family, society, marriage, technology, art, warfare, economic production and exchange, civilisation, politics and religion.

Underlying human adaptation is perforce the natural environment adapted to; so real material CONSTRAINTS on socio-cultural forms loom large in cogent anthropology. Simultaneously, the socio-cultural environment, constituted by the concomitant EMERGENCE of human capacities for REFERENTIAL DETACHMENT, language and praxis, is itself so complex and essential to individuals that it, too, becomes a catalyst for adaptation. Thus CULTURE, a primary adaptive mode, also becomes a structural condition of further adaptation. This promotes a REFLEXIVITY driving increasingly rapid adaptive transformations.

Referential detachability means, among other things, that human discourse can be studied in isolation from its extra-discursive referents. This has provided the underlying rationale for a contemporary ethnological school, derived from French POST-MODERNISM, which actually denies the importance of any extra-discursive referent, and discusses representation apart from other real strata. This approach builds on the idealism of cultural RELATIVISM, prominent since early ethnology sought an academic identity distinct from biology, psychology and sociology, and relied on the culture concept for it. However, the self-contradiction – that cultures can be relative to each other yet unrelated to the world, and the correlative postmodernist contradiction that politics does not pertain to real relations – is now duly undermining enthusiasm for this form of IRREALISM.

Superseding it is a range of reality-based approaches, in development since the 1980s, which at least tacitly recognise that both the MATERIAL and IDEAL realms are real. These approaches include PHENOMENOLOGY as the base description of experience in a socio-environmental context; an older cultural ecology now enhanced by computerised mapping, modelling and monitoring; landscape anthropology relating people's local values and understandings to their immediate surround; and practice theory, dialectically framing agency with its structural conditions, as manifest environmentally and socially. Practice dialectics shows the effects STRUCTURE and AGENCY each has on emerging social and environmental forms. For example, the idea that open spaces benefit both people and ecosystems can be appropriated by strategically motivated agents, who work to create the philosophical, legislative and economic conditions for open-space creation and preservation; in turn, if protected open-spaces actualise, they become new structural conditions to which agents must cognitively, behaviourally and socially adapt.

CR undergirds all such reality-based ethnological approaches by showing the world to be ontologically stratified as physical, chemical, biological and conscious, each layer emerging from, but not reducible to, prior ones. This grounds anthropology's distinct but compatible physical and meaning-based investigations of humanness. CR shows knowledge to be both crucial and fallible, thus invoking ontic moral constraints enjoining development of increasingly accurate models of reality. Humans are complexly part of natural and cultural strata – which, through human involvement, also dialectically inform each other. Simultaneously, humans are aware of their

anthropology, philosophical   43

participation in each stratum severally, as well as the resulting system, this awareness itself having real consequences at every level.

Anthropology's shortcomings include its failure to (1) convey the fascination of such investigations to the public, (2) induce designers of socio-economic developmental schemes to accord due importance to local knowledge and values, and (3) settle on a practicable definition of culture. However, CR dialectics recognises culture as intelligibilia with material implications, a view largely in accord with ethnologist Brad Shore's (1996) concept of analogical schematisation, which denotes culture's linked cognitive and material manifestations. See also PHILOSOPHICAL ANTHROPOLOGY, Brereton 2004b.

DEREK P. BRERETON

anthropology, philosophical. See PHILOSOPHICAL ANTHROPOLOGY.
anthropomorphism. See ANTHROPISM.
anthroporealism, anthroporealist exchanges. See ANTHROPISM, IRREALISM.
anti-naturalism. See HERMENEUTIS; NATURALISM.
anti-naturalistic fallacy. See NATURALISTIC FALLACY.
anti-realism. See IRREALISM; ONTOLOGY; REALISM; TRANSCENDENTAL REALISM.
antinomial critique. See CRITIQUE.
antinomianism, adj. antinomian (Gr. *anti*, over against, in opposition to + *nomos*, law). The view that one is not bound by moral law; in a wider sense, opposition to the institutionalised forms of law, religion and politics. There is a pronounced antinomianism in this latter sense in the thought of Bhaskar.
antinomy, **antinomialism**, adj. antinomial (Gr. *anti*, over against, in opposition to + *nomos*, law). Literally, contradiction in law. One of 'the myriad figures that OPPOSITIONALITY in philosophy takes', signifying CONTRADICTION or opposition between conclusions that seem equally logical, reasonable or necessary, which if accepted entail rejection of the law of non-contradiction (i.e. acceptance of 'both A and –A') (*P.* 242). Close to, but distinct from, DILEMMA. *Antinomy of transcendental pragmatism*: see RATIONALISM.
apodeictic or **apodictic/deictic** (f. Gr. *apodeiktikos*, demonstrative, both in the sense of pointing out and proving, f. *apo-*, away from, *deiknumai*, to point out). A *deictic* (or TOKEN-*reflexive*), like an INDEXICAL, expression is one whose REFERENCE is determined by the context of utterance, e.g., 'that one there'. An apodeictic proposition asserts what must be the case and, if it is a valid description of the conclusion of a transcendental argument (a transcendental necessity), cannot reasonably be doubted. See also INFERENCE; JUDGEMENT; MODALITY.
aporetic. See APORIA.
aporia (Gr. difficulty, loss, want, lack, or need), pl. *aporiai* or its Anglicised version, *aporias*, adj. *aporetic*, n. *aporeticity*, the quality of being aporetic. A 2E figure of OPPOSITIONALITY signifying 'interminably insoluble indeterminacy' (*P.* 242). The *aporetic method*, pioneered by Aristotle, is 'very close to the method of immanent CRITIQUE' (*P.* 8). Philosophical *aporetics* attempt to resolve aporiai. See also PROBLEM.
a priori/a posteriori (L., what comes before/what comes after). A distinction in epistemology between knowledge, propositions or concepts that are prior to or

independent of any (further) experience and those based on experience. While all ANALYTIC propositions are a priori (knowable without empirical evidence) and logically necessary (they cannot be false), for CR not all a priori propositions are analytic, i.e., there may be synthetic a priori knowledge, established by transcendental argument and retroduction from historically relative premises (see INFERENCE). Similarly, while all synthetic propositions are a posteriori (based on empirical evidence) and logically contingent (they can be empirically false), not all a posteriori propositions are synthetic, i.e., there may be analytic (necessary) a posteriori knowledge, i.e., knowledge of natural necessity, as when scientists deduce the necessary tendency of a thing from its intrinsic structure and arrive at a real definition of it as a NATURAL KIND (*RS2*: 174, 209 ff.). 'Water tends to boil when heated' is an analytic a posteriori truth because the predicate is contained in the concept of the subject in virtue of the molecular and atomic structure of water, known to science. This defeats Hume's 'drastic disjunction' between (analytic) 'truths of reason' and 'matters of fact' (see TRANSCENDENTAL REALISM). The concept of a *false* analytic a posteriori – 'an idea as shocking to the pretensions of analytical reason' as that of a FALSE NECESSITY (*D*: 199 n) – is entailed by the fact that when science locates deeper levels of structure it corrects real definitions. See also PHILOSOPHY.

arch of knowledge tradition. The great *arch of knowledge* (Olroyd 1986) is comprised of (1) upward curve or limb: induction (also/or retroduction [Peirce], conjectured hypothesis [Popper]); (2) keystone: general laws or principles; (3) downward curve: deduction of lower-order laws and particular facts. The arch of knowledge *tradition* dates back to Aristotle and was not fundamentally challenged until the 1970s, when its collapse provided the context for the genesis of CR. It presupposes ACTUALISM. When CAUSAL LAWS are construed as empirical regularities or constant conjunctions, it is impossible to put the keystone in place, for such laws must necessarily be true for all particular cases. 'The sceptic in theory (cf. the upward limb) must be a dogmatist in practice (cf. the downward limb)' (*P*: 41, cf. 17–23, 35–6). The dominant modern form of the tradition is positivism. Historically, EMPIRICISTS have sought to induce (2) from experience, thereby eliminating natural necessity (the problem of induction); RATIONALISTS have sought to deduce (2) a priori, thus by-passing empirically grounded scientific theory. This is PRIMAL SQUEEZE, produced by the epistemic–ontic fallacy (which has the consequence that rationalism and empiricism seem to exhaust the alternatives), and ultimately by ontological monovalence. See also CAUSAL LAW; EXPLANATION; INFERENCE; UNHOLY TRINITY.

archē (Gr., pl. *archai*). A beginning, starting point, first principle (also first cause or primal reality; supreme power, sovereignty, office – as in 'Archbishopric'). Not to be confused with 'arch' (f. L. *arcus*) in the sense of 'a curved structure', deployed metaphorically in 'ARCH OF KNOWLEDGE TRADITION', though the notion of an indubitable *archē* is intrinsic to the latter.

Archimedean (vantage-) point. The notion, rejected by postmodernism and CR alike, that philosophy can start out from some utterly secure vantage-point, with a view of TOTALITY and immune from EPISTEMIC RELATIVITY (see, e.g., Descartes 1641: 102). Named after the Greek scientist Archimedes (*c*. 287–212 BCE), who is reported to have claimed that he could move the world with a long enough lever and a firm place to stand on. It has no etymological connection with ARCHĒ, but has come to mean much the same thing. See also FOUNDATIONALISM.

argument forms. See DIALECTICAL ARGUMENT; INFERENCE.

Aristotelian mean, doctrine or principle of. See INTEREST.

Aristotelian propaedeuctics. See PROPAEDEUCTIC.

A-series, B-series, C-series. In his classic argument for the unreality of time, McTaggart (1908, 1927) distinguished the A-serial 'ordination of events by the explicitly tensed *past, present* and *future* {. . .} from the B-serial *earlier, simultaneous* and *later* or the spatio-temporally *indefinite* C-series' (*DG*: 393). CR defends the irreducibility of the A-series to the B-series, i.e., 'the *reality of TENSE* and the *irreducibility of SPACE–TIME* on any WORLD LINE both for the transitive observer and for the intransitively observed' (*D*: 210). This means that time is not moving through us but we through it 'as in effect time-travellers on our historical world-lines (embedded in world-history), producing not the future {. . .}, but its content' (*SR*: 219). This is a non-anthropic position (or the A-series is non-anthropically real) in that, unlike most defences of A-serial tense, it does not rely exclusively on an argument that it is necessary for the intelligibility of human experience and action (*D*: 301). Being is characterised by *A-series processuality* quite independently of human being. A universe stripped of its A-serial properties would be a BLOCK universe, incompatible with embodied agency and modern cosmological theory. (The specific sense of apprehension of past and future, on the other hand, is probably a product of the human mind, and from the perspective of meta-Reality, which 'lift{s} the veil of time' (*MR*: 104), past and future are [real] illusions, i.e. have no real objects.) Because it is impossible to predict the future, CR is also committed to the open C-series. See also ANTHROPISM; PROCESS; SPACE–TIME.

assertoric, assertoric imperative, assertorically imperatival sensitised solidarity. See DCR; EMANCIPATORY AXIOLOGY.

asymmetry. See PRIMACY.

atomism (f. Gr. *atomos*, not able to be cut, indivisible). 'Literally meaning without interior space. Argued in C2.1 to be conceptually incoherent. But associated, as *corpuscularianism*, with the Cartesian paradigm of action-by-contiguous-contact, and the achievement of Newtonian locally celestial closure, it provides a source of philosophy's analogical GRAMMAR {*ontological atomism*}. It is complicit with, and provides a model for, bourgeois individualism {*social* or *sociological atomism*}. It is entailed by EMPIRICIST justificationism, the dominant twentieth-century form of the quest for incorrigibility or search for an unhypothetical starting point {*epistemological atomism*}. As such, it implies the reductionist demand for an autonomised or empty mind' (*DG*: 394–5). Social atomism is more immediately presupposed by the abstract UNIVERSALITY of the empiricist theory of CAUSAL LAWS and in turn entails spatio-temporal PUNCTUALISM. Indeed, it provides the clue (already indicated in *RS*) to the definition of the 'central polarising, constitutive couple, the overarching DUALITY {. . .} of western thought: the individual, ego or homunculus, the more or less punctual, atomistic tacitly gendered man and complete uniformity, abstract universality, the ACTUALIST constant conjunction of the same. And the shocking thing is that both terms of this couple are completely false, and indeed rationally groundless' (*RM*: 75). *Atomicity* is the quality of being atomic. Social atomism contrasts with social FUSIONISM. See also EMPIRICAL REALISM; EVENT; EXTENSIONALISM; FOUNDATIONALISM; KNOWLEDGE; PDM.

attachment. See DETACHMENT.

A-type, **B-type**, and **C-type determinations**. Natural, mixed and social determinations, respectively. Human agency, 'moored *socially* in a complex of social relations and *physically* at determinate locations in space and time', is B-type (*SR*: 116, 129). This entails that its constitution can never be wholly either social or physical (see also SUBJECTIVITY).

Aufhebung. See SUBLATION.

autogenesis or **autogeny** (f. Gr., self + origin, source, birth). Self-creation or spontaneous generation (in biology, of living organisms from non-living matter; in cosmotheogeny, self-creation by a creator ex nihilo). *Autogenetics* refers to an allegedly self-generating system of thought or social forms, e.g., 'linear Hegelian autogenetics' (*D*: 240), rejected by CR in virtue of the multiple and contingent determination of most results entailed by the openness of the world. Not to be confused with AUTOPOIESIS. See also CREATIVITY.

autology. See FREEDOM, ETC.

autonomisation. An *autonomised mind*, as conceived in EMPIRICISM and PHENOMENALISM, is empty of extraneous influences – a 'blank slate', capable of apprehending 'the facts'. *Autonomised sense experience* is that which, according to these approaches, secures a relation of either identity or one-to-one correspondence between propositions and that to which they refer. It is fundamental to the positivist theory of knowledge, closely bound up with the EPISTEMIC–ONTIC FALLACY, ontological ATOMISM, the REIFICATION of facts, and the FETISHISM of CLOSED SYSTEMS and CONSTANT CONJUNCTIONS (see *SR*: 254–6).

autonomy. See FREEDOM, ETC.

autoplastic. See ALLOPLASTIC.

autopoiesis, adj. *autopoietic* (f. Gr. *auto-*, self + *poiēsis*, a making, forming or creating). Literally, pertaining to SYSTEMS that are *self-producing* or *self-constructing*. In traditional systems theory, systems were seen as open, transforming inputs into outputs. Biologists Maturana and Varela (1980, 1987) developed the concept of autopoiesis to explain the special nature of living as opposed to non-living systems. Autopoietic systems are closed and self-referential – they do not primarily transform inputs into outputs, instead they transform *themselves into themselves*. The components of an autopoietic system enter into processes of production or construction to produce more of the same as necessary for the continuation of the system. The output of the system, that which it produces, is its own internal components, and the inputs it uses are again its own components. They are said to be organisationally closed but interactively open (Mingers 1995). The paradigm example is a single-celled organism such as amoeba.

Attempts have been made to apply the concept to social systems (Luhmann 1986), but this has been criticised (Mingers 2002a). Bhaskar uses the idea in several ways: in discussing the nature of science and scientific knowledge as generated from previous knowledge (*SR*: 54, 83); more fundamentally, as the basic mechanism of societal (re)production (*D*: 156–7); and finally he characterises EMERGENCE itself as autopoietic (*D*: 49). See also AGENCY; FUNCTIONAL EXPLANATION; RECURSIVITY.

JOHN MINGERS

auto-subversion. See ABSENCE; SUBJECTIVITY.

avidya, avidya-Tina formation. See ALIENATION; DEMI-REALITY; IDEOLOGY; PROCESS; TINA SYNDROME.
axiological imperative. See AGENCY; AXIOLOGY; CRITICAL REALISM; HOLY TRINITY.
axiological interest. See AXIOLOGY.
axiological necessity, axiological need. See ALETHIA; ALIENATION; AXIOLOGY; CRITICAL NATURALISM; DENEGATION; DETACHMENT; EMANCIPATORY AXIOLOGY; NEEDS; PHILOSOPHY; REALITY PRINCIPLE; RECOGNITION; REFERENTIAL DETACHMENT; TINA SYNDROME.
axiology (f. Gr. *axia*, worth, value + *logos*). A branch of philosophy concerned with the study of the nature of value (worth) and what kinds of things have it. It can embrace aesthetics, epistemic values, etc., but is normally confined to ETHICS. Commonly thought to contrast with deontology (the theory of moral duty) and/or praxiology (the theory of action [see EXPLANATORY CRITIQUE]), in CR it, rather, complements them. Collier's *Being and Worth* (1999) is an exercise in *ethical axiology* or *axiological ethics*. It defends an objective (ontocentric) theory of value, that being is intrinsically good, by contrast with the dominant modern subjective (ANTHROPOCENTRIC) view that value depends on human interest. In Bhaskar, axiology is primarily both *the theory* (in the TD) *of agentive (human) AGENCY* and that to which the theory refers (in the ID), i.e., agency that is its own reward (intrinsically satisfying or valuable) and efficacious, 'concerned precisely with the absenting of absences' (*D*: 83) and tendentially productive of the good life for all, i.e., EMANCIPATORY AXIOLOGY or the *axiology of emancipation*. Irrealist philosophy lacks an axiology in this sense, which pertains to irreducible intentional causality at 4D; or its axiology is de-agentifying (reifying and/or disembodying). Because of this irreducibility, a 4D phenomenon in the ID is sometimes referred to as 'ontological (or axiological)', etc., and 1M–4D as the *ontological–axiological* chain. Likewise, the realm of human subjectivity in the TD may be referred to as 'epistemological and axiological'.

The *axiological imperative* (introduced at *PN3*: 87) says that (1) the human agent is by nature spontaneously active or 'always already acting' (*SR*: 188): given that she exists and intentionality is irreducible, she must act ('one can no more abstain from activity than from life' [*SR*: 32], and refraining from action in a specific context is itself a form of action), including *appraise beliefs* from the standpoint of their suitability for acting upon; and (2) nobody can act for her. In an always already moralised world, (1) 'imperiously demands a grounding or critique of our values' (*P*: 110); while (2) furnishes the cardinal principle underlying the late Bhaskarian doctrine of the PRIMACY of SELF- or subject-REFERENTIALITY in collective emancipation, entailing and being entailed by an *axiological commitment* implicit in AGENCY and the JUDGEMENT FORM to 'the establishment and the prefigurative existence of a society in which each being's welfare counts as much as my own' (*RM*: 14). The *axiological moment* is 'the spot from which we must act' (*D*: 9), i.e., the 3L concrete context. *Axiological necessity* (or *practical necessity*) is another name for the REALITY PRINCIPLE or ALETHIC truth. It is what the always already acting human agent encounters and must address, the necessity of the world in an open context, which may come to be known as such. It constellationally embraces – but is sometimes used as a synonym for – *axiological need* (or *transcendental–axiological need*), which denotes the agent's own basic human NEEDS, i.e. any need the satisfaction of

48 axiology

which is transcendentally necessary for agentive agency (such as the need for food, truth, autonomy or non-alienation), and the same holds for an *axiological interest* (any interest transcendentally necessary for agentive agency). When denied overt expression, thereby ALIENATING agents from what is intrinsic to their well-being, axiological needs entrain pathologies and TINA COMPROMISE FORMATIONS; axiological necessities/needs may consequently be *Tina necessities*. More formally, axiological necessity is co-extensive with natural necessity > transcendental necessity > dialectical necessity 'when epistemologically mediated' (*D*: 103). *Axiological inconsistencies* are practical (practice–practice, or, given the DUALITY of theory and practice, theory–practice) INCONSISTENCIES or CONTRADICTIONS. Thus, for example, the Hegelian constellational closure of history is axiologically inconsistent with 'the transformative potential implicit in the praxis necessary to reproduce {it}' (*D*: 341). The axiological imperative means that, when faced with logical and/or theory–practice contradictions or with the epistemological undecidability of some issues (e.g., When is a thing no longer itself? When to allow a premature baby to die? What is the extent of the present, or of this locale? When and in what degree is the dialectical suspension of analytical reason appropriate? Which is the most satisfying 'description, action, way of life or social system'? [*D*: 81]), our situation is one of *axiological indeterminacy* (absence of sufficient, sometimes of any, grounds for rational autonomy of action – we must act, but we do not know what to do); or we are in a *problematic axiological choice situation*. Such unresolved indeterminacy in the intrinsic or intentional aspect of social life is quite consistent with a determinate outcome, as consistency and grounded beliefs are not the only causal factors at work. The most creative response to it is to redescribe the alternatives in the light of a transformed theory–practice ensemble which either abolishes the grounds of the apparently mutually exclusive alternatives or locates them within a deeper context of understanding of moments of TRANSITION and change – 'the non-VALENT Socratic strategy' (*D*: 72, 81 f.).

Axiological indeterminacy is to be distinguished from *axiological underdetermination*, which is entailed by the depth-openness of the world and is a necessary condition for freedom of choice, hence AUTONOMY as self-DETERMINATION, and free will. Indeterminacy tends to make rational agency impossible, underdetermination is a condition of its possibility. Although theory–practice inconsistency is a source of axiological underdetermination, the latter does not entail 'either (a) that there are no rational criteria for action or (b) that in a world without T/P inconsistency axiology will be determined, as distinct from autonomous. T/P consistency leaves options open. A world without T/P inconsistency would not be, as the Stoic-Spinozan-Hegelian tradition has thought, a (rationally) determined one. It would be *open* and up to the agents to do what they wanted within their causal powers' (*D*: 247).

A *concrete axiological judgement* (CAJ) prescribes a course of action appropriate for the specific circumstances of a particular situation. In the expanded schema for the transition from facts to values, it helps to bridge the gap occupied by the CP clause in (4) and the initiation of action: (1) theoretical critique (demonstrating falsity of P); (2) explanatory critique (explaining the structural source of P); (3) value judgement (negative evaluation, CP, of the structural source of P); (4) practical judgement (positive evaluation, CP, of action rationally directed at removing the source of P); (5) *concrete axiological judgement* (this is what we should do in this context to remove it); (6)

transformation in the agent's praxis; (7) emancipatory action, i.e., praxis oriented to emancipation; (8) transformative praxis; (9) emancipated (free) action. It is of course consistent with a decision not to act, in which case the schema would stop at (5) (adapted from *SR*: 188; see also EXPLANATORY CRITIQUE; EMANCIPATORY AXIOLOGY).

An *axiological standpoint* (see PERSPECTIVE) refers to our practical (axiological) commitments and interests which provide the immediate context for CAJs. The *axiology of freedom* or EMANCIPATORY AXIOLOGY, entailed by EXPLANATORY CRITIQUE, is the shortest real DEFINITION of dialectic in the realm of geo-history (see also DIALECTIC OF FREEDOM). *Axiological asymmetry* and the *asymmetry of (axiology and) emancipation*: see PRIMACY.

axiology of freedom. See EMANCIPATORY AXIOLOGY.

B

backwards causation. See CAUSALITY; SPACE–TIME.
bad (dialectic, infinite, totality, etc.). See the qualified concept.
balance. See SOPHROSYNE.
base/superstructure. See INFRA-/INTRA-/SUPER-STRUCTURATION.
base-level. See ZERO-LEVEL.
basic acts. See AGENCY; MEDIATE; PERCEPTION.
Beautiful Soul, Stoic, Sceptic, Unhappy Consciousness. Dialectical figures in Hegel's *Phenomenology of Spirit* (1807 [*PS*]) which function as substantive social concepts consequent upon the inauguration of lordship and bondage (see MASTER–SLAVE), each of the last three of which explicates the implicit logic of its predecessor in the drive of self-consciousness or human 'sentient socialised self-awareness' (*P*: 174) to overcome the subject–object opposition (see DIRECTIONALITY) – Stoic → Sceptic → Unhappy Consciousness – and hence is underpinned by ALIENATION, of which the Beautiful Soul is for Hegel the archetypal figure.

Stoicism (named after a school of classical Greek philosophy whose leader taught in a *stoa* [covered gallery or cloister]) acknowledges, but adopts a posture of indifference to, reality, including master–slave-type relations; hence the freedom it enjoys in thought is only an abstract freedom, 'not the living reality of freedom itself' (*PS*: 122). Stoicism is thus SCOTOMATIC, afflicted with blind spots. Scepticism (likewise named, after the Skeptics [f. Gr. *skeptesthai*, to look out, consider], who refused to grant that any knowledge was possible) is schizoid or 'UNSERIOUS', denying the existence of reality in theory but affirming it in practice, so committing performative contradiction: 'it pronounces an absolute vanishing, but the pronouncement *is*, and this consciousness is the vanishing that is pronounced' (*PS*: 125). Hence scepticism and *dogmatism* are 'mutually complicit DIALECTICAL COUNTERPARTS (*P*: 193), closely bound up with FIDEISM and FUNDAMENTALISM. Serious scepticism results in silence and SOLIPSISM. The Unhappy Consciousness is Scepticism aware of its theory–practice INCONSISTENCY but unable to overcome it, and so wretched and 'agonisingly self-divided' (*PS*: 131) taking refuge 'in asceticism or OTHER-WORLDLINESS and in respect of power$_2$ relations in {. . .} introjective identification with the master's ideology or projective absorption in a fantasy world made for slaves' (*DG*: 406). On Kojève's reading of Hegel, which Bhaskar endorses on this point, the real basis of scepticism is private property (presupposing GENERATIVE SEPARATION): 'only those who do not need to sell their labour-power can afford to be sceptics' (*D*: 331). The Beautiful Soul, for its part, vainly seeks to overcome the dividedness of self and world by withdrawing, alienated, from society.

These dialectical figures loosely correspond to the moments of the ontological–axiological chain, as shown in Table 7.

Bhaskar accordingly treats these figures as 'the archetypical ideological "attitudes"' to master–slave-type relations, and deploys them as a diagnostic guide to the modern history of philosophy and the irrealist, split consciousness of the DEMI-REAL, commencing with Hegel himself, who begins his intellectual odyssey as a Beautiful Soul striving to overcome alienation, but never transcends Stoicism, Scepticism and Unhappy Consciousness because he comes to terms with power$_2$-stricken society, resolving its contradictions in an endist TINA COMPROMISE, when they can be resolved only in practice. The figures function as 'ideologies of bondage and legitimation {. . .} in a duplex fashion', i.e., involve a 'double alienation' (from reality as such and in the implicit endorsement of the alienation underpinning power$_2$ relations) which 'haunts IRREALIST philosophy' (*D*: 362, e.a.). The heterologous Unhappy Consciousness, split between de-ontologised enlightenment and faith of one kind or another, between rationality and rationalisation, has been pre-eminent in the philosophy common rooms of the West ever since. The *predicament of the Beautiful Soul* can only be resolved by engaging in totalising depth-struggle to abolish the sources of alienation.

being. See ONTOLOGY; REALITY.

Bestand, das Bestehende (Ger., existence, the actually existing). Used by Bhaskar for the established order or the status quo, brimming with 'commonsense and uncommon nonsense' alike. Positivism is 'the philosophy of *das Bestehende*' (*SR*: 305).

binary opposites. See DUALITY and DUALISM.

bind (vb, n.). Has three closely related meanings in CR: to constrain or CONTRADICT (to fasten, fetter), often used together with *block*; to cause to cohere, set *bounds* to; and obligatory, mandatory (e.g., Kantian ethical rules, which are *binding*). In master–slave

Table 7 Irrealism and master–slave relations

Causal–Axiological chain	1M Non-Identity	2E Negativity	3L Totality	4D Transformative Agency
Irrealist errors	actualism	ontological monovalence	ontological extensionalism	reification/ disembodiment
Irrealist attitudes to master–slave relations	Stoicism (indifference)	Scepticism (denegation)	Beautiful Soul (alienation)	Unhappy Consciousness (introjection/ projection)
Performative consequence	scotomatism	unseriousness	effeteness	split
Forms of alienated (split) consciousness	extra-jection (splits the world from thought about it)	retro-jection (splits geo-historical change from the present)	intro-jection (splits the psyche)	pro-jection (splits minds from bodies)

society, the present is bound by the massive presence of the past; emancipatory praxis *unbinds* it. See also MIND; PRINCIPLE OF INDIFFERENCE.

biota (New Latin, f. Gr. *biotē*, life). All the plant and animal life of a particular region or period.

bipolar, bipolar dual, bipolarity. See ABSENCE; DCR; DUALITY.

bivalence. See VALENCE.

blockism/punctualism. IRREALIST theories of SPACE–TIME, entailed by the Humean account of causal laws, which deny the reality of TENSE and PROCESS, hence cannot sustain an adequate concept of CAUSALITY. *Blockism* (short for *block universalism*) is entailed by the ACTUALISM of the Humean account, hence may be thought of as its spatio-temporal DUAL (necessary correlative); *punctualism* is entailed by its ATOMISM, of which it is likewise the spatio-temporal dual. Blockism postulates the simultaneous co-existence of all times and events, i.e., spatio-temporal closure – the absence of an open universe; one implication is that, since the future already exists, having been predetermined, directional change is impossible. Punctualism, 'an atomistic conception of space–time, refuted by the DISTANCIATED spatio-temporality necessary for time-consciousness and more generally any non-contiguous (or even contiguous) causality' (*DG*: 402), is the converse fallacy that only the here-now exists. An ANTHROPIC version of each may be distinguished: *God's-eye view* or *Laplacean blockism*, according to which, once science discovers the laws governing the present (which contains the future and the past, such that events are pre-ordained before they are caused), 'nothing would be uncertain [to science], and the future, as the past, would be present to its eyes' (Laplace 1819: 4, cit. *RS2*: 67–8); and *ego-present-centric* (*INDEXICALIST*) *punctualism*. 'Squeeze {indexicalism} on to here and it becomes punctual: the present is reduced to a punctual now here' (*P*: 71).

Neither blockism nor punctualism 'is compatible with either embodied agency {which entails transformative change} or cosmological theory, which posits a rapidly expanding divergent and, therefore, as yet unmade universe increasingly unknowable for any given state of technology, in which causes bring about qualitative and quantitative changes, that is to say (to be laborious about it), events do not happen before they are caused to occur'. Their failure to come to grips with the implications of the theory of relativity may be explained by their ONTOLOGICAL MONOVALENCE: 'the past and the future are both negative, but in different causal directionalities – one is gone (spatially behind us), the other is ahead (not yet)' (*D*: 253–4). See also A-SERIES; CAUSALITY; CENTRISM → TRIUMPHALISM → ENDISM; DETERMINISM; FIXISM/FLUXISM.

body. See AGENCY; COGNITIVE SCIENCE; CULTURE; DE-AGENTIFICATION; DISABILITY; META-REALITY; MIND; PHENOMENOLOGY; SELF; SOCIAL CONSTRUCTIONISM; SOCIOLOGY OF THE BODY; SOCIOLOGY OF HEALTH.

border, boundary. See SYSTEM; TRANSITION.

bourgeois triumphalism. See CENTRISM → TRIUMPHALISM → ENDISM; PDM.

bracketing, dialectical bracketing. See EPISTEMOLOGICAL DIALECTIC; PHENOMENOLOGY; PROPAEDEUTIC.

brain. Two insights from CR have strong relevance to the study of the brain and the history of neuroscience. These insights are (1) a materialist notion of EMERGENCE, and (2) the role of experimentation in the physical and non-physical sciences.

The realist theory of science supports a materialist notion of emergence. The properties φ of *x* are just as real as the constituent objects which enable them (RS). Reality is stratified into different levels through successive iterations of this ontological principle. The complexity of the brain is unique in the number of emergent levels it spans (Bunge 1980), from the biological macro-molecules involved in neuronal activation and synaptic learning, to neurons and neuronal circuits, to functionally differentiated areas of cortex and other brain structures, to whole nervous system interactions.

The role of experimentation is to untangle from the flow of actual events the causal laws that underwrite these events. Only under experimental closure can the universal nature of a scientific law be demonstrated. Neuroscience shares with the social sciences the impossibility of experimental CLOSURE; the different levels of emergent structures are only analytically separable even during exacting experimental scrutiny. When an experimental effect is observed, it necessarily involves interactions at the molecular level, whole systems of the brain and intervening levels. When the brain belongs to a social being, we may add social context to this list of interacting structures.

Neuroscience has followed a largely reductionist path, perhaps best exemplified by Barlow's (1972) *single neuron doctrine*, the equivalent of methodological individualism in the social sciences. In recent decades, the importance of studying brain circuits at a range of scales has become acknowledged, but tends to be plagued with false dichotomies. The goal has become discovering which circuit or region underlies a particular effect. Truly integrative approaches, pioneered by neuroscientists such as Freeman (1975), recognise the importance of multiple and interdependent dynamical processes and structures that span a range of temporal and spatial scales in the brain. See also MIND.

DAVID ALEXANDER

branch, branch point. See TRANSITION.
break point. See TRANSITION.
B-series. See A-SERIES

C

capitalism. A social system in which the social relation between capital and wage-labour is the pivotal organising feature, and capital accumulation the primary driving force. The concept is central to MARXISM and is derived from Marx's critical analyses of capital, since the actual concept capitalism is not utilised by Marx in major texts like *Capital*. Its precise definition and elaboration is subject to much debate and disagreement within Marxism(s), and likewise between Marxists and non-Marxists, since the concept is also more broadly rooted in political economy and social science. There is a strong tendency, however, for Marxists to prioritise the concept with or without second-order specification such as 'Fordist', 'disorganised', or 'informational' capitalism, whereas non-Marxists are rather inclined to use and develop other concepts such as industrial society, economy, modernity or network society. For anti-Marxists, such as mainstream economists and NEO-LIBERALS, the concept is redolent with Marxism and so is viewed as something to be either stripped of any critical content and connotations, or avoided at all costs and replaced with euphemisms like market economy or welfare society.

In CR the significance of the concept is relative, first to the importance attributed to its relation to Marxism, and second to the level of abstraction. Bhaskar frequently uses it, and refers to its Marxian and Marxist origins, but leaves it fairly unspecified and, indeed, under-theorised, since his interests are more philosophical and formal than substantive. The same applies to other CR philosophers with Marxist leanings, such as Andrew Collier. For Tony Lawson (2003b) and others working on and with CR (in economics) but without close affiliation to Marx and Marxism, the concept is either non-existent or insignificant – no matter whether the primary interest is methodological or theoretical. Therefore, since capitalism is primarily a theoretical concept particularly favoured by Marxists, it is mainly critical realists of this persuasion and working theoretically who have contributed to our critical understanding of the complex and enigmatic social system it refers to, which continues to intrigue and challenge scholars and activists alike. It can indeed be argued that much of what has been produced on capitalism(s) by Marx, within Marxism and in other related traditions is implicitly CR, but we have also witnessed important contributions from scholars explicitly utilising and advancing philosophical insights from CR. Bob Jessop (1990, 2002a, 2002b) reconstructs and develops the regulation approach to the study of postwar and contemporary capitalism, and critically incorporates various other influences in his continuous endeavour to shed light on capitalism and the capitalist state. Richard

categorema, categoremata, categorematic 55

Marsden (1999) rereads Marx and Foucault in order to argue for a typical CR third-way stance beyond traditional Marxism and postmodernism, and theorises postmodernity as the contemporary condition for the expansion of capital. Kathryn Dean (2003) analyses capitalism and citizenship by entering into open and critical dialogue with classical and recent theorists about the nature and recent transformations of capitalism, but also pushes the EXPLANATORY CRITIQUE of capitalism forward by incorporating a utopian moment. Such work is certainly in line with the spirit of CR. Critical realists both have made, and may be expected to continue to make, important contributions to philosophical underlabouring for critical theories of capitalism and the midwifing of new and enriched accounts of this all-important social system of the present, especially accounts which articulate and mediate the aspirations of all who understand the way in which capitalism both suppresses and occludes human potential to flourish and the historical importance of moving beyond it. See also FETISHISM; PDM; POLITICAL ECONOMY; POLITICAL THEORY; REIFICATION; Brown et al. (eds) 2002; Ray and Sayer (eds) 1999.

PETER NIELSEN

categorema, categoremata, categorematic. See CATEGORY.
categorial error. See ALIENATION; CATEGORY; EVIL; IDEOLOGY; REALISM; TINA SYNDROME.
categorial realism. See CATEGORY; REALISM.
categorial structure (of the world). See CATEGORY.
categorical imperative. See UNIVERSALISABILITY.
category (f. Gr. *katēgorein*, to speak against, accuse, f. *kata*, against + *agora* assembly, forum). A fundamental class or constitutive mode of being, e.g., CAUSALITY, ABSENCE, EMERGENCE. Under the influence of Kant, the dominant modern account of the categories is subjectivist: they are interpretative/classificatory schemas or forms which the human mind (in the TD) imposes upon the world. While such a view has adherents within CR (Dean 2003; Nellhaus 2004), who, however, stress that categories arise through our embodied interaction with the world, Bhaskar himself espouses *categorial realism*, whereby categories are constitutive features of the world (in the ID) – albeit 'very abstract or skeletal' ones – which may be described or misdescribed in the TD, hence have a TD/ID BIVALENCE; or the world itself has a *categorial structure*. Thus, for Bhaskar, while 'causality' is an object in thought, it refers, in the purely natural world, to a real object independent of thought; to 'hold that causal laws existed and acted independently of human beings but not causality or natural lawfulness would be akin to being a realist about knives, forks and spoons but not about cutlery' (*EW*: 34). While a social object is of course dependent on thought (though not exhausted by it), it is existentially intransitive in relation to thought about it; or 'social reality, like natural reality, is really pre-categorised (in the ID) independently of any *account* of its categorial constitution or categorisation (in the TD)' (*EW*: 35). Bhaskar accordingly *geo-historicises* the categories of being to include *social categories* such as money, capital, wages, housing, higher education, religious worship and war, referring to such categorisation as 'fundamental constitutive structure' (*D*: 108; *EW*: 34–5). Put another way, he is saying that there are determinate structures and real stratification in the world (including the social world), real ESSENCES and natural kinds; from this perspective, Kant involuted

ontological structure within the human mind (*P*: 204). There is nothing about social constructs which preclude them from having essential characteristics (Groff 2004). Categories of any kind, '*if valid*, are constitutive of reality as such, irrespective of their categorisation by observers or thought' (Bhaskar 1997a: 140–1, e.a.). If invalid, they *miscategorise* reality as such and are in part constitutive of (social) reality, though dependent on valid categories.

The difference within CR is perhaps partly terminological. Most would doubtless agree that social objects, like natural objects, may have essential characteristics, and that the social is always already pre-categorised, but some baulk at the thought of extending the latter notion to the natural world – categories are held to be essentially ideational or semiotic, hence to pertain exclusively to the epistemic order, except where, as in the case of the social, they are partly constitutive of being; nothing is gained by referring to real distinctions in the natural world as categories, and, indeed, a version of the epistemic–ontic fallacy is avoided. But to use the category 'causality' to refer both to the transitive object and the real object – as Bhaskar has done from the outset, in the footsteps of Hegel and Marx (*D*: 104) – is no more to commit the epistemic–ontic fallacy than to refer to 'housing' in the ID as 'housing', and arguably a more consistent (DIALECTICAL CRITICAL) *NATURALISM* results. This relates to the differences within CR concerning the concept of ontological and ALETHIC TRUTH, where the latter is the epistemologically mediated 'real reason or dialectical grounds for the phenomena of the world', whether social or natural (*P*: 163; see also *RM*: 49; ONTOLOGY).

Another view has it that there are two irreconcilable kinds of categorial realism operative in Bhaskar's later work in the field of ethics, one pertaining to real socio-historical categories and the other to categories having to do with the essential nature of human beings as a species which are in fact irreal (idealist and illusory) and intended to guarantee the possibility of an ideal eudaimonian telos (Hostettler and Norrie 2003). Such a view arguably is sociologically reductionist or HISTORICIST and ACTUALIST, conflating 'the deep content of the judgement form and the latent immanent TELEOLOGY of praxis' (*P*: 154) with the actual, overlooks Bhaskar's own claim to transcend the old dispute between metaphysical IDEALISM VS. MATERIALISM, and foists a DUALISM on Bhaskar where he only acknowledges real distinctions within a constellational identity (see also EMANCIPATORY AXIOLOGY; HUMAN NATURE).

Category mistakes con-fuse (see FISSION/FUSION) one category with another, e.g., epistemology with ontology, or place an entity in the wrong category, e.g., 'Saturday is a bed'. Categorial error figures centrally in IDEOLOGY and TINA FORMATIONS, and more fundamentally in ALIENATION and (in PMR) the DEMI-REAL. The term was introduced by Gilbert Ryle (1949).

Categorematic (n. *categorema*, pl. *categoremata*) denotes any word or symbol that has independent meaning (e.g., 'red', 'animal', 'Aristotle'), i.e., that names things in categories. *Syncategorematic* (n. *syncategorema*, pl. *syncategoremata*) denotes any word or symbol that has no independent meaning (e.g., 'and', 'all', 'some'), i.e., words or phrases that somehow link or go together with categories. In a wider usage, it may refer to any product/process that lacks all but relative independence from others: e.g., philosophy (science, etc.) are, on the principle of epistemic relativity, syncategorematic with 'the contingent actuality of social practices' which supply 'premises for its arguments and potential referents for its conclusions' (*SR*: 12–13).

cathexis (Gr., a holding, keeping hold of). Conscious or unconscious attachment of emotional energy on to an idea, object or person. In Freudian psychoanalysis, the libidinal energy (in the ID), analogous to an electrical charge, invested in something (person, idea, etc.), signifying 'culture's "occupation" of the biological by way of the subject's more or less active response to an "invitation" to attend to a part of the world in a particular manner' (Dean 2003: 19); and the process (TD) whereby this is effected. See also CULTURE; REALITY PRINCIPLE; SUBJECTIVITY.

causal–axiological chain, causal chain. See CAUSALITY; MELD.

causality. Most basically, the power to bring about change: the characteristic way(s) of acting possessed by things (whether particulars/NATURAL KINDS or SYSTEMS) in virtue of their intrinsic STRUCTURES (relations between internal elements); the irreducible property whereby entities generate or prevent (offset), *enable* or *CONSTRAIN* effects. Structures, including SOCIAL STRUCTURES, possess *CAUSAL POWERS/liabilities* and *TENDENCIES*. Whereas causal powers are 'potentialities which may or may not be exercised', tendencies are potentialities which are exercised but may be unactualised and/or unmanifest to people – powers 'dynamised or set in motion' and TRANSFACTUALLY efficacious (acting in open as well as closed systems), hence normically qualified (their effects may be offset by the operation of other powers). Causal power is thus the wider concept: it may not only be possessed unexercised, but, qua tendency, be exercised without being actualised or manifest to people: causal power > tendency. The two concepts come together in the concept of *generative mechanism*, which may be either or both. The distinction between powers and tendencies is entailed by the consideration, derived from the analysis of experimental activity, that generative mechanisms 'not only persist but are efficacious in open systems'. This means that phenomena are to be explained 'by reference not just to enduring powers but {also to} the unrealised {and/or unmanifest} activities of things'. *CAUSAL LAWS* must therefore be analysed as tendencies (ID) or tendency statements (TD), not simply as powers, and their analysis entails 'a new kind of conditional': the *normic*, as distinct from subjunctive, *conditional*, expressed in TRANSFACTUAL (tendencies) as distinct from COUNTERFACTUAL (powers) statements. A powers statement 'says that A would ψ in appropriate circumstances', a transfactual or tendencies statement that it really is ψ-ing. Powers and liabilities are the 'active' and 'passive' sides of the same coin. For a power to have an effect on a thing, the entity must have a propensity to be affected or be 'complicit' with it. Powers are thus the capacity to do or become, liabilities the capacity to suffer or be affected ('what Hobbes called a "passive power"' (see CONTRADICTION) (*RS2*: 50–1, 87 n).

When stimulated, released or enabled, then, powers and generative mechanisms are tendencies, and as such act 'with NATURAL NECESSITY and UNIVERSALITY (within their range) so as to codetermine the manifest phenomena of the world, which occur for the most part in open systems' (*P*: 23, e.a.). Where a thing just is its powers and tendencies (mechanisms), these are the same as structure. This is the case with ULTIMATA and with PERSONS and some other geo-historical entities, which are DISPOSITIONALLY identical with their (changing) causal powers (see IDENTITY), exemplifying, at moments of TRANSITION, the coincidence of identity and change. Otherwise mechanisms and structures are distinct, i.e., mechanisms (powers and tendencies) are *of* (instantiated in) structured things. This allows for the possibilities that 'the same mechanism (for example, the market) may sustain (or undermine) a multiplicity of distinct structures {...}; the same structure {...} may be reproduced (or transformed) by the joint activity of a number of different mechanisms; the same kind of mechanism may sustain alternative structures; and the same structure may be reproduced by a variety of different types of mechanism'. While in its primary meaning a tendency just is a transfactually efficacious causal power, there are a range of distinctions to be made within this base concept which are registered in the entry for it. The polysemic and open-textured nature of the concepts clustering around the domain of the real is quite deliberate, designed to chime with different PERSPECTIVES (*PN3*: 169–70).

Causality thus embraces not only powers and mechanisms, but also their stimulating, releasing and enabling conditions. Indeed, in CR it is implicitly a 'multicomponential' and polysemic concept, which DCR draws out and makes explicit. Analytically, '*to cause is to change is to negate is to absent*' a pre-existing state of affairs; this is one of the senses in which, *pace* Hegel, Spinoza may be said to have been right that 'all DETERMINATION is negation'. Causation may occur 'in any of the four senses, and four modes, of absenting' corresponding to the moments of the CONCRETE UNIVERSAL ↔ SINGULAR (*P*: 20). This is displayed in Table 8, which highlights the role of structures – the proper object of science – in each mode. (It should be borne in mind, however, that structures themselves are emergent and subject to transformation). Each of the four modes or levels of causality is irreducible.

A *causal chain* (causation) for CR thus comprises, or may comprise – not a sequence of events, as in the regularity view of causation (see CAUSAL LAW) – but (1) the transfactual efficacy of generative mechanisms; (2) the RHYTHMIC (spatio-temporal) exercise of their powers; (3) multiple MEDIATION by HOLISTIC CAUSALITY; (4) (in the human sphere) intentional AGENCY; (5) a concretely singularised outcome ('e.g., a change in a state of affairs') co-determined by (1)–(4), 'with consequences and possible feedback effects') (*P*. 83). It is vital not to reify the distinctions made here. Most senses and modes of ABSENCE will be involved in the causation of most outcomes, and in the human sphere will typically all be involved, each mediating the other. Thus an agent involved in transformative praxis exemplifying product-in-process (4D) will be engaged in reflexive self-monitoring as well as social intra-action (3L) and multiple rhythmics involving transformative negation (2E) with their own transfactual efficacy in addition to that of her own multiform causal powers (processes-entified-in-products) (1M). All four kinds of causality, which correspond to Aristotle's CAUSES in the manner indicated, are accordingly implicated in the TMSA.

The dialecticisation of 1M thus brings out what was already implicit in it. 2E causality is 'the dynamisation, which is also the spatialisation' of causal powers in 'some more or less determinate and mediated field' (*D*: 240). This involves *directional absenting*, which is entailed by the impossibility of *backwards causation*: causes cannot come after their effects (see BLOCKISM; DIRECTIONALITY), because 'a power cannot, analytically, be possessed subsequent to, although it may be simultaneous with, its effect)' (*P*. 71). Since all absentings are in SPACE–TIME, directional absenting is *tensed spatialising PROCESS* or rhythmic, which may have supervenient causal powers of its own, including those deriving from mediation and intra-RELATIONALITY. The reality (causal efficacy) of space–time issues in a concept of the *existential constitution* of things by their

Table 8 'To cause is to change is to absent is to transform and so redetermine' (D: 240)

Causal–Axiological Chain	1M Non-Identity	2E Negativity	3L Totality	4D Transformative Agency
Fourfold polysemy of absence	*product* (simple absence, the result of some absenting process)	*process* (directional absenting) *process-in-product* (existentially constitutive geo-history)	*process-in-product* (existentially constitutive intra-relationality)	*product-in-process* (the actual or potential exercise of a product's powers, in the human world reproductive or transformative praxis)
Modes of absence	*transfactual causality* (causal efficacy) of the generative mechanisms of structures	*rhythmic causality* (exercise of the causal powers of structures [spatio-temporalising causality])	*intra-active holistic causality* of structures (totalities)	*intentional causality* of structures possessed by people as laminated structurata (stratified subjects) (reasons as causes)
Concrete universal ↔ singular	universality	processuality (rhythmicity)	mediation	concrete singularity
Aristotle's causes	material formal	efficient final	formal material	final efficient
Types of tendency	tendency$_a$	tendency$_b$	tendency$_c$	tendency$_d$

geo-historical processes of formation – manifest in the presence of the past and outside – where GEO-HISTORY embraces the whole of being, not just human being. Existential constitution thus signifies not just causal formation by, but constitutive dependency on, the past and outside. 'Geo-historical entities, such as social individuals or the cosmos, are *processes-in-products* (in process)' which must be thought 'under the aspect of their identity with their continually changing causal powers' (*P*: 61 n). In this way space (classically the dimension of difference or ALTERITY), time (classically the dimension of change) and causality are unified under the sign of the fourfold polysemy of absence, such that there is a '*constellational identity of causality* {. . .} *and space–time*' (*D*: 77).

While antecedence is indispensable to causality as becoming at 2E, causes may sometimes be simultaneous with their effects. Here we are in the realm of 3L *HOLISTIC CAUSALITY* or constitutive intra-relationality, whereby an entity may be at least partly constituted by another entity's causal powers (see also IDENTITY; REFLECTION; RELATIONALITY). Holistic causality 'may be said to operate when a complex coheres in such a way that (a) the TOTALITY, i.e., the form or structure of the combination, causally codetermines the elements; and (b) the form and structure of the elements causally codetermine each other, and so causally codetermine the whole' (*DG*: 399). Holistic causality is thus *HETEROCOSMIC*, i.e., radically alters the combining elements. It is compatible with both EXPRESSIVE and (asymmetrically) structured totalities whose elements may be tendentially mutually exclusive or dialectically CONTRADICTORY (see DCR; STRUCTURE), which are by far the more pervasive in geo-history, and which must likewise be thought as processes-in-product.

At 4D we are in the zone of *INTENTIONAL causality*, made possible by the emergent powers of MIND. Here causality is 'fundamentally absenting {. . .} by potentially rational praxis' (*P*: 214), the bearer of the possibility of a eudaimonian society. Both reproductive/transformative agency and social forms are *products-in-process*. In intentional causality, mind affects matter, i.e., psychological states affect neurophysiological states. This is a transcendentally necessary condition of human praxis, but, because we do not know what the mediating mechanism is, Bhaskar, in *PN*, introduced a concept of *transcategorial causality* (*PN3*: 90, 101 ff.), a manner of speaking which he later dropped in view of the robustness of the theory of EMERGENCE and of reasons as causes (*P*: 103).

Causality is thus a category that embraces and unifies all four of the CR stadia, which form a *causal–* or *ontological–axiological chain*. Since to cause is to absent, its home, however, is at 2E, the abode of negativity. Moreover, if at 1M causality qua absence is 'the hub of existence' (being), i.e., has ontological PRIORITY over presence, and is the most important criterion for ascribing REALITY to beings ('whatever is capable of producing a physical effect is real', *RS2*: 113), and at 3L is 'the whole in the heap, the void at the heart of being' (*D*: 229), at 2E (and 4D) it comes into its own as (intrinsically tensed, spatio-temporalising) *absenting process*, or the *unfolding of being*. See also CREATIVITY; POTENTIA; REALISM.

causal law. On the received Humean-empiricist view, there is no necessity in nature: causal laws are invariable sequences of event-types, i.e., cosmic regularities. This *regularity view of laws* is implied by the *regularity view* of CAUSATION, which ties causation as it is in the world to the presence of regularities in nature: to call a sequence of events *c* and *e* causal is to say that this sequence instantiates a regularity, namely an invariable succession between event-types *C* and *E*. Accordingly, causal laws are regularities, and in particular those regularities that embody a certain time-asymmetry. But not all regularities constitute causal laws. Nor can all regularities be deemed laws of nature. So Humeans have to draw a distinction between those regularities that constitute *laws of nature* and those that are merely accidental. Only the former can underpin causation and play a role in explanation. Among the several empiricist attempts to draw this distinction, two stand out. The first takes it that the difference between laws and accidentally true generalisations is merely a difference in our epistemic attitudes towards them. For instance, we are supposed to take as laws those generalisations that are confirmed by their positive instances (this is a version of the NICOD-CRITERION). The second, known as the Mill–Ramsey–Lewis view, claims that the regularities that constitute laws of nature are expressed by the axioms and theorems of an ideal deductive system of our knowledge of the world, which strikes the best balance between simplicity and strength. Whatever regularity is not part of this best system is merely accidental: it fails to be a genuine law of nature. But neither of these views offers a purely objective account of laws of nature. In any case, according to the Humean view, laws of nature are not necessary, since there is no necessity in nature. But laws are supposed to have MODAL force in that they support COUNTERFACTUAL conditionals.

Non-Humean conceptions take it that lawhood cannot be reduced to regularity. At best, regularities can be symptoms of the presence of laws. Some non-Humeans claim that laws are necessitating relations among natural properties. An attraction of this view is that it makes clear how laws can cause anything to happen: they do so because they embody causal relations among properties. But the necessitating relation is taken to be *contingent*: the laws that obtain in the world could have been different. Other non-Humeans take laws to hold necessarily. One central claim here is that laws of nature are ontologically dependent on the intrinsic natures (ESSENCES) of NATURAL KINDS: given that the natural kinds are essentially what they are, and given that their members are thus intrinsically disposed to behave in certain ways, the causal laws they give rise to are fixed and invariable. The distinctive CR view is that causal laws are objective and necessary: they stem from the causal powers either of particulars or of systems. CR objects to the event ontology that underlies the Humean conception of causal laws. It takes causal laws to be independent of the patterns of events that might be used as symptoms for the presence of laws. This independence is captured by the fact that laws are underpinned by generative mechanisms, which are active, enduring and causally efficacious in open systems, that is outside the controlled environment of the laboratory. Since causal laws embody claims about generative mechanisms and their TENDENCIES, they support normic (or TRANSFACTUAL) conditionals: if object A has the tendency to φ, then it is true that A *is* φ-ing, even if the φ-ing is not observed or is counteracted by other tendencies. See also CONSTANT CONJUNCTION; *RS*3; Armstrong 1983; Harré and Madden (1975); Lewis 1986a; Psillos 2002a.

STATHIS PSILLOS

causal power. See CAUSALITY; NATURAL KIND; POWER; TENDENCY.

causes: material, efficient, formal and final. Kinds of cause (explanatory factor) distinguished by Aristotle. The *material cause* of a statue, for example, is the matter from which it is made; its *formal cause* (*EIDOS*) is the implicit potential and capacities both of

62 central state materialism (CSM)

the marble and of the sculptor, tools, etc. (their essences); its *efficient cause* is the sculptor's agency, i.e., the exercise of her capacities or potentialities; and its *final cause* is its purpose or end (L. *finis*, Gr. *TELOS*): creation of a statue. Thus, as Table 9 depicts, in any creative process the material cause is what was transformed (the immanent ground at 3L and 1M in DCR terms), the formal cause is what was implicit (at 1M and 3L), the efficient cause is the external impulse or agency (2E and 4D), and the final cause the intention (4D and 2E) (cf. *SR*: 54–5; *P*: 163–4, 182 n; *MR*: 105 n).

Material causes are central to the conception of the person–society connection in the TMSA, according to which people's activities necessarily reproduce and/or change pre-given social forms, and to the materialist conception of history, where 'material circumstances' refers to social relations and ideas as well as to material objects (cf. *D*: 273). Formal causes have a CR counterpart in the causal powers and/or shaping role of structures and concepts, and receive explicit recognition in SOCIAL FORM. Efficient causes have their CR analogue in the transcendent cause or irruption of the new in the CREATIVE moment proper, and in the exercise of causal powers generally, at 2E and 4D (e.g., a belief that is acted upon). Final causes are exemplified in intentional human activity, where alone, for CR, unlike Aristotle, TELEOLOGICAL explanation is valid (as an account of what is actually done). Note that 'final' does not have its normal connotation of 'last' or 'ultimate' in this context. *Finality*: the relation or quality of being a final cause. *Counterfinality* refers to the unintended consequences which thwart or CONTRADICT the finality (purposes) of agents' intentional action. Hence it is conceptualised in PMR as unfulfilled intentionality or *karma*. See also CAUSALITY; EMERGENCE.

central state materialism (CSM). See MIND.

centre–periphery. See ABSENCE; CO-PRESENCE; DIALECTIC OF FORM TO CONTENT; FISSION; GROUND; PERSPECTIVE; REFLECTION.

centrism → triumphalism → endism. Errors endemic to IRREALISM which conceptually entail each other in the order indicated. The concepts are implicit in Marx's critique of Hegel (see HEGEL–MARX CRITIQUE), and central to the Bhaskarian critique of irrealism generally, the main elements of which are displayed in Table 10.

Centrism refers to any view which takes human being or aspects of human being (the atomistic ego, a way of knowing, a language, a politics, a party, a nation) as the centre or goal of the/its universe. The chief centrism identified by CR is *anthropocentrism* (see ANTHROPISM). In Hegel it goes hand in hand with EXPRESSIVISM (often written *centrism-expressivism*), and is rooted in the epistemic fallacy and actualism. *Decentring* or

Table 9 Aristotle's causes mapped to the concrete universal and the ontological–axiological chain

Causal–Axiological Chain	1M Non-Identity	2E Negativity	3L Totality	4D Transformative Agency
Concrete universal ↔ singular	universality	processuality (rhythmicity)	mediation	concrete singularity
Aristotle's causes	formal material	efficient final	material formal	final efficient

centrism → triumphalism → endism

Table 10 Centrism → triumphalism → endism and the realist/materialist critique

IRREALIST TEMET	*Underpinning error (unholy trinity)*	*Form of realist/ materialist critique*	*Substantive critique*
Centrism	epistemic fallacy/ actualism	transcendental realism/epistemological materialism	denegation of autonomy of nature
Triumphalism	speculative/ positivistic illusion	ontological or emergent powers materialism	cognitivism
Endism	ontological monovalence	practical materialism (cf. TMSA)	denegation of geo-historicity

Key source: P: 126

decentrification refers to a movement away from centrism, as when Darwin demonstrated that humans are not the telos of the cosmos. *Triumphalism* is the overweening exorbitation of human powers (or those of a class, group, etc.) to know, to possess and to control, entrained by the SPECULATIVE/POSITIVISTIC ILLUSION. CR takes its distance in particular from *epistemological* or *cognitive triumphalism* or *cognitivism*, as exemplified, e.g., in Hegel, 'which identifies what is (and is not) with what lies within the bounds of human cognitive competence', pointing out that 'reality is a potentially infinite totality of which we know something but not how much' (*D*: 15). Cognitive triumphalism is thus falsified by 'the depth-openness of nature' (*DG*: 401). 'There is no necessity that the world should be knowable to us, even if it is understandable why it should seem to be so. We are contingent temporary flotsam on a sea of being' (*P*: 82). While PMR is committed to the possibility of universal self-realisation in which 'at the limit we {. . .} become one with totality' (*MR*: 324), it should be noted that this depends on rejection, not acceptance, of cognitive triumphalism. Both centrism and triumphalism are rooted in the anthropic fallacy, which underpins, and is masked by, the EPISTEMIC FALLACY.

Endism – the view that while there once was history it has now come to a halt or *plateau* (adj. *plateau-nic*) – by contrast depends mainly on ONTOLOGICAL MONOVALENCE, and in fact is sometimes equated with it (e.g., *MR*: 128); in a purely positive world there could be no change, hence history would indeed be at an end. There would of course continue to be a future, but there would be no more qualitative social and institutional change or ideologies of change: the future would in this sense be constellationally englobed within the present. Such denial of the ongoing nature of GEO-HISTORICITY is characteristic of master-classes, because its continuation must sooner or later spell the end of their own rule, and is 'a potent motif of the philosophy tradition' (*D*: 64) which assists in legitimising that rule. *Post-dated endism* is the view that history has not yet come to an end, but will at some time in the future, as in some Marxist understandings of communism. *Epistemological endism* or *absolutism* is the belief that we already infallibly know all that is important to know. Endism is falsified by the irreducibly transformative nature of human praxis which absents and creates, even as it reproduces, the given. In eudaimonian society, far from historicity having ended, it would be embraced and shift up a gear as implicit and stymied human creative potentialities were unlocked and unfolded in an in-principle never-ending open evolutionary process, with an emphasis on being and letting be rather than having and controlling. If 'negative incompleteness' were ever eliminated, i.e., illusion and major social contradictions and problems, 'positive incompleteness' (absence of total development) would always remain. See also BLOCKISM/PUNCTUALISM; DEMI-REALITY; PDM.

change. See ALTERITY/CHANGE; PROCESS.

chaos/complexity theory (**CCT**), like CR, was born of the social and scientific turmoil of the 1960s. Though elements of both problematics predate that decade, the sense that it marked a turning point emboldened many scholars to experiment with intellectual syntheses. Thus, it is not surprising that CR and CCT have overlapping agendas. Two points of that convergence are germane: both sought a third, ameliorative path between the stalemated orthodoxies that dominated their respective fields; and both reconceptualised the relation between the natural sciences and the humanities, while preserving their respective objects of inquiry. Hence, CR critiqued both positivism and cultural hermeneutics and dialectically united them, while CCT sublated classical Newtonian science and the non-linear paradigms then illuminating evolutionary and historical dynamics.

The relative youth of both problematics can be seen in their shifting nomenclature. The orthographic awkwardness of the term 'chaos/complexity theory', for example, signals the factional disputes that greeted the fledgling worldview. Originally described agnostically as 'non-linear analysis', its premises and methods quickly disseminated across many disciplines. While there is some dispute over who first coined the term 'chaos', the rapidity of its adoption is now part of popular lore accompanying its history. The very idea of a 'science of chaos' communicated the same performative whimsy then inspiring the naming of sixties rock groups. As the non-linear perspective proved itself, however, whimsy gave way to the 'routinisation of chaos'. The name 'chaos theory' was replaced in some quarters with the more respectable cognomen 'deterministic chaos', though hints of the former absurdity still lingered. Later, 'dissipative systems theory' added an even more sober patina, while its routinisation seemed assured with the emergence of the lapidary title 'complexity theory'. Today, only the persistence of the orthographic oddity 'chaos/complexity' remains to suggest the unsettled factional issues still dividing the scientific community.

As for CCT itself, it is the study of SYSTEMS, i.e., bounded entities which are internally differentiated and holistically integrated. These systems can usefully be partitioned into 'equilibrated conservative systems' and 'far-from-equilibrium' (FFE) evolutionary systems. Conservative systems preserve the boundary parameters of time and space, and of matter and energy, so that they remain analytically constant across the history of the structures being examined. This parametric invariance permits formulation of law-like propositions that have predictive powers. Hence, once a constellation's initial conditions (in the case of mechanics, its position and momentum) and its dynamic principles are known, both their future and past states can be mapped analytically. A crucial requirement for such prediction is that 'superposition' operates – i.e., that the total effect of multiple causes is represented by the sum of those causal powers and can be discretely analysed into the components associated with each causal power. The predictive power of Newtonian mechanics is paradigmatic and is considered by many to be tantamount to science itself.

FFE systems, by contrast, are paragons of chaos/complexity. They are historical entities in that, over the course of their development, the bounding parameters delimiting their evolution themselves evolve. As a result of this 'evolution of evolutionary space', FFE systems are 'irreversibly' and EMERGENTLY structured entities, are 'sensitively dependent on the initial conditions' from which they originate, and follow trajectories that are 'path-dependent' in the sense that their trajectories are seldom strictly replicable. These characteristics deviate from the conservative paradigm so much that analytic mapping of potential states and outcomes becomes problematic. CAUSALITY for complex systems is complex and interactive. Superposition does not hold. This corresponds exactly with CR's understanding of the nature of causality as manifested at the level of the actual. In that FFE systems undergo irreversible changes at critical points in their temporal evolution, new system properties and boundary parameters emerge. Similarly, once set in motion, sensitivity to initial conditions allows FFE constellations to follow divergent trajectories and to establish new evolutionary end states. Hence, the predictive protocols used by Newtonian science to validate its propositions are difficult to achieve. Predictions are not, however, impossible. Predictions of future states and historical retrodiction become increasingly problematic, however, the further FFE systems move away from their initial conditions. Predictive protocols in non-linear systems are thus often said to be 'asymptotic'.

This asymptotic proviso springs from the reproductive dynamics typifying FFE systems. FFE change is non-linear, recursively structured, and progressively cumulative over the life of the system. FFE change is incommensurate in that, at critical points in their evolution, small initial differences can mushroom into disproportionately large outcomes. This recursive reproduction is markedly different from the additive and integrable nature of change in the conservative Newtonian systems. In the latter, small changes in initial conditions produce small results, whereas large changes generate large results. Finally, the iterative dynamics of FFE reproduction assume an emergent character so that FFE evolution produces irreducibly complex and ontologically layered structures.

Although the two terms are used almost as synonyms, there is a useful distinction to be drawn between chaotic systems and complex ordered systems. Chaotic trajectories are unstable, while complex ordered systems are remarkably robust. That is to say, complex ordered systems remain much the same for long periods of time but can and do undergo sudden and profound changes of kind – metamorphoses. They continue to exist but in new forms. This corresponds to CR's sense of structure as relatively permanent but inherently mutable.

As with all newborn problematics, CCT has been selectively appropriated by its competitors, many of which were objects of CCT's original critique. So-called 'big science', for example, has appropriated CCT's modelling techniques and has used them to tackle its own emergent anomalies while simultaneously retaining its hegemonic grip on scientific praxis. There are those scientists and laymen alike who interpret the mathematics of chaos/complexity from a mystical neo-Pythagorean worldview, using pre-Socratic premises to construct a teleological vision of world history. In humanist circles, advocates of postmodernism have been drawn to CCT's indeterminate ontology and its challenge to causal canons of explanation. Finally, a worldview has emerged from non-equilibrium thermodynamics that attempts to unify the sciences, the arts and the humanities under chaos/complexity itself. This later synthesis resonates closely with CR and punctuated equilibria theories of evolution. See also MEDIATE.

DAVID BYRNE, DAVID L. HARVEY

charity, principle of. See CRITIQUE.

chiasmus (f. Gr. *chiazein*, to mark with or like a *chi* [χ]). (1) A figure by which the order of words in one clause or sentence (or of syllables in a word) is inverted in another to alter the accustomed meaning. Thus Bhaskar occasionally deploys *outwith* (Scots for 'without') to subvert dichotomous or dualistic thinking by suggesting that that which is outside is also, by valid PERSPECTIVAL switch, inside; in the concluding sentences of the transcendental argument for the primacy of absence, for example – 'Within the world as we know it, non-being is at least on a par with being. Outwith it the negative has ontological primacy'(*D*: 47; cf. *SR*: 77) – 'outwith' has the effect of reminding us that epistemology is contained within ontology and suggesting that the negative has ontological primacy *within* being as such (Totality), for the unknown must be deemed (an immensely vaster) part of it (than the known). (2) A 2E figure of OPPOSITIONALITY signifying any kind of inversion or 'juxtaposition of the terms of a polarity' (*P*. 242).

choice, axiological. See AXIOLOGY; DILEMMA; PROBLEM; VALENCE.

class, social. See COLONIALISM, ETC.; EMANCIPATORY MOVEMENTS; ETHICAL IDEOLOGIES; FEMINIST THEORY; GENERATIVE SEPARATION; IDEOLOGY; LEGAL STUDIES; MASTER–SLAVE; OPPOSITIONALITY; POLITICAL ECONOMY; POWER; SOCIAL FORM, SOCIOLOGY OF KNOWLEDGE; SOLIPSISM, ETC.

classification. See NATURAL KIND; RESEMBLANCE.

closed and **open systems**. See ACTUALISM; AUTOPOIESIS; CAUSAL LAW; CAUSALITY; CENTRISM, ETC.; CLOSURE; CONSTANT CONJUNCTION; CONTRAST EXPLANATION; COUNTERFACTUAL/TRANSFACTUAL; ECONOMICS; EMERGENCE; EMPIRICAL REALISM; EXPLANATION; FINITE/INFINITE; FUTURES STUDIES; IDENTITY THEORY; POLITICAL ECONOMY; SYSTEM; TENDENCY; TOTALITY.

closed-systems (constant conjunction) form. See CONSTANT CONJUNCTION; CRITIQUE.

closure. The distinction between *open* and *closed SYSTEMS* plays a key role in CR thought. Following *RS*, a closed system is defined as any in which a strict (possibly stochastic) *CONSTANT CONJUNCTION* of events, or *event regularity*, of the sort 'whenever event x then event y' occurs. Systems in which such regularities are absent are termed open. A closure is then simply a system in which strict event regularities occur.

The terms *closure of concomitance* and *closure of causal sequence* are introduced by Lawson (2003b) to distinguish two types of closed system. A closure of causal sequence is a system in which a strict event regularity obtains and where the event, y, is causally conditioned by some earlier event, x. A closure of concomitance is a system in which a strict event regularity arises and in which the events x and y are correlated but where neither causally conditions the other (where, for instance, the movements in some third factor affect both x and y). The significance of the distinction lies in the fact that it is closures of causal sequence that are presupposed by *deductivism* (see EXPLANATION; INFERENCE). While closures of concomitance are plentiful in both the natural and social realms, closures of causal sequence are found to be rare. This latter observation is rendered intelligible by noting that closures of causal sequence presuppose the existence of a MECHANISM that is both stable and insulated from other mechanisms, a situation that from a CR perspective is relatively unlikely to occur (as least spontaneously) in a reality that is conceived as structured, emergent and so on.

PHILIP FAULKNER

cognitive science. The scientific study of the thought process. Originally cognitive science was conceptualised solely in terms of conscious processes such as symbol manipulation, memory formation, and learning. Now, however, it encompasses neurobiology, cognitive functions ranging from phonemes to concepts, and a 'neurocomputational' level connecting the two. These levels are understood as emergent layers: thought cannot be reduced to biology, nor can thought be treated as disembodied. The research supporting cognitive science often involves neuroscience, psychology, linguistics and ethnography.

According to this research, embodied experience generates iconic MODELS such as 'container', 'source-path-goal', 'force', 'link', 'cycle', 'part-whole', 'up-down', 'centre-periphery', 'hot-cold', and so forth. These provide the basic conceptual templates or 'image schemas' underlying thought. Image schemas permeate cognition in large part by being extended metaphorically to abstractions at various levels. Such activities occur almost entirely unconsciously, and so we are normally unaware that phrases such as 'person A is in a relationship with person B' or even 'Polaris is in Ursa Minor' involve a metaphor of containment ('in'). Image schemas provide the means by which we are able to conceptualise objects of investigation that we cannot perceive directly, particularly relationships among things.

Cognitive science supports the claim that there is no theory-free perception, because image schemas are involved in conscious categorisation and conceptualisation. According to CR, images and metaphors are essential to scientific theorisation; cognitive science not only verifies this point, but also demonstrates that it holds at the most fundamental levels of thought. Thus images are integral to epistemological RELATIVISM, yet because they emerge from embodied experience they do not undermine judgemental rationality.

Cognitive science has identified many other processes demonstrating that human cognition depends on numerous imaginational activities. As a whole, cognitive science has a wide range of implications for CR, such as its theories of the MIND–BODY relationship, reasons as causes, and human AGENCY. Cognitive science is also of interest because some of its researchers have drawn philosophical implications that closely match CR. See also Fauconnier and Turner 2002; Lakoff and Johnson 1999; Nellhaus 2004.

TOBIN NELLHAUS

cognitive relativism. Another name for JUDGEMENTAL RELATIVISM.
cognitive theory of ethics, cognitivism. See ETHICS.

cognitivism or **cognitive triumphalism**. The TRIUMPHALIST view, denied by CR, that reality is necessarily knowable to us. We can fallibly know (something of) reality, but this is entirely contingent. Cognitivism also denotes a non-triumphalist approach to ETHICS.

coherence theory of truth. See TRUTH.

coincidence (of distinctions and connections, identity and difference, identity and change). See CONSTELLATION; CO-PRESENCE; DIALECTIC; DISTANCIATION; DUALITY; FISSION/FUSION; TABLE OF OPPOSITION.

co-inclusion. See CO-PRESENCE.

cold/hot societies. See GEO-HISTORICITY.

collective agency. See AGENCY; INTERESTS; POLITICS.

Colletti contradiction. See CONTRADICTION; CRITIQUE.

colligation. See EXPLANATION.

colonialism, neo-colonialism, post-colonialism. *Colonialism* refers to asymmetrical relationships of domination/subordination between societies, in the last five hundred years generally European and non-European. The modern meaning of colonialism has come a long way from its origins in *colonia* (a network of Latin settlements) in early Roman history. Colonialism comprises economic, political, military, technological, moral/ethical, cultural and legal relations between societies. Its forms of governance and institutional mechanisms have undergone changes and reconstitution during the different phases of capitalism: mercantile, industrial and monopoly-finance capitalism. Colonial rule may be direct or indirect, and historically it has led to settler and non-settler forms of state (Hettne 1978; Mamdani 1996).

After the Second World War, the formal independence of colonised societies in international law gave rise to the term *neo-colonialism*. This emphasises the continuation of economic structures and relations of exploitation, the continued drain of resources and transfer of wealth from the (neo-)colonies/periphery to the imperial centre/core, and the mutual entailment of development (of the countries of the 'North') and underdevelopment (of the countries of the 'South'). The 'linguistic turn' in social theory, with its emphasis on linguisticism, discourse, relativism, deconstruction, particularism, identity and difference over structures, totalities and causality, and, most important, inattention to emancipation of the oppressed as a real possibility, produced *post-colonialism*, the colonial variant of POSTMODERNISM (Dirlik 1997).

Conceptually there is considerable overlap between the terms colonialism and imperialism. Generally, imperialism refers to the external, expansionist dimension of capitalism. It is useful to see it as driven by a generative mechanism constituted by the institutions, ideologies, cultures, practices, laws, discourses (economic, political and moral/ethical) that cumulatively give capitalist societies their capacities to dominate other societies, and their articulation with colonial societies, the SOCIAL FORMATION or structure constituted by imperialism.

Although the philosophical UNDERLABOURING of CR holds out possibilities of radical and exciting theoretical advancements, colonialism, with or without the prefixes 'neo' and/or 'post', remains an underdeveloped area of study within CR.

The challenge of colonialism is to transcend problems arising from DUALISTIC approaches prompted by the asymmetrical relationships between societies and peoples under capitalism and simultaneously to sustain ideas of UNIVERSAL values and a

colonialism, neo-colonialism, post-colonialism 69

common human destiny. The dualistic approaches are entailed in theories like dependency theory, world systems theory and critical development theory on the one hand and, on the other, identity theories underpinned by the subaltern 'Other' (see PDM).

Typically, CR interrogates the common grounds that sustain dualistic positions. The pre-eminence of ONTOLOGY in CR, taken with the geo-historical contingency of social structures, makes it possible to postulate a distinctive ontological status for colonial societies, qua societies constituted through distinctive geo-historical processes, i.e., by imperialism, the expansionist, external dimension of capitalism. Identifying the external and internal relations constitutive of colonial societies can then delineate the attributes of a colonial society. Colonial societies can therefore be known and understood by synthesising the objective (structural), subjective (agency), internal (national) and external (international) dimensions. Sociological categories of analysis like class may then be tempered by GEO-HISTORICITY without sliding into a deconstructionist project that makes it difficult to sustain theory or a mechanical structuralist project that subsumes geo-historical differences for theoretical consistency.

A CR approach to colonialism in future research programmes could help address a number of problems in social theory with regard to (depending on theoretical orientations) the neo-/post-colonial societies/'Third World'/'South'/ 'underdeveloped' societies. The most important of these is the conflation of the juridical state with the sociological state and the ramifications of the conflation for how colonialism is envisioned in the postwar era. More recently, debates about 'Empire' and 'recolonisation' in the wake of 'GLOBALISATION' highlight the tensions arising from the conflation, viz., the interrelatedness of societies on the one hand and the continued causal powers of some states over others, on the other.

The inclusiveness of the concept MASTER–SLAVE-TYPE RELATIONS means that asymmetrical relations between societies, qua societies, may be analysed, understood and explained without conflating or reducing them to categories like class relations, another persistent problem in social theory in relation to colonialism.

Equally, the diversity of possible positions and differences within CR, especially within the trajectory of (T)(D)CR/PMR, could present a CR approach to colonialism with potentially contentious questions. The most important of these relates to geo-historical DIRECTIONALITY. Historically, teleological deterministic, evolutionary theories, e.g., 'social Darwinism', were problematic for colonised peoples everywhere owing to the indeterminacy of their place in the evolutionary schema, the limitations it imposed on their own destinies, and the implicit curtailment of their right to self-determination. Such positions are generally discredited even in mainstream theory today. While the CR concept of directionality connotes a TENDENCY (towards the unity of theory and practice in practice at the level of the whole species) whose actualisation is contingent, it none the less raises issues for colonised people which must be addressed in any future research programme.

Tentatively, it is possible to speculate that attempts to resolve problems of geo-historical directionality within 'Western' intellectual traditions of scientific methods and biological and physiological analogies could see a return of the old critique of 'Eurocentrism' (see Dickens 2000). Seeking answers in inter-paradigmatic dialogues between 'Western' and 'Eastern' intellectual traditions risks detaching the

70 comment, dialectical (dc′)

moral/ethical and cultural dimensions of the colonial critique and the agency of the colonised peoples themselves from its mooring in radical, often rich, structural critique of capitalism and imperialism/colonialism.

Notwithstanding the challenges, by clearing the philosophical grounds for integrating nature-society-social structures-human lives (people), CR holds out the possibility of intertwining the humanistic inclinations of people in capitalist societies with the aspirations for self-determination of the peoples in colonised societies in theoretically consistent ways. By transcending the duality entailed in 'neo-' and 'post-' colonial critique and by tying the colonial critique to problems of human emancipation, future research programmes could pave the way for transcending the real structural divisions and conflicts of interests imposed by history and geography on people by imperialism and capitalism. See also CAPITALISM; NEOLIBERALISM; POLITICAL ECONOMY; Bhaskar *RM*, ch. 1, and (with Laclau) 1998f; D'Souza 2006; Gruffydd Jones 2001; Morgan 2003a.

RADHA D'SOUZA

comment, dialectical (dc′). See DIALECTICAL COMMENT.
commitment, axiological. See AXIOLOGY.
commodification. The process, fundamental to capitalism, whereby human capacities, products and services (e.g., intellectual property rights, parking spaces) are progressively turned into commodities or things that can be bought and sold issuing in expansion of the sphere of private property. The commodification of the human capacity to work (labour-power) is originary and paradigmatic. Closely bound up with REIFICATION and FETISHISM.
comparative method. See CONTRAST EXPLANATION.
complicity. A 2E figure of OPPOSITIONALITY signifying 'acceptance of or dependence upon an incompatible position' (*P*: 243). See also DIALECTICAL ANTAGONISTS.
compromise formation. See TINA SYNDROME.
conation, conative. See CONATUS; DESIRE; VOLITION; WANTS.
conatus (L., an attempt, effort, endeavour; impulse, inclination, f. L. *conari*, to endeavour, strive). A TENDENCY for a thing to persist in its being and to grow and develop (tendency$_b$). For Spinoza – an important influence on the ethical thinking of Collier (1999) and Bhaskar – all beings possess such a tendency, hence are valuable in themselves (have objective value). In humans the conatus takes the form of a drive, grounded in the nature of our agency and discourse (see EMANCIPATORY AXIOLOGY), to rationality (defined as the capacity to take into account the nature of the object) and to FREEDOM; this is the *pulse* of freedom. At the level of social structure, by far the most powerful conatus at work in the world today is the drive to capital accumulation (*RM*: 172 n). A *conand* is a way of speaking, commonly deployed in CR, designed to *stretch* our ordinary notions of what is normal or possible; e.g., 'you have eyes at the back of your head' (*MR*: xlvi) stretches our ordinary notion of the visual to suggest that we see and feel with our whole body-field. The *conative domain* (n. *conation*) is one of the five bases of human ACTION (see also DESIRE; VOLITION). See also DISTANCIATION; ENCHANTMENT.
concept-dependence. See CRITICAL NATURALISM.

conceptual emergence. See CREATIVITY; EPISTEMOLOGICAL DIALECTIC; INFRA-/INTRA-/SUPER-STRUCTURE.

conceptual realism. See CONVENTIONALISM; EMPIRICAL REALISM; RATIONALISM.

conceptualism. See UNIVERSALS/PARTICULARS.

concrete/abstract (L., to grow together and to draw apart, respectively). Necessary (internally related) correlatives: concrete is not-abstract, and vice versa. In so far as concrete has 'a positive meaning of its own, its nearest synonym is "well rounded", in the sense of balanced, appropriate and complete for the purposes at hand' (*D*: 128); a 3L figure closely related to TOTALITY, it satisfies all four moments of the CONCRETE UNIVERSAL and (in the human world) all four planes of SOCIAL BEING. Abstract refers to only one aspect of the concrete: 1M universality (transfactuality) (of causal laws, categories, etc.). 'It is the world itself, not our thought of it, that is abstract and ideal' (*RS2*: 237). Licit *abstraction* is thus both the (normally) mental process of isolating out and focusing on the (putatively) universal or general, omitting or absenting other aspects of the phenomenon being studied, and its end result, which is registered in *concepts* (where concept is understood in its most general sense as a principle of classification), and, within that class, definitions of real TYPES and NATURAL KINDS. Abstraction may be made at various levels of generality and geo-historical specificity and from different PERSPECTIVES, ranging, in the case of the human social world, for example, from the concretely singular person through to historically specific as well as general social relations (e.g.,, capital–labour relations and master–slave-type relations, including from the perspective of one or other pole of the relation), the human species (e.g., needs, powers), and what it shares with other animals (e.g., bodies, instincts) and with the material world (weight, extension, movement, etc.) (Ollman 2003: ch. 5). Although indispensable for science and praxis generally, abstractions are necessarily one-sided and partial, though their arbitrariness may famously be more or less 'rational'.

While concrete and abstract are empirical-relational terms (as in parent–child) rather than purely relational (as in east–west), neither concrete nor abstract is to be *equated* with the empirical or actual; i.e., mechanisms, powers and dispositions at the level of the real, as well as events or experiences, and their descriptions, are aspects of the concrete and are only abstract if treated in isolation. For CR the *fallacy of misplaced concreteness* is an error of ACTUALISM, which reduces the real to the actual, treating only the actual as concrete or real; or it identifies 'a real tendency or aspect in being with a particular manifestation in actuality' (*RM*: 170 n). Abstract and concrete have a TD/ID bivalency: Humean empiricism, for example, abstracts from the world in the TD, but is itself a concrete (well-rounded) case of empiricism in the world (ID), and its description in the TD may be concrete (or abstract). It is in fact a common mistake to equate the concrete with the empirical by comparison with the theoretical, which is deemed to be necessarily and/or uniquely abstract. It is the case that the higher the level of generality, i.e., the more removed from the data of sense, the more abstract/less concrete an analysis may be said to be, e.g., ontology is more abstract than social ontology, which is more abstract than social theory, which is more abstract than geo-history, and so on; but any theory or interpretation that successfully 'follows the object' in a well-rounded way is concrete by comparison with a theory that does not (a theoretical science is only abstract if – as is often the case – it does not retotalise its

72 concrete/abstract

conceptions to satisfy all four moments of the concrete universal), and practices and social forms as well as theories – reflecting the DUALITY of theory and practice – may be abstract (e.g., commodity exchange and money abstract from use-value; this is *material abstraction* [Arthur 2002: 79]). Thus Bhaskar opines that Marx sought to establish a 'concrete science of human geo-history' which has as its intransitive object 'the *pan-concrete* totality (of totalities)' (*D*: 128, citing T. Smith [1990]), i.e., 'the generative mechanisms and causal structures which account in all their complex and multiple determinations for the concrete phenomena of human history' (Bhaskar 1998a: xvi), and that if the description of capitalism in *Capital* is accurate its ultimate intransitive object is *capital-in-concretion*, i.e., the fundamental processual dynamic or rhythmic whereby capital tends to make the world over in its own image. *Concretion* (correlative of abstraction) is the (re-)totalising movement from the abstract to the more concrete via a dialectic of the abstract and concrete, as when philosophical theses are recast as scientific ones to assist in identifying 'their insights, aporiai, tensions and effects' and in seeing the '*complex in the simple*', and then reincorporated into philosophy (*D*: 14, 212).

Since it splits or fragments what is whole, abstraction often issues in ill effects: ALIENATION, FETISHISM, REIFICATION and WORLDLESSNESS. Bhaskar derives two sets of four *modes of illicit abstraction* (note the implication that *some* modes of abstraction are entirely *licit*), (1) from the moments of the concrete universal: *destratification*, *deprocessualisation*, *demediation* and *desingularisation*; and (2) from the moments of the ontological–axiological chain: *destratification* (again), *denegativisation* (positivisation), *detotalisation* and *de-agentification*. These modes are instances of two category mistakes which, while apparently opposed, actually complement each other: FUSION (destratification at 1M) and FISSION (the other modes at 2E–4D). Together these modes generate the PROBLEMS of philosophy and so provide a key for the metacritics of each level of DCR. They all involve an ABSENCE – 'of the concepts of structure and HETEROLOGY, concretion, relationality and totality, agentive agency, and, above all, absence itself' (*D*: 43) – which must, for the rational resolution of the problems of philosophy, enabling it to play an emancipatory role, be repaired through a process of *retotalisation*, i.e., absenting the relevant absences (*re-agentification* at 4D = retotalisation at the level of the agent). (a) Destratification or *homology* (fusion) sequesters *essential* questions of being. It is implied by the epistemic fallacy, mediated by subject–object IDENTITY theory and its generalised form ACTUALISM, and is underpinned by the logic of commodification and FETISHISM (which turns on the reduction of powers [labour-power] to their exercise [labour]). (b) Deprocessualisation and denegativisation sequester *existential* questions, positivising knowledge and being. (c) (i) Demediation absents the MEDIATION that occurs in most determination. It is an essential component of scientific experiment, where it is licit if accompanied by retotalisation. (ii) Detotalisation is a 3L figure which also occurs in the other three modes of illicit abstraction, and so is regarded by Bhaskar as affording 'an umbrella response to the problems of philosophy'. It sequesters or splits off a part of being, e.g., facts from values, reflexivity from philosophy, and is 'symptomatic of an alienated world' (*D*: 321) which the *retotalising* DIALECTICS OF FREEDOM aspire to transcend. Finally, (d) (i) desingularisation absents or ignores the concrete singularity of any thing; (ii) de-agentification is its human form, which is itself of two main kinds: (dualistic)

DISEMBODIMENT or *disintentionality*, and (reductionist) reification, productive of radically de-centred, impotent and empty selves. The dialectics of freedom therefore incorporate *re-agentification* (*re-centrification* and *re-empowerment*) as an essential component.

In philosophy and science and, indeed, human praxis generally, however, abstraction is a necessary means, given 1M transfactuality and the universality of causal laws, of arriving at a more adequate and ultimately concrete theory. It reaches its zenith, perhaps, in the directly experimental sciences, where there seems to be 'a kind of implacable inverse-law by which the values of precision and generality are achieved only by abstracting away from the conditions of real-world applicability' (Norris 2003: 10–11). For science to promote human flourishing, however, what has been broken up or split must then be retotalised in a richer synthesis (the concrete in thought). This duplex movement from the concrete to the abstract and the abstract to the concrete is admirably exemplified in the materialist DIFFRACTION and retotalisation of the Hegelian dialectic by Marx and Bhaskar, and hopefully in the logic of this dictionary. See also DIACHRONY/SYNCHRONY; EPISTEMOLOGICAL DIALECTIC; IRREALISM; MELD; RELATIONALITY; TOTALITY; VALENCE (ontological monovalence); Sayer 1984; *SR*: 301–2.

concrete singular. See CONCRETE UNIVERSAL ↔ SINGULAR.

concrete totality. See CONCRETE; CONCRETE UNIVERSAL; TOTALITY.

concrete universal ↔ singular. The moments of CR's *concrete universal* comprise a *multiple quadruplicity* corresponding to the 1M–4D modes of being–becoming: *universality* (1M non-identity), *processuality* (2E negativity), *mediation* (3L totality), (concrete) *singularity* (4D transformative agency, but also betokening result or outcome generally) (*UPMS*), as illustrated in Table 11. 'Concrete' does duty here in its positive meaning of 'well-rounded'. However, it is not essential to the CR concept, unlike Hegel's, 'that it, or its components, comprise TOTALITIES' (*D*: 128). An extended materialist version of Hegel's, and arguably implicit in Marx, it eschews Hegel's 'realised idealism', viewing universals as properties of particular things, and adds PROCESSUALITY (the spatio-temporalisation of structural effects – or their geo-historical trajectory in the human world – which Hegel's actualism cannot accommodate) to Hegel's *UPS* schema (universality, particularity, singularity [*Allgemeinheit, Besonderheit, Einzelheit*] – where 'particularity' corresponds to [particular] 'mediation[s]' in the UPMS schema). Like Hegel, however, it rejects the abstract universality of modernity in favour of dialectical universality; the former presupposes the external RELATIONALITY of things, the latter the constellational containment of external within internal relations. Given the materialist DIFFRACTION of dialectic, the CR concrete universal comprises a *multiplicity* of quadruplicities, hence is dubbed a 'multiple quadruplicity', all of whose elements are in principle separable (i.e., are real distinctions within a dialectical unity, neither to be collapsed into oneness nor treated as autonomous) and subject to multiple determination. Separability is attested by the *demediation* that occurs in scientific experiment, and by the fact that structures may manifest universalities (qua unactualised possibility) without singularities. Multiple determination is a necessary feature of open SYSTEMS.

What Bhaskar calls the *concrete singular* thus combines all four moments of the concrete universal, or the concrete universal = concrete singular (which might therefore also be called a *concrete DUALITY* [*D*: 178, 199] or totality). A human being, for example, consists

74 concrete utopianism

Table 11 The concrete universal ↔ singular

Causal–Axiological Chain	1M Non-Identity	2E Negativity	3L Totality	4D Transformative Agency
Concrete universal ↔ singular (UPMS)	universality	processuality	(particular) mediation(s)	(concrete) singularity (not just human)
Hegel (UPS)	universality		particularity (mediations)	singularity
Qualities of alethic truth	grounded	dynamic	totalising	context-sensitive

in a core universal HUMAN NATURE, particular mediations and the RHYTHMICS of her world-line, 'uniquely individuating her {...} as in effect a natural kind sui generis' (*DG*: 395). It thus displaces the regnant, constitutive and false duality of the PHILOSOPHICAL DISCOURSE OF MODERNITY: abstract universality and the atomistic ego (*RM*: 78–9). The logic of concrete singularity is to treat all individuals, though sharing a common humanity, as ethically different, not – as in Kantian ethics – the same; 'ought presupposes, not implies, can' (*P*: 146 n, 148). In this way, the concrete universal emphasises the interconnectedness and becoming of being, rendering PHILOSOPHY 'concrete-dependent' (*D*: 108); and concrete singularity is 'truly the key to the realm of FREEDOM' (*D*: 264). Concrete universal ↔ singular grounds the concept of dialectical UNIVERSALISABILITY, which is a fortiori radically different from abstract universalisability, and which plays a central role in the DIALECTICS OF FREEDOM. Within the CR system, concrete universal ↔ singular functions as a generative grammar, a criterion of truth, and a test for consistency and wholeness. Thus its four moments are broadly aligned with all the correspondences of the MELD schema, including the qualities of alethic truth; and it provides criteria for EMANCIPATORY AXIOLOGY. Hostettler and Norrie (2003) correctly suggest its importance for ETHICS, but perhaps emphasise the unity of its moments at the expense of their distinctness. See also EMANCIPATORY AXIOLOGY; MEDIATE; PDM.

concrete utopianism. Introduced to the world as a concept by 'the philosopher of utopia', Ernst Bloch (1959), who contrasted it with *abstract utopianism*. The Bhaskarian concept is in some respects similar, as are the systems in which the concept is located, and Bhaskar himself may be entitled to join Bloch in Adorno's list of 'the very few philosophers who {do} not recoil in fear from the idea of a world without domination and hierarchy' (cit. Geoghegan 1996: 162). What has above all inspired the Bhaskarian concept, however, is the theory and practice of William Morris (1891, 1973; see also Beaumont 2005; Thompson 1959).

For Bhaskar, concrete utopianism 'consists in the exercise of constructing MODELS of alternative ways of living on the basis of some assumed set of resources, counterbalancing ACTUALISM and informing hope' (*DG*: 395). This is by no means the exclusive prerogative of (emancipatory) social science; it is a freely flowing energy inscribed in the processes of transformation themselves (and indeed is consciously established practice in sections of the current Green, Red and Black movements) (cf. Holloway 2002). The Bhaskarian concept is introduced in the course of elaborating the

DIALECTIC OF FREEDOM, where it is a vital component of EMANCIPATORY AXIOLOGY; and it reverberates through the later philosophy. As in all model-building and thought-experiments, 'creative fantasy' (cf. Bloch) plays 'a constitutive role', 'identif{ying} "the positive in the negative"', yielding 'at once hope and possibility to {. . .} totalising depth-praxis' (*D*: 209, 294) – a process in which praxis educates fantasy, and fantasy praxis (and which thus has nothing in common with actualist blueprints or inflexible planning). Gramsci's slogan is amended to read, 'optimism of the will and *realism*, informed by concrete utopianism, not pessimism, of the intellect' (*P*: 215). CONCRETE is intended in its positive meaning of well-rounded and appropriate for the purposes in hand, and links the concept to the CONCRETE UNIVERSAL; if it is not concrete, utopianism is taken in a pejorative sense as not 'naturalistically grounded in a fully FOUR-PLANAR analysis of human being' (*D*: 350) and so not satisfying principles of ACTIONABILITY and PREFIGURATIONALITY. (In Blochian terms, this distinguishes the 'objectively-real possible' [concrete] from the merely 'formally' possible [abstract] [Geoghegan 1996: 4, 32].) Its function is precisely to 'pinpoint the real, but non-actualised, possibilities inherent in a situation, thus inspiring grounded hope to inform emancipatory praxis' (*P*: 112 n). Concrete *utopianism* is grounded philosophically in, among other things, a keen sense of the reality, ontological primacy and boundlessness of unactualised possibility; and the theory of the transcendentally real SELF in the later philosophy stresses that in many areas of our everyday lives we already act in 'the way social utopians have believed we could act' – in terms of reciprocity, non-egoism, trust and reconciliation – or, 'spirituality is a concrete reality, here and now' (*RM*: 15–16). At a meta-philosophical level, Bhaskar engages in an exercise of '*metacritical (metatheoretical) concrete utopianism*' which attempts to articulate the tendential rational DIRECTIONALITY of the historical process (*D*: 279).

CR's demonstration of the openness of the world and the insistence of the pulse of FREEDOM, which has a positive as well as a negative moment, leads it to reject the endist postmodern pessimism of writers like Jameson (2004: 46) who defend the 'essential reasonableness' of the view that the function of utopianism 'lies not in helping us to imagine a better future but rather in demonstrating our utter incapacity to imagine such a future – our imprisonment in a non-utopian present without historicity or futurity – so as to reveal the ideological closure of the system in which we are somehow trapped and confined'. Jameson's (qualified) ENDISM is itself a negative utopianism. As Eagleton avers (2000: 174), 'the truly starry-eyed utopian {. . .} is he who imagines that the future will be pretty much like the present'; for the stupendous dynamic of capital accumulation is running up against absolute ecological constraints and other CONTRADICTIONS. The common aspersion that utopian thinking leads to the slaying of millions is ideological nonsense: the 'slaughter-bench of history' (Hegel) is owned and operated by power$_2$ and its slaves (cf. *D*: 339, 366–7). See also EUDAIMONIA; PERSPECTIVE; UNIVERSALISABILITY; Beaumont 2005; Patomäki 2004.

conditionality requirement. The condition in the principle for the highest form of FREEDOM, universal flourishing: 'the free development of each as a *condition* for the free development of all'. It states that it is 'only if I recognise the concrete singularity of each and every other individual that I am not guilty of theory/practice inconsistency and heterology in asserting my own' (*P*: 149). Freedom is accordingly *indivisible*.

confirmation. See THEORY CHOICE.

76 conflation, conflationism

conflation, conflationism. See SOCIAL STRUCTURE; SOCIAL THEORY.

conjuncture, adj. conjunctural. Calls attention to the fact that DETERMINATION in open systems is multiple. Its mainstream English usage goes back to the seventeenth century, where it means (1) any combination of events and circumstances, and (2) one that is critical or betokens a crisis. Althusserian Marxist usage exploits (2) to denote 'the exact balance of forces, state of OVERDETERMINATION {...} of the contradictions at any given moment to which political tactics must be applied' (Althusser and Balibar 1970: 311).

connection, connectionism, connectivity. See CAUSALITY; CONSTELLATION; CONTRADICTION; DCR; DIALECTIC; DUALITY; MEDIATE; RELATIONALITY; TOTALITY; TRANSITION.

connector point. See TRANSITION.

conscience. See EMANCIPATORY AXIOLOGY; SELF.

consciousness. To be distinguished from the conscious MIND, which it constellationally embraces and which is, on the other hand, only the 'tip of the iceberg' of mind as a whole, as Freud (*PN3*: 111) and COGNITIVE SCIENCE have taught us. (1) Embodied (capacities for) 'subjective states of sentience or awareness' (Searle 1995, cit. Moll 2004: 54) common to many life-forms. It includes the 'supramental' or 'unthought' (consciousness unmediated by thought), as in basic human ACTIONS and understandings or spontaneous practice drawing on tacit knowledge (*D*: 147 n.). (2) Embodied (capacities for) *self-consciousness*, reflexive self-awareness, REFLEXIVITY or INTENTIONALITY, involving the capacity to be aware of what we are doing, and that we are doing it (second-order monitoring) (*RS2*: 48 n32; Archer 2003b: 95). This presupposes a capacity for REFERENTIAL DETACHMENT, entailing a self-world, TD/ID, distinction and the capacity to have knowledge, via the production of concepts, models, etc., of objects in the ID, including of our own activities and nature, which is a condition of possibility of SUBJECTIVITY, and vice versa (*D*: 325). (2) is peculiar to humans, presupposing possession of language phylogenetically (*PN3*: 81–2). In Archer's (2000) ontogeny, self-consciousness or reflexivity is the (prime) EMERGENT personal POWER (PEP) to deliberate internally upon what is to be done in the natural, practical and social circumstances that people encounter. (3) Understanding of the world, whether at the level of person (whether token or type), group (e.g., Christian consciousness), SOCIAL FORM (*forms of consciousness*, as in the UNHAPPY CONSCIOUSNESS), or human species (e.g., its 'moral consciousness', which is 'in principle open', or today's 'globalising self-consciousness' made possible by developments in electronic technology [*D*: 211, 270, e.a.]), and whether theoretical (*discursive*) or *practical consciousness* (cf. FORM OF LIFE). In PMR, the ALIENATION of self-consciousness (absence of reflexive consciousness of self) underpins all the forms of alienation and division in the demi-real (*EW*: 38). *False* or *inadequate consciousness*: see IDEOLOGY. *Transcendental identity* (*in*) *consciousness*: see META-REALITY. See also DIALECTIC OF CONSCIOUSNESS AND SELF-CONSCIOUSNESS IN REASON; EVOLUTION OF SOCIETY.

consensus theory of truth. See TRUTH.

consequence explanation/law. See FUNCTIONAL EXPLANATION.

consistency/inconsistency (f. L. *con*-, together + *sistere*, to put, place). Consistency is among the necessary conditions for social life (*PN3*: 62). CR sets great store by it – in the context of a critical and emancipatory rationality that constellationally

consistency/inconsistency

overreaches the formally logical and merely instrumental – ultimately because human agents have an interest in it as indispensable for discourse, discovery, praxis, autonomy and flourishing. Bhaskar (*SR*: 200) cites Nietzsche with approval: 'From the beginning we are unlogical and therefore unjust beings and we can know this: this is one of the greatest and most insoluble disharmonies of existence.' For Bhaskar, justice, emancipation and enlightenment consist ultimately in theory–practice consistency, which is fundamentally consistency with our transcendentally real selves (see EMANCIPATORY AXIOLOGY).

It is important to distinguish between the (1) *formal LOGICAL* consistency of statements, (2) *dialectical, developmental* or *DIRECTIONAL* consistency of a PROCESS, exemplifying *developmental negation*; (3) *theory–practice* (T/P) or *practico-epistemological* consistency of agents in a process; and corresponding forms of *inconsistency*. REASON or RATIONALITY embraces all three forms of consistency.

(1) A set of statements are logically consistent if they are free of logical CONTRADICTIONS, i.e., if they do not breach the law of non-contradiction. We may speak of intra- and inter-theoretical consistency (*theory–theory consistency*) in this sense.

(2) There are no general formulas for developmental consistency, which is 'redeemable only in the course of, and at the end of, the day' (*D*: 1): criteria of progressive import are necessarily specific to the process concerned, e.g., the growth of an organism or a research programme. The latter necessarily breaches the law of non-contradiction at vital points (see EPISTEMOLOGICAL DIALECTIC) and depends upon it at others. 'To be developmentally consistent is to know when to be inconsistent, when to grow, when to mature, when to apply a DIALECTICAL COMMENT on dialectical comments, when to stop and when to wait until the agents concerned have made up their own minds into what their freedom consists. Dialectical processual consistency recognises the authenticity of every concretely singular agent's own narrative or story no less than the rights of her being' (*D*: 170). Developmental consistency is exemplified by the Hegelian 'speculative proposition', which describes a subject (society) – the site of necessary contradictions – in a complex process of formation; a process designated by Bhaskar *subject negation* or *subject-developmental negation* (see FIXISM/FLUXISM).

(3) T/P consistency in a praxis in a process embraces both (1) and (2). It is imposed, in a world in which we cannot abstain from acting, by the REALITY PRINCIPLE or alethic truth. Because all theory is also practical and all practice quasi-propositional, *theory–practice* constitutes a DUALITY, i.e., one may be seen under the aspect of the other. Therefore (a) either a theory or a practice may be T/P consistent or inconsistent; and (b) formal logical contradictions qua intra- or inter-theoretical (*theory–theory*) contradictions may be a form of T/P inconsistency (or of *practice–practice* inconsistency [see AXIOLOGY]). Since all four moments of the CONCRETE UNIVERSAL are involved in process, T/P consistency is formally 'transfactual, processual, concrete, agent-specific and transformative'. While criteria covering every particular case cannot be given for it any more than for (2), such that 'only detailed skilled judgement will do' (*D*: 170), no counter-hegemonic praxis would be likely to succeed without (i) being ACTIONABLE (capable of being accomplished by *these* agents), (ii) specifying a plausibly attainable end which the praxis (practical reason) PREFIGURES (entailing *means–ends consistency*), and (iii) elaborating explanatory theories (theoretical reason) of the current

78 constant conjunction

conjuncture, of the goal aspired to and of the TRANSITION from one to the other. As a test for consistency and a criterion of their truth, both the praxis and the theories should be dialectically UNIVERSALISABLE. Lack of universalisability signifies T/P incompleteness (*P*: 243).

Inconsistency or *incoherence* is intimately related to incompleteness and lack of wholeness. As a form of ABSENCE, 2E T/P inconsistency or contradiction (also *PERFORMATIVE* contradiction, *REFLEXIVE* inconsistency, *transcategorial* inconsistency) – the denial in theory or practice of what is necessary in practice – and more generally any lack or inadequacy in a theory, practice or social form, may manifest itself as 3L split or de-totalisation, 4D auto-deconstruction and 1M non-identity, where a theory may require something alterior to itself for completion, e.g., a TINA supplement. Its archetype is given in SCEPTICISM. The dialectic of agency and discourse (see EMANCIPATORY AXIOLOGY) shows that, for acting and speaking social beings, theory–practice inconsistency is a form of heterology or ALIENATION. It 'leads to pathologies of action, from repression through compromise formation to ad hoc grafting' (*DG*: 395). The demonstration and resolution of theory–practice inconsistency in a (from Hegel) *dialectic of inconsistency and incompleteness* within the dialectic of dialectical and analytical reason (*D*: 84; see also GÖDELIAN DIALECTIC) is the fundamental motor of immanent CRITIQUE and dialectic generally, and of emancipatory axiology. Though dependent on its correlative (consistency), inconsistency must accordingly (*pace* formal logic) 'be conceded a value in its own right' (*D*: 280). See also PHILOSOPHY; PROPAEDEUTIC.

constant conjunction (of events). Hume's EMPIRICAL REALIST term for a CAUSAL LAW, otherwise known as an *empirical invariance* or universal empirical generalisation ('whenever x, then y'). Constant conjunctions presuppose closed SYSTEMS and the truth of *regularity DETERMINISM*; their specifically CR name is $TENDENCY_7$. However, in our open-systemic world, closed systems exist only rarely outside the artificially closed conditions of experimental laboratories. Because they are produced by humans, constant conjunctions cannot be either a sufficient or even a necessary condition for a causal law; regularities are the empirical grounds for laws, not the laws themselves, which have 'a real basis independent of events' (*RS2*: 45). The illusion that constant conjunctions are all we have to go on in science – that causal laws *are* constant conjunctions which constitute the world – is a species of FETISHISM, based on the EPISTEMIC FALLACY, which may also be viewed as 'constant conjunctivitis' (*D*: 117); this is the *constant conjunction* or *closed systems form*, the social meaning of which is concealment of the reality of (social) structural causation (*R*: 9). 'There is no statement which is universal and empirical {. . .} the way in which true universality operates in nature {. . .} is behind the surface of events which it governs and co-determines' (*RM*: 79). See also ACTUALISM; DETERMINISM.

constative. See PERFORMATIVE.

constellation. Adapted by Adorno from Walter Benjamin ('ideas are related to objects as constellations to stars': cit, Jarvis 1998: 175), and Bhaskar in turn made it his own term of art (*D*: 114; Norrie 2004a) as 'a figure of containment within an overreaching term {. . .} from which the overreached term may be diachronically or synchronically EMERGENT' (*DG*: 395–6, e.a.); it is used in both 'materialist non-pejorative and idealist pejorative ways' (*D*: 115). Thus epistemology (TD) is

constellational identity

constellationally contained within (or *connected* with, or *embraced* or *overreached* by) ontology (ID), the present-future within the past, reasons within causes, structure within agency; power$_2$ within power$_1$, theory within practice, etc.; or ontology *constellationally contains* (embraces, overreaches) epistemology, etc.: ontology > epistemology; past > present-future; causes > reasons; agency > structure, power$_1$ > power$_2$, practice > theory. The two together constitute a *constellation*, which may be a species either of IDENTITY or of DUALITY or UNITY; they are, then, *constellationally relative* terms (*P*: 151). Strictly speaking, constellational *containment* '(in the sense of being part of)' signifies constellational *IDENTITY* (of which all the above are examples) and constellational *connection* '(in the sense of being bound together)' *constellational UNITY* or DUALITY (*D*: 114–15), e.g., the unity of the ethical and the political within the political. However, identity and unity are sometimes used interchangeably. (Constellational containment/connection is closely related to preservative SUBLATION, of which it is the genus.)

Constellation is thus a 3L concept, incorporating the figure of emergence and closely related to TOTALITY – indeed, another name for it – and *constellationality* names (1) the relationality of its elements, conveying the idea of their internal and necessary connectedness, and (2) their distinctness. Constellationality is accordingly *the coincidence of real distinctions and connections* in the world, or *the CO-PRESENCE of non-identities* (see ALTERITY/CHANGE) *within an overarching IDENTITY or UNITY (identity-in-difference or unity-in-diversity)*, the thinking of which requires the art of dialectic. After the dialectics of co-inclusion or co-presence, which embrace it (co-presence > constellationality), constellationality is the key figure of MEDIATION and the key to resolving the problem of OPPOSITES. Thus the problem of the oppositionality of subject and object is resolved in the figure of the 'constellational unity of the unity of subject and object {. . .} within subjectivity {. . .} within objectivity' (*D*: 272), that of dialectical and analytical reason in the 'constellational unity of analytical and dialectical reason within dialectical reason for the sake of absolute reason' (*P*: 99, e.a.). Constellations and totalities will in turn be recursively embedded within others. In a world characterised by irreducible change and difference, they are partial and in open PROCESS, intra- and inter-connected by all manner of contingent and external as well as internal determinations. *Constellational realism* holds that to exclude anything from being is to detotalise or alienate it (*EW*: 34); it is thus REALISM about everything.

In a *constellation of concepts* (*PN3*: 169), such as is articulated in this dictionary, the elements are both interdefined and distinct, and interarticulated with other conceptual and experiential constellations, as well as in movement. A *constellation of social structures* constitutes a SOCIAL FORMATION. See also ALTERITY/CHANGE; CENTRISM, ETC.; DEMI-REALITY; DISTANCIATION (constellational embedding); GEO-HISTORICITY (constellational closure); HOLY TRINITY; constellational MONISM; constellational PRIMACY or priority.

constellational identity. Bhaskar's term of art for a figure found in Hegel and redeployed in non-actualist manner. See ALETHIA; ALTERITY/CHANGE; CAUSALITY; CONSTELLATION; DISTANCIATION; HOLY TRINITY; IDENTITY; INTRANSITIVE, ETC.; SPACE–TIME; ULTIMATA.

constellational unity. See ALTERITY/CHANGE; CONSTELLATION; DUALITY; EPISTEMOLOGICAL DIALECTIC; INTRINSIC/EXTRINSIC; UNITY.

constitution, existential. See CAUSALITY; RELATIONALITY.

constitutionality. The *principle of constitutionality* calls attention to the existential constitution of any present (distanciated stretch-spread) by the past and outside (*P*: 143). It is a significant consideration in EMANCIPATORY AXIOLOGY. Contrasts with the principle of PREFIGURATIONALITY, which inhabits the present-future.

constitutive falsity (false necessity). See ALETHIA; ALIENATION; CRITIQUE; DEMI-REALITY; DIALECTICAL ARGUMENT; IDEOLOGY.

constitutive perspectivity or **perspectivality**. See PERSPECTIVE.

constitutively false morality. EXPLANATORY CRITIQUE may demonstrate the *constitutive morality* of a social system to be both false and necessary, e.g., that capitalist society is founded on the false notion, necessary for its reproduction, that wages represent full and fair payment for labour performed, or that master–slave-type societies in general are founded on the false notion, necessary for their functioning, that GENERATIVE SEPARATION is objectively moral (*D*: 264–5, *P*: 112–13). See also DCR.

constraint/enablement. A *constraint* is 'an absolute or relative prohibition, whether natural or social and remediable or not' (*DG*: 396). An *enablement* by contrast makes something possible. Constraint is a negative, enablement (or *enabling condition*) a positive, form of CAUSALITY, which like any other form can be possessed unactualised. And, since (it is argued – but to see it requires a series of PERSPECTIVAL SWITCHES) to constrain is to contradict is to negate is to cause, constraint is closely linked to the central category of ABSENCE and absenting (underpinning change and difference alike). When subscripts are used to distinguish constraints that are operative in the human world from other forms of constraint, *constraint*$_1$ (corresponding to POWER$_1$ and SUBJECTS$_1$) refers to the capacity to constrain as such, whether by a social form or not; *constraint*$_2$ (corresponding to power$_2$ and subjects$_2$) refers to social constraints, and in particular the constraining power of MASTER–SLAVE-type relations. Likewise, Bhaskar distinguishes enablement from *empowerment*. The former is more inclusive, with empowerment reserved for human agency. Emancipatory praxis proceeds in part via a dialectic of empowerment and constraint.

Both constraint and enablement play a key role in the TMSA, where they appear as modes in which social structure *conditions* agency. They arguably entail an ontological distinction between structure and agency, society and people: there must be something that is constrained and enabled and something that does the constraining and enabling. In DCR, constraint is generalised to include any kind of *ill* in the social sphere, for ills may be viewed as absences which function as constraints on human causal powers to flourish; this is constraint as the *normatively negative* form of absence. It is then in turn *NEGATIVELY GENERALISED* (accompanied by a positive generalisation of the concept of FREEDOM) 'so that to constrain an ill is to disempower, contradict, overthrow or otherwise *constrain a constraint*' (*D*: 277, e.a.). Constraining constraints thus figures prominently in the real DEFINITION of geo-historical dialectic as 'absenting constraints on absenting ills (or absences) – or, in effect, *the axiology of freedom*' (*D*: 238). By valid perspective switch, any constraint on an agent's flourishing may be seen as ALIENATION from what is intrinsic to her well-being. See also CONTRADICTION; EMANCIPATORY AXIOLOGY; ETHICS; NEEDS; PRIMACY (of constraints over enablements in the demi-real).

constructionism (constructivism). See SOCIAL CONSTRUCTIONISM.

containment. See CONSTELLATION; CO-PRESENCE.

content–form. See ABSENCE; CO-PRESENCE; CRITICAL NATURALISM; DIALECTIC OF FORM TO CONTENT; EMANCIPATORY AXIOLOGY; LEARNING PROCESS; PERSPECTIVE; REFLECTION.

context of discovery, context of justification. See ONTOLOGY; PHILOSOPHY OF SCIENCE; TRANSCENDENTAL REALISM.

contingency. See MODALITY.

contradiction (f. L. *contra*, against + *dicere*, to speak) may refer to 'any kind of dissonance, strain or tension', but is most clearly specifiable in relation to human action, where it refers to 'a situation which permits the satisfaction of one end or {. . .} result only at the expense of another: that is, a bind or CONSTRAINT. An *internal contradiction* is then a double-bind or *self-constraint*' (*D*: 56, e.a.). This meaning may be generalised to any form of action. *External contradictions* (constraints) are cosmically pervasive, 'established by the mere fact of determinate spatio-temporal being' (*D*: 57), which is subject to constraining laws of nature. Internal contradictions probably are pervasive, too, for change would not be possible unless ongoing things possessed both a CONATUS to persist in their being and a conatus to desist (either internal 'COMPLICITY' with exogenous sources of change or an internal counter-conative tendency). This establishes their FINITUDE, which is the most basic form of *existential contradiction*. Contradiction is thus a 2E figure *par excellence*.

Formal logical contradictions ('both A and $-A$') are a species of internal contradiction which obtain between propositions. When committed, they are constituents of the social world, *pace* Hegel, who LOGICISES the whole of being, and Kant, who extrudes them from being. They include *theory–theory* INCONSISTENCIES. In virtue of the DUALITY of theory and praxis, *theory–practice* inconsistencies are *quasi-logical contradictions*. (Practice–practice inconsistencies are AXIOLOGICAL inconsistencies or *contradictions*.) The norm or *law of non-contradiction* – the centre-piece of the ANALYTICAL PROBLEMATIC – states that we cannot truly affirm of a given subject *both A and $-A$* (it is *either A or $-A$*) (this should not be confused with the law of excluded middle, which prohibits *neither A nor $-A$*). A necessary condition of intentional action, it applies transhistorically (Archer 2000) but is subject to the principle of epistemic relativity or geo-historicity: the characterisation of 'A and $-A$' will be context- and meaning-dependent. Further, it does not hold in regard to points of TRANSITION and intersection, arbitrarily rules out epistemologically undecidable phenomena, and is necessarily violated by any science at critical points (see EPISTEMOLOGICAL DIALECTIC; TABLE OF OPPOSITION).

Dialectical contradictions, which are also internal, are a species of *dialectical connection* (of opposing elements) – itself a species of *necessary connection* and *connection*: connections ≥ necessary connections ≥ dialectical connections ≥ dialectical contradictions. In dialectical connections, the elements of a TOTALITY are distinct yet inseparable in that they existentially presuppose (constitute or causally intra-act with) each other. In dialectical contradictions, the elements are in addition tendentially mutually exclusive, hence potentially or actually transformative: dialectical contradictions ≥ antagonisms ≥ conflicts ≥ overt struggles (*D*: 59). They contrast with *non-dialectical oppositions*, e.g., 'a Newtonian conflict of forces or a Kantian *Realrepugnanz*', in which the elements stand in an external (although perhaps necessarily connected) relationship to one another (*PN3*: 71; *P*: 87–8). Dialectical are not as such logical contradictions, but the latter are dialectical contradictions when grounded in a common mistake. However, we can designate them as such only when – as 'Hegel intends to do, and sometimes succeeds (brilliantly) in doing' (*D*: 59) – we can isolate the mistaken ground and remedy its contradictoriness. Instead of fearing contradictions, therefore, we should exploit them for their *dialectical fertility*. Power$_2$ social relations exemplify antagonistic dialectical contradictions, constituting a *unity* (not identity) *of opposites* – the elements are connected, not contained, within a whole. These are the mainstay of materialist analysis and critique. A *Colletti contradiction* is a specific form of dialectical contradiction involving 'the necessary co-existence in social reality of an object and a categorially false presentation of it', e.g., the wage-form and the social relations that generate it (*PN3*: 70; see also CRITIQUE). Bhaskar follows Marx in distinguishing also the following. (1) *Structural dialectical contradictions*. These are synchronic or loco-periodised, intrinsic to a particular social form (e.g., between wage-labour and capital). (2) *Geo-historically specific dialectical contradictions* that contribute diachronically to bringing new social forms into being through the operation of opposing TENDENCIES$_2$ (e.g., of the forces and relations of production), and are also referred to as *RHYTHMIC contradictions*. These are grounded in (1), structural contradictions. And (3) an original dialectical GENERATIVE SEPARATION (contradiction) or ALIENATION, which is a necessary condition of both (1) and (2) (*D*: 70–1, see also CO-PRESENCE). The materialist dialectic differs from the Hegelian in that it is *simultaneously* dialectical and contradictory, involving 'a simultaneity of grounded (transfactual) presence and (actual) absence, of practical (existential) inclusion and mutual (tendentially transformative) exclusion. It is this that makes it genuinely dialectically contradictory in a stratified ontology that pre-exists the discourse that describes it' (*D*: 62–3). This is the basis of the materialist DIFFRACTION of dialectic (see also TABLE OF OPPOSITES). The resolution of such contradictions can only be practical – the non-preservative transformative negation or sublation of their ground in a non-antagonistic social form through what Bhaskar theorises as transformed, transformative, totalising, transformist praxis (TTTTΦ).

contraries. See DIALECTICAL ANTAGONISTS; NEGATIVE PRESENCE; TABLE OF OPPOSITION.

contrast explanation. Critical realists argue that social reality is quintessentially open, structured, processual and interconnected. Attention has accordingly been given to uncovering explanatory methods appropriate to such conditions. One such (advanced by Lawson 1997 and 2003b, drawing on the tradition of the comparative method) is contrast EXPLANATION. The essence of this approach is, not to explain some x, but rather to explain why x rather than y in conditions where y was expected, given that a process thought to be the same as that producing x has produced y. For example, the quest is not to explain crop yield (which involves knowing all the factors responsible) but why it is much higher at one end of the field. The point here is that by asking why x rather than y, that is, why at one end crop yield is higher (x) rather than the same as elsewhere (y), it can with reason be assumed that all factors affecting yield are fairly constant throughout the field over time except the one (set) making the difference to the yield. The application of the method of contrast explanation requires, then, merely that (1) over some region, referred to as the *contrast space*, good reasons are available to encourage researchers to expect that two outcomes of a certain kind have the same or a similar causal history; and (2) that a posteriori the researchers are surprised by outcomes that diverge from those anticipated. The controlled experiment can be seen as constituting a special case of the method of contrast explanation. However, closure is not a necessary condition for the success of projects in contrast causal explanation. The conditions necessary for contrast explanation can be shown to hold for the social as well as for the natural realm.

One practical implication of the method of contrast explanation is that, if further understanding of any open-systemic, multiply determined event is sought, a productive strategy involves comparing the given phenomenon with an array of different contrasts or foils. More broadly, and in line with certain themes in feminist theory, contrast explanation not only highlights the situated nature of the researcher, but also insists that this situatedness or standpoint (see PERSPECTIVE) constitutes a resource rather than a constraint. Here interested standpoints are valuable, even indispensable, aids to the explanatory process. The detection of previously unknown causal mechanisms via this method presupposes that people are capable of forming relevant contrasts, are liable to perceive some of them as puzzling, unusual, undesirable, etc., and disposed to act on their surprise. Since it is the interests of the researcher that determine which causal mechanism is likely to be pursued, it follows that science can benefit if undertaken by individuals who are situated differently. See also Lipton 2004.

STEPHEN B. PRATTEN

contrast space. See CONTRAST EXPLANATION.

control (dual, multiple). See DEPTH; DETERMINATION; FREEDOM, ETC.; INFRA-/INTRA-/SUPER-STRUCTURE.

conventionalism. The IRREALIST (anthroporealist + transcendent realist) doctrine, formulated by Henri Poincaré (1854–1912) and Pierre Duhem (1861–1916), that logical and mathematical truth are created entirely by linguistic convention, not imposed by the world; more generally, that scientific laws and theories are likewise matters of convention and cannot express empirically ascertained truths about intransitive reality. It is thus a 'hybrid of VOLUNTARISM and REIFICATION' (*SR*: 28): the scientific community decides what is to count as fact, but once established the convention tends to be reified and NATURALISED. Such a position is a form of conservative FOUNDATIONALISM in that it rests on 'a bedrock of established usage' (*SR*: 88) that can only be FIDEISTICALLY adhered to, because conventionalism lacks concepts of depth, natural necessity and alethic truth. *Sociological conventionalism* (e.g., Kuhn) holds that scientific neophytes come to accept scientific theories on the basis of authority. The conventionalist thus tends to be a conformist who, like Hume, 'upholds law, order, private property and the prevailing order of things, epistemic and social alike' (*D*: 359). When foundationalism is rejected and epistemic relativity espoused, fideism leads inexorably to a SUPERIDEALIST ontology of subjective conceptual realism, or the view that when our theories change the world changes with them. *Social conventionalism* holds that social forms have their origins in consent or agreement among people (e.g., social contract theory); it is a collective form of voluntarism. See also SOCIAL CONSTRUCTIONISM.

conversation, inner or **internal**. See AGENCY; CRITICAL NATURALISM; PHILOSOPHICAL ANTHROPOLOGY; SUBJECTIVITY; Archer 2003b.

co-presence or **co-inclusion**. The co-occurrence of the absence and presence of something ('what is present from one perspective, at one level, in some region may be absent from, at or in another' [*D*: 48]) (1) at different times within a DISTANCIATED temporal stretch (e.g., positive contraries and negative subcontraries in an EPISTEMOLOGICAL DIALECTIC, more generally, the co-existence of the existentially constitutive past and of the future as increasingly shaped possibility within the present); and/or (2) simultaneously at different levels within a spatial spread (e.g., the actual absence and real presence of opposites in a multiply determined result), more generally, the presence or absence of the real and possible in the actual and of the outside in the inside (the intrinsic outside) (inner complicity with impacting object, the penumbra in the figure, the periphery in the centre, infrastructure/superstructure within or on structure). For Bhaskar, co-presence is the 'golden nugget' of Hegelian dialectic, and, within that, the epistemological dialectic its 'rational kernel' (see HEGEL–MARX CRITIQUE).

The leading figure of OPPOSITIONALITY at 2E, at 3L co-presence is loosely affined to CONSTELLATIONALITY, which it embraces, hence is the key CR figure of MEDIATION interlinking all four moments of the ontological–axiological chain. In the *dialectics of co-presence* or *co-inclusion* (of absence and presence) it mediates and unifies 2E *dialectics of PROCESS* and TRANSITION with the *dialectics of opposition*, including reversal of power$_2$ relations, as illustrated in Figure 3. It embraces the coincidence and intermingling of identity and change (non-identity) and of totality (e.g., CR) and its aspects (e.g., the CR stadia) within any dialectical process (2E–3L); of identity and difference (non-identity) in structural generation (2E–1M); of material and efficient causes in human activity, of the praxis of the dead within that of the living, of solidarising intersubjectivity within atomising globalisation (1M–4D). It completes, in the guise of dialectical opposites, the traditional TABLE OF OPPOSITION, and is the key to resolving 'the PROBLEMS and problematics of philosophy' (*P*: 245), including the social CONTRADICTIONS (rooted in opposites) underpinning them, embracing such concepts as GENERATIVE SEPARATION or ALIENATION (real presence but actual absence of our essential nature), MASTER–SLAVE-type relations (actual slavery, essential non-slavery), TINA FORMATION (absence of a necessary concept, presence of a false concept), the co-presence of the duplicitous counterpart within DIALECTICAL ANTAGONISM, and the totalising depth-praxis of EMANCIPATORY AXIOLOGY. Epistemologically, it necessitates dialectic as the art of thinking the co-presence of distinctions and connections.

The philosophy of META-REALITY generalises co-presence such that everything is implicitly enfolded qua possibility, synchronically, within everything else without the reduction of externality and difference. This is effected by transcendental argument from the experience of transcendental identity consciousness in science and elsewhere (see CREATIVITY). We do not have to see ourselves in the other and the other in ourselves (Sartre, see also RECOGNITION) because we are already 'in' each other (profoundly interconnected in virtue of ingredient cosmic 'implicate order'). This conception also draws on the quantum phenomenon of non-locality or action-at-a-distance thematised in *Dialectic* (*RM*: 90, 184–5, 274).

core universal human nature. See ALIENATION; CONCRETE UNIVERSAL; ETHICS; HUMAN NATURE; SUBJECTIVITY.

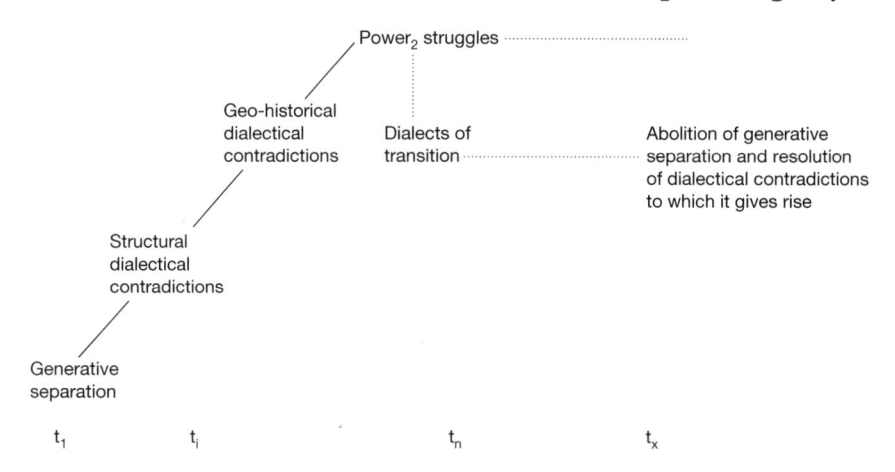

Figure 3 Dialectics of co-presence or co-inclusion, departing from generative separation: a possible schema
Source P: 240.

corporate agency. See AGENCY.
corpuscularianism. See ATOMISM.
correspondence. Unless otherwise indicated, used throughout this work as a metaphor for 'answering to', 'being congruous with' or 'expressing' rather than as designating 'one-to-oneness' or 'likeness'. *Correspondence theory of truth*: see ALETHIA; HOLY TRINITY (epistemic relativity); TRUTH.
corrigibility. See ATOMISM; CRITICAL NATURALISM; FOUNDATIONALISM; HOLY TRINITY.
cosmic body. See META-REALITY.
cosmic envelope. See GROUND; META-REALITY; ULTIMATA.
cosmic incapacity (of science). See REALITY; ULTIMATA.
cosmogeny, cosmogony (f. Gr. *cosmos*, order, ornament, the world + *genesis* and *gonē* respectively, both of which derive from a common root meaning birth, origin). Synonyms for (1) the generation or evolution of the universe, and (2) an account or theory of the same.
counterfactual/transfactual. A CR construal of CAUSAL LAWS supports both *counterfactual* or *nomic* (universal) conditionals in the subjunctive mood at the level of the actual (events) ('were this match struck in appropriate circumstances, it would light') and *transfactual* or *normic* conditionals in the indicative mood at the level of the real (mechanisms) ('matches tend to strike when struck'). A counterfactual conditional presupposes the falsity of the antecedent (it is not instantiated), a transfactual conditional that the consequent may be unrealised: the lighting of a match indicates that there is a TENDENCY in play that is not invariably manifest at the level of events. *Nomic* (f. Gr. *nomos*, law), law-like; *nomological*, the study of laws. *Normic* (f. L. *norma*, a rule, precept, pattern), rule-like. *Counterfactual*, contrary to fact. *Transfactual* (also *transphenomenal, transsituational*), operating 'across' both closed and open situations; universal – the concept takes us 'to a level at which things are really going on

86 counterfinality

irrespective of the actual outcome' (*RS2*: 51). Construed in this way, counterfactual truth is underpinned by transfactual TRUTH and, properly understood, nomic UNIVERSALS are not empirical invariances or CONSTANT CONJUNCTIONS of events, but 'transfactual or normic statements with factual instances in the laboratory (and perhaps a few other effectively closed contexts) that constitute their empirical grounds; they need not, and in general will not, be reflected in an invariant pattern or regularly recurring sequence of events. {...} Once we allow for open systems then laws can only be universal if they are interpreted in a non-empirical (trans-factual) way, i.e., as designating the activity of generative mechanisms and structures independently of any particular sequence or pattern of events' (*RS2*: 14). *Nomic realism* is realism that sustains the universality of laws within their range and a concept of natural necessity. *Nomic DIRECTIONALITY* is tendential directionality. See also CAUSALITY; NATURAL KIND (natural necessity).

counterfinality. See CAUSES (final causes).

counter-hegemonic struggle. See CRITIQUE; DEPTH; EMANCIPATORY AXIOLOGY; HEGEMONY; HERMENEUTICS; IDEOLOGY; RECOGNITION; REVERSAL; SOCIAL CUBE.

covering law model. See EXPLANATION; SOCIAL STRUCTURE.

covert/overt effects. See TINA SYNDROME.

Cratylus*, reductio ad *Cratylus. See SOLIPSISTIC, ETC.

creativity. The capacity to produce something new and valuable (see CAUSALITY); in dialectical terms, a positive bipolar of ABSENCE. When exercised it results in EMERGENCE, which is 'the explication of an implicit or enfolded potential' (*RM*: 253), resulting in something hitherto absent from actuality. Creation thus betokens the CO-PRESENCE of absence and presence. CR has always rejected the VOLUNTARIST notion that change within space–time can be purely creation ex nihilo. While it is impossible to prove that such cases can never occur, the 'cosmic incapacity' of science entails that it can never know that they do: what appears to be a miracle at t_1 may be shown to have conventionally understood causes at t_2 (*RS2*: 205). So far as we know, change always involves the transformation of pre-existing material causes, and the material continuity of fundamental elements through change (e.g., of a gene pool in species-change) is transcendentally necessary for science to be possible (*RS2*: 84–5, 205). In the sociosphere we never create or make, but only change, our knowledge and our social forms. In Bhaskar's early works, where the emphasis is – against voluntarism – on the material continuity of cognitive change, insistence on this goes hand in hand with an emphasis on the creative power of the human mind and imagination in the dialectic of scientific discovery and more generally of the self-organising power of nature in emergence.

In and since *Dialectic*, Bhaskar argues that emergent change always involves an *element* of creation ex nihilo, i.e., *pace* Hostettler and Norrie (2003), who claim that Bhaskar elides the distinction between de novo and ex nihilo creation, emergence is always both ontologically immanent (de novo, i.e., has a material cause) and epistemologically TRANSCENDENT (ex nihilo), because it is impossible to predict or infer the new gestalt. Change does not reduce ontologically to pre-existing elements, though it depends on them; rather, the novel element is added by (1) the creative power of matter in material emergence, which makes quantum leaps 'of (one feels like saying) the materialised imagination – or even, with Hegel, reason {...}. This is matter as creative, as AUTOPOIETIC' [*D*: 49]); and (2) the (related) creative power of human praxis (see AGENCY), involving thought/unthought, imagination, intuition and tacit know-how as well as practical effort, in conceptual, social and material cultural emergence (which provides an analogue for extra-human material emergence) (see EPISTEMOLOGICAL DIALECTIC; SUBLATION). However, since transformation of material causes is a necessary condition for emergence, it remains the case that we never create, but only change, our knowledge or social forms; but emphasis is now placed on the fact that in changing them we may add something new, such that an emergent level is constituted. In causal terms, in emergence an epistemologically transcendent cause (an implicit potential or DISPOSITION) at the level of the real erupts on to an immanently prepared ground at the level of the actual (cf. Nellhaus 2004). The pre-existing formation is transformatively negated, but *what is negated does not include the implicit*, which *emerges* (*MR*: 105). This conception seems at one with Marx's account of the labour process, whereby living labour does not just transform existing materials, but adds something new. In Bhaskar, as in Marx, the totality of master–slave relations depends entirely upon the spontaneous creativity of the slaves, the abolition of which in the transition to eudaimonia could inaugurate an epoch of the free development of human creativity as an end in itself (cf. *RM*: 85).

In conceptual emergence – the move to a more comprehensive theoretical formation – in science (or any LEARNING PROCESS), the moment of insight comes 'out of the blue'. As depicted in Table 12, the philosophy of META-REALITY argues that such moments are examples of *transcendental identity consciousness*, whereby humans imaginatively 'become one' with what they are trying to understand. This is possible because, at the level of ULTIMATA, there is an underlying identity in all things, including humans. Although the ground for the creative insight must be thoroughly prepared by an enormous amount of work and *thought* – itself a creative power – the creative moment proper requires a state of *unthought*, a gap, space or clearing within normal thought patterns – a relaxed 'letting be' – which allows what was hidden to become manifest (or to manifest *itself* as the eruption of 'a moment of supra-mental consciousness'). PMR generalises the concept of creativity to the whole ontological–axiological chain, whose moments are now explicitly seen as embracing 'cycles' or 'circles' of the human creative process, moving from implicit impulse, will and raw materials (1M) to thought/unthought and transcendence (2E), emotion and formation (3L), making and objectification (4D) and reflection of objectification to its maker (5A–7A), corresponding to the human capacities for freedom, creativity, love, right action and fulfilment, which are so many forms of manifestation of the ground-state and modes of creative transcendence, and mirroring the cycle of cosmogenesis. At 4D all human action is now seen as involving an element of creativity ex nihilo on the grounds that every authentic human act produces something new or is spontaneously creative, in mimesis of cosmic creation. At the level of cosmogenesis, PMR holds that if there was a unique beginning to everything, as posited by Big Bang, it could only have been a creator's act of radical autogenesis out of pure nothingness. This notion is canvassed in *Dialectic* (46, 47 n, 77), but, albeit it plays a role in the deduction of the ontological priority of absence, Bhaskar there appears to reject it. (See *MR*: 105 ff.). In PMR the overall intent is, inter alia, to stretch our understanding

Table 12 Elements of the creative process in PMR > CR

Causal–Axiological Chain	1M Non-Identity	2E Negativity	3L Totality	4D Transformative Agency	5A Reflexivity (TDCR)	6R (Re-) Enchantment	7A/Z Non-Duality
Basic human capacities	freedom (negative completion)	creativity	love	right-action	fulfilment (self-realisation, enlightenment) (positive completion)		
Elements of the creative process	*raw materials*, incl. relational, semiosic (material causality)	*transcendence* (epistemological, causal) (final, efficient causality)	*formation* (formal causality)	*objectification* (final causality)	*reflexivity* or reflection of objectification to the maker (formal causality)		
Components of action	will	thought/unthought	emotion	making (action and objectification)	fulfilment (or not)		
Non-dual components of action (modes of transcendence)	*transcendental consciousness* of real self (loss of ego)	*transcendental identification* in consciousness (away from subjectivity)	*transcendental teamwork* or holism	*transcendental agency* (absorption in activity)	*transcendental retreat* into self-identity (away from objectivity)		
Transcendental morphogenesis of action	ground (implicit potential)	emergence (creativity)	identification, unification (love, solidarity)	agency (capacity for right-action)	reflection or fulfilment	perception (apperception, intuition, and insight)	awakening
Cosmogeny	polyvalent foundational impulse (from implicit potential)	creativity (transcendental emergence)	love (transcendental identification)	action (transcendental agency)	fulfilled intentionality of the foundational impulse (reflexivity or reflection of objectification back to the creator)		
Cosmotheogeny (cycle of cosmic creation)	self-creation of the creator ex nihilo†	emergence of realm of duality becoming and time	emergence of realm of demi-reality and of binding nature of love	(commencement of return cycle from alienation) individual self-realisation	individual and universal self-realisation, then god-realisation (theosis)‡; reflexivity		

† Corresponding to the descent of consciousness in traditional cosmotheogenies, and to Big Bang in modern scientific theory
‡ Corresponding to the ascent of consciousness in traditional cosmotheogenies

crisis 89

of what is 'natural' and to put in question our stubborn belief, underpinned by 'the ILLUSION of physical being', in the brute physicality of being (see PDM).
crisis. See TRANSITION.
criteria for rational agency. See FREEDOM, ETC.; SOLIDARITY.
critical dialectics. Dialectics which practise CRITIQUE in any of its forms, especially immanent critique of other problematics. To be distinguished from PRESENTATIONAL and SYSTEMATIC DIALECTICS.
critical discourse analysis (CDA) subsumes a number of rather diverse approaches (differentiated in Fairclough and Wodak 1997) to analysing relations between *discourse* in an abstract sense, or *semiosis*, and other elements or moments of social processes (see also SIGNS, where Bhaskar's few comments on semiosis are discussed). Semiosis is the *making of meaning*, and the means or resources for making meaning include language but also other semiotic modes such as 'body language' and visual images. CDA is consistent with CR in claiming that discourse has *causal powers* and liabilities whose actualisation is, however, contingent upon non-discoursal context (the interconnection of discourse with other causal powers), and in resisting 'discourse-imperialism' and the conflation of discourse and material practices, which undermines the TRANSITIVE/INTRANSITIVE distinction and leads to a version of the EPISTEMIC FALLACY. CR has operated with 'an insufficiently concrete and complex analysis' of discourse to the detriment of social analyses informed by it (Fairclough, Jessop and Sayer 2004), and CDA is a congenial approach which it can draw upon as well as add to. Given the understanding of semiosis as meaning-making, the question of semiosis/discourse is closely associated with the question of CULTURE for CR.

We can base an account of how discourse produces causal effects which is consistent with CDA accounts (e.g., Chouliaraki and Fairclough 1999) on some key features of CR. Discourse is a facet (along with non-discoursal facets) of the *real*, the *actual*, and the *empirical*. Languages and other semiotic systems and 'orders of discourse' (see below) are elements of the real with causal powers which may be activated under certain conditions in *texts* in a broad sense, including written and spoken texts and the 'multi-modal' texts of television and the internet. Texts are elements of the actual and of (actors' experiences of) the empirical. This contrasts with Nellhaus' proposal (1998) that *semiosis* rather than *the empirical* is the third ontological domain (along with the real and the actual). Nellhaus recognises, however, that the empirical (in conventional CR terms) has non-semiotic facets and both the real and the actual have semiotic facets, and his proposal requires what might be seen as a somewhat cumbersome complication of CR categories to deal with this. See also Carter and Sealey's (2000) proposed restructuring of the field of sociolinguistics on the basis of CR perspectives, specifically Margaret Archer's ANALYTICAL DUALISM and Derek Layder's theory of 'social domains'.

In the analysis of concrete social processes and action, and their relationship to the reproduction and transformation of social structures, a CDA based upon CR principles proceeds from a recognition of the positioning of discourse (texts) in complex *dialectical RELATIONS* with other moments (including facets of the material world, social institutions and social relations, the identities, natures and habitus of social agents, and features of practical contexts). In the concrete instance, each of these moments is an actualisation from the potentials defined by structures – in the case of discourse, these include *semiotic systems* (most obviously languages) but also, at a lower level of abstraction, mediating *orders of discourse*, forms of social structuring of semiotic variation associated with specific social fields or institutions (e.g., Jessop 2004 suggests that the 'knowledge-based economy' is an order of discourse, or 'semiotic order'). The dialectical relationship between moments is such that on the one hand discourse (meaning-making) may contingently have constructive effects on other moments and their relations, and cumulatively on their associated structures and relations between structures, but on the other hand the meaning-making and causal effects of discourse are 'contextually' constrained by the relationships of other moments to discourse. With regard to the causal effects of discourse, it is important to differentiate, as Sayer (2000a) argues, between *construal* and *construction*: discourse may construe aspects of the non-discoursal in particular ways, but what constructive effects such construals have depends upon various non-discoursal as well as discoursal conditions.

This difference and relationship can be clarified in terms of the evolutionary mechanisms of *variation*, *selection* and *retention* which shape the relationship between discourse and social structuration (Jessop 2004). This can best be elaborated in terms of particular *discourses* ('discourse' in the more concrete sense of divergent and positioned ways of representing aspects of the world). Meaning-making adds to the variety of discourses which construe the world in different ways, but the selection of particular discourses (privileging them over others for interpreting events, legitimising actions, etc.) and the subsequent retention of some of these (in hegemonic projects, institutions, habituses) depend on both discoursal and non-discoursal conditions, including for example their resonance with other discourses (discoursal) and institutionalised power relations (non-discoursal).

These relations are part of the complex dialectic alluded to above and extend into the *operationalisation* of discourses, comprising *enactment*, *inculcation* and *objectification*, which is contingent upon selection/retention. Discourses may be enacted as ways of acting, inculcated as ways of being (identities), and objectified in built environment, technology and 'bodily hexis' (Bourdieu). The first two of these modes of operationalisation may themselves be either semiotic or non-semiotic in character: discourses may be enacted as *genres* (ways of interacting communicatively) or as, for instance, organisational procedures which will include genres but also non-semiotic elements; discourses may be inculcated as *styles* (ways of interacting associated with particular identities, e.g., contemporary business or political managers) or as bodily dispositions. *Orders of discourse* (see above) are constituted as particular configurations or articulations of different discourses, different genres and different styles.

CDA brings to transdisciplinary or postdisciplinary social research based on CR principles resources for analysing the EMERGENCE of meaning in texts through analysis of *texturing* as a productive and potentially creative activity of social agents working given (structurally available) semiotic resources – orders of discourse, discourses, genres, styles, and grammatical, semantic and lexical forms and relations. *Intertextuality* is a central property of semiosis with respect to emergence, both the concrete intertextuality of relations (e.g., 'reported speech') between particular texts and the more abstract relations between different orders of discourse, discourses, genres or styles.

Different versions of CDA and its precursors show differences in theory and methodology which bear upon their compatibility with CR, as well as being applied in different sorts of social research. The discussion above relates especially to the *dialectical approach* distinguished below. We conclude with a summary differentiation of these versions, which are more fully discussed in Fairclough and Wodak (1997). *French discourse analysis* from the 1960s to the 1980s gave a discursive turn to Althusser's theory of ideology, emphasising the ideological effects of discursive formations in positioning people as social subjects, with a focus on political discourse in France. The *critical linguistics* which developed in Britain in the 1970s drew from French discourse analysis but gave particular prominence to the ideological significance of linguistic choices (e.g., using passive rather than active sentences), working with a linguistic theory known as Systemic Functional Linguistics. The term *critical discourse analysis* was adopted for a number of distinct approaches developed from the 1980s of which we mention four. A *socio-cognitive approach*, which has been applied especially to the reproduction of racism and ethnic prejudice in discourse, argues that there is no direct relation between discourse structures and social structures, that they are mediated by personal and social cognition. A *discourse-historical approach* seeks to maximise knowledge about the historical sources and background of social and political fields in which discursive 'events' are embedded, focusing upon sexism, racism and anti-Semitism in their discourse aspects. A *social semiotic approach* draws attention to the 'multi-semiotic' character of most contemporary texts, exploring ways of critically analysing visual images and the relationships between them and language. Finally, a *dialectical approach* which conceptualises discourse as a moment of social processes dialectically related to others has been applied especially in transdisciplinary research on contemporary social change.

NORMAN FAIRCLOUGH

critical legal studies. See LEGAL STUDIES.

critical naturalism (CN). The second moment in the CR system after TRANSCENDENTAL REALISM (TR). When its implicitly dialectical character is brought out, it is *dialectical critical naturalism* (DCN). DCN embraces all the major social concepts of DCR, explicitly highlighting that DCR embraces NATURALISM in all three senses indicated below 'as a special theory of the social sciences' (*D*: 259). Because human emancipation presupposes genuine knowledge of the social forces that chain us, an overriding concern has been to establish and promote the possibility of such knowledge, in the context of an intellectual milieu where this possibility has been radically called into question. It is for this reason that the CR system historically begins with philosophy of science (*RS*), then asks whether and how scientific knowledge of society is possible (*PN*). TR is deduced from the human activities of scientific experimentation and application. CN is independently deduced from human INTENTIONALITY as such, which it views as 'criterial' for the sociosphere (*SR*: 122), and via immanent critique of the antinomies of social theory. The fact that it is nevertheless consistent with the TR theory of science strengthens its case – hence its long title: *transcendental realist critical naturalism* (*PN3*: 137). At the most abstract and general level – in terms of the subsequently elaborated ontological–axiological chain – whereas TR thinks being as structured and differentiated (1M), CN thinks it (additionally) as constellationally containing MIND and concepts as EMERGENT *products* (1M), and negativity (transformative AGENCY), CONTRADICTION (between the way objects are manifested in human experience and the way they really are) and emergence

critical naturalism (CN)

as *PROCESS* (2E), where process is the way in which the causal powers both of social structures and of the human AGENCY which reproduces them are exercised and their effects realised. 'This is the general form of the spatio-temporality of social life' (*D*: 156), encapsulated in the TMSA and SOCIAL CUBE.

Mainstream social science, social theory and philosophy are riven by dichotomies, ultimately reflecting social splits and contradictions (see PROBLEM), which are debilitating for social science and emancipation alike and which CN therefore seeks to transcend and heal through a process of immanent critique that, by elaborating a fuller, richer conceptual formation, remedies the constitutive incompletenesses or absences in the rival theories that give rise to the oppositions. This procedure was later formalised in the EPISTEMOLOGICAL DIALECTIC. Table 13 sets out the most important of these dualisms and their resolution by CN. They are all species of the dualism of OBJECTIVITY/SUBJECTIVITY.

The resolution of (1) espouses naturalism in all three of its meanings as ontological or emergent powers materialism, methodological unity-in-diversity, and ethical naturalism. On it, the social sciences can be sciences in the same sense as the experimental sciences of nature but in ways that are specific to their different subject-matters (all science follows its object). Both POSITIVISM and HERMENEUTICS operate with a false Humean account of a CAUSAL LAW, entailing that positivism is otiose in relation to both natural and social science, and that hermeneutics reassesses or abandons its DUALISM. All science is concerned with the discovery, not of CONSTANT CONJUNCTIONS, but of intelligible connections in its subject-matter (see EXPLANATION). In contrast to positivism, CN sustains 'the *transfactuality* of social structures, while insisting on their *conceptuality* (or concept-dependence). And in contrast to hermeneutics, it can sustain the *intransitivity* of both beliefs and meanings, while insisting on their susceptibility to scientific explanation and hence *CRITIQUE*, in a spiral (rather than circle) which reflexively implicates social science as a moment in the process that it explains' (*PN3*: 21–2). On the solution to (2), all social life is embedded in networks (emergent TOTALITIES) of social relations, which are the paradigmatic subject-matter of social science rather than individuals or collectivities. Social identities are constituted RELATIONALLY. Transformationalism (3) is specified by the transformational model of social activity (TMSA) in CN, whose elaboration in

Table 13 Dichotomies in mainstream social thought and their resolution by critical naturalism

Prevailing antinomy	CN resolution
(1) naturalism/anti-naturalism (positivism/ hermeneutics: nature/society)	critical naturalism (non-positivist) methodological unity-in-diversity
(2) collectivism (holism)/individualism	relationism, emergentism
(3) reification/voluntarism (structure/agency)	transformationalism (TMSA)
(4) causes/reason	reasons, when acted upon, are causes
(5) body/mind	synchronic emergent powers materialism (SEPM), ontological materialism
(6) facts/values	explanatory critique, ethical naturalism

Source Bhaskar 1998: xiii–xiv

critical naturalism (CN)

DCN is four, planar social being. On this, social structure is the condition and outcome (t_1) of intentional agency which reproduces and changes it (t_2). This implicates four kinds of DUALITY: *duality of praxis* (intentional action is also largely unintentional reproduction of social forms); *duality of structure* (social structure is both the temporally prior condition and the outcome of praxis); *duality of agency and structure* (they are existentially interdependent but essentially distinct); and *duality of mediation and transformation* (where mediation is process as defined above and the HIATUS-in-the-duality situates the possibility of non-substantial process and disembedding [see DISTANCIATION]). Archer's *analytical dualism* (see MORPHOGENESIS) is designed to respect the irreducibility of the elements of these dualities. She has also (Archer 2003: ch. 4) theorised the mechanism whereby structure is mediated by agency as the REFLEXIVE deliberations or *internal conversation* of AGENTS. On the resolutions of (4) and (5), MIND is a sui generis real emergent power of matter, and reasons, when acted upon, are (materially conditioned) causes. Together they establish the relative autonomy of PSYCHOLOGY from SOCIOLOGY. Human discourse, reflexivity and depth-stratification of the person presuppose a distinction between real reasons and deliberated possible reasons, and between real reasons as underlying motivation (WANTS and beliefs), which may be more or less unconscious and ideologically constituted, and reasons that are mere rationalisations. On the resolution of (6), values can be derived from facts. This is discussed below.

CN is a *qualified* naturalism. While acknowledging that naturalism has limits (set out, together with their derivation, in Table 14, see also SOCIAL THEORY), CN argues (1) that, while the mode of INFERENCE in all science is broadly retroductive-analogical (of which transcendental and DIALECTICAL ARGUMENTS are species and sub-species, respectively), any science must have METHODS and procedures specific (appropriate)

Table 14 Limits to naturalism

Type	Derivation	Limits
ontological	transformational model of scientific activity (TMSA) (CN) social cube (DCN)	social, unlike natural, structures are (1) concept-dependent, entailing a hermeneutical starting point for social science; (2) activity-dependent, entailing (quasi-)autopoiesis; (3) more space–time specific, entailing relative transience (geo-historicity)
relational	TMSA (relational character of social life entails causal interdependency of social science and its subject-matter)	social-relational dependence: social, unlike natural, structures are causally impacted by science
epistemological or methodological	TMSA (openness of social systems, impossibility of closure)	absence of decisive test situations for social sciences
critical	TMSA (unacknowledged conditions, unintended consequences, tacit skills, unconscious motivation of agency)	social, unlike natural, objects include beliefs about themselves; situates the possibility of explanatory critique of consciousness and social forms

94 critical naturalism (CN)

to its subject-matter; and (2) there are means whereby hypotheses in the social sciences can adequately be grounded empirically.

The list incorporates changes made in response to criticism since the publication of *PN*. Within CR, only the last two – absence of decisive test situations and the possibility of explanatory critique – now seem at all controversial. In DCN, activity-dependence is NEGATIVELY GENERALISED to include past activity, obtaining the result that social structures may persist without any (present) human agency or even despite it (e.g., demographic structures or the thawing icecaps, which, while not social structures as such, are socially produced phenomena), and that 'all the *decisive* moments of social life are negative' (*D*: 160; see also ABSENCE). The (quasi-)AUTOPOIETIC nature of social forms is entailed by the fact that social causes, mediated by intentional agency, enter into their own reproduction, or society is itself a social product. In relation to the epistemological or methodological limit, Bhaskar's essential point is that the social, unlike the natural, world is 'ineradicably open'; social science is therefore denied a decisive predictive test (closed experiments) for the development and replacement of theory, which must rely exclusively on tests in terms of explanatory power. However, well-controlled direct experiment is available only to some of the natural sciences, otherwise the contrast is one of degree only and the problems of confirmation and falsification this poses can arguably be offset to a large extent (see CONTRAST EXPLANATION; DEMI-REG; THEORY CHOICE).

Although the theory of explanatory critique (EC), or ethical naturalism, was elaborated fully only after CN, and is often regarded as its successor moment, its possibility was already situated in *PN* by the thesis of the limited nature (owing to unintended consequences, unconscious motives, etc.) and contingent corrigibility of agents' accounts and conceptualisations, which do not exhaust or constitute the social, and may indeed obscure aspects of it; and its substance outlined. DCN, moreover, explicitly embraces both EC and EMANCIPATORY AXIOLOGY. EC holds that, contrary to Hume, ethical values can logically be inferred from purely factual premises or scientific results, so providing a grounding for ETHICS. If social science can demonstrate that a belief is false and give a true account of its causes, then it immediately issues in a negative valuation of those causes and any structure, states of affairs, etc., sustaining them, CP, and a positive valuation of action directed at removing the causes, CP. Note that, if successfully carried through, this achieves a transition not just from facts to values but also from (explanatory) theory to practice, form to content, demonstrating their dialectical unity and entailing that social science is intrinsically emancipatory. The action-guiding character of morality then poses no problems for CR moral realism.

The correctness of EC as defined here is presupposed by the dialectics of freedom and EMANCIPATORY AXIOLOGY, which it entails; indeed, for the middle and later Bhaskar, social science as such *is* essentially EC in that such critique is transcendentally necessary for it (see CRITIQUE) and 'without parallel in the field of the natural sciences' (*SR*: 134), albeit it is 'conditioned by human dependency on the natural order', i.e., naturalism as ontological materialism. Though few would dispute that knowledge of the social causes of false beliefs and social ills is essential to their remedy, EC as refuting Hume's law and entailing that social science can play a significant role in the formation of emancipatory values has, however, been questioned strongly by some critical realists

critical naturalism (CN)

(see EXPLANATORY CRITIQUE). Since so much hinges on it, some comment is called for here. EC has been criticised essentially on the grounds that (1) the premises of any EC include hidden value premises; (2) conclusions derived from EC are fallible and so might be mistaken; (3) the inferences hold only in virtue of casually or complacently deployed CP clauses which (3.1) cloak the fact that there are other goods and ills than truth and falsity, and hence that the conclusions of ECs may be overridden by other considerations; (3.2) 'effectively admit that no logical deduction from factual premises alone can produce practical value judgements about what is right or wrong, and about what should be done, in particular circumstances' (Hammersley 2002: 42); and (3.3) overlook the probability that powerful vested interests will resist the proposed changes and worse consequences ensue, i.e. beg the question of feasibility.

(1) itself overlooks that, since the subject-matter of social science includes not just social objects but beliefs about them, social science has a mandate to explain and *criticise* social beliefs – they clearly may be logically faulty or inadequate or false in terms of what they are *about*, i.e. on strictly cognitive criteria. The only value that *necessarily enters into* such explanation and criticism is commitment to truth and consistency, but that is a condition of any factual discourse (and any human communication, and therefore any human being) whatsoever, and thus participates in the very constitution of what a fact is – i.e., considerations of truth are intrinsic to facts. Non-cognitive values and interests do play a motivating and orientating role in science and in the acceptance or rejection of its conclusions – *science is not value-free* – but it remains the case that no value judgements other than those already involved in the assessment of the cognitive power of an EC are *necessary* for the derivation of its evaluations (*SR*: 200). It is Bhaskar's claim that the development of his own philosophy has been 'powered by the single norm of truth' (*RM*: 175), and that is in principle possible. (2) overlooks that the fallibility or corrigibility of knowledge-claims (entailed by epistemic relativity) is presupposed by EC, applies universally (including to any defence of the status quo) in a world in which we none the less must act (see AXIOLOGY), and is consistent with JUDGEMENTAL RATIONALISM. (3) is beside the point. First, that there are goods other than truth and ills other than falsity, which might be deemed overriding in particular situations, does not gainsay that, CP, truth is good and falsity ill, any more than 'gravity is undermined by the existence of double-decker buses' (*SR*: 187). Second, EC stipulates only *that* something should be done, CP, to remove the causes of falsity, which is undeniable; *what* (if anything) should and can feasibly be done is for social movements articulated with social science and philosophy to discover and effect, and this might, precisely, involve counter-hegemonic struggle. Third, in an open world CP clauses are a condition for moving, not just from facts to values, but from facts to facts: to invoke a law of nature is not to say what will happen but what would happen, CP (Bhaskar 1998a: xix). Far from being casual, their deployment in the argument for EC is standard scientific (and philosophical) procedure.

While this still leaves a very long way to go on the road to generalised free flourishing, the significance of the formal demolition of Hume's dichotomy – a monument to ALIENATION, heterology and heteronomy that serves as an intellectual pillar of the status quo – should not be underestimated. It strikes a mighty blow *at the level of philosophy* against the notion that the oppressed only have their arbitrary subjective preferences to go by, valorising *critical explanatory* social science and morality, entailing the

96   critical realism

remoralisation of the world, and emancipatory praxis. *Values are not science-free.* Ethical naturalism grounds a distinction, within moral realism, between the domains of actually existing morality (dm$_a$) and moral reality or alethia (dm$_r$) which explanatory critical research may discover. Most important, its transition from facts to values and theories to practices 'can be *generalised* to cover the failure to satisfy other AXIOLOGICAL NEEDS, necessities and interests besides truth, including those which are necessary conditions for truth, such as basic health, education and ergonic efficiency' (*D*: 262; cf. *SR*: 191–3). Hammersley (2002) suggests that this smuggles in a value premise: that there *are* needs can be established factually, but it does not follow that they *ought* to be met, let alone always and generally. This overlooks that meeting basic needs is a condition for science itself to flourish, the more universally the better from the perspective of truth; indeed, it is a universalisably necessary condition for successful action as such (*D*: 287; Gewirth 1979). Further, in a world of universal human free flourishing, truth would prosper as never before, or so it can plausibly be argued. Freedom is a condition of truth (and vice versa) (cf. *D*: 292). Coupled with the totalising DEPTH-investigation and depth-praxis it logically entails for creatures who must act and speak, truth *can* make us free, and in making us free liberate our powers to know. Benton and Craib (2001), Collier (1994) and Potter (2000) are or provide excellent critical introductions to CN and TR. See also PHILOSOPHY OF SOCIAL SCIENCE.

critical realism. A movement in philosophy, social theory and cognate practices that seeks to underlabour for science and other ways of knowing in order to promote the cause of TRUTH and FREEDOM, hence the transformation of social structures and other constraints that impede that cause and their replacement with wanted and needed ones, or emancipation. It is in no way a predominantly personal vision; it is rather, in broad outline, the dialectically necessary worldview supervenient, by processes of immanent critique, on the philosophical discourse of modernity, pinpointing its weaknesses, embracing its strengths, and drawing on the great pre-modern discourses to adumbrate an outlook fitting for a sustainable social order of free flourishing. The single best essay-length introduction and overview is Bhaskar's 'General introduction' to *Critical Realism: Essential Readings* (1998a). Dean et al. (2005) is an excellent recent critical overview and assessment. Any philosophy worthy of the name must be able to situate itself reflexively and coherently sustain itself in relation to the philosophical thought of the geo-historical epoch in which it appears. This is attempted in relation to CR in PHILOSOPHICAL DISCOURSE OF MODERNITY (PDM). Here there is space only to locate CR briefly within its more immediate philosophical and socio-historical context, and sketch the features of the overall system that unify the four main moments of its development: TRANSCENDENTAL REALISM (TR), CRITICAL NATURALISM (CN), EXPLANATORY CRITIQUE (EC) and DIALECTICAL CRITICAL REALISM (DCR).

CR is a child of the social and political turmoil that accompanied the end of the postwar boom in the mid-1960s and the beginning of a recessionary epoch in the West, producing a strong sense of the possibility of new eudaimonian beginnings. Within the Western Marxist tradition the crisis fuelled a veritable renaissance, as intellectuals, often organically linked to social movements, abandoned the straitjacket of disciplinary boundaries and sought the deep structural roots of the crisis, imperialism, revolution and counter-revolution in the South, the impasse in the Soviet bloc, the 'cultural revolution' in China, and the reappearance in a stark form after 1968 of the world-historical problem of agency. The hallmark of much of this work was the understanding, shared by CR, that the oppressed could, with the aid of science, fallibly come to apprehend the real causes of their suffering and act to transform them. Metacritically, such an approach is in the tradition of the drive to freedom linking the philosophies of Descartes, Kant, Hegel, Marx and DCR (*D*: 335). Working in the opposite direction were and are profound tendencies, which CR seeks to counteract, to cognitive relativism and irrationalism or superidealism, which inexorably result from abandonment of foundationalism and acceptance of the correct thesis of epistemic relativity (itself causally bound up with physical relativity and the relativising power of money and the market) in the absence of a robust ontology. In the PHILOSOPHY OF SCIENCE, the opportunity to elaborate such an ontology was provided by the collapse of the ARCH OF KNOWLEDGE TRADITION in its modern positivist incarnation in the 1960s and 1970s. The positivist orthodoxy came under sustained attack from a range of quarters, calling into question the monistic theory of scientific development and the deductivist theory of scientific structure on which it hinged. As Bhaskar has recounted in a number of contexts, TR was developed out of both these strands, the former entering into his account of the transitive dimension (TD) and the latter contributing to his account of the intransitive dimension (ID) (see *RS*: 8 f.; *SR*: chs 1.1, 1.5; *R*: ch. 9; *D*: 224 f.; Bhaskar 1993b: 481–2, 1998a: x–xii; REALISM).

While Bhaskarian critical realists were not the first to deploy it (see below), the term 'critical realism' evolved independently in the UK in the 1980s when there was a need to refer to transcendental *realism* and *critical* naturalism as a unit. People began substituting 'critical' for 'transcendental' in the former, the resulting hybrid was accepted by Bhaskar on the grounds that ' "critical", like "transcendental", suggested affinities with Kant's philosophy, while "realism" indicated the differences from it' (*R*: 190; Bhaskar 1998a: ix), and the term stuck. It was later extended to include explanatory critique (already proleptically contained in CN) and DCR, but continues to have a restricted meaning that embraces TR and CN (as they have developed and are developing) only.

The relation and correspondences of CR to other critical realisms, and to other REALISMS that hold that objects of knowledge (perception, etc.) are relatively or absolutely independent of our knowledge of them, awaits in-depth research. This core realist notion is an old one but underwent a revival in the guise of *new realism* (a form of direct or naïve realism) in the UK and North America in the early twentieth century in reaction against idealism (*CDP*: 610). In German, 'critical realism' (*der kritische Realismus*) has a wide range of uses dating back at least to J.H. Herbart (1776–1841), the successor to Kant at Köningsberg, and his followers (Sullivan 2006). The English term appears to have first been used in the late nineteenth century to translate the 'transcendental realism' of Eduard von Hartmann (1875, see Scot Henderson 1876) – a form of absolute idealism asserting categorial realism – and '*der realistische Kriticismus*' (literally, 'realist [Kantian] critique'), deployed in the late nineteenth century by the Austrian and German neo-Kantians Alois Riehl (1844–1924) and Wilhelm Wundt (1832–1920) to designate their philosophical outlook (Morelli 2003; Paggmore 1957 ch. 12). The English term was taken up by the American philosopher Roy Wood Sellars (1880–1967) as the title of his work on epistemology (1916) and by the English

critical realism

philosopher Dawes Hicks (1917, fl. 1862–1941), and in the 1920s and 1930s designated something of a philosophical movement that also included A. O. Lovejoy, George Santayana and C. D. Broad (see especially Drake et al. 1920; Dawes Hicks 1938; R. W. Sellars 1924, 1929, 1937, 1939); its non-British wing is sometimes referred to as *American critical realism*. Epistemologically, it was intended to provide a third way between new realism and idealism by assigning a mediating role to 'the mental'; we do not directly perceive external objects but their effects in the form of an image, idea or impression that more or less accurately represents the objects. It is thus a form of REPRESENTATIVE REALISM. Metaphysically it was more diverse, embracing forms of dualism as well as a materialism, espoused by Sellars and Santayana, that 'saw cognition as simply a function of conscious biological systems' (*CDP*: 194–5, 656, see also Chisholm 1982). Sellars's position was to be of most lasting influence, because it underwent sophisticated development at the hands of his son, Wilfrid Sellars (1912–89), a key figure in the revival of realism, especially in the PHILOSOPHY OF SCIENCE, in the last forty years (see, e.g., Niiniluoto 1999; Psillos 1999). Bhaskar, whilst acknowledging that Sellars's position was near to transcendental realism (*RS*: 26), has taken his distance from this latest wave of 'new realism', on the grounds that it has empiricist or empirical realist and politically mainstream proclivities (*R*: 190–1). The 'scientific materialism' or 'scientific realism' of the Argentinian-Canadian physicist-philosopher Mario Bunge (1959, 1963, 1977, 1979) derived an emergentist, stratified, complexly interconnected ontology directly from science, largely by-passing explicit transcendental procedure. Maurice Mandelbaum's (1964, fl. 1908–87) 'radical critical realism' insisted on the independence of object from percept and the sui generis reality of the social, and criticised the covering law model of explanation (Lloyd 1986b; Verstegen 2000), but went hand in hand with a reificatory conception of 'societal facts' or sociological reductionism that, like Durkheim's, is incompatible with the TMSA (Mandelbaum 1955). In Australia, the realist tradition inaugurated by the Scottish-Australian philosopher John Anderson (1893–1962) explicitly saw itself as entailing empiricism and positivism (Anderson 1962; Baker 1986); and, while an emergentist strand, best exemplified in the scientific essentialism of Brian Ellis (2001), has in part grown out of that tradition, it is an emergentism that is reductionist in relation to the sociosphere. In another neck of the realist philosophical woods, the 'critical realism' of the Canadian Jesuit philosopher-theologian Bernard Lonergan (1904–84) appears to commit the epistemic fallacy in viewing the real as that which is known (Dunne 2006; Lonergan 1957), as does that of the French Thomist philosopher, Jacques Maritain (1882–1973, see Sweet 2004). The current movement in theology known as 'critical realism' (see PHILOSOPHY OF RELIGION; Shipway 2000) was significantly influenced by the tradition of representative realism, but is currently being taken in a strongly Bhaskarian CR direction by Alister McGrath (2002). By and large, while there are important commonalities between all these various kinds of realisms, one can say that Bhaskarian CR is sharply distinguished from the others by its robustly transcendental and immanently critical method, its outright rejection of empiricism and positivism, its thoroughgoing emergentism, its understanding of social science as necessarily explanatory critical (entailing rejection of the fact–value and theory–practice dichotomies) and its explicitly emancipatory stance. In short, it is a mistake to focus – as so often happens – on the 'realism' of critical realism at the expense of the 'critical'.

critical realism

CR is more in the line of descent of CRITICAL THEORY, which in its classical programmatic statement (Horkheimer 1937) had as its goal not 'simply an increase of knowledge as such' but 'emancipation from slavery', than of the realist tradition in mainstream philosophy.

That said, there is of course a sense in which CR is a development *within* mainstream thought (cf. McGrath 2002: 208), because it takes its departure from within the analytical tradition and develops by critique of mainline tendencies in philosophy and social theory. Bhaskar has called attention to three key features of its overall development:

(1) Its immanently critical character
(2) Its constituting a dialectical process
(3) Its constituting a progressive movement from questions to do with science and objectivity through to normative and subjective questions of freedom and ultimately universal self-realisation, a development powered by the single norm of truth. (*RM*: 175)

(1) CR has developed by a process of double immanent CRITIQUE of the PDM and of its own phases, a process that is ongoing and in principle without end. Once foundationalism is abandoned, the characteristic procedure of PHILOSOPHY necessarily assumes an immanently critical form. It is important to grasp that, for CR, PHILOSOPHY is neither autonomous from nor reducible to science and other ways of knowing, but distinct and relatively autonomous yet interrelated, constituting a dialectical unity with them. Immanent critique proceeds, via a process of transcendental argument, essentially by identifying theory–practice inconsistencies, contradictions and anomalies in rival discursive formations and remedying the constitutive absences or incompletenesses that give rise to them, thereby effecting a move to a fuller, richer conceptual TOTALITY; CR thus both critiques and reflects recent realist (1M), Red (2E), Green and feminist (3L) and reflexive (4D) turns in thought. Hence it is that the category of ABSENCE unifies the development of CR, as well as the 1M–4D moments of the ontological–axiological chain (see also MELD). Thus the great inconsistency in the discourse of modernity that TR seized upon was that between the view that statements about being are to be analysed in terms of our knowledge of being (the EPISTEMIC FALLACY) and the necessary secretion simultaneously of an implicit ontology. The absence generating this contradiction was that of the possibility of ontology, i.e. there was a dogmatic embargo in place, aptly dubbed by Maki (2001) 'ontology avoidance': one could not refer to the real world – the possibility of which had been ALIENATED in the sense of banished from our minds – only to our descriptions of it. Its remedy was the revindication of ONTOLOGY and the elaboration of TR's defining concepts: INTRANSITIVITY, TRANSFACTUALITY and stratification (see DEPTH). Today ONTOLOGY is the flavour of the month in the Western academy.

TR thus broke a taboo imposed by conventional or irrealist (non-transcendental realist) philosophy on what it is rational for philosophy to concern itself with, and the method of immanent critique entailed that the system of CR continued to develop in this vein. And so CN absents a banishment of rational discourse about duality, contradiction and split. EC lifts an embargo on rational discourse about values, and

DCR on rational discourse about negativity, dialectic, process, change and emancipatory axiology.

Embedded within the critique of PDM is a more detailed critique of Hegel, Marx and the Marxist tradition (see HEGEL–MARX CRITIQUE; Bhaskar 1983b; *R*: ch. 7, *PF*: app. 2). This has gone hand in hand with a reassessment of the mature Marx as a critical realist – indeed, as the 'comet' who anticipated many of its leading tenets (*SR*: 120). It is worth registering in summary form here that CR is significantly indebted to Marx for, inter alia, implicit concepts of the transitive and intransitive dimensions; the domains of the real and the actual; transfactuality and tendency; the relational conception of social forms; the holy trinity of judgemental rationalism, epistemic relativity and ontological realism and their converse in the unholy trinity; the diffraction of dialectic; the nature of moral alethia; the speculative illusion, monovalence and endism, fetishism, eternisation and naturalisation, and – far less implicitly – ideology-critique, depth-explanation and judgemental rationalism practical materialism; and generative separation/alienation and four-planar social being.

CR has developed equally by isolating and remedying incompletenesses in its own phases, as displayed in Table 15.

(2) The method of immanent critique thus ensures that CR is a developing or *dialectical process*. Nothing can be taken as 'limited or fixed', over and done, in this philosophy, including its own prior phases. Thus the diachronically earliest phase, TR, is very much alive and developing as a research programme in the philosophy of science that feeds into 'the arts and practices {. . .} of life' quite generally (*RM*: 179 n) and then back into TR, and the same goes for the successor phases. Nor is there any implication that research in the area of earlier phases is somehow less valuable than research in the later; on the contrary, since the later are implicit in the earlier moments, it is vital from their point of view that research in the earlier flourishes. Your true critical realist starts with the framework that seems most appropriate to her object and follows the logic of discovery wherever it takes her.

(3) The progressive movement of the developing system from issues of objectivity to 'normative and subjective questions of FREEDOM and ultimately universal SELF-realisation' is at one with its REVERSAL from anti-ANTHROPISM at 1M to HUMANISM at 4D. (It should be noted that 'self-realisation' does not here displace 'freedom' but rather names its positive dimension: the fulfilment of human needs and potentialities.) Closely related to this is an ongoing attempt to transcend and heal the rift in conventional thinking and problem-fields between subjectivity and objectivity and related antinomies and contradictions, symptomatic of the fivefold ALIENATION of human being, by the same method of absenting constitutive absences through moves to more complete conceptual totalities, and to demonstrate the practical prerequisites and proclivities for transcending the social contradictions such conceptual antinomies are causally bound up with. This, underpinned by commitment to discovery of truth, is the main driving force of the system.

Thus CN thematises mind and the concept-dependent nature of social forms as emergent products (1M) and processes (2E), entailing concepts of social relationism and transformationalism (see TMSA), and continues the task of transcending the subjectivity/objectivity split already begun by TR. The 'LOGIC OF SCIENTIFIC DISCOVERY' as spelled out in TR had shown that science can fallibly achieve objective

Table 15 Phases of development of CR mapped to its stadia

Stadion / phase	1M Non-Identity	2E Negativity	3L Totality	4D Transformative Agency
CR as a whole: thinking being	as such and in general	processually + as for 1M	as a totality + as for 2E	as incorporating human praxis and reflexivity + as for 3L
Form of reflexivity – immanent critique of	philosophical discourse of modernity (PDM)	PDM + 1M	PDM + 1M, 2E	PDM + 1M, 2E, 3L
TR: thinking being as (Absents banishment of rational discourse about being)	structured and differentiated, entailing concepts of intransitivity, transfactuality and stratification			
CN inflection: thinking being as (Absents banishment of rational discourse re contradiction)	constellationally containing mind (SEPM) and concepts (conceptual emergence qua product)	negativity, contradiction, emergence qua process, entailing concepts of social relationism and transformationalism		
EC inflection: thinking being as (Absents banishment of rational discourse re values)	intrinsically valuable	negativity qua absenting constraints, understood as maximising totality	totality, understood as including values (retotalisation of being)	
DCR inflection: thinking being as (Absents banishment of rational discourse re change)	alethic truth (reality principle)	negativity qua (determinate) absence, generalised to the whole of being as real and essential to change	totality, understood as maximised by praxis (which absents incompleteness)	transformative praxis and reflexivity (emancipatory axiology)

Key source *RM*: 165–90.

Note Stadia go horizontally, phases vertically. TR = transcendental realism; CN = critical naturalism; EC = explanatory critique; DCR = dialectical critical realism. Reading horizontally and bracketing the second and third row, cells (including blank ones) should be interpreted (1) as implicitly or potentially containing the content of the cells in the vertically successor phases (thus 'inflection' refers to the specific emphasis given by the new phase of CR to what was already implicit in its previous moment); and (2) as actually containing ditto in the predecessor phases. In the topmost row, each stadion should be understood as presupposed or preservatively sublated by its successor (nothing is annulled), such that 4D > 3L > 2E > 1M. The same holds (bracketing the second and third row) reading vertically in the leftmost column: DCR > EC > CN > TR, but here the sublations are not entirely preservative in that the earlier phases are deepened and cross-fertilised by dialecticisation; i.e. they are *essentially* preservative (cf. *D*: 9 n).

knowledge, thereby bridging the TD and ID. CN enables us to see, by absenting the absence of an adequate account of, e.g., a law of nature, such that laws and causal explanation are thinkable for the sociosphere, that the apparent opposition between, e.g., positivism and hermeneutics is illusory. From the perspective of each other they are positive contraries or absences (i.e. lacking in relation to each other, and so in conflict), but from the perspective of the new, fuller conceptual totality of CN which identifies and remedies the common ground of their lack they appear as negative sub-contraries or presences, CO-INCLUDED within a distanciated present together with their positive contraries: the theoretical (though not the social) basis for their antagonism has been overcome (see DIALECTICAL ANTAGONISTS). This is the characteristic Hegel-derived therapeutic procedure deployed by CN in the mode of distanciated time and elaborated and formalised in DCR's EPISTEMOLOGICAL DIALECTIC, the logic of which (already implicit in TR as the logic of scientific discovery) points in the direction of the removal of the social (not just conceptual) causes of such oppositionality via depth-praxis and emancipatory axiology.

EC is the application of the same logic to the field of ethics (a form of *ethical naturalism*), made possible by the fact that 'social objects include beliefs about themselves' (Bhaskar 1998a: xvii) and by the thematisation of *moral realism*. Ethical naturalism, contra the Humean tradition, licenses a move from facts to values, theory to practice, providing a grounding for ETHICS. Again, such a move is already implicit in TR. Commitment to truth and consistency is a condition, not just of science and human flourishing, but of any human discourse whatsoever and therefore of any human being, and enters into the very constitution of the nature of a fact. While other values do play a role in various aspects of science and philosophy, if the researcher really wants to know they do not necessarily enter into the cognitive assessment of factual claims and theories. It is thus in principle possible for the development of a theory or system to be driven by 'the single norm of truth'. For human being, then, truth is a good (the highest good of all, higher than freedom [Bhaskar 1999]) and falsity (or misleading, inadequate, etc., ideas or social forms) an ill, CP. To criticise a belief as false (e.g., the TR critique of the positivist conception of a causal law already mentioned) is therefore ipso facto to criticise any belief or action informed by it as ethically wrong, and positively to evaluate their rejection and the acceptance of the conceptual schema that provides a more adequate account, CP. In discovering a falsity, philosophy has discovered a value, just as in discovering the falsity of, e.g., racism science has discovered and valorised a value: anti-racism. The belief in each case is shown to be incompatible with a true account of what it purports to be about. EC extends this to the social causes of false beliefs, as was already adumbrated in CN. Thus, if the propositions 'racism is false' and 'social structure (S) causally generates racism' are true, it follows that S should be changed, CP, and that we should take measures to effect this, CP. Here the belief is 'incompatible with its own true explanation' (ibid.: xviii, e.a.). Where false ideas are IDEOLOGIES in that they play a causal role in maintaining S, as in the case of racism, there is a double mandate to change S, CP. In this way EC extends the logic of identifying and remedying absences to the whole ethical sphere, and imperiously demands the unity of theory and practice in practice, which is theorised in the EMANCIPATORY AXIOLOGY elaborated in DCR.

critical realism 103

On moral realism, 'morality, like knowledge, has an intransitive object/ive' (*P.* 151), which we may come to know, and being as such is intrinsically valuable, such that antianthropism applies in ethics too. The basic line of transcendental reasoning establishing this is as follows. Human intentional agency, as the power to bring about a state of affairs that would not otherwise have obtained, presupposes the openness of the world. In an open world, the irreducibility of intentionality necessitates that we *appraise* beliefs from the standpoint of their suitability for acting upon, and *select* a course of action from among a range of possibilities (for we cannot refrain from acting, and so appraising – this is the AXIOLOGICAL imperative) (*PN3:* 87). Appraisal and selection presuppose criteria, and criteria presuppose categories of 'better' and 'worse' as well as 'true' and 'false' in respect of beliefs about how the world is. Like ideas in general, therefore, human values and morality are 'parts of the natural world, products of the naturalised process of thought' (Bhaskar 1997a: 143), an ontic or emergent property of the world rather than 'a cultural add-on' (Brereton 2005; see also his 2000; cf. Archer 2000: 318–19). In other words, human intentionality presupposes a dialectical unity of theory and practice, facts and values (which GENERATIVE SEPARATION geo-historically shatters and CR aspires to heal and retotalise). Furthermore, as emergent properties of the world, values must have been implicit in being in some sense qua possibility prior to human being, hence values are intrinsic to being as such at the level of the real (cf. Collier 1999). In this way, EC re-totalises being by including values within it, and human being by thematising the possibility of a social order based on theory–practice CONSISTENCY.

The same line of reasoning establishes the real existence of *dialectically* universal human rights to truth (learning) and freedom at the level of the real, where freedom is understood as embracing not only liberties and equities (the absence of master–slave-type relations), but also needs and potentialities (capacities, opportunities, resources) for development, presupposing de-alienation; and a CONATUS or TENDENCY$_b$ to realise them (see DIRECTIONALITY; HUMAN NATURE). That is, it is argued that it is intrinsic to the nature of human beings in any conceivable social form to need and to strive to achieve truth and freedom in this sense, which are a condition of each other and properly carried through issue in the understanding of moral truth or ALETHIA (morality's 'intransitive object/ive') as universal free flourishing, in which the flourishing of each is a condition of the flourishing of all. Our standpoint as intentional embodied agents necessarily includes the satisfaction of DESIRE, which entails the removal of constraints on its satisfaction, entraining a 'positive freedom to', and entailing recursively – in a dialectic of SOLIDARITY implicated by the expressively veracious and fiduciary aspects of human speech and action and proceeding by the unity of immanent critique and dialectical UNIVERSALISABILITY (for not to solidarise in the struggle for freedom is to commit theory–practice contradiction) – a meta-desire to remove constraints on the removal of constraints, and to attain the explanatory critical knowledge about the world that is indispensable for the totalising depth-praxis necessary to effect this. TRUST (our primary existential) (3L), explanatory critical theory (2E), depth-praxis (substantive emancipatory axiology) (4D) and freedom as the realisation of moral alethia (1M) thus comprise an ETHICAL TETRAPOLITY. Transformative praxis at 4D, which absents incompleteness, is understood as maximising totality via the struggle for a higher, more richly differentiated global social unity practising an ethic of people- and world-care.

In this way, DCR develops 'a meta-theory for the social sciences, on the basis of which they will be capable of functioning as agencies of human self-emancipation'. As a necessary condition of this, it develops 'a general theory of dialectic – or better, a dialectic' (*D*: 2) (in effect a totalising philosophy of geo-history), at the centre of which is an account of determinate absence, generalised to the whole of being as real and essential to change, and issuing in the 1M–4D system of DCR. And it rounds this off with the elements of a totalising critique, chiefly in terms of the lack of a concept of determinate absence, of IRREALIST tendencies in philosophy that militate against the dialectics of truth and freedom. Being as such is now conceptualised as an open totality that endlessly unfolds via creative leaps of, as it were, 'the materialised imagination' (*D*: 49), which transcend contradictions and constraints at ever more complex emergent levels. In the contingent transition to eudaimonia, in which emancipatory philosophy and social science play an indispensable, albeit by no means the most important role, we overcome the dualism of capitalist modernity and know ourselves at home in the endless creative unfolding of being.

Such at-oneness with each other and the world is the hallmark of the spirituality thematised by the philosophy of META-REALITY, which continues the drive to totality by thematising being, though overlaid by the disenchantment of the DEMI-REAL, as always already enchanted or valuable, and constellationally embracing, besides the world of duality, of which CR provides the best account, an underlying non-duality which people access in their daily lives via the power of transcendence – precisely the capacity to make creative moves to greater coherence and completeness – that they share with the rest of being.

Of course, no such comprehensive account of human-being-in-nature could fail to arouse controversy – which, however, if it is immanently critical, is the very lifeblood of CR philosophy and social theory. Among the issues currently subject to lively discussion and debate are the status of transcendental arguments; the precise nature of TD/ID distinction; the marriage of depth-stratification with entity relationism; propositional versus alethic truth; the derivation of values from facts; the whole area of falsity, false consciousness and ideology; the ontological primacy of absence and the asymmetry of being; rational geo-historical directionality as a real tendency versus an ideal goal; the relation between meta-ethics and normative ethics, objective morality and historically specific morality; the concept of the demi-real; and the spiritual turn and its take on the world-historical problem of agency.

The great bulk of CR substantive research carried out to date has been within a TR/CN, rather than within an EC or DCR, framework. This is doubtless at least in part because the latter are in the relatively early stages of their reception; there is no reason why they should not lend themselves equally to exciting empirically based research – on the contrary, like the ontologies they articulate, they seethe with possibility that may be expected to stimulate a great deal of such work, and there are signs that this is happening. Hugh Lacey (2000, 2002a, b, 2003b, 2006; Lacey et al. 2002) has pioneered substantive research in the spirit of EC, Heikki Patomäki (2006a, b) draws on DCR and emancipatory axiology in relation to FUTURES STUDIES, and there is lively work in hand in a number of other areas within a DCR framework, which it is hoped *DCR* will assist in stimulating.

critical (or limit) situation   105

While politically CR adopts an openly (anti-elitist, anti-substitutionist) emancipatory stance, it is not affiliated or aligned with any particular political or social movement. It seeks rather to assist in promoting a triplex emancipatory helix encompassing philosophy and social theory, empirically based research and a rich diversity of social movements on a planetary scale.

critical (or **limit**) **situation**. See TRANSITION.

critical theory (CT). In its original meaning, a school of thought also known as the Frankfurt School because of the affiliation of key figures with the University of Frankfurt. Although CT started out in the 1920s at the famous Institute for Social Research, the term was only coined by Horkheimer (1937) in a pivotal piece where the CT of particularly Marx in his dialectical critique of political economy is opposed to the traditional theory of Descartes and mainstream scientists. The first generation included Adorno, Marcuse and Fromm, and was distinguished by attempts to combine and advance insights from Marx, Weber and Freud in critiques of instrumental reason in philosophy, science and society. Habermas became the leading theorist of a second generation that propagates a novel version of CT focusing on communicative action, and within recent years a third generation is rising around the work of Honneth concerning RECOGNITION. Honneth furthers a strategic move from negative critique and deconstruction to positive normativity and a constructive spirit that was initiated by Habermas. As a highly influential philosophical tradition rooted in unorthodox Marxism, CT is a continuous source of inspiration and engagement for critical realists. The philosophical groundwork of the early critical theorists is indispensable, and the original critiques of instrumental reason constitute a lively alternative to RATIONAL CHOICE THEORY in mainstream social science. Likewise the consequent developments and reorientations of CT enforce progressive agendas and pose important challenges to critical realists and indeed anyone else involved in social critique and emancipation. Nowadays CT also has a wider meaning which embraces any critical, theoretical approach with liberatory intent, including feminist theory and CR itself. See also FETISHISM; ENCHANTMENT; PDM; REIFICATION; Engelskirchen 2004; Norrie 2004a; Outhwaite 1987.

PETER NIELSEN

critique is the 'critical' in CRITICAL REALISM, oriented to demystification, hence emancipation, and ultimately grounded in the human capacity reflexively to accept or reject (transcend) the received socio-cultural tradition, a necessary condition for which is the depth-stratification and intransitivity of the world, including of its actually existing discourses and moralities. Closely related to Marxian critique, it differs from Kantian critique in viewing the 'forms which structure and inebriate experience', not as a priori contributions of the human mind, but as 'objective systems of constraints, historically produced, reproduced and potentially transformable', which critical explanatory theory may show to be false or systematically misleading (*SR*: 198). Its various forms are set out in Figure 4.

Critique is 'paradigmatically distinguished from *criticism* in that it isolates the source or ground of the imputed error' (*DG*: 396, e.a.), whereas criticism 'knows how to judge and condemn the present, but not how to comprehend it' (Marx, cit. *PN3*: 65), i.e., it applies valuations without any kind of causal grounding. The fundamental form of

critique

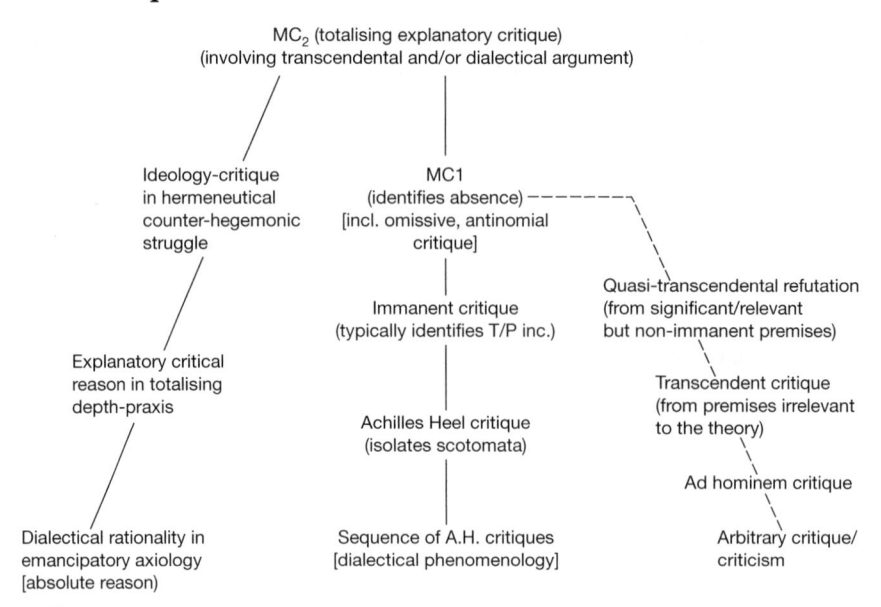

```
                    MC₂ (totalising explanatory critique)
              (involving transcendental and/or dialectical argument)
                   ╱              │
                  ╱               │
         Ideology-critique       MC1
         in hermeneutical        (identifies absence) ─ ─ ─ ─ ─ ╮
         counter-hegemonic       [incl. omissive, antinomial             ╲
         struggle                 critique]                                ╲
            ╱                      │                         Quasi-transcendental refutation
           ╱                Immanent critique                (from significant/relevant
          ╱                 (typically identifies T/P inc.)   but non-immanent premises)
     Explanatory critical          │                                ╲
     reason in totalising          │                          Transcendent critique
     depth-praxis                  │                          (from premises irrelevant
       ╱                     Achilles Heel critique            to the theory)
      ╱                      (isolates scotomata)                   ╲
     ╱                              │                           Ad hominem critique
    ╱                               │                                ╲
  Dialectical rationality in    Sequence of A.H. critiques      Arbitrary critique/
  emancipatory axiology         [dialectical phenomenology]     criticism
  [absolute reason]
```

Figure 4 A topology of critiques
Adapted from D: 242

critique, whether in PHILOSOPHY, science, everyday life or counter-hegemonic struggle, is *immanent critique*. Its hallmark is the avoidance of the 'bad circularity' or arbitrariness implicit in external criteria of knowledge (e.g., judging Socrates by Rorty's criteria) by taking its departure from *within* the accounts it seeks to situate, correct or replace – abandoning all pretence of an ahistorical Archimedean starting point and deploying a process of transcendental argument (see INFERENCE) to demonstrate either that an account is theory–practice inconsistent or, if consistent, beset with APORIAI or problems that are insoluble in its own terms. Thus the CR critique of the Humean theory of a causal law demonstrates that it is inconsistent with the experimental practice that identifies laws. Since there is little point in demolishing a straw figure, the fundamental interpretive principle of immanent critique is HERMENEUTIC *adequacy*. This will usually entrain a *principle of charity*: before criticising a position, cast it in its strongest possible light and even correct and improve it.

Bhaskar distinguishes immanent from *transcendental* and EXPLANATORY CRITIQUE (*PN3*: 120). The conclusion of a transcendental argument may function as an immanent critique of a theory (e.g., the deduction of EMERGENCE functions as a critique of reductive materialism), but a critique is transcendental if it demonstrates that an account is inconsistent with the possibility of science (or of human intentional agency as such) and shows what its conditions of possibility are, issuing in a *transcendental refutation*. It is explanatory if, in addition, it explains the persistence of the inadequate account sociologically, licensing a negative evaluation of its social causes, CP, and its designation, especially where it involves CATEGORIAL error, as *IDEOLOGY*. This will typically involve DIALECTICAL ARGUMENT, and so may also be characterised as *dialectical explanatory critique*. Latterly, Bhaskar has emphasised what these critiques have in common – that 'transcendental argument, dialectical argument, immanent critique and retroductive-analogical explanation in science are all roughly the same in form: they say we have a certain phenomenon or a position which someone is holding, let's see what must be the case for that phenomenon or position to be possible' (Bhaskar 2003b: 97). Noting that explanatory critique is encompassed within *metacritique₂*, we can say that the main forms of critique – all of which involve transcendental argument – are *metacritique₁* (MC₁) (which identifies an absence or incompleteness entailing the real NEGATION of a theory); *metacritique₂* (MC₂) (which in addition explains the absence as well as the social reproduction of the defective theory); *immanent critique* (which identifies either theory–practice inconsistencies or aporiai together with their source), and *Achilles heel* (AH) *critique* (which identifies a weakness in a theory at what was taken to be its strongest point). Thus we have MC₂ > MC₁ > immanent critique > AH critique (cf. *D*: 62 n, 185). A cumulative sequence of AH critiques constitutes either a dialectical PHENOMENOLOGY, magnificently exemplified in Hegelian DIALECTIC and witheringly deployed by Bhaskar to critique the irrealist tradition in Western philosophy, or (deployed upon one's developing self) a *dialectical life* (*D*: 105; cf. *EW*: pt II).

Within this ordination, *omissive critiques* and ANTINOMIAL *critiques* are species of metacritique₁ which, respectively, identify the absence of a transcendentally necessary category, and exhibit interrelated structures of CONTRADICTION between conclusions which seem equally necessary. IDEOLOGY-*critique* is a species of metacritique₂ (totalising explanatory critique) oriented to transformative praxis. The basic form of explanatory critique provides the minimal necessary criteria for designating a belief or theory as *ideological*: it demonstrates both the *falsity* or inadequacy of a theory and its *necessity*, i.e., provides a superior explanation of the phenomena addressed by the theory, together with a range of others, and shows why the inadequate account is believed, issuing in a mandate to change the social cause of falsity, CP. Thus if T = scientific theory, P = proto-scientific theory, I = ideology, V = valuation, S = structure or system, and φ = axiological commitment, then (*P*: 147):

$$T > P \; T \exp I(P) \rightarrow -V[S \rightarrow I(P)] \rightarrow V\varphi_{(-S)}$$

Ideology-critique characteristically goes further, however. Its formal specifications are set out in detail in *PN*: ch. 2 and *SR*: chs. 2.6 and 3. Briefly, (1) it shows not just a theory, but the theory–practice ensemble within which it is embedded, to be (characteristically categorially) false or misleading; that is, it shows that falsity pertains to social forms – which are quasi-propositional – (e.g., the wage-form, which collapses powers to their exercise), not just to the theory of them: $(-(-T_1[P_1])$. In such a case, it is not so much that the subject deceives herself (makes cognitive mistakes) but that the structure of reality produces the deception in her, analogously to the effect of a mirage or of a good actor in the role of a real historical character. (2) It explains the falsity as a necessary misrepresentation of the essential structures (social relations) generating it, to which it is internally related and which it therefore functions to conceal. (1) and (2) together constitute a dialectical CONTRADICTION of a specific kind, a *Colletti contradiction* or *false necessity*: 'the necessary coexistence in social reality of an object and a (categorially) false presentation of it, where it is the inner (or essential) structure of the object which generates the categorially false presentation (or appearance)' (*SR*: 195) (this is *constitutive* or ALETHIC *falsity*), and their demonstration constitutes a DIALECTICAL EXPLANATORY ARGUMENT (da') or *dialectical explanation*. (*Pace* Hammersley [2002] this does *not* commit ideology-critique to functionalism [see FUNCTIONAL EXPLANATION].) Such mystification characteristically involves fundamental category mistakes, e.g., the presentation of the social as natural and ahistorical in FETISHISM. *First-order* and *second-order* (ideology-)*critiques* expose category mistakes, characteristically involving either illicit FISSION or FUSION. A *second-order critique* shows a social form to be 'true but systematically {...} self-misrepresenting, in being validly applicable to experience but only *within* certain {... often geo-historically specific} limits, contrary to the form's own self-presentation', e.g., the value-form, or the FACT-FORM, or the closed-systems or CONSTANT CONJUNCTION form produced within laboratories. A *first-order critique* demonstrates the social form itself (e.g., the wage-form) to be false, or more formally that the categories informing it are 'not properly {...} applicable to experience at all' (*SR*: 199, 304–5). Ideology-critique, and more generally explanatory critique, may thus ground a threefold criticism, (1) of theories *(theoretical ideologies)*, (2) of social practices *(practical ideologies)* and (3) of the generative social structures that underpin them, issuing in a mandate to change them, CP (see also ALIGNATION). Explanatory critique is a transcendentally necessary, not optional, explanation-form – a condition of every rational praxis: for there can be no action without judgements as to the truth or falsity of beliefs and no rational change of beliefs without explanation.

It is the internal relationality of the social sciences and their subject-matter, i.e., the concept-dependence of social forms, that renders critical explanatory theory in general possible, for it entails that criticality is '*intrinsic* to {their} explanatory function {...} If criticism without explanation is impotent, explanation without criticism will often just be simply false' (*SR*: 193–4). This is '*the DUALITY OF THEORY AND CRITIQUE*' (*D*: 260, e.a.). Social science itself is accordingly conceptualised *as* explanatory critique (and thence, in the DIALECTIC OF SOCIAL SCIENCE, as EMANCIPATORY AXIOLOGY and DIALECTICAL REASON). But its social situatedness further entails that it is *conditioned critique* – that it is itself subject to 'the same possibilities of unreflected determination and historical supersession it situates. Hence continuing self-reflexive *auto-critique* is the *sine qua non* of any critical explanatory theory' (*SR*: 170, 210, e.a.). The possibility of the effectivity of such theory in human history (bearing in mind that social science need not be restricted to the academy) 'comprises perhaps the only chance of non-barbaric, i.e., civilised, survival for the human species' (*SR*: 180).

Among the forms of non-immanent and non-explanatory criticism, which are far weaker, are *quasi-transcendental refutation*, which departs from significant or relevant, but non-immanent, premises; *transcendent critique*, which operates from premises that are irrelevant to the theory; *infinite* or *vicious regress*, which demonstrates the impossibility of a non-circular resolution; *reductio ad absurdum*, which demonstrates incoherence; *ad hominem critique*, which goes for the person instead of the theory; *indirect* criticism of a theory on grounds of its social effects; and *arbitrary critique*, or any criticism which is not immanent (see *PN3*: 120–3; *D*: 80, 242). See also CONSISTENCY; CRITICAL NATURALISM; INTRANSITIVE, ETC.; IRREALISM; PHILOSOPHY; REASON; TINA SYNDROME.

C-series. See A-SERIES.

CSM (central state materialism). See MIND.
cube, social. See SOCIAL CUBE.
cultural relativism. See HERMENEUTICS; HOLY TRINITY.
cultural studies. See CULTURE.
culture, cultural analysis. *Culture* is an extraordinarily complex term. Among its current senses are 'civilisation', 'ethnicity or heritage', 'beliefs and customs', 'style, taste or refinement', and 'development of thought and expression'. For CR this last sense is primary, provided that 'development' is understood as involving structures, practices and products that are social and embodied. However, views differ on what that definition entails, and, all told, culture is as yet undertheorised within CR. This is unfortunate, because having such a theory allows one to analyse personal and social identity, feelings, arts, ideology, and other cultural forces that make life experiences meaningful.

One broad thesis does ground all CR approaches to cultural analysis: culture is *RELATIVELY AUTONOMOUS* from other major social structures. In other words, cultural activities have aspects linking them to social structures such as economic systems and gender relations, and to the body, yet culture has its own characteristics which make it irreducible to any other structure: ideas and structures are connected yet not completely bound together. This basic position leaves plenty of room for a variety of approaches, but it also distinguishes a CR approach from deterministic views which make culture purely epiphenomenal, and from post-structuralist analyses asserting that social relations collapse into language. At the same time it allows CR to be receptive to insights that other approaches may offer.

POST-STRUCTURALISM (or, more broadly, POSTMODERNISM) undoubtedly dominates the field of cultural analysis at present, although its stranglehold seems to have weakened lately. One of its distinctive theses is that a crucial entity in cultural production is 'structured like a language', such as the psyche for Lacan, or society for Giddens. The analogy with language generally becomes a conflation with language, sometimes called the *linguistic fallacy*, which is a version of the EPISTEMIC FALLACY. At the heart of this fallacy is the Saussurean concept of SIGNS, which lacks referential connections outside the mind and hence lacks a way for language to be subject to an 'outside': instead, the 'outside' is wholly subject to language and/or its mode of operation. Nevertheless, just as CR shares certain views with postmodernism, the work of some postmodern thinkers (such as Derrida and Foucault) can contribute to CR.

There are other predecessors to CR cultural theory. Many draw from Marxism, but mix in other ingredients. One was Raymond Williams (1977), who examined culture according to Marxist terms (e.g., forces of production, HEGEMONY), but also developed concepts such as 'structures of feeling', which allowed supple analyses that could be sensitive to the qualities of lived experience. The Bakhtin circle's philosophy of language and theory of the dialogic nature of discourse melded Marxism and phenomenology in ways that give them strong kinship with CR. Another seminal thinker is Brecht, who discussed both the interpretation and the purpose of cultural production, and defined artistic realism not as verisimilitude, but rather as the revelation of society's causal structures. Interestingly, Brecht was a playwright, Bakhtin's key concept was dialogism, and some of Williams's major writing is on drama: it seems dialectics has been a part of realist cultural theory from the start.

Several recent cultural theories have adopted an avowedly realist stance, apparently without any knowledge of CR, but solidly in its stream nevertheless. Mohanty (1997), for example, develops a realist concept of the sign, and from there follows an argument leading from a social (and dialogic) theory of language to a defence of multiculturalism (as an aspect of social justice) based, not on relativism, but on the very fallibility of theory: 'the realist will favour cultural diversity as *the best social condition* in which objective knowledge about human flourishing might be sought' (243).

There have been several significant efforts towards a specifically CR cultural theory, but they are not all in agreement. For Archer (1996), culture consists of 'all intelligibilia' constituting a single 'Cultural System'. Her analysis is more a sociology of ideas than a fully fleshed cultural theory; nevertheless, she provides major components for such a theory. Two of her positions stand foremost. First, there is an ontological and analytical distinction between culture and agency (and implicitly between culture and social structure), because culture pertains to logical relations among propositions whereas agency involves causal relations. Consequently, culture is not an epiphenomenon of structures, nor are structures simple cultural (conceptual, linguistic) products, nor are structure and culture so intertwined as to be even conceptually inextricable. Second, culture develops in a cyclical process starting from cultural conditioning, through socio-cultural interactions, to cultural elaboration, which sets the stage for the next cycle. The dynamics of this cycle depend in large part on four possible situational logics among ideas, depending on whether particular ideas are logically compatible or contradictory, and whether they are (historically or socially) independent or inescapably connected to the other. Each of these situations involves social and logical processes that may lead to various types of group formations and disintegrations.

Archer's social ontology can be refined into a view that society possesses three distinct strata, namely structures, agents and discourses, and that each stratum has material, sociological and meaningful aspects (in different degrees and roles) which interact in various ways. But Archer's assertion that there is only one 'Cultural System' is essentially rationalist: COGNITIVE SCIENCE has demonstrated that analogical relations are at least as important as logical ones (see SIGNS), and that culture is linked to embodiment. The argument has various ramifications for theories about the mind/body connection, the process of cultural creation, agency, and culture as a totality. In particular, the embodied practices involved in a particular social structure – communication – generate discursive strategies pervading much thought and art. Communication structures also provide much of culture's relative autonomy from other social structures. The role of embodiment and materiality is such that there is not one universal Cultural System, but instead many cultures, often possessing subcultures, which are 'porous' to each other: in CR terms, they are open, partial TOTALITIES. The point leads straight to Mohanty's defence of multiculturalism. Archer's own later work on agency (2000) stresses embodiment and materiality but without fully taking on board their implications for the theory of Cultural System.

Analysing culture in terms of partial totalities is a promising route because it obliges the critical realist to consider both the internal and the external relations of a cultural object (large or small). Consequently, many different analytic threads can be pursued and integrated. These include: (1) the object's internal semiotic or discursive structures (by way of close readings) and its meaning in both its time and ours; (2) the micro-history of cultural agents (producers and audiences) such as their biography, circumstances and purposes of production, relationships with other agents, dialogic activities and reception practices; and (3) macro-historical analysis of the structural conditions of production and reception, including their connection to larger social structures, the material forces of production, cultural formations, discursive/ performance strategies, social positions, institutions, governmental policies, etc.

Crucially for the critical realist, these approaches are not independent. One must know about the larger culture in order to interpret a particular cultural product, and thought is highly mediated by social relations, practices and institutions. Certainly, artists' intentions are significant, since the agent is the most immediate causal power producing the cultural object. But artists act under material, social and discursive preconditions and with given resources. The audience can interpret in its own manner and may even wrest the object from its original context and use it quite differently. Intermediaries such as corporations may disseminate cultural works for their own purposes. In short, the creation and reception of cultural works must ultimately be understood in terms of discourses, agents and structures. Moreover, cultural works have connection to many different discourses (religious, political, philosophical, etc.), agents (individuals, groups and institutions) and social structures (economics, gender, communication, etc.). However, although all of these approaches are needed for a *complete* analysis, to a certain extent individual studies may focus on one aspect or another.

But why study culture? Perhaps the concept of partial totality can be deployed here as well. There are intrinsic merits and pleasures in experiencing and interpreting cultural works, and in understanding their place within the larger currents of biography and history. Interpreting a work's meaning-system and explaining the forces that led to the work's creation and affected its reception can enrich our experience of it. Reversing perspectives, cultural products can offer insights into the world of their creation or reception. Similarly, one may interpret and explain the overall system of a culture or sub-culture, which may entail issues of ideology, hegemony and counter-hegemony. And one may glean from other cultures knowledge of the world unavailable to oneself, to understand more deeply socio-cultural structures and dynamics, and thereby expand our view of the reasons for changing the world, ways of doing it, and what might be achieved in the process. See also Archer 1988/1996; Bakhtin 1981; Bernard-Donals 1994; Brecht 1964; Chouliaraki and Fairclough 1999; Mohanty 1997; Nellhaus 1998, 2000, 2005; Voloshinov 1929; Williams 1977.

TOBIN NELLHAUS

D

4D (fourth dimension). See MELD.
δ node or **δ moment**. See TRANSITION.
das Bestehende. See BESTAND.
Dasein (Ger., to be there, to exist). Used as a noun by Hegel and Heidegger to signify determinate being and human being in time respectively. (The later) Bhaskar's usage is similar to Hegel's but, whereas in Hegel it is the intrinsic cosmic structure of logical necessity that comes to self-realisation in *Dasein*, in Bhaskar it is the open process of potentiality that contingently does so.
DCR. Acronym for DIALECTICAL CRITICAL REALISM.
DEA and **DET models**. Moral reasoning is a species of practical reasoning or problem-solving. In the *DEA model of practical reasoning* we *d*iagnose a problem, *e*xplain it and *a*ct to resolve it. In the *DET model of normative change* we *d*escribe, *e*xplain and *t*ransform actually existing morality. Both will depend upon the pure and applied models of EXPLANATION of theoretical reason (DREI[C] and RRREI[C]) (*D*: 260–1).
de-agentification. See AGENTIFICATION.
de-alienation. See ALIENATION; DIALECTIC OF DE-ALIENATION; EMANCIPATORY AXIOLOGY; PROBLEM OF ALIENATION.
decentrification, decentring. See CENTRISM, ETC.; PROBLEM OF AGENCY.
decisionism. See ETHICAL IDEOLOGIES.
deconstruction. See DIALECTICAL COMMENT; HERMENEUTICS; POST-STRUCTURALISM; RATIONAL RECONSTRUCTION.
deduction, deductivism. See EXPLANATION; INFERENCE.
deductive–nomological (D–N) model of explanation. See EXPLANATION; SOCIAL STRUCTURE.
de-evaluation of reality. See ALIENATION; DIALECTIC OF MORALITY; NATURALISTIC AKA ANTI-NATURALISTIC FALLACY.
defence mechanism. See METAPHYSICAL LAMBDA (λ) CLAUSE; TINA SYNDROME.
definition. A *nominal definition* (not to be confused with *nominal essence* [see NATURAL KIND]) specifies the meaning of a linguistic expression, typically in terms of actualistically understood characteristics or manifest appearance: e.g., 'yellow malleable metal' for 'gold'. Realists commonly use it to mean a provisional working definition at the outset of an inquiry – the object as conceptualised in the existing literature. If the inquiry is successful, the nominal definition is recast, at the Leibnizian

de-geo-historicisation 113

stage of the EPISTEMOLOGICAL DIALECTIC, to provide a *real definition* capturing the essential structure or ALETHIC truth of the thing: e.g., 'element with atomic number 79' for 'gold'. It specifies a reason distinct from, but explanatory of, the corrected initial description, which instantiates *natural necessity* and may provide the basis for its definition as a *natural kind*. On the CR account real definition plays a crucial role in the epistemological dialectic of any science, including social science. The real definitions of a science constitute its fundamental theoretical principles.

de-geo-historicisation. See GEO-HISTORY.
deictic. See APODEICTIC.
de-materialisation. See VIRTUALITY.
demediation. See CONCRETE/ABSTRACT; CONCRETE UNIVERSAL ↔ SINGULAR; MEDIATE.
demi-actual(ism). See ACTUALISM; DEMI-REALITY.
demi-reality (L. *dimidius*, halved, semi-, curtailed) names and theorises the nightmare of error and illusion – in which falsity prevails over truth, evil over good, heteronomy over autonomy, war and repression over love and peace – from which we are trying to awake. Within the CR system, the concept has its origins in the (Marxian and) Bhaskarian critique of Hegel, specifically Hegel's notion that 'what is rational is actual and what is actual is rational': the *demi-ACTUAL* is the name given by Bhaskar to that part of actuality which Hegel's system cannot rationally transfigure or explain; the *demi-absolute* is the part that it can. (This borrows from Hegel's own thematisation of the figure of 'demi-', e.g., in respect of women, who achieve full actuality only via their relationship to men.) The *demi-present* is the Hegelian present, which is irrational because it is glaciated, englobing a constellationally closed future (see also ENDISM; GEO-HISTORICITY). 'Demi-' thereby acquires from Hegel the connotation of 'irrational existent', and all that is required for the concept of the *demi-real* to be born is for the CR 'real' to be subtended to the CR understanding of the 'demi-actual' (*D*: 64, 73 n, 91, 115, 194; *DG*: 393; *P*: 123 n).

Its concept is first explicitly elaborated in *EW*, but it builds extensively on the earlier theorisation of IDEOLOGY (which draws heavily on Marx, who 'anticipated Nietzsche's great insight that "among the conditions of life might be error"' [*SR*: 119]) and the TINA SYNDROME; indeed, the demi-real *is* an objective multiply compounded '*avidya-tina* formation' (*EW*: 34–5, 38 n), where 'avidya' signifies ignorance, based on CATEGORIAL error, of our true selves and the totality they inhabit (our inner and outer SELVES), hence ALIENATION from them – such that demi-reality might also be called *Tina-reality*. Metacritically, then, the demi-real is ultimately constituted by ABSENCE of a normatively negative kind (the absent self and totality, which it occludes) – i.e., by alienation. While error and illusion are of course also subjective, ontological monovalence and the epistemic–ontic and other irrealist fallacies are embodied *in* social reality (*D*: 90) as *lived illusions*. This complex four-dimensional maze of *objective false* (or mis-) CATEGORISATION is '(1) *irrealist* in character (i.e., not realist); (2) *demi-real* in truth-value (i.e., false), but (3) *real* in causal aefficacy (and hence being), although dependently so' (*EW*: 6 n). 'False' here conveys the notion of an illusion, i.e. lacking a real object; it would perhaps be better to say, however, that the demi-real, like the Tina-reality it is, is '*half* false and half true' (*RM*: 84, e.a.), resulting as it does from denial in theory or practice of what is necessary in practice. As illustrated in Table 16, it involves three kinds of (emergent, causally efficacious) *untruth* or falsity (constitutive falsity), corresponding to three of the components of the truth tetrapolity (see ALETHIA): epistemic untruth *about* an object or being, epistemic-ontic (expressive-referential) untruth *in* an object or being (e.g., the fact-form, the wage-form), and ontological untruth *of* an object or being *to* its essential nature (e.g., any persons who have been alienated from their potentialities [*D*: 282], and social relations that are untrue to the essential nature of human being – what has been called STRUCTURAL SIN [cf. *EW*: 37]); all four moments of the causal chain are satisfied if we add the untrustworthiness implied by (2) and (3).

The emphasis on categorial error appears to make the human condition depend fundamentally upon (objectively sedimented) philosophical mistakes (cf. *EW*: 39), but, since PMR largely preservatively sublates DCR, it is arguably consistent with a historical materialist account – if such does not extrude the robust causal power of ideas and ideology from the geo-historical process – of the emergence and persistence of power$_2$ relations, and of struggles to change them. Objectified categorial error both underwrites and expresses alienation, which presupposes Marxian primary accumulation and the recursive piling of a mountain of debris at the feet of the angel of history (Benjamin 1940: 259–60). While, at a meta-theoretical level, ignorance is dialectically necessary for enlightenment – there could be no enlightenment without a prior forgetting (*EW*: 38 n), that is, a condition that has been well and truly met in the barbarous demi-real of today. We live our lives to considerable extent within – not a fly-bottle, as Wittgenstein has it – but a seemingly endless Kafkaesque cubic stretch-flow *web* (*maya*) of ignorance, illusion and alienation with intersecting levels and mazes. Our tendential striving towards the moral ALETHIA of the species necessitates that we extricate ourselves (*shed* the demi-real), a task which presupposes the totalising depth-praxis of the earlier philosophy (see EMANCIPATORY AXIOLOGY). Shedding is a form of *DISEMERGENCE* (e.g., de-commodification, dereification, de-alienation), presupposing a 'simultaneous self-TRANCENDENCE in a deeper {...} and wider {...} totality' (*EW*: 11–12).

Table 17 relates demi-reality (fourth column) to the key moments and figures of PMR, mapped to the (loosely corresponding) CR domains of reality; correspondence is loose because relative and demi-reality are themselves structured and differentiated.

Table 16 Modes of untruth in the demi-real

Causal–Axiological Chain	1M Non-Identity	2E Negativity	3L Totality	4D Transformative Agency
Components of theory of alethic truth	ontological, alethic (ID)	adequating (warrantedly assertible) or epistemic (2E and 1M) (TD)	normative-fiduciary (IA)	expressive-referential (4D and 1M) (TD/ID)
Modes of untruth	(3) *of* an object or being *to* its essential nature (ID)	(1) *about* an object or being (at any one level of reality) (TD)	untrustworthiness	(2) *in* an object or being (at that level of reality) (TD/ID)

Table 17 Key moments and figures of PMR mapped to the CR domains of reality

Domains of Reality	**Real** *experiences, concepts and signs* events *mechanisms*	**Actual** *experiences, concepts and signs* events [mechanisms]	**Empirical**, more generally **Subjective** *experiences, concepts and signs* [events] [mechanisms]
Realms of Reality	demi-reality relative reality *absolute reality* the enfolded, the implicit (necessity, possibility)	demi-reality *relative reality* [absolute reality] the unfolded, the explicit	*demi-reality* [relative reality] [absolute reality] the falsely unfolded
Meta-Philosophical Principle	dualism duality *identity* (non-duality, non-alienation)	dualism *duality* (without alienation) [identity]	*dualism* (alienation) [duality] [identity]
Philosophy	irrealism critical realism *meta-Reality*	irrealism *critical realism* [meta-Reality]	*irrealism* [critical realism] [meta-Reality]
Ontological Principle	irrealism realism *truth*	irrealism *realism* [truth]	*irrealism* [realism] [truth]
Orientation to Being	evading being thinking being *being being*	evading being *thinking being* [being being]	*evading being* [thinking being] [being being]
Subjectivity–Objectivity Relation	diremption (alienation) expressive unity *unity-in-diversity*	diremption (alienation) *expressive unity* [unity-in-diversity]	*diremption* (alienation) [expressive unity] [unity-in-diversity]
Social Principle	war struggle *love and peace*	war *struggle* [love and peace]	*war* [struggle] [love and peace]
Modes of Freedom/ Unfreedom	alienation unity *autonomy* (identity)	alienation *unity* [autonomy]	*alienation* [unity] [autonomy]
Modes of Truth	adequating (epistemic) expressive-referential *alethic*	adequating (epistemic) *expressive-referential* [alethic]	*adequating (epistemic)* [expressive-referential] [alethic]
Modes of Untruth	about something in something *of something to its essential nature*	about something in something [of something to its essential nature]	about something [in something] [of something to its essential nature]
Modes of Knowing	ideology science *dialectical reason*	ideology science [dialectical reason]	*ideology* [science] [dialectical reason]
Modes of Reason	analytic, instrumental dialectical *unthought*	analytic, instrumental *dialectical* [unthought]	*analytical, instrumental* [dialectical] [unthought]
Modes of Intellection	discursive intellect intuitive intellect *ground-state* or *identity consciousness* (supramental)	discursive intellect *intuitive intellect* [ground-state consciousness]	*discursive intellect* [intuitive intellect] [ground-state consciousness]
Modes of Self	ego embodied self (person) *transcendental* or *alethic self*	ego *embodied self* [transcendental self]	*ego* (a real illusion) [embodied self] [transcendental self]

Note. Lower-most (primary) levels constellationally embrace upper (secondary) levels, hence have ontological, epistemological and logical priority over them – the priority of the enfolded over the unfolded, the possible over the actual. Upper levels, which thus presuppose primary levels, are either limited or false in relation to them, and function to occlude them. Bracketed levels are not given in the concept of unbracketed levels but are presupposed by it.

It is important to note that the theory of the demi-real is *not* claiming that we inhabit only a shadowy half or third of reality: demi-reality is constellationally embraced by and dependent on ABSOLUTE and relative reality, which it occludes; we inhabit all three, and they us. The necessary, the true and the autonomous are CO-PRESENT with the unnecessary, the false and the heteronomous. The false is a moment of the true (and not vice versa, as Debord [1967] would have it).

demi-reg (demi-regularity). An *event regularity* (or CONSTANT CONJUNCTION) that holds imperfectly over a restricted region of space–time. The significance of such rough-and-ready event regularities to CR analysis lies in their providing a prima facie indication of an occasional, but less than universal, actualisation of a *mechanism* or *TENDENCY*. The term originates in the work of Tony Lawson (1997, 2003b), who argues that contrastive demi-regs provide an important source of directionality to social scientific investigation, particularly those employing the method of CONTRAST EXPLANATION.

PHILIP FAULKNER

denegation (f. L. *denegare*, to deny, refuse, reject). A 2E figure of OPPOSITIONALITY signifying 'denial in theory, affirmation in practice' (*P*: 242). It is the inverse of *unseriousness*, or affirmation in theory, denial in practice. Both are a form of theory–practice CONTRADICTION or INCONSISTENCY. In either case, immanent CRITIQUE shows that the critiqued theory unwittingly expresses and presupposes 'the very content it would deny'. The ground of this contradiction 'lies in the *transcendental–AXIOLOGICAL necessity* of the denied position' (*SR*: 298; see also *EW*: 10 n, REALITY PRINCIPLE). Denegation thus plays a crucial role in the constitution of TINA FORMATIONS.

denegativisation (positivisation). See CONCRETE/ABSTRACT.
de novo creation. See CREATIVITY.
de-ont. See ONT/DE-ONT.
deontic logic. See LOGIC.
deontic modality. See MODALITY.
deontological, deontology. See AXIOLOGY; ETHICS; ONT/DE-ONT.
dependence. See PRIMACY (priority).
deprocessualise, deprocessualisation. See CONCRETE/ABSTRACT; PROCESS.
depth or **ontological depth** together with INTRANSITIVITY and TRANSFACTUALITY, is the central 1M category or dimension of being. It captures 'the idea of the real *stratification* of being apart from our knowledge of being' (*SR*: 63, e.a.), which is made explicit in the concept of *depth-stratification* (or the *multi-tiered stratification* of being, etc.). By the same token it also captures the idea of the real stratification of our knowledge of being, which – since the logic of discovery proceeds in the opposite direction to the unfolding of being (see below) – is no historical accident but necessarily inversely reflects ontological depth. The stratification of being is thus a transcendentally necessary condition of the intelligibility of the stratification of the sciences and of scientific change and growth. More generally, it is a condition of the intelligibility of human intentional action as such, i.e., of the capacity reflexively to produce more adequate accounts of the world (see CRITIQUE), closely affined to ALETHIC TRUTH. As such it is the basic principle of theoretical science, which – since it cannot know what

depth, or ontological depth

lies beyond its farthest frontier – necessarily assumes that depth is in principle limitless; and it is the key to the resolution of the 1M PROBLEMS of philosophy, and in particular the problem of induction. In the case of concretely singular individuals, including people, their depth-stratification is also captured in the notion of a *laminated structuratum* (see STRUCTURE) or *system*. Ontological depth or stratification contrasts with the ontological *depthlessness* or flatness characteristic of *de-stratifying* actualist approaches. A vertical dimension of being, it is to be distinguished from ontological *differentiation*, which applies horizontally and is entailed by the *open SYSTEMIC* nature of the world, or, the transfactual efficacy of its mechanisms. The two concepts are captured in the concepts of *vertical, theoretical* or *depth-realism* and *horizontal* or *transfactual realism* respectively, and come together in the concept of *depth-open* (adj.) and *depth-openness* (*P*: 163). (It should be noted, however, that, together with intransitivity, transfactuality can itself be regarded as a kind of depth, so that there are three kinds of depth in CR: intransitivity, transfactuality and stratification [Bhaskar 1998a: xi–xii]).

Since epistemology is constellationally contained within ontology, *depth-realism* is thus synonymous with ontological realism at 1M in its vertical aspect, and ontological depth embraces the key concept of (EMERGENT) *level* or *stratum* (pl. *strata*), whether in being as such (including human being, viz. the depth-stratification of the SUBJECT and social stratification [see SOCIAL STRUCTURE]) or in our knowledge of being. Things have 'their own level of being' (*RS2*: 78) or of STRUCTURE, and it is this in-itselfness that the various corresponding sciences seek above all to explore, rather than the complex circumstances in which things exist and act (see also NATURAL KIND). This provides science with a non-arbitrary principle of stratification. While *higher* or *higher-order* entities or levels and strata are emergent from and thus unilaterally dependent on *lower* or *lower-order* ones (see PRIMACY) in that they could not exist without them, 'the historical order of the development of our knowledge of strata is opposite to the causal order of their dependence in being' (*RS2*: 169). Because the order of epistemological discovery thus necessarily reverses the ontological order (which is then matched in the order of EXPLANATION), what is higher or more superficial from an ontological is lower or deeper from an epistemological PERSPECTIVE, and vice versa. Thus from an ontological perspective biological are higher than physical entities, but from an epistemological perspective the reverse is the case. This means that the EPISTEMOLOGICAL DIALECTIC, no less than the unfolding of being, goes up, metaphorically speaking, not down. Bhaskar's usage in regard to the natural world and being as a whole accordingly usually reflects this, but sometimes departs from it to remind us (1) that what is ontologically lower (deeper) can also be considered ontologically higher in the sense of fundamentally ingredient (see ULTIMATA); and relatedly (2) that the distinction between epistemology and ontology is constellationally contained within ontology. However, in the social world what is ontologically is also usually epistemologically higher for Bhaskar: thus social forms are higher in both senses than intentional agency, reflecting the fact that he sees sociality ontologically as supervenient on 4D, 'since all the concepts necessary for a philosophical characterisation of sociality can be derived from the sole premise of intentional agency' (*PG*: 250), and epistemologically as having explanatory PRIMACY over it (within the demi-real). In the ethical or practical sphere, 'higher' and 'lower' mean 'more' and 'less important' respectively. What is important is, however, not unrelated to ontological

118 depth or ontological depth

depth and/or totality. Thus (alethic) truth is the highest norm, higher even than freedom, because ultimately it just is (the epistemologically mediated essential causal powers of) reality, the understanding of which is of utmost importance to human flourishing. 'Level' is also used in the more general sense of level of analysis or discourse, together with its referent (e.g., 1M or 4D, abstract or concrete, real or actual, 'Leibnizian' or 'Humean'), and is thus sometimes synonymous with domain, stadion, moment, etc.

The notions of real stratification and transfactuality are brought together in the concept of the *depth-openness* of the world, which acts as a bridging concept to 2E and 3L. Ontological depth is consistent with process and totality, with the fluidity and intra-activity of the world as understood by modern biology and physics. The world is (very complexly) both hierarchically structured and fluidly unfolding and interacting, i.e., REDUCTIONISM cannot provide an adequate account and only the conceptualisation of emergent levels of being subject to dual or multiple control can halt the vicious regresses it entails.

Depth-investigation or *depth-inquiry* (DI) refers (1) to the general process whereby human inquiry probes deeper (or moves higher) into the complexly related levels of being; and (2) to the practical application of the theory of EXPLANATORY CRITIQUE to particular cases in the social world, whether historically reconstructed or living, and thence to the empirical confirmation, correction and further development of that theory, for which it is indispensable; the objective of the whole exercise being EMANCIPATION (replacing unwanted, unnecessary and oppressive sources of determination with rationally wanted and empowering ones). In this more technical sense, a DI is 'any co-operative enquiry, which includes the frustrated agent(s) concerned, into the structure of some presumed set of mechanisms, constituting for that agent an unwanted source of determination, with a view to initiating, preserving or restoring the agent's well-being, including her capacity to think, speak, feel and act rationally' (*SR*: 202). Its concept and formal structure are elaborated in Bhaskar 1980, 1982, and *SR*: 200–11 (see also Rogers 2004). It presupposes depth-stratification 'within the constitution of our theoretical and practical agency' (*SR*: 206), a (corrigible but grounded) theory of human nature (human capacities and needs) and that willingness to learn on the part of the investigators (including the frustrated agent) and continually to develop their theories/practices and their own self-understanding that Marx calls attention to in the third *Thesis on Feuerbach*. It is characterised neither by the pure movement of thought, as in idealism, nor by an external relationship between 'expert' and client, as in empiricism, but by the internal relationality of theory to practice in practice.

What constitutes well-being and rationality cannot be stipulated in advance of DI; it is discovered during its course. This is captured in the concepts of *depth-rationality* or *depth-explanatory critical RATIONALITY* and *depth-praxis* (see also EMANCIPATORY AXIOLOGY). An array of DIs linked into explanatory critique articulated in an emancipatory movement deploying the logic of dialectical UNIVERSALISABILITY constitute *totalising depth-inquiry* and *-praxis* (TTTTφ). Participants in such emancipatory movements are (totalising) *depth-strugglers*. A *meta-DI* is the philosophical investigation of the socially mediated emancipatory drive or pulse of freedom as such, which both underlabours for depth-praxis and is in turn informed by it. The DIALECTIC OF FREEDOM constitutes a *depth-dialectic*.

depth explanation 119

A *depth-explanation* provides a causal account, arrived at during the course of DI (or more generally explanatory critique), of whatever is preventing the agents from acting rationally and flourishing, i.e., it identifies the blocking mechanism(s) and issues a mandate to transform them, CP. CR aspires to provide a general conceptual framework and methodology for *depth-explanatory social science*. A *depth-critique* is synonymous with explanatory critique. *Critical depth-semantics* is a species of metacritique which identifies agents' unawareness, misdescriptions or misunderstandings of their actions, explains them in terms of social causes and redescribes them in the light of the explanatory critique (*SR*: 167).

The possibility of DI is 'a transcendental condition of any human science, and hence at a remove of any science or philosophy and of every rational practice or act of self-understanding at all', which is 'necessitated by the existential intransitivity and enabled by the causal interdependency of the phenomena of sociality', whereby critique is 'internal to (and conditioned by) its objects' (*SR*: 206, 210, e.a.; *RR*: 113). It is these features of the human situation that make it possible for us to identify the real reasons or grounds why we are mistaken or our needs are not being satisfied, and act to change them. In this double movement, in which both the causes of beliefs and actions and the causes of those causes are investigated and criticised, the human sciences are 'part of the programme, paramorphic (but non-identical) to that of Kepler, Galileo and Newton, of the investigation of the underlying structures producing the manifest phenomena of social life' (*SR*: 207).

depth-explanation, depth-inquiry, depth-investigation. See DEPTH.
depth-practice, depth-praxis. See AGENCY; DEPTH; EMANCIPATORY AXIOLOGY; TTTTφ.
depth-rationality. See DEPTH, RATIONALITY.
depth-realism. See DEPTH.
depth-semantics. See DEPTH.
derealisation of reality. See EPISTEMIC–ONTIC FALLACY.
dereification. See REIFICATION.
Derridean supplement, supplementarity. See NEO-LIBERALISM; TINA SYNDROME.
descriptive adequacy. See HERMENEUTICS.
descriptivism. See ETHICAL IDEOLOGIES.
desingularisation. See CONCRETE/ABSTRACT; DE-AGENTIFICATION.
desire (f. L. *desiderare*, to long for some person or thing that is absent or lost; to require, need; to miss, lack; to lose). Presupposing and impelled by ABSENCE (as its etymology suggests), desire is fundamental to intentional ACTION, albeit only the most elemental component in one of its five 'componental springs' or bases: the *conative domain* (f. L. *conari*, to endeavour, strive). The other domains are the *cognitive* (providing discursive knowledge that), *affective*, *expressive* and *performative* – which must be supplemented by *competencies* (practical knowledge how) and *facilities* (resources). 'Pure or unimpeded' desire may be regarded as WILL, and 'desire correspondingly as will, which, encountering some obstacle, requires beliefs about the manner as well as the object of its satisfaction' (*PN3*: 96).

Action typically involves, as central (conscious or unconscious) elements, a cognitive as well as a conative component: *beliefs* about the way things are and the likely effects

120 desire

of the action on them, and *desire* to change (absent) the way things are (see also WANTS). Since this presupposes a *meta-desire* to remove constraints on the fulfilment of desire, thence a series of *meta-meta desires* to absent constraints, recursively, the concept is fundamental to the AXIOLOGY OF FREEDOM. It is 'informed desire that drives {emancipatory} praxis on' (*D*: 169), and it is elemental desire manifest in the PRIMAL SCREAM that sets it going and inaugurates REFERENTIAL DETACHMENT and the TRANSITIVE DIMENSION, leading to acknowledgement of the REALITY PRINCIPLE and ALETHIC TRUTH (see *D*: 303), such that 'the good society is implicit in elemental desire' (*P*: 161). Desire also includes the desire to be desired. Bhaskar agrees with the young Hegel that this cannot be reduced to the desire for RECOGNITION; it is, rather, the desire to be loved and united with other people in a state of non-ALIENATION (see also POLYAD, TRUST).

Since desire is oriented, not to the action, but to its result, it can, however, itself lead to a form of alienation or divided intentionality, such that, instead of focusing on the doing itself, we focus on the result. In market-driven societies this becomes an insatiable striving for more and more, the *desire for desire*, underwritten by exploitation, instrumental rationality (which enslaves reason to desire), transactional conditionality, and reductive or mechanical (irrealist) materialism (on which the human world is comprised of atomistic egos standing over against each other and the rest of the world) (see PDM), and systematically promoted by power$_2$. This sets up a bad infinite, as Shakespeare, standing at the cusp of post/modernity, so clearly saw:

> Then everything include itself in power,
> Power into will, will into appetite;
> And appetite, an universal wolf,
> So doubly seconded with will and power,
> Must make perforce an universal prey,
> And last eat up himself.
> (*Troilus and Cressida*)

Today, flourishing therefore requires the (DIALECTIC OF THE) *EDUCATION OF DESIRE* to only collectively attainable wants and needs within the bounds of global ecological constraints and a shift to an emergent powers orientation and dialectical rationality. The desire for desire is transmuted in this directional process into freedom from such desire and, indeed, freedom from desire as such, i.e., having and wanting are subsumed under being; when the desire for non-alienation is satisfied, it is no longer desire but being. There is thus no split in CR ethics, unlike in Kantian, between duty and desire.

In PMR, desire tends to be treated as one of the purely negative emotions – the bipolar opposite of fear – which are parasitic on the positive emotions of love, care, solidarity and trust. However, this should be construed as desire constituted in and by the DEMI-REAL, grounded in ignorance of and alienation from our true selves and the totality they inhabit, hence in ontological insecurity deriving from 'lack of integration and totality' (*EW*: 106): the *desire-form* (cf. the FACT-FORM), which must be shed for eudaimonia to be possible (see ESTEEM; SELF). Beneath this, the desire to absent any constraints on freedom and flourishing and for de-alienation beats on. But, in keeping with the primacy of SELF-REFERENTIALITY, *if* we all gave up on desire as such, now, and lived the good for human being, we would *ipso facto* free ourselves from constraints other than those imposed by the laws of the physical world (*RM*: 62–4). See also DIRECTIONALITY.

desocialisation of science. See EPISTEMIC–ONTIC FALLACY.

destratification. See CONCRETE/ABSTRACT.

DET model of normative change. See DEA and DET MODELS.

detachment/attachment. CR distinguishes five kinds of *detachment*, all turning on ABSENCE or absenting: *material, referential, transcendental, dialectical* and '*logical*'. The last four presuppose human INTENTIONALITY, the last three pertain to various forms of argument.

(1) *Material detachment* occurs in EMERGENCE with the constitution of a new, sui generis real, level of being. Here reducibility to lower-order levels is absented. In the next three variants, humans detach or absent themselves and their acts of reference from (2) what they are about (*REFERENTIAL DETACHMENT*); (3) the (subjective) premises of an argument (*transcendental detachment*); (4) refuted or sublated premises/conclusions (*dialectical detachment*). (Conversely, they detach these referents from their own acts of reference.) (2) is the generalised form of REFERENCE, of which the other human variants are species. We referentially detach '*for* the sake of AXIOLOGICAL need' (*D*: 105). In (3), by contrast, we detach ourselves or our thought *from* axiological need, i.e., from premises having to do with our subjectivity. Transcendental arguments that are true to the spirit of Kant proceed from premises that 'retain some essential relation to human subjectivity'. CR transcendental detachment is from these premises, such that we see that 'there are {transcendental and} dialectical necessities {. . .} just as there are natural necessities, quite intransitive and independently of our knowing them or any necessary relation to human subjectivity' (*D*: 104; see also MODALITY). In this process, we detach 'our conclusions from our premises, just as scientists and logicians do {. . .} so establishing ontological propositions sui generis' (*P*: 54). Transcendental detachment is brilliantly deployed in the dialectical argument for the concept of *a meta-reflexively totalising (self-)situation* or the constellational identity of the identity of subject and object within subjectivity within an overarching objectivity (*D*: 149–50, 271–2). (4) is simply the jettisoning or abandonment of refuted premises or conclusions in order to move on in any LEARNING PROCESS: 'once a phase is overcome in the history of philosophy or social life, such as the epistemic fallacy or the fear of death, we can discard it as an outmoded way of belief or being' (*D*: 104). Finally, (5) '*logical*' detachment (which overlaps transcendental detachment) also detaches conclusions from premises. Thus in deductive logic one can get from 'If P then Q and P' to 'Q'. In DIALECTICAL ARGUMENTS conducted within an irrealist or ontologically actualist framework, Q cannot be detached from P – a deficiency which, in realist dialectics, the concepts of ontological stratification and transfactuality, turning on reference to (referential detachment of) the absent, or NEGATIVE REFERRAL, make good (see *D*: 105, 137–9). This connects with the completion of the TABLE OF OPPOSITION, which Hegel achieves, also by referentially detaching absence, in the mode of distanciated time rather than ontological depth. In dialectics, 'logical' is more properly 'ontological' detachment (or its failure); when successful, it enables 'reference to the structural and transfactual reasons for, or truth of, phenomena' (*D*: 105; see also ALETHIA).

Failure of detachment is also termed *attachment*, as in the failure of material detachment of a disease in the presence of traces on a body or of the past generally in the present (see TENSE), or personal attachment to an object in DESIRE. The varying degrees of attachment or detachment of the elements of a TOTALITY constitute their RELATIVE AUTONOMY. In PMR, attachment is (positively) desire and (negatively) fear and aversion, grounded and pursued in ignorance, in the DEMI-REAL, of our transcendentally real SELVES. It is manifested in conditionality, instrumental RATIONALITY, craving, exploitation, FETISHISM, *AMOUR-PROPRE*, and IRREALIST philosophy. In a state of *non-attachment* we are true to our innermost being.

determination, determinism. CR sustains a concept of *ubiquity determinism* or *the principle of CAUSALITY* (a version of the *principle of sufficient reason*), which is necessarily presupposed by science: everything has a cause, there must be a reason in the nature of things for changes and differences, identities and equilibria. (It entails the *principle of sufficient practical reason*, which specifies that there must be grounds for social differences, e.g., inequity, and is of cardinal importance for EMANCIPATORY AXIOLOGY.) But it rejects determinism as it is usually understood, viz., the ACTUALIST thesis 'that for everything that happens there are conditions such that, given them, nothing else could have happened' (Bhaskar 2003a: 154) – a thesis that BLOCKISM extends to embrace the future: everything that happens or will happen is wholly predetermined. Such a view is incompatible with the openness of the world, in which laws operate as tendencies and *determination* (i.e., causation) is multiple and plural, established by CR. Thus the past is *determinate* (i.e., is not just nothing but an absent structured process), although in some respects not *determinable* (capable of being authoritatively accounted for), but it is not *determined* (it could have been different) in an actualist sense; or it is determined, but only in the sense of ubiquity determinism; the future is *indeterminate* (not yet 'somehow, sometime' [Morgan 2005a]). Determination is the 'para-temporal' correlative of the 'para-spatial' concept *MOMENT* (*P*: 83).

In EMPIRICISM and positivism, determinism takes the form of '*regularity determinism*, viz. that for every event x there is a set of events $y_1...y_n$ such that they are regularly conjoined under some set of descriptions' (Bhaskar 2003a: 154), i.e., x and $y_1...y_n$ are CONSTANT CONJUNCTIONS, and their ubiquity in nature is presupposed. Regularity determinism commits the EPISTEMIC FALLACY, conflating the actual with the necessary and possible, (epistemological) predeterminism with (ontological) determination, and overlooks that 'relations of natural generation are not, as a general rule, logically transitive. Thus it is not the case that, because S_1 produced S_2 and S_2 produced S_3, S_1 produced S_3 – if, for example, S_2 possessed emergent powers {...} or the system in which S_3 is formed is open or if some of the processes at work are STOCHASTIC' (*D*: 235).

Hence CR borrows the concept of *overdetermination* from Althusser (who got it from Freud), signifying 'the *multiple determination* of events, structures and totalities, and of the contradictions which constitute, reproduce and transform them' and a methodological injunction 'to search for internal and interconnectedly generated as well as external and analytically separable causes' (*RR*: 181, e.a.). Multiple (or complex or CONJUNCTURAL) determination is related to, but distinct from, *multiple* (or dual) *control*. Multiple control is multi-level determination; it pertains to higher agencies setting boundary conditions for the operation of lower-level laws, and lower-level strata providing the framework conditions of possibility of higher-order ones (see INFRA-/INTRA-/SUPER-STRUCTURE); it makes *self-determination* possible (*RS2*: 111). Multiple determination (overdetermination) is the totality of causal forces co-operating to produce a result, such that multiple determination > multiple control. Overdetermination carries the additional connotation that more causes (internal or external or both) than are necessary to produce a result are operative, e.g., execution by firing squad, or systems in which functional relations obtain (see FUNCTIONAL EXPLANATION). However, in the multiply contradictory and open world of master–slave relations it is entirely consistent with the possibility of functional disconnections leading to opportunities for systemic change (Dean 2005). It is also consistent with the quite general *underdetermination* of higher-order by lower-order phenomena entailed by EMERGENCE (or the *underdeterminacy* of higher-order phenomena), of which *epistemic indeterminacy* (the underdetermination of theories by evidence, hence the formal underdetermination of rational choice of theories) and *AXIOLOGICAL indeterminacy* (resulting from epistemic undecidability or the experience of contradiction) are species (see also HOLY TRINITY [judgemental rationality], PHILOSOPHY). (Note. While underdeterminacy and indeterminacy, together with cognate terms, are sometimes used synonymously, there are subtle differences: underdeterminacy may pertain to the already fully determined, indeterminacy may not; indeterminacy is consistent with the complete absence of determination, underdetermination is not; AXIOLOGICALLY, complete absence of determination stymies rational agency, underdetermination is a condition of its possibility.)

Regularity determinism is further undermined by quantum theory, which suggests that, at the level of elementary particles, reality is comprised of seas of POTENTIA; and by CHAOS THEORY, 'illustrating that non-linear dynamic systems can yield highly irregular (chaotic and unpredictable) results' (Bhaskar 2003a: 154). The so-called *problem of free will* vs. determinism arises from the reduction of the necessary and possible to the actual, issuing in *historical justificationism* (affined to epistemological justificationism or FOUNDATIONALISM; e.g., 'given the conditions, Stalinism [or Fascism, etc.] was inevitable'). When ACTUALISM is abandoned, the problem dissolves, and 'the possibility of a naturalistic revindication of human agency, of the causality of reasons and the potential applicability of the predicate "free" to agents, their ACTIONS and their situations, is once more opened up' (Bhaskar 2003a: 154). 'Freedom is irrespective of the arrow's flight, no regarder of the river's flow' (*SR*: 142). This suggests a solution to the problem of evil: freedom presupposes the possibility of making mistakes, of acting otherwise than in accordance with our true natures, hence of the emergence of baneful (but unnecessary and dispensable) social structures (cf. *EW*: 37, 89; EVIL).

Psychic ubiquity determinism is the principle that, for every belief or action, there is a set of real reasons, constituting its rationale, which, together with other causes, explains it (*PN3*: 96). (Reasons that are acted on are causes.) It is arguably a necessary presupposition of investigation in the human sciences. It is neither to suppose that reasons completely explain every action, nor that they are true, nor that the relevant processes of reasoning are valid. See also Norris (2000).

detotalisation. See CONCRETE/ABSTRACT; FISSION; MEDIATE (DEMEDIATION); TOTALITY.

development. See CONSISTENCY; DIRECTIONALITY; PROCESS.

developmental or **dialectical consistency**. See CONSISTENCY.

developmental negation. See CONSISTENCY; FIXISM/FLUXISM.

diachrony/synchrony (f. Gr. through time, with or across time, respectively). The question of the relationship of the *diachronic* and *synchronic* is one of the perennial problems of social theory. They are often thought of as opposites, diachrony pertaining to change and process and synchrony to stability and system, or, at the limit, temporality and a-temporality (a purely analytical moment), respectively; but for CR they are distinct yet interconnected, exemplifying DUALITY or dialectical interconnection. Their distinctness is ultimately given in the mutual irreducibility of 1M DIFFERENCE (space, spatial position and organisation) and 2E CHANGE (time, PROCESS) (see LOCO-PERIODISATION; SPACE–TIME). In terms of the distinction between the domains of the real and the actual/empirical, synchrony pertains vertically to structures and causal powers (1M, 3L) (the proper object of theory), diachrony horizontally to their processual exercise (2E, 4D) (the domain of practical/applied explanations). At the level of structure or system, synchrony pertains to *reproduction*, diachrony to GEO-HISTORICAL *transformation* (cf. morphostasis and *MORPHOGENESIS; TRANSITION*). As such, the distinction is necessary for science, in particular because, in an EMERGENTIST world, deep structural change or stability will be explanatorily more significant than other kinds. Fundamental interconnection is attested above all by the unifying concept of ABSENCE, which can equally be derived from difference as well as from change, and the fourfold meaning of which can denote synchronic difference (1M product, 3L process-in-product) as readily as diachronic change (2E process, 4D product-in-process). In relation to the different concepts of TENDENCY aligned to this fourfold polysemy, Bhaskar comments 'one might well ask why cannot causal powers not just be processually exercised, but there be powers to process (consider "maturing", "growing", "developing")?' (*D*: 146). Clearly they can be, and structures may themselves change (or there may be changes within unchanging structures), powers may acquire powers, and effects constitute structures. Such considerations indicate that the diachrony/synchrony distinction should not be hypostatised. Both synchronically and diachronically, the CR totality is open (*D*: 273).

Indeed, while there are passages in Bhaskar which indicate that diachrony and synchrony are on a par or 'ontologically co-equal' (*SR*: 246), his general sense in *Dialectic* seems to be that diachrony is *at least* on a par with synchrony and probably primary in that it overreaches the latter in a manner analogous to that in which dialectic overreaches analytics in the dialectic of analytical and dialectical reason. In fact, since dialectic (ID) at bottom *is* the geo-historical process, and analytics is the philosophy of ontological stasis, there is a sense in which diachronic is perhaps synonymous with dialectical overreach. Synchronic analysis cannot handle inter-systemic transitions, hence cannot grasp geo-history as TOTALITY. This is at one with Bhaskar's call for the ('long overdue' [*D*: 146]) geo-historicisation of social theory. While it is true that no transformation of structure is ever total, reproduction is always also transformation, and reproduction/transformation together must be seen as 'a space–time *flow*' (*SR*: 213, e.a.). On the other hand, the weight attached to the massive inertial presence of the past suggests co-equality. The emphasis in SYSTEMATIC DIALECTICS, SYSTEMS theory and STRUCTURALISM, by contrast, is on synchrony. See also CAUSALITY; CHAOS/COMPLEXITY THEORY.

dialectic. At once *argument, change* and *freedom*, each of which 'rationally presupposes its predecessor' (*D*: 377), unified by the concept of ABSENCE. Thus argument involves

dialectic from anthropism to anti-humanism   125

the absenting of mistakes, change absents states of affairs, and the process of freedom absents constraints or ills, which can be seen as absences. Hence the real DEFINITION of geo-historical dialectic as (1), formally, *'absenting absences*, or applied recursively, as *absenting constraints on absenting absences'*, and (2), substantively, the axiology of freedom (*D*: 377). Like all CR concepts, it has a TD/ID bivalence, so that it is both an intransitive process or processes and the transitive attempt to express it/them in thought. Dialectic is thus much more than the method it is often taken to be, though it is that, too: as argument, it 'depends upon the art of thinking the coincidence of distinctions and connections' (*D*: 180).

In *Dialectic*, the real definition is arrived at via (1) a retroductive-explanatory account of dialectic from its genealogy, which goes back to ancient Greek philosophy via Marx, Hegel, Kant, Schiller and Plotinus; (2) a process of transcendental deduction within a CR framework from premises of human agency and discourse; and (3) critique of the ANALYTICAL PROBLEMATIC (see *D*: 175 ff.). See also DCR; DIALECTICS; DIRECTIONALITY; LEARNING PROCESS.

dialectic from anthropism to anti-humanism. See ANTHROPISM, DIALECTICS.
dialectic from anti-anthropism to humanism. See HUMANISM, DIALECTICS.
dialectic of agency. See EMANCIPATORY AXIOLOGY.
dialectic of cognition. One of the twelve routes, sketched in *Dialectic*, taken by the DIALECTIC OF DESIRE TO FREEDOM in the transition from form to content, commencing from constraints on desire (*D*: 287). See also LEARNING PROCESS.
dialectic(s) of co-inclusion or **co-presence**. See CO-PRESENCE.
dialectic of consciousness and self-consciousness in reason or the **dialectic of consciousness via self-consciousness to reason** echoes Hegel's transitions from consciousness to self-consciousness and reason. It incorporates a metacritical *dialectic of instrumental → explanatory critical → dialectical reason* (see RATIONALITY) or *dialectic of social science*, which moves from a conception of social science as neutral to one of it as EXPLANATORY CRITICAL and EMANCIPATORY AXIOLOGICAL in a two-way dialectic of form and content, thence as dialectic or DIALECTICAL REASON. In the current context of 'post-Fordist' changes in the labour process and globalisation, it envisages the possibility of 'a dialectic of globalising self-consciousness' (*D*: 270).
dialectic of de-alienation or **autonomy**. One of the twelve routes, sketched in *Dialectic*, taken by the DIALECTIC OF DESIRE TO FREEDOM in the transition from form to content. In so far as anyone is alienated (separated) from something essential to her needs, the realisation of her potentialities or well-being, and does not want to remain in that condition, she will be committed to its restoration, CP. This makes 'a prima facie case for *socialism*, in so far as it rationally portends the sublation of the generative separation of the immediate producers from the means and materials of production and *their* rational regulation of their use' (*D*: 288). The dialectic of de-alienation has its immediate origins in Marx, but is extended by Bhaskar to encompass (abolition of) the totality of MASTER–SLAVE-type relations, including internalised or intra-psychic forms of ALIENATION. It is a paramorph of the desire, in love, to be united with the loved one which was movingly thematised by the early Hegel but later glossed by him as the desire for RECOGNITION.
dialectic of desire (generic). One of the twelve ways of effecting the transition from form to content in the dialectic of desire to freedom, it broadens the DIALECTIC

126 **dialectic of desire for freedom**

OF DE-ALIENATION via the consideration that desire includes desire to be recognised. By the logic of dialectical universalisability, this will involve (1) RECOGNITION of 'the capacity to enjoy rights and liberties', entailing 'the *real enjoyment* of equal and universally reciprocally recognised rights and liberties, including the right to de-alienation and the enjoyment of health, education, access to resources and other liberties'; thence (2) the right of all slaves to be free of master–slave-type relations; and (3) acquisition of the cognitive and empowering conditions for effecting this and for universal free flourishing, etc. (*D*: 288).

dialectic of desire for freedom. See DIALECTIC OF DESIRE TO FREEDOM.

dialectic of desire to freedom and the **dialectic of desire for freedom** are two sides of the *DIALECTIC OF FREEDOM* (see EMANCIPATORY AXIOLOGY). The two concepts highlight that the dialectic of freedom is equally a dialectic of (1) DESIRE and of (2) FREEDOM. In (1) we move from desire to the *desire for freedom*; in (2) from *desire to freedom* (*D*: 230). (2) presupposes the desire to remove constraints on desire; (1) that 'human beings, by and large, want to be free, under some (sets of) description(s)' (*D*: 289, e.a.).

dialectic of dialectical and analytical reasoning. See DCR; EPISTEMOLOGICAL DIALECTIC; PROPAEDEUTIC.

dialectic of dialectical universalisability and immanent critique. See UNIVERSALISABILITY.

dialectic of discourse. The *dialectic of speech action*, and, within that, of *judgement*. See EMANCIPATORY AXIOLOGY.

dialectic of the education of desire or **dialectic of the education of desire for freedom**. The process whereby desire is, or may be, rendered compatible with autonomy and the responsibilities of good citizenship. One of the twelve routes, sketched in *Dialectic*, taken by the DIALECTIC OF DESIRE TO FREEDOM in the transition from form to content, it takes its departure from the trustworthiness of the primary POLYAD, which establishes the reality of the social bond, endowing the infant with 'the existential security that at once silences its scream, nurtures its self-ESTEEM and lays the basis for the *amour de soi* which underpins solidarity and altruism alike'. The DIRECTIONAL logic of this process is towards the absenting of power$_2$ relations and the presencing of a normative order 'based on the traditionally feminist virtues of care, sensitivity to the suffering of others {. . .}, solidarity and *TRUST*'. Desire is both satisfied and, as the child grows, shaped by the REALITY PRINCIPLE into only collectively attainable WANTS and NEEDS. 'The logic of desire and of interest', mediated by the EPISTEMOLOGICAL DIALECTIC, 'point in the same direction' (*D*: 286).

dialectic of equity. See EQUITY.

dialectic of figure to ground or **figure and ground**. A dialectic that 'occurs in PERCEPTION, REFERENCE and PERSPECTIVAL SWITCHES (which can also be used to highlight the plight of a marginalised minority – or majority)' (*D*: 237).

dialectic of form to content. Effects the TRANSITION from form to content in the dialectic of freedom (see EMANCIPATORY AXIOLOGY). It is often grouped with the *dialectics of figure to ground* and *centre to periphery*, because figure (e.g., capital) and centre are a kind of form for which ground (e.g., labour) and the periphery supply content.

dialectic of freedom. Embraces (1) the DIALECTIC OF DESIRE TO FREEDOM and the *dialectic of desire for freedom* as two poles of the one dialectic; and (2) in an extended

dialectic of inconsistency and incompleteness 127

sense all the 4D dialectics listed in Table 18. It is an 'extension, generalisation and radicalisation of the dialectic of freedom' in Hegel and Marx. For details, see EMANCIPATORY AXIOLOGY.

dialectic of inconsistency and incompleteness. See CONSISTENCY.

dialectic of instrumental → explanatory critical → dialectical reason. See DIALECTIC OF CONSCIOUSNESS AND SELF-CONSCIOUSNESS IN REASON.

dialectic of interests (generic). See INTERESTS.

dialectic of judgement. See EMANCIPATORY AXIOLOGY.

dialectic of learning processes. See LEARNING PROCESS.

dialectic of the logic and practice of dialectical universalisability. See UNIVERSALISABILITY.

dialectic of malaise. One of the twelve routes, sketched in *Dialectic*, taken by the DIALECTIC OF DESIRE TO FREEDOM in the transition from form to content. It commences from a malaise or ill as such, which is a constraint on desire, and proceeds, via the logic of dialectical UNIVERSALISABILITY, to the absenting of all ills as such and their identified causes, and so to 'the free flourishing of each and all' (*D*: 287).

dialectic of material interests. See INTERESTS.

dialectic of master and slave. See MASTER–SLAVE.

dialectic of morality. Departs from rejection of the alienating de-evaluation of reality accomplished by the ANTI-NATURALISTIC FALLACY. Making a distinction between *transitive* or relational and *intransitive* (actually existing) *morality*, its de-alienating praxis goes from (1) description of actually existing morality → (2) redescriptive morality → (3) explanatory critical morality ↔ (4) totalising depth-praxis, in principle without end, entailing that the moral evolution of the species, like geo-history as such, is open and unfinished. Since (2) is achieved via immanent critique, and (3) ↔ (4) via explanatory critical theory[†] and depth-praxis, coupled with the logic of dialectical universalisability, the dialectic of morality is an integral strand in EMANCIPATORY AXIOLOGY (interwoven with the DIALECTIC OF SOCIAL SCIENCE).

dialectic of the real and the actual. An ontological dialectic that occurs in EMERGENCE.

dialectic of real geo-historical processes. See UNIVERSALISABILITY.

dialectic of recognition. See EMANCIPATORY AXIOLOGY; RECOGNITION.

dialectic of scientific discovery. See EPISTEMOLOGICAL DIALECTIC.

dialectic of self and solidarity. See SOLIDARITY.

dialectic of the 7 E's. See ESTEEM.

dialectic of social science. Proceeds from social science conceived (1) as value neutral, to (2) as EXPLANATORY CRITIQUE, to (3) as EMANCIPATORY AXIOLOGY, and finally to (4) as DIALECTICAL REASON. It goes hand in hand with the DIALECTIC OF MORALITY.

dialectic of solidarity. See SOLIDARITY.

dialectic of speech action. See EMANCIPATORY AXIOLOGY.

dialectic(s) of structure and agency concern the interplay of the elements of the dislocated duality of social structure and agency. See AGENCY; MORPHOGENESIS; SOCIAL STRUCTURE; TMSA.

dialectic of subjectivity. The developmental logic of changing four-planar HUMAN NATURE, incorporating the DIALECTIC OF MORALITY (cf. *D*: 349–50). It is contingent whether it issues in the eudaimonian society.

128 dialectic(s) of transition

dialectic(s) of transition. One of the twelve routes, sketched in *Dialectic*, taken by the DIALECTIC OF DESIRE TO FREEDOM in the transition from form to content, involving 'two-way traffic between truth and freedom, form and content' (*D*: 288) and incorporating (1) EXPLANATORY CRITIQUE†, concrete utopias and theories of TRANSITION (explanatory critical theory† [EC]) and (2) totalising depth-praxis (see EMANCIPATORY AXIOLOGY).

dialectic of truth. See ALETHIA; DIALECTICAL REASON; EMANCIPATORY AXIOLOGY; EPISTEMOLOGICAL DIALECTIC.

dialectic of truth and freedom (or **truth ↔ freedom**), or **dialectic of content from truth to freedom**. See EMANCIPATORY AXIOLOGY.

dialectic of unity-in-diversity. See UNITY.

dialectic of universalisability. See UNIVERSALISABILITY.

dialectic of wants and needs. See WANTS.

dialectical antagonists (also **complements, contraries, counterparts, couples** or **pairs**) are in conflict over relatively superficial matters but necessary to each other in that they are grounded in (a) common category mistake(s), hence tacitly complicit. So their antagonism or opposition is ultimately phoney, and appears as such when overreached by a fuller conceptual formation which makes good the category mistake (see NEGATIVE PRESENCE). The figure first appears in Marx, though not by that name (cf. *D*: 88). Though related, the antagonism is not that of *dialectical opposites*, which is necessary in that it is rooted in the common causal ground of a dialectical CONTRADICTION, conveying tendencies to change manifest socially in conflict and overt struggle. Or dialectical counterparts are illusory (mutually inclusive), not real (tendentially mutually exclusive), oppositions, though both kinds of opposition figure in the explanatory critique (metacritique$_2$) of philosophical and social theoretic problematics. Thus positivism and the hermeneutical tradition are a dialectical pair, grounded in acceptance of the categorially false Humean account of natural science and the natural world, which rests on the illicit FUSION effected by the epistemic–ontic fallacy. The resulting DUPLICITY is itself ultimately grounded in acceptance of, or indifference to, real dialectical contradictions embodied in master–slave-type society, underpinned by illicit FISSION (generative separation or ALIENATION). Dialectical pairs are therefore pervasive in power$_2$-stricken society. They include (philosophically, where they constitute or are part of *theory problem–field solution sets*) analytical irrealism/dialectical irrealism; anthroporealism/transcendent realism; classical (subjective) empiricism/objective idealism; dualism/monism, reductionism; (metaphysical) idealism/materialism; disembodiment, voluntarism/reification of agency; expressivism/extensionalism; social holism/individualism; Derridean deconstruction/Habermasian speech action theory; and (substantively) a reformist evolutionism abstract utopianism Democrats/Republicans, New Labour/Tories, Social Democrats/Christian Democrats, etc. Such dialectical twosomes are well caught in the literary figures of Tweedledum and Tweedledee. They should be distinguished from *DUALS*, which embrace them ([co-presence > constellation] > dual > dialectical pair) but may be complementary, not ostensibly antagonistic, e.g., the epistemic–ontic fallacies. A 2E figure, dialectical antagonists falls within the *dialectics of co-inclusion* or *CO-PRESENCE* (of absence and presence). The absence of the concept of the non-actual real in positivism and interpretivism, for example, produces – via philosophical aporetics, as reality bites – the presence within each of a counterfeit concept, 'the empirical world', and so on. By remedying such absences in a fuller, richer conceptual formation, CR heals and moves beyond their antagonism (see CRITICAL NATURALISM; EPISTEMOLOGICAL DIALECTIC).

dialectical argument. Broadly, anything ranging from an 'expansive sense of interplay' in argument to 'any systematic interconnection that unites a body of thought, such as absolute idealism' (*D*: 103). To be distinguished from dialectic *as* argument or conversation, which is one of the oldest traditions of dialectic, nowadays continued by HERMENEUTICS. In a more restricted and rigorous sense, dialectical arguments are 'a species of the genus of transcendental arguments of the kind of retroductive-explanatory arguments familiar to science': retroductive explanatory arguments > transcendental arguments > dialectical arguments (see EXPLANATION; INFERENCE). As such they play a vital role in the EPISTEMOLOGICAL DIALECTIC. Turning on 2E notions of negativity and/or CONTRADICTION and licensing negative evaluations (see CRITIQUE), dialectical arguments 'establish false necessities as ontological conclusions' (*D*: 199), where 'false' should be construed as 'categorially false' (or limited, incomplete, incoherent, etc.), i.e., based on a CATEGORY mistake (see also ALIENATION; IDEOLOGY), and 'necessary' as 'caused' (see DETERMINATION). This is *dialectical EXPLANATORY CRITIQUE*. False necessities are a species of *DIALECTICAL NECESSITY*.

If *transcendental arguments* establish the ontological conditions of possibility of some widely accepted premise, *dialectical arguments* are transcendental arguments that demonstrate 'the conditions of possibility (dr′) of the conditions of impossibility (dc′)' of such a premise (*D*: 46), where dr′ stands for DIALECTICAL REASON, in the sense of ALETHIC truth of, and dc′ for DIALECTICAL COMMENT. The dialectical comment typically shows that the initial premise has both a degree of truth and a degree of falsity, i.e., is 'true in one respect but not in another' (*D*: 105). Thus in the argument for the ontological primacy of ABSENCE over presence, the initial premise is positive existence – being(s) (ONTS) exist. A transcendental argument establishes that the condition of possibility of their existing is a world containing non-being (negative existents or de-onts) as well as onts. Dialectical DETACHMENT from the initial premise then permits a dialectical argument showing the possibility of the impossibility of onts existing, i.e., if there were a total void, in the past or the future. The initial premise is then seen to be true in the sense that onts do actually exist, but false in the sense that it omits any reference to de-onts; to that extent it is inadequate or limited, based on a category mistake (being = onts). It will sooner or later be called to account by the REALITY PRINCIPLE, necessitating (if it is to be saved) TINA COMPROMISES. A condition of possibility for the persistence of false beliefs is social causation, so the falsity is necessary.

Such an argument might be thought not to establish the reality of false *necessities* in *nature*, as distinct from the sociosphere, to which (false) beliefs pertain. However, Bhaskar insists that there may be natural as well as social false necessities (*D*: 107). This is because in his view ideas are 'parts of the natural world, products of the naturalised process of thought' (1997a: 143). Wherever the natural world as such as well as the social world (e.g., the wage-form, which collapses powers to their exercise) is falsely categorised, beliefs – which are constellationally contained within ontology – will be both false and necessary.

A *dialectical explanatory argument* or *dialectical explanation* is a dialectical argument that substantively explains 'contradictory forces in terms of a structured common ground'

130 dialectical bracketing

(*D*: 87). In philosophy, a dialectical explanation explains the opposing factions constituting a PROBLEM-field, e.g., hermeneuticism vs. positivism, in terms of a common mistake; a METACRITICALLY$_2$ dialectical explanation in addition explains its social causes. A dialectical explanation of a rival theory, showing that it is false and why it persists, constitutes its *dialectical refutation*. Dialectical arguments and explanations enable the traditional TABLE OF OPPOSITION to be completed. If valid, they effect dialectical PERSPECTIVAL SWITCHES. They are very different from Nietzschean and Heideggerian ERASURE, which likewise establishes the conditions of possibility of the conditions of impossibility of some premise, only to equate them, whereas CR, deploying the concept of a META-REFLEXIVELY TOTALISING (SELF-) SITUATION, moves on to expand the conceptual field and to totalising depth-praxis. The DIALECTIC OF FREEDOM exemplifies dialectical argument. See also DIRECTIONALITY.

dialectical bracketing. See EPISTEMOLOGICAL DIALECTIC; PHENOMENOLOGY; PROPAEDEUTIC.

dialectical comment (dc′). Paradigmatically reports theory–practice inconsistencies in a theory that has been subjected to immanent CRITIQUE (see also NEGATION OF THE NEGATION; SUBLATION). Allowing finer distinctions, it is a metacritical comment on a preceding comment, theory, phase, etc. that it 'has a degree of truth, but also a degree of falsity, that is, that it is true in one respect but not in another' (*D*: 105). It may call attention to 2E inconsistency or contradiction, 3L split, 4D auto-subversiveness, or 1M non-identity whereby a theory requires something other than itself for completion, each of which may be the site of invocation of a TINA λ clause by those attempting to save a theory. Marx's critique of, and dialectical comment on, political economy and capitalist society provide a good example. Together, or generalised, these kinds of comments comprise a metacritical *dialectical remark* (drk′ or drk†): drk′≥ dc′. There are clear affinities between 'the generalised dialectical remark and civilised conversation oriented to trust-worthy or fiduciary communications' (*D*: 175), as in the Habermasian ideal-speech situation and theory of communicative action. *Dialectical distinction* (dd′) (from Aristotle) and *dialectical limit* (dl′) (from Kant) are species of dialectical comment or remark. The latter may be regarded as an '(anti-)dialectical remark' (*P*: 132), with a modern equivalent in Derridean deconstruction. In Kant, it is designed to show that there are unsurpassable limits to human intelligence, which are given in EMPIRICAL REALISM and which, if transgressed, produce irresolvable ANTINOMIES. CR accepts that we do not know all the world or how much of it we know, that human reason is potentially limited, and that transitive or epistemic relativity is real, but rejects Kant's EMPIRICAL REALISM and the conceptual realist aspiration to GROUND the conditioned in the unconditioned, resting content to ground the more in the less conditioned (*D*: 121–2).

dialectical connection. See CONTRADICTION; DUALITY; TRANSITION.
dialectical or **developmental consistency**. See CONSISTENCY.
dialectical contradiction. See CONTRADICTION; DIALECTIC.
dialectical counterparts. See DIALECTICAL ANTAGONISTS; DUPLICITY.
dialectical critical naturalism (DCN). See CRITICAL NATURALISM.
dialectical critical realism (DCR). Prior to 1993, Roy Bhaskar was best known for his realist accounts of the natural and social sciences (see CRITICAL NATURALISM;

dialectical critical realism (DCR)

TRANSCENDENTAL REALISM) and, stemming from the latter, his concept of EXPLANATORY and EMANCIPATORY CRITIQUES. Less well known in this period was his interest in the concept of dialectic, which was pursued in a number of brief articles concerned with the character of Marxist thought and the Hegel–Marx nexus. In his *Dialectic: The Pulse of Freedom*, Bhaskar combines these two interests to provide a new dialectical account of CR which recognises the importance of the dialectical tradition in Western thought, and, on the basis of established CR concepts, reveals its limits and moves beyond them. DCR also underlabours further for the modern social (and natural) sciences and, in particular, provides a philosophical basis for Marxist social theory consistent with Marx's own insights. Most ambitiously, it establishes the groundwork for a new ethical theory which, building upon emancipatory critique, could resolve the problematic theory–practice dichotomy associated with radical forms of social science.

In the preface to *Dialectic*, Bhaskar describes his work as involving 'a preservative generalisation and enrichment of hitherto existing critical realisms that is 'a non-preservative sublation of Hegelian dialectic'. To Hegel's dialectics of identity, negativity and totality, Bhaskar will offer the four terms of non-identity, negativity, totality and transformative agency, and he adds that where he and Hegel share common terms DCR will provide radically different interpretations of the concepts in question. The central task of *Dialectic* is to synthesise dialectical methods with existing realist ideas of ontological depth and the differentiation of the ontic from the epistemic in order to provide a meta-framework for the understanding of real geo-historical processes. This pushes CR further into the fields of 'REFERENCE and TRUTH, SPATIO-TEMPORALITY, TENSE and PROCESS, the logic of dialectical UNIVERSALISABILITY and on to the plane of ETHICS'. There, a combination of moral realism and ethical naturalism opens up the possibility of moving from 'the form of judgements to the content of a freely flourishing society' (*D*: xiii).

The dialecticisation of critical realism

This complex and profound development of CR involves a broad and multiform treatment of dialectic: in its historical and systematic forms; *epistemically*, as the logic of argument and the method of immanent CRITIQUE; *ontologically*, as the dynamic of conflict and the mechanism of change; and *normatively-practically*, as the AXIOLOGY of FREEDOM. Central to dialectic is the concept of ABSENCE, for dialectic is defined as 'the absenting of CONSTRAINTS on absenting absences or ills (which may also be regarded as constraints)' (*D*: xiv) and this applies whether it be an epistemological matter of remedying an argument, an ontological matter of socio-historical change, or an ethical question of human freedom. This standpoint contrasts with an entire tradition from Plato which has prioritised (and positivised) the positive, producing the characteristic error that Bhaskar calls *ONTOLOGICAL MONOVALENCE*: the reliance on a 'purely positive, complementing a purely actual, notion of reality' (*D*: 4–5). Argument, the socio-historic development of human being and ethics are all marked by what they lack. There is a fundamental bipolarity of absence and presence, so that negativity is a condition of positive being; it is this essential relationship that is dialectic, a process that is 'the pulse of freedom' (*D*: 385).

132   dialectical critical realism (DCR)

The dialecticisation of CR is itself a dialectical *PROCESS* that has four moments (see MELD). Bhaskar schematises this theoretical development as from a first moment (1M) to a second edge (2E), then to a third level (3L), and on to a fourth dimension (4D). 1M primarily involves existing CR concepts like structure, differentiation, change, alterity (for example epistemic/ontic non-identity within ontology), transfactual efficacy, EMERGENCE and systemic openness. These are reworked and enhanced at 2E in the light of dialectical categories such as negativity, NEGATION, becoming, CONTRADICTION, process, development and decline, MEDIATION and reciprocity. 1M concepts might be seen as implicitly calling for (as explicitly lacking) the dialecticisation they receive at 2E. 1M concepts suffice 'for, e.g., an adequate account of science which abstracts from space, time and the process of change' (*D*: 8); but which science can afford to rest content with abstraction from the spatio-temporal or from change? At 2E, the marriage of CR and dialectics opens up the prospect of an understanding of the way sciences must be if they are to be adequate to 'cosmology {. . .} human geo-history {. . .} personal biography, laborious or routinised work but also joyful or idle play' (*D*: 8–9). 1M concepts taken to the 2E dialectical edge are opened up to the world of the *CONCRETE UNIVERSAL*, the unity of being in its subjection to, and expression of, 'CAUSALITY, space and time in tensed RHYTHMIC spatialising process' (*D*: 392). The move from 1M to 2E can perhaps be summarised as an attempt to avoid taking CR categories in a static or rigid way and to build a sense of GEO-HISTORICITY, fluidity, blurring and incompleteness into their use (see also DISTANCIATION).

At 2E, however, a third level (3L) is already invoked. Most simply, an absence or omission, say an incompleteness in a scientific theory, generates a contradiction (manifest, for example, in science in a band of significant anomalies or aporiai), split or alienation, which can be remedied only by resort to a greater totality, e.g., a fuller, deeper, wider or more complete scientific theory. The internal and intrinsic connectedness of phenomena deduced from the dialecticisation of 1M at 2E reveals the implicit need to theorise TOTALITY and CONSTELLATIONALITY as well as their internal forms: CONNECTIVITY, RELATIONALITY, REFLEXIVITY, concrete universality, SUBJECTIVITY and OBJECTIVITY, autonomy-within-duality and hiatus. What is more, if we then recognise the pro-activity and the constitutionality of these internal forms within constellations, sub-totalities or totalities, as we must once we acknowledge the possibilities of knowledge and agency, then 3L is also the level which produces new philosophical accounts of REASON, rationality and PHRONESIS (practical wisdom). This in turn will lead to the fourth dimension (4D) of 'the unity of theory and practice *in practice*' (*D*: 9) (see DIALECTICAL REASON).

The bipolarity of absence and presence

At the core of DCR is an argument about absence, emergence and dialectical connection and contradiction, which leads to the development of DCR concepts of mediation, totality, relationality and what Bhaskar labels TINA FORMATIONS. Chapter 2 of *Dialectic* begins with two sections on absence and emergence, and these can be seen as standing for the essential bipolarity of the negative and the positive that is

dialectical critical realism (DCR)

entailed by placing absence at the heart of positivity. For DCR, 'non-being is a condition of possibility of being' (*D*: 46) and dialectic 'just is, in its essence, the process of *absenting absence*' (*D*: 43). Absenting processes lead to the emergence of the new. They are crucial to the ONTO-LOGIC of change, while argument involves the epistemic absenting of mistakes, and absenting absences which block NEEDS is essential to moral freedom. Importantly, if absence (negativity) is one pole of the positive, then the positive cannot be successfully positiv*ised*. Absence opens up the critique of the fixity of the subject in the traditional subject–predicate propositional form (see FIXISM/FLUXISM). Dialectic becomes the 'great "loosener", permitting empirical "open texture" {...} and structural fluidity and interconnectedness' (*D*: 44).

The positive correlate of absence is emergence. A 1M CR category already well known in the idea of 'synchronic emergent powers materialism' (see MIND), emergence involves the generation of new beings (entities, structures, totalities, concepts) 'out of pre-existing material from which they could have been neither induced nor deduced' (*D*: 49) – 'something new', a quantum leap: matter as creative or AUTOPOIETIC. Taking emergence to 2E, Bhaskar now links new entities with their own causal powers to space–time as a 'relational property of the meshwork of material beings' (*D*: 53) and to the possibility of emergent spatio-temporalities with their own processual rhythmics. This gives rise to a 2E insistence that the here and now embody the presence of the outside and the past, so that 'emergent social things {...} not only presuppose {...} but also are *existentially constituted* by {...} or merely contain {...} their geo-histories' (*D*: 54). Neither emergence nor absence should be seen as an abstract metaphysical category: they are linked to real processes in the socio-historic and natural development of human being in the world.

Contradiction and dialectical motifs

Holding absence or real negation and positivity together as a bipolar dual generates contradiction in argument, in history and in practice. Contradictions may be external, internal, formal-logical or dialectical. Dialectical contradictions involve 'connections between entities or aspects of a totality such that they are in principle *distinct* but *inseparable*' (*D*: 58) so that there is both existential presupposition *and* opposition. They establish tendential mutual exclusion at the nub of socio-history, KNOWLEDGE and the pursuit of freedom.

As regards knowledge, the idea of dialectic as the 'great loosener' suggests an important nexus between dialectical and ANALYTICAL reasoning. Formal-logical contradictions within analytical thinking may indicate the site of real dialectical contradictions, for the former are understood as 'real constituents of the *Lebenswelt*' that cannot be resolved, *pace* Hegel, through LOGICISING being. To the contrary, DCR, with its emphasis on ontological depth and structural causation identifies the common ground in contradictory propositions, not sublating them, but situating them in structural and causal contexts and in the rhythmics of geo-history. This engenders a *dialectic of dialectical and analytical reasoning* in which dialectical reasoning 'overreaches' (rather than transcends) analytical reasoning. Logical contradictions may be the sign of real contradictions in the world, to be located and explained, thereby pointing the way forward to the understandings – and the practices – required to overcome them (see EPISTEMOLOGICAL DIALECTIC).

A variety of dialectical motifs rest upon a realist understanding of absence, emergence and contradiction. Amongst these are *MEDIATION*, the idea of an 'intermediary or means of some sort' (*D*: 113), which is central to the idea of the bipolarity of being and absence, and which is necessitated by the ideas of internal connection and particularity within totality. Linked with mediation is the idea of *DUALITY*, described broadly as 'the combination of existential interdependence {. . .} and essential {. . .} distinction' (*D*: 115). Duality locates the specific within the general, agency within structure, freedom within the conditioned, and it is marked by two closely linked dialectical motifs: those of *hiatus-in-the-duality*, which defends autonomy against either reificatory or voluntaristic collapse, as well as locating the possibility of dislocation; and *PERSPECTIVAL shifts*, such as that required by the duality of agency and structure in sociological contexts. Similarly, *CONSTELLATIONALITY* signifies the necessary connectedness of things, such as the dialectical unity of dialectical and analytical reason described above: dialectical reason builds on the analytical, overreaching but not transcending it, while the latter lacks without the former.

What, however, if, either in theory or in practice, the dialectical overreach is denied and an analytical proposition 'turns its back' on a dialectical conclusion? The lack or absence that dialectic would have made good is suppressed, and this suppression requires defence, supplementation or compromise. This is the realm of the ironically titled *Tina* ('There is no alternative') *formation*. The phrase was infamously coined by Margaret Thatcher when prime minister of the UK. The irony lies in the fact that there usually are alternatives, so that Tina statements are false closures enforced in an area of various possibilities, requiring various forms of supplementary and contradictory control to maintain them. A practical and literal illustration would be the closure of the British mining industry effected by the temporary establishment of a virtual police state – in the name of freedom. *Tina formations* are 'internally contradictory, more or less systemic, efficacious, syntonic {. . .} ensembles {. . .} displaying DUPLICITY, equivocation, extreme plasticity {. . .} and rational indeterminacy (facilitating their ideological and manipulative use)' (*D*: 117).

Of course, Tina formations are only exposed as such in so far as their limits and obfuscations have been revealed through their systematic grounding in more complex and conflicted contexts and understandings. Such understandings are no more than the 'drive to totality' (*D*: 123) endemic in scientific activity, and *TOTALITY* is the final dialectical concept to consider here. As we have already said, the 2E treatment of 1M concepts invokes a move beyond 2E and on to a third level occupied by totalising dialectical concepts. Totality is the key 3L concept, just as absence plays that part at 2E. The domain of totality is that of 'intra-actively changing embedded ensembles, constituted by their geo-histories {. . .} and their contexts, in open potentially disjointed process' (*D*: 126). It involves, from the point of view of the whole, and thereby complementing absence's effect on the part, a 'break with our ordinary notions of identity, causality, space and time', requiring us to see things as they are '*existentially constituted*, and permeated, *by their relations with others*' (*D*: 125). In the ontic realm of totality, identity thought cannot be adequate and there is a need 'to conceptualise *entity relationism*', thus underlining the need once more for a move beyond analytical thinking.

dialectical critical realism (DCR) 135

Prioritising ontology

Throughout these moves, questions of epistemology are embedded within questions of ontology. Underlying DCR is a complex ontology of human being that insists on its unfinished, POWER$_2$-dominated character. The ontology is seen in the depiction of human nature as *four-planar social being* or the *social tetrapolity* (see PHILOSOPHICAL ANTHROPOLOGY), while the role of power is seen in the significance of *power$_2$ relations* for human society and philosophy. Social life qua totality is constituted by four dialectically interdependent planes: of material transactions with nature, inter-personal action, social relations and intra-subjectivity. Four-planar social being must be conceived in terms of depth and stratification, and the elements of each plane are 'subject to multiple and conflicting determinations and mediations' (*D*: 160) in determining what Bhaskar calls the CONCRETE UNIVERSAL.

This depiction of the multiplicity and structuring of human social being is aligned with the important theme of power$_2$, or *generalised MASTER–SLAVE-type, relations*. This links DCR to both Hegel's master–slave dialectic and Marx's analysis of wage-slavery, IDEOLOGY and the FETISHISM of commodities, but it is also generalisable to all socially structured power relations, such as those of gender, race and age, through which agents, groups or classes get their way 'against either {...} the overt wishes and/or {...} the real interests of others (grounded in their concrete singularities)' (*D*: 153). This is an important substantive argument about the nature of modern and historical societies and the multi-form *constraints$_2$* to which they give rise. Sanctioned by DCR's realist social scientific grounding, it explains what dialectical ethics and practice orient themselves to and against, as well as licensing modern critical phenomenological readings of, for example, 'the co-existence of Disneyfication/McDonaldisation, poverty and waste' (*D*: 162) in the 'New World Order'.

But more than this, DCR relates the significance of historical power$_2$ relations to the forms of Western philosophy. This intrinsic philosophical significance is seen, for example, in Bhaskar's linking of analytical philosophy to the expression of an *ontology of stasis*, so that analytics unselfconsciously assumes the role of 'normalisation of past changes and freedoms, and the denegation of present and future ones' (*D*: 177). Since a philosophy of ontological stasis is closely linked to the cardinal philosophical errors that characterise Western philosophy, such as ontological monovalence, DCR raises fundamental – appropriately dialectical and realist – questions about the relationship between power relations and knowledge in the 'Great Arch' (see ARCH OF KNOWLEDGE TRADITION) deriving from Plato and Aristotle. Master–slave relations, money, instrumental reasoning, REIFICATION and ALIENATION may all be systematically connected to IRREALISM, actualism, monovalence, analytics and theory–practice split.

Ethics in DCR

The foregoing describes how CR concepts (at 1M) are dialecticised (at 2E) and brought together in a larger picture (at 3L). What follows considers the conceptions of ethics and practice entailed by DCR (at 4D). 'My project', says Bhaskar, 'is normative' (*D*: 279),

136 dialectical critical realism (DCR)

and *Dialectic* builds an ethics based upon ontology, the nature of dialectic as the absenting of absence, dialectics of truth, desire and freedom, and explanatory critique. These lead to a moral theory that moves 'from primal scream to universal human flourishing' (*D*: 180), in which 'concrete singularity {the free flourishing of each} is the relational condition of concrete universality {the free flourishing of all}', and where this ethical conclusion is concretised as 'an immanent and tendential possibility {. . .} necessitated by structural conditions', but one that is 'held in check by global discursively moralised power$_2$ relations' (*D*: 202) (see DIRECTIONALITY).

Ontology has always played the major part in CR, and it is the foundation stone of Bhaskar's ethics. It grounds what he calls alethic truth, defined as 'dialectical reason and ground in theory and the absence of HETEROLOGY {unreconciled otherness}; it is true to, for, in and of itself'. With regard to the epistemic, it 'furnishes the non-arbitrary principle of ontological STRATIFICATION that powers the dialectic of scientific discovery'. With regard to the ethical, which unites what we know with what we are, 'the true = the moral good = freedom, in the sense of universal human emancipation' (*D*: 219–20). It is the unifying ontic ground for the cognitive and normative aspects of human being, understood in four-planar form.

As ethical basis, ALETHIA is centrally linked to dialectic. Bhaskar writes that, 'at the outset', 'the most important thing to appreciate' is that any ill 'can be looked upon, or dialectically transposed, as an absence, and any absence can be viewed as a constraint' (*D*: 259), and this is tied to the alethic, for in 'the moral realm, alethic truth, the good, is freedom, [which] depend[s] on the absenting of constraints on absenting ills' (*D*: 212). It is this grounding of the ethical in the ontological that distinguishes DCR from other contemporary philosophies. The *TRANSFACTUAL* character of moral truth means that Bhaskar's ethics depends on, for example, neither a (neo-Kantian) ideal-speech situation, *pace* Habermas, nor a (neo-contractarian) original position, *pace* Rawls (*D*: 221).

Dialectical universalisability

In turn, dialectic goes hand in hand with the idea of *dialectical universalisability*. This involves a characteristic form of argument in which a commitment to the negation of x entails further commitment to the negation of those things that x itself entails. Thus the ontological starting point of the absenting of constraints on being entails a progressive commitment to the absenting of all such constraints. Any truth statement 'can be seen to imply a commitment to the project of universal human emancipation, involving the abolition of the *totality of master–slave relations*'; and the claim that the 'desire to overcome constraints (including constraints$_2$) on the satisfaction of desires, wants {. . .} and needs {. . .} implies a CONATUS {. . .} to knowledge of all four planes of the social tetrapolity' (*D*: 180).

In addition to being universalisable, such a judgement has to be 'oriented to the concrete singularity of the addressee {. . .} and universalisable to any other concrete singular so situated'. This orientation to concrete singularity relies, in Kantian terms, on an 'assertoric' – there is no surreptitious categorical – imperative, on what an agent ought to do in her situation, but there is still a possible generalisation from the addressed agent to the addressee. If the addressee experiences constraints on her needs, the addressor, through his orientation to her concrete singularity as expressed in his judgement, stands in her place, implying solidarity and commitment to the critical explanation of her situation. That in turn entails a location of her concrete singularity within a 'theory of human nature-(needs and interests)-in-society-in-nature'. This dialectical generalisation then involves a further universalisation since any social ill discovered through solidarisation will be seen as a constraint on freedom, so that any truth judgement 'implies a commitment to universal human emancipation and a society in which the concrete singularity of each and all is realised' (*D*: 178–9).

This argument for 'assertorically imperatival sensitised solidarity' (*D*: 262) is linked to the DIALECTIC OF *DESIRE TO FREEDOM* that ultimately drives, and demands, the search for truth. The starting point of this dialectic is the ontology of absence, the experience of it, and the desire to absent that experience. Absence is experienced as constraint, as 'unfulfilled needs, lacks, wants or, in the setting of primary POLYADISATION {i.e., 'PRIMAL SCREAM'}, elemental desire' (*D*: 285). Desire involves the RECOGNITION of *difference* (between desirer and the desired), which entails REFERENTIAL DETACHMENT (the separation of the act of desiring from the desired object), which in turn entails the recognition of the intransitive world of causality, ontological stratification and alethic truth. These in turn involve the recognition of power$_2$ relations, the absences they create and the desire to absent those absences.

Moral realism and ethical naturalism

The existence of such relations leads Bhaskar to distinguish between *moral realism* and *ethical naturalism*. Moral realism entails that morality is an 'objective real property' (*D*: 259), i.e., alethic, but a distinction has to be made between morality as practical, relational and explanatory-critical and as it actually exists within an already moralised world. The distinction is the basis for a *critical* moral realism that is not reducible to a sociology of ethics. Ethical naturalism entails that moral properties can be suitable objects of study for the social sciences, and the combination of realism and naturalism leads to the conclusion that the constitutive morality of a society can be shown to be essentially limited or false in its ethical claims. It is an appropriate function of a social science to be involved in examining beliefs about social objects, beliefs which it may show to be false. If the social science can also explain the falsity, then, subject to a *CP* clause, 'one can move without further ado to a negative evaluation' of the belief and 'a positive evaluation of any action rationally designed to absent it' (*D*: 262).

Some see a tension between these two standpoints. On the one hand, DCR's prioritisation of ontology embeds ethics, like epistemology, within processes of sociohistoric development of human being. It suggests the unfinished and contained nature of ethical categories (hence ethical naturalism). On the other hand, arguments based on dialectical universalisability or truth-finding seem to indicate a direct line to the good society, if only humankind would grasp it (hence moral realism). Yet, if this were so, it would also suggest that ethics transcends the socio-historic, de-emphasising (or radically changing) the underlying ontology.

138 dialectical detachment

A key point to note concerns the highly generalised and abstract character of the categories used to establish the alethic state. The dialectics of truth and freedom, or the eudaimonic society, are hard to deny, but they are equally hard to concretise. They require embedding in existing social relations and practices, which could specify in the here and now what they actually mean. This, Bhaskar stresses, is the task of *totalising depth-praxis*. But, once in the here and now, how would these general and abstract categories guide us? We might all believe in world peace, but how do we actually achieve it? One way to resolve the tension between the ABSTRACT and the CONCRETE, the ethically natural and the morally real, is to envisage actually existing morality as expressing by dialectical implication the deeper moral truths that Bhaskar expresses in abstract form in his moral realism. These exist as inchoate ethical potentialities in the morality of the here and now which could be brought to fruition through processes of social transformation away from existing power relations. Thus, liberal society develops ideas of freedom as agentive capacity, as negative and positive liberty, and these disclose a deeper conception of freedom as emancipation. The full contours, as opposed to the abstract outlines, of such a concept cannot, however, be grasped ahead of the fundamental processes of social change needed to bring it about.

This suggests for DCR a morality which is critical of modern forms, socio-historical and processual in its orientation, and directed to the lived experience of those experiencing constraints, especially constraints$_2$. Reflecting the experience of resistance to oppression, dialectic cannot itself be 'in the business of telling people {. . .} what to do'. It is best conceived as an 'inner urge that flows universally from the logic of elemental absence', manifesting itself wherever lack, need, want or desire are blocked. It is most especially experienced where power$_2$ relations hold sway. It can be conceived as ontic pulse, both real and moral, as 'the heartbeat of a positively generalised concept of freedom'. As such, it is the 'irrepressible' (*D*: 299) guide to how far humankind has come, and how far it has yet to go.

ALAN NORRIE

dialectical detachment. See DETACHMENT.
dialectical development. See CONSISTENCY; DIRECTIONALITY; PROCESS.
dialectical distinction (dd′). See DIALECTICAL COMMENT.
dialectical duplicity. See DIALECTICAL ANTAGONISTS; DUPLICITY.
dialectical explanation, dialectical explanatory argument. See DIALECTICAL ARGUMENT.
dialectical fertility. See CONTRADICTION; EPISTEMOLOGICAL DIALECTIC.
dialectical freedom. See FREEDOM, ETC.
dialectical generative separation. See GENERATIVE SEPARATION.
dialectical ground (*DG′*). See GROUND.
dialectical identity. Another term for CONSTELLATIONAL IDENTITY.
dialectical indivisibility of freedom. See CONDITIONALITY; FREEDOM, ETC.
dialectical intelligibility. 'In Hegel this depends upon the teleologically generated presentation, comment on (i.e., immanent CRITIQUE of) and preservative supersession of conceptual and socio-cultural forms. In Marx {and Bhaskar} it is transformed to comprise the EXPLANATORY CRITIQUE of the causally generated production of social phenomena – from crises to categories – in terms of their underlying causal grounds'

dialectical irrealism 139

(*D*: 102). Because reasons are causes, TELEOLOGICAL explanation in Hegel is in fact causal, but it is not grounded in underlying social causes.
dialectical irrealism. See IRREALISM.
(a) dialectical life. See CRITIQUE.
dialectical limit (dl′). See DIALECTICAL COMMENT.
dialectical logic. See LOGIC; TRANSITION.
dialectical necessity. See AXIOLOGY; DIALECTICAL ARGUMENT; MODALITY.
dialectical opposites. See DIALECTICAL ANTAGONISTS; TABLE OF OPPOSITES.
dialectical overreaching. See CONSTELLATIONALITY, DCR; DIACHRONY.
dialectical pair. See DIALECTICAL ANTAGONISTS.
dialectical perspectival switch. See PERSPECTIVE.
dialectical praxis (dφ) is practically oriented transformative negation of contradictory states of affairs. It has two ratchets: transformed transformative totalising transformist praxis (dφ′ or TTTTφ); and transformed, transformative, trustworthy, totalising, transformist, transitional praxis (dφ′ or TTTTTTφ). See also AGENCY; DIALECTICAL REASON; EMANCIPATORY AXIOLOGY.
dialectical process (dp′). See PROCESS.
dialectical propaedeutics. See PROPAEDEUTIC.
dialectical rationality. See DIALECTICAL REASON.
dialectical reason (rationality) or **absolute reason (dr′** or **dr†)**. (1) epistemologically has connotations ranging from enlightenment and demystification to the creative, flexible processual thinking necessary for the EPISTEMOLOGICAL DIALECTIC and totalising depth-praxis; (2) ontologically, the real or *alethic reason* for or *ALETHIC truth* of a thing. (1) and (2) come together in (3) 'The unity – or better, coherence – of theory and practice in practice' (*DG*: 393), or *dialectical reason*; i.e., the coherence of *theoretical reason* (dr_t') and *practical reason* (dr_p'), or EXPLANATORY CRITIQUE ↔ EMANCIPATORY AXIOLOGY, in a two-way dialectic of truth ↔ freedom: dr_a' in dφ, where 'a'' stands for absolute reason and φ for the DIALECTICAL PRAXIS that transformatively negates the prior (contradictory) state of affairs, constituting an 'unachieved but realisable' *practical TOTALITY* (*D*: 122). *Dialectical rationality* is sometimes used synonymously with dialectical reason, but also has a distinctive meaning which includes, in addition, the mediation of practical wisdom or PHRONESIS (*P*: 66). Theoretical reason seeks to 'adjust our beliefs to the world', practical reason to 'adjust the world to our will' (*D*: 65). For creatures who have no choice but to act (see ACTION; AXIOLOGY), the one presupposes the other; and since practice is quasi-propositional and theory quasi-practical they constitute a DUALITY. Dialectical reason is thus closely related to REFLEXIVITY in one of its meanings. The concept derives from Plato and more immediately from Hegel's theory of truth (to which Hegel himself was untrue) 'which would vindicate the unity of theory and practice in practice, that is, in transformation of (socialised) reality to comport it to a rationally grounded notion of it' (*D*: 340). (If the notion is scientifically grounded, this does not contradict the realist methodological prescription to 'follow the object'.) REASON is said to be ABSOLUTE, not because epistemological absolutism (ENDISM) and FOUNDATIONALISM are in any way endorsed, or relativity resiled from, but because the coherence of theory and practice in practice is its highest (optimal) form, which contingently maximises human freedom and flourishing in an open, always

140 dialectical refutation

unfinished process. In PMR the concept is generalised to take in the totality of being (*cosmic consciousness*); at the level of the concretely singular person, it is most immediately manifested in *spontaneous right-action* and more mediately in *self-realisation* or the positive fulfilment of needs and potentials. 'Coherence' is substituted for 'unity' because 'unities can be of many (e.g., antagonistic) types' (*D*: 378). Ultimately this coherence is nothing other than dialectic itself, or dialectic is (the process of) the coherence of theory and practice which tendentially approximates dialectically rational geo-historical DIRECTIONALITY. Dialectic is driven onward by absence (incompleteness, inconsistency) and its repair in an ever more complete totality, in principle without end. In philosophy and cognate practices this process takes the form of immanent CRITIQUE both of rival discourses and of its own past phases. See also CONSISTENCY; DIALECTIC; REASON.

dialectical refutation. See DIALECTICAL ARGUMENT.

dialectical remark (drk′ or drk†). See DIALECTICAL COMMENT.

dialectical result. (1) The outcome (dr^o) or result of a *dialectical PROCESS*, dp′; (2) a result ($dr^†$) that resolves the contradiction in dp′; (3) a result (dr') that provides a *rational* resolution of the contradiction in dp′ (and signifies in addition DIALECTICAL REASON itself understood ontologically as ALETHIC TRUTH); (4) a result (dr'') that additionally conforms to Hegelian radical preservative determinate negation; (5) and affords us reconciliation to life (dr'''); (6) and encourages mutual recognition in a free society (dr''''). The finer discriminations are used in the critique of Hegelian dialectic (*D*: 28).

dialectical sublation. See SUBLATION.

dialectical suspension or bracketing. See PROPAEDEUCTICS.

dialectical totality. See TOTALITY.

dialectical unity. Another term for CONSTELLATIONAL UNITY.

dialectical universalisability, universalisation, universality. See CONCRETE UNIVERSAL; UNIVERSALISABILITY.

dialectics. Since DIALECTIC is fundamentally PROCESS (absenting absences) and the attempt to grasp it, there are in principle as many dialectics as there are processes – an indefinite array. These fall into four broad classes: (1) *ontological* ('dialectical ontologies which may operate at different levels', i.e. dialectics of emergence and disemergence, etc.); (2) *epistemological* (relating to our knowledge of [1]); (3) *relational* (concerning the dialectical interplay of our knowledge between (1) and (2); and (4) *practical* or *ethical* (concerning what we should do) (*D*: 3, 97–8). Within these, a wide variety of categorisations can be made. Thus in the domain of concepts there are at least five classes of dialectics: *discursive, conversational, hermeneutical, methodological* or *heuristic*, and *philosophical*; and the twelve routes taken by the dialectic of desire to freedom, e.g., may in turn be grouped into four: *systematic, intrinsic, intrinsic extrinsic* and *extrinsic intrinsic* dialectics (see EMANCIPATORY AXIOLOGY), and so on. However, it should be noted that, in the sociosphere, what constitutes a *significant* dialectic is fundamentally related to the criteriology of the good society, deduced from human intentionality as currently understood, such that 'above all there is the dialectic of freedom' (*D*: 202). This entry tabulates (Table 18) the main dialectics specified in the CR system under the different moments of the ontological–axiological chain, only a small proportion of which have been assigned separate entries. It is important to bear in mind (1) that these – as ID/TD dimensions of being/concepts for thinking being – constitute an interrelated unity.

dialectic(s), Schillerian   141

Thus, as absence in one of its four modes, the dialectics of 2E encompass all the other stadia; and likewise those of 3L as totalising, those of 1M as the outcome of absenting and mediation, and those of 4D as a specific kind of 2E absenting. And relatedly, (2), that there are dialectics within and between the various moments. Where a dialectic is located in Table 18 accordingly depends on what its *primary* moment is taken to be. Thus the epistemological dialectic crucially involves all four stadia, but appears under 1M because, without grounding in causal mechanisms, it is ultimately impotent. Likewise the dialectics of co-presence appear at 2E because that is the home of absence. Since 1M and 3L, on the one hand, and 2E and 4D, on the other, have a special affinity (totality is a structure and praxis transformative negation), many of the dialectics appearing under the one 'belong' equally to the other. The realist dialectic from *anti-anthropism to HUMANISM*, and its irrealist counterpart from *ANTHROPISM to anti-humanism*, go from 1M to 4D, and so are not included in the table.

dialectic(s), Schillerian. See PROCESS.

dialectics of DCR. The dialectics pertaining to each of its four stadia (*D*: 306–7). See DIALECTICS; MELD.

dialectics of infra-/intra-/superstructuration. See INFRA-/INTRA-/SUPERSTRUCTURATION.

dialectics of opposition. See CO-PRESENCE, REVERSAL.

dialectics of reversal. See REVERSAL.

dialectics of transition. See CO-PRESENCE; TRANSITION.

dialogue, dialogicality, dialogism. See CULTURE; ELENCHUS; HERMENEUTICS; PHENOMENOLOGY; PHILOSOPHY; SIGNS.

dichotomy (f. Gr. *dichotoma*, a cutting in two). A split, DUALISM.

diesseitig. See THIS-WORLDLY.

difference. See ALTERITY.

differentiation/stratification. 1M concepts grounded in the distinctions between CLOSED and open systems (evincing *differentiation*) and structures and the events they generate (issuing in *stratification*) (*SR*: 27). The former pertains to *horizontal* or *transfactual realism*, the latter to *vertical, theoretical* or *DEPTH-realism*. See also ALTERITY/CHANGE.

diffraction. Used by Bhaskar in respect of (1) the liberation of DIALECTIC from its 'Hegelian moorings', and (2) the process of CRITIQUE whereby, following Marx, he effects this before 'retotalis{ing} it under the sign of absence' (*D*: 370). Diffraction connotes not only pluralisation and differentiation, but also the play of ABSENCE and presence: in optics it refers to the fringes of light and dark bands produced by the deflection and decomposition of light when it passes through slits or past sharp edges. This underlines that we are dealing with a unity-in-diversity, not a fragmentation or splitting. The diffraction of dialectic is often referred to as the *MATERIALIST* diffraction because it is carried through from the standpoint of epistemological materialism (TRANSCENDENTAL REALISM), ontological materialism (in the sense that the social is seen as EMERGENT from the natural world), and practical materialism (i.e., the TMSA).

dilemma. A 2E figure of OPPOSITIONALITY signifying a logical problem whereby we are faced with an either/or choice between two equally unacceptable positions (cf. FORK), entailing for its resolution rejection of either the principle of BIVALENCE or the law of EXCLUDED MIDDLE, or both. An example is the *dilemma of de-geo-historicisation*

Table 18 Key dialectics of critical realism

Causal Chain	1M Non-Identity	2E Negativity	3L Totality	4D Transformative Agency
Formal principle	non-identity	negativity	totality	transformative agency (praxis)
Substantive content: realist	*alterity*	*absence and presence* (co-presence)	*inter-/intra-relationality*	*freedom* (absence/desire/need/lack) (agency)
	identity-in-difference and alterity	*process*, change	holistic causality	
	infra-intra-/super-structuration (product) stratification	emergence (process)	constellationality	truth ↔ freedom, via wisdom
		process and product	unity-in-diversity	explanatory critique†;
	emergence	development	duality (and non-duality)	counter-hegemonic
	the real and the actual	directionality	perspectival switch	reversal, struggle
	ground	contradiction	recursive embedding	(reality principle)
	dialectical reason and dialectical ground	opposition, incl. reversal	retotalisation (concretion)	emancipatory axiology;
	abstract and concrete	decay	emergence (process-in-product)	totalising depth-praxis
	essence and existence	delay (lapsed time)	disemergence	(ethical tetrapolity)
	mechanisms and events	lag	immediate and mediate	dialectical universalisability
	signifier and trace	spatio-temporality	figure and ground	and immanent critique
	form to content	differentiated rhythmics	needs and wants	form to content; figure to
	juxtaposition and chiasmus	transition	centre and periphery	ground; centre to periphery
	simple and complex	limit	inner and outer; part and	7 E's (re-agentification:
	metaphor and metonymy	boundary	whole inclusion and exclusion	recentrification;
	the real and the virtually real	frontier	globalisation and localisation	re-empowerment;
	epistemological d (scientific discovery)	node	present and past	*incorporating*:
		disjuncture and constitutive	*reflexivity*	*emancipation*
	dialectical and analytical reason	geo-history	dialectic of consciousness and self-consciousness to reason	consciousness + self-consciousness → reason
	incompleteness and inconsistency	islocation and distanciation	*ethical dialectics*	instrumental → explanatory
	explanation	d *generative separation*	(the dialectics of freedom)	critical → dialectical reason
	truth	+ 4D dialectics	practical reason	morality
			universalisability	*desire to freedom* (12 routes:)
				1. agency and discourse

de-totalisation; alienation split			self and solidarity; recognition 2. education of desire 3. malaise 4. cognition 5. equity 6. de-alienation 7. desire (generic) 8. transition 9. desire for freedom 10. universalisability 11. interests (generic) 12. material interests politics of totalising depth-praxis democracy life; movement; representative; emancipatory politics
Substantive content: irrealist or ideological; categories of realist meta-critical dialectics	explanatory and taxonomic knowledge natural necessity and natural kinds destratification; identity expressive unity (fusion) homology; regress; vicious circle; stasis supplementarity and grafting mediatisation, virtualisation and hyperrealisation representation of sectional interests as universal	de-negativisation; ideological oppositions; antinomies aporial (omissive critique)	*metacritics* master and slave philosophy and science de-agentification (reification, fetishism, hypostatisation; disembodiment, voluntarism; ideological mystification)
		compartmentalisation (fission); representation of universal interests as sectional	

Note All items should be read as if prefaced by 'dialectic(s) of' except where 'd' (= dialectic[s]) appears.

144 dilemma of de-geo-historicisation

(see GEO-HISTORY). Close to, but distinct from, ANTINOMY. Like T/P contradictions and epistemological undecidability, such logical contradictions place us in a *problematic AXIOLOGICAL choice-situation*. Since dilemmas have presuppositions, the most creative response to them is to redescribe the alternatives in the light of a transformed problematic that typically invokes a deeper level of structure.

dilemma of de-geo-historicisation. See GEO-HISTORICITY.
dilemma of dualism. See DUALISM.
dilemma of reductionism. See REDUCTIONISM.
direct realism. See REALISM.
directionality pertains to 2E PROCESS (see also CAUSALITY), which is directional change or absenting ('development'), or 'the mode of spatio-temporalising structure', necessary conditions for which are the irreversibility (directional asymmetry) of time and the reality of TENSE. Directional change is a condition of the intelligibility of our embodied AGENCY, 'for to act is to bring about a state of affairs which (unless it were overdetermined) would not otherwise have occurred' (*D*: 252), and is presupposed by the theory of an expanding, divergent universe. In our multiply contradictory and open world, directional absenting is rarely, even in its simplest manifestations, unilinear. It is rather an uneven and multiply divergent unfolding and enfolding, punctuated by fold-back and retrogression (strikingly exemplified in the process of biological evolution). Just as there is 'no uniquely determined or predetermined path linking amoebae to humans' (*PN3*: 100–1), so there is none from hunter-foraging modes of production to eudaimonia – '*social life is not a success story* and *things could have been* (just as they might be) *different*' (*SR*: 140). Nor does directional change have a conceivable END (see also GEO-HISTORICITY). CR is accordingly strongly critical of all evolutionary HISTORICISMS (see also PDM).

In this 'non-centrist-expressivist-triumphalist-endist' (non-modernist, non-bourgeois, non-Whiggish, nor Social Democratic non-Fukuyaman) sense, we may none the less speak of *progress* in the human realm (and probably also within the process of biological evolution), providing that, in accordance with the principle of EPISTEMIC RELATIVITY, it is recognised that all judgements of progressive import are necessarily intrinsic to the process concerned and made from a particular position in theoretical 'time', be it a research programme, an emancipatory struggle, a historical epoch or the overall process of geo-history as such (see CONSISTENCY; cf. Minnerup 2003). Progress may be negative, consisting merely in the slowing down of degeneration or ENTROPY (e.g., from an emancipatory point of view, of the current erosion of civil liberties in the context of the 'war on terror'), rather than the positive speeding up of negentropy. (It may of course also be more absolutely negative, i.e., consist in movement away from the desired direction altogether, as in the progressive degeneration of a research programme.) Diversity and plurality are among its necessary conditions.

In the case of science generally and social science specifically, positive progress is embedded and registered in the EPISTEMOLOGICAL DIALECTIC (cf. Norris 2003; Norris and Papastephanou 2002) and EXPLANATORY CRITIQUE respectively (cf. Lawson 2003b: 61 n). Geo-history for its part is argued by Bhaskar to exhibit *tendential rational directionality* overall. This concept has its immediate origins in Hegel's and Marx's understanding of human agency as involving both separation of self from nature (see REFERENTIAL DETACHMENT) with a potential for ALIENATION and a fundamental

directionality

drive to overcome it through self-realisation (Sayers 2003), which finds an echo in Adorno's notion of an underlying impulse to freedom in human history (Norrie 2004a) and is exemplified in the Bhaskarian notion of 'the deep content of the judgement form and the immanent latent teleology of praxis' (*P*: 154; see also EMANCIPATORY AXIOLOGY). It 'pushes history from behind' ('*teleonomic push*') as well as 'pulling it from in front' ('*teleological pull*') (*D*: 22, which has 'from in front *and outside*' in relation to Hegel because it is *epistemologically* but not ontologically transcendent and so illicit in Hegel) (see TELEOLOGY). It does this in Hegel at the level of the actual, in Marx and Bhaskar at the level of the real by the aspiration and struggle to repair the lack of the good society that is immanent in our praxis. 'Directionality' is imparted to this process also by the constraints on free flourishing, whether natural or social, which desiring agents have an interest in gaining the necessary knowledge and power to absent. Thus we can say that directionality is 'imposed on the education of desire by the REALITY PRINCIPLE, i.e., ALETHIC truth, mediated in practice by the meta-ethical virtue of wisdom' (*D*: 176). By 'rational' is meant that the directionality is towards the moral ALETHIA of the species via DIALECTICAL REASON, or the coherence of theory and practice in practice, entailing the abolition of master–slave-type relations as such and in general and transition to EUDAIMONIA (marking the end, not of history, but of its power$_2$-stricken phase, and the commencement of an epoch of authentic free development and diversity). 'Tendential' highlights that it pertains to the domain of the real and the possible, not to the actual or to the 'ideal': it is a real (causal) TENDENCY (tendency$_b$), and while it is as such necessary and inexorable so long as there is human being, and issues in real effects unless prevented by countervailing causes, it is highly contingent whether and how it will be actualised, because this depends both on human praxis (hegemonic as well as counter-hegemonic) and on accidents. This is a fundamental point that POSTMODERNISM overlooks in maintaining that narratives of emancipation necessarily rest on unilinear views of history (the end of which is ironically proclaimed to be already secured in postmodernity itself, one-sidedly characterised as the era of 'fragmentation, multiplicity and pluralism' [Woodward 2002]). As a normic universal, rational directionality cannot be falsified by the actual course of history, 'only by the provision of a better, nobler, norm more fitting to the needs and propensities of developing four-planar socialised humanity' (*D*: 280). Its analysis is itself causally efficacious, offering a meaningful interpretation of geo-history productive of hope (cf. Patomäki 2006b, who, however, controversially holds that Bhaskar is concerned largely with the logic of the 'ideal', rather than real, process of emancipation).

Such a view seems fully compatible with the historical materialist thesis concerning the tendency$_b$ of human productive powers to develop across history (cf. Creaven 2003; Nolan 2002a, b), providing the thesis is not interpreted actualistically and the process of freedom is not reduced to the economy of time and the absenting of material scarcity, or seen as spearheaded by the West or any other vanguard or elite, or as coming to an end in communism. As already noted, tendential rational directionality is grounded in human agency as such, wherever it is – 'a species-specific ineliminable *fact*', whether in the mode of mowing a lawn or TTTTφ (*D*: 210) – which knows no East nor West nor South nor North. While its robustness has on balance been vastly enhanced by the development of capitalism, whose historical origins contingently lay in the West, it is

146 **diremption**

no more West- or North-centric than the globalising logic of capital itself. Unlike the latter, however, it is genuinely dialectically universalising, thrusting across and beyond all power$_2$ divisions of class, gender, ethnicity, nationality, religion, and COLONIALISM. The DIALECTIC OF (desire to) FREEDOM within which it is embedded necessarily moves from 'form to content, centre to periphery, and figure to ground' (*D*: 208). This is the fundamental logic of *alter-globalisation*.

At the end of the day (and during its course), tendential rational directionality is none other than DIALECTIC (maximally defined as the process of absenting constraints on absenting ills) or the PULSE OF FREEDOM itself, which beats most strongly in the breasts of the downtrodden, the exploited, the oppressed, who may increasingly come to understand, in conditions of post/modernity conducive to 'a dialectic of globalising self-consciousness', that the freedom of each in an interconnected world is a condition of the freedom of all, so enabling 'geo-historical directionality {. . . to} catch up with dialectical rationality' (*D*: 270) and redeeming all the crushed CONCRETELY UTOPIAN strivings of the past (Benjamin 1940).

diremption. See ALIENATION; GENERATIVE SEPARATION; PROCESS (dialectical process).

disability. The concept of disability has a long and varied history, although for many decades in the twentieth century it was seen as biomedical in origin and import. One of the most influential 'frames' for defining disability was the *International Classification of Impairments, Disabilities and Handicaps* (ICIDH), published by the World Health Organisation (WHO) in 1980. Bodily impairment was presented in the ICIDH as 'first cause' in a causal chain, often leading to functional disadvantage (disability), and sometimes also to psychosocial hindrance (handicaps) (G. Williams 2001). There was initially little critical interrogation of this frame within medical sociology. Medical sociologists focused on conditions that were often taken as biomedical 'givens', albeit from the perspective of those with the condition rather than that of physicians.

It was disability theorists who challenged this frame vigorously enough to promote a rethink. For them the prime mover, causally, was not the biomedical condition, nor even the 'tragic victim', but rather the oppressive society in which people with putative disabilities live (Oliver 1990). According to the 'social model', disability and dependency are caused by society. Although this form of disability theory/activism undoubtedly occasioned a shift in thinking about disability, not all commentators accept its wholesale dismissal of the thinking behind the ICIDH. G. Williams (2001), for example, contends that (1) most disability emerges over time from chronic illness; (2) people are only temporarily able-bodied, since disability can, and is likely to, affect most of us sooner or later; and (3) disability undeniably has to do with the pain or discomfort of bodies.

There have been some attempts to develop a CR perspective on disease and disability. S. Williams (1999), for example, insists that the body is a real entity, no matter what we call it or how we observe it (in short, notwithstanding epistemological relativism). Moreover, it has its own mind-independent generative structures and causal mechanisms. Disease labels are merely descriptive, not constitutive of disease itself. Williams draws on Archer's (1995) morphogenetic framework to maintain that disability is the sole product of neither the impaired body nor the oppressive society. It is rather an emergent property, located, temporally speaking, in terms of the *interplay*

discourse, dialectic of 147

between the biological reality of physiological impairment, structural conditioning and socio-cultural interaction leading to structural elaboration/reproduction.

Developing Williams's analysis via an account of juvenile Batten disease, Scambler and Scambler (2003) have advanced an 'ontological differentiation thesis'. This holds that the individual is a complex (dynamic, dialectical) mix of psycho-organic and socio-cultural properties. In studying disability, attention must be paid both to capacities transmitted 'upstream' (that is, from the biological to the socio-cultural) *and* to those transmitted 'downstream' (that is, from the socio-cultural to the biological) (see Creaven 2000). It is to be hoped that CR studies will make headway in applying such an approach. See also Bhaskar 2006 (with Danermark); Danermark 2002, 2003.

GRAHAM SCAMBLER

discourse, dialectic of. See EMANCIPATORY AXIOLOGY.
discourse analysis. See CRITICAL DISCOURSE ANALYSIS.
disembedding. See DISTANCIATION.
disembodiment. See ACTUALISM; CONCRETE/ABSTRACT; DE-AGENTIFICATION; RATIONALISM; VOLUNTARISM.
disemergence. See ALIENATION; DEMI-REALITY; DISTANCIATION (disembedding), EMERGENCE; ENTROPY.
disenchantment. See ENCHANTMENT.
disintentionality. See AGENCY; CONCRETE/ABSTRACT.
disposition is used in the sense of 'basically a POWER' throughout Bhaskar's work. *Dispositional REALISM*, a concept introduced in Bhaskar (1997a), is to that extent synonymous with *causal powers realism* (see CAUSALITY). However, *disposition* highlights the logical, epistemological and ontological PRIORITY of the *possible* and implicit over the actual and explicit, which entails *MODAL realism* and 'a three-tier (possibility, exercise, actualisation), in contrast to a two-tier (grounding, manifestation) analysis of dispositions' (*EW*: 28). Moreover, PMR works with a very broad concept of disposition which embraces, besides powers and liabilities, 'properties {...} affordances, {...} fields, rights, duties, and so on' (*EW*: 28), i.e., with the breakdown of the fact/value dichotomy, powers have an ethical as well as a conventionally causal dimension. For *dispositional identity*, a concept introduced in *RS*, see IDENTITY and ULTIMATA. See also TENDENCY.

distanciation or **distantiation** has currency in a wide range of fields, including philosophy, social theory, human geography, globalisation studies, information theory, and drama and film theory. Bhaskar builds on Giddens's usage to connote 'the play of ABSENCE and presence' as between 'a) stretching {(time) or spreading (space)} (and thereby extending presence or *embedding*) and {...} b) distancing (and thereby absenting and possibly *disembedding*)' (*D*: 7 n, e.a.).

Disembedding may involve shrinkage or compression as well as distancing. Embedding has connotations of (RECURSIVE) insertion and nesting, and as such is linked to 3L. Distanciation refers to either or both, i.e., constellationally contains both as distinctions within an identity; each can be seen under the aspect of the other. Thus the social order is both disembedded (EMERGENT) from the natural order and embedded (nested) within it. The concept *constellational embedding* thus refers to both moments of distanciation, which incorporates 'a notional moment of the coincidence

148 distraction

of IDENTITY and change' (*D*: 273). Because of the irreversibility of time, it is asymmetrical (see PRIMACY).

Distanciation is above all a (key) 2E category or form of absence/absenting which, since it pertains to 'the transcendental parameters of any conceivable social life' (*D*: 13), is ineliminable. Indeed, it is closely connected to and may be involved in transformative NEGATION, which itself may be stretched out spatio-temporally (embedded), or split off and dislocated (disembedded), e.g., in a geo-historical epoch and the transitions to and from it. Such stretching may eventually issue in sheer difference, as in the *conceptual distanciation* (indispensable to scientific development) whereby erstwhile commensurable theories may sometimes become incommensurable. However, it may also involve mere spatio-temporal distancing (separation) without transformation (real negation without transformative negation). This illustrates the difference between difference and change, i.e., their irreducibility, and links distanciation (and negativity) to 1M ALTERITY (*D*: 257). Distanciation is also closely related to PROCESS, RHYTHMICS and EMERGENCE/disemergence (where embedding/disembedding find counterparts in INTRASTRUCTURATION/SUPERSTRUCTURATION). Indeed, at its most general level it *is* the process that produces, within ontological stratification and spatio-temporality, the CO-PRESENCE or co-inclusion of 'presence and absence, past and present, inside and outside, essence and appearance, transfactual and actual' (*D*: 63). Thus we have the concepts of *distanciated stratification* (e.g., of an embodied personality) and *distanciated spatio-temporality* (developmental process). The latter is a condition of the intelligibility of time-consciousness and non-contiguous CAUSALITY. Both may involve multiple and uneven vertical (depth) as well as horizontal (temporal) and lateral stretching or spreading and dislocation or divergence, i.e., stratification has a processual dimension. They come together in the notion of *multiply emergent distanciated spatio-temporalities* (*D*: 299) and in the concept of a *meta-reflexively totalising situation* (which they make possible) (see AGENCY; DETACHMENT/ATTACHMENT; REFLECTION).

From the point of view of human flourishing, distanciation may be benign or malign. In GLOBALISING post/modernity it is increasingly in evidence in the disembedding (involving absenting, shrinking or compressing) or EMERGENCE of space from time (a telephone conversation, mass mediatisation, 'the end of history'); of time from space (an aeroplane flight, conceptualisation); of space from place (a hypermarket, a virtual community, the spatialisation of culture and the city); and time from tense (abstract homogenised labour-time, such that 'time is money'). While the shrinking of SPACE–TIME confers some obvious potential benefits, it is fraught with problems for the human NEED for community, sociability and intelligibility. Today four-planar human being is distanciated both spatially (globalised) and – notwithstanding ENDISM – temporally (the massive presence of the past) as never before, presenting opportunities (as well as risks) for humanity to move beyond its $power_2$-stricken phase, including for the vast stretching of the moral imagination (*moral distanciation*) that this will require.

distraction (f. L. *distrahere*, to draw apart) pertains to degrees of spatial spread, e.g., those between local, regional, continental and global analysis (*D*: 107). Cf. DISTANCIATION, which may be temporal as well as spatial.

diversity. See ALTERITY/CHANGE; UNITY.

dφ. See DIALECTICAL PRAXIS; TTTTφ.

domains of the empirical (or **subjective** or **semiosic**), **the actual and the real**. See REALITY.

double aspect theories hold that mind and matter are attributes of a single substance (e.g., Spinoza, Strawson). They 'suffer from an ambivalent ontological commitment. If the mental attributes are real (irreducibly causally efficacious) then they amount to SEPM; if not to CSM' (Bhaskar 2003a: 387).

double tendentiality. See TENDENCY.

doxa (adj. doxastic). Used by some philosophers since Plato to distinguish mere belief or opinion (Gr. *doxa*) from genuine knowledge (Gr. *epistēmē*). For Foucault, an *episteme* (adj. epistemic) is 'the underlying structure of a discourse, the fundamental concepts within which all thinking in {a} particular period is determined' (Benton and Craib 2001: 180). It is to be distinguished both from *epistemology* (adj. epistemological, see KNOWLEDGE, THEORY OF) and *problematic* and *PROBLEM-field*. See also IDEOLOGY.

doxastic modality. See MODALITY.

DREI(C) model of pure scientific explanation. See EXPLANATION.

dual, duals. See DUALITY.

dual control, **dual determination**. See DEPTH; DETERMINATION; FREEDOM, ETC.; INFRA-/INTRA-/SUPERSTRUCTURE.

duality and **dualism** (f. L. *dualis*, of or pertaining to two). Related but distinct concepts pertaining to 3L. A duality is a combination, exemplifying non-arbitrary *dialectical connection* and *dialectical distinction* (dd′) of existentially interdependent and essentially distinct correlatives, e.g., *theory and practice*, 'so that typically, but not always, one may be seen under the aspect of the other' by PERSPECTIVAL SWITCH (*DG*: 397). Thus practice may be seen as theory because it is quasi-propositional, and theory as practice because it is work on and with material causes. This is the *duality of theory and practice*. Duality is thus closely affined to CONSTELLATION (cf. TOTALITY) and CO-PRESENCE: co-presence ≥ constellationality ≥ duality. The elements of a duality are normally also elements of a constellation, but a constellation is often more far-flung. Thus subject and object constitute a duality (subject–object) located within a constellation, viz., there is a constellational UNITY of (1) the unity of subject–object (2) within subjectivity (level of ontogeny) (3) within an overarching objectivity (level of ontology). The greater reach of constellationality is not gainsaid by the fact that (1) and (2) together also constitute a duality or unity, as do (2) and (3). Moreover, constellationality embraces identity in addition to unity. Like constellationality, duality is deployed in both materialist/realist non-pejorative and idealist/irrealist pejorative ways; and its elements may be non-antagonistic or antagonistic. Non-pejorative uses include dualities of: absence or absence–presence (what is present at or from one ontological level, spatio-temporal stretch or perspective may be absent at another); agency–structure, beliefs–meaning; intentionality (at once epistemological and practical, focused on objects and objectives); concrete universal ↔ singular (a *concrete duality*); structure (outcome and condition of agency); praxis (conscious production, unconscious reproduction/transformation); praxis–poiesis; theory–critique; theory–practice; subject–predicate; TD–ID; and truth qua expressive-referential dual. Pejorative uses include the epistemic–ontic fallacy; the induction–transduction problem (within a Humean problematic); fetishism of conjunctions–reification of facts; and fetishism/reification–de-agentification of the knower.

150 duality of theory and critique

The irreducibility of the distinctness of the elements of a duality is called a *hiatus* (L., a gap, break in continuity, etc.), constituted by lapsed time (either past activity or inaction) or spatial distanciation (see also NEGATIVE GENERALISATION). This feature of dualities is underlined by the concept *duality-with-a-hiatus* or *hiatus-in-the-duality*, e.g., the *tensed hiatus* or quantum leap from agency to structure. A *bipolarity* (L. *bi-*, twice, two + *polus*, pole, the quality of having two poles or opposite extremities) or *polarity* is a duality in which the two elements are opposites, e.g., EMERGENCE is the *bipolar opposite* of absence, or its *positive bipolar*; the next step conceptually is SPLIT, with which bipolarity is sometimes identified. Thus we have: co-presence > constellationality > duality > bipolarity > split. In terms of the TD/ID distinction, duality is aligned with INTRANSITIVITY, split and ALIENATION (or rather, their *possibility*) with transitivity and REFERENTIAL DETACHMENT (see also *EW*: 24). Bivalence and *HOMONYMY* are synonymous with duality when its elements are not opposed. *Ambivalence* (see VALENCE) is also sometimes synonymous with duality, but is normally used in a sense in which the elements are incompatible; it is more closely related to DUPLICITY. *Dual* (n.) in its non-pejorative sense either substitutes for duality (as in 'truth as an expressive-referential dual') or refers to the elements of a duality, which are then duals of each other. A *bipolar dual* is a bipolarity. *Irrealist duals*, like DIALECTICAL ANTAGONISTS or pairs, which they embrace (dual > dialectical pair), are grounded in a common category mistake, but unlike them are complementary rather than ostensibly antagonistic.

Whereas in its non-pejorative meaning duality encompasses a valid categorial distinction, *dualism* embodies a false one resting on the categorial error of FISSION or split; it thus constitutes a DIALECTICAL PAIR with MONISM and REDUCTIONISM respectively. (1) Any view (*epistemological dualism*, *ontological dualism*), prevalent in IRREALIST thinking, especially from the onset of modernity, that sees the world ANTINOMIALLY or in terms of two disparate parts (*binary opposites*, *dichotomies*), e.g., nature/culture, nature/spirit, reason/affect, reasons/causes, fact/value, material/mental, body/mind, noumena/phenomena, waves/particles. An apparently unbridgeable gulf, not a hiatus-in-duality, is posited between the two different orders. (2) The world so divided, e.g., any dichotomous conceptual structure, a master–slave-type social structure or, more generally, DEMI-REALITY. At the 4D level of agency, dualism's denial of intentional causality is vulnerable to the DILEMMA that its very statement is 'an intentional causal act with causal effects' (*P.* 244). In Archer (1995, etc.), *analytical dualism* is a mode of analysis that respects the hiatus-in-duality of agency–structure and other *dualities*.

In PMR, dualism pertains to demi-real relative being, duality to relative being, identity to absolute being. Demi-reality is a constellational unity of all three, relative reality of the last two. *META-REALITY* constitutes the constellational identity of duality (relative reality) and identity (absolute reality). There would be no dualisms or antagonistic dualities in a eudaimonian society. In PMR duality in its antagonistic or oppositional sense is often synonymous with dualism.

duality of theory and critique. Applies wherever the subject-matter of theory is already conceptualised under some description, e.g., a social form or a scientific theory. The explanatory function of theory then entails criticism of the social form as inadequately conceptualised. This means that SOCIAL SCIENCE as such is EXPLANATORY CRITIQUE, which natural science will also practice in so far as it gives

duality of theory and practice

causal accounts of inadequate theories. In virtue of the *duality of theory and practice*, what holds for theoretical reasoning also applies to practical reasoning arising from the frustration of some desire, want or interest: it logically presupposes both explanation and critique of the causes of the frustration, together with a theory of how to remove them (see EMANCIPATORY AXIOLOGY).

duality of theory and practice. See CONSISTENCY; DUALITY; DUALITY OF THEORY AND CRITIQUE; EMANCIPATORY AXIOLOGY; PERSPECTIVE.

duality of truth. See DUALITY; HOLY TRINITY; PRIMACY.

duplicity (f. L. *duplicare*, to fold in two, double). In ordinary English the quality of being double in action or conduct (double-dealing, deceitful); and the quality of being double simpliciter (doubleness). It is closely affined to – indeed often synonymous with – *equivocation* (f. L. *aequivocare* [*aequs*, equal, the same + *vocare*, to call], to call by the same name). (1) The use of words in more than one sense, especially ambiguously or with a double meaning; and (2) doing this in order to mislead. (To be distinguished from *equivocity* [f. L. *aequs* + *vocare*], which in Bhaskar normally means having two or more senses, equally appropriate, as in 'there is an *equivocity* of *freedom from* and *freedom to*' [*D*: 260].) Mostly Bhaskar exploits both senses of duplicity/equivocation to refer in particular to the point of identity or con-FUSION in subject–object [epistemic–ontic] IDENTITY (or equivalent) THEORY which becomes a point of internal duplicity as the anthropocentric ANTHROPIC and EPISTEMIC FALLACIES define the world in terms of human knowledge, and the world so defined returns in the reciprocating anthropomorphic ontic fallacy to naturalise knowledge, e.g., in the fetishism of facts and their constant conjunctions. In this *duplicitous* anthroporealist exchange, the natural world is humanised and the human world naturalised. Duplicity thus contrasts with DUALITY in its non-pejorative meaning – which does not collapse but sustains the distinction between epistemology and ontology, the TD and ID; particular instances of duplicity are, however, *duals* in the pejorative sense. Formally, duplicity is manifest as ANTINOMY. Substantively, since a complete identity theory is practically (transcendentally and axiologically) unachievable, duplicity passes over into *ideological plasticity* (which may be thought of as duplicity generalised or multiplied – 'susceptibility to a multiplicity of incompatible positions' [*P*: 243]), generating TINA COMPROMISE FORMS as reality causally impinges. (The related figure of *pliability* normally pertains to structures [products], and *elasticity* to structures in process [geo-historicity] [cf. stretching, DISTANCIATION].) Ideological plasticity is basically secured in a non-virtuous circle or bad infinite: illicit fusion produces an internal duplicity which generates fission or split; this in turn necessitates external *grafting* or supplementation, hence illicit fusion and duplicity, leading to further split. The internal structure of any identity theory (e.g., positivism) will therefore depend upon equivocation. While they may coincide, simple internal duplicity and external complementarity or grafting are to be distinguished from the *dialectical duplicity* of antagonistic but tacitly *complicit* counterparts, e.g., positivism and interpretivism, which are themselves likewise grounded in illicit fusion/fission (see DIALECTICAL ANTAGONISTS). Duplicity and equivocation are related to *ambivalence* (proneness to, rather than equivocation between, two or more incompatible positions [see VALENCE]); occasionally they are used synonymously. See also OPPOSITE.

dyad, dyadisation (primary). See POLYAD.

E

ecofeminism. See FEMINIST THEORY.

ecological asymmetry or the **relational asymmetry**. So called because it is vividly exemplified in the relation of organism to environment, which may include individual to society or more generally subject to object, part to whole: when the two clash 'it is the species, subject or part, not the environment, object or whole which "gives" or goes under'. It is thus a 3L figure. Embedded within ENTROPY, it supports the postulate of the *methodological PRIMACY of the pathological*, espoused also by Marx, Durkheim, Freud and others, which brings out 'just those features of successful action or adaptation which the very success of action tends to elide or obscure', thereby throwing into relief or rendering transparent what needs to be explained (*SR*: 140–1) (cf. CONTRAST EXPLANATION). Mature post-Darwinian understanding of the ecological asymmetry involves what the human species, in the hunter-forager moment of its existence, has long since known: that we belong to/were made for the world rather than that the world belongs to/was made for us, 'and that we survive as a species only in so far as second nature respects the overriding constraints imposed upon it by first nature' (*SR*: 221).

ecology. Introduced in 1866 by the German biologist Ernst Hackel (1834–1919), the concept of ecology refers to the scientific study of the distribution and abundance of living organisms and how these properties are affected by interactions between the organisms and their environment. These organisms and environments can be studied at different levels, ranging from *physiological ecology* and *behavioural ecology* (the adaptations of individual organisms to their environment), via *population ecology* (the dynamics of the populations of a single species), *community ecology* (interactions between species within a single community), *ecosystem ecology* (flows of energy and matter through the biotic and abiotic components of ecosystems), to *landscape ecology* (processes and relationships across multiple ecosystems or very large geographical areas). Complexity, understood as the number of entities and processes in the systems studied, is most pronounced at the latter, higher levels. Ecology can also be sub-divided according to the species focused on, or to the type of ecological formation studied (e.g., arctic ecology, desert ecology, etc.).

Historically, the ecology concept has, however, been applied in many other contexts. In reductionist natural sciences there have been traditions that consider ecology only as a subordinate discipline within biology. With the advent of modern environmental discourse in the 1960s it became more common to consider ecology as an overarching concept (Commoner 1971), but with the later strong focus on genes and genetics within biology certain reductionist traditions again sidelined ecology as a minor discipline. Ecology has also been applied purely as a social science concept. The so-called Chicago School of sociology used the term urban ecology, and applied ecological principles in their studies of urban people and life. Subsequently the concept of ecology has also been used to express complex social interrelations at the level of social structure. Such uses are less frequent today. It has become more common also in the social sciences to apply the concept of ecology – and even such a term as ecopolitics – in the same way as in the dominant environmental discourse.

A main principle of ecology is that every living organism has an ongoing and continual relationship with every other element that makes up its environment. An *ecosystem* can be defined as any situation where there is interaction between organisms and their environment. Humans are no exception to this. Along with humanity's increasing interferences with the natural environment, resulting from technological development, population growth as well as social mechanisms tending to include increasingly large proportions of nature into the circuit of commodity production, human ecology has emerged as a sub-field within ecology focusing in particular on the interactions between human societies and the natural environment. Such interactions can again be studied at different levels of aggregation, e.g., individual households, cities (cf. the research field of urban ecology), regions, or the global population. Because of the increasing use of non-local resources and the regional and even global impacts of some local human activities, the contexts at higher geographical levels must, however, be included also when studying the ecology of households and local communities.

As evident from the above, ecology is a holistic science, drawing heavily on and overarching a number of more 'narrow' sciences such as biology, geology, physics, meteorology and – in the case of human ecology – also the social and human sciences. It depends on INTERDISCIPLINARY integration of insights from a number of specialised fields. It involves several of (and in the case of human ecology arguably all) the strata of which – according to critical realist ontology – reality is comprised. The complexity of relationships and interactions implies that ecological SYSTEMS, in particular those at higher levels, must be considered as OPEN SYSTEMS where the multiplicity of causal powers and liabilities precludes one from making certain and precise predictions about the changes in the distribution and abundance of living organisms in an ecosystem resulting from a change in one of its internally or externally related components. This does not mean that likely impacts on ecosystems owing to changes in some causal factors cannot be anticipated, for example in the composition of vegetation and animal life within a given region due to a given increase in the annual mean temperature. However, such predictions can never be very precise. Moreover, owing to the multitude and complexity of causal powers involved in an ecosystem, a large proportion of these factors remain unknown: there is, in the case of ecosystems, a considerable gap between the intransitive and the transitive dimension of reality. Environmentalists have therefore advocated the so-called *precautionary principle*, according to which lack of full scientific knowledge should not be used as a reason for abstaining from or postponing measures to avoid environmental degradation where a danger of serious and irreversible damage exists.

As a scientific discipline, ecology does not in itself say anything about which types of ecological situation should be considered desirable or undesirable. Combined with ethical and political values, ecological knowledge is, however, a base for environmental policies. The concepts of environmental degradation and ecological crisis refer to ecological situations conflicting significantly with such values. The philosophical sub-discipline of *environmental philosophy* (or ecophilosophy) deals in particular with ontological and ethical aspects of the relationship between humans and nature, but also the topic of a fair distribution of limited natural and environmental resources between humans across and within generations (A. Næss 1989).

To a great extent, environmental degradation and ecological crisis have been the unintended side-effects of human activities and technologies aiming to improve human welfare and economic growth. Many technologies for repairing or reducing negative ecological consequences have also turned out to produce other, unintended ecological damages, typically by relocating impacts (temporally, spatially and topically). The creation of unintended environmental problems, as well as the belief in 'technological fixes' as a solution to ecological problems, is partly a result of a reductionist conception of reality, where the complexity of causal powers and mechanisms operating in open systems is largely ignored. In particular, such fragmented analyses are typical for neoclassical environmental economics, where both the ecological impacts and their causes are looked at in separation from their environmental and social context. (Such analyses could also be characterised as 'IDEOLOGICAL' in the sense that their set of ideas are false and at the same time more or less contingently necessary in order to legitimate the belief that economic growth can be 'decoupled' from ecological degradation [P. Næss 2006b].) Distinct from this fragmented approach, CR considers that physical, biological, psychological, psycho-social, socio-economic, cultural and normative kinds of mechanisms, types of context and characteristic effects are all essential to the understanding of the phenomena in fields such as the study of ecological problems (Bhaskar 2006). See also HOLISTIC CAUSALITY; REFLECTION; RELATIONALITY; SUSTAINABILITY.

Petter Næss, Karl Georg Høyer

economic history became established as a distinct area of specialisation in the early twentieth century, and up to the 1950s what is now termed 'old' economic history dominated. This involved a narrative style, along the lines of traditional historical scholarship, providing an interpretive approach that took into account the wider socio-economic context while concentrating on economic phenomena and placing a greater emphasis on straightforward quantitative evidence and descriptive statistics relating to economic phenomena.

From the 1950s on, however, 'new' economic history came increasingly to dominate the discipline. Cliometrics, as it is sometimes known, focuses on data measurement and explaining the development of economic phenomena over time through the use of econometric modelling techniques. The classic statement of the New Economic History is provided by Fogel (1966), who emphasises the need for a 'scientific' approach based on logical positivism and the nomological-deductive methodology. Mathematical modelling techniques require strictly functional relationships between variables. In practice this means that the discipline lends itself to neo-classical assumptions about human nature, such as self-interested utility maximisation, whereby human behaviour can be simplified down to strict regularities of the 'whenever a then b' type. Modelling requirements for 'hard' quantitative data also mean that the discipline tends to downplay, or simply ignore, qualitative evidence and non-economic causal mechanisms and processes.

The regularity determinism at the core of mainstream work in economic history leads to two widespread and related errors. First is the regularity error of thinking that a law must always reveal itself at the level of the empirical. So when economic historians think they are dealing with a well-grounded regularity, such as the idea that the direction and volume of migration are determined mainly by wage differences, then they tend to believe that that must prove to be the case in all circumstances and rarely if ever consider that non-economic factors could be dominant in significant instances.

The corollary to the regularity error is the error of pseudo-falsification (see *RS*). This is to assume that if a set of circumstances occurs in which a law does not apparently hold, then that law is invalidated *in toto* for all circumstances. For example, in debates about the Prebisch–Singer thesis (that the terms of trade for primary products tend to decline relative to those for manufactures) the citing of particular (brief) historical periods when the terms of trade for primary products have risen is taken as all that is required to prove that the tendency does not exist (Pinkstone 2002). For critical realists, however, the absence of evidence for a tendency that is thought to be well grounded need not be the end of research but, rather, the trigger to a search for countervailing forces that may have disrupted or blocked the tendency on the occasion in question.

A CR approach to economic history may well hark back to the 'old' economic history, in that it would be more likely to take on board the role of non-economic causal mechanisms in historical explanation. In contradistinction to the 'new' economic history, it would be less likely to use modelling techniques that exclude evidence which is difficult to quantify. In addition, by explicitly recognising the open nature of the social world, a critical realist economic history would also make a clear distinction between the role of persistent structural forces and historically specific factors in the explanation of particular historical questions. See also Pinkstone 2003a.

BRIAN PINKSTONE

economics demands mathematical or econometric modelling as prerequisite for serious 'science' and hence can be seen as the home of positivism in the social sciences. Textbook or core price theory (microeconomics) remains based upon marginalism, (ordinal) utility theory, and perfect competition. In this respect the discipline remains linked with the founders of neo-classical economics, Menger (1871), Walras (1874) and Jevons (1871). However, the contemporary discipline has developed away from certain core theoretical assumptions, making it in some respects amorphous at the level of substantive theory (in contrast to its uniform insistence on mathematics and econometrics, at the level of method). For example, whereas once isolated from other social sciences through assuming perfect competition, economics is now 'colonising' other social sciences by 'theorising' institutions as rational responses to market imperfections (e.g., information asymmetries). This 'economics imperialism' is associated with 'new institutional economics', 'new political economy', 'new economic sociology', and so on (Fine 2002). Macroeconomics, which became a separate half of the discipline under the influence of Keynes (1936), has been thoroughly incorporated within the dominant individualist framework. The original emphasis given by Keynes to the role of aggregate demand has virtually disappeared from mainstream macroeconomics.

There are several dissenting schools of economic thought, though these remain on the very margins of the discipline. Post-Keynesian economics stresses aggregate demand, history and uncertainty as central to Keynes and to the comprehension of the economy. Austrian economics, founded by Menger, retains utility theory but rejects the static, closed system nature of mainstream economics. Feminist economics embraces diverse perspectives often highly critical of mainstream economics. (Old) Institutional and Evolutionary economics, founded on the work of Veblen (1899) amongst other authors, stresses the importance of social context. Marx (1867) and Sraffa (1960) respectively lend their name to two further schools of thought.

Since the late 1980s, Tony Lawson, followed by a group of critical realists originally based at Cambridge University, has developed CR within economics. The 'Cambridge group' have had a strong impact within heterodox economics and within the sub-discipline of economic methodology. Lawson (1997, 2003b) *defines* mainstream economics by its 'deductivist' insistence on closed-system mathematical and econometric modelling. Drawing upon the CR analysis of closure conditions, Lawson argues that deductivism leads to the absurd individualist fictions of mainstream economics.

Lawson and other members of the Cambridge group argue that heterodox schools of thought generally employ an open-system approach (with the exception of Sraffian economics) and hence can unite under the banner of CR. There has been a general, though cautious, acceptance of CR within post-Keynesian economics. Within Marxist economics, as within Marxism more broadly, CR has become an important contributor to debate (Brown, Fleetwood and Roberts 2002). The CR engagement with Institutional, Feminist and Austrian economics has generated ongoing debate within each of these schools. Certain CR perspectives within economics have developed that are distinct from the Cambridge group. For example, some CR economists argue that the Cambridge group goes too far in its critique of the use of mathematics and econometrics. These economists emphasise the fluidity, as opposed to rigid dualism, of the closed/open system distinction (Downward [ed.] 2003).

If the prospects for CR within heterodox economics appear to be bright, the prospects for survival of heterodox economics within the economics discipline are, by contrast, uncertain. The increasingly broad substantive scope of the discipline has not led to a more tolerant or enlightened methodology; rather it has tightened the positivist straitjacket. See also Downward, Dow and Fleetwood 2006; Fleetwood 1999a; Lewis 2004.

ANDREW BROWN

ecstasis (f. Gr. *ek* + *histanein*, to place outside). Displacement; (of the mind) distraction, astonishment; a trance (hence Eng. ecstasy). Used by Bhaskar, in a probable appropriation of Heidegger's *ek-stasis* as humanity's (*Dasein*) standing outside itself stretched towards the future, to mean (a particular kind of) 'orientation to': 'The past brings possibilities and openings for the future and the best "ecstasis" or orientation to

education 157

it is one of the *creative transformative* use of these' (*D*: 135). See also AGENCY; CREATIVITY; REFLECTION.

education. See STUDIES OF EDUCATION.

eduction. See INFERENCE.

efficient cause. See CAUSES.

ego, egoism. See ANTHROPISM; CENTRISM; DEMI-REALITY; ESTEEM; PDM; SELF; SOLIPSISM.

ego-(ethno-anthro-)present-centrism. See ANTHROPISM; BLOCKISM/ PUNCTUALISM; INDEXICALISM.

eidos (Gr., f. *eidein*, to see: that which is seen, a form, shape, figure; a kind or species), adj. eidetic. In Aristotle, *eidos* denominates 'formal cause', in Husserl 'essence'. In *eidetic ACTUALISM* (or *eidetic identity* theory), the actual is either a direct manifestation of transcendent forms (Plato) or things are ultimately identified with the actualisation of their potentialities (Aristotle, Hegel). The FOUNDATIONALISM which underpins the Platonic–Aristotelian fault line and ultimately the UNHOLY TRINITY is thus an *eidetic foundationalism*. *Eidetic eternity* (*D*: 48, 96) refers to the ETERNISATION of conceptual forms. An *eidetic memory* is a 'photographic' one. CR is an immanent *eidetic realism*.

elasticity. See DUPLICITY.

elenchus. Socratic progenitor, modified and systematised by Aristotle, of *DIALECTICAL ARGUMENT*, which proceeded dialogically and immanently by exposing and repairing inconsistencies in a position, and is differentiated from Sophistic *eristic*, which aimed at rhetorical success, by its orientation towards the disinterested pursuit of truth and groundable ideals (*D*: 16, 100). As an adjective, eristic usually means 'controversial' or 'debatable'. See also INFERENCE (dialectical argument).

eliminative materialism. See EMPIRICISM; MATERIALISM; MIND.

emancipation. The transformation from unwanted, unneeded and/or oppressive sources of determination to wanted, needed and/or liberating ones (*SR*: 171), both carried through by AGENCY and establishing the conditions for its fuller development (cf. Lacey 2002b). It is paradigmatically structural: 'transformation *in* structures rather than a marginal adjustment of states of affairs' and 'transformation *to* other (needed, wanted and empowering) structures rather than to a realm which somehow magically escapes determination' (*R*: 187). In PMR it is conceived, centrally, as DISEMERGENCE or shedding of sources of determination which are false to our essential natures. See also EMANCIPATORY AXIOLOGY; FREEDOM, ETC.

emancipatory axiology (EA) or the **axiology of emancipation**. The *axiology of freedom*, incorporating both moments of the real definition or alethia of dialectic as, formally, 'the absenting of absence' and, socio-subtantively, 'the process of the development of freedom' (*D*: 176). More fully, it is (1) *formally*, the theory of the possibility of a planetary society of unity-in-diversity in which the free flourishing of each is a condition of the free flourishing of all (the moral ALETHIA for humans); (2) *substantively*, totalising depth-praxis, including research, tending in that direction and implied by the theory; and (3) the dialectical cross-fertilisation of (1) and (2), issuing in DIALECTICAL REASON or the coherence of theory and practice in practice. It embraces the *dialectic(s) of emancipation* or *freedom*, which, along with metacritics, is the main class of 4D dialectic (involving, however, all the other moments of the causal

158 emancipatory axiology (EA) or the axiology of emancipation

chain, too, such that the 'most natural' way to present the whole DCR system of categories is in terms of the dialectic of freedom [*D*: 301–3; *P*: 168–9]). The archetypal form of this dialectic is given in the sequence:

> elemental desire – referential detachment – constraint – understanding of the causes of the constraint – dialectic of solidarity (immanent critique and dialectical universalisability) – totalising depth-praxis – emancipatory axiology (*P*: 166–7).

Metacritically, Bhaskar locates it within 'a conative drive to freedom linking Descartes to Kant to Hegel to Marx to dialectical critical realism', each remedying omissions in the predecessor philosophy (*D*: 335). He sees it as a 'radical extension, generalisation and deepening of Hegel's dialectic of freedom, and {of} Marx's own radicalisation of Hegel's dialectic of reconciliation, mutual recognition or forgiveness into a DIALECTIC OF DE-ALIENATION' (*D*: 377; see also MASTER–SLAVE) and as a meta-theoretical exercise in CONCRETE UTOPIANISM which attempts, not to predict the future in historicist fashion, but to inform it by articulating the tendential rational DIRECTIONALITY of geo-history (*D*: 279). Socio-substantively it constitutes a DIALECTICAL (explanatory) ARGUMENT (*D*: 307).

EA is entailed by the theory of explanatory critique (EC), or the CR form of ETHICAL NATURALISM, and therefore presupposes that EC is correct. This holds that, contrary to Hume, it is logically possible to get from purely factual premises to value conclusions, so providing a grounding for ETHICS. While its validity has been questioned strongly by some critical realists (see EXPLANATORY CRITIQUE), its correctness is here assumed. For a response to the main criticisms that have been made, see CRITICAL NATURALISM.

The dialectics of agency and discourse

Once we abandon Humean fact-value Manichaeism in favour of the dialectical unity or coherence of theory and practice, it is easy to see that there are, in fact, more direct transitions from facts to values and praxis than that afforded by EC. (1) To criticise a theory or belief, even at the elementary level of technical rationality, is *ipso facto* to criticise any *action* informed by it as wrong, CP (see CRITIQUE). (2) Human action (AGENCY) is itself paradigmatically driven by DESIRE, and 'analytic to the concept of desire' is that we have an interest in removing the constraints on it, including constraints$_2$, i.e., in being 'autonomous or self-determining in some relevant respect' (*P*: 141) – constraints which, from this perspective, are ills, as well as 'falsehoods to concretely singularised human nature' (*D*: 281). (This is the universal moment of the *pulse of freedom* – 'the latent immanent teleology of praxis' [*P*: 154]. It is in this sense that people *as such* are free: it is in their nature to desire freedom – a theorem without which EA fails.) (3) A transition from facts to values, theory to practice, is intrinsic to any judgement, more generally speech act, e.g., a truth judgement: '"X is true" entails "act on X (in appropriate circumstances) CP"'. It is just this that makes applied science, and indeed any rational or even intentional (speech or other) action possible' (*SR*: 183). Values are *immanent to practices* '(as latent or partially manifested tendencies) {...}, or normative discourse is utopian and idle' (*SR*: 210).

emancipatory axiology (EA) or the axiology of emancipation 159

The logical implications of (2) and (3) provide us with a *formal* account of the nature of moral truth or alethia, which is 'universal concretely singularised human flourishing in nature' [*D*: 145]), and entrain the two basic dialectics of transition that formally drive EA and enframe the many other dialectics of freedom (which travel the same route): the *dialectic of agency* and the *dialectic of discourse*. *Substantively*, the general contours provided by this account have to be fleshed out by emancipatory movements and an *explanatory critical theory*[†] *complex* (EC†), consisting of ECs, exercises in CONCRETE UTOPIAS, and theories of TRANSITION and providing empirically grounded theories of the immanent possibilities of developing four-planar social being. It should be borne in mind, however, that the formal account itself embodies transitions from form to content, theory to practice (see below); that, in relation to emancipatory science and practical programmes, it is an exercise in both metaphysics α and metaphysics β – and in just the same way that CR PHILOSOPHY is in general – both *meshing in with* and *underlabouring* for them and elucidating and deconstructing/reconstructing the conceptual fields informing them; and that it is premised on the possibility of, and an invitation to, productive dialogue between social movements, the human sciences and philosophy in a triplex emancipatory helix. In view of this, complaints about the high level of generality and formality of the dialectics of freedom (e.g., Hostettler and Norrie 2003) seem misplaced. '[I]t is not an objection to a criterion that it is formal if it can inform a substantive one' (*D*: 170), and both types of metaphysics have an indispensable role to play in any conceivable social transition, or, philosophy must itself be transformed. EA expresses the *unity* or coherence of formal and substantive theory and practice in practice, dφ' or TTTTφ: '*transformed* (autoplastic), *transformative* (ALLOPLASTIC), *totalising, transformist* (oriented to deep-structure global and dialectically universal change) *praxis*' (*D*: 156, e.a.). More fully, TTTTTTφ: transformed, transformative, *trustworthy*, totalising, transformist, *transitional* praxis (*D*: 266). (*Autoplastic* is self-changing praxis, *alloplastic* other-changing; they are entailed by the other moments of this directional process.) Far from 'pulling global salvation out of the critical realist hat' (Sayer 2000a: 170 n), the theory of EA highlights that its realisation is down to us. In specifying wherein moral alethia consists, that the good society is possible and that depth-strugglers can come to know how to realise it – that 'every second of time {is} the gate through which the Messiah might enter' (Benjamin 1940: 226), – it constitutes itself as an integral rhythmic in this process. If it is objected that EA depends, not only on the logic of the dialectics of agency and discourse, but also on the entire scaffolding of dialectical CRITICAL NATURALISM, Bhaskar's answer is that both flow alike from the conditions of possibility of intentional agency (*D*: 303).

The dialectic of discourse

Analytically, any judgement or speech act has four internally related components or dimensions, the PERFORMATIVE qualities of being (1) *expressively veracious*, (2) *imperatival-fiduciary* (trustworthy), (3) *descriptive*, and (4) *evidential*. These are derivable from what is involved in (but not limited to) a truth judgement or 'truth talk' (deciding and saying that a proposition is true), which is axiologically necessary in that it 'satisfies a transcendental-axiological need, acting as a steering mechanism for language-users to find their way about the world' (*P*: 62–3), and entails REFERENTIAL DETACHMENT

160 emancipatory axiology (EA) or the axiology of emancipation

or the language-user's *absenting* herself or her speech act from the referent: (1) giving one's assent to it; (2) implying that it is reliable and can be acted on in appropriate circumstances; (3) claiming that it indicates the way things are; and (4) is in principle empirically grounded. These comprise the *judgement form*. They are loosely aligned to the moments of the causal–axiological chain, the theory of alethic truth, and the concrete universal ↔ singular, and to the logics of the *ethical tetrapolity* (see below) and of social transitions, as set out in Table 19. This is because they REFLECT the moments of the ontological–axiological chain, such that, in thinking the judgement form, we are thinking aspects of the moments of being–becoming.

Each of these moments is dialectically *UNIVERSALISABLE* as follows:

(a) *expressive veracity*: 'if I had to act in these circumstances, this is what I would act on';
(b) *fiduciariness*: 'in exactly your circumstances, this is the best thing to do'; [fiduciary-imperatival aspect]
(c) *descriptive*: 'in exactly the same circumstances, the same result would ensue';
(d) *evidential*: 'in exactly the same circumstances, the reasons would be the same'.
(*D*: 221, e.a.)

Providing 'the same circumstances' are understood in realist causal terms, i.e., normically and concretely, such that (c) is optimally grounded in (d), (c) and (d) are entailed by UBIQUITY DETERMINISM, or the principle of sufficient reason. (a) is implied by (b) – my advice to you is what I would do myself – and vice versa. Because of the concrete singularity of agents, the *imperatival* aspect of (b) is *assertoric*, not categorical in the Kantian sense (applying independently of the agent's situation); i.e., it is *sensitive* to the agent's particular wants, needs and interests, issuing in judgements that are not patronising and inimical to autonomy. The imperative and descriptive-evidential aspects together entail that the duality of theory–practice is built into or intrinsic to the judgement form. Given (a) and (b) and the internal relatedness of all four components, this means that in every act of judgement (and by extension 'every speech act' [*D*: 222]) we make an *AXIOLOGICAL commitment* to assist and empower the addressee as best we can wherever she is constrained in meeting her wanted needs: (a) stipulates and, in implying (b), presupposes, not only how the addressor would act in the addressee's situation, but also that she show *SOLIDARITY* with the addressee – more specifically, as explained above, 'assertorically imperatival sensitised solidarity' (*D*: 262). This entrains a *dialectic of solidarity* (or of *self and solidarity*) (presupposing a *dialectic of RECOGNITION* and commitment to the content of the EC† of her situation) to remove constraints on the addressee's flourishing, or, in other words, to absent her ALIENATION from anything intrinsic to her well-being; this dialectic is nicely exemplified currently within, e.g., Iraq Veterans against the War. Given that saying how the world is to offer implicit advice to act on that information, not to solidarise (consistently with other priorities of a balanced life, *amour de soi* and anti-substitutionism) is to commit performative contradiction and heterology (so immanent critique must be assiduously practised). There is no a priori limit to such solidarity, which logically must range from mere advice, to moral and material support, to engagement in *totalising DEPTH-praxis* (TTTTφ) to achieve the desired end, implying a conatus or tendency$_b$ to

Table 19 Components of the judgement form, with some correspondences

Causal Chain	1M Non-Identity	2E Negativity	3L Totality	4D Transformative Praxis
Judgement form	evidential (1M in so far as our judgement is optimally grounded)	descriptive (2E in so far as our object is changing)	fiduciary-imperatival (trustworthy)	expressively veracious
Theory of alethic truth – truth as	ontological, alethic (ID)	adequating (warrantedly assertible) or epistemic (2E and 1M) (TD)	normative-fiduciary (IA)	expressive-referential (4D and 1M) (TD/ID)
Concrete universal ↔ singular	universality	processuality	mediation	concrete singularity
Logic of ethical tetrapolity	(4) freedom qua universal human emancipation, the realisation of moral truth (alethia)	(2) explanatory critical theory† complex	(1) fiduciariness	(3) totalising depth-praxis
Logic of transition (social)	(4) emergence	(2) dialectical contradiction	(1) dialectical connection	(3) negation of the negation (geo-historical transformation of geo-historical products)

Note that human agents are themselves concrete universal ↔ singularities.

Source P: 168–9, D: 218–20.

knowledge of four-planar social being via EC†. Furthermore, the logic of concretised dialectical universalisability commits both addressor and addressee to removing all dialectically similar constraints, 'and, then, by a further step, to getting rid of all unwanted, unnecessary and remedial constraints as such, precisely insofar as they are dialectically similar, viz. as constraints' (*P*: 111–12). Combined with EC†, which it implies, the judgement form thus exemplifies the alignment of dialectical reason (the coherence of theory and practice in practice), ALETHIC TRUTH, theory–practice CONSISTENCY, dialectical UNIVERSALISABILITY and AUTONOMY in the sense of true to, for and in oneself and the other. In this way, the goal of concretely singularised universal emancipation – a society in which 'the free development of each is a condition of the free development of all' (Marx) – is implicit in every expressively veracious judgement:

> [expressive veracity → axiological commitment] → (1) fiduciariness → (2) content of the explanatory critical theory complex [= explanatory critique + concrete utopianism + theory of transition] ↔ (3) totalising depth-praxis of emancipatory axiology {= TTTTφ} → (4) freedom qua universal human emancipation (*DG*: 397; cf. *D*: 179)

(1) – (3) comprise EA in so far as it embodies *the practical presuppositions of discourse*; (2) ↔ (3) constitute a substantive two-way *dialectic of truth and freedom* (or *truth* ↔ *freedom*, for each is a condition of the other), mediated by phronesis, or the coherence of theory and practice in practice: dialectical reason; (1) – (4) comprise *the ethical tetrapolity*. (4) is the practical enactment of emancipatory axiology, or of dialectic as the pulse of freedom.

As indicated, the *dialectic of judgement* is a form of the more general *dialectic of discourse* or *speech action*. On this basis, Bhaskar suggests that a rapprochement with Habermasian discourse ethics is possible, but points out that the latter lacks an 'evidential' dimension, i.e., is not grounded transfactually at 1M, treating ethics as 'autonomous from science and history' (*SR*: 210; *P*: 142–3, 160; cf. *R*: 189). This can be made good within a CR framework. It by no means entails that the dialectic of speech action is the only component of CR dialectics of freedom.

The dialectic of agency

Just as discourse has practical presuppositions, so *praxis* has corresponding *discursive presuppositions*. Like that of discourse, the dialectic of agency is set off by (experience of) absence; or it is a *dialectic from absence to freedom* (*D*: 180), or of *desire to freedom*, more briefly a *dialectic of negativity* (*D*: 208). Thus we get it going by noting that, in seeking to absent an ill that is functioning as a constraint on the desire that animates our praxis, we are logically committed to remove all dialectically similar ills, CP. This entails absenting their causes, hence EC† and totalising depth-praxis. Theory–practice consistency in this process then requires that our actions and the theories informing them be both 'directionally progressive {2E} and universalisably accountable' in the sense that they are grounded transfactually at 1M, concretely at 3L, and agent-specifically or actionably at 4D (i.e., have regard for what other agents are disposed to and can do)

emancipatory axiology (EA) or the axiology of emancipation

(*P*: 141–2; *D*: 170). By the logic of dialectical universalisability we then move to the absenting of all remediable ills as such, which implies 'a commitment to absenting the absence of the society which will remedy them' (*P*: 147). The good society is thus implicit in every deed, commencing with an infant's PRIMAL SCREAM – 'a presupposition of the most elemental desire, the first initiating act of referential detachment, induced by negativity in the guise of absence' (*P*: 210).

The dialectic of agency thus has a fundamentally similar overall logic to that of the dialectic of discourse, such that we could rewrite the schema given above as:

> [desire to absent constraint (ill)] → logical commitment to absent dialectically similar ills] → (1) absent all ills as such → (2) content of the explanatory critical theory complex [= explanatory critique + concrete utopianism + theory of transition] ↔ (3) totalising depth praxis of emancipatory axiology {= TTTTφ} → (4) freedom qua universal human emancipation

This is because each dialectic expresses the other pole of the coherence of theory and practice – and, ultimately, of the concept of absence as both absent being and ill-being, mediated by the logic of dialectical universalisability – effecting a *transition*, or *ethical dialectic, from form to content* in two important ways: to the action entailed by solidarity and totalising depth-praxis and to the free flourishing of all as the content of the good society (*D*: 179–80). As acting and speaking social beings we cannot be true to ourselves (non-heterologous, autonomous) unless we work for such a society, nor can we truly flourish in a DEMI-REAL social context founded on a contrary ethic of spinning, twisting, stealing, lying and cheating, where to speak truth is to 'simplify' or 'rant' and public figures who profess honesty set examples for crooks. The deep content of these dialectics is later thematised as the *transcendentally real* or *alethic self*, the existence of which is in principle perfectly testable empirically; it is that which announces to the spinner that he is a fraud, suggesting a secular explanation for the phenomenon of *conscience*. Emancipation is thus a coming home to our selves, or SELF-realisation. Both dialectics entrain an array of principles that need to be observed, constituting limits to universalisability. These include cognitive non-triumphalism or awareness of the limits of our knowledge of the immanent possibilities of social being; principles of *ACTIONABILITY* or *feasibility* and of *PREFIGURATIONALITY*, incorporating a principle of means–ends consistency (these are entailed by the requirement for theory–practice consistency); a principle of *self-emancipation* or *SELF-REFERENTIALITY* (in virtue inter alia of the goal of autonomy); and, via the logic of dialectical universalisation itself, a principle of the indivisibility of FREEDOM.

Other routes of the dialectic of freedom

As already indicated, the dialectic of freedom embraces many other dialectics besides the dialectics of agency and discourse. Many are listed in the entry for DIALECTICS (Table 18); the twelve dialectics appearing under the dialectic of desire to freedom are so many different ways (among others) of arriving at the same result. They can be divided into four types. (1) *SYSTEMATIC DIALECTICS* – so called because they concern

164 emancipatory critique

the overall logic of EA and the ethical tetrapolity – embrace the dialectic of agency/discourse and its sub-dialectics of solidarity and recognition. (2) *Intrinsic dialectics* – so called because they are logically compelling (INTRINSIC or normative aspect) but not causally grounded (extrinsic or causal aspect) – are exemplified by the DIALECTIC OF EQUITY, which in its own right augments the argument for the dialectic of freedom. (3) *Intrinsic extrinsic dialectics* are *dialectics of LEARNING PROCESSES* which are both logically compelling and provide causal grounds for their implementation. They include the DIALECTIC OF MORALITY (see also ETHICS) which describes and redescribes actually existing morality, a *DIALECTIC OF THE EDUCATION OF DESIRE FOR FREEDOM* by the REALITY PRINCIPLE, a *DIALECTIC OF DE-ALIENATION* which restores wholeness or retotalises what has been detotalised by alienation and split, and a *DIALECTIC OF SOCIAL SCIENCE* itself 'from neutrality to explanatory critical rationality to emancipatory axiology to dialectic' (*D*: 289). (4) *Extrinsic intrinsic dialectics* are also dialectics of learning processes, but they are rationally and reflexively persuasive rather than logically compelling, and their leading edge is social structural. They are exemplified by the *DIALECTICS OF INTERESTS*, including *material interests*, which Bhaskar sees as a fundamental motor of totalising depth-praxis. The whole of EA may be seen as part of a more general *dialectical LEARNING PROCESS* or *dialectic of learning processes*. See also DIRECTIONALITY.

emancipatory critique. See CRITIQUE; EMANCIPATORY AXIOLOGY.

emancipatory movements. Collective actions by groups/people in society directed at transforming one or more aspects of social relations perceived by them to be oppressive, undesirable or unwanted, and/or create alternatives that are perceived to be more desirable. Movements for social change may or may not be emancipatory, depending on their orientations to the social structures within which they arise and the alternative values and visions that inform the movements. The PULSE OF FREEDOM, to greater or lesser degree, is the motor of emancipatory movements, whether or not it is articulated as such.

EMANCIPATION is the lynchpin that holds together a range of ideas, concepts and positions in CR, inter alia: STRUCTURE/AGENCY, MASTER–SLAVE-TYPE RELATIONS, EXPLANATORY CRITIQUE, EMANCIPATORY AXIOLOGY (praxiology), HEGEMONIC and counter-hegemonic struggles, EUDAIMONIAS and CONCRETE UTOPIAS, ETHICS, ALIENATION, and the emotional and SPIRITUAL dimensions of social life and HISTORY. The diversity of views on emancipation is evidence of its importance to CR (Collier 1998; Creaven 2003; Lacey 1997; Roberts 2002). Its application to specific questions in social science is an important emerging area of study within CR (Cruickshank, ed., 2003b, ch. 6; Dean 2003, Morén and Blom 2003b; Willmott 2003). However, emancipatory movements bring the concept of emancipation in philosophy and social theory down to the murkier terrain of politics and social practices.

It is generally accepted that CR does not take particular political positions; is compatible with a number of diverse political positions; and has a broad affinity with socialism in a general sense as the heir/legatee of emancipatory movements historically (*SE*: 190). While philosophy and theory entail abstractions, emancipatory movements are invariably context-specific. Evaluation of emancipatory movements is possible only when it is empirically informed; and this takes the inquiry to a different type and level

emancipatory movements 165

of engagement. None the less, it may be useful to identify some questions that emancipatory movements pose for CR.

Emancipatory movements embody a number of pulls and pressures. First, under capitalism, communities are formed around specific interests structured around market relations and instrumental rationality exercised through politics as the means for addressing conflicts of interests. Imperialism exacerbates the tensions by spatialising and racialising conflicts of interests (Collier 1992) and by counterposing positive values/practices in the 'West' to negative values/practices in the 'East' in the way social questions, e.g., human rights, are framed (Sen 1997). Consequently there is a hiatus between the philosophy and theory that affirms emancipation and human freedom on the one hand and political practices within the wider architecture of capitalism on the other (D'Souza 2002, 2004).

Second, the format of society under capitalism based on the formal and institutional separation of economy-state-civil society relations imposes a structure on social movements. The challenge for emancipatory movements is to transcend that structure in context-specific ways but informed by universal philosophical and theoretical insights, i.e., sustain diversity *and* unity in social practices. That in turn raises difficult and complex organisational questions for emancipatory movements.

Third, social movements embody tensions between immediate defensive actions and long-term actions for conscious structural change. Defensive actions are often necessitated by problems presented by a situation and solutions that are constrained by the situational dynamics. Long-term objectives of structural change require not only sharing common visions and values about alternatives, but also tailoring the short-term/immediate actions and goals in tandem with long-term objectives.

Fourth, the dialectics of emancipatory movements and social change entail periods of 'incubation' when oppression/structural constraints may undergo accommodation and toleration and periods of open revolutionary upheavals (Marx 1852). Perceptions and assessments of *both* periods call for judgements and evaluations based on participation in and engagement with social actions.

Fifth, emancipatory movements presuppose a spiritual disposition to act in certain ways and transcend the self in order to attain emancipation/liberation, e.g., as martyrs do in liberation struggles (Uberoi 1999). Emphasising one aspect, viz., structural, discursive, moral, etc., over the others leads to very different approaches to social movements (see, e.g., Giri 2001; Lacey 2002b).

Sixth, contrary to conventional socialist movements, the shift away from reductionist and instrumentalist readings of class and proletarian leadership in social change in CR calls for recognising that participants in emancipatory movements act as spokespersons for others. It becomes necessary therefore explicitly to acknowledge and debate the ethical and moral responsibilities entailed in that role, a task that the ethical/moral aspects of CR underlabours for.

Seventh, the institutional embeddedness of most CR research within the academy has ramifications for the values, approaches, priorities and objectives that inform research. In the postwar era, the boundaries between social science, an integral part of the academy designed primarily to produce knowledge for reifying and reproducing existing structures and relations in society, and praxiology as a distinct type of theorising informed by radical philosophy and theory but oriented towards social change, has been blurred considerably for reasons that can be explained sociologically. Consequently, praxiological questions are constrained by institutional limitations of the academy, on the one hand, and developments in social theory become inaccessible and de-linked from emancipatory movements and practices, on the other. The question becomes significant in an international context where material, social and intellectual disparities are stark, especially between societies of the 'North' and 'South', and at the same time interactions between privileged sections of those societies are extensive and facilitative.

The disparities and aforesaid questions warrant another tier of underlabouring so that, just as philosophy underlabours for the social sciences, the social sciences can in turn underlabour for praxiology by defining their roles in ways that make the developments in CR relevant to those engaged in emancipatory movements (D'Souza 2005, ch. 17). See also Collier 1999; Hartwig 2001; Hodgson 1999a, b, 2000.

RADHA D'SOUZA

emancipatory praxis is any praxis tending to EMANCIPATION. In its broadest sense it is synonymous with EMANCIPATORY AXIOLOGY ($d\varphi'$ or $\tau\tau\tau\tau\varphi$).
emancipatory social science. See IDEOLOGY; SOCIAL SCIENCE.
embedding/disembedding. See DISTANCIATION.
embodiment, embodied intentional agency, embodied person(ality). See AGENCY; PHENOMENOLOGY; SUBJECTIVITY; SOCIOLOGY OF THE BODY.
emergence. A concept currently defined in philosophy in terms of three characteristics. (1) that some substance, entity, property or system β is dependent for its existence upon some other substance, entity, property or system α; (2) that dependency implies some form of co-variance where fundamental changes in α mean fundamental changes in β; and (3) that the form, operation and consequences of β cannot be reduced to α. Thus, though (1) and (2) imply some form of relation that may perhaps be conceptualised as non-constant conjunction, or irregular, and/or multiply realisable causation, (3) makes the form of that relation conceptually problematic because irreducibility implies some form of disjuncture between α and β such that β cannot be translated, explained or predicted from α alone.

Though problematic, the concept is also highly attractive because it mediates between long-standing contradictory positions and problems in debates in philosophy of science, the nature–society relation and the philosophy of MIND. In *RS*, Bhaskar carries through the Copernican revolution in ontology by suggesting that the intelligibility of scientific knowledge and practice implies a DEPTH-reality of generative mechanisms with particular causal powers actualised in events and experienced and interpreted by people who, particularly from the artificial closures of laboratory experiment, construct scientific knowledge of what are otherwise open systems. Science conducts experiment and research whose effectiveness is related to its appropriateness to its domain (the ecological, the biological, the geological, the chemical, the physical/subatomic, the astrophysical/cosmological, etc.). Thus scientific knowledge (with obvious overlap) is differentiated by domain, but the distinction, in so far as it is appropriate, relates to the objects and processes. Some objects and processes are necessary to others and thus more basic. Accordingly, reality is stratified and emergent (*RS2*: 119), and this is what makes sense of the distinctiveness of domains of scientific knowledge, though distinctions are always provisional since they are subject to the possibility that new connections or even real reductions to more basic causal phenomena may be found. Bhaskar, therefore, makes a virtue out of the relation – dependency ambiguity of emergence by using it to make sense of our experience of the difference between domains, where in earlier applications it produced basic contradictions with the dominant empiricist–positivist framework of science. Mill, for example, in his *System of Logic* (1843), termed the concept – in relation to synergistic causal effects of organic chemical processes – *heteropathy*, but could not reconcile it to his empiricist model of induction. Once the concept of emergence is placed in terms of the ontology of a stratified depth-reality of open systems it gains a greater coherence. As part of a CR ontology it contributes to breaking down the explaining–understanding divide between natural and social science by providing a plausible basis for a naturalised argument for causal social being and for explanatory critique, and analytical distinctions between agents and structures and the partial TOTALITIES of societies. Reason, planning, devising and so on are emergent properties that must be studied in a way appropriate to them. An emergent account of consciousness itself is also highly plausible since it avoids the mysticism of dualism and the basic intentionality contradiction of reductionism. Again, within an overall ontology, basic framework contradictions associated with the work of earlier proponents are avoided, such as Broad (1925), who argues for a differentiation of the sciences on the basis that different objects are subject to different trans-ordinal laws, but persists in identifying higher-ordinal objects in physicalist terms, producing a materialist contradiction (see MATERIALISM). In DCR, while fundamentally a 1M and 2E concept, emergence satisfies all four moments of the ontological–axiological chain and the fourfold polysemy of ABSENCE as product (1M), process (2E), process in product (3L) and product in process (4D).

Disemergence is the inverse of emergence: 'the decay, demise or disjoint detachment of the higher-order level' (*D*: 51). In the sociosphere, it is a desideratum of the DIALECTIC OF DE-ALIENATION and EMANCIPATORY AXIOLOGY generally; in PMR such disemergence goes by the name of *shedding*.

JAMIE MORGAN

emergent falsity. See ALETHIA; CRITIQUE; DEMI-REALITY; DIALECTICAL ARGUMENT; IDEOLOGY.
emergent powers materialism. See EMERGENCE; MATERIALISM; MIND.
emergent totalities, principle of. See INTERESTS; PRINCIPLE OF TOTALITY.
emotion. See REASON.
emotivism. See ETHICAL IDEOLOGIES.
empirical, actual, real – domains of (d_e, d_a, d_r). See REALITY.
empirical invariance. SEE CAUSAL LAWS; CONSTANT CONJUNCTION; EMPIRICISM; EXPLANATION.
empirical realism (ER). The view 'that the world is constituted by the objects of actual (and sometimes possible) experiences' (*P*: 6, e.a.). First named by Kant, it has an extended meaning in CR. Every epistemology presupposes some view of what the world must be like for knowledge under its descriptions to be possible, hence some kind of REALISM. ER is the ANTHROPOREALIST ontology – 'of experience and

168 empirical realism (ER)

atomistic events constantly conjoined' (*RS2*: 221–2), hence of closed systems – presupposed by mainstream philosophy of science and social science, whether in the tradition of Humean empiricism/positivism or of Kantian and neo-Kantian transcendental idealism and more broadly the hermeneutical tradition (which adopts the Humean concept of a causal law for the natural world). It posits an IDENTITY or at least a correspondence between experiences and events, and events and causal laws, hence epistemology and ontology, such that $d_r = dr_a = d_e$ for being as such (positivism) or for the natural world only (other traditions), thereby inter alia making REFERENTIAL DETACHMENT (theoretically) impossible. From a CR perspective, where typically $d_r > d_a > d_s$ (where 's' is 'subjective', i.e. experience and concepts/semiosis) which may vary independently of each other, it thus involves intertwined category mistakes, combining an EMPIRICIST epistemology with an ACTUALIST ontology: the reduction of causal laws to CONSTANT CONJUNCTIONS of events and states of affairs (the *actualist fallacy*); and the reduction of events (hence real causal powers and laws) to experience (the EPISTEMIC FALLACY, issuing in the concept of the *empirical world* [*SR*: 45]). Both fallacies are horizontally as well as vertically uniformitarian, eliminating open systems (hence differentiation) and ontological stratification or DEPTH alike, i.e., issuing in a 'flat undifferentiated ontology' (*SR*: 250) and BLOCKISM and FIXISM. Since each is involved in the other, and the epistemic (together with the anthropic) fallacy is the more primordial mistake, 'epistemic fallacy' often does duty for both.

Since ontology cannot be reduced to epistemology, the epistemic fallacy 'merely covers the generation of an implicit ontology based on the category of experience {hence 'empirical'}; and an implicit realism based on the presumed characteristics of the objects of experience, viz. atomistic events, and their relations, viz. constant conjunctions' (*RS2*: 16) (hence 'realism') – presumptions strongly bound up with the quest for certain FOUNDATIONS for knowledge and above all epistemological individualism (*SR*: 242–3) (which attempts 'to reduce knowledge to an individual acquisition in sense-experience' seen 'as the neutral ground of knowledge that [literally] defines the world' [*RS2*: 18]). Empirical realism thus involves three key interrelated assumptions: the ubiquity of closed systems, entailing regularity DETERMINISM; a mechanistic model of action by contiguous contact, presupposing ontological EXTENSIONALISM; and an ATOMISTIC conception of the human subject (sociological individualism) based on 'an epistemological individualism in which men are regarded as passive recipients of given facts and recorders of their given conjunctions' (*RS2*: 16). While they were rendered plausible in the heyday of positivism by Newtonian cosmology and physics and the rise of possessive individualism, these assumptions have since been radically called into question by the Einsteinian revolution and the turn to intersubjectivity and ENTITY RELATIONISM, such that they now constitute a lapsed analogical GRAMMAR. ER is accordingly a conservative ideology for science which, unable to come to terms with the universe of modern physics, but 'spontaneously generated' by the social forms of market society (*R*: 192), cannot give an adequate account of scientific EXPLANATION or confirmation (and falsification), or think science in the TD as an ongoing social praxis in which knowledge is *produced* by means of knowledge, hence cannot sustain the rationality of scientific growth and change.

It is, of course, possible to reject empirical realism and still commit the epistemic fallacy, i.e., 'to define the world in terms of the possibility of *non*-empirical knowledge of it' (*RS2*: 38, e.a.). ER accordingly has a number of close anthroporealist relatives – apparent contraries which it often mutually presupposes (and vice versa). *Conceptual realism*, where $d_r=dr_a=d_{concepts/signs}$, is associated historically with the RATIONALIST tradition. It comes in two main forms. *Objective conceptual realism* (e.g., Plato, Descartes) holds that real objects are constituted, hence deducible, by thought or reason which, while presented as independent of human minds, is in reality modelled on human reason. In *subjective conceptual realism*, also known as *superidealism*, real objects are explicitly viewed as the products of human thought or discourse. For CR, concepts are of course real, but they do not exhaust or constitute social let alone natural objects. In the romantic strain of *intuitional realism*, real objects are 'identified wholly or partially in terms of human intuition, sensibility or affect' ($d_r=dr_a=d_{intuition}$) (*SR*: 8; see also ENCHANTMENT), and in *voluntarist* or Nietzschean *realism* in terms of human will ($d_r=dr_a=d_{will\ to\ power}$). In that all these realisms collapse the domain of the real to the actual, they are also, or incorporate, ACTUALISMS. While *TRANSCENDENT* or *ineffable realism* is not an anthroporealism in that it acknowledges a realm of pure other-being (e.g., Kant's thing-in-itself), deeming it inaccessible to human being, it must presuppose in practice empirical, conceptual or some other form of anthroporealism in relation to that part of being that is accessible. Moreover, all anthroporealisms (empirical, conceptual, etc.) presuppose transcendent realism, in that 'neither a purely transcendent nor a purely immanent analysis is sustainable' on anthroporealist assumptions, resulting in TINA COMPROMISE FORMATIONS which themselves must in part accommodate to reality understood in (re-described) TRANSCENDENTAL REALIST terms (*SR*: 9 f.); your anthroporealist in theory is necessarily a realist in practice (*RM*: 84–5).

empiricism is the position in epistemology that knowledge ultimately derives from sense experience. Its key early proponents include Locke (1690), Berkeley (1710), Hume (1777) and Mill (1843). Early empiricism rejects innate knowledge or knowledge a priori but is otherwise split between MATERIALISM and IDEALISM in matters of metaphysics on the basis of further argument concerning the nature of that sense experience. According to Locke, the properties of an independent reality of objects provides reliable sense experience to a mind that begins as a 'white paper'. Sense data provide that basis but are subject to the distinction between nominal essences, or the abstract idea of the object arising from what the senses observe, and real essences, or the object's insensible constitution (see NATURAL KIND). From this distinction Locke restricts knowledge to nominal essences. Berkeley radicalises this position on the basis that there can be no ideas without the mind. He defines properties or qualities such as colour, shape and tactility in sense experience as a set of ideas, and concludes that there is no reason to infer that material objects provide the basis of properties and qualities.

Critical realists (*RS2*: 1; *PN1*: ch. 4; Sayer 1984; Collier 1994) have identified numerous incoherencies in the implicit ontology of empiricism. The early empiricist focus is on the rejection of the legitimacy of knowledge about the possible mechanics of, or structures facilitating, sense experience according to the criterion that these cannot be experienced and thus cannot be known (only that which can be experienced through the senses can be known). The epistemology of empiricism is, therefore, ACTUALISTIC. At the same time, in both the materialist and idealist empiricist

170 empiricism

metaphysic, experience is essentially passive since it is provided to the senses. In epistemology, empiricism creates a criterion contradiction, i.e., how can one know that all that can be known is that which is experienced? If this is not itself an experience, does not the criterion contradict its own content?

In ontology the criterion not only commits the ontic and EPISTEMIC FALLACIES by simultaneously reducing experience to the passive givens of reality and reality to a theory of knowledge (of forms of nominal essences), but also creates further incoherence because empiricists selectively violate the criterion of that theory. In doing so, they refer to intransitive structures, practices or sustaining possibilities that fall outside actualism. Berkeley's idealist empiricism is constructed in terms of a prior theistic commitment and is essentially an attack on materialist atheism. He uses it to infer an unobserved supreme mind or God that provides and coordinates properties and qualities amongst its sentient creations. Locke's and Hume's violations take the form of failed resolutions of conceptual problems and acknowledgements. Locke's empiricism differentiates sense experience from the subsequent analysis of it to allow the further differentiation of knowledge and sense experience, and thus make sense of critical reason and the possibility of perceptual error. But he cannot persuasively reconcile how a blank-slate mind can critically construct progressive knowledge from given experience. Passivity thus gives way to structuring in terms of an implicit interaction that undermines his rejection of innate knowledge. This leads to inferences about the structure of the mind and to what was to become the Kantian problematic. Hume meanwhile reduces experience to particular atomistic sensory snapshots and claims these as the only facts. According to Hume, it is from these atomistic experiences that disposition, custom or habit constructs causal relations of two kinds – from regular CONSTANT CONJUNCTION of events (whenever fire then heat) and from expectations of usual subsequent events (physical violence followed by bruising) (see also CAUSAL LAW). This was to bequeath problems of a CLOSED-SYSTEM conception of reality to future accounts of scientific METHOD. Moreover, for Hume, disposition, custom or habit is no more an indication that the two events genuinely are conjoined, raising the PROBLEM OF INDUCTION, than of how they are conjoined, neither of which can be known. Yet Hume seems compelled to acknowledge that science has been effective in postulating unobserved entities, in hypothesising successively deeper causal principles and mechanisms, such as gravity, and in changing the very nature of disposition, custom and habit as they pertain to sense experience, which cannot then be passive, but must be a socially structured practice. From a CR perspective, therefore, he violates the empiricist criterion by acknowledging that science is not limited by the original epistemological criterion, and by acknowledging that reality is more than actualism and atomism, and that science has been effective in identifying this.

In the early twentieth century *logical empiricism* or *logical positivism* emerged, based on the early work of Wittgenstein (1922) and of key members of the Vienna Circle such as Schlick, Carnap, Neurath and Waismann, and populisers such as Ayer (1936). Positivism is broadly 'characterised by a *MONISTIC* theory of scientific development and a *DEDUCTIVIST* theory of scientific structure' (*P.* 18). While it has an ancient lineage, it was 'in its essentials fully developed by Hume' (Kolakowski 1966). The term itself was coined by Comte (1853). Logical positivism was *positivism* in the sense that science is the highest and only genuine form of knowledge. It was *logical* in that the new logic developed by Frege and others provided the analytical tool for showing both that science has universal methodological meta-characteristics and that non-science is meaningless. It was *empiricism* in that scientific content, and thus genuine knowledge, derives from sense experience. Following Wittgenstein, the contradiction that the original epistemological criterion itself was not a sense experience was squared by the argument that logic was an analytical necessary truth based on convention, but not itself an element of the world or synthetic truth (a form of eliminative materialism). The world consists of states of affairs of which the object-language provides logical corresponding pictures via observation. An observation provides a corresponding protocol or atomic sentence ('this grass is green'), and accumulations of them provide the basis for hypotheses ('grass is green'). A hypothesis is either verified or subsequently refuted by the accumulation. 'Grass is green' can thus be true or false but not, in this form of empiricism, meaningless. 'God is great', meanwhile, is literally nonsensical. The focus of philosophy thus became philosophy of science with an agenda to reduce the scope of philosophy by rejecting as genuine knowledge metaphysics, evaluative aesthetics, and ethics. Attention also shifted to the problem of distinguishing between models of scientific method, based on problems of deduction and induction, and between different scientific theories. Much of Carnap's and Hempel's seminal work in this area was published in the early volumes of the *Minnesota Studies in the Philosophy of Science*. In Bhaskar's view 'the supreme "logical positivist", *avant la lettre*' is in many respects, ironically, Hegel (*D*: 90). *Ultra-positivism* (or *ultra-pragmatism*) is the view that theory is the direct expression of practice (*SR*: 174).

Empiricism is extremely important for a number of reasons. Its forms have provided a major strand in modern Western philosophy. From a CR perspective, logical empiricism shares the basic deficiencies of early empiricism. It is actualistic, atomistic (ontologically EXTENSIONALIST), overly focused on regularity and closure as the basis of CAUSATION, and provides no clear conception of how experience and knowledge-seeking are contingent and socially situated. Though logical empiricism has declined since the 1960s, its legacy is felt in philosophy and philosophy of science in the influence of its modifications in Quine's naturalism (1977) and Popper's falsificationism (1934) as well as Kuhn's and Feyerabend's alternatives. Its legacy is felt in the SCIENTISTIC ideology of modernity, particularly the grip of behaviourism and positive sociology and positive economics; in the explaining/understanding division of the natural and the social, which CR opposes (see CRITICAL NATURALISM); and also in the reactions to and critiques of scientistic ideology such as Frankfurt School CRITICAL THEORY and forms of POSTMODERNISM. DCR situates empiricism as one element in ONTOLOGICAL MONOVALENCE in a chain of 1M to 4D errors. See also Collier 1994; Danermark et al. 2002; Keat and Urry 1975; Layder 1994; Outhwaite 1987.

JAMIE MORGAN

empowerment, re-empowerment. See AGENCY; AGENTIFICATION; CONCRETE/ABSTRACT; CONSTRAINT/ENABLEMENT; EMANCIPATORY AXIOLOGY; ESTEEM; FREEDOM, ETC.; PROBLEM OF AGENCY.

enablement, enabling conditions. See CONSTRAINT/ENABLEMENT; TENDENCY.

enchantment. A leading concept (together with *disenchantment* and *re-enchantment* or *re-spiritualisation*) in the philosophy of META-REALITY, signifying a world in which human being is at one with nature deemed to be intrinsically valuable and meaningful. Re-enchantment is already opined in *SR*: 97 to be a perhaps 'necessary' dream (of intuitional realism). Strictly speaking, PMR is not aimed at re-enchanting the world, for it is always already enchanted, but at critiquing the disenchanted philosophies and forms of life that overlay and occlude its enchantment. Disenchantment is a form of ALIENATION from reality. See also DIRECTIONALITY, PDM.

endism. See CENTRISM → TRIUMPHALISM → ENDISM.

endosmotic refutation. *Osmosis* (f. Gr. *ōsmos*, push, impulse), in biology, medicine and physics the movement and intermingling of fluids or gas through a permeable membrane; more generally, a metaphor for any kind of transition and intermingling. *Endosmosis* (Gr. *endon*, within), the inward flow from without of fluids or gas through a permeable membrane; a metaphor for any kind of inward flow, e.g., of meaning into a tradition, the divine into the world (antonym *exosmosis*). *Endosmotic refutation* is Bhaskar's term of art (*D*: 342) for the inevitable (in virtue of the transformative nature of praxis) ex post facto refutation of a (constellationally) closed philosophical system such as Hegel's by real-world changes seeping back into its system of categories (e.g., changes in concepts of space–time, causality, class and women since Hegel's day).

enlightenment. CR, as distinct from DCR, is sometimes located (e.g., Collier 1994) within the tradition of the modern enlightenment. In *Dialectic* and elsewhere Bhaskar draws on the pre-bourgeois tradition of dialectical and spiritual enlightenment to essay a (partially preservative) SUBLATION of the philosophy of modern enlightenment in the cause of a new eudaimonian enlightenment. The great failing of the modern enlightenment is that it is *de-ontologised* and *de-negativised* (monovalent), productive of heterologous Unhappy Consciousnesses, split between reason and faith of one kind or another and promoting an ethic of having and world-control rather than being and world-care. Its strength is commitment to scientific knowledge as a means to a better life and to freedom. At the level of the embodied person, enlightenment in CR is *SELF-realisation* or the non-heterologous (true-to-self) freely flourishing fulfilment of one's needs and potentials. This is the core of its meaning, and of the concept of *fulfilled intentionality*, in PMR. See also BEAUTIFUL SOUL, ETC.; CONSISTENCY; DEMI-REALITY; DIALECTICAL REASON; EMANCIPATORY AXIOLOGY; EXPLANATORY CRITIQUE; EXPRESSIVISM; META-REALITY; PHRONESIS, PROCESS; SOPHROSYNE.

entelechy (f. Gr. *entelecheia*, in a state of perfect being, f. *en-*, in, *telos*, a complete state, *exein*, to have or possess). Used by Aristotle to denote the full actualisation or realisation of potentiality; and by Bhaskar to refer to Hegel's ACTUALIST (constellational) closure of his system and of history, more generally any view of history which sees it as 'the progressive realisation of a present which appears as its end, at once telos and finis' (*SR*: 289). See also CENTRISM → TRIUMPHALISM → ENDISM; DEMI-REALITY; DIRECTIONALITY; GEO-HISTORICITY; TELEOLOGY.

enthymeme, n., adj. enthymatic (f. Gr. *enthyma*, a rhetorical argument). An argument in which a premise and sometimes the conclusion is suppressed, often for rhetorical effect.

entity relationism (ontological relationism). See ALTERITY/CHANGE; DCR; ONTOLOGICAL RELATIVISM; RELATIONALITY; TOTALITY.

entropy/negentropy. A concept based on the second law of thermodynamics, which states that everything in the universe, i.e., relative reality in the terminology of PMR, is inevitably subject to decay and dissipation (*disemergence*) via the conversion of 'free' (open) possibility into 'bound' products, *entropy* constellationally embraces *negentropy* or a principle of potential EMERGENCE; i.e., SPACE–TIME is (inter alia) *both* a generally entropic process *and* has the property of potential emergence. Closely related to CREATIVITY, negentropy, or the *negentropic transform*, especially in PMR, connotes the converse of entropy: the conversion of 'bound' into 'free' or 'unbound' energy (*MR*: xlvi). Bhaskar sometimes writes (*neg*)*entropy* for entropy/negentropy to highlight the dialectical unity of the two concepts as forms of change. The concept of negentropy was introduced to the world by Schrödinger (1944) to refer to the ability of living systems not only to avoid the effects of entropy pro tem but to increase in organisation, and is now widely used in the life-sciences. In POSTHUMANIST discourse, *extropy* is by contrast the undialectical opposite of entropy which assumes that inertia and decay can in effect be abolished by science, technology, rationality and self-change. Because it embraces forms of change, (neg)entropy is a 2E concept: '1M suffices for, e.g., an adequate account of science which abstracts from space, time and the process of change, which posits "principles of indifference" or "metaphysical inertia". At 2E {...} the very principles of indifference are called into question and difference, and we have "metaphysical (neg)entropy". This is the moment of cosmology, of human geo-history, of personal biography, laborious or routinised work but also of joyful or idle play' (*D*: 8; *ER*: 574) – the moment of creativity unbound. See also CONSTELLATION, FINITE/INFINITE.

environmental ethics, politics. See ECOLOGY; SUSTAINABILITY.

epiphenomenalism. The doctrine that only physical states have causal power and that mental states are entirely caused by them (i.e., by brain and neurophysiological states). An *epiphenomenon* is a mere by-product of some process with no significant or relevant causal power of its own. See also MIND; REDUCTIONISM.

episteme. See DOXA.

epistemic. See EPISTEMOLOGY, ETC.

epistemic–ontic fallacy. After ONTOLOGICAL MONOVALENCE, the primordial failing of Western philosophy, generative of the PROBLEM of the one and the many and the attendant problems of universals and induction, the *epistemic fallacy* is 'the analysis or definition of statements about being in terms of statements about our knowledge (of being)' (*DG*: 397), or the reduction of ONTOLOGY to epistemology, being to knowledge of being – the *epistemologisation of being*; for example, in Humean empiricism, or 'Dummett's verificationist view to the effect that the past only exists if there is evidence which would make it a fact' (Norris 2003: 10). This commits the category mistake of conflating two ALTERIOR modes of being, things and thoughts; or it is an illicit FUSION.

Closely related to epistemologisation is the *logicisation* of being (*epistemo-logicisation* [*D*: 67, 291, 311] refers to both). Logicisation regards the principles of formal logic (identity, non-contradiction) as axiomatic of reality as a whole, either by (1) *transposing* the principles to reality (i.e., ontologically), thereby *pan-logicising* being (Aristotle [for whom being is necessarily non-contradictory], Hegel [for whom it is necessarily contradictory, albeit in an incessantly self-cancelling way]), exemplifying fusion; and (2) *contra-posing*

174   epistemic–ontic fallacy

the principles to reality, such that logical contradictions are extruded from it, thereby detotalising or *illogicising* being (Kant, logical positivism), exemplifying fission, ultimately underpinned by ontological monovalence. CR rejects both views. Against (2), it holds that once a logical contradiction is committed it is part of (social) reality – this is '*not illogicising* being' (*D*: 81, e.a.); against (1), that formal logical contradictions obtain only between propositions; and against both, that logic 'does not determine the nature of being, but at best establishes what the world must be like if we are to perform certain operations successfully' (*D*: 74, e.a.; see also MODALITY [the modal fallacy]). An important contextual factor in the rise of CR historically was the failure of the formal and *logicist* programmes in the philosophy of mathematics and science. Conceptual or cognitive formations must not be regarded, any more than the world as such, as exhausted by logic or concepts, but as material causes of knowledge reproduced and changed '*in use*' (*SR*: 302–3).

Since ontology is in fact irreducible to epistemology, the epistemic fallacy merely masks the generation of an implicit ontology (ACTUALISM, thence EMPIRICAL REALISM), establishing a drive to subject–object IDENTITY theory, where it entails 'the converse *ontic fallacy*, viz. the definition, or assumption of the compulsive determination, of knowledge by being' (*DG*: 397, e.a.), or the *ontification* or *ontologisation of* (current) *knowledge*, hence its NATURALISATION and ETERNISATION, e.g., in the FETISHISM of facts and their conjunctions, underpinned by the fetishism of closed systems. The ontic fallacy thus is, or reveals, the social meaning of the epistemic fallacy (*R*: 181). If the epistemic fallacy collapses the intransitive dimension, effecting the *derealisation of reality*, the ontic fallacy collapses the transitive dimension, effecting the *desocialisation of science* and other ways of knowing (*SR*: 253). Nowadays – whether in analytic, Marxist, post-structuralist or postmodernist philosophy – the epistemic fallacy most usually takes the form of the *linguistic fallacy*: 'the analysis of being as our discourse about being' (*D*: 206), which finds classic expression in Wittgenstein's *Tractatus*: 'To give the essence of propositions means to give the essence of all description, therefore the essence of the world' (cit. *PN3*: 163 n).

The epistemic (together with its logicising variant) and ontic fallacies are dialectical counterparts or DUALS which, while apparent antagonists, in reality mutually presuppose and support each other – a fact registered in the concept *epistemo-onto-logicisation* (*D*: 248) – transmitting positivity from knowledge to being and reflecting it back, thus stymieing criticism and critique. 'The epistemic fallacy merely masks the dogmatic complacency that *our* knowledge is determined by being, i.e., the ontic fallacy, in the achievement of generalised subject–object identity theory, i.e., paradigmatically knowledge as universal and necessarily certain' (*D*: 111). Because it entails an empirical realist framework, continuing commitment to the epistemic fallacy in the contemporary context, now that fundamentalism has been rejected and epistemic relativism accepted, leads inexorably 'from positivism via conventionalism, pragmatism {and} constructivism to {subjectivist} superidealism' (*D*: 228). The epistemic fallacy is at the heart of the ANALYTICAL PROBLEMATIC, and is both underpinned by and functions to conceal the ingrained ANTHROPISM of Western thought. It is also buttressed by, and buttresses, ontological monovalence, and next to it plays the leading role in the IRREALIST ensemble which positivises, reifies and eternises 'current knowledge and more generally the status quo' (*D*: 184), such that the course of nature itself becomes identified with intellectual and socio-political conformity (indeed, the epistemic fallacy = irrealism in its core sense, i.e., non-transcendental realism); historically its origins, like that of ontological monovalence, unsurprisingly lie in the quest for an unhypothetical starting point as 'a response to worrying change and diversity' (*D*: 309). The fact that it generates an implicit ontology exposes it as a sham; or it constitutes its reductio ad absurdum. It is definitively refuted by the transcendental deduction of epistemic/ontic non-identity or alterity within ontology (the TRANSITIVE and INTRANSITIVE DIMENSIONS). See also REFERENTIAL DETACHMENT; SPECULATIVE/POSITIVIST(IC) ILLUSION; UNHOLY TRINITY.

epistemic perception, epistemic stance. See PERCEPTION.

epistemic relativity, epistemic relativism. See HOLY TRINITY; KNOWLEDGE; PERSPECTIVE; POST-STRUCTURALISM; SAPIR–WHORF; SOCIOLOGY OF KNOWLEDGE; TRANSCENDENTAL REALISM.

epistemological absolutism. See ABSOLUTISM; CENTRISM, ETC.

epistemological circle. See HERMENEUTICS.

epistemological dialectic. Embraces (1) the TRANSITIVE DIMENSION as such, and (2), more specifically, the process of production of knowledge by means of knowledge in science, philosophy, etc. The present entry is concerned mainly with (2). Both conceptions draw on an understanding of 'the ONTO-LOGICAL structure of human activity or PRAXIS as essentially transformative or poietic – as consisting in the transformation of pre-given material (natural and social) causes' (*SR*: 122) – and indeed their early formulation by Bhaskar is the prototype of the TMSA. The epistemological dialectic incorporates the *dialectic of EXPLANATION* or *explanatory dialectic*, and in fact may be seen as essentially the form assumed by that dialectic at the level of epistemology. Because it is adapted from Hegel – whose epistemological dialectic was for Marx (and is for Bhaskar, exemplifying the dialectics of CO-PRESENCE) the 'rational kernel' in his system – and crucially incorporates GÖDELIAN DIALECTICS, it is sometimes referred to as the *Hegelian/Gödelian* or the *Hegelianesque* epistemological dialectic. Unlike the Hegelian version, however, it proceeds in the mode of ontological depth as well as distanciated time. It seeks to capture at an abstract level *the logic of scientific discovery* (which may be regarded as constituting its core) – a logic imparted by the stratification of reality – which will of course be highly subject- and context-specific.

The essential, triadic movement of the dialectic of explanation from knowledge of manifest phenomena to a structural account of what generates them, and criticism and correction of the initial hypothesis or theory in its light, can be seen, at the level of epistemology, to be one in which INCONSISTENCY (CONTRADICTION, APORIA, ANOMALY), caused by incompleteness (some relevant conceptual or empirical ABSENCE) (2E), generates a move to greater completeness (TOTALITY) (3L) – the moment of *TRANSCENDENCE* – via the postulation and identification of causal mechanisms (1M) and the attendant reconceptualisation or transformation (transformative negation) (2E, 4D) of our theories and research programmes, and so on recursively. Both dialectics incorporate a *dialectic of truth*. Thus we have

> epistemological dialectic, incl. Gödelian dialectics > dialectic of explanation > dialectic of truth

176   epistemological dialectic

The epistemological dialectic is thus broadly similar in form to – indeed, sometimes another name for – immanent CRITIQUE (cf. *D*: 34, 61), the procedure par excellence of (critical) philosophy as well as of science. More broadly yet, it is synonymous with the process of CR *dialectical reasoning*, embracing (as can be seen above) all the 'modes of reasoning and practice' appropriate to 1M–4D 'and their unity' (*D*: 13) (see DIALECTICAL ARGUMENT; INFERENCE). In particular, it incorporates a *dialectic of dialectical and analytical reasoning*, in which formal logical analysis (including deduction and induction), though constellationally overreached by dialectical reason in a dialectical UNITY, plays a vital role; however, the living creative process of science often breaks the laws of formal logic. In this dialectic, which finesses Hegelian dialectic (HD) (see Figure 5), qualitative change is effected via a number of imaginative quantum leaps or TRANSITIONS, designated by Bhaskar *transforms*. These exemplify (1) preservative determinate negation in Hegel, or *conceptual SUPERSTRUCTURATION* or EMERGENCE; and (2) transformative negation, which is only partly preservative, in Bhaskar. In either case, the transition effected exemplifies *conceptual SUPER-STRUCTURATION* or *EMERGENCE*, and provides an analogue of material emergence and a non-arbitrary principle of stratification for science. In PMR, something like the epistemological dialectic is fundamental to all learning processes and arguably indeed the fundamental developmental processes of nature as such (*RM*: 43).

The ρ (*rhō*) *transform* is the transition in HD from the pre-reflective thought of everyday life (cf. Wittgenstein) to the Understanding ('U') (which represents the principle of identity and the moment of 'reflective' or analytical thought). In the CR dialectic it stands for the extra-scientific inputs (the relational dialectics into science), including scientific training (or philosophical, etc.), and U is taken to refer to something like Kuhnian *normal science*.

The σ (*sigma*) *transform* is the leap in HD of Dialectical reason ('D') (representing the principle of negativity) required to identify the contradictions, anomalies, etc., in our conceptualisations and experience. In CR dialectic it similarly stands for the emergence within normal science (philosophy, etc.) of major anomalies and contradictions, suggestive of some lack or inadequacy in its practice and some theory–practice inconsistency on the part of the scientific community. It issues in a DIALECTICAL (negative) comment – dc′ – on the practice of the community which lays bare these lacks and inconsistencies.

The τ (*tau*) *transform* is the leap in both dialectics of positive (speculative) Reason ('R') (representing the principle of rational totality) required to resolve these contradictions, etc., via an expansion of the conceptual field and issuing in a dialectical (negative) comment – dc′ – on the practice of the pre-existing community which lays bare its lacks and inconsistencies.

In both the σ and τ transforms 'the dialectical fertility of contradictions' is seen to depend 'upon their analytical unacceptability' (*D*: 20), as 'D' (dc′) reports and 'R' (dr′)

$$\xrightarrow{\rho} U \xrightarrow{\sigma} D = dc' \xrightarrow{\tau} R = dr' \xrightarrow{\upsilon} \varphi$$

Figure 5 The CR epistemological dialectic
Adapted from *D*: chs. 1.6, 1.7

epistemological dualism

remedies an absence in our conceptualisations. However, in CR dialectic, logical contradictions are suspended or bracketed rather than transcended, as in Hegel. Further, in CR dialectic both these transforms incorporate moments of indeterminate and underdeterminate negation, i.e., they are not instances of linear radical negation, as in Hegel; nor are they radically preservative, as in Hegel.

The overall result of HD is our return, via the ν *(upsilon) transform*, reconciled, to the now rationally comprehended and transfigured actuality we left in the ρ transform. The contradictions (theory–practice inconsistencies) identified in the σ transform are resolved in thought via 'analytical reinstatement' in 'an allegedly closed consistency' (*D*: 340), not in practice.

The CR dialectic stretches the ν *transform* to incorporate the extra-scientific outputs (the relational dialectics out of science) – applied science and technology – and the (re-)appropriation by the lay community of the skills and knowledge formed within science, and adds the φ *(phi) transform*, or the resolution (real negation or absenting) of contradictions in practice (dφ). Whereas the HD is circular in structure (*D*: 22, fig. 1.1), the CR dialectic can be represented as an open-ended process, with a new phase of discovery beginning after dφ.

For CR 'D' ushers in, and 'R' completes, an epoch of *scientific revolution*. Retroductive-analogical thinking, paramorphic model-building, and the absenting, distanciation or transformation of pre-existing knowledge all play vital roles in this epistemological dialectic. 'The determinate result of this labour of transformative negation {...} is the identification of a new level of ontological structure (S_2), described in a new theory (T_2) capable of explaining most of the significant phenomena explained by T_1 (at U) plus the anomalies at D, albeit in its own (T_2's) terms' (*D*: 34). The *phenomena* identified by T_1 at S_1 are for the most part 'saved' or preserved, but the *theory* itself is negated, falsified ('lost') and replaced by (transformed into) a new theory which could not have been predicted but only worked for and won in the transitive process of science. In view of the depth-openness of the world, this result in principle initiates the possibility of a new round of epistemological (incorporating explanatory) dialectic leading to the discovery of a new level of ontological structure, and so on indefinitely. Given that S_1 explains *x*, when S_2 has been referentially detached, there can be no practical doubt that an objective truth (the ALETHIA of *x*) has been discovered. The epistemological dialectic has an important analogue or counterpart in the domain of falsity: the logic of false consciousness whereby TINA SYNDROMES are constituted. *Epistemological dialectics*: see ONTOLOGICAL DIALECTICS. See also PROPAEDEUTIC; TRANSITION.

epistemological dualism. See DUALISM.
epistemological endism. See ABSOLUTISM; CENTRISM, ETC.
epistemological materialism. See DIFFRACTION; KNOWLEDGE; IDEALISM VS. MATERIALISM; META-REALITY.
epistemological problematic. See FOUNDATIONALISM.
epistemological realism. See IDEALISM VS. MATERIALISM; KNOWLEDGE; TRANSCENDENTAL REALISM.
epistemological (or **cognitive**) **triumphalism**. See CENTRISM → TRIUMPHALISM → ENDISM.
epistemologisation, epistemo-logicisation, epistemo-onto-logicisation. See EPISTEMIC–ONTIC FALLACY.

epistemology. See EPISTEMOLOGICAL DIALECTIC; EPISTEMOLOGY, ETC.; KNOWLEDGE.

epistemology, etc. In *SR* (36–8, from which much of what follows quotes and paraphrases), Bhaskar makes a series of 'seemingly cumbersome distinctions', to be invoked only when 'absolutely necessary', the essentials of which it will be helpful to set out separately here, correcting for printing errors.

Philosophical ONTOLOGY delineates 'the *general categorial form* of the world' presupposed by science and other human activities (the province of *metaphysics* α [see PHILOSOPHY]). S*cientific ontology* articulates 'the *specific contents* of the world' established or postulated by some substantive scientific theory (which it is the task of *metaphysics* β to elucidate) (*D*: 107). The distinction follows from the argument establishing the relative autonomy of PHILOSOPHY in relation to science and other practices. *Ontology* (ontological, etc.) refers to both *ontology$_1$*, propositions in the general philosophical theory of being; and *ontology$_2$*, propositions in the transcendental theory constituted by reflection on the presuppositions of scientific and other activities. *Ontic* (ontical, etc.) refers to both *ontic$_1$*, whatever pertains to being generally, rather than some distinctively philosophical (or scientific) theory of it (ontology), e.g., the ontic presuppositions of a work of art; and *ontic$_2$*, the INTRANSITIVE objects of some specific, historically determinate, scientific or other empirically based investigation. *Ontics* refers to the class of intransitive objects of specific epistemic inquiries. A world without human beings would have an ontology$_2$ (although obviously there would be no one in such a world to articulate it), encompassing the intransitive objects of non-actualised (and perhaps humanly impossible) scientific inquiries, but not an ontology$_1$. This relates to Bhaskar's views that (1) 'there can be no philosophy as such and in general, only the philosophy of particular historically determinate social forms' (*SR*: 12); and (2) while 'both the premises and conclusions of {transcendental} arguments remain contingent facts, the former (*but not the latter*) {are} necessarily social, and hence historically transient' (*PN3*: 5, e.a.). The concepts of the 'ontological' and the 'ontic' are not equivalent (or theoretically indebted) to Heidegger's, but there is some overlap. Ontic$_1$ encompasses Heidegger's 'ontological', the realm of everyday pre-understood being, and ontic$_2$ overlaps with Heidegger's 'entities'. *Ontologisation* or *ontification* (of knowledge) results from committing the ONTIC FALLACY.

Epistemology pairs with 'ontology', *epistemic* with 'ontic'. Consequently, *epistemology* refers to both *epistemology$_1$*, propositions in the general theory of KNOWLEDGE; and *epistemology$_2$*, propositions in the transcendental theory of knowledge constituted by reflection on the presuppositions of scientific and other activities. *Epistemic* refers to both *epistemic$_1$*, whatever pertains to knowledge generally, rather than some distinctively philosophical (or scientific) theory of it (epistemology); and *epistemic$_2$*, the transitive process or product of some specific, historically determinate, scientific or other empirically based investigation. Epistemic also has a related but distinct meaning as the adjective of episteme (see DOXA). *Epistemologisation* (of being) results from committing the EPISTEMIC FALLACY. See also INTRINSIC/EXTRINSIC; LOGICISM; ONT/DE-ONT.

epochē (Gr. stoppage, pause, suspension). Used by the Greek sceptics to indicate refusal to confirm or deny; by Husserl to signify the bracketing of common-sense assumptions; and by Bhaskar to refer to the suspension of thought at the moment of CREATIVE discovery.

equality. See EQUITY.

equity. A specific kind of *equality*, viz., concretely singularised equality, or equality that takes into account differences in needs, potentialities and responsibilities. The *dialectic of equity* or of *concretely singularised equality of autonomy* (*P*: 148) is one of the twelve routes, sketched in *Dialectic*, taken by the DIALECTIC OF DESIRE TO FREEDOM in the transition from form to content, which may itself be regarded as 'an extended explication of the *principle of equity*'. This holds that the good society, as the basis for a thoroughgoing equity, must tend to a core equality in resources, opportunities, rights and freedoms, inclusive of everyone in virtue of our common HUMAN NATURE, i.e., in virtue of the PRINCIPLE OF EQUIVALENCE. There are two related basic elaborating arguments. (1) As Gewirth (1979) has shown, freedom and well-being, including absence of gross inequality, are 'universalisably necessary conditions for successful action'. The dialectic of equity develops and generalises this position to 'an argument for the *realisation* of the *potential* of all agents to perform dialectically similar acts; and from there to an argument for the *development* of all dialectically similar potentials; and from there it is but a short step to argue for the development of all *potentials qua potentials* and we are at {. . .} the eudaimonistic society' (*D*: 287–8, e.a.). Note that on this model equality does not imply sameness but the richly differentiated development of potential. (2) Core equality is necessary because gross inequality is ALIENATING or detotalising – hence false to our human nature – based in that generative separation that divides us from ourselves, the means of producing a livelihood, the product of our labour, other people, and/or the system of social relations that valorises such arrangements; i.e., on this score, too, a basic equality is an aspect of the moral good (see ALETHIA; ETHICS) for creatures of our kind (see also FISSION/FUSION).

Far from equity being necessarily incompatible with FREEDOM, its absence impedes it (constraints$_2$, entailing inequities, are constraints on freedom) and its presence extends the scope of dialectical UNIVERSALISABILITY beyond the boundaries of sectional interests to the species as a whole.

equivalence. See PRINCIPLE OF EQUIVALENCE.
equivocation, equivocity. See DUPLICITY.
erasure. In Heidegger and Derrida, the formula, deriving from Nietzsche, that 'the condition of the possibility of x = the condition of the impossibility of x', e.g., knowledge of the real world. There is nothing to be said for it 'as a solution to philosophical problems' (*D*: 250, 272). It is very different from a DIALECTICAL ARGUMENT, which also establishes the conditions of possibility of the conditions of impossibility of x, not to equate them, but to move on to expansion of the conceptual field and to transformative praxis via the fundamental concept, inter alia, of a *meta-reflexively totalising* (*self-*)*situation* (see REFLECTION). See also SCANDAL OF PHILOSOPHY.
ergodic (f. Gr. *ergein*, to shut in or out, include or exclude). Pertaining to CLOSED systems.
ergonic (f. Gr. *ergon*, work). *Ergonic efficiency* concerns the efficiency (economy of energy, time, resources, etc.) with which any task is performed.
eristic. See ELENCHUS.
error. 'There is nothing wrong with being in error. The only thing wrong is to live in a state where you know you are in error. That is the real compromise' (*MR*: 98). It is lived by Unhappy Consciousnesses everywhere. In a complex, open world, error is

inevitable and a necessary part of every LEARNING PROCESS. The experience of error is probably fundamental to the emergence of human reflexivity and self-consciousness both phylogenetically and ontogenetically (Nellhaus 2004).

error, categorial. See ALIENATION; CATEGORY; EVIL; IDEOLOGY; KNOWLEDGE; TINA SYNDROME.

escape clause. See METAPHYSICAL LAMBDA (λ) cLause.

Essence Logic. See GENERATIVE SEPARATION; INFRA-/INTRA-/SUPER-STRUCTURE; NATURAL KIND.

essentialism. An argument can be classified as essentialist if it holds that an essential property yields explanatory knowledge of how individuals, groups, institutions, structures, etc., operate. Such an argument is reductionist and determinist, because it reduces the level of causal explanation to one essential property, and holds that this determines all other aspects of the objects in question. A classic example is the view that the social world is to be explained in terms of a fixed universal HUMAN NATURE: knowledge of human nature *qua* essential property yields knowledge of the social world because the essential property determines individuals' behaviour. To know the essential property is to know *the* moving force behind the behaviour of people. Essentialist arguments about human nature may be criticised for taking a reductionist and determinist approach to complex phenomena which are more than mere expressions of some essential property; and for defining human nature to fit the prevailing social and political order, in an attempt to justify that order. For example, human nature may be defined as competitive in an attempt to justify 'free market' liberal capitalism.

Social science, or at least SOCIOLOGY, arose in reaction to such essentialism. Sociologists sought to argue that society was a moving force in its own right and not an epiphenomenal expression of human nature. The degree to which individual agents were regarded as having the capacity to influence social structural factors depended on the philosophical, methodological and political inclinations of the sociologist theorising social reality. This issue of defining a social ontology to say what social reality is opens up the 'STRUCTURE–AGENCY problem' in sociology (and social science more broadly). Critical realists attempt to resolve the problem without falling into structural determinism or individualist voluntarism, by arguing that social structures are emergent properties that are open to change by groups of agents over time (Archer 1995).

With the cultural turn and rise of relativism, the charge of essentialism has been levelled at social scientists who seek to explain social reality by reference to structures or structures and agency. It is argued such people are guilty of succumbing to a bad Western metaphysical urge to explain the given in terms of some unobservable stratum of ultimate reality. The politics associated with reference to structures may often be taken to be radical, with such social scientists arguing for a change to existing structures. However, critics would maintain that no group can assume a privileged position outside a language-game or discourse, and that any attempt to do so results in an authoritarian politics that seeks to impose a naïve vision of the good society, failing to recognise and embrace difference and plurality.

Defending the CR ontology from the charge of essentialism, Sayer (1997b) argues that we need to distinguish between erroneous forms of essentialism that are reductionist, determinist and often used to justify right-wing political ideologies, and acceptable forms of essentialism that are relational. Examples of the former are essentialist views of human nature and gender differences, and an example of the latter the view that the need to make a profit is an essential feature of capitalist social relations. Sayer accepts the connotation of fixity that the term essentialism has, and argues that it can be replaced by a reference to structural/causal relations that operate in open systems. For Sayer we need to accept the existence of structural relations, which have features essential to them, but this does not lead to a reductionist and determinist ontology.

Nevertheless, CR is still open to the charge of essentialism because arguments pertaining to the status of the ontology are subject to a tension. An emphasis on the ontology being a fallible interpretation of reality that is open to revision sits uncomfortably next to claims about the ontology being the definitively correct definition of reality (Cruickshank 2004). There is a slide from what Bhaskar (*RS3*) refers to as 'metaphysics as a conceptual science', which simply clarifies ontological presumptions within the sciences, to metaphysics as a dialogue-stopping claim to have a privileged access to Being. This slide occurs when critical realists seek to escape the epistemic fallacy by moving beyond fallible knowledge claims about being, to treat ontology as a theory of being that transcends the fallibility of knowledge claims. The latter rendering of ontology can be described as essentialist because the ontology would be taken to be a direct expression of the essential moving force of social reality rather than a fallible meta-theory. The problem of essentialism remains despite the move away from anti-sociological notions of human nature determining individuals. See also NATURAL KINDS; PHILOSOPHICAL ANTHROPOLOGY.

JUSTIN CRUICKSHANK

esteem. Bhaskar follows Rousseau (1754: 182 n) in distinguishing sharply between two kinds of self-love or self-respect, *amour-propre* and *amour de soi*. *Amour-propre* is actualist self-love with an eye to the opinion of others and narrowly construed self-interest. The 'baneful' manifestation of the 'AGONISTIC and expressive' aspects of developing human nature (*P.* 150), it is in love with being loved, detests being disliked, and hungers for power and status. Conceited and intrinsically egoistic, lacking in truthfulness and sincerity, it is premised on the ideological illusion that human SUBJECTS are atomistic and opposed, disconnected from social and natural being. It is a creature of ALIENATION, hence common to all master–slave-type societies, but reaches its zenith in capitalist post/modernity (cf. Dean 2003).

Amour de soi is love of our transcendentally real or essential selves as human beings, or SELF-*esteem*. Grounded in and sustained by relations of care, SOLIDARITY and TRUST, it is an inner resource that provides an indispensable basis for altruism: 'only the empowered can empower' (*D*: 222). It prizes theory–practice CONSISTENCY, truth and free flourishing. It can exist without *amour-propre*, but not vice versa: even as it denies and occludes *amour de soi*, *amour propre* depends on it and knows itself for a fraud – a form of self-loathing.

Amour de soi is what Anna Karenina had, but lost; what Konstantin Levin did not have, but found. *Amour-propre* is what George W. Bush has in abundance: 'America's most saleable commodity' (Pinter 2005). However, the moral alethia of humankind – 'the free development of each as a condition for the free development of all' – presupposes that we all lose it. In this, the *dialectic of the 7 E's*, underpinned by the trust

182 eternisation (eternalisation)

and solidarity of totalising depth-praxis and politics (TTTTTTφ) – the best remedy (repeatedly attested in myriad social movements) for narcissistic, fragmented selves – may assist:

> self-*e*steem ↔ mutual *e*steem (where the intra-dependence of action itself reflects both the fiduciary nature of the social bond and the reality of oppressive social relations) ↔ *e*xistential security ↔ *e*rgonic efficiency ↔ (individual → collective → totalising) *e*mpowerment ↔ universal *e*mancipation ↔ *e*udaimonia (*D*: 265).

eternisation (eternalisation). The act or process of *eternising*, of causing to continue indefinitely, often used in respect of someone's fame. Adapted by Marx to refer to the process, rooted in ALIENATION, whereby transitory and often illusory or IDEOLOGICAL contemporary phenomena come to be viewed as permanent, and used in that sense in CR. Thus the modern illusion that people have, not embodied personalities interconnected with each other and the world, but egos sharply divided from and opposed to each other and the world, comes to be regarded as a permanent, real feature of human nature conceived of as trans-human-historical (as 'natural'). This not only *dehistoricises* the historically specific, but *detemporalises* it (*SR*: 199). Eternisation can embrace whole forms of life: 'in transcendentising actuality, Hegel eternalises it, transmuting it sub specie aeternitatis' (*D*: 336). Grounded in the EPISTEMIC–ONTIC FALLACY (R: 181), it commits the converse of the GENETIC FALLACY, and is synonymous with NATURALISATION in its strongest sense.

ethical hyper-naturalism. See NATURALISM.

ethical ideologies of post/modernity include *emotivism, decisionism, personalism, prescriptivism, descriptivism*, and *sociological reductionism* (of ethics), all of which are forms of moral IRREALISM premised on the ANTI-NATURALISTIC FALLACY, which denies any possible moral truth grounds (there can be no 'ought' from 'is'), and are hence (with the exception of descriptivism) *non-cognitivist* ethical outlooks. Notwithstanding their differences, the uneasy co-existence of which is secured by a TINA compromise formation, all are accordingly tacitly complicit in sustaining existing power$_2$ relations, which they *discursively moralise*.

Emotivism, whose post/modern proponents include Hume and Nietzsche, reduces morality to arbitrary personal preference: ethical statements are neither true nor false but expressions of emotions, desires and attitudes, and moral disagreement is rationally irresolvable. Historically, emotivism thus smoothed the path for cynicism and *nihilism*, which holds that the only value is the power to establish values. It is the dominant ethical ideology in globalising capitalist society, systemically generated by it and profoundly embedded in its culture and habitus, its key assumption informing many otherwise opposed ethical theories. *Decisionism*, espoused by the early Heidegger and Sartre, is the closely related view that, because there are no objective rational grounds for reaching a decision in ethical matters, human subjects are radically free and so must ultimately choose or decide. *Personalism* 'is characterised by the attribution of responsibility to the isolated individual in an abstract, desocialised, deprocessualised, unmediated way, with blame, reinforced by punishment (rather than the failure to satisfy needs), as the sanction for default' (*D*: 265). Its *abstractly universalised* contrasts with CR's *concretely singularised* ethics. *Kantian prescriptivism* holds that moral

ethical naturalism

rules are universally binding regardless of the particular context. *Prescriptivism*, in its modern version, a theory proposed by R. M. Hare in response to emotivism, holds, like CR, that moral judgements are necessarily prescriptive or universalisable: if you say that someone should φ, then you are committed to φ-ing in relevantly similar circumstances. However, unlike CR, it accepts the descriptive/evaluative dichotomy and that adopting an ethical principle is ultimately a matter of personal commitment, thereby conceding the substance of emotivism. *Descriptivism* is the cognitivist but actualist undialectical opposite of prescriptivism, holding that the meaning of any moral statement is purely descriptive or factual, i.e., determined by its truth conditions. It thus reduces values to facts, the possible to the actual. *Sociological reductionism* reduces ethics to EPIPHENOMENA of the social. All these ethical ideologies are forms of *JUDGEMENTAL RELATIVISM* which fails to see the constellational identity of the possibility of judgemental rationalism (in the IA of the TD) within the actuality of ethical relativity (in the EA of the TD) within the necessity of moral realism (in the ID).

Since morality ultimately reduces, on most of these accounts, to arbitrary will, the question that arises sociologically is: *whose* will (cf. MacIntyre 1981: 104). Bhaskar tellingly aligns the three leading theories to the figures of the master and the slave, legitimation and bondage: 'Emotivism is the moral ideology of those who do not need to work {i.e., of the master-class proper, the predominant owners of wealth}, decisionism is the ethics of the slaves of masters who are themselves masters of slaves {i.e., managers and bureaucrats}, personalism is the philosophy of slaves who accept the description {and who therefore do not struggle for emancipation}' (*D*: 366). Nihilism, we might add, is the moral ideology of the overmen of the global corporations and imperial state apparatuses. See also Bull 2000; Collier 1999, 2003b.

ethical naturalism in CR is the theory and practice of explanatory critique. See also CRITICAL NATURALISM; DCR; NATURALISM; NATURALISTIC FALLACY.

ethical relativism or **relativity**. See ETHICAL IDEOLOGIES; HOLY TRINITY; NEEDS; PRAGMATISM; UNHOLY TRINITY; WORLD-LINE.

ethical tetrapolity. The developmental logic implicit in human agency and discourse that goes from '[expressive veracity → axiological commitment] → (1) fiduciariness {3L} → (2) content of the explanatory critical theory complex [= EXPLANATORY CRITIQUE + CONCRETE UTOPIANISM + theory of TRANSITION] {2E} ↔ (3) totalising depth praxis of EMANCIPATORY AXIOLOGY {4D} → (4) FREEDOM qua universal human emancipation {1M, (*DG*: 397).

ethics. Mainstream ethical theory has come to be divided into the 'meta-', the 'normative', and the 'applied' – concerned, respectively, with the derivation, content and practical implications of the principles deemed appropriate for the evaluation of human life and actions. CR treatments have tended to dwell upon the meta-ethical: on the nature and sources of the evaluative in social life and its critique. Yet the 'take' on the ethical definitive of CR work itself gives the lie to a strict severing of ethics from other modes of inquiry, and indeed to a rigid construal of the internal tripartite division. This is partly because of the rejection in CR of the Humean (and otherwise predominant) fact/value dichotomy. But it is also because of the particular *way* in which the dichotomy is put into question. The notion of EXPLANATORY CRITIQUE itself implies a view of the relation between the evaluative and the valued which shifts

184 ethics

the register of ethical discussion to a terrain distinct from that most trodden in mainstream ethical debate. Even so, elements in CR treatments of the ethical reflect themes familiar from both ancient and modern ethical traditions.

What is it for ethics to be 'realist'? The conventional definition of 'moral realism' is pitted against key ETHICAL IDEOLOGIES of post/modernity, most notably emotivism and decisionism, but also the relativism and perspectivism typical of purportedly radical challenges to established meta-ethical traditions. It insists that there are moral properties, such that it is (as it were) either *true* that rape is wrong, or it is not, and that the answer to this is independent of the way this issue is mediated by any particular perspective, tradition or process of discursive constitution. This meta-ethical position has been coupled with various normative frameworks, from Platonism to fundamentalist religion to Kantian Socialism. But this transferability begs questions about the relation between the form of moral realism and the wider normative commitments to which it may give rise. For CR, given its rejection of the fact/value dichotomy, the negotiation of this relation is key. It has been conducted in various ways; here, for the sake of contrast, we cover three.

In *PE* (151) Bhaskar articulates two points central to any CR approach: morality has a 'properly ontological employment', and 'like knowledge, has an intransitive object/ive'. These claims connote a further two fundamental commitments: a rejection of ethical ANTI-NATURALISM (i.e., of the notion that 'the fact/value divide leaves ethics ungrounded'), and of moral IRREALISM. CR ethics is thus committed to defending the cognitivism inherent in the conventional definition of moral realism given above, and particularly its rejection of the idea that moral properties might somehow reduce to creations either of the will or of affective experience. Where it goes from here, however, has been cause for debate. There are perhaps two main areas of focus. First: the nature of value. *How* – and to what extent – are moral values intrinsic, or objective? Second: the scope of value judgement. To what extent are norms historically specific, and which, if any, transcend their mediation through particular social conjunctures?

In the work of Bhaskar himself, the question of ethics emerges most strongly from the late 1980s onwards. Up until that stage, though an ethical impetus had clearly been core in CR's development, the relation between meta- and normative ethics had yet to be staked out. In *SR* (187), considering alternatives to normative actualism, Bhaskar argues that the historicity of action-situations 'vitiates the universality of norms, placing them under the sign of an actual or possible scope restriction'. Between here and *Dialectic*, there is a shift in register; so that when Bhaskar begins to talk more substantially about the normative dimension of his project it appears in more universalist, formalised terms. It still has an anchor in the historically specific; it is 'concretely singularised' (*P*: 160). But the 'balance' in attention, on the one hand, to 'concrete singularity and transfactuality', and on the other to 'absence and actionability', has seemed to many to shift over to the latter. Because freedom is as much a condition of truth as vice versa, and because of the 'fiduciary nature' of the expressively veracious remark, Bhaskar argues – in a re-tuned echo of Habermas – that the ideal of human solidarity is implicit even in the most mundane communicative actions (*D*: 291–2). All actions imply a commitment to the goal of universal human flourishing. This universalistic moment is extended, in the later work, where it is presented with a more Eastern inflection; talk of a 'core universal human nature' (itself reminiscent of Marx's 'species-being') is replaced by talk of a 'ground-state', itself understood in terms of an essential creativity. With the 'spiritual turn' thus taken, the relation between the historical and the ethical appears less intimately couched.

Andrew Collier, in *Being and Worth*, defends a realism about value which draws on Spinoza and Augustine in developing three main claims. The first concerns the question of ethical motivation. For Collier, invoking Macmurray, REASON is the capacity to behave in terms of the nature of the object (morality's 'properly ontological employment'). Reason has 'shoving-power'. But it is also affective; emotions, far from being severable from reason on the Humean model, are inevitably informed by reason. The second claim is that all being, as being, is good. The third is that evil, rather than existing in itself, consists in the privation of being. For Collier (1998), his own position avoids the formalism he finds in Bhaskar's appeal to a freedom implicit in every human desire or action. Yet the appeal to intrinsic worth itself brings well-rehearsed problems, about the nature of the worth itself, and about the range of 'things' considered good (Benton 2004). Collier's understanding of evil as privation of being points towards the significance of ABSENCE, but in a separate register from Bhaskar's own emphasis on the priority of negativity to positivity. Indeed, it might seem to signal a reverse direction, in which absence is an aberration: the absenting of evil is a making present of the good, 'good' and 'being' understood here as convertible. Not all lacks need be evil, just because evil is. But in Collier's version absence itself (aside from the – arguably quite distinct – sense of the *absenting* of ills or constraints on freedom) is subject to the prior presumption that it is the sign of a deviation from the good.

Third, comes a more explicitly historicist variant, rejecting the residual essentialism of Bhaskar and Collier, through an implication that this commits an unwarranted support for foundationalism. For Nick Hostettler and Alan Norrie (2003), Bhaskar's 1980s work is split between a commitment to a corroboration of the normative force of a common, transhistorical human nature, and the historicist momentum of his own recognition of the differentiation and stratification characteristic of social being. The case for a tension hinges on the identification of that split, and its avoidability: this latter, on these terms, depending on the avoidance of Habermas's (Kantian) 'autonomisation' of ethics vis-à-vis science and history. The case's avoidability itself depends on the possibility of realising the ethical force of a realist emancipatory project on the basis of the 'concrete' universal, which itself is here conceived as free from any 'ontic' commitment to an essential human nature.

These three perspectives have a heuristic value in configuring possible trajectories of CR dealings with the ethical. It does not seem reductive to argue that the story of CR treatments of ethics plays out in a way reminiscent of the sometimes tortuous relations of Marxism to its ethical underpinnings and implications. Marx's often latent normativity has been read in Hegelian, Aristotelian and Kantian terms: prioritising, respectively, the historicity of values; the generalised conditions of human flourishing; and the universality implicit in its own promise of a just society. Working this out in the field of CR promises a nuanced reconciliation of these three trajectories. This promise stems precisely from the scope offered by a stratified conception of reality, including both social structures and agents, in which each might offer a source both of value and of (in Bloch's terms) the 'not-yet' of the promise of improved social relations.

At the heart of CR ethics, as an ethics of emancipation, is a rejection of abstract formalism and a process of normative engagement with the conditions of our own flourishing; to this extent, virtue ethics seems a more proximate corollary than the other two great modern traditions of Kantian deontology and utilitarianism.

This is perhaps best exemplified in the appeal – most explicit in Collier – to an ethics which transcends anthropocentrism and embraces the worth of the non-human. At any rate, a key aspect of realist ethics has to be a broader sense of ethical relationality than those afforded by the Kantian or utilitarian traditions – one capable of encompassing the natural environment, including non-human species. It amounts to a rejection of both modern and postmodern ethics in their elementary forms, but one which extends the emancipatory themes of both. This is universalist in a particular sense – not one which requires for its identification a God's-eye view, but one which starts from, and seeks also to end with, a basic presumption of the equal worth of human beings as such, and the inherent value of other nature besides – i.e., not simply as bestowed by human agency. Reflecting the coalescence of epistemological relativism and ontological realism, there may be some space for a descriptive particularism, but with the refusal of its normative counterpart which would itself run counter to a genuinely inclusive emancipatory politics. Horizons of knowing about ethical norms/values are not to be equated with their extension. Were this a common denominator among CR accounts of the relation between meta-ethics and normative ethics, the field of applied ethics (delineating rights and wrongs in euthanasia, the sex industry, surrogate motherhood, punishment, animal experimentation) would still remain ripe for contestation. Here, for all the ethical impetus inherent across the gamut of practical CR theorising, there is work to be done. See also DEA and DET MODELS; DIALECTIC OF MORALITY; EMANCIPATORY AXIOLOGY; Sayer 2005.

GIDEON CALDER

ethnicity, ethnocentrism. See ANTHROPISM; MASTER–SLAVE; RACE.

ethnography. A method of conducting social research concerned with the study of human behaviour within specific contexts. It attempts to study behaviour within a defined context as opposed to abstractly or in isolation, and the ethnographic research process aims to provide social researchers with an empirical sensitivity. As with many research METHODS, ethnography has been coupled with both positivist and interpretivist philosophies. CR researchers have also been able to utilise ethnographic research as a starting point for their investigations. The explanatory commitment of many ethnographic research projects, and subsequent attempts to articulate deep causal structures following ethnographic observation, leads to an implicitly realist philosophy detectable in much ethnographic work. Brewer, for example, argues that Paul Willis's *Learning to Labour* (1977) is essentially critical realist (Brewer 2000: 51). Explicitly CR ethnographies focus upon the generation of theoretical explanation of empirically observed phenomena such as racism in a medical setting (Porter 2000), or the development of practitioner research (McEvoy and Richards 2003; Carspecken 1995). Whilst it is encouraging that examples of ethnographic research exist which make use of CR, a fundamental treatment of the relationship between the two might ultimately lead to the foundation of a theoretically informed methodology for ethnography.

MATTHEW BRANNAN

ethnology. See ANTHROPOLOGY.

eudaimonia (Gr., happiness, prosperity, flourishing, f. Gr. *eu*, well + *daimon*, god/goddess, destiny, fortune, one's genius or lot). In ancient Greek ethics, the ultimate justification of morality, achieved through the cultivation and exercise of the virtues. Nowadays it refers most generally to any conception of ETHICS (*eudaimonism*) that sees human happiness and flourishing as the highest good.

In CR it has the specific meaning of the moral ALETHIA of the human species, entailed by and expressing 'both the deep content of the judgement form and the latent immanent teleology of praxis' (*P*: 154): the free flourishing of each as a condition of the free flourishing of all, which is the good society for human being. Adoption of the term by CR thus does not betoken endorsement of any particular Greek ethics, but rather willingness to draw on pre-modern dialectical and spiritual traditions in order to articulate an ethics fitting for a post-power$_2$ global order. The argument for eudaimonia as moral alethia is set out in EMANCIPATORY AXIOLOGY; indicated here are the substantive social principles this argument suggests, subject to what ongoing depth-investigation and praxis might discover about the possibilities of four-planar social being.

Universal free flourishing presupposes the inversion and transformation of the Hegelian triad of the family, civil society and the state into the domains of (1) *universal civic duty*, (2) *social virtue* and (3) *individual freedom*. This is set out in Table 20, with some correspondences. It is posited on the abolition of master–slave-type relations *in toto*, including the social relations of commodity production and exchange and of the power$_2$ state (which are designed to exclude people from the social determination of their lives). When it is taken into account that such a society will necessarily be in a permanent state of transition or *open process*, the moments of eudaimonia can be seen to match all four moments of the concrete universal and the causal–axiological chain.

The domain of (1) universal civic duty enshrines the ethical core of eudaimonia oriented to concrete singularity (cf. *P*: 154) – 'a normative order informed by the values of trust, solidarity, sensitivity to suffering, nurturing and care' – in universally reciprocally recognised rights (freedoms) and duties, such that one may be seen under the aspect of the other; thus duty is 'conceived precisely as the right to be subject to universal rights and participate in globalised democracy' (*D*: 296). The *supreme good* is moral alethia. The *highest good* is the supreme good 'aesthetically enjoyed in creative flourishing' (*P*: 154). Such a society does not entail that people would speak the truth all the time: speaking less than the truth out of sensitivity to the other (white lies) falls entirely within its compass. Among other principles it would be necessary to recognise are:

- constitutional democracy organised around people's councils or assemblies forged to articulate self-determination, with as much local autonomy and participatory democracy as possible;
- 'the massive redistribution, transformation and limitation of resource use dictated by considerations of EQUITY and ECOLOGY', including the socialisation of knowledge;
- the co-operative organisation of production and services by inter-linked autonomous associations at a local and regional level;

Table 20 The moments of eudaimonia, with some correspondences

Eudaimonia	(1) universal civic duty (rights, responsibilities)	open process (permanent transition)	(2) social virtue (enterprise, participation)	(3) individual freedom (unity-in-diversity)
Hegel	(3) state	{end of history}	(2) civil society	(1) family
Marx	(1) realm of necessity	open process	(2) civil society	(3) realm of freedom
CR concrete universal	universality	processuality	mediation	concrete singularity
Causal-Axiological chain	1M non-identity	2E negativity	3L totality	4D transformative agency
Logic of ethical tetrapolity	(4) freedom qua universal human emancipation, the realisation of moral truth (alethia)	(2) explanatory critical theory complex	(1) fiduciariness	(3) totalising depth-praxis
Logic of transition (social)	(4) emergence	(2) dialectical contradiction	(1) dialectical connection	(3) negation of the negation (geo-historical transformation of geo-historical products)
Politics (democratic, participatory)	emancipatory/transformative movement		representative/syndicalist	life

- distributive principles along the lines of 'from each in accordance with their wants, abilities and needs' and 'to each according to their essential needs and innovative enterprise', such that minimally no one is forced to sell their labour-power or work for a master;
- recognition of the indispensability of diversity and pluralism for political, scientific, educational, etc., creativity (*D*: 267–8, 296)

A domain of (3) individual freedom (unquestioned choice) is the sine qua non for the realisation of (1) and is thus the crowning glory of eudaimonia. Such a realm is transcendentally necessary because autonomy and creativity presuppose AXIOLOGICAL underdetermination. Individual freedom should accordingly be unconstrained, 'save by respect for global constraints and universal rights', allowing for the manifold development of human creativity as an end in itself. The good life for each would be a unity-in-diversity – coherence within a rich all-roundedness.

Finally, a domain of (2) social virtue is required as a mediating zone between necessary duties and individual freedom to provide the practical, democratic, political, agonistic and expressive diversity and plurality, duly recognitive of individual differences, that are a necessary condition for totality and progress respectively (*P*. 150). Initiative and enterprise in all these spheres would be materially rewarded if necessary, i.e., where they were not seen as their own reward. See also CONCRETE UTOPIANISM; CENTRISM; DIRECTIONALITY; ENLIGHTENMENT; FREEDOM ETC.; GEO-HISTORY; POLITICS; PROBLEM(S) OF AGENCY; SOCIAL CUBE; SOPHROSYNE; TRANSITION; UNIVERSALISABILITY.

event. The counterpart in the non-human sphere of an act or ACTION in the human, both of which pertain to the domain of the actual, but may function as mechanisms (see REALITY). In keeping with much modern science and the theory of the CONCRETE UNIVERSAL ↔ SINGULAR, CR moves decisively away from a conception of events as ATOMISTIC and punctiform instantiations of abstract universality, and of mass events as atomistic or additive collectivities (mere sums of their parts), to one which embraces the 'holistic or quantum concept of an *event itself* as a mass, collectivity or TOTALITY {..., for} example, as a distribution or spread in space or a succession or flow in time, or both – a RHYTHMIC matrix in SPACE–TIME' (*EW*: 27). See also DISTANCIATION; NEXUS; PROCESS.

evidentialism. See FIDEISM.

evil in Bhaskar's later philosophy is the (categorial) error of acting contrary to our transcendentally real SELF or essential nature (self-alienation), which in an open world we are able to do in virtue of the reality of FREE WILL (which also makes self-realisation possible). Considered intransitively and more basically in an explanatory sense, it is the objectified result of such action – a real emergent false being, purveying mystification and ignorance – conceptualised in the Catholic tradition as 'structural sin' (Pope John Paul II, *Solicitudo Rei Socialis* [*On Social Concern*], 1987; cf. *EW*: 37 et passim). It is thus not just privation, lack or absence of being (though it depends on these), but 'an emergent power in its own right' (*EW*: 89). However, consistently with the theory of emergence, it is unilaterally dependent for its existence (parasitical) on the more fundamental ground-state realities of freedom, creativity, love and spontaneous right-action, and so is unnecessary. The *problem of evil* is thus resolved by the figure of

190 evolution of society

CO-PRESENCE (cf. *EW*: 105–6). While it is thematised explicitly only in PMR, such a conception is arguably strongly implicit in DCR, especially in the theories of ALIENATION, IDEOLOGY and the TINA SYNDROME, hence prefigured also in CN. See also DEMI-REALITY; EMANCIPATORY AXIOLOGY; ETHICS; HUMAN NATURE; META-REALITY; POLYAD.

evolution of society. This account of social evolution stresses that human adaptation generally is a response to real conditions. The model is therefore dialectical. Attempted coping creates conditions that in turn call for revised modes of coping. One result of this dialectic has been evolving social arrangements. These, too, become novel conditions requiring responses. Consequent real conditions eventually determine the degree to which any response has been adaptive. Humanness is thus founded on a dialectic of CONSCIOUSNESS and TECHNOLOGY, generating emergent social forms.

The first hominids that relied on technology are dubbed *Homo habilis*, 'Man the capable'. This species, and with it the genus *Homo*, emerged some two million years ago from bipedal primates in eastern Africa. Crude, unhafted, intentionally made stone tools were likely first used to process carrion. This adaptation marks the transition to humanness, and began a series of changes leading to today's globally distributed social and technological networks.

The earliest human social forms derived from the sorts of multi-adult troops that characterise terrestrial apes. Resource distribution and extraction variously constrains bonobo and chimpanzee social organisation, but in general these species exhibit coalitions, hierarchies, social boundaries, mate selection and other complex social arrangements.

To adapt to an open environment replete with predacious carnivores, early hominids thus also would have relied on complex social dynamics, as well as on anatomical features such as opposable thumbs, flexible shoulder-girdles, and an upright bipedal gait. The first glimmer of consciousness, the awareness of being aware of real conditions, would have been through inference. In nature, certain perceptible phenomena indicate the existence and location of other phenomena. Animal tracks indicate animals. The ability to infer one dimension of reality from another imparted to early humans a vast increase in potency. Consciousness transformed the environment from a mere container into a potential creation. It could be manipulated to meet human needs.

The emergence of consciousness was not rapid. *Homo habilis*, living in small nomadic bands, made and used only one type of crude stone tool for half a million years. Their successor, *Homo erectus*, left Africa perhaps a million and a half years ago and colonised most habitats of Eurasia. Yet for another million years *erectus* used only one type of hand axe and probably, in tropical Asia, bamboo sharps. Crucial to *erectus*' success was inference enhanced by a cranial capacity about 80 per cent of ours; they were likely capable of learning not only from direct experience but also through rudimentary information transmission. This would not have required language, as games, sports, dance, pantomime and much ritual do not.

None the less, when language evolved, probably between two hundred thousand and one hundred thousand years ago, it exponentially expanded knowledge accumulation and sharing. This is the evolutionary problem solved by human sociality. With the emergence of complex, culturally shaped social forms underwritten by a robust psyche, learning itself became the human mode of adaptation. Humans occupied the knowledge niche in nature. And, because humans perforce must make principled selections among alternate strategic possibilities, morality, in the abstract, is embodied in human nature (See also ETHICS). It is not imposed supernaturally. And, because we are by nature learners, open pathways to learning are an incontrovertible human RIGHT.

The developed psyche is capable of diverse affect, reality-based inference, a tacit theory of other minds, learning through both personal experience and cultural transmission, language, the capacity to devise and test virtual models of alternative possible realities, projection of thoughts onto reality, and social as well as landscape mapping. These capacities are summed as CULTURE, the Lamarckian transmission of acquired information, occurring in a selective environment.

In two ways selection theory is pertinent for sociality. It accounts for some social phenomena such as the lengthy infancy required by the maturation of a large brain. And cultural forces selectively retain social models. Social forms are part of the human phenotype, amenable to selection and evolution because they exhibit the requisite features: variation, transmissibility and at least tacit competition.

Human society, therefore, is a more or less organised, enduring system of interpersonal relationships. Beginning with the existentially primary dyads of male–female and mother–offspring, these include relations between siblings and more distant kin; relationships fostered by proximity and/or delayed reciprocity; inter-family alliances; task groups and coalitions; strangers and enemies. This social complex characterises the tribal social organisation that succeeded bands when expanding populations stimulated competition for resources.

About ten thousand years ago, domesticates in southwest Asia led to the first storage and associated material culture: pits, baskets, pots and granaries. These material structures in turn constrained developments in social structure. Sedentism arose, leading to satellite villages, roads, trade, rulers, allegiance, tribute, distribution, territory, rights, gifting, ethnic markers, rivals and war. This pattern developed wherever agriculture and sedentism did, namely, southwest Asia, northeast Africa, the Indus Valley, southwest China, and Mesoamerica.

The storage of material and knowledge resources constitutes the cultural appropriation of time and space, which began when stone tools first extended hominid prowess beyond the present. See also ANTHROPOLOGY, EVOLUTIONARY PSYCHOLOGY.

DEREK P. BRERETON

evolutionary psychology uses neo-Darwinian evolutionary theory to explain human cognitive and emotional phenomena. Neo-Darwinism adds to Darwinian processes of natural and sexual selection the post-Darwinian concepts of kin selection and reciprocal altruism. Natural selection is the unguided process that none the less leads to an ordered biota because individuals fitter in a given environment differentially pass genes supporting their manifestly more adequate adaptations on to offspring. Sexual selection complements natural selection as the sexes choose mates based on perceived desirable qualities commensurate with their respective reproductive strategies. Bearers of offspring, usually females, perceive an advantage in having mates willing and able to provide resources. Males in such circumstances infer benefit from inseminating many females, and infer progeny's benefit from the mother's health, in turn inferred, in humans, via youth, beauty and absence of visible deformity. Each sex selects for some of its opposite's important traits. Such 'perceptions' and 'inferences' are selectively evolved algorithms, and need not be fully conscious or intentional.

The supplementary, twentieth-century concept of kin selection has built on Mendelian genetics to solve an evolutionary problem. How could altruistic behaviour, requisite for sociality, evolve if it lessened the altruist's own reproductive success? The answer is altruists' genes are transmitted by kin the altruism benefits. Kin recognition is not necessary, as early hominid bands would have been largely kin anyway; kin recognition is possible, however, merely by a juvenile noting the individuals to whom its mother provides care. Reciprocal altruism extends altruism from the close-knit kin group to larger social forms, as social complexity increases. A social, hence adaptive, advantage would accrue to one perceived as cooperative; cheaters would be duly apprehended via compensatory algorithms – using facial expressions to infer emotional states accurately – given the several epochs such systems had to evolve.

Evolutionary psychology posits such evolved algorithms as grounding much emotional and cognitive phenomena. Fear is behaviourally manifest in desire to distance oneself from perceived threats. It originates in pre-cortical brain areas in service of survival, as selected in the environment of evolutionary adaptation (EEA). Disgust triggers eliminative responses designed to avert danger from poisoning. Most phobias are directly traceable to environmental threats from extreme heights, tight enclosures, vast spaces, venomous creatures, fear of non-kin, etc. Depression, a psycho-emotional withdrawal, was likely beneficial in some stressful EEA conditions; immobilisation can sometimes trump any sort of activity. Conversely, so-called 'attention deficit disorder' likely afforded the individual and his group advantages from the ability simultaneously to attend to multiple stimuli.

PSYCHOLOGY has struggled to supersede non-evolutionary models devised by Freud and behaviourism; both posited a psyche moulded primarily by lifetime experience. Intra-family conflict grounded the neurosis-prone Freudian unconscious, while behaviourism attributed patterned behaviour to conditioning by immediate environmental stimuli. Only Darwinism, however, provides the time, depth and intergenerational transmission process required to explain fully why certain experiences do in fact register as salient. The reason is our psyches have been differentially prepared to notice types of stimuli that impacted reproduction in the EEA.

The criticism of EP, that it reduces socio-cultural phenomena to determination by genes, is refuted by noting that many human responses depend for their activation on particular environmental circumstances, and even in the presence of particular circumstances remain highly variable. EP accounts for manifest species-level cognitive and emotional constraints, and is not rigidly deterministic. All humans fear fire because such fear was adaptive in the EEA; but we can still immolate ourselves in socio-cultural protest. Se also ANTHROPOLOGY; EVOLUTION OF SOCIETY.

DEREK P. BRERETON

exchange of non-equivalents. See ANTHROPISM; FISSION/FUSION; PRINCIPLE OF EQUIVALENCE.

excluded middle, law or principle of. See ANALYTICAL PROBLEMATIC; CONTRADICTION; TABLE OF OPPOSITIONS.

existential connection, existential constitution, existential permeation. 2E figures of OPPOSITIONALITY whereby one internally related element is affected by another. Their distinguishing features are set out in RELATIONALITY and TOTALITY. See also CAUSALITY.

existential intransitivity. See INTRANSITIVE, ETC.

ex nihilo creation. See CREATIVITY.

explanandum, **explanans**. See EXPLANATION.

explanation. The process of making some initial phenomenon intelligible. The result achieved is the *explanans* (L., that which explains) and the phenomenon – statement, event, process, etc. – is called the *explanandum* (L., the thing to be explained). Explanations are multi-form. They can be lay or scientific, natural or social scientific, applied or theoretical, full or partial, etc. The CR treatment of explanation can be distinguished by its emphasis on elaborating its metaphysical presuppositions. This allows a series of problems to be identified with the previously dominant *deductivist* account of explanation. It also enables the deleterious effect on certain (particularly social science) disciplines of the acceptance of the view that such deductivist accounts of explanation are universally applicable to be explored and catalogued. More positively, it facilitates a firmer ontological grounding for a causal account of explanation, a better means of examining the prospects for causal explanation in different domains, the analysis of strategies that may be adopted to pursue such causal projects, a clarification of the distinct mode of INFERENCE implied by causal explanation and a clear distinction to be drawn between applied and theoretical explanations.

Contemporary deductivism derives immediately from the theory of explanation implicit in Hume's analysis of CAUSALITY. This theory was articulated explicitly by J. S. Mill and has been fully elaborated by Popper and, most extensively, Hempel. It is variously known as the deductive-nomological (D–N), covering law or Popper–Hempel theory (or model) of explanation. In the case of 'statistical laws' the D–N description is often replaced by the 'inductive-probabilistic' (I-P) label. According to deductivist explanation, the explanandum must be deduced from a set of initial and boundary conditions plus universal laws of the form 'whenever event x then event y'. The deductivist model of explanation can be illustrated with a simple example. To use it to explain the event that a car radiator contained ice this morning, a set of initial conditions and at least one universal 'covering law' which logically entails this event need to be identified. One possible explanation therefore comprises the following set of initial conditions: (1) the car radiator contained water yesterday, (2) the radiator does not leak, (3) the temperature fell below 0°C last night, along with the empirical law: water freezes at 0°C. On this deductivist conception, explanation and prediction amount to much the same thing (this is the symmetry thesis). The former entails the deduction of an event after it has (or is known to have) occurred, the latter prior to (knowledge of) its occurrence. According to deductivism, the explanation of laws, theories and sciences similarly proceeds by deductive subsumption.

There is now a substantial literature that has subjected the deductivist account to near-fatal criticism. CR has added an ontological dimension to these critiques.

explanation

Deductivism stands squarely upon the Humean principle of empirical invariance; viz., that laws are or depend upon empirical regularities of the form 'whenever event x then event y'. That is to say, a presupposition of the universality, or wide applicability, of deductivism is that reality is characterised by a ubiquity of spontaneously CLOSED SYSTEMS. Pointing out that CONSTANT CONJUNCTIONS of events are rare and must (outside a few naturally closed contexts) be artificially produced, Bhaskar (*RS*) has argued that in order to understand the significance of experimental and applied scientific activity it is necessary to abandon the view that the generalisations of nature consist of event regularities and accept an account of the objects of the world, including science, as *structured*, i.e., irreducible to the events of experience, and *INTRANSITIVE*, i.e., existing and acting independently of their being identified. As the causal agency of scientists is a necessary condition for experimentally produced and controlled invariances, if laws are identified with those event conjunctions the logical implication is that scientists themselves, in their experimental activity, cause the laws of nature. Further, if laws are identified with constant conjunctions, then this leaves unaddressed the question of what governs phenomena in open systems (where few such strict conjunctions obtain). Clearly, what scientists produce in the laboratory are not laws of nature, but their empirical grounds. The generalisable features of the results achieved within the experimental context relate to the underlying mechanisms and their mode of operation. In the experimental situation an event invariance is sometimes achieved between the mechanism (experimentally insulated, and so empirically manifest) effects and the conditions of its being triggered. The legitimate object of generalisation is the experiment-invariant mechanism and its mode of operation that have temporarily been experimentally isolated and empirically identified. Deductivism could only be of relevance in a highly restricted set of circumstances; as a general account of scientific explanation these ontological considerations imply that it must be regarded as entirely implausible. Critical realists have further rigorously demonstrated how this outdated and mistaken deductivist account of explanation, when uncritically adopted, serves severely to constrain the forms of theory advanced and viewed as acceptable in certain social science disciplines (see Lawson 1997, 2003b). As a mode of inference, *deduction* is of course 'unobjectionable if it specifies the deducibility of a transfactually efficacious tendency', thereby licensing the attribution of natural necessity, but 'false and pernicious if applied actualistically' (*D*: 235).

From a CR perspective, the world is constituted, not only by events and states of affairs and our experiences or perceptions of those actualities, but also by structures, powers, mechanisms and their tendencies that, although perhaps not directly observable, nevertheless exist, whether or not detected, and govern the actual events, etc., that we do – or may – experience. Laws are analysed as the TENDENCIES of mechanisms which may be exercised without being manifest in particular outcomes. Characteristically they set limits, impose constraints and co-determine outcomes rather than generate uniquely fixed results. The satisfaction of the CP clause is regarded as a condition for the predictive success of law statements but not for the causal efficacy of laws: it is a condition for the actualisation of the tendency designated in the statement; effectively, it acts as a reminder that the system in question may not be closed and that the tendency postulated in a law statement may not act in isolation. This points, contra deductivism, to an asymmetry between explanation and prediction.

explanation 195

To provide a causal explanation of some phenomenon is to identify aspects of, or otherwise elucidate, its causal history. From a CR view, fundamental here is the uncovering of structures, powers, mechanisms and tendencies that facilitate or produce the phenomenon to be explained. While CR does not preclude the activity of uncovering events standing in the causal history of others, a knowledge merely of events and their conjunctions allows at best a very partial and limited explanation and understanding of any outcome. The same may be said for knowledge – though it is indispensable for social explanation – merely of the reasons agents have for acting (*rational explanation*) or their own conceptualisations of their activities, emphasised in the interpretive tradition (see HERMENEUTICS). The ontology sustained in CR makes it clear that explanation of phenomena at one level in terms of powers, structures and other factors lying at a deeper level is always a potentially fruitful direction of research.

The construction of a causal explanation typically involves employing antecedently existing cognitive resources and operating on the basis of something like analogy or metaphor, to develop a MODEL of the mechanism which, *if* it were to exist and act in the postulated way, would account for the phenomenon concerned. The reality of the postulated mechanism (and usually there will be several competing hypotheses) must then be empirically assessed. Once this is achieved any explanation accepted must itself in principle be explained; and its explanation in turn explained and so on. This is the *dialectic of explanation* incorporating *a dialectic of explanatory and taxonomic knowledge* which captures the essential logic of the transitive process of science as it moves recursively from empirical result or regularity (knowledge corresponding to the HUMEAN LEVEL of natural necessity) to imagined mechanism via model building (KANTIAN LEVEL) to real explanatory mechanism via empirical testing (LOCKEAN LEVEL, licensing the attribution of natural necessity expressed in a statement of CAUSAL LAW), and sometimes to a real DEFINITION of the explanatory structure as a NATURAL KIND (LEIBNIZIAN or taxonomic LEVEL, the level of necessary truth [analytic a posteriori knowledge]). To explain something, then, is neither to deduce nor to induce it (though both will often be involved [Wuisman 2005]), but paradigmatically to collect or *colligate* (n. *colligation*) it under a new scheme of concepts, designating the structures, mechanisms, powers, etc., producing it. To pursue causal explanation a mode of INFERENCE is required that takes us behind surface phenomena to their causes, or more generally from phenomena lying at one level to causes lying at a different, deeper one. This is *retroduction* or *abduction*.

On the structured and differentiated ontology of CR two basic modes or broad schema of explanation can be identified. The first, referred to as pure (or theoretical or abstract) explanation or the DREI(C) model, proceeds in a few basic steps. First, a regularity in some phenomenon, typically anomalous for existing theory (e.g., the invariance of an experimental result), is described; second, some explanatory mechanism is retroduced, antecedently available cognitive resources are used to make plausible models of unknown mechanisms; ('Feyerabend's moment') in a multiplicity of research programmes ('Lakatos' moment'); third, competing explanations are elaborated and some are eliminated on the grounds of their inferior empirical adequacy (Popper's moment); so that, fourth, the identification of the causal mechanism at work is hopefully achieved, whereupon the latter becomes the phenomenon to be explained and the initial theory is corrected in the light of the new knowledge. In that the statements of the tendencies producing the phenomenon are retrospectively deducible from the explanatory structure, which may itself come to be defined as a natural kind, this model gives us the best possible grounds for attributing natural necessity and necessary truth. The second mode of explanation, referred to as applied (or practical or concrete) explanation, or the RRREI(C) model – a form that is essential when conditions are fundamentally open – proceeds in a manner that is somewhat different. First, a complex event or situation of interest is *r*esolved into its separate components, i.e., into the effects of its separate determinants; second, these components are then *r*edescribed in theoretically significant terms; third, a knowledge of independently validated tendency statements is utilised in the *r*etrodiction of possible antecedent conditions, which involves working out the way in which known causes may have been triggered and interacted with one another such as to give rise to the concrete phenomenon under investigation; whereupon, fourth, alternative accounts of possible causes are *e*liminated on evidential grounds. This may be followed by *i*dentification and *c*orrection, as in the pure model.

Subject to qualification, both models of pure (or theoretical) and applied (or practical) explanation are operative in the social domain. But CR does not license the blind imposition of results derived from reflection on the conditions of the natural sciences to the social sphere. Rather, it is only in virtue of an independent analysis of the subject-matter of the social (psychological, etc.) sciences (see *PN*), where it is shown that there are knowable structures at work in the social domain partially analogous, but irreducible, to those identified in nature, that a position is reached from which it is possible to see that the characteristic modalities of explanation may apply equally well in the social as in the natural sphere. Critical realists have thus supported a qualified brand of *naturalism* stressing the specificity of the ways in which the movement from manifest phenomena to explanatory structures is achieved in different domains. The issue of how causal explanation can proceed in the context of a social realm characterised as open, holistically constituted and dynamic is pursued by those critical realists who elaborate upon the method of CONTRAST EXPLANATION. See also CAUSAL LAW; CRITICAL NATURALISM; EXPLANATORY CRITIQUE; Ruben 1990.

STEPHEN B. PRATTEN

explanationism. See INFERENCE.

explanatory critical reason. See EXPLANATORY CRITIQUE; RATIONALITY.

explanatory critical theory† **complex** or **explanatory critique**† **(EC**†**)**. Comprises EXPLANATORY CRITIQUES, exercises in CONCRETE UTOPIAS, and theories of TRANSITION. Organically linked and articulated with EMANCIPATORY MOVEMENTS (see also EMANCIPATORY AXIOLOGY), it provides empirically grounded theories of the immanent possibilities of developing four-planar social being.

explanatory critique (EC). CRITIQUE of a phenomenon that follows from diagnosing that it is part of the explanation of why a false belief is held (*cognitive EC*), or why some social or personal ill persists (*NEEDS-based EC*); *explanatory critical reason* is the theory and practice of EC, a form of *ethical naturalism*. A (cognitive) EC is doubly critical; it involves cognitive critique (falsity) of a belief and ethical critique of the cause of the false belief being held. Bhaskar has repeatedly urged the importance of one kind of EC: ethical critique of a SOCIAL STRUCTURE that follows, CP, from a soundly established

explanatory critique (EC)

social science theory, in which are confirmed both the falsity of a widely accepted belief, and the causal role of the structure in explaining why the belief is widely accepted. His most detailed argument is in *SR*: ch. 2, § 6, 7; an earlier version is in ch. 2, § 6, and a later development in *D*: ch. 3, § 7 (see also *PF*: app. 1; Collier 1994: ch. 6).

Bhaskar argues that, in the social sciences, if a theory, which confirms (1) that some widely held belief is false and (2) that a prevailing social structure is an important causal factor in sustaining the prevalence of this false belief, becomes soundly established – in accordance with such uncontroversial cognitive criteria as empirical adequacy and explanatory power – then (3) a negative ethical valuation of the structure follows, CP, from the theory. Moreover, if a soundly established theory (the same or another) confirms (4) that a certain activity may contribute to displacing the structure, then (5) a positive ethical evaluation of the activity follows, CP, also. In some cases, the false belief, e.g., that the structure is not causally implicated in the persistence of social ills, may be an IDEOLOGICAL belief, one whose being held widely is a condition for the structure's maintenance. Then, the inference to (5) amounts to a positive valuation of EMANCIPATORY activities aimed at removing these ills.

EC has affinities with the modern ENLIGHTENMENT idea that scientific knowledge informs human EMANCIPATION (Hammersley 2002) and, relatedly, with the quest of early modern philosophers (e.g., Bacon and Descartes) to identify, for the sake of eliminating them from scientific practices, the causal sources of the widespread errors inherited from the past. These ideas, however, were developed in a way that depended on a sharp separation of fact and value, generally associated with 'Hume's law', that 'ought' cannot be derived from 'is', or that value cannot be derived from fact (see NATURALISTIC FALLACY).

In contrast, Bhaskar's argument maintains that from 'factual' proposals expressed in the kinds of theories he describes (i.e., that certain beliefs are false and that holding them is explained in a certain way), CP, value judgements follow. Clearly, knowledge of the causes of false beliefs being held (especially ideological ones) – and, more generally, knowledge of the causes of persistence of social ills – can inform emancipatory practices in important ways, for an ill cannot be removed without removing its causes. Over and above this, however, Bhaskar claims that EC also has far-reaching philosophical implications; that the possibility of EC refutes Hume's law, and that the social sciences intrinsically bear an emancipatory impulse (*SR*: 169). These latter claims have generated considerable controversy and they will be the principal focus of this entry.

Bhaskar considers that, if the factual propositions, (1)/(2)/(4), are confirmed in theories that are soundly established in accordance with uncontroversial cognitive criteria, then the inferences from (1)/(2) to (3), and from (1)/(2)/(4) to (5), refute Hume's law. On a common interpretation of Hume's law, that 'ought' is not *logically entailed* by 'is', however, the role of the CP clause ensures that it does not (Hammersley 2002). But Collier (1994: 170) interprets the inferences as 'more like evidential or scientific than deductive arguments', and I think that they are best expressed in the pragmatic, rather than formal, mode: if one accepts that the factual proposals are confirmed in appropriate theories, then, CP, it is unintelligible to deny the value judgements. On these interpretations, refuting Hume's law amounts to demonstrating that the inference from the factual proposals to the value judgement does not involve the MEDIATION

198 explanatory critique (EC)

of any value judgements, in particular, that the CP clause does not hide any such mediation.

Bhaskar's argument is that EC is possible in the social (and related) sciences. He identifies instances – 'templates' (*SR*: 179) – of the two inference patterns in 'Marxist analysis of IDEOLOGY, Nietzsche's analysis of the genealogy of morals and Freud's analysis of repression and rationalisation' (Bhaskar and Collier 1998c: 386), and his argument is intended to articulate their general features. Nevertheless, he seems to stop short of endorsing that these analyses are confirmed in theories that are soundly established in accordance with uncontroversial cognitive criteria; and, otherwise, he offers no successful examples of EC. This points to a general feature of Bhaskar's mode of argument. It proceeds, not from a critical analysis of actual, on-going research practices in the social sciences and their established results, but rather from a (TRANSCENDENTAL) ARGUMENT for conducting research in accordance with the TMSA. This model proposes (among other things) that social phenomena, including what beliefs are held, are partly causal products of (non-observable) generative mechanisms, which include relatively enduring social structures, and which require DEPTH-investigation for their identification. For Bhaskar, the possibility of EC, derived from the TMSA, suffices to refute Hume's law, and to underlie one of the central tenets of CR, that there is a rich interplay between fact and value rather than a sharp separation between them. But, in the absence of critical reflection on successful examples, it remains unspecified how EC might actually function in social scientific inquiry, and what value judgements and specific directions for action might thereby be supported. Sayer (2000a: 160) comments that the argument, claiming to refute Hume's law, is 'strangely at odds with the experience of practical instances'.

Writers who are broadly sympathetic to CR have criticised Bhaskar's argument in a number of different ways. Some of them, e.g., have rejected the TMSA as a framework for social scientific inquiry (Chalmers 1988; Harré 2002; see also Rogers 2004); then, the only way to recover the argument would be to base it on the analysis of actual successful examples, in which the inference patterns identified by Bhaskar could be defended.

The lack of successful examples and, more generally, considerations pertaining to actual social scientific practices and their relationships to actual emancipatory movements, rather than the rejection of TMSA, are also key to the arguments of some other critics (Lacey 1997, 2002b; Sayer 2000a). These critics do not doubt that a rich interplay – 'entanglement' (Putnam 2002) – between fact and value, and between explanation and ethical critique, is essential in the social sciences. Nevertheless, they question that Bhaskar's argument adequately captures this interplay, which may, e.g., have more to do with value judgements having factual presuppositions than with logically deriving value judgements from facts (Lacey 1997). They maintain that the interplay between fact and value in the social sciences needs to be unpacked in detailed and specific ways before the emancipatory potential of the social sciences can be discerned. This does not depend on the refutation of Hume's narrow logical claim (which does not preclude value judgements having factual presuppositions), although it does presuppose rejecting the Humean-inspired view that fact and value can be kept separate and values kept out of the sciences (see also Lacey 1999).

Before turning to the criticisms in detail, some clarification will be helpful. Bhaskar, appealing to the two inference patterns, affirms that the social sciences can be sources

explanatory critique (EC)

of novel value judgements. But, while he countenances no logical moves from values to facts, he also recognises clearly that values are presupposed at the outset of inquiry, motivating and directing it. Values play a role not only in the choice of problems to investigate (see also Sayer 2000a), but also in the adoption of strategies that specify the kinds of theories to pursue, the kinds of concepts they will deploy, the kinds of possibilities they are capable of identifying, and various methodological matters that concern the procuring of relevant evidence (Lacey 2002b). Even the adoption of strategies that incorporate the TMSA, which anticipates the possibility of structural change, draws upon a view of human nature and the possibility of human transformative agency, which underlies adopting emancipatory values (*SR*: 207). What specific values are actually presupposed, and so what possibilities are considered worth investigating (including the possible barriers to their actualisation and the possible consequences of their further actualisation), may be drawn from 'a particular area of existing progressive practice, which provides a model which we would like to generalise' (Sayer 2000a: 163) or from actual emancipatory movements like those that constitute the World Social Forum (Lacey 2002b). No facts may be inferred from these values, not even the factual claim that there are genuinely actualisable possibilities of the kind considered worth investigating. Factual claims must be appraised in the light of available empirical evidence and strictly cognitive criteria (e.g., as mentioned above, empirical adequacy and explanatory power), which are distinct from ethical and social values. Although a theory, with components instantiating (1) and (2), may not be developed and tested outside of the motivating context provided by particular values, this does not imply that these values serve among the criteria of cognitive appraisal of the theory. It is indispensable to Bhaskar's argument that no ethical values have been presupposed in establishing the theories with respect to which he wishes to deploy his inference patterns. Otherwise, at the outset, his claim to have derived a value judgement from established facts would be trivialised.

Critics maintain that, in each of the inference patterns put forward by Bhaskar, value judgements do play a mediating role in the inference from (hypothetically) confirmed factual premises to the value judgement that constitutes the conclusion – a role that is disguised by the casual use of the CP clause – 'subject to a ceteris paribus clause {...} one can move without further ado to [3]' (*D*: 261–2). Consider the inference made from (1) and (2), CP, to (3). When does the CP clause obtain? According to the critics, only if the following conditions (among others) are satisfied: (a) an established theory confirms that the social structure, causally implicated in the false belief being widely held, is not also a significant causal factor in sustaining a range of positively valued phenomena; and (b) there are sound reasons to believe that there is a feasible alternative structure (Sayer 2000a: 162, 168), which would not sustain the same, comparable or worse social ills (including the cognitive ones) to those sustained under the actual structure (Lacey 1997). Interestingly, Bhaskar notes that conditions like these must be satisfied (*SR*: 185) but he quickly glosses over them, as if their obtaining can readily be relied on. Yet value judgements are involved in accepting both (a) and (b).

With respect to (a) this is immediately apparent, for valuations of a range of effects of the social structure are involved. Moreover, those who hold the values that are highly embodied in the social structure are not likely to endorse the value judgements needed to mediate Bhaskar's inference. Sayer (2000a) calls Bhaskar's use of the CP clause 'complacent' (158), in part because it glosses over condition (a), oversimplifies by considering one factor at a time, and provides no grounds for considering the inference to have normative force in the face of disagreements about value judgements. For Sayer, this points to the need for CR to attend to developing a theory of valuative and normative issues, and he does not think that inferences drawn from established results in the social sciences will play a very significant role in that theory.

With respect to (b), when we limit it so that it refers only to cognitive ills, it is more difficult to discern the involvement of value judgements. The proposition, (6), that an alternative structure, which would not also sustain the widespread holding of false beliefs, is feasible, is itself open to investigation in the social sciences. At the same time, the infeasibility of alternatives, the negation of (6) [($\sim$6)], is likely to be one of the beliefs on which the stability of the structure depends, and which are presuppositions of holding the values embodied in it – i.e., the structure is likely to be a causal factor in sustaining that the belief in ($\sim$6) is widely held. Thus the normative force of the inference depends on providing evidence for (6), a proposition of comparable status to (1) and with a comparable role in making the inference to (3). Minimally, the CP clause hides that there is greater complexity to the inference than first meets the eye.

Those who hold the values embodied in the structure will be predisposed to reject (6), as well as to question the strength of the cognitive credentials of any alleged confirmation of it and the significance of research conducted under the strategies that supposedly produced it. We might put it: CP, they will deny that proposition (b), which is implicit in the CP clause of Bhaskar's inference, can be upheld – where their own CP clause signifies 'unless compelling evidence to the contrary is provided', but where the values they hold nourish little motivation to conduct investigation on (b).

Bhaskar does discuss propositions of the type (6); their negations – 'There *is no* alternative' – function within the *TINA SYNDROME* (*D*: 116), which is routinely deployed in legitimations of predominant structures. This serves to bring attention to the structure's causal role in maintaining the widely held belief in ($\sim$6). But having such a causal source does not imply that it is false; evidence would need to be gained in inquiry to show that. Bhaskar makes clear that evidence for (or against) (6) could be obtained in empirical inquiries conducted under strategies that reflect the TMSA, but (as already pointed out) he offers no examples of successful inferences of the two types and, therefore, in view of the implicit role of (b) in the CP clause, no examples of propositions of type (6) that are well supported empirically. Even so, as already indicated, he deploys his CP clause with little commentary, almost as if the genuine plausibility (and not just logical possibility) of alternatives is a consequence of the TMSA. It was pointed out earlier that adopting strategies that reflect the TMSA is linked with (by way of the view of human nature that it draws upon) holding emancipatory values. Among the (hypothetically) factual presuppositions of holding these values is the feasibility of alternatives (Lacey 1997) – as the World Social Forum proclaims: 'Another world is possible'; if there were none, attempting to further the embodiment of these values would represent only an illusion (Lacey 2002b). Of course, that is not an argument for the truth of (b), but it is one to adopt the stance: CP, accept (and act in the light of the assumption) that (b) is true – where this CP clause, like the one deployed by those who hold the values embodied in the structure, signifies 'unless compelling evidence to the contrary is provided'. In this case, contrary to the earlier one, the values adopted provide motivation to engage in investigation that, in principle, could confirm (b) (although it could also disconfirm it). Rationally holding a set of values is always grounded in factual presuppositions (Lacey 1999: ch. 2); (a) and (b) are among the presuppositions of holding emancipatory values.

That is the context – where emancipatory values are held – in which Bhaskar's CP clause may be deployed soundly 'without further ado'. This is not dogmatic, since research conducted under the same strategies that leads to accepting (1) and (2), in principle, might lead to the evidentially informed confirmation or undermining of (a) and (b). Neither is it trivial, for the inferences might lead those who hold emancipatory values to make new or modified value judgements, or to consolidate their commitment to them, so that EC could have an important role in informing the activities of emancipatory movements. But it does not refute Hume's law, and it does not show that the social sciences have a significant role in the formation of values or the potential seriously to challenge hegemonic value judgements.

When dealing with hegemonic social structures, it is unlikely that the CP clause will ever be considered satisfied – except in the value-laden context described – and, thus, it is unlikely that specific instances of Bhaskar's inferences will gain normative force that cuts across value outlooks. Perhaps that explains why it is so difficult to produce actual successful examples of EC. When we consider theories with a narrower focus, perhaps successful examples will be forthcoming. Rogers (2004) has argued that the 'theory of action' – a theory concerned with the effectiveness of professionals in dealing with issues that require interpersonal competence and with effective interventions to improve their effectiveness – provides a sound instance of Bhaskar's inference that has general normative force or, more realistically, normative force for a wider range of value outlooks. Here we do not always encounter the same kind of investment in the values of the *status quo*, and the very context of seeking effective interventions provides openness to alternatives, so that we can anticipate that any impasses concerning the counterparts of (a) and (b) may be more easily resolved, although that does not mean that they do not involve value judgements. If Rogers's analysis survives criticism, it will suffice to show that EC, which carries normative force across a range of value outlooks, is indeed possible. Then critical reflection on it and further instances that may be offered should enable us to explore both the future horizons and the limits of EC. This is of considerable importance and serves emancipatory interests, whether or not EC serves to refute Hume's law or to provide a significant source of emancipatory values. See also Edgley (1976).

HUGH LACEY

explanatory dialectic (dialectic of explanation). See EPISTEMOLOGICAL DIALECTIC.

expressive totality, expressive unity. See EXPRESSIVISM; TOTALITY.

expressively veracious component. See EMANCIPATORY AXIOLOGY (judgement form).

expressive-referential. See ALETHIA; HOLY TRINITY; TRUTH.

expressivism (intensionalism)/extensionalism. Contrasting forms of detotalisation or REDUCTIONISM, the former depending on illicit FUSION, the latter on illicit FISSION. Denegation of TOTALITY 'encourages a tendency towards {either} analytical extensionalism or romantic expressionism, the non-dialectical or undifferentiated restriction of REASON to purely analytical or else expressivist modes of thought' (*EW*: 11), respectively.

The main philosophical error pinpointed by CR at 3L, (*ontological*) *extensionalism* is the denial of internal RELATIONALITY or intra-action, the void in irrealist thought constituted by the absence of a concept of the internally real; correspondingly, 'the division of a totality into discrete, separable, externally related parts, manifest {...} in, for example, the extrusion of thought, or contradiction, or morality, from reality – for instance, in the fact/value divide' (*DG*: 398). It could equally well be called ontological externalism or separatism, because it holds that all things are discrete and externally related, and is both manifest in and underpinned by ALIENATION or split. In the West, it dates back to Aristotle and Plato but 'its canonical, and also extreme version, is Hume's famous dictum that things "seem conjoined but never connected". (This is an extreme formulation because it denies {not only internal relations but} even necessary relationships between externally related things.)' (*D*: 10). Its watchword is disconnect; in detotalising and compartmentalising, it divides and rules. 'Ontological extensionalism is made for empire-builders, manipulators and the masters of subjects$_2$ who want to distract their eyes from the top of the power$_2$ relations on which they sit' (*D*: 305). *Logical* or *analytical extensionalism*, which is at the heart of the ANALYTICAL PROBLEMATIC, where it 'converts the love of wisdom into the fetishism of technique' (*D*: 360), is its manifestation in formal logic, which none the less has an indispensable (but properly subordinate) role in the EPISTEMOLOGICAL DIALECTIC. Because as embodied persons we are physical beings spatially separated from each other for whom physical entities but not relations are immediately visible, it will always seem that people and things are externally related and disconnected; this is the (real) 'illusion of physical being' (*RM*: 95), which, however, we may come to understand *as* an illusion (cf. the FACT-FORM). Ontological extensionalism is brilliantly lampooned by Ted Hughes in the figure of Crow (representing four-planar human being), who sees things only in separation:

> He could see the bread he cut
> He could see the letters of words he read
> He could see the wrinkles on handskin he looked at
> Or one eye of a person
> Or an ear, or a foot, or the other foot
> But somehow he could not quite see
> {...} luckily his camera worked OK
>
> (*Crow*)

Expressivism is a tradition of thought originating in the European *Sturm und Drang* of the late eighteenth century, sometimes treated as synonymous with the Romantic tradition, sometimes as a strand within it; in either event, it is in protest against the dominant objectivationist and extensionalising modes of thought of capitalist modernity (to which it counterposes forms of semantic and logical *intensionalism* focusing on the sense or meaning of expressions), demanding union and wholeness among humans and with nature, rejecting the dichotomy of meaning and being, and seeing authentic human life as an expression of an essential inner nature. Where in error, it thus makes the opposite mistake to extensionalism: it either fuses real differentiations within a TOTALITY into a mélange, thereby screening conflict and contradiction, and giving rise to a rhetoric of organicism (e.g., in fascism) or collectivism (e.g., in the erstwhile Soviet bloc), or (as *expressivist CENTRISM*) inflates one aspect of totality having to do with human being into the whole (e.g., the identity of being and thought in thought in Hegelian philosophy). It lacks 'a sufficiently strong concept of alterity necessary for true totality' (*D*: 270). It is not that there are not, or may not be, expressive unities or totalities. 'Hermeneutics provides a good initial heuristic for understanding what it is to think in this dialectical mode {of totality}. In a painting it is not only that the parts cannot be understood except in relation to the whole and vice versa but – and this is the clue to the Hegelian totality – they mutually "infect" each other – the whole is in the part, as my body is in my writing hand. This is what Althusser meant by "expressive totality".' This is 'not necessarily wrong – it is merely a particular kind of totality. Montage, and pastiche generally, and entities like the British Working Class in February 1992 provide examples of very different sorts of totalities' (*D*: 12). In general, for CR, totalities must be thought of as complex and open, recursively embedded in other totalities, 'punctuated by alterities, shot through with spaces, criss-crossed by traces and connected by all manner of negative, external and contingent as well as positive, internal and necessary determinations, the exact form of which it is up to science to fathom' (*D*: 56). In such a universe, in which external relations are real but constellationally contained within overarching internal relations, there is no role for a philosophy of pervasive internal relations in which everything is an expression of everything else (e.g., Ollman 1971, 1993, 2003), for DIFFERENCE as well as CHANGE is irreducible; the world *is* internally related overall, but in a highly differentiated, disjoint and punctuated way.

CR thus no more denies the reality of (open and partial) expressive totalities – let alone of the vitally important expressive dimensions of human life, including the expressive-referential moment in TRUTH – than it denies a role for analytical extensionalism within the overall dialectic of dialectical and analytical reason. Instead of rejecting these dichotomous positions holus-bolus, it takes what is valid in each – as in the broader INTERPRETIVIST and POSITIVIST traditions within which they are respectively embedded, or in IRREALISM, which embraces them both – and sublates them in a richer, fuller theoretical determination whose watchword at 3L is neither simple expressive unity nor disconnection, but identity-in-difference and unity-in-diversity, requiring dialectic, qua 'the art of thinking the coincidence of {real} distinctions and connections' (*D*: 180, e.a.), if it is to be thought at all.

However, there can be little doubt that Bhaskar has always looked more kindly on expressivism than on extensionalism, regarding the latter as by far the greater error. There is a powerful expressivist current in his later thought, manifest, for example, in the denunciation of possession at the expense of expression, and in the theory of generalised CO-PRESENCE. The later Bhaskarian totality remains irreducibly differentiated and open, however, and it could be argued of the Bhaskarian philosophy as a whole that, like the thought of Marx on some interpretations (Daly 1996; Taylor 1977 pp. 203, 252), it draws on earlier dialectical and spiritual traditions to marry expressivism with the radical bourgeois enlightenment in its attempt to forge an outlook fitting for a sustainable post-slave order in which diversity and creativity flourish. The notion of ALIENATION, central to both thinkers, arguably 'belongs intrinsically to an expressivist structure of thought' (Taylor 1975: 548).

extension, extensionalism. See EXPRESSIVISM/EXTENSIONALISM.
extensive and **intensive margins of inquiry**. See TOTALITY.
external relations. See RELATIONALITY.
extrajection. See FISSION.
extrinsic. See INTRINSIC/EXTRINSIC.
extrinsic intrinsic dialectics. See DIALECTICS; EMANCIPATORY AXIOLOGY.

F

φ **(phi) transform**. See EPISTEMOLOGICAL DIALECTIC.
fact, fact-form. See ALIENATION; CRITIQUE; FISSION/FUSION; IDEOLOGY; ONT/DE-ONT; PERCEPTION; REIFICATION.
fact–value. See CRITICAL NATURALISM; EXPLANATORY CRITIQUE; EMANCIPATORY AXIOLOGY; ETHICS; NATURALISTIC FALLACY.
Fall, the. See ALIENATION; PROCESS.
fallacy of misplaced concreteness. See CONCRETE.
fallibilism. See HOLY TRINITY.
falsifiability, falsification, falsificationism. See HOLY TRINITY; PARADOXES OF FALSIFICATION; PHILOSOPHY OF SOCIAL SCIENCE; THEORY CHOICE.
falsity, false being, false consciousness, false necessity (constitutive falsity). See ALETHIA; ALIENATION; CRITIQUE; DCR; DEMI-REALITY; DIALECTICAL ARGUMENT; EVIL; IDEOLOGY; ILL; NEEDS; PHILOSOPHY OF SOCIAL SCIENCE; REALISM; REASON; TRUTH.
family resemblance. See RESEMBLANCE.
fantasy, phantasy. See ALIENATION; BEAUTIFUL SOUL, ETC.; CONCRETE UTOPIA; DE-AGENTIFICATION; FISSION/FUSION; INVERSION; VIRTUALITY.
fault-line, Platonic–Aristotelian. See PRIMAL SQUEEZE; UNHOLY TRINITY.
feasibility principle. See ACTIONABILITY.
feminist theory (FT) underpins feminism as a political movement of women uniting to transform the gender order, and thus the entire social world, by ending male dominance and female subordination. From the mid-1960s to early 1980s feminist theory was unreflectively realist, using versions of EXPLANATORY CRITIQUE to draw attention to the discrepancy between the real capacities and needs of women and those attributed to them by official ideology and policy. In the 1970s feminists (1) criticised misogyny and the sexual double standard in Western literature and in PSYCHOANALYSIS; (2) argued that the oppression of women affected class-privileged and white women as well as working-class, minority and 'third world' women, so that female political solidarity ('sisterhood') had a rational basis; (3) distinguished 'sex' (physical sexual difference, seen as enduring) from 'gender' (normative characteristics and relations, and sense of identity, seen as socially constructed).

FT prior to, or outside of, the linguistic turn may be roughly divided into *liberal*, *radical* and *Marxist/socialist* sub-types, with differing ontologies and ethics. In all three trends there are elements of ontological STRATIFICATION. Liberal FT emphasised the actualaly or potentialaly equal capabilities of women and men, and the consequent injustice of social arrangements which assume relevant differences or, through discriminatory policies, construct them. Rejecting gender hierarchy, liberal FT implicitly accepts other dimensions of inequality. Radical FT developed the structural concept of 'patriarchy', usually seen as emergent from men's (but not from women's) real nature. Its adherents called for a woman-led cultural revolution to reverse the patriarchal preference for men over women, culture over nature, reason over emotion, conflict over nurturance, and put an end to 'compulsory' heterosexuality. (For ecofeminists, the very future of the planet depended on this reversal.) Separatism was either a tactical necessity or a permanent goal. Socialist feminists discussed whether 'patriarchy' was an aspect of capitalism or an analytically separable social structure (Hartmann 1993). In 'the domestic labour debate' Marxist feminists analysed the unpaid work done in households in terms of the labour theory of value. Both theorised power in the public and private realms, and the relationship between the liberation of women and of the working class.

Sisterhood was grounded in an epistemology of women's experience, formalised as *feminist STANDPOINT theory*. Drawing an analogy with class, Hartsock (1983) suggested that 'subjugated knowledges' available to women because of their work and oppressed role offered more accurate perspectives on the social world. In later versions the standpoint was worked for rather than automatic, and incorporated women's diversity in a sort of 'convergent realism' (e.g., Collins 1991). Postmodern critics welcomed the PERSPECTIVALISM but rejected the idea that any view of the social world is 'more accurate' than another. Critical realists have defended the concept of a standpoint as enabling (Lawson 1999; New 1998).

For Harding and others standpoint theory was a method: (women's) experience must become the starting point for research, and researchers must acknowledge their own positioning to achieve 'strong objectivity' (1993). The older feminist critique of science here converged with newer postmodern critiques, e.g., Haraway's (1991) critique of the 'God-trick' of objectivity and insistence that knowledge is situated. CR critics have agreed with Harding's and Haraway's epistemic relativism, but insisted on the distinction between epistemology and ontology and the possibility of rational judgements between theories (see HOLY TRINITY).

The cultural turn in FT has produced much interesting work, but critical realists have argued that in its REDUCTIONISM and IRREALISM it has been detrimental to the feminist political project. Explanatory critiques are rejected, since claims to truth are considered bids for power. Causal accounts of women's oppression are rejected, both a priori as 'universalising', and (using Humean notions of cause) because of counter-instances. Sexuality is theorised, usually in an idealist way, but other material forms of oppression are neglected. Since the social world has been reduced to texts, deconstruction and parody are the recommended forms of political action. By the 1980s a correct insistence on women's diversity was, in an ACTUALIST error, seen as undermining the category of 'woman', so that feminism itself became VOLUNTARISTIC, a matter of 'groundless solidarity' (Elam 1994). Strong SOCIAL CONSTRUCTIVISTS (especially Butler 1990, 1993) have gone further, rejecting the sex-gender distinction on the grounds that gender is discursively prior and sexual difference a mere 'constitutive construction'. Yet the posited gender dichotomy of discourse is either taken for granted or explained in terms of 'sexual difference', citing post-Lacan feminist psychoanalytic theory – as grand a narrative as they come. In response, critical realists have defended the ontology of sex, opposing the 'radical contingency' of strong social constructionism's anti-ESSENTIALISM, which 'loses all critical purchase on any oppressive exercise of power' (Sayer 2000b: 98). Ethical naturalism offers grounds to oppose women's oppression because it injures the causal powers of women and impedes their flourishing – a CR position also consonant with the ecological feminist ethics of Cuomo (1998). See also Hull 2006; Lawson 2003a; New 1996b; 2003, Soper 1990; Walby 1990, 2001.

CAROLINE NEW

fetishism. The process whereby human beings and social relations/systems are perceived to be natural, closed and immutable whereas they are really geo-historical, open and reflexive, and whereby dehumanisation and naturalisation are typically aggravated by construing inanimate things as endowed with human and social qualities. The authoritative statement is in the final section of the first chapter in *Capital*, Vol. I, where Marx develops an account of the fetishism of commodities. Commodity fetishism, together with fetishism as a broader notion of mystification and INVERSION in capitalist societies, is a distinctive feature of Marx's critique of political economy and one of the all-important points of divergence separating Marxism from mainstream economics and social science, and is consequently a crucial issue in CR. Bhaskar refers sympathetically to Marx's analysis of commodity fetishism on several occasions (*PN2*: 52; *D*: 61) and various analogical uses of the concept of fetishism abound in his work – such as the fetishism of systems (*R*: 22), the fetishism of FACTS (*PF*: 98) and the fetishism of (CONSTANT) CONJUNCTIONS (*D*: 205) – which are seen to stand in mutually supportive relation to the fetishism of commodities. Bhaskar calls attention to this borrowing when he summarises the relationship between Marxism and CR (*PF*: 143); and, as one would expect, notions of fetishism are particularly salient in CR texts that are clearly associated with Marx and Marxism. The most elaborate and explicit CR account of fetishism is articulated by Dean (2003). See also ANTHROPISM; CONCRETE/ABSTRACT; CRITIQUE; IDEOLOGY; FISSION/FUSION; REIFICATION; VIRTUALITY.

PETER NIELSEN

Feyerabend's moment, Feyerabendian moment. The MOMENT in the dialectic of EXPLANATION when a multiplicity of hypotheses is considered for testing. See also EPISTEMOLOGICAL DIALECTIC; Feyerabend 1975.
fict. See ONT/DE-ONT.
fictionality. See AESTHETICS.
fideism. In PHILOSOPHY OF RELIGION, the view that religious belief is based on faith alone since it cannot be justified by evidence or orthodox reason; faith therefore stands apart from reason, and can never be reconciled with it. It contrasts with *evidentialism*, the view that religious belief is rationally acceptable only if supported by the totality of other beliefs one knows to be true. A third view is what William Hasker has called *experientialism*, which holds that some religious beliefs 'may be directly grounded in religious experience without the mediation of other beliefs, and may be rationally

warranted on that account' (*CDP*: 294; see also Alston 1991). Bhaskar generalises the concept of fideism to include any belief or complex of beliefs that cannot rationally be justified, i.e., that is accepted on the basis of faith, authority, conformity to social rules, or caprice and will-to-power, constituting a *transcendent realism* (see SOLIPSISTIC, ETC.). See also SCEPTICISM; UNHOLY TRINITY.

fiduciariness, fiduciary. See TRUST.

figure–ground. See ABSENCE; CO-PRESENCE; PERSPECTIVE; REFLECTION.

final, final cause, finality. See CAUSES.

finite/infinite (L. *finitus*, having an end or limit, *infinitus*, without end or limit, unbounded). In Bhaskar, as in Hegel (see *dialectic*, lead quote) and Marx, correlative terms, not opposites. The finite is a moment of the infinite, constellationally contained within it. A *good* or *true infinite* is further distinguished from a *bad infinite*. The latter refers to an endless advance or regress from one thing to another. While in Hegel this is represented by a straight line indefinitely extended at either end, in Bhaskar it is more of the same, however represented. It is thus a *vicious infinite regress* resulting from 'the absence of a non-HOMOLOGOUS resolution' (*D*: 80). Capital's limitless drive to quantitative profit and the multiplication of wants are social examples proffered by all three thinkers. The current 'war on terror', which fails to address the social causes of what it seeks to repress, is another.

Bhaskar, however, radically parts company with Hegel in respect of true infinity. For Hegel, a good infinite is 'any relatively self-contained reciprocal or circular structure' (Inwood 1992: 141). Like a circle or sphere, it is thus finite yet beyond beginning or end, an AUTOPOIETIC closed TOTALITY, as in Hegel's ENDIST constellational closure of being and knowledge. Bhaskar, taking his cue from Marx's critique of Hegel, arrives at the opposite conclusion: the concept of an *open* totality is more complete than a closed one because it contains the latter as a special case (*D*: 25–6). In announcing closure, Hegel commits himself, for fear of change, to a bad, not a good infinite: a wearisome endless reinstatement of the same, which in the sociosphere is impossible in virtue of the transformative character of human AGENCY, and not coincidentally is also the logic of capital (Arthur 2002; Browning 1996). The true infinite is thus an open (absent) totality (cf. *EW*: 42 n.), unfolding and unfinished. Its central application is to the cosmos as a whole; it pertains to ABSOLUTE being, which constellationally contains the zone of finite being: to be is ultimately to be able to become, and being is an endless creative unfolding. It does not in the main apply to the totalities recursively embedded within the TOTALITY of totalities, including geo-history (relative being): real partial totalities are 'finite, limited and conditioned' (*D*: 126).

When we experience the *sublime* – for example, at the moment of scientific discovery of a new stratum of being – we 'comprehend the awesome aspects of our ignorance', that 'reality is a potentially infinite totality of which we know something but not how much' (*D*: 15). This experience is simultaneously 'a recognition of our *finitude*, projected outwards and reflected back on to us' (*SR*: 98). Finitude (or transitoriness) is an absolute constraint imposed by nature on any concrete singular, 'the simple existential ALETHIA of absence' (*D*: 200). Finite human beings – contingent coagulations of possibility – are none the less capable of infinitude (see RECURSIVITY), and short of annihilation there is no conceivable end to the physical, social and moral evolution of the species. The achievement of a society of universal social individuals, the flourishing of each of whom

208   **first moment** (also **prime moment**) (**1M**)

is the condition for the flourishing of all, would mark, not the closure of geo-history, but the radical openness of a new beginning.

The Bhaskarian concept of infinity is accordingly represented by an open spiral or helix, not by a circle, the symbol of closure. In today's world it is the totalising drive of capital that constantly seeks to effect the bad infinite of a closed circuit, such that it controls all the conditions of its own reproduction (Arthur 2002; Dean 2003). Bhaskar agrees with Marx that it can never completely succeed, ultimately because the capital-totality is parasitically grounded in, and in denial of, that which is ontologically not-capital and prior to it: the inalienable CREATIVE capacities of living labour as distinct from wage-labour, or, labour capacity as distinct from the commodity labour-power (see PRIMACY, Arthur 2003), which – as 'the content obscured by the form, the living masked by the dead' (*D*: 241) – can very well do without capital but not vice versa.

Transfinity, a term introduced into mathematics by Georg Cantor in 1874, refers to an infinity (or infinite set) of infinities. Thus in Bhaskar $power_2$ relations constitute a 'potential transfinity' (*D*: 153). See also ALIENATION; ALTERITY/CHANGE; DIFFRACTION; ENTROPY.

first moment (also **prime moment**) (**1M**). See MELD.

fission/fusion. When illicit, two fundamental forms of CATEGORIAL ERROR grounded in the EPISTEMIC–ONTIC FALLACY, which, while apparently opposed, tacitly complement each other, as can be seen in the following examples: the *fusion* of world and experience in EMPIRICAL REALISM, and the *fission* of being in the ACTUALIST account of laws. The former involves the ABSENCE of a *distinction* (between the ID and the TD), producing a *false* totality and rendering the 1M concept of alterity impossible, the latter involves the absence of a *connection* (between laws in closed and open systems), producing a *split* or *detotalised* totality. ('Dialectic', Bhaskar reminds us when broaching this topic, 'depends upon the art of thinking the coincidence of *distinctions* and *connections*' [*D*: 180, e.a.].) Fission and fusion are also in evidence as the ontological presuppositions of methodological individualism (social ATOMISM – fission) and holism (social fusionism, resulting in an 'oversocialised' conception of agents).

Fission and fusion are not, however, confined to IRREALIST philosophy and social theory: they are indispensable to the functioning of the mode of production itself in modernity/postmodernity, which irrealist philosophy thus helps to reproduce (and vice versa). Thus one form of illicit fusion is *the exchange of non-equivalents* (labour for labour-power), which is pivotal to the wage-labour/capital contract, and without which capitalism could not operate. This is in turn connected with the *representation of sectional interests as universal*, which is characteristic of free-market ideologies, as when sectional Anglo-American interests are identified with those of 'the international community'. And a prevalent form of illicit fission at work in capitalism (which itself may be sourced to *GENERATIVE SEPARATION* or *fission*) is the *non-parity of equivalents* which is evident when women and immigrants are paid less for the same work than native males, which is also connected with an ideological mechanism: *the representation of universal interests as sectional* (as when global ecological crisis is presented as a sectional, purely Green concern). Fission and fusion may thus be put critically both to vital 'politico-ethical' and (in philosophy) to 'systematic-diagnostic use' (*D*: 181) (see also PRINCIPLE OF EQUIVALENCE; EQUITY). As regards the former, the illicit fission at work in the non-parity of equivalents, which contravenes the norm of truth, provides the basis for an argument for 'a core basic human equality in resources and opportunities', which (with suitable CP provisos) 'immediately {...} entails an argument for the abolition of {...} power$_2$ differentials, at least of the present gross kind', and 'augments our argument for freedom and shows the sense in which freedom and equality {...} are not necessarily at loggerheads' (*D*: 181). Regarding the latter, CR seeks to show 'how the initial illicit fusion of the EPISTEMIC FALLACY at work in empiricism and rationalism coheres with the fissions caused by ONTOLOGICAL MONOVALENCE and the PRIMAL SQUEEZE {...} to produce the dominant – and analytic – irrealist problematic within a potentially infinite array' (*D*: 182). The dialectics of fission and fusion are thus central to METACRITIQUE and IDEOLOGY-CRITIQUE and underline the primary role of the concept of absence in CR.

Split is often used synonymously with fission. A 2E figure of OPPOSITIONALITY which does its detotalising work at 3L, it is also close to DICHOTOMY and to ALIENATION (estrangement from self), and so is central to ETHICS: 'properly speaking split and wrong are the same thing' (*MR*: 142 n). *Gulf* is its spatial counterpart. A *split-off* is 'a dualistic dichotomy' (*DG*: 396), i.e., a split that is productive of DUALISM. It may take the form of *extra-jection* at 1M (e.g., of the concept of ontological stratification from irrealist thought by PRIMAL SQUEEZE, splitting being and thought about being); *retro-jection* at 2E (e.g., reading our own spontaneous, factualising consciousness – the FACT-FORM – back into history [ETERNISATION], thus splitting [dehistoricising, detemporalising] the present); *intro-jection* at 3L (e.g., of the master's viewpoint within the psyche of the slave, creating internal split); or *pro-jection* at 4D (e.g., slaves identifying with the fantasy world made for them by masters, splitting minds from bodies). In split-off, the divided whole or any resulting 'fragment' (e.g., the periphery split from the centre) is marginalised, isolated or rendered impotent. *Split intentionality* is unfulfilled intentionality, because it issues in AXIOLOGICAL indeterminacy; the UNHAPPY CONSCIOUSNESS is its archetypal figure. More generally, 'detotalisation at 3L is symptomatic of an ALIENATED world. Its sign is split, and is clearly connected, as cause and effect, to 2E opposition', including the 'generally dichotomous character' of the social sciences (*D*: 321, 370, see also OPPOSITES). In an alienated world, fission and fusion constitute a bad dialectic or bad infinite. Illicit fission necessitates external grafting, hence illicit fusion, whose DUPLICITY generates further split ad infinitum. The only way out of such a dialectic is the abolition of its social ground. See also BEAUTIFUL SOUL, ETC.; DIALECTICAL ANTAGONISTS; EXPRESSIVISM/EXTENSIONALISM; FETISHISM; IDEOLOGY; TINA SYNDROME; TOTALITY.

fixism/fluxism. A form of *identity-thinking* informing the ANALYTICAL PROBLEMATIC, *fixism* is (1) the ontological notion that particulars are constant (TOKEN–token IDENTITY); and (2) the corresponding epistemological notion that subjects in the subject–predicate propositional form are fixed and can therefore be caught in rigid definitions – negation and change are always in the predicate, and accidental rather than necessary. *Fluxism* is the DIALECTICAL OPPOSITE of fixism, transmitting the flux in thought to things. Everything is flowing, any definition is too rigid.

CR espouses, by contrast to fixism, which is a form of REIFICATION, (a) the PROCESSUALITY of particulars (*developmental negation*) and (b) *subject-developmental negation* or self-transformation, a species of radical negation (see *D*: 139, 249–50; *RM*: 87–8;

210 **flourishing**

ABSENCE). The latter builds on the Hegelian 'speculative proposition' whose subject (society) is in complex process of formation, involving the idea of a necessary CONTRADICTION in the subject (*contradictio in subjecto* issuing in *subject negation*) developed by neo-Kantian philosopher Bruno Fischer (1824–1907). From this perspective fixism's apparently precise definitions are in fact arbitrary, and prevent, in particular, adequate conceptualisation of science as a process (see EPISTEMOLOGICAL DIALECTIC, KNOWLEDGE).

Though Bhaskar's terms of art, introduced in *Dialectic*, fixism and fluxism were generated de facto by Parmenidean MONISM, making their first appearance in the guise of the PROBLEMS of the one and the other and of the one and the many. They both overlook that, metacritically, change and difference (see ALTERITY/CHANGE) (i) are mutually irreducible, and (ii) presuppose each other, exemplifying dialectical connection; i.e., that we are situated within *metacritical limits* – 'in a world between the absence and ubiquity of change' (*D*: 255). For CR, a dialectical or developmental PROCESS involves both (1) stages of development (entailing difference, as seen, e.g., in the development of a chrysalis into a butterfly), and (2) process (change). Fixism in effect denies process (change is reducible to difference); fluxism in effect denies development (difference is reducible to change). They are thus both forms of illicit FUSION, lacking a concept of ontological depth. They are accordingly sublated in the notion of 'structured open systemic flux' (*D*: 126). See also ACTUALISM; BLOCKISM; CONSISTENCY; IRREALISM; RHYTHMIC.

flourishing. See ETHICS; EUDAIMONIA; FREEDOM, ETC.; EMANCIPATORY AXIOLOGY.

foldback (fold-back), principle of. See INTERESTS.

fork. An argument pointing more than one way; containing a DILEMMA. Those employing the notion usually hope to show that an opponent's argument entails a fork, every 'branch' of which involves problematic or intolerable consequences for the position espoused. *Hume's fork*: alleged dichotomy between relations of ideas and matters of fact. See also OPPOSITES.

form analysis. See SOCIAL FORM.

form–content. See ABSENCE; CO-PRESENCE; CRITICAL NATURALISM; DIALECTIC OF FORM TO CONTENT; EMANCIPATORY AXIOLOGY; PERSPECTIVE; REFLECTION.

form of life (Ger. *Lebensform*). Wittgenstein's term for practical, as distinct from theoretical, consciousness; more broadly, for the always already systems of social practices, necessary for language acquisition and use, which (for him) constitute the human world. The complementary notion of a *language-game* highlights that language is simultaneously a social activity (part of a form of life), and the means whereby that activity and the 'world' is constituted as something determinate and identifiable. The terms suggest the indissoluble unity of linguistic rules, social activity, and objective situations. Language-games/forms of life are governed by rules internal to themselves, such that criteria for deciding what is to count as meaningful, true, rational, etc., are internal. They are the ur-stuff of the social world, constituting its limits, the ULTIMATA beyond which it is impossible for human being to go. While the concepts must be rejected as embodying the LINGUISTIC-ONTIC FALLACY and, if pushed to their logical conclusion, as incompatible with the possibility of communication itself (see *PN3*: ch.

4), the insight that social practice is a necessary condition of language-learning and use is important. Archer (2000) has emphasised the role of the embodied and non-semiotic dimensions of social practices in this. In CR, 'form of life' is sometimes used also in the same broad sense as *society*. See also (by no means the same) SOCIAL FORM.

formal causes. See CAUSES.

formal logic. See ANALYTICAL PROBLEMATIC; CONSISTENCY; EPISTEMIC–ONTIC FALLACY; EPISTEMOLOGICAL DIALECTIC; EXPRESSIVISM/EXTENSIONALISM; IRREALISM; LOGIC; PROPAEDEUTIC.

formalism. See PDM.

foundationalism and **fundamentalism.** Epistemological concepts often used synonymously to capture the notion that one's KNOWLEDGE, whether in epistemology proper or in economics, politics, religion, etc., is or can be certain and incorrigible in starting out from indubitable foundations, indemnified from scepticism, error and change. These include, in the EMPIRICIST tradition, theory-independent or brute facts and basic statements, and in the HERMENEUTICAL tradition incorrigible or brute lay interpretations ('*interpretative fundamentalism*') (*SR*: 160 f.). The project to identify certain foundations is the *epistemological problematic* (*PF*: 31, 33). On this view, those who are unable to establish and set out from non-inferential or unassailable premises can know nothing at all (cf. Descartes). The quest for certain foundations is expressive of fear of change, which 'manifests itself in the normative guise of the fear of error' (*SR*: 23). It 'inevitably splits reality into two (viz., that which conforms to its criterion and that which does not)'. Historically it thus gave rise to the first great problem of philosophy, the PROBLEM of OPPOSITES or of the one and the other, which in turn gave rise to the second, that of CHANGE or of the one and the many. What fundamentalism ignores is that we can never start from scratch or an indubitable starting point because we are always 'THROWN' into an already existing EPISTEMOLOGICAL DIALECTIC or process, entailing epistemological relativism and the possibility of CRITIQUE. It thus arrives at the opposite conclusion to that of its dialectical twin, (epistemological) ENDISM or *absolutism*, which assumes that we are, or can be, left with nothing to do. Modern foundationalism, however, typically holds that foundational beliefs need not provide certainty or deductively (rather than probabilistically) support non-foundational beliefs. Its dominant twentieth-century epistemic form is empiricist *justificationism*. Foundationalism and fundamentalism are closely related to REDUCTIONISM (in that 'an autonomised or empty mind' [*DG*: 395] is presupposed) and MONISM (which uniquely satisfies 'the demand for the unconditioned, the starting point which is the ending point' [*D*: 357]) as one of a cluster of primal errors generated by and generative of the EPISTEMIC FALLACY and ONTOLOGICAL MONOVALENCE. The most powerful fundamentalism in the world today is the ideology of the free market.

Within CR, the position of the later Bhaskar in ETHICS has been argued, not uncontroversially, to be foundationalist in an *ontological* sense, i.e., grounded in an ahistorical, idealist conception of the human ESSENCE (Hostettler and Norrie 2003). Of course, if 'foundational' means 'grounded in (existentially and/or causally intransitive) reality', as it often does in postmodernist contexts, then CR is ontologically foundationalist. See also ARCHIMEDEAN POINT; HOLY TRINITY.

212 four-planar social being

four-planar social being. See DCR; HUMAN NATURE; PHILOSOPHICAL ANTHROPOLOGY; SOCIAL CUBE.
fourth dimension (4D). See MELD.
free will. See ALIENATION; DETERMINATION; NATURAL KIND; SOCIAL STRUCTURE.
frontier. See TRANSITION.
freedom, etc. 'To be free {...} is to know and to possess the power and disposition to act in or towards our real individual, social, species and natural interests' (*R*: 187). Though closely related, *freedom* should be distinguished from *autology*, *homology* and *autonomy* (or the negative of the corresponding contrary meanings of *heterology* and *heteronomy*) as illustrated in Table 21.

Autology (Gr. *autos*, self + *logos*, word, thought, reason), *true of*, or applicable to, *itself* (contrary: *heterology* in sense [a]). Closely related to *homology* (f. Gr. *homologein* [*homos*, the same + *logos*], to agree), *the same as itself* (contrary: heterology in sense [b]) (*DG*: 398). Ontologically, homology presupposes closure and depthlessness. 'The avoidance of homology, intimated by Hegelian dialectic {...} and refined by critical realist ALETHIC TRUTH {...} is the key to the resolution of most of the textbook PROBLEMS of philosophy' (*DG*: 400, e.a.), i.e., genuine explanation of a phenomenon must be in terms of a mechanism other than its instances, thus avoiding infinite regress. *Autonomy* (Gr. *autos* + *nomos*, law), *true for* and/or *to* and/or (if it is de-ALIENATED) in *itself* (contrary: heterology in sense [c], or *heteronomy*); self-DETERMINATION (within constraints). Autonomy has three components: (i) cognitive – one must know what is in one's real interests; (ii) empowered – one must have the requisite skills, resources and opportunities and know how to act on them; and (iii) dispositional – one must be motivated to do so (*D*: 260). If we use our autonomy rationally and wisely, we shall tend to be able to act in our real interests. This is *rational autonomy* or *rational agency*. It is thus a 'theoretico-practical bridge concept linking truth to freedom mediated by wisdom in a two-way dialectic' (*DG*: 395), closely connected to DIALECTICAL REASON or the coherency of theory and practice in practice (*D*: 281). One can only act in one's real interests if one knows the alethic truth of one's situation – the constraints on one's autonomy and the possibilities afforded – and allows oneself to be guided by PHRONESIS.

Autonomy is the 'root notion' of the broader concept of *freedom*, which constellationally contains it, ranging from (1) *agentive freedom* ('the capacity to do otherwise analytic to the concept of agency') via (2) *formal legal freedom*, (3) *negative freedom*

Table 21 Modes of freedom and unfreedom

autology	**homology**	*freedom* → *emancipation* → **autonomy (self-determination)** → *dialectical freedom*	**heterology**	**heteronomy**
(a) true of itself	(b) the same as itself	(c) true for, to and in itself	not (a) true of, (b) the same as, or (c) true for, to, or in itself	not (c) true for, to and in itself, i.e. not autonomous or free

from constraints, (4) *positive freedom* (to do, become), (5) *emancipation from specific constraints*, to *universal human emancipation*, (6) *autonomy* (self-determination) in varying degrees all the way to universal concretely singularised human autonomy in nature, (7) *well-being* ('oriented to the satisfaction of NEEDS and the absence of remediable ills') and universal concretely singularised well-being, (8) *flourishing*, and (9) universal concretely singularised human flourishing in nature – *dialectical freedom* or *the EUDAIMONISTIC society* (in ETHICAL terms, the supreme good or moral ALETHIA) (*P*: 145; *D*: 102).

Underlying this conception of freedom – which presupposes axiological underdetermination and multiple control, hence the depth-openness of the world – is the 'naturalistic vision that *wo/men as such are free* {. . .}, but that to win their freedom they must absent the constraints$_2$ on it' (*D*: 291, e.a.). *Emancipation* (emancipatory practice) 'consists in the *transformation*, in self-emancipation by the agents concerned, *from an unwanted and unneeded to a wanted and needed source of determination*'. It spans the 'logical gap between "knowing" and "doing", which can only be bridged by "being able and wanting to do in suitable circumstances"'; while 'causally presaged and logically entailed' by EXPLANATORY CRITIQUE, 'it can only be effected in *practice*' (*SR*: 171), i.e., it presupposes an emancipatory POLITICS. Since the theme of absenting constraints is sometimes taken to indicate that the Bhaskarian conception of freedom is purely negative, it is important to underline that it is equally positive: 'there is an *equivocity* of *freedom from* and *freedom to* {. . .} as two poles of ultimately the one concept' (*D*: 260). By perspectival switch, 'the absence of a capacity to do x can always be viewed as a constraint on x', hence negative freedom from specific constraints as positive freedom to do x (*P*: 145). Freedom is never a purely cognitive matter, either, so can no more be the simple recognition of necessity (Engels) than escape from it (Sartre) (*SR*: 171). Indeed, both formally and substantively the 'negative generalisation' of the concept of constraint to include inequities, alienations and master–slave-type relations is accompanied by a '*positive generalisation of the concept of freedom* so that this includes not only rights, democracies and equities (including the absenting of constraints$_2$) but also needs and possibilities or potentialities for development' (*D*: 277–8; *P*: 144). The consideration that you cannot advise someone to do something and not be prepared to do it yourself in exactly similar circumstances without theory–practice inconsistency (see EMANCIPATORY AXIOLOGY) grounds the *dialectical indivisibility of freedom*. For 'an increase in your concretely singularised freedom means, ceteris paribus, an increase in my freedom, universalised out as an increase in my right to freedom and ultimately in every agent's freedom'; conversely, 'not to concede to others the rights and benefits one appropriates to oneself' is also to commit performative contradiction. It is this (principle of) *the indivisibility of freedom* that 'provides the transcendental deduction of the CONDITIONALITY REQUIREMENT' in the formula for eudaimonia, that is, which makes 'the free development of each {. . .} a *condition* of the free development of all') (*P*: 146; see also EMANCIPATORY AXIOLOGY). The positive side of freedom is also conceptualised as the *fulfilment* of one's needs and potentials, which in PMR goes by the name of SELF-*realisation*.

Heterology (Gr. *heteros*, other + *logos*) '(a) *not true of*, or applicable to, *itself* (contrary: autology); (b) *not the same as itself* {i.e., other than itself} (contrary: homology); and/or (c) *not true for* and/or to *itself* (contrary: autonomy which, if it includes de-alienation, entails true in itself also)' (*DG*: 398, e.a.). (a) and (c) are its most prevalent sense: untrue

214 fulfilment/unfulfilment, fulfilled intentionality

of or to itself. (c) also goes by the name of *alterology*. *Heteronomy* (*heteros*, other + *nomos*) is subjection to the rule of another being or power, or to an external law (contrary: autonomy), hence *not true for*, to or in itself (cf. ALIENATION); it is synonymous with heterology in sense (c). Thus the autonomy of people consists in 'reflexive self-direction', their heteronomy in 'passive adaptation' (Dean 2004: 117).

fulfilment/unfulfilment, fulfilled intentionality. SEE AGENCY; CRITICAL NATURALISM; DESIRE; DIALECTICAL REASON; ENLIGHTENMENT; FISSION/FUSION; FREEDOM; META-REALITY; REASON; SELF.

functional explanation (FE). A species of the genus *consequence explanation* (CE) of the family causal EXPLANATION; it may be *TELEONOMIC* or (in the case of individual action and certain tendencies it may ground at the level of the social real) *TELEOLOGICAL*. CE explains something in terms of its consequences, FE in terms of its beneficial consequences for its own continuing existence (Figure 6).

When unpacked, this can be seen to involve causal explanation as normally understood, where causal laws are construed as tendencies and powers, not empirical regularities, albeit the form of a consequence law differs from a statement of an ordinary causal law (see *SR*: 142–54); Cohen's useful account (1978), from which the examples are taken, is vitiated by actualism. That is, causal explanation > consequence explanation > functional explanation, whether teleonomic or teleological.

Analytically, on a CR construal, there are four components of FE: (1) *dispositional power (tendency) claim*, P (e.g., hollow bones, A, tend to facilitate flight, B, in virtue of some mechanism, M_1); (2) *functional gloss* on P: it is beneficial or functional for some end, e.g., species survival, i.e., B is a function of A; (3) *consequence claim*: it is because of P that A comes or came about, i.e., if A tends to do B, then A exists; (4) theoretical elaboration of the *consequence-producing* or *adaptive mechanism(s)*, M_2, implied in (3), whereby P tends to produce A.

Whereas an ordinary causal law may be glossed (in this example) as 'if P then A', i.e., P → A, i.e., in terms of our example, (A → B) → A, a *consequence law* (since B provides the context or environment) has the form B. (A → B) → A. This does not involve explaining causes by their effects: the dispositional power cited in (1) pre-exists or is concurrent with the establishment of functional relations. *Ascribing* a function is not in itself explanatory (for example, 'the function of the heart is to pump blood'); we must be able to demonstrate the existence of a consequence mechanism. This is the main mistake of *functionalism*: it assumes that, just because something plays a useful role, that explains it. In the biological example we can specify the consequence-producing mechanisms: chance variation and natural selection; these mechanisms drive the *teleonomic* process of biological evolution in the direction of increasing emergent complexity. In the social world they arguably have their counterparts in intentional agency and social selection. All explanations of individual action can be cast in

(A) hollow bones in birds   Growth of Protestantism

↓ ↑   ↓ ↑

(B) facilitate flight   Development of capitalism

Figure 6 Functional explanation

functional or consequence form: the mechanism is the belief that the action will have the intended effect. Social outcomes, however, are never wholly intended, so social FE, if legitimate, cannot be reducible to intentional explanation; it is therefore not assimilable to *teleological explanation*, which is in terms of ends or purposes and is valid only in respect of individual action (the functional gloss gives the *appearance* of social teleology only), though human praxes may ground teleological tendencies at the level of the social real (see also EMANCIPATORY AXIOLOGY). However, intended social outcomes may play a role through trial and error. It is important to note the differences from biological evolution: social change is neither 'automatic' (it depends on intentions) nor 'blind' (functionality becomes a human project [see GEO-HISTORICITY]), and it is 'Lamarckian' in that we pass on 'adaptive' learning. FE of social phenomena can rarely if ever be complete, because social processes are much more open than biological ones and there is chronic contestation and conflict. It does not entail *functionalism*, the view that all elements of social life are interrelated in a mutually reinforcing or beneficial way; though functionalism, if true, would entail FE. The relation between functionally necessary mystificatory phenomenal forms and essential relations in the CR account of IDEOLOGY is, *pace* Hammersley (2002), only quasi-functional (*SR*: 195 ff.). See also PRVS MODEL.

fundamentalism. See FOUNDATIONALISM.

fusion. See FISSION/FUSION.

fusion of horizons. See DIALOGUE; HERMENEUTICS.

future. See SPACE–TIME; TENSE.

futures studies. A new concept of progressive time emerged in eighteenth- and early nineteenth-century Europe. Social theorists and political philosophers working in the traditions of Kant, Marx or Spencer projected development and progress into the future, for instance in terms of 'perpetual peace', or 'socialism' and 'communism', or 'evolutionary development'. However, the idea of studying the future systematically in its own right emerged decades later. In 1902, H. G. Wells, who failed to share the popular nineteenth-century faith in inevitable progress, argued that the future of humankind is a no less suitable subject for scientific inquiry than the past and the present. Futurology should not be just a branch of sociology but rather an attempt to bring together all relevant disciplines 'for whatever they could tell about the future of earth and especially its human fauna' (cit. Wagar 2004: 37).

In spite of Wells, it took some sixty more years before futurology became a practical endeavour. The World Futures Studies Federation (WFSF) – an organisation of some 500 individuals and 60 institutions around the world whose mission is to promote futures education and research – was established in 1967. It emerged in the 1960s from the ideas and pioneering work of such persons as Igor Bestuzhev-Lada, Bertrand de Jouvenel, Johan Galtung, Robert Jungk, John McHale and others who conceived of the concept of futures studies, often at the global level. Since the early 1960s, futurologists have developed their own (usually rather small) institutes and networks among those who share an orientation towards the future, but understand their task, not in terms of predicting the future, but in terms of analysing trends and possibilities and building scenarios. Although important for planning activities in modern organisations ranging from corporations to nation states, futures studies have remained relatively marginal in the human sciences.

futures studies

A quick examination of contemporary practices and discussions concerning methods of futures studies (see, e.g., Bell 2000 and recent issues of *Futures*) reveals that futurology is not an independent science, although its wholism and disrespect for established disciplinary boundaries can be fruitful when studying local and global problems. Futurology is more a temporal orientation. Futures studies are dependent on the knowledge of experts and lay actors in various fields. Futures studios, future barometers and the so-called Delphi-technique are all based on the idea of putting a group of experts from the relevant fields together and asking them to assess possible futures collectively in order to overcome the subjective biases of any particular analyst. Very often, these types of futures studies have focused on the values of particular quantifiable variables such as the price and availability of oil at some future point in time or on the date of particular technological innovations such as a commercially viable fusion reactor.

Different estimates are usually grounded on the assumption that certain empirical regularities remain fairly constant across time and space. In retrospect, many earlier expectations concerning future possibilities have proved either wrong or inflated. There are many famous examples of failed predictions (see Bobrow 1999: 5–6). Sometimes technological 'impossibilities' have proved wrong almost instantaneously. In 1895 the president of the British Royal Society asserted that 'heavier-than-air flying machines are impossible'; in 1958 the head of IBM thought that 'there is a world market for about five computers'. Economic predictions based on 'business as usual' have also failed miserably. In 1929 one of the best-known of all American economists, Irving Fisher, proclaimed that 'stocks have reached what looks like a permanently high plateau' and, acting on this assumption, lost between US$8 and US$10 million in net worth. The end of the Cold War is a similar example in world politics, illustrating how faith in the continuity of past experiences can be seriously misleading.

It is true that some of the pioneers of futurology were preoccupied with prediction. Outside economics, however, the idea of predicting the future has not survived the test of time. Even in mainstream economics, the notion of prediction is ambiguous. Milton Friedman argued in the early 1950s that the 'relevant test of the *validity* of a hypothesis is comparison of its predictions with experience' (Friedman 1953: 8–9). However, the meaning of the term 'prediction' in this sentence is far from clear. In practice, it usually means postdiction, i.e. 'predicting' the past – all those things that have already happened – rather than any kind of foresight about aspects of the future. It is now widely acknowledged that even in the most favourable circumstances – business as usual – the real predictive power of econometric models is not only contingent but also rather limited in terms of both time and space. Usually, reliable and precise economic predictions cover only the next few months.

The standard CR explanation is that both Friedman and most mainstream economists have not noticed that they are attempting to apply deductivist ideas and methodological tools that are only suitable for use in CLOSED SYSTEMS of directly observable phenomena. As Bhaskar has maintained (*R*: 20–36), outside astronomy, these systems are only to be found in the laboratories of classical, Newtonian physicists. Outside artificially created existential and causal closures, there are only open systems. The point of artificial closures is to isolate the mechanisms, laws and explanatory structures that are also causally efficacious – i.e. capable of transfactually producing outcomes – in the open systems that exist in nature. In open systems, which typically also include non-observable components and layers, explanation and prediction are asymmetrical; and meaningful, precise scientific predictions are, in general, impossible (which does not, however, mean that nothing can be anticipated). Apparently, this argument would seem to indicate that CR sides with prevailing practice in most of the social sciences: one should remain silent about possible futures because the future is unknowable. A few critical realists have protested, knowing that all planning directly involves the consideration of different futures (Næss 2004); and arguing against the clear-cut dichotomy between open and closed systems and making the reasonable point that almost all systems – including our solar system – are in fact situated somewhere between absolutely open and absolutely closed systems, i.e. they are in fact closed to a varying degree (Töttö 2004: 269–84 et passim).

The study of possible futures is indeed an essential task for the social sciences (Patomäki 2006a, b). Futurology is first and foremost a temporal orientation. The ontology of CR provides a starting point for studying possible futures. Realist ontology posits that the future – which is real but not yet determined and therefore consists of a multiplicity of different possibilities (see TENSE) – unfolds through various transforming events and nodal points, themselves presupposing particular concept- and action-dependent historical social structures. Since everything takes place in relatively open systems, forecasts are necessarily contingent on a number of uncertain conditions: multifarious geo-historical processes and mechanisms (such as homeostatic causal loops), and modes of responsiveness linked with learning and self-regulation in layered systems. On the other hand, this conditionality is also the purpose behind futures studies. The idea is not to adjust ourselves to given future realities, but to try to shape worlds that are yet to come. Furthermore, in recognition of the interpretive nature of scenario-construction, studying possible futures is also a self-reflexive exercise in cultural studies, moral philosophy and creative ability.

Heikki Patomäki

G

gender (f. Gr. *gen-*, to produce, give birth). Originally a term referring to grammatical categories and other kinds, was appropriated by Anglophone second-wave feminists to distinguish between (1) sexual difference/dimorphism ('sex'), and (2) socially constructed identities, rules, beliefs and practices ('gender') through which sexual difference is given meaning and organised in social life. Early postwar gender theorising focused on gender identity as a psychological property acquired through learning, but psychoanalytic and sociological criticism led to the development of more complex stratified ideas of gendered subjectivity. The distinction between sex and gender historicised the social relations associated with sex, seeing them as real but changeable structures of social practice. It arguably retains political clout, since at the Fourth UN Conference on Women in 1995 several Catholic and Muslim states formed an alliance to replace all references to gender in the conference documents with the term 'sex'.

In European writing the terms 'sexual difference' or 'sexed subject position' are more often used where Anglophones would say 'gender'. Gayle Rubin (1975), drawing on Marx, Lacan and Lévi-Strauss, described sex/gender systems as sets of social relations and practices through which biological sexuality is given social and cultural form and meaning. These practices include the sexual divisions of labour, incest taboos, the exchange of women and the institutionalised dominance of heterosexuality. Poststructuralists have since argued that the sex/gender binary is a false DUALISM, derived from other couples such as nature/human and nature/culture. 'Materialist feminism' (where discourse is the 'material') emphasises the domination inherent in the social production of women. Post-Lacanian 'sexual difference theory' insists on the discursive, imaginary, yet enduring and compelling nature of sexual difference. For both, the denaturalisation of 'woman' is a central project. The supposed naturalness of sexual difference is, in these views, a socially constructed regulatory fiction requiring exposure. 'Early critiques [of the binary logic of the nature/culture distinction]' did not focus on historicising and culturally relativising the 'passive' categories of sex or nature. Thus, formulations of an essential identity as a woman or a man were left 'analytically untouched and politically dangerous' (Haraway 1991: 132).

Critical realists, in contrast, do see sexual difference as 'natural', in that its ontology is not primarily socially caused. The vast majority of humans have either 'male' or 'female' clusters of characteristics (and there is a finite range of variation on the various dimensions within these clusters). Male and female are NATURAL KINDS constellationally contained within the human species (which is itself a natural kind).

Their different properties, their clustering and structure, result from various underlying causal processes theorised in evolutionary biology. Males and females share most of the causal propensities of humans, but each sex has its own powers and liabilities (in certain circumstances possession of a womb permits pregnancy, in others it makes its owner liable to uterine cancer). As for intersexuality, the incidence of which is exaggerated by post-structuralist gender theorists, it does not undermine sexual dimorphism. It, too, '*is a reflection of the causal structure of sex*, a real structure that intimately links male and female bodies while explaining differences. It is not the effect of a "resistance" to discourse or a lawless chaos or even a continuum' (Hull 2006).

To be a man is more than being male, and similarly to be a woman is always to be a member of particular socio-cultural groups that attribute certain meanings and obligations to femaleness. Gender is among the crucial cultural and social EMERGENT properties of persons, affecting their agency in terms both of their constructed capacities, opportunities and motivations. Some SOCIAL STRUCTURES are intrinsically, others contingently, gendered. A CR theory of gender would theorise the construction and effects of gendered structures at all levels. Gender orders are legitimised in terms of local beliefs about sexed bodies, their capacities and liabilities, and about the nature of the social world. Where post-structuralists commit the epistemic fallacy by confusing what we believe and know with what is the case, critical realists recognise that local beliefs about the forms and implications of sexual difference may be better or worse representations of reality – leaving a space for critique and social change. Gender orders arguably prevent NEEDS of both women and men being met, and thus impede their flourishing. New (2001a) has argued that these are complementary, dialectically related oppressions. See also CULTURE; MASTER–SLAVE; RACE; New 1996b, 2003, 2005; Sayer 2000a, 2000b; Walby 1990.

CAROLINE NEW

general conceptual scheme. See INTERDISCIPLINARITY, ETC.; PHILOSOPHY; THEORY, ETC.

generalisability. See UNIVERSALISABILITY.

generalisation. See CONTRAST EXPLANATION; COUNTERFACTUAL/TRANSFACTUAL; EXPLANATION.

generalisation, negative and **positive**. See NEGATIVE GENERALISATION; TMSA.

generative grammar. See GRAMMAR.

generative mechanism. See CAUSAL LAW; CAUSALITY; CONCRETE/ABSTRACT; COUNTERFACTUAL/TRANSFACTUAL; TENDENCY.

generative separation or *dialectical generative separation* or *FISSION* or *diremption* (see PROCESS [dialectical process]) is Bhaskar's term of art for the Marxian notion of a primary or original separation or ALIENATION of the immediate producers from their labour, which violates a primary dialectical unity of theory and practice (think of a gardener just gardening) and generates a fivefold alienation of producers from – besides (1) their labour and its product – (2) the means and materials of production, (3) each other, (4) the nexus of social relations within which their production takes place, and (5) ultimately themselves. They are thus alienated at all four planes of their social being or HUMAN NATURE, a condition which underpins the whole gamut of

220 generic dialectic of desire

MASTER–SLAVE-type relations and their ideological legitimations. Generative separation produces a gash in four-planar social being inaugurating power$_2$ relations (*P*: 241), i.e. is transcendentally presupposed by them, and grounds the 'Essence Logic' (*D*: 333) or fundamental RHYTHMIC of social systems. Its potential is given in the emergent power of REFERENTIAL DETACHMENT. The concept is redolent with 'the slaughter-bench of history' (Hegel), speaking of the violent sundering of that which should be whole and, in our own distanciated geo-historical stretch-spread, of a process set in motion in England in the seventeenth century and still reverberating around the world – in India, China, Brazil, Kosovo, Russia and the Middle East – with bloody butterfly effect. Violent separation or split is the literal meaning of diremption (f. L. *dis-* + *emere*, to take apart). In PMR primal generative separation is seen ultimately as alienation from the divine, encompassing – what is encapsulated in the Marxian concept – alienation from our true selves and the totality they inhabit (*EW*: 10).

generic dialectic of desire. See DIALECTIC OF DESIRE.

generic dialectic of interests. See INTERESTS.

genetic fallacy. The mistake of arguing that, because something was originally x, it must still be x; e.g., people evolved from apes, therefore they must still be a form of ape. This violates the principle that 'x cannot be the sole source of, or {be} completely explained in terms of y, if x is relatively unchanging and y is not'. It is related to the normative *transhistorical justificationist fallacy*: e.g., because x was right for our ancestors, it must be right for us, which violates the principle that 'y cannot be fully justified or comprehensively criticised in terms of x, if x is relatively unchanging and y is not'. The general point inscribed in these principles is that 'in a world characterised by inconsistency, diversity or change (or, in the normative mood, error, evil, unhappiness, etc.), the invocation of alleged or real invariants (such as the given, human nature, sense-experience, reason, or God) can neither adequately explain the occurrence nor underwrite the rationality of one rather than another set of historically actualised (or possible) practices' (*SR*: 46). Such arguments abstract illicitly from dialectical development or PROCESS, making the converse mistake to that involved in ETERNISATION. In the philosophy and history of ideas they are often bound up with the EPISTEMIC FALLACY: when being is reduced to our means of knowing being, the only criteria for deciding issues of truth are internal to the paradigm, and so questions of origin within the tradition assume an irreal importance. The two fallacies are often confused, to the point where it is sometimes convenient to refer to both (the resulting condensate) as the genetic fallacy.

geography of religion, (GOR) A subfield of human geography concerned with the geographic phenomena of religions, beliefs, practices, experiences, sacred spaces and places, and the diffusion and contestation thereof. Approaches to (GOR) include ecclesiastical geography, empiricism, positivism, Marxism, social constructionism, humanism, phenomenology and post-structuralism. Empiricism and positivism, frequently described as the 'poetics' of GOR as founded in the work of Mircea Eliade, represent the core approaches dominating the work published even in the last decade. Marxism and social constructionism, which are gaining in popularity, accentuate the social and political conflicts which surround religious sites and boundaries and are often expressed as the 'politics' of GOR, an example being the work of Chidester and Linenthal. CR has, to this contributor's knowledge, not yet been applied in GOR. CR could enlighten epistemic gaps, help incorporate poetics and politics, provide fertile soil for methodological pluralism, and open new doors for discussion among geographers of various traditions. CR is especially relevant to GOR in light of its recent 'spiritual turn', with examples including Roman Catholic applications (Porpora), Bhaskar's commitment to the possibility of universal self-realisation in TDCR and the philosophy of META-REALITY, and even an attempt philosophically to reconcile Christianity and Marxism (Collier 2001). The application of Bhaskar's TMSA could also provide rich opportunity for study. See also PHILOSOPHY OF RELIGION; Chidester and Linenthal, (eds), 1995; Eliade 1968; Kong 1990, 2001; Morgan 2003b; Park 1994; Porpora 2001a; Sopher 1967.

MICHAEL P. FERBER

geo-historical directionality. See DIRECTIONALITY.
geo-historicity (**historicity**), **geo-history**. Added to 'history', 'geo-' highlights that space and time, ALTERITY and CHANGE, are ontologically irreducible. In any inquiry, neither should be assumed in advance to be the more important. The term refers either to specifically human geo-history or to the whole process of formation of the cosmos. However, human geo-history differs from SPACE–TIME in that change is always substantial, whereas space–time incorporates non-substantial change in the form of action-at-a-distance. Human geo-history is the process of production of the contents of our future, mediated by the past and what we understand of it (*SR*: 149).

Lévi-Strauss distinguishes the *cold*, concrete societies of 'primitive' peoples which have no sense of history, i.e., in which the present is experienced as the enduring perpetuation of the past (a common view of non-master–slave peoples), from *hot* historicised societies in which the present is seen as radically different from the past, a product of change over time. (*Geo-*)*historicity* is both (1) the quality of having such a self-reflexive consciousness of history; and (2) that to which the consciousness refers, the quality of spatialising diachronic change which is exhibited by all societies, though at widely varying rates. At a meta-level, (3), it is conceptually a derivative of ABSENCE, embracing the whole of being, not just human being. Geo-historicity (3) is thus any PROCESS of DIRECTIONAL change; cf., 'There certainly is historicity and geographicity in nature (e.g., in biology and geology) and probably – through cosmology (and especially cosmogeny) – extending into the domain covered by chemistry and physics too' (*PN3*: 176). Geo-historicity (1) is a relatively recent, emergent phenomenon (which anthropologists themselves helped to impose on 'cold' societies). However, as Bhaskar suggests, hunter-foragers 'certainly had a past (and outside)' (*MR*: 103) in the sense that they lived in the full presence of their ancestor-creators and myths, with a keen sense of their origins and distinct identity within the cosmic order; and 'hot' societies have it in common with them that, according to the ruling ideologies, there is no future, in the sense that qualitative social change is unthinkable: from the perspective of total history, even as historicity (1) emerged in hot societies, it was frozen by ENDISM and ONTOLOGICAL MONOVALENCE, such that there was a history once, but not now (Marx). 'Hot' societies are thus only 'half' geo-historicised, split between past and future – the dominant outlook lacks a sense of the '*futuricity* of PRAXIS' (*D*: 251, e.a.) – and rampant ETERNISATION *de-geo-historicises*. De-geo-historicisation is productive of a philosophical DILEMMA: ' "There is no history" – but the very statement of this, consisting in transformative praxis, constitutes a performative contradiction' (*P.* 244). A eudaimonian society would be fully historicised, combining an awareness, both (a) of radical departure from the past (which itself made that departure possible) and of redemption of the crushed utopian aspirations and possibilities of the past (the presence of the future within the past), and (b) of future change as necessary and desirable for human flourishing, with a re-ENCHANTED view of the cosmos as the unfolding of being and of its own continuity and connectedness with this process – 'embracing process and change, openness to the future, as an essential part of our being' (*MR*: 128). We would live our lives in the moment but in the full presence (in the sense indicated) both of the past and of the future, such that the distanciated present was present to itself – not absent, as in the 'fast-twitch' PUNCTUALIST here-now of market societies –, cradling us 'as if on a boat rocking on a lake' (Hölderlin). *World historicity* (2), whereby all societies are caught up in the globalising process of capital, constituting 'a new RHYTHMIC' (*MR*: 103), was inaugurated in the dawn of modernity. See also DEMI-REALITY; HISTORIOGRAPHY; PDM; TENSE.

Gibsonian perception theory. See PERCEPTION.

globalisation has become part of the *lingua franca*. Commentators tend to agree that it entails socio-spatial relations that are significant beyond the traditional conception of the nation state – that there is:

- an incomplete but globally expanding infrastructure of travel, communication, trade, finance, diplomacy and law, some of it operative through large corporations trading, retailing, or producing in networks within and across several states and regions; but also entailing additional operation and regulation by nation state members of various global institutions such as the UN, the World Bank, the IMF and the WTO, constituting a network of global governance initiatives – themselves monitored and held imperfectly to account by a variety of NGOs such as Amnesty International.
- a degree of economic, political and cultural interconnectedness around the human world that cannot simply be defined in terms of relations between nation states as discrete units. Decisions and events in one locality (wages in Beijing) affect others (unemployment in Birmingham). Such interconnectedness is reciprocal or interpenetrative (McDonald's moves East, Tandoori West), and, empirically speaking, is tending to increase in density, particularly through the medium of technological change (global share and foreign currency trades have increased year on year since the linked computerisation of exchange networks). As a result, interconnectedness seems to be shrinking social or conceptual space through its relation to time – the Roman world was two months across whilst the industrialised world is a matter of hours in person and seconds by media (see SPACE–TIME; DISTANCIATION).

Beyond these acknowledgements, conflict concerning the nature and significance of particular characteristics of globalisation has been a principal feature of debate. Is globalisation something radically new in terms of new technologies and possibilities, or more a matter of continuity, the latest incarnation of longstanding forms of social relations? What is its political and social significance – is it progressive or oppressive, something to be lauded or overcome? Positions on these questions span a spectrum from the positive accounts of neo-liberalism, to the negative accounts of neo-imperialism. The former associates globalisation with emancipation through unfettered market development and democratisation. The latter associates it with the primitive abuses of sophisticated capitalism. A variety of critical realists working in international and area studies have contributed to the analysis of globalisation. For example, Patomäki (2002) has done influential work on global financial instability based on an analysis of structures as mechanisms with causal powers. He, like the philosopher Hugh Lacey (2002b), and in keeping with the emancipatory emphasis in CR, is active in the World Social Forum. Lacey has highlighted the significance of Bhaskar's pre-DCR EXPLANATORY CRITIQUE argument such that the empirical establishment of theory T that shows belief P about object O to be false should lead to the transformation of O (*SR*: 184). This might be further extended in terms of dialectics and power$_1$ and power$_2$. Patomäki (2004) has produced, in cooperation with a series of international lawyers, academics and activists, blueprints for new international institutional forms aimed at a genuine democratisation and accountability in global relations.

JAMIE MORGAN

gnoseology (gnosiology), adj. gnoseological. Another name for epistemology or theory of KNOWLEDGE. *Gnosis* was the common word for knowledge in classical Greece and only later, with the rise of *Gnosticism*, acquired the special meaning of higher insight that leads to salvation, construed as the freeing of the essentially divine spiritual part of ourselves from imprisoning matter. There are arguably (Collier 2001c) gnostic tendencies in this special sense in the later thought of Bhaskar, who, however, stresses that 'the way of gnostic being' is to seek emancipation through practical, THIS-WORLDLY engagement ('practical mysticism') (*RM*: 221).

Gödelian dialectic. Gödel's incompleteness theorems show 'very roughly that no formal system can be shown to be both consistent and complete. Extra-formal systems are usually, and I have argued necessarily, neither' (*P*: 27 n). *SR* builds on Gödel to argue that 'the determinant outcomes of internally inconsistent (and so formally indeterminate) systems can only be explained by reference to factors, uses or principles not contained within the system {. . .} The resolution of contradictions is never just a matter of immanent or internal necessity, but always presupposes external constraints upon or selection of outcomes' (*SR*: 262; see also TINA SYNDROME). In *Dialectic*, Bhaskar demonstrates that contradiction and openness are necessarily pervasive extra-formally. A *Gödelian dialectic* thus moves extra-formally as well as formally, in a triadic pattern whereby incompleteness (ABSENCE) generates inconsistency (CONTRADIC-TION) in a theory, leading to greater completeness (TOTALITY); as such it is a vital component of the EPISTEMOLOGICAL DIALECTIC and the DIALECTIC OF FREEDOM. There is no conceivable end to such dialectics.

good (dialectic, infinite, society, totality, etc.). See the concept qualified in this way.

graft, grafting. See CONSISTENCY; DUPLICITY; TINA SYNDROME.

grammar. A *generative grammar* is Chomsky's term of art for a set of structural elements and rules that generates the well-formed sentences of a language in a manner comparable to the way in which the elements and rules of chess generate possible situations. A philosophical or scientific ONTOLOGY may likewise be viewed as a grammar: an *ontological grammar* generative of possible situations on the chessboard of the ACTUAL. An *analogical grammar* is the indispensable stock of metaphors, analogies, models, etc., available to a field of inquiry or theoretical approach which helps to generate hypotheses and solutions (whether scientific or ideological, manifestly or latently) and lend the approach plausibility. The term is G. Buchdahl's (see *RS2*: 242). See also CAUSAL LAW; THEORY, ETC.

grand narratives, grand theories, etc. See DIRECTIONALITY; FEMINIST THEORY; NARRATIVE; PDM; PHILOSOPHY OF SCIENCE.

great arch tradition. See ARCH OF KNOWLEDGE TRADITION.

great chain of being. See PRINCIPLE OF PLENITUDE.

ground, grounding. 1M figures deployed in CR chiefly in the conventional sense of a reason, principle, basis or explanation for something, especially a fundamental or adequate one (the underlying *causal grounds*) (n.); and (vb.) to supply such (cf. to *earth* something). Hence a theory, opinion, etc., may be well *grounded* or not. In DCR, ground in the sense of the real or dialectical reason (dr') for things is their ALETHIC TRUTH. Grounding thus admits of several degrees, ranging from the mere detection of a possibly causal correlation to the referential detachment of a new level of ontological structure and its definition as a natural kind (see EPISTEMOLOGICAL DIALECTIC; EXPLANATION).

A *dialectical ground* (*DG'*) underpins (1) real dialectically contradictory or opposing forces (e.g., GENERATIVE SEPARATION underpins the master–slave social opposition); and (2) the apparent OPPOSITION, but real COMPLICITY, of DIALECTICAL ANTAGONISTS. Such oppositions can only be abolished by transformation or abandonment of their common ground(s) (alienated social relations, categorial error[s]) (see ALIENATION; CRITIQUE; IDEOLOGY). The *ungrounded* has no basis in reality; e.g., if there is no route from facts to values, ethics is ungrounded or *groundless*. In *figure and (dialectical) ground*, ground signifies the 3L CO-PRESENCE of the outside and/or past within an entity. Thus present in the figure of the centre is the periphery that helps to sustain or ground it (see also DIALECTIC OF FORM TO CONTENT; RELATIONALITY).

In PMR, the *ground-state*/ cosmic envelope is the ultimate, implicitly conscious, field(s) of possibility ingredient in, and necessary for, the whole of being, including human being (see ULTIMATA); the ground and alethic truth of being, its *META-REALITY* (*MR*: xiv, 249). It has ethical as well as conventionally causal dispositional properties – a conception entrained by the breakdown of the fact/value divide, such that being is intrinsically valuable, and by the principle of the priority of the possible over the actual (what is actual must have existed independently as implicit possibility).

ground-state. See ALETHIA; CREATIVITY; ETHICS; GROUND; META-REALITY; SELF-REFERENTIALITY; SUBJECTIVITY; TELEOLOGY; ULTIMATA.

gulf. See FISSION/FUSION.

H

habitus (L. *habitus*, form or appearance; nature or disposition). Often thought of as an exclusively Bourdieuian concept, but actually with origins in the classical tradition of sociology, explicitly in Weber and implicitly in Durkheim and the Frankfurt School. It is deployed by a range of CR authors to capture the notion of historical depth-PSYCHOLOGY, referring to the historically emergent structure of embodied dispositions to act habitually and according to current styles that informs people's practice, 'ensuring the heavy weight of the past in the present' (Lawson 2003b: 45–6). Social structures, though distinct, are not external to people: they *in-here* within and through them. See also IDEOLOGY; Camic 1986.

haecceity. See NATURAL KIND.

having. See ALETHIA; DESIRE; ENLIGHTENMENT; EXPRESSIVISM.

health, medical sociology. See SOCIOLOGY OF HEALTH AND MEDICINE.

Hegel–Marx critique. DCR is importantly derived by complex metacritique of Hegel and Marx. As an aid to understanding the relevant arguments in *Dialectic*, Tables 22 and 23 set out the contours of this critique.

hegemony. Generally, the way leadership or dominance is exercised through the attainment of consent. While the balance of force and consent is important, the term also implies cohesion. Hegemony therefore also relates to the way a project can be constructed and held together. This means examining the conditions under which different social groups and their interests can be brought together into a bloc. Gramsci (1971) relates such a project to underlying social conditions through his notion of the historical bloc.

The concept of hegemony as described is not really present in Bhaskar's work. However, he does use the idea of hegemonic struggle when outlining a project of human emancipation. These struggles are understood as *HERMENEUTIC hegemonic/counter-hegemonic struggles* in the context of generalised master–slave POWER$_2$ relations (*D*: 62). By power$_2$ relations Bhaskar means the capacity to get one's way against the overt wishes or real interests of others in virtue of structures of exploitation, domination, subjugation and control. These relations are what Bhaskar calls generalised MASTER–SLAVE-type relations, and it is around these that hegemonic struggles are waged. Strangely, though, Bhaskar usually talks of these struggles as hermeneutic, only several times mentioning other more material but still conceptualised hegemonic/counter-hegemonic struggles (e.g., *D*: 269; *DG*: 402). Thus hegemony is mainly conceived of as a struggle over description and re-description.

Table 22 The fine structure of the Bhaskarian critique of Hegel

Hegel aspires to	Hegel's criteria	Form of critique	Overall critique	Immanent critique	Omissive critique	Antinomial critique
(α) realise the traditional goals of philosophy within an immanent metaphysics of experience	(a) rationality and seriousness (corresponds to Being in *Logic*, and to Understanding in the epistemological dialectic)	transcendental realism/ epistemological materialism	(α′) 'realised' idealism (principle of identity) centrism-expressivism, denegation of autonomy of nature *loses intransitivity*	unseriousness: faulty transitions; actualism, empirical realism, eternisation: substitutes a logical present for a tensed geo-history, and spirit for structure, rendering both science and change impossible. Untrue to his theory of truth which entails the transformation of actuality to conform it to its notion. Analytical reinstatement.		
(β) unity in diversity	(b) totality (corresponds to Essence in *Logic*, and to Dialectic in	ontological realism/ materialism	(β′) constellational spiritual monism ('logical mysticism'): speculative illusion,		replete with absences – of key 1M–4D concepts, incl. determinate	

	the epistemological dialectic)		cognitivism, triumphalism *loses transfactuality*	absence, open totality and TTTTφ. Failure to transcend transcendental idealism re science; subordinates causality to teleology.
(γ) avoid the fate of the **B**eautiful Soul without **U**nhappy **C**onsciousness	(c) clarity (corresponds to *N*otion in *Logic*, and to **S**peculative reason in the epistemological dialectic)	practical materialism	(γ') preservative dialectical sublation (immanent teleology), endism ontol. monovalence *loses transformative praxis*	interpretive antinomies, e.g., his philosophy of history is no consolation for the slaughter-bench of history

Table 23 Hegel–Marx critique

Form of critique (c.f. holy trinity)	Underpinning error (c.f. unholy trinity)	Marx on Hegel Formal	Marx on Hegel Substantive	Bhaskar on Hegel Formal	Bhaskar on Hegel Substantive	Bhaskar on Marx Formal	Bhaskar on Marx Substantive
transcendental realism/ epistemological materialism	epistemic–ontic fallacy (irrealism)	principle of identity	denegation of autonomy of nature – dialectical complicity idealism/ empirical realism	'realised' idealism (subject–object identity, actualism)	centrism, sociological reductionism of ethics	underdevelopment of (i) scientific realism; (ii) critique of empiricism (actualist residues); (iii) ID and objectivity; (iv) normativity; (v) discourse/communication; (vi) necessary mediations re individuality; (vii) historical materialism vs. critique of political economy	presentational linearity, ethical sociological reductionism, political negativism, lack of concrete utopianism (provides formal criteria for the good society only) and prefigurationality
ontological or emergent powers materialism	speculative/ positivist illusion, involving primal squeeze	logical mysticism	cognitivism/ theoreticism (the only labour is abstract mental labour)	constellational spiritual monism	triumphalism		triumphalism qua Promethean-ism, class$_2$-unidimensionality, functionalism, teleology, evolutionism
practical materialism (cf. TMSA)	ontological monovalence	triple inversions: absolute idealism/ universals as properties of particular things; epistemological rationalism/	denegation of historicity of social forms	preservative sublation (underpinned by immanent spiritual teleology)	endism	principle of transformative sublation leads to underestimation of material presence of past	post-dated endism, programmatic expressivism

	knowledge as empirical; idealist sociology/civil society. No distinction objectification and alienation; alienation of spirit and of labour		proclivities to actualism, monism, demoralisation, utopianism in pejorative sense
Mystical shell (*in nuce*)	[as for Bhaskar, the terms alone differing]	ontological monovalence (absence of determinate absence – uncancelled contradiction)	
Rational kernel	(a) above all, epistemological dialectic, incl. dialectical explanation, entailing quasi-ontological stratification and its diffraction via transformed concepts of (b) alienation, labour, and social totality in process (Essence Logic, inner dynamic); (c) negativity, contradiction, praxis and dialectical reason. Also, (d) the exoteric form of presentational dialectics	epistemological dialectic, incl. dialectical explanation, etc. [as for Marx]; generalisable to any learning process; entailing transcendence in a greater totality.	(i) critical dialectics: epistemological dialectic, depth-explanation, (implicit scientific realism) and explanatory and ideology critique, entailing the holy trinity of **CR** (ii) systematic or presentational dialectics (inscribed within [i]). (iii) practical dialectics – transformative negation of the causes of falsity.
Golden nugget		dialectics of co-presence	moral alethia: the free development of each as a condition of the free development of all
Platinum plate		diagnostic value of philosophy: irrealism reflects irrealist surface categorial structure of society, realist deep structure	

Note Marx's critique of Hegel has been cast largely in CR terms. Conversely, the Bhaskarian critique of Hegel is essentially the Marxian critique, e.g., 'logical mysticism' = 'constellational spiritual monism', 'denegation of historicity' = 'endism').

If we look at what a more material concept of hegemony might look like, we can draw on Bhaskar's TMSA to relate hegemony to the production and reproduction of social structures. Thus structures are reproduced and occasionally transformed by human agents. To this we might add conserved, so that hegemony now becomes the struggle over the reproduction, conservation and transformation of social structures. This reveals the political moment of structure and the structural aspect of hegemony running alongside the more agential conception of hegemony as leadership and consent. Indeed, the structural and agential aspects of hegemony have to be understood as related, in that agential projects depend upon the underlying conditions of structural reproduction, while this structural reproduction in turn can only be realised through agential activity (see SOCIAL STRUCTURE). Therefore hegemonic projects might be said to be emergent out of underlying structural conditions while developing their own degree of specificity. This specificity means that hegemonic projects can always act back upon the conditions out of which they emerge in unfavourable ways, thus generating a crisis of hegemony. Under these circumstances an alternative counter-hegemony may vie for power (Joseph 2001, 2002, 2003a).

Such a position and its reliance on the TMSA model is criticised by Jessop (2003b). The TMSA model is said to be DUALIST, and in its place Jessop offers a STRATEGIC-RELATIONAL APPROACH that examines various mechanisms that produce greater or lesser social coherence. Rejecting the retroduction from hegemonic projects to structural hegemony, Jessop argues that specific structures selectively reinforce specific modes of action and discourage others (strategic selectivity) while actors orient their strategies to take account of this situation. See also HOBBESIAN PROBLEM OF ORDER; RECOGNITION; REVERSAL; SIGNS.

JONATHAN JOSEPH

Hempel's paradox. See NICOD'S CRITERION.
hermeneutic adequacy. See HERMENEUTICS.
hermeneutic hegemonic/counter-hegemonic struggles. See HEGEMONY; HERMENEUTICS.
hermeneutical circle. See HERMENEUTICS; MENO'S PARADOX.
hermeneutics or **interpretivism** (Gr. *hermēneuein*, to interpret, explain, make clear, f. *Hermes*, the messenger of the gods). The art, science or philosophy of interpretation or elucidation of meaning, first in ancient and/or sacred texts, by the nineteenth century in texts generally, and thereafter, by analogy, in all human activity and socio-cultural contexts. As a philosophical tradition it has defined itself historically in opposition to POSITIVISM, positing a dichotomy or DUALISM of method between the natural and the human sciences, reflecting a putative absolute difference in subject-matter. On this account, the subject-matter of the social sciences is uniquely meaningful, conceptual and rule-governed, such that the fundamental aim of social science must be to elucidate its meaning and trace conceptual connections, either by (logocentric) *immersion* in it (e.g., Winch) or by (dialogical) *fusion* of horizons or meaning-frames (e.g., Gadamer); for meaningful action can be understood but not causally explained (reasons are not causes, RATIONAL EXPLANATION is not causal explanation) – it is the natural sciences that are concerned with causes and empirical method. Interpretivism is thus fundamentally anti-NATURALIST.

hermeneutics or **interpretivism**   231

Such a contrast goes back crucially to Kant, whose EMPIRICAL REALISM denied space to science which his transcendental idealism used 'to make room for human freedom from causal determination' in the transcendent or ineffable realism of a noumenal realm (*SR*: 287 n; Kant 1787, *preface*). This was echoed in the Hegelian split between nature/spirit. In the hands of Wilhelm Dilthey, Heinrich Rickert, Max Weber and others, this dualism of body and mind subsequently generated dichotomies between the physical/social, phenomena or sensibilia/intelligibilia, nomothetic/ideographic, causal explanation (*Erklären*)/interpretive understanding (*Verstehen*), etc. Within the tradition, *neo-Kantians* (e.g., Weber and Habermas) try to fuse hermeneutics and positivism into a unified approach to social theory, whereas *dualists* (e.g., Winch and Gadamer, from the Wittgensteinian and Heideggerian traditions, respectively) hold that positivism has no relevance to the social world. Interpretivism also informs PHENOMENOLOGY, ethnomethodology and the Nietzschean PERSPECTIVISM of most POST-STRUCTURALISTS and POSTMODERNISTS ('there are no facts, only interpretations'), which likewise often go hand in hand with a positivist ontology of the natural world. It should, however, be seen as only one, though the main, current of the 'anti-scientific romantic reaction that has always co-existed in symbiosis with positivism' (*PN3*: 126) and will be with us until the 'blessed end' of the social system with which they both resonate (Marx 1939, cit. *R*: 208 n).

While at bitter odds with positivism in many ways, hermeneuticism tacitly presupposes positivism by taking its account of science and the natural world for granted, in particular its perceptual criterion for ascribing reality and construal of causal laws as empirical regularities, thereby ensuring that it dances ultimately to the tune of its enemy, i.e., it is the (junior) dialectical partner of positivism. This DIALECTICAL ANTAGONISM of the 'champions of meaning and of law' is resolved in CRITICAL NATURALISM, which sustains both the conceptuality of social reality and the transfactuality (generality) of its causal laws (*PN3*: 123–4). Interpretivism is thus committed to EMPIRICAL REALISM and ACTUALISM (empirical for the natural, conceptual for the social world – a split leading to the reification of agency in the extrinsic aspect [i.e., physicalist reductionism] and its disembodiment in the intrinsic aspect), underpinned by an UNHOLY TRINITY in which the epistemic fallacy often appears as the linguistic fallacy, whereby the limits of language are the limits of the world. This is its chronic failing. The identification of social being and thought in conceptual actualism must be rejected on the grounds that all conceptualisations are corrigible and that social being has an irreducible material dimension, both in the physical sense (the human body, buildings, roads, etc.) and in the sense of other material causes (social structures, etc.).

Tacit complicity is further evidenced by the reproduction of positivist themes in transposed form in the interpretivist account. Thus it often espouses a VOLUNTARISM which is 'the simple inverse of positivism's blanket determinism', reflecting 'in the last instance the same mistaken notion of a causal law' (*PN3*: 20). Likewise, its sensible/intelligible distinction depends upon the Humean notion that there are no intelligible connections in nature. Its conceptual actualism and sociological reductionism, whereby concepts or people's own understandings of what they are doing in their activities exhaust or constitute the social, is the counterpart for the social world of its empirical realism for the natural. This commits it to the view that lay

232   hermeneutics or interpretivism

conceptualisations are incorrigible, ruling out CRITIQUE, hence rationally defensible conceptual change; for, once we grasp lay conceptualisations, that is all there is to know – a notion which ignores unintended consequences, tacit knowledge and the unconscious, as well as that conceptualisations may be inadequate and/or distort what is going on, or that aspects of social reality may not have been conceptualised at all, and renders social science neutral or indifferent to social practices. Notwithstanding the hermeneutical critique of Cartesian foundationalism, this tends to an *'interpretative FUNDAMENTALISM'*, mirroring the empiricist's 'basic statements' as the foundation of knowledge, and to 'a displaced hermeneuticised SCIENTISM'. The *'brute data'* of the social world become *'brute interpretations'* (*SR*: 160–1). Likewise, the hermeneuticist social ontology of actions (informed by concepts) is the mirror image of positivism's ontology of events, and Winchian 'rule-uniformity' can be viewed as a 'metonymic hermeneutic displacement of the PROBLEM OF INDUCTION' (how can we know we are carrying on in the right way when there are a potential infinity of rules governing any activity, e.g., going for a walk?) (*D*: 318; cf. *P*: 197).

The interpretivist phenomenal/intelligible dichotomy is reinforced by an essentially twofold argument that reasons cannot be analysed as causes. First, it is claimed that they are not logically independent of the actions they govern. But 'natural events can likewise be explained in terms of their causes (e.g., toast as burnt)'. Second, they are said to 'operate at a different language level (F. Waismann) or belong to a different LANGUAGE-GAME (Wittgenstein) from causes' (Bhaskar 2003a: 483). But, if reasons are not causally efficacious, what grounds can there be for preferring one reason-explanation to another?

Notwithstanding these shortcomings, from a CR perspective the hermeneutical tradition has a number of very important interrelated achievements to its credit (cf. *PN3*: 152). It usefully highlights:

(1) The *pre-interpreted* character of social reality. Social reality is largely always already interpreted (conceptualised, linguistified, valued) under some or other descriptions, i.e., *'already brought under the same kind of material in terms of which it is to be grasped'* (*PN3*: 21). This provides the indispensable starting point for social science and the humanities (there is no other), though only a starting point, as depth-inquiry seeks to uncover generative mechanisms which may lead to criticism and correction of agents' understandings. That is, in the process of description, explanation and redescription which is science, the social scientific investigator must first achieve *hermeneutic adequacy* or *Verstehen* in relation to her subject-matter, via a mediation of horizons, by reading meaning out of it rather than into it, and by thinking in the mode of TOTALITY, as a condition for generating *descriptive adequacy*. In general, the most adequate (re)descriptions will be those entailed by the theory with most explanatory power. Thus, theoretically contextualised, the statement that 'police executed an innocent Brazilian as a suspected terrorist in the Stockwell tube, London, in July 2005' might provide a more adequate description of the event than a statement to the effect that they 'shot' him. Descriptive adequacy, while respecting the authenticity of lay accounts, does not leave them unchanged; 'not to call a spade a spade, in any human society, is to misdescribe it' (*PN3*: 61). Social science thus requires a *triple hermeneutic*, just as it may involve a triple critique: into the social world of the scientist, the scientific practice within which she is working, and the subject-matter she investigates. (Since its subject-

hermeneutics or interpretivism

matter is not concept-dependent, natural science requires a double hermeneutic: *P*: 105.) 'In short, just as natural science has no FOUNDATIONS, there are no foundations of social knowledge – scientific or lay' (*PN3*: 60). The hermeneutic paradigm is thus consistent with realist metatheory of science. Properly understood and practised, it complements both deconstruction (it is committed to reconstruction) and SEMIOTIC inquiry into the production of meaning, leading to the possibility of metacritique of a DISCOURSE as incapable of producing an adequate description of a phenomenon (e.g., of global warming by neo-liberalism).

(2) The *non-presuppositionless* character of social life and inquiry. Because of our vehicular THROWNNESS into a context of beliefs and values, social inquiry, like its subject-matter, never starts from scratch (always has presuppositions) and never comes to an end; it is both pre-existing and ongoing. However, precisely because of this, and because of a core human nature shared by all and our practical transactions with natural objects, we normally have some prior understanding of the situations we face (see POLYAD; SUBJECTIVITY).

(3) The *indexicalised character of expressions*. In both social reality and social science, the meaning of expressions depends on the specific (changing) context of utterance. This entails EPISTEMOLOGICAL RELATIVITY, but not the judgemental relativism or the incommensurability of solipsistic forms of life that ensues from the collapse of the intransitive dimension in the hermeneutic paradigm.

(4) The *dialogicality* of social life and inquiry. Since this is also emphasised in the older tradition of dialectic as argument, we may say that (critical) hermeneutics is *a form of dialectic*, or there are *hermeneutical dialectics* (*D*: 139, 375). In both traditions, openness to dialogue or the fusion of meaning-frames, in which prejudices are challenged and outlooks broadened, is a precondition for access of wisdom and virtue, involving (CR notes) the identification and elimination of mistakes, hence absenting absences. By the same token, the metaphor of fusion must not be hypostatised such that we forget that in practice, in keeping with the thesis of epistemic relativity, 'the melting is always in the here and now' (*PN3*: 157). Moreover, arguments presuppose 'mutual RECOGNITION of the participants, which may involve a struggle, as Hegel famously realised' (*D*: 101, e.a.) – hence the CR concept of *hermeneutic* HEGEMONIC/counter-hegemonic *struggles*, in addition to more 'material' kinds of struggle, in power$_2$-stricken society. Hermeneutical dialogicality is at the heart of the dialectical critique of analytical reason (see also ANALYTICAL PROBLEMATIC; PROPAEDEUTIC):

> Any piece of deductive reason presupposes the establishment of a common context of utterance. And its existence presupposes the possibility of its re-establishment. We are immediately involved in a meta-hermeneutics dependent upon transformative praxis or absenting agency, and a meta-semiology implicating the changing intra-relationality of a nexus of signs. {...} From the most elementary to the most recondite, analytic reasoning is entirely *dialectically dependent* upon the processes of transformative negation necessary to ascertain, in an open-ended hermeneutic, 'what x means' or when two instances of A are to count as the same. Such hermeneutics is a constant, like the speed of light, or the 'UNIVERSAL WHORE', necessary for any inter-subjective exchange, communication, measure or intelligibility and presupposing our 'vehicular' thrownness. (*D*: 190)

234 heterocosmic

The *hermeneutical circle* (which is also an *epistemological circle*), whereby the parts cannot be understood unless there is some prior understanding of the whole, and vice versa, is a condition of all communication. (And a condition of that is the intransitivity, hence detachability and susceptibility to causal inquiry, of beliefs and meanings, which the hermeneutical tradition cannot sustain.) Bhaskar distinguishes four hermeneutical circles, each involving the four features itemised above. The first two are universal, the last two distinctive of and necessary (but not sufficient) for the social sciences. (C1), the circle of *inquiry* (C of I), highlights the non-presuppositionless character of any inquiry as a precondition for it to occur (see MENO'S PARADOX). Here we are concerned with the clarification and exploitation of presuppositions. (C2), the circle of *communication*, is a precondition for any dialogue or act of communication. It stipulates that communication is impossible unless the parties share some presuppositions, and unnecessary unless there is the possibility of disagreement (some non-identity in what they want to say to each other). This demonstrates that the (logocentric) ideal of understanding by empathy or immersion, such that there is an identity (or correspondence) between scientific and lay accounts, is impossible, i.e., inconsistent with the possibility of communication. Communication will always involve re-description. (C3), the circle of *inquiry-into-communication* (C of I [C]), obtains where inquiry is into other cultures, traditions, etc. Whereas the circle of communication presupposes a common context of utterance or shared horizon, here one must be established. It highlights the dialogic nature of inquiry, and its paradigmatic method is listening and translating. Again, a cardinal rule here is to select 'that translation which is *explanatorily most adequate* (whether or not it is most charitable) in the context of what is already known about the organisation of the particular society in question (and of societies in general)' (*SR*: 168). (C4), the circle of *inquiry-into-texts* (C of I [T]), applies where inquiry is into products or objects rather than subjects. Its paradigmatic method is 'reading', and it calls attention to the inadequacy of any (logocentric) theory of reading which focuses exclusively on recovery of authorial intentions.

These circles are 'circular' just in virtue of their dialogicality, their indexicalisation to some place-time, their presupposing some prior understanding, the preinterpreted character of social reality, and their susceptibility to a meta-hermeneutic if the base hermeneutic fails. They can all be expressed in terms of part–whole relationships because meanings are holistic or systemic, 'existing as they do only in virtue of (the use of) systems of difference' (*PN3*: 154). When conjoined with EXPLANATORY CRITIQUE and EMANCIPATORY AXIOLOGY, they may transmute into emancipatory *spirals*. Since it is TOTALITY that closes these circles, ensuring that communication and objects of inquiry appear as 'at least potentially intelligible to us' (*D*: 273), the home of hermeneutics in DCR is at 3L. Hermeneutics provides 'a good initial heuristic for understanding what it is to think' in the mode of totality (*D*: 12). However, only when conjoined with concepts of intransitivity, transfactuality, dialectical PROCESS, agentive agency – which presupposes that reasons (that are acted on) are causes – and struggle can it be regarded unreservedly as a friend of emancipatory science.

heterocosmic (Gr. *heteros*, other + *cosmos*, pertaining to another cosmos or world). *Heterocosmic affinity* is a concept Bhaskar borrows from Rosen (1982) to refer to an affinity between (some of) the concepts of radically different theories arising from the incorporation in one theory of transformed concepts deriving from another, whether

heterology, heteronomy   235

actual or implied. Thus Marx's concept 'ALIENATION of labour', adopted by Bhaskar, has heterocosmic affinity with Hegel's 'alienation of spirit', which it transforms. For his part, Hegel in effect posits a heterocosmic affinity between 'a self-creating god' and his own 'radically closed autogenetics' which is 'impossible' (*D*: 78) because each concept logically rules out the other. HOLISTIC CAUSALITY is *heterocosmic* because it transforms the radically different 'worlds' of the combining elements (*EW*: 130).

heterology, heteronomy. See FREEDOM, ETC.

hiatus, hiatus-in-the-duality. See CRITICAL NATURALISM; DCR; DISTANCIATION; DUALITY; INTRANSITIVE; NEGATIVE GENERALISATION; STRUCTURALISM; TENSE.

higher/lower, higher order/lower order. See DEPTH; PRIMACY.

highest good. See EUDAIMONIA.

historical justificationism. See DETERMINATION.

historical materialism. See DEMI-REALITY; DIRECTIONALITY; IDEALISM VS. MATERIALISM; PRIMACY.

historicism/scientism. 'Scientism: the dislocation of knowledge from the socio-historical realm and a consequent lack of historical REFLEXIVITY. Historicism is the opposite: the reduction of science to an expression of the historical process and a consequent judgemental RELATIVISM' (Bhaskar 1983b: 256). Both are species of ACTUALISM. Scientism is often associated with a positivistic NATURALISM that denies that there are any significant differences between the methods appropriate to social sciences, on the one hand, and the various natural sciences, on the other, and holds that such methods yield objective and impartial knowledge (values do not enter into the determination of facts); it thus identifies the domain of truth with what can be known by science (*R*: 183). Clearly, CR is not scientistic, as is sometimes claimed, in this sense; nor is it science-centred, as it can be derived without recourse to science (*D*: 229 ff.; *P*: 162). Historicism often goes hand in hand with HERMENEUTICISM; it fails to distinguish between the *transitive* and *INTRANSITIVE* dimensions, and the *INTRINSIC* and *extrinsic* aspects of science and other ways of knowing, viewing ways of knowing as EPIPHENOMENA, and its method of empathy or immersion in the past inevitably results in empathising with the victors, thus benefiting the current master-classes (Benjamin 1940: 258) of history. It is thus 'the enemy of novelty {. . .} locked into the forms of the past {. . .} "the sterile incarceration of history"' (Geoghegan 1996: 50, citing Ernst Bloch). For Popper (1944), historicism is the distinct but related view that history is governed by (actualistically conceived) inexorable laws such that deductively justified predictability is warranted (the future, yet to be caused, is unconditionally predictable), implying that history is a closed system. Popper's view that Marx was a historicist in this sense is challenged by CR, which reassesses the mature Marx as a scientific realist committed (unlike Popper himself) to a non-positivist conception of laws as TENDENCIES operative in open SYSTEMS; Hegel, who announced the end of history, is more vulnerable to Popper's critique (*D*: 351, 367–8). The falsity of Popperian historicism does not entail that the social sciences cannot make conditional predictions, subject to a CP clause. *Historicise* (n. *historicisation*) is often used unobjectionably to refer to acts or processes of de-NATURALISATION and de-REIFICATION, e.g., grasping the status quo as part of a historical process rather than as immutable. See also EMPIRICISM; ENDISM; HUMANISM; IDEOLOGY; REDUCTIONISM.

236 historicity

historicity. See GEO-HISTORICITY.

historiography. The theory and practice of the writing of history, that is, in CR terms, 'history' in the TD as distinct from 'history' in the ID (e.g., 'the historiography of France' and 'the history of France'), and therefore largely synonymous with the practices of the academic discipline of history as established in most European and North American universities from the mid-nineteenth century onwards. That discipline in its mainstream has been, and continues to be, resolutely and self-consciously empiricist, working with a 'historical method' defined (after Leopold von Ranke) as the reconstruction of past events 'as they actually happened' from the evidence of documentary and other sources. The classical form in which mainstream historiography presents the results of its research is the chronological historical NARRATIVE.

As a discipline, historiography self-consciously separates itself from the social sciences by making its proper domain historical AGENCY – the unique actions of individuals and groups – while leaving the investigation of STRUCTURE to social sciences such as sociology and economics. Although every historian will have to make assumptions concerning the structural nature of society in order to be able to narrate historical agency at all (because all such narrative requires some statements about what caused the events narrated), these assumptions usually remain implicit, unexamined, and frequently disguised as mere 'common sense'.

The essential arbitrariness with which historians string together their 'FACTS' into narratives has led postmodernist critics to conclude that the stories told by narrative historiography 'are as much invented as they are found' (Hayden White 1978). This is true to the extent that the 'facts' do not in and of themselves contain a narrative, but the postmodern 'linguistic turn' – a considerable influence on a younger generation of historians – has only made matters worse by suggesting that history cannot be more than storytelling and abandoning historical realism for representationalism, even replacing the concept of 'history' itself with that of 'memory'. In CR terms, both the empiricist mainstream of historiography and its postmodernist critics share the EPISTEMIC FALLACY of reducing the real world to our knowledge of it, which in the case of the latter is compounded by the LINGUISTIC FALLACY of reducing knowledge itself to the language we use to describe it.

Both the reconstruction of historical events from the empirical evidence and their description in narrative form are, of course, entirely justified and necessary activities. In its mainstream, empiricist ('Rankean' or 'neo-Rankean') form, however, historiography is pre-scientific in the same sense that, for example, a meteorology confining itself to a narrative of past climatic events rather than explaining the mechanisms producing cloud formations, precipitation and climate change would be pre-scientific: as a narrative that does not acknowledge the ontological depth of reality and fails to investigate the unobservable generative mechanisms which produce historical events (or, in the above example, climatic events). A CR-informed narrative historiography would be what Margaret Archer (1995) has called an 'analytic narrative of emergence'. Such a narrative would not only focus on 'the interplay between the real, the actual and the empirical to explain precise outcomes' (Archer) but also as a 'narrative of emergence' liberate historiography from its self-imposed artificial straitjacket of a 'past' bounded by 'pre-history' (left to anthropopaleontology, anthropology and archaeology), on the one hand, and the 'present' (left to the social sciences) on the other. The main properties of such a historiography can be outlined as follows.

It would understand the historicity of human societies as emerging from the historicity of the human species and the historicity of the human species as in turn emergent from the historicity of nature (see GEO-HISTORICITY), thereby transcending the anthropocentric assumptions of the conventional opposition between the 'humanities' (the realm of 'free will' and contingency) and the 'natural sciences' (as about inert matter governed by laws) and enabling itself to ask the same *kind of* questions about human history as those posed by science about the natural world.

It would abandon the rigid separation of questions about agency from questions about structure implied in the historical discipline's isolation from the social sciences, and become part of the long-overdue project of unifying (or 'geo-historicising', to use Bhaskar's expression) the social science disciplines. This does not preclude, of course, the pluralism of specialisations and methods that can also be found within the natural sciences.

It would acknowledge that, just as the natural and social sciences are not simply driven by 'curiosity' but by the desire to understand generative mechanisms and their interaction in order to be able to apply this knowledge for some practical benefit, the quest for an understanding of human history cannot be 'disinterested' (see EXPLANATORY CRITIQUE; PERSPECTIVE). Past events were generated by mechanisms, tendencies and structures which may not at all have receded into 'the past' – as highlighted in Brecht's 1955 warning about fascism: 'but the womb from which it was born/is still fertile'. Recalling the social ills of the past, therefore, entails a practical commitment to absenting their causes, in so far as they persist, from the future (see EMANCIPATORY AXIOLOGY; ENDISM). See also Lloyd 1986a; Minnerup 2003.

GÜNTER MINNERUP

history, philosophy of. See FUTURES STUDIES; GEO-HISTORICITY; HISTORIOGRAPHY; and *DCR passim*.

Hobbesian problem of order. This so-called problem, '*at the very least as thematised in sociological theory*, is a HOMEOMORPH of the Humean problem of induction' (*P:* 31, e.a.); that is, it has the same structure, grounded in a common (social) cause. The problem is thematised in mainstream sociological theory, especially by Parsons (1937), as the view that, 'in the absence of external constraint, the pursuit of private interests and desires leads inevitably to both social and individual disintegration' (Dawe 1970, cit. Van Krieken 1997), where the individual is conceived of as 'natural' or 'presocial'. On this conception there is no way that the abstract social universality of the rule of law can be grounded in pre-social atomistic individuals, just as causal laws qua constant conjunctions can never be grounded in (the observation of) atomistic states of affairs. This is the homeomorphism, underpinned by an atomistic conception of the world.

This understanding of Hobbes is, however, arguably radically mistaken – hence the Bhaskarian proviso. First, the state of nature in Hobbes 'is a deduction from the appetites and faculties not of man as such but of civilised men' (Macpherson 1962: 29), i.e., *socialised* possessive individuals. Second, Hobbes does not see the centralised authority of Leviathan as 'externally imposed' but as socially founded in morality and education. Sociology's celebrated 'normative solution' to the problem of order is therefore already contained in Hobbes, although he formulated the problem differently.

Bhaskar's fundamental point is that this is a pseudo-solution because the problem is misconceived in either case. Just as it is in the nature of strontium to burn with a red flame, so it is in the nature of a form of life structurally constituted by GENERATIVE SEPARATION to possess a relational dimension with a monopoly over internal and external use of force. Power$_2$-stricken society is necessarily held together by IDEOLOGY backed by the threat of force (see also HEGEMONY). When power$_2$ relations are abolished, society will cohere in virtue of an achieved rational understanding of the objective interests of each and all.

holism. SEE CRITICAL NATURALISM; DIALECTICAL ANTAGONISTS; FISSION/FUSION; MONISM (*ontological holism*); SEMANTIC HOLISM; SOCIAL STRUCTURE (*methodological holism, sociological holism*); TMSA; TOTALITY.

holistic causality. The 'characteristic mode of operation of a TOTALITY', which occurs when a complex coheres such that '(a) the totality, i.e. the form or structure of the combination, causally determines the elements; and (b) the form or structure of the elements causally codetermine each other and so causally codetermine the whole' (*P*: 77; cf. *D*: 127). It may take milder forms than DETERMINATION, e.g., conditioning, sustaining. See also CAUSALITY; EMERGENCE; HETEROCOSMIC; MEDIATE; REFLECTION; RELATIONALITY; SYSTEM; TELEOLOGY/TELEONOMY; TOTALITY.

holy trinity. Like the UNHOLY TRINITY of IRREALISM, the holy trinity of CR is a trinity because it exhibits a triplex constellational identity-in-difference, or one-in-threeness: (1) the possibility of *judgemental rationalism* (in the INTRINSIC ASPECT of the TD), presupposing (2) the actuality of *epistemic relativism* (in the extrinsic aspect of the TD), presupposing (3) the necessity of *ontological* depth-*REALISM* (in the ID). (2) is also entailed by (3), such that (3) ↔ (2) → (1). There is thus in CR an 'absence of strain between {. . .} realism and relativism, relativity and rationality' (*SR*: 98). 'Holy' puns on 'holes', (underlining the cardinal presence of the concept of ABSENCE). Table 24 indicates how the moments of the holy trinity articulate with other key constellationally distinct concepts. Both the holy and the unholy trinities are implicit in Marx, the elements of the former corresponding to practical, epistemological and ontological materialism, respectively (see also HEGEL–MARX CRITIQUE).

The derivation and conceptualisation of *ontological realism* is set out at many places in this work, including DEPTH; INTRANSITIVE, ETC.; REFERENTIAL DETACHMENT; ONTOLOGY; and TRANSCENDENTAL REALISM.

The principle of *epistemic relativism* or *relativity* 'asserts that all beliefs are socially produced, so that all KNOWLEDGE is transient, and neither truth-values nor criteria of rationality exist outside historical time' (*PN3*: 83); or there is real epistemic GEO-HISTORICITY and diversity: 'there is no way of knowing the world except under particular, more or less historically transient descriptions' (*SR*: 99), no neutral languages or invariant standards or principles that can be applied mechanically. (The law of non-contradiction – presupposed as it is by intentionality – is transhistorical, but what are deemed to be tokens of 'A' and '–A' will very much depend on the context of meaning.) Epistemic relativity is thus synonymous with *PERSPECTIVAL relativity* (*D*: 77): we always view the world from within some epistemic framework located within an epistemic

Table 24 The holy trinity and its key correspondences

HOLY TRINITY	Transcendentally	Dimensions of any cognitive process	Key categories	Distinctions within being	Human being in being	Specifically human being	UNHOLY TRINITY
judgemental rationalism	possible	transitive dimension† – intrinsic aspect of	dialectical reason	ethics	subjectivity	rational nature	ontological monovalence (judgemental relativism)
epistemic/moral relativism	actual	transitive dimension† – extrinsic aspect	absence (change) work (on)	epistemology	sociality	social nature	epistemic–ontic fallacy (epistemic absolutism)
ontological/moral realism	necessary	intransitive dimension	1M–4D	ontology	being	biological nature	primal squeeze on natural necessity and empirical science (ontological irrealism)

† Including the meta-critical dimension of philosophical inquiry.

240   holy trinity

WORLD-LINE within geo-history. It underlines the essential transformability of all human cognitive achievements. This means, inter alia, that some thoughts that were available to, say, Darwin were not in principle available to, say, Aristotle. And it entails, Bhaskar argues, the rejection of any correspondence theory of TRUTH (see also ALETHIA), except as a metaphor. 'There is no correspondence, no conformity, no similarity between objects and thought', except where the objects of thought are thoughts; there is just 'the general relativity of our knowledge': the changing expression of the absolutely or relatively independently changing world in thought (*RS2*: 249–50). Perspectival relativity should be distinguished from *perspectivalism*, such as Nietzsche's or Weber's, premised on the idea of the infinite or undifferentiated complexity of the (empirical) world itself, necessitating its conceptual simplification from specific viewpoints, an idea that is at the heart of the Weberian ideal type (see NATURAL KIND). Perspectival or epistemic relativity is premised, to the contrary, on the notion of an objectively differentiated (processual) world order.

Just as epistemology is constellationally contained within ontology, so epistemic is contained within *ontological*, including *social-relational, relativity,* diversity and change (see PERSPECTIVE; RELATIONALITY). Epistemic relativity is explicitly modelled metaphorically on physical relativity and, as already noted, the concomitant notion of divergent WORLD-LINES; and its containment within the constellational identity of the holy trinity allows philosophy 'to internalise the consequences of the twentieth-century revenge', in science and society alike, of 'the repressed fact of change'(*D*: 320). The changing process of science (and other ways of knowing) is relative to the comparatively enduring (more slowly changing) structures of the world and the constancy of trans-cosmic laws. As it follows the flight of time's arrow, it moves both vertically into the depths, and horizontally (latitudinally and longitudinally) over the face, of being–becoming – from the closed systems to which science aspires, to its application in the processual openness of the world.

Epistemic relativity can be derived in two ways. (1) From *ontological* (DEPTH-)*realism*, which – respecting a distinction between the sense and reference of propositions – entails that science must change to accommodate the progressively deeper strata it discovers and the reconceptualisation of more superficial strata in the light of this. Epistemic relativity (the geo-historicity of knowledge) (in the TD) thus does not refute, but actually depends upon, ontological realism (the otherness of the objects of knowledge) (in the ID); indeed, they mutually presuppose each other. Together they capture the two aspects – epistemic and ontic, expressive and referential – of what Bhaskar in *SR* (99 f.) called the *duality of TRUTH*, later incorporated in the TRUTH TETRAPOLITY (as its third moment) and dialectically developed into the concept of truth as ontological and ALETHIC. (2) From the necessary pre-existence of social forms for human intentional action to be possible (the argument for REFERENTIAL DETACHMENT, intransitivity and the TMSA). On this construal, human activity must be *work* on pre-existing (in the case of mental work, cognitive) materials, i.e., has a material CAUSE. Scientists (and philosophers, etc.) are always 'THROWN' into a moving, already, made EPISTEMOLOGICAL DIALECTIC or *PROCESS of production of knowledge by means of knowledge*, as into the social process more generally, which their activity reproduces/changes; they can never start from scratch, as *epistemological FOUNDATIONALISM* presumes. Epistemic relativism thus depends fundamentally upon

holy trinity 241

ABSENCE, involving as it does transformative negation and absenting process. It embraces change and diversity – unlike absolutism, whose drive to overcome the 'problem of relativism' has its seat in fear of change.

Since it presupposes depth-realism, epistemic relativism entails the *corrigibility* of knowledge-claims or epistemic *fallibilism*: all our theories are destined to be revised as depth-investigation proceeds and mistakes and incompletenesses (absences) are remedied (see CRITIQUE). Fallibilism thus refers, not just to human weakness (proneness to error) – the usual interpretation – but also to strength (the capacity for rational criticism of beliefs). 'It is only if the working scientist possesses the concept of an ontological realm, distinct from his current claims to knowledge of it, that he can philosophically think out the possibility of a rational criticism of these claims. To be a fallibilist about knowledge, it is necessary to be a realist about things. Conversely, to be a sceptic about things is to be a dogmatist about knowledge' (*RS2*: 42). Epistemic relativity entrains the apparent paradox that we can be sure that our work contains errors, without being able to tell which and where (this is a version of the *paradox of the preface*). Error in this sense is as intrinsic to the nature of the epistemic process as dying is to living things. However, fallibilism is also the name given by C. S. Peirce, and sometimes so used by Bhaskar, to the related but erroneous doctrine of *falsificationism* (Popper). This holds that all knowledge-claims are conjectural and that science progresses by falsification, not confirmation, of such conjectured empirical invariances – not by discovering the true but by eliminating the false. This overlooks that the problem of induction is not resolvable within an actualist ontological framework, and that in order to demonstrate a mistake epistemologically some proposition within some theoretical framework must be accepted, i.e., be regarded as non-conjectural, and depends upon the unconditional predictions condemned by Popper in his attack on HISTORICISM.

Material causes may be transformed, however, such that science displays discontinuity within continuity. Epistemic geo-historicity raises problems of *incommensurability* and *meaning-variation* in relation to diachronic *meaning-change* within science ('*paradigm*' *relativism*, including the issue of theory-dependence). These are discussed in detail in *SR* (ch. 1.6). They have affinities with issues of *cultural relativism* arising from diachronic social change and from synchronic *meaning-diversity* within cultures. Incommensurability is the thesis that it sometimes transpires that rival theories have no meaning in common (their terms are not inter-translatable), thereby apparently rendering a (judgementally) rational choice between them impossible, i.e., entailing a form of *judgemental* or *cognitive relativism* or *irrationalism* (the view that 'all beliefs [statements] are equally valid, in the sense that there can be no [rational] grounds for preferring one to another': *PN3*: 57), and encouraging the ontological relativist view (*subjective superidealism*) that there is no theory-independent world, such that when our theories change the world changes with them. CR allows the *possibility* of incommensurability as a limiting case (however, cf. *P*: 50) but rejects the view that it entails judgemental relativism, drawing on the distinction between the sense, REFERENCE and referent of expressions to argue that 'difference of meaning does not preclude identity (or at least commonality) of reference' (*SR*: 73). Indeed, the very intelligibility of incommensurability presupposes that 'there exists a field of real objects', which does not change with our theories, 'with respect to which *rival* theories are

242   homeomorphic (homoeomorphic)

incommensurable'. One can thus accept the incommensurability of theories and yet make judgementally rational choice between them (see THEORY CHOICE). There may be theories with no referential overlap but, then, they are not deemed incommensurable: 'no-one bothers to say that the rules of cricket and football are incommensurable' (*RS2*: 258). Further, the transhistorical existence of the causal powers of the natural world and of common human potentials and needs (at the level of the real) provides the basis of a (changing) '*mensurating material object language*' which serves as a 'bridgehead' between radically different theories and cultures (*SR*: 89–91). Superidealism cannot sustain the intelligibility of the major conceptual transformations, and especially of the notion of incommensurability that it calls attention to, and so is immanently refuted.

Epistemic relativity gives no grounds for *judgemental* relativism (the oft-heard argument that epistemic relativity is self-refuting confuses the two). On the contrary, if we are to act at all (and we cannot not act – the *AXIOLOGICAL imperative*) 'there must be grounds for preferring one belief (about some domain) to another' (*PN3*: 58). Judgemental *rationalism* or *rationality* presupposes epistemic relativity, especially epistemic GEO-HISTORICITY, because the context and descriptions of the world within which judgements are made are always necessarily historically specific, and issues requiring judgement (e.g., what is to be done?) only ever arise with the historical emergence of alternatives. Judgemental rationality may thus be said to occur *within* epistemic relativity; its exercise is always historically situated. Thus, in terms of constellational (ontological) containment, we have ontological realism > epistemic relativity > judgemental rationalism. This does not entail, however, that we get nowhere; on the contrary, it presupposes that science is a *developmental* process (see EPISTEMOLOGICAL DIALECTIC). Where it makes genuine depth-discoveries, the vertical movement of science sustains the (judgemental) rationality of its geo-historical transformations (scientific change and revolutions). The primary criterion for THEORY-CHOICE is therefore explanatory power (see also CRITIQUE; HERMENEUTICS). Epistemic indeterminacy is in practice resolved by the fact that there are usually very few theories both consistent with the data and promisingly progressive.

CR rejects judgemental relativism in ETHICS as well as epistemology. Indeed, the holy trinity as such has its counterpart in CR ETHICS, where judgemental rationalism presupposes *ethical relativity* (entailed by ethical naturalism) which presupposes *moral realism*. Judgemental relativism fails to see the constellational identity of these moments (see ETHICAL IDEOLOGIES). Understanding the full implications of judgemental rationalism involves grasping CR as a whole, and especially its drive to EXPLANATORY CRITIQUE (including the metacritique of philosophical ideologies) and EMANCIPATORY AXIOLOGY. See also HISTORICISM; ONTOLOGY; SAPIR–WHORF; SOCIOLOGY OF KNOWLEDGE; Agar 2005.

homeomorphic (homoeomorphic). See -MORPH, MORPHO-.
homology. See FREEDOM, ETC.
homonomous, homonomy (f. Gr. *homos*, same + *nomos*, law). Subject to, or having, the same law or mode of growth.
homonym (f. Gr. *homos*, same + *onoma*, name). The same word in pronunciation or spelling used to denote different things. In Bhaskar *homonymy* designates to the quality of some object or process whereby it possesses different yet *interrelated* aspects, e.g.,

negation has a process/product *homonymy* or BIVALENCE (see ABSENCE; CAUSALITY). It is thus closely affined to DUALITY, thence TOTALITY.

horizontal realism. See DEPTH; DIFFERENTIATION.

hot/cold societies. See GEO-HISTORICITY.

human nature. A theory of human nature is a condition of any moral discourse. Bhaskar argues that an adequate theory will recognise at least three distinct but interrelated components. (1) A *core universal nature*, grounded in shared genetic constitution, the ALETHIC SELF (DCR) and the GROUND-STATE/COSMIC ENVELOPE (PMR), and manifested in certain species-wide distinctive powers (e.g., to acquire and use language), NEEDS (e.g., for autonomy), 'and very probably WANTS (and so INTERESTS)' (cf. Archer 2000). (2) A highly differentiated, *historically specific nature* whose development is initiated at the time and place of our birth and is shared with other people 'who are subject to the same general determinations'. (3) A 'more or less *unique individuality*' (*SR*: 207–9). Since all three aspects are changing (and changeable), (1) far more slowly than the others, we can add (4) *PROCESSUALITY* to this list and note that it then corresponds to the four moments of the CONCRETE UNIVERSAL ↔ SINGULAR and the ontological–axiological chain, as illustrated in Table 25. Such a concept of human nature has nothing in common with the fixist, eternised views such as Hume's that 'mankind is {. . .} much the same in all times and places' (cit. *SR*: 289). Its three changing components together comprise *four-planar social*

Table 25 Four-planar social being, or human nature

Causal–Axiological Chain	1M Non-Identity	2E Negativity	3L Totality	4D Transformative Agency
Concrete universal ↔ singular	universality	processuality	mediation	(concrete) singularity
Components of human nature	(1) core universal	(4) changing	(2) historically specific	(3) unique individuality
Four-planar social being or human nature (social tetrapolity)	d. (intra-) subjectivity (the stratified person)	c. social relations (concept-dependent, site of social oppositionality)	b. inter-/intra-subjective (personal) relations – transactions with ourselves and others	a. material transactions with nature (making)
Alienation (fivefold) from	d. ourselves (5)	c. the nexus of social relations (4)	b. each other (3)	a. the labour process: labour and its product (-in-process), (1), material and means of production; (2) more broadly, nature and material objects

being or *socio-spatio-temporal being* or the *social tetrapolity* or *human nature* (see also PHILOSOPHICAL ANTHROPOLOGY). The four (dialectically distinct but interdependent) planes of the social tetrapolity bring greater specificity to the 'general determinations' of (2) and (3), which include (1). They are material transactions with nature; inter-intra/-personal action; social relations; and intra-subjectivity, i.e., the stratified person. Social relations are 2E because they are concept-dependent and the locus of social CONTRADICTION. The concept of four-planar social being generalises the TMSA, a 1M figure, to the other stadia of the ontological–axiological chain: negatively (2E, see NEGATIVE GENERALISATION), holistically (3L, addressing feminist and ecological themes) and transformatively (4D). It should be thought of as a DISTANCIATED four-dimensional time–space flow (*D*: 94, 153, 158–61, 276). At planes (b) and (c) it constellationally contains another cubic stretch-flow, the SOCIAL CUBE, comprising power$_2$, discursive/communicative and normative/moral relations, intersecting in ideology. Because it is itself a cube it is itself sometimes referred to as 'the social cube', but the two concepts are distinct. Each is a cube because it is open at either end diachronically considered, giving it two additional planes, denoting the not-yet and the intransitive determined.

Like any philosophical concept of a nature, four-planar human nature should not be hypostatised; it has no independent meaning, 'acting as a placeholder for, and awaiting filling by, substantive scientific descriptions'. While 'four-planar social being' and 'human nature' are often used interchangeably, Bhaskar also keeps them distinct so that 'human nature' can also be deployed as (1), core universal human nature, for if 'an independent meaning can be put on human nature then it may function as a norm against which, for example, social institutions, can be judged' (*D*: 283 n). As an ensemble of powers and needs, (1) can only be exercised or realised via (2) and (3), but it is not reducible to them; to collapse powers and needs to the conditions of their exercise or realisation in this way would be to commit an ethical version of the epistemic fallacy. (This is questioned by Hostettler and Norrie 2003.) Thus (1) grounds 'the *existence* of rights (and goods) for all human beings *qua* human beings, in virtue of their possession of a common (although always historically mediated) nature {. . .} even though these rights (and goods) can only come to be formulated as demands, recognised as legitimate and exercised as rights under very definite historical conditions' (*SR*: 209); or the human species at any one time is united by a 'chain of identities-in-difference (or, if you prefer, non-equivalent equivalences)' (*D*: 122). It also makes emancipation, which presupposes the prior existence of the powers of the emancipated in an undeveloped or unactualised form, possible. While this commits Bhaskar to the reality of universal human rights, it should be noted that the UNIVERSALITY concerned is dialectical, not abstract. PMR links (3) to ULTIMATA (the qualities of the GROUND-STATE/cosmic envelope). See also ETHICS; NEEDS.

human rights. See RIGHTS.

humanism. An outlook or worldview, rather than a body of doctrine, centring on one or more of the following ideas. (1) *A shared HUMAN NATURE* which unites us as a species in terms of basic potentialities and needs and distinguishes us from other animals; understood non-actualistically, it is consistent with a rich diversity of socio-cultural forms. (2) *Human history as a rationally DIRECTIONAL process* of freedom or emancipation, understood either (a) actualistically and deterministically, e.g., the Whig view of history, modernisation theory (see PDM), HISTORICISM in Popper's sense; or (b) in realist terms as a TENDENCY structurally induced on the basis of human striving to flourish, the outcome of which is highly contingent. (3) *VOLUNTARISM* in its CR connotation. This accords explanatory primacy either to individual or to collective agency in history. It is often associated with a HERMENEUTICIST orientation and an anti-naturalist history/nature DUALISM which denies (1), e.g., existentialism. (4) *Moral humanism*. The view that 'human well-being or flourishing {. . .} is the primary, sole, or ultimate object of proper moral concern' (Benton 2001: 3). Historically, this has provided the main secular alternative to religious ethical orientations (cf. RATIONALISM).

Most varieties of humanism are ANTHROPIC. For CR, to be human is, among other things, precisely to be able to be non-anthropic (see also DETACHMENT; cf. Archer 2000: 147 f.). CR moves from *anti-anthropism* at 1M to *humanism* in senses (1) and (2b) at 4D, a crucial basis for which is EMERGENCE (specified by non-reductive naturalism), which is indispensable for EXPLANATORY CRITIQUE and EMANCIPATORY AXIOLOGY (*SR*: 211): 'a world without humanity is a condition of the possibility of everything we call "human"' (*D*: 150). (It deploys a negative or pejorative concept of humanism in relation to its other senses.) IRREALISM moves the opposite way: from *anthropism* at 1M to *anti-humanism* in the CR sense at 4D (de-agentification through reification and disembodiment). Because CR humanism is developed dialectically from a non-anthropic position, it does not entail giving priority to human flourishing over that of non-human beings, as in (4). On the contrary, 'human well-being is recognised to be itself dependent on coexistence and interrelationship with flourishing non-human nature' (Benton 2001: 4).

Transhumanism is the belief that we can move beyond the human, progressively transforming what we are, especially via biotechnology, IT and AI. *Posthumanism*, especially in North America, is the realisation of the transhuman project. The posthuman is a self-programming, self-constituting, potentially unlimited being. The discourse of the posthuman faces the difficulty that it necessarily both excludes humanism (thereby undermining its claim to universality) and tacitly includes it (thereby losing the category of 'post'). In European thought, posthumanism signifies more a move beyond humanism which is non-anthropic (does not accept [4]) and which stresses the limitations of human ingenuity. See also ENTROPY/NEGENTROPY; GEO-HISTORICITY; PHILOSOPHICAL ANTHROPOLOGY; PRIMACY; STRUCTURALISM.

Humean level. The establishment of an empirical result or regularity in the DIALECTIC OF EXPLANATION preliminary to ascertaining the generative mechanisms that produce it. See also EPISTEMOLOGICAL DIALECTIC.

Hume's fork. See FORK.

Hume's law. See CRITICAL NATURALISM; EXPLANATORY CRITIQUE.

hylomorphism (Gr. *hylo-*, matter + *morphē*, form). The view, first advanced by Aristotle, that material objects consist of a unity of matter and form. Matter by itself is only potential, requiring form for its actualisation. On Bhaskar's construal, it is a species of IDENTITY THEORY or ACTUALISM.

hyper-naturalism. See NATURALISM.

hyper-rationalism. See RATIONALISM.

hyperreal, hyperreality, hyperrealisation. See VIRTUALITY.

hypostatisation. The 'illicit substantification' of concepts, conceptual forms and conceptual relations, treating them as if they were substances (*SR*: 97). See RATIONALISM; REIFICATION.

I

icon. See COGNITIVE SCIENCE; SIGN.
ideal type. See HOLY TRINITY; NATURAL KIND.
idealism does not stand in simple opposition to REALISM, because it is not a denial of existence, but is rather a family of ontological claims about the nature of reality. The core of idealism is that reality is not mind-independent. The subsequent issue is on what mind, and in what sense, reality is dependent? *Absolute idealism* posits a pervasive entity, consciousness, spirit or cognate system that can be immanent, transcendent or progressively realised in or as reality. In the Western tradition this has ambiguous roots in classical Greek philosophy. For Plato, in *Timaeus* and *Phaedo*, material objects are imperfect copies of transcendent mind-independent eternal forms. However, following Plotinus and Dionysius the Areopagite, in the Neo-Platonism of the Renaissance they are God's transcendent ideas and are thus dependent. Such absolute idealism can segue into either objective or subjective forms of idealism. *Objective idealism* focuses on either the imposition of idealist reality on the individual or on the common cognitive conditions of that idealist reality for conscious beings. *Subjective idealism* focuses on the role of human agents and/or social forms in the constitution of an idealist reality. The two are rarely entirely distinct because idealist philosophers typically explore an interrelationship between the subject and some form of proof of an absolute idealist being or some condition of its possibility, and, though sceptical concerning mind-independent external things in themselves, do not necessarily reject outright that there are *also* such external things in themselves. As a result of this metaphysical ambivalence, there is also a nominalist strain in idealism. Nominalism argues that there may be mind-independent external objects, but it is not the properties of those things which determine how they are distinguished by the mind. The major modern innovators of forms of idealism include Descartes, Berkeley, Kant, Fichte and Hegel.

Descartes (1637) tends towards a highly subjective or individual form of idealism verging on nihilistic solipsism. His search for a definite Archimedean point on which to base his system leads to Cartesian dualism and *cogito ergo sum* where everything beyond his own thought, including his own body, may be illusion. But the concept of perfection within his thought is used to infer a perfect or absolute being as the source of that concept. Berkeley's empiricist idealism (1710) argues that knowledge is founded in sense experience and that the qualities and properties perceived require no additional material object (though he is ambivalent about whether there are material objects) beyond the sensation of those qualities and properties. This *esse est percipi* (to be is to be perceived) is subjective in the sense that it is based on the sense experience of the individual, but there is an objectivist strand to Berkeley since, unlike Descartes, other minds experiencing the world in a similar way are taken as a given rather than simply left to sceptical doubt. If other minds experience the world in similar fashion, he argues, either there are real material objects that somehow make a difference to the properties of experience to produce this objective reproducible experience in common, or there is an absolute being that coordinates the properties or qualities that are experienced. For Berkeley, since we have no access beyond the properties of experience and talk of properties is meaningless without a mind that has them, talk of material objects in addition to those properties of mind is superfluous. A theistic interpretation of Occam's razor thus leads him to confirm a coordinating absolute being.

Kant's transcendental idealism (1781) continues the progression towards objective idealism by reversing the dominant Lockean empiricist presumption of his time that properties of material objects dictate the content of experience to the passive senses. Kant asks what must be the case for experience and argues that a universal reason or knowledge a priori imposes the structure of time, space and causation on experience rather than the other way round, creating a split between phenomena and noumena, one where the thing in itself is acknowledged, but remains as unknowable as the unobservable in Humean empiricism. Since experience, including causation, is structured by the a priori of reason, Kant was never able plausibly to address what difference things in themselves make to our knowledge of them. Accordingly, Fichte (1794–1800) criticised Kant's concept of the thing in itself as superfluously mind-independent in a way that is inconsistent with Kant's own transcendental precepts. For Fichte, the thing in itself represents a vestige of materialist empiricism that he terms dogmatism. He takes Kantianism towards a more subjective idealism where self-consciousness 'produces' objectivity, but one whose moral order of freedom is proof of God – a God that has neither consciousness nor person. Hegel, as the last of idealism's major innovators, develops this notion of the self-conscious production of objectivity, arguing that the conditions of the production of objectivity by the self-conscious subject are themselves objective, in the sense of being a teleological expression of an absolute idealist spirit that is progressively realised through stages of development of that self-consciousness. Accordingly, his critique of the philosophical systems of his predecessors (1807), including Fichte, identifies them as moments in this development. Self-actualisation and genuine understanding are thus a mutual historical process, a simultaneous subjective–objective reconciliation, through a dialectic, where higher stages of knowledge, culture and the state emerge from the real tensions of the dynamics of absolute spirit that structure the process. Whether the realisation of absolute spirit is theistic remains a matter of debate within Hegelian scholarship.

Idealism, particularly that of Kant and Hegel, has provided a major source for immanent CRITIQUE in the development of CR ONTOLOGY (*RS1*: ch. 1; *D*; *P*. 116–18) and also in the internal critique of CR itself (Groff 2004: ch. 2). Critical realists argue that idealism commits the EPISTEMIC FALLACY by reducing being to knowledge of being. The basic idealist commitment that reality is not mind-independent has tended to fail to account for what difference reality might make to cognition, even if sense experience is not simply a passive given. As such, critical realists tend to reverse idealism partially, whilst refusing both an exclusively MATERIALIST understanding and

248 idealism vs. materialism

EMPIRICISM. In early CR, Bhaskar argues that science explores the tendencies of underlying mechanisms or structures rather than the conjunction of events because laboratory conditions of artificial closure are usually required to create such conjunctions. The fallibility of perception and the fallibility of knowledge only make sense in terms of the properties of such relatively intransitive mind-independent mechanisms. The capacity for ideas/practices to affect reality, i.e., the human causal power to isolate operative mechanisms in given events in laboratory experiment, and also to manipulate mechanisms through conscious activity and technologies based on our understanding of them, presupposes that those mechanisms are prior to and separate from the intervention and that the mind is able to grasp aspects of those mechanisms adequately. The effectiveness of mind as a constituted part of reality, rather than mind-dependency, then becomes an emergent causal power of conscious creatures in a stratified depth-realist reality. Whilst this cannot be a refutation of extreme subjective idealism or SOLIPSISM, it is a more persuasive response to other forms of idealism than Moore's influential realism (1939), which argues that our common-sense intuitions based in experience are usually correct; for example, I experience my own hands, therefore I have hands. This is because the empirical truth of the experience of hands cannot address whether the reality of hands is ideational. CR precisely focuses on the plausibility of ideas on the traditional ground of argument concerning the existence of putatively independent external objects, but does so in a way that makes sense of degrees of mind effectiveness, moving on to link these to theories of embodied and positioned practice (*PN1*: 51). As such it addresses the central philosophical problem of the interaction between the human and the world. Thus the refutation of idealism in the philosophy of science leads also to ontological arguments about society, the meaning of social structure and the capacities of the socialised human, including the possibility of freedom and emancipation. Here immanent critiques of postmodernism and the cultural turn in general (Archer 1988/1996; Norris 1997; Sayer 2000a) and specific modern incarnations of idealism (many of which would refuse the label) in the form of Berger and Luckmann's reduction of social structure to reified agency (*PN1*: 40–2), the Winchian reduction of reality to culture and knowledge to belief (*PN1*: 169–95), and the Rortian rejection of formal philosophy in the name of human emancipation (*PF*) have played an important role. In *Dialectic*, an immanent critique of Hegel and Marx provides the grounding for a DCR that tries to move beyond the problematics of idealism and materialism. Some critical realists, however, identify idealist (confusingly termed irrealist) elements in DCR and Bhaskar's subsequent work (Hostettler and Norrie 2003; Norrie 2004a). It should be noted that idealism is still a vibrant element in disciplinary philosophy. The neo-Kantianism of T. H. Green and the ethical idealism of F. H. Bradley stood as a counterpoint to logical empiricism in the early twentieth century. More recently, Davidson's work in epistemology and various works in the philosophy of mind, ethics, rights and language have kept a mainly atheistic idealism alive (Coates and Hutto 1996).

JAMIE MORGAN

idealism vs. materialism. A key figure of OPPOSITIONALITY in irrealist thought, underpinning the dichotomies of mind and body, history and nature, hermeneuticism and positivism, values and facts, dualism and reductionism/monism (see also

idealism vs. materialism 249

PROBLEM). The issues at stake in this ancient debate are set out in IDEALISM and MATERIALISM. For CR (subjective) idealism and (mechanical and reductive, or irrealist) materialism are DIALECTICAL ANTAGONISTS, grounded in the common mistake of denial of the causal efficacy of ideas; emancipation depends upon their untruth. CR sublates this dichotomy by viewing ideas (including the ideal in the normative sense), not as opposed to, but as emergent 'parts of the natural world, products of the naturalised process of thought' (Bhaskar 1997a: 143). Corresponding to the moments of the HOLY TRINITY, as indicated in Table 26, CR upholds (1) *practical materialism*, asserting 'the constitutive role of human transformative agency' in the reproduction and transformation of social forms (entailed by the TMSA); (2) *epistemological materialism* (i.e., TRANSCENDENTAL REALISM), asserting the existential intransitivity and transfactual efficacy of the objects of scientific thought; and (3) *ontological* or *emergent powers materialism*, in the sense of 'the unilateral dependence of social upon biological (and more generally physical) being and the EMERGENCE of the former from the latter' as causally and taxonomically irreducible, with powers to react back on it (*DG*: 400; Bhaskar 2003a: 386). (1) is entailed by (2) and (3), which mutually entail each other: ontological depth-realism ↔ epistemological materialsim (transcendental realism) → practical materialism. In terms of constellational (ontological) containment, (3) > (2) > (1). *Historical materialism* is a substantive elaboration of (1), asserting the hypothesis of the causal PRIMACY of the form of labour process in human history. It is regarded by Bhaskar as heuristically valuable. In *Dialectic*, Bhaskar endorses the general thrust of Marx's analysis of the fundamental rhythmics of the capitalist mode of production. In a conception owing something both to Hegel and to Vedic thought, the philosophy of META-REALITY qualifies and elaborates (3) by arguing that consciousness is implicitly, i.e., qua inchoate possibility, enfolded in all being, but that 'matter is not enfolded in all consciousness' (ideas have 'emergent relations and connections', which, albeit existentially dependent on a physical substrate, 'are not mediated in any way by physical causality'), and that this suggests that 'matter is a *synchronic* {para-temporally} implicit, implicate or enfolded power of consciousness,

Table 26 Forms of CR realism/materialism, with some correspondences

Forms of CR realism/ materialism	Distinctions within being	Key categories	Holy trinity	Human being in being	Specifically human being
(1) practical materialism	ethics	praxis, dialectical reason	judgemental rationalism (possible) (IA of TD)	subjectivity	rational nature
(2) transcendental realism/ epistemological materialism	epistemology	absence (change), work	epistemic relativism (actual) (EA of TD)	sociality	social nature
(3) ontological or metaphysical realism/materialism	ontology	1M–4D	ontological realism (necessary) (ID)	being	biological nature

250 identity, law or principle of

and {in accordance with the modern scientific account, human} consciousness is a *diachronically* and synchronically emergent power of matter' – a conception intended to transcend idealism/materialism dualism in a new realist materialism (*MR*: 70, 330; *RM*: 92, e.a.). In PMR reductive and mechanical materialism is explicitly thematised as a form of IRREALISM (e.g., *EW*: 25 n; see also PDM).

identity, law or principle of. See ANALYTICAL PROBLEMATIC.

identity of indiscernibles/indiscernibility of identicals. Converse ACTUALIST principles first formulated by Leibniz and sometimes jointly referred to as 'Leibniz's law(s)'. The former states that, if there is no discernible difference between two entities, they are one and the same entity. But difference may be indiscernible at the level of the empirical and actual, yet real and undiscerned at the level of intrinsic structure. The latter states in effect that no two entities are exactly alike. But, unless there were things with identical structures, general knowledge would be impossible (see NATURAL KIND; TOKEN/TYPE). Leibniz's laws fail to think the coincidence or CO-PRESENCE of IDENTITY and DIFFERENCE and of identity and CHANGE (see *P*: 34).

identity theory, or **subject–object identity theory,** or **identity-thinking** is best illustrated in terms of Leibniz's argument in the *Monadology* that every proposition can be expressed in a subject–predicate form and that every true proposition is a statement of identity of one of two kinds. (1) Truths of reason where statements have (or can be reduced to) an internal self-evident truth quality inherent in the relation of the terms, taxonomy, use of language, etc. Thus, in 'all widows are women whose husband has died', 'widow' is substitutable by what follows (i.e., α is α); whilst in 'all equilateral triangles are triangles', 'equilateral triangles' are a sub-form of the general form (i.e., $\alpha\beta$ is β). Denial of the truth of these identities involves contradiction. (2) Truths of fact refer to propositions where identity is contingently assumed because statements are existential or empirical. Leibniz thus causes the theory of identity to be bound up with (1) the problem of knowledge a priori, which he argues can only be a rationalist truth or reason; and (2) what reduces to an essentially empiricist correspondence theory of truth (of fact, of experience, of individual events). The theory of identity, especially of truth of reason, has undergone a great deal of subsequent development and critique based on problems of reduction and the difference between identity and association. However, the rationalist/empiricist split on matters of identity has continued to be a basic philosophical problem.

Forms of RATIONALISM and EMPIRICISM in the philosophical tradition provide the grounds for critique of aspects of identity theory in *Dialectic* (e.g., 72–3). Non-dialectical identity theories tend to deny CONTRADICTION and privilege analytic method over coherent ontology (itself a contradiction), resulting in a LOGICISING of being which is reduced to tests of conformity such as logical propositional forms, whether of reason or of fact (or analytic or synthetic). Since being is reduced to individual non-contradictory and contingent propositional statements, both reason and fact are constructed in an ACTUALISTIC and atomistic way. This is an EPISTEMIC FALLACY, whereby being is de-totalised and reduced to falsely construed superficialities of either internal relations of statements or single perceived conjoined event-types. The implicit ontology secreted in such identity theorisation cannot account for the occurrence of those internal relations or those events, or changes in either, because (1) method seeks what is fixed and defines this as what is true, producing a static ontology where lack or real absence or potential cannot be incorporated because the focus is not on the structures that give rise to the actual; and (2) in using contradiction as a measure of falsity, such identity theories are unable to distinguish real contradictions or tensions. The implicit ontology is thus non-dialectical and MONOVALENT.

Consequently, Bhaskar counterposes a series of ontologically coherent concepts of identity (*D*: 20, 76–8; *DG*: 399). *Constellational identity* refers to the containment or co-inclusion of one thing within another. For example, knowledge is an aspect of reality. Accordingly, there are grounds of being towards which knowing is directed and that make knowledge possible (epistemology is within ontology). The ontological basis of epistemology is the fallibility and contingency of KNOWLEDGE (see also HOLY TRINITY). Within constellational identity, therefore, the epistemic warrant cannot itself be a form of identity or isomorphism without producing an unwarranted form of contradiction. Judgementally rational and continually critically appraised concepts and measures of adequacy within and between recognised discourses provide a conditional and contingent basis for currently justified belief and varying degrees of confidence. In accordance with depth-realism, the main focus of knowledge is on generative mechanisms and their tendencies. Since change is an aspect of reality, any concept of the identity of fundamentals of reality (ULTIMATA) must be compatible with change and must allow for process, transformation, potential, real absence as potential, and real contradictions or tensions.

This pushes one not only beyond non-dialectical monovalent philosophy but also, as part of the overall immanent critique of dialectics in *Dialectic*, beyond Hegelian dialectic. According to Bhaskar, Hegel's dialectic is self-contradictory in an unintended way: since all Hegel's contradictions are internal his dialectic suffers from a defect of *achieved identity* (*D*: 24) or a basic inability to deal coherently with its own problematic of change. It is constellationally closed and thus can be no more than an expression of a fixed process, i.e., a linear dialectic to some definite end or closed totality whose mechanism, in addition to the epistemic fallacy that it occurs in and through knowledge (see IDEALISM), operates according to a preservative sublation (*DG*: 404), achieving a self-identity of Spirit and understanding at each stage indicative of an underlying privileging of a positivity: a philosophical monovalence that cannot coherently deal with absence or real negation and the genuine openness of reality.

On the basis of this critique, Bhaskar emphasises the non-preservative potential of dialectics in an open TOTALITY where ultimata are themselves not fixed. Causal powers and the exercise of causal powers may change, precisely because real negation or the absenting of absence may be radically transformative. In order to highlight the incompleteness of real dialectical processes and their moments of resolution in new states, Bhaskar therefore introduces the concepts (*D*: 77–8) of *dispositional identity* (of a thing with its changing causal powers) and RHYTHMIC *identity* (of a thing with the exercise of its changing causal powers) within an overall process of constellational identity. *Identity-in-difference*: see ALTERITY/CHANGE; CONSTELLATION; INTRANSITIVE, ETC.; META-REALITY; SPACE–TIME. See also ANALYTICAL PROBLEMATIC; FIXISM; MONISM; TOKEN/TYPE.

JAMIE MORGAN

identity theory of mind. See MIND.

ideology. Just as the main strands of modern social theory have their counterparts within CR, so do the different approaches to ideology. These fall into two broad classes oriented around (1) a 'negative' view associated with the classical Marxist account of ideology as false (mystified and mystifying) consciousness abstracted from the real processes of geo-history (e.g., Ollman 1971); and (2) a 'positive' view whereby ideology either refers to the political ideas or worldview associated with different standpoints, groups or social classes, or more generally embraces the entire sphere of CULTURE understood as more or less opaque, a universal and necessary dimension of the human condition (e.g., Althusser, Habermas and in CR, Dean 2003; Joseph 2002).

Because it is difficult to reconcile the positive view (whereby – at least in its broader version – ideology is ETERNISED and CR itself must be ideological) with the possibility of EXPLANATORY CRITIQUE of consciousness, argued to be a transcendentally necessary condition of intentional agency (*PN3*: 64), Bhaskar locates himself squarely within the first approach, operating both with a general concept of ideologies as lived systems of false or inadequate ideas which can be explained or situated in terms of the historically specific $power_2$ relations they function to screen and legitimate, and with a restricted concept (designated *ideology†*) whereby the ideas involved are fundamental CATEGORY mistakes, e.g., viewing war as a game or confusing powers with their exercise. Since they contravene transcendental-axiological necessities or the REALITY PRINCIPLE, ideologies in the restricted sense necessarily give rise to, or themselves constitute, TINA COMPROMISE FORMATIONS. In either case the consciousness involved is false in that it is misleading, inadequate or untrue in terms of what it is about. It should be noted that there is nothing out of the ordinary about the notion of false consciousness: every time we have an argument with someone, we assume its possibility, to deny which is thus to commit performative contradiction (Sayer 2000a: 48). While Bhaskar concedes to position (2) that the spontaneously occurring mystification pertaining to the fact-form is necessary to any conceivable society, he points out that false consciousness *of* this form is not necessary (*SR*: 283–5).

On this conception, ideology is correlative to science, which demonstrates it to be (1) *false* or inadequate; (2) *necessary* (i.e. caused), so long as its social conditions of generation persist; and (3) beneficial in its effects on the reproduction of the social relations that generate it, and thus susceptible of quasi-FUNCTIONAL EXPLANATION (see also CRITIQUE; DIALECTICAL ARGUMENT; EMANCIPATORY AXIOLOGY). Such science is emancipatory, and must be distinguished both from *ideologies for science* (which rationalise the practice of Kuhnian 'normal' science) and *ideologies of science* (SCIENTISM). Produced and reproduced at the intersection of the $POWER_2$ with the discursive/communicative (cognitive) and normative/moral sub-dimensions of the SOCIAL CUBE (the *ideological intersect*), ideology functions in general to secure social cohesion and moral legitimacy in the context of generalised MASTER–SLAVE-type relations, in which class is seen to play a pivotal role. Fundamentally generated by oppressive and exploitative social structures, and reproductive of them, it is in no way intrinsic to $power_1$, or universal or necessary. It accordingly falls into two main classes: *ideologies of legitimation*, which function to secure the position of dominant groups, and *ideologies of bondage* whereby dominated groups internalise the viewpoint of the master (see BEAUTIFUL SOUL, ETC.; ETHICAL IDEOLOGIES). This last produces an *ideological INVERSION*, falsely turning power$_2$ relations upside-down in consciousness, as do commodification, reification and fetishism.

In keeping with Bhaskar's emphasis on the importance of ideas in geo-history, neglected by many Marxists (but not by Marx), ideology is a more fundamental concept than – though related to – 'the political' and the concept of HEGEMONY, and indeed is the 'site' of hermeneutic and other counter-hegemonic struggles over power$_2$ relations. This is not to say that ideology can 'dispense with' the coercive power of the state; rather that, if it ceased to grip the masses, they themselves would dispense with it together with the power$_2$ of the state. By no means confined to a 'superstructural' level, it pervades human practices at all levels, including science, as noted above, and philosophy, in multiple and contradictory ways. The concept of *resonance* or *REFLECTION* is designed to capture such diffracted pervasiveness. Meta-theoretically, ideology is both a fundamental cause and a consequence of human ALIENATION, with which it is closely bound up; indeed, ideology *is* conceptual and practical alienation, underpinned by self-alienation, entailing categorial error. In terms of the overall process of geo-history, the *first ideological reflection* corresponds to the epoch of 'simple undifferentiated expressive unity' which, in an illicit FUSION, abstracted from the rich diversity of the world, during which alienation was only a potential consequent upon the capacity for REFERENTIAL DETACHMENT. The *second ideological reflection* is the epoch of alienation proper in which the effects of illicit FISSION are generalised throughout society, reaching their apogee in the commodification, reification and fetishism of post/modernity (*D*: 306). The *alienation from reality* which haunts the philosophical *ideology of irrealism* is underpinned by 'primal GENERATIVE SEPARATION' or the divorce of people from the means of producing their livelihood (*D*: 362). Ideology as false consciousness is thus in no sense, as often alleged, an elitist or condescending theory whereby the upholders of the theory are enlightened and everyone else is a dupe. Everyone is embroiled within ideology in master–slave-type societies, and the pulse of freedom beats in the breasts of all.

In PMR, humans in master–slave-type society come to be seen as denizens of an emergent zone of being, the DEMI-REAL, constituted by four-dimensional labyrinthine layered webs of ideological ignorance (*avidya*) and illusion (*maya*) not of their own making but the acervative deposit of social practices informed by categorial error and construed, in accordance with conceptual emergentism, as *false being* or *constitutive falsity* (see ALETHIA). In this account of ideology/alienation, the falsity of theories and social forms is now explained ultimately in terms of an underlying falsity to the essential nature of human beings and the world; or *conceptual alienation* is explained by *real practical alienation* which in turn is explained by *self-alienation* (alienation from self and TOTALITY) (*EW*: 33–9). The seeds of this conception were sown, inter alia, in the earlier detailed analysis of the POSITIVISTIC ILLUSION (*SR*: ch. 3) and of the TINA SYNDROME, and in the implicit theorisation of the transcendentally real SELF in the dialectics of agency and discourse (see EMANCIPATORY AXIOLOGY). See also CONTRADICTION; Larrain 1979.

ideology critique. See CRITIQUE; EXPLANATORY CRITIQUE; IDEOLOGY; PDM.

ill, ill-being. Ills that are 'unnecessary and unwanted {. . .} can be considered as absences, and so constraints, but also as falsehoods to concretely singularised human nature' (*D*: 281); in the latter sense, ill-being is *false being*. See also ABSENCE; ALIENATION; CONSTRAINT; EMANCIPATORY AXIOLOGY; ETHICS; EVIL.

illicit fission, illicit fusion. See FISSION/FUSION.

illogicism, illogicise, (not) illogicising being. See EPISTEMIC–ONTIC FALLACY.

illusion. An illusion is real, but false, lacking a real object. Examples include logical CONTRADICTIONS, CATEGORY mistakes, the CONSTANT CONJUNCTION form, the atomistic ego, a mirage, the illusion of physical being (see EXTENSIONALISM), and the POSITIVISTIC/SPECULATIVE ILLUSION. See also ALIENATION; A-SERIES; DEMI-REALITY; IDEOLOGY; TINA SYNDROME.

immanence/transcendence. Correlative terms whose general meaning is indwelling or inherent (internal), and moving or being beyond (external), respectively. Thus God is often believed to be either transcendent (existing outside) or immanent in (existing within) the world, or a combination of both. A criticism or CRITIQUE is immanent if it is on the basis of a theory's own assumptions, and transcendent if it deploys external criteria.

In CR, transcendent and transcendence have highly pejorative connotations in certain contexts. (1) A *transcendent critique* is arbitrary. (2) *Transcendent* or *ineffable REALISM* is belief in the reality of things that cannot rationally be justified, hence that is accepted on the basis of faith, authority, convention, or caprice (see EMPIRICAL REALISM; FIDEISM; IRREALISM; SOLIPSISTIC, ETC.). In other contexts, however, they are perfectly kosher expressions, both epistemologically and ontologically. (1.1) The irruption of the new in EMERGENCE (see also CREATIVITY) is *epistemically transcendent* (beyond any means of inferring or predicting at that moment), and in the human sphere transcendence is the capacity to go beyond the limits of existing beliefs and practices, and/or the process whereby we do so. (1.2) Relatedly, *contradictions* may be *transcended* in a more adequate conceptual framework, or form of life (see EPISTEMOLOGICAL DIALECTIC), or being; this is an aspect of SUBLATION. Both consist in a move to *greater TOTALITY* or *completeness*, surmounting the old state of affairs. (2) *Verification-transcendent* (or *recognition-transcendent*) *truth* refers to the objective existence of an order of things and range of truth-values beyond the limits of our current knowledge (not any possible knowledge, as in Kant, though there may be realms that are in principle unknowable), i.e., the objective existence of *transcendent beings* (see MODALITY; TRANSCENDENTAL REALISM). It is entailed by the unfinished and open character of science and, more generally, the PRIMACY of the possible over the actual (see also ALETHIA).

The philosophy of META-REALITY generalises transcendence as the move to greater completeness in all learning processes and indeed all aspects of our lives (*self-transcendence*), which we regularly achieve in identity-consciousness (*transcendental identity consciousness* or *transcendental identification*), whether with our real self, an object or person, our agency (absorption in what we are doing), or others in collective work, thereby losing our false sense of ourselves as separate and isolated. Transcendence in this sense has nothing to do with *transcendent realism*; it is THIS-WORLDLY and indispensable for moving beyond ALIENATION and MASTER–SLAVERY, and indeed for any form of social life.

Transcendental has the same derivation as transcendent. Its various CR usages are defined and discussed under CRITIQUE (*transcendental critique*), DIALECTICAL ARGUMENT, INFERENCE (*transcendental argument*), PERSPECTIVE (*transcendental perspectival switch*), and TRANSCENDENTAL REALISM. See also UNIVERSALS/PARTICULARS.

immanent critique. See CONSISTENCY; CRITIQUE; EPISTEMOLOGICAL DIALECTIC; IMMANENCE; PDM; PHILOSOPHY.
immediacy, immediate. See MEDIATE; PERCEPTION.
imperative, axiological. See AXIOLOGY.
imperialism. See COLONIALISM.
imperialism, social or **sociological**. See SOCIOLOGICAL IMPERIALISM.
in, for and **of itself**. See FREEDOM, ETC.
inaction, inactivity. See NEGATIVE GENERALISATION; ZERO-LEVEL.
incommensurability. See HOLY TRINITY.
inconsistency. See CONSISTENCY.
incorrigibility. See ATOMISM; CRITICAL NATURALISM; FOUNDATIONALISM; HOLY TRINITY.
indeterminacy, axiological. See AXIOLOGY; CONTRADICTION.
indeterminacy, epistemological. See DETERMINATION; PHILOSOPHY.
indeterminacy, ontological. See ABSENCE; DETERMINATION.
indeterminacy of explanation. See PARADOX OF THE INDETERMINACY OF EXPLANATION.
indeterminate negation. See ABSENCE; TRANSITION.
indexicality. In semantics, an *indexical* is an expression whose REFERENCE is determined by the context of utterance, e.g., 'now', 'here', 'this', 'she'. *Indexicality* may thus be synonymous with self-reference, or more broadly with the dependence of the meaning of expressions on the context of utterance (*PN3*: 113, 152). An '*indexicalised* observer' (*D*: 38) occupies some determinate here/now. *Indexicalism* in the philosophy of SPACE–TIME and Bhaskar is the ego-present-centric view that only the 'now' (the present) exists (is real) – there is no past or future. In non-Bhaskarian usage it may also refer to the notion that the properties objects have are PERSPECTIVE-indexed. See also BLOCKISM/PUNCTUALISM.
indifference, principle of. See PRINCIPLE OF INDIFFERENCE.
individualism. See ATOMISM (*social atomism, sociological individualism*); EMPIRICAL REALISM (*epistemological* and *sociological individualism*); SOCIAL STRUCTURE; VOLUNTARISM (*methodological individualism*).
individuation. See POLYAD; PROBLEM OF INDIVIDUATION; RELATIONALITY; TOTALITY.
indivisibility of freedom. See CONDITIONALITY; DCR; FREEDOM, ETC.
induction. See INFERENCE; METHOD; NICOD'S CRITERION; PROBLEM OF INDUCTION; SOLIPSISTIC, ETC.
industrial relations. Commitment to CR meta-theory is rare in the discipline of industrial relations (IR), with Edwards (2005), Fleetwood (1999b) and Godard (1993) the only examples to be found. Edwards (2005) surveyed three leading IR journals over a five-year period (1997–2001) and concluded that: 'Of the 353 papers reviewed, none used the term critical realism but 27 might be seen as clearly consistent with the approach on the criterion that they are interested in causal mechanisms and underlying causal factors'. Yet Edwards also sees CR as 'giving a grounding to the instincts of many IR researchers'. To see why this might be the case, let us take the standard definition of IR, namely, the 'study of the rules governing employment' (Clegg 1979: 1). From the perspective of CR, social rules can be interpreted as specific forms of social structure.

The latter can, in turn, be interpreted as governing the employment relation in virtue of the human agents (especially, but not restricted to, employers and employees) that interact with them. In this case, we have a commitment to something like an ontology of agency and structure. And arguably the most developed agency-structure framework is Bhaskar's TMSA and, more recently, Archer's morphogenetic perspective. The aim of studying the rules governing employment can be interpreted as a focus, not upon outcomes or consequences, such as patterns of events or law-like regularities, but rather upon the conditions that make action possible. That is, upon the structures or rules that govern the actions of employers and employees. In this case, we have a commitment to something like the causal–explanatory model where the task is to explain, and explanation comes in terms of an analysis of causal structures. If such interpretations are reasonable, then IR could benefit in future from CR meta-theory.

STEVE FLEETWOOD

ineffable realism. See EMPIRICAL REALISM; HERMENEUTIC; IMMANENCE/TRANSCENDENCE; REALISM; SOLIPSISTIC, ETC.; TINA SYNDROME.

inference. A cognitive process in virtue of which a conclusion is drawn from a set of premises. It is meant to capture both the psychological process of drawing conclusions and the logical or formal rules that entitle (or justify) the subject to draw conclusions from certain premises. A rule of inference is an argument-pattern, i.e., a linguistic construction consisting of a set of premises and a conclusion and an (often implicit) claim that the conclusion is suitably connected to the premises (i.e., it logically follows from them, or is made plausible, probable or justified by them). Arguments can be divided into deductive (or demonstrative) and non-deductive (or non-demonstrative).

Deductive arguments are logically valid: the premises are inconsistent with the negation of the conclusion. That is, a deductively valid argument is such that, if the premises are true, the conclusion has to be true. This essential property of valid deductive argument is known as truth-transmission. Yet it comes at a price: deductive arguments are not content-increasing (the information contained in the conclusion is already present – albeit in an implicit form – in the premises). Though deductive inference has been taken to be justified in a straightforward manner, its justification depends on the meaning of logical connectives and on the status of logical laws. Those who deny that the laws of logic are a priori true argue that deductive inference is justified on broadly empirical grounds.

Non-deductive arguments are ampliative in the sense that the content of the conclusion exceeds the content of the premises. In its simplest form (enumerative) *induction* licenses an inference from the premise that all observed A's have been B to the conclusion that (probably) *all* A's are B. Induction is the mode of inference that establishes generalisations, whereas *eduction* refers to a process of initial clarification which licences only next-instance predictions. Traditionally, ampliative inferences have been taken to be problematic in their justification, the reason being that any attempt to justify them will have to presuppose some ampliative rules of inference or some premises (e.g., that nature is uniform) whose own truth can only be established by ampliative rules. Several attempts have been made to meet this sceptical challenge. One of them is to think of ampliative inferences (especially induction) as a form of probabilistic reasoning, whereby the reasoner follows the axioms of probability calculus in order to specify the probability of the conclusion (typically a universal generalisation) in relation to the premises. But the challenge to the justifiability of ampliative inference has been based on the deeply problematic assumption that a rule of inference is not justified unless the truth of the premises *guarantees* the truth of the conclusion. Ampliative inference, however, can have probative force: it licenses the acceptance of a conclusion as reasonable on the basis that the premises confer enough warrant on it. *Deductivism*, the view that the only reasonable inferences are deductive, is wrong both as a descriptive and as a normative thesis. The distinctive view of CR on this matter is that the justification of induction is a largely a posteriori matter and depends on the causal structure of the world. In particular, inductive inferences are justified when there are generative mechanisms that connect particulars in virtue of their causal powers.

Abduction has been another much-discussed mode of ampliative inference. In a famous characterisation, Charles Peirce described abduction as the inferential process which proceeds as follows: the surprising fact C is observed; but, if A were true, C would be a matter of course; hence, there is reason to suspect that A is true. Currently, 'abduction' is used as a code name for inference to the best explanation, whereby a hypothesis is reasonably accepted if it is the best explanation of some phenomena or evidence that needs to be explained. The proper application of inference to the best explanation requires the ranking of competing potential explanations in terms of their explanatory quality. *Explanationism* is the view that explanatory considerations should guide inference and reasonable belief. *Retroduction*, viz., the inference from effects to explanatory structures, is a species of inference to the best explanation. As deployed by Bhaskar, *transcendental arguments* are species of retroductive argument 'in which the premises – the undisputed understanding of some human activity or of human agency given in the theory/theories being subjected to immanent CRITIQUE – embody some categorical necessity' (*DG*: 405) which the analysis brings to light. Their nature and role in CR are discussed in *RS2* (257–9); Callinicos 1994, 2001, 2006; Bhaskar (with Callinicos) 2003b; Kaidesoja 2005, 2006; Morgan 2004b, 2006; see also TRANSCENDENTAL REALISM. *DIALECTICAL ARGUMENTS* are transcendental arguments that 'establish false necessities as ontological conclusions' (*D*: 199, e.a.). Retroduction and *retrodiction* (inference from effects to causes or from later to earlier states of systems via retroduced explanatory structures, e.g., when a doctor infers from a symptom in a patient that one of the generative mechanisms involved is an influenza virus) are grouped together with induction and *transduction* (inference from closed to open systems) under the umbrella concept *transdiction* (inference from the observed to the unobserved) or the *transdictive complex*. See also Dummett 1974; Harman 1995; Lipton 2004; Mellor 1988; Psillos 2002b.

STATHIS PSILLOS

infinite. See FINITE.

infra-/intra-/super-structure. *Superstructuration* or *superstructure-formation* is 'the characteristic movement of the Hegelian dialectic' (*DG*: 404), whereby conceptual EMERGENCE occurs (see EPISTEMOLOGICAL DIALECTIC). By transcendental perspectival switch, it can be seen also to involve *intrastructuration*, such that a new conceptual form is either superimposed on (Model A) or intra-posed within (Model B) an existing emergent level. Thus superstructuration and intrastructuration are aspects of the same process of emergence, which is sometimes signified by the concept of *structuration* (issuing in *stratification*) (e.g., *D*: 11) (not to be confused with the structuration

258   injunctive paradox

theory of Giddens). *Dialectics of intra-/super-structuration* provide the main substantive content of 1M dialectics of non-identity (*P*: 167), but because emergence partakes of the fourfold polysemy and causal modes of absence they are equally at home in all four CR stadia.

Emergence situates the phenomenon of MULTIPLE CONTROL in open systems, whereby 'higher-order agencies set the boundary conditions for the operation of lower-order laws', e.g., economic considerations in capitalist systems tend to govern the application of the physical principles of mechanics. This principle offers 'keys to the unravelling of the old Marxian conundrum of the "superstructures"' (*D*: 53, also 161–3, 353; *P*: 75, 97). The new level of structure can be seen *both* as setting boundary conditions for the base level (Model A) *and* as being formed within the base level which supplies *it* with framework principles or conditions of possibility (Model B). Thus the 'politics of the new world disorder', for example, both regulate/promote and occur within, as their very condition of existence, globalising capitalist commodification (see also FUNCTIONAL EXPLANATION).

Bhaskar promotes, as an exemplifying schema for substantive research, an *infra-/intra-/super-structural model* of, at any rate, capitalist society, whereby the mode of material production or infrastructure ('a unity of forces and relations of production' [*SR*: 245]) conditions the rest of social life more strongly than vice versa, such that the contemporary world is 'dominated by the logic of commodification' or capitalism's 'Essence Logic' (*D*: 98, 333), but with no commitment to the crude base/superstructure models which lump together all ideas in a superstructure: (different) ideas and ideologies are associated with (different) social practices at every level of the sociosphere, i.e., social practices and relations are concept-dependent and exhibit varying degrees of RELATIVE AUTONOMY (*PN3*: 66). A relational conception of the political also entails that it cannot be assigned exclusively to a superstructural level. The emphasis on structure in 'infra-/intra-/super-structure' highlights that realism's main interest is in identifying and explaining the real (interrelated) structures which generate 'different aspects of the concrete flux of social life' (*SR*: 158).

In PMR, the infrastructure/superstructure metaphor is extended to embrace the notion of realist deep structure, in particular human potentiality or our transcendentally real selves (at the level of non-duality) overlaid and occluded by an irrealist categorial surface structure sustaining and sustained by power$_2$ relations, together constituting the DUALISM of the DEMI-REAL. This is true to the emphasis in Marx on the creative powers of people in the labour process (the 'forces of production') overlaid by mystificatory structures of exploitation and control.

injunctive paradox derives from the conservative injunction to reproduce the status quo. It is autosubversive because it relies on human transformative AGENCY or praxis which is inherently world-changing (cf. *D*: 210). It is 'an ethical displacement of the problem of induction' (*D*: 119), which vainly seeks an actualist warrant that the course of nature will not change. Bhaskar deploys it with telling effect against Hegel's ENDISM.

inner or **internal conversation** or **dialogue**. See AGENCY; CRITICAL NATURALISM; PHILOSOPHICAL ANTHROPOLOGY; SUBJECTIVITY; Archer 2003b.

inorganic body. See META-REALITY; SELF.

instance-confirmation. See NICOD'S CRITERION.

instrumental rationality or **reason**. See RATIONALITY; REASON.

instrumentalism. The anti-realist view, shared by many PRAGMATISTS (e.g., Peirce, James) and some positivists (e.g., Mach, Schlick), that theories and concepts are neither true nor false, but rather instruments for moving from one set of data to a predicted set, or more generally for integrating, predicting and controlling, by conferral of 'warranted assertibility' (Dewey), our practical interaction with our environment. Some theories and concepts have greater utility in this respect than others, and it is this – rather than their putative truth or reference – which makes them justifiably acceptable. See also CONVENTIONALISM; PHILOSOPHY OF SCIENCE; TRANSCENDENTAL REALISM.

integrative pluralism (IP), also **structured pluralism**, and **developing** (or **dialectical**) **integrative pluralism (DIP)**. Names given by Bhaskar in *SR* (101 ff.) to his conception of philosophical ONTOLOGY as asymmetrically stratified and differentiated, dynamic and interconnected, in which both ALTERITY and CHANGE are irreducible and external relations are both real and constellationally contained within the internal RELATIONALITY of open (totalities nested within) TOTALITY. DIP contrasts with, and is intended to sublate, the dialectical couple of ontological *DUALISM* and *MONISM* or *holism*, recognising real distinctions as well as connections between the hierarchies of levels, the DIALECTICAL art of thinking which humans must acquire if they are to understand much at all. The concept appears but rarely (e.g., *D*: 350) in Bhaskar's subsequent works, probably because it does not lend itself to putting 2E at the hub of the ontological–axiological chain and can readily be interpreted as metaphysical dualism or DOUBLE ASPECT THEORY. *SR* acknowledges that the ineluctable social change entailed by the TMSA may affect 'even perhaps the feasibility of an orientation such as that of IP' (159). The philosophy of META-REALITY espouses a differentiated and stratified monism.

intension, intensionalism. See EXPRESSIVISM.

intention, intentional causality, intentionality. See AGENCY; CAUSALITY; INTRINSIC; MIND; REASON; REFLEXIVITY.

inter-action/intra-action. See RELATIONALITY.

interdisciplinarity, etc. *Multidisciplinary* research draws on more than one discipline, without challenging disciplinary identities, in order to study an object that transcends disciplinary boundaries (e.g., the city). *Crossdisciplinary* work illuminates one discipline or its object from the perspective of another, as when a philosopher critiques music theory. *Interdisciplinary* research is constituted on the basis of the integration of a number of disciplines into a research cluster which provides, or purports to provide, a new framework of understanding (e.g., COGNITIVE SCIENCE). *Transdisciplinarity* refers to the integration of intellectual frameworks at a higher level than disciplinary boundaries, providing a research model that may be transportable from one discipline to another. Here, too, disciplines remain distinct (you cannot have inter-, cross-, multi- or transdisciplinarity without disciplines). *Postdisciplinarity*, by contrast, envisages the demise of disciplines as we know them and revels in the freedom, eclecticism and new sense of unity this is supposed to bring.

The challenge to disciplinary boundaries in the human sciences over the last forty years or so, which included a phase of *antidisciplinarity*, is related to the reaction against scientism and the modernisation phase of globalising capital which saw the rise of POSTMODERNISM, with its actualistic emphasis on difference and identity at the expense of universality and totality (see PDM). This has resulted in a crisis of identity in the humanities and many social sciences. A fundamental enabling condition of this is that conventional boundaries were often arbitrary in that they lacked an adequate underlying philosophical rationale or scientific ontology, i.e., they were not aligned with real differences in social objects, reflecting the dominance of EMPIRICAL REALISM (*SR*: 104–5). 'The trouble with social science {. . .} is not that it has no (or too many) paradigms or research programmes; but rather that it lacks an adequate general conceptual scheme', which critical naturalism aspires to provide (*RS2*: 195; see also THEORY, ETC.). Only thus can the requisite continuity for progressive research programmes be secured. The increasing interest in CR across the whole gamut of the humanities and social sciences suggests that this conceptual underlabouring project is bearing fruit.

CR is by no means antidisciplinary and, indeed, leaving aside the various life and natural sciences, vindicates the relative autonomy of some traditional social disciplines, including PSYCHOLOGY, SOCIOLOGY, HISTORY and PHILOSOPHY, as well as some that are not so traditional (e.g., CULTURAL ANALYSIS), by establishing the relative autonomy of their object of study as an emergent level within the sociosphere. This is in accordance with its epistemological and methodological prescription to 'follow the object'. But, given the fluidity and interconnectedness of social reality – including of conceptual and analogical formations, which are often sociospherical in their higher reaches (López 2003a, b) – and the laminated nature of many social systems, this principle equally validates interdisciplinarity in certain areas as defined above, as well as multidisciplinary, crossdisciplinary and transdisciplinary emphases as appropriate, with disciplinary boundaries open and semi-permeable rather than closed; ontologically, we are never faced with a mono-disciplinary situation. Another principle – that of the explanatory reducibility or not of the elements of one science to another – provides an objective basis for interdisciplinarity (see REDUCTIONISM). On this principle, 'quantum mechanics and chemistry would belong to the same branch. But electromagnetism and mechanics, neurophysiology and psychology and {. . .} psychology and sociology would belong to different branches' (*RS2*: 181–2). Postdisciplinarity – the current buzz – is more contentious. Sayer has argued a CR case for a *postdisciplinary social science* on the grounds that disciplines are intrinsically both parochial and imperialist, hence prone to illicit 'reductionism, blinkered interpretations, and misattributions of causality' (Sayer 2000a: 7, see also his 2005). However, Cruickshank (2006) has plausibly suggested that Sayer's own practice is in effect multidisciplinary or crossdisciplinary as defined here, rather than postdisciplinary. Of course, if it takes the CR injunction to 'maximise explanatory power' seriously, any discipline – social scientific or otherwise – will tend to totalise, both synchronically and diachronically. CR sees this as no bad thing, as the multi-faceted nature of the pluriverse, especially the sociosphere, calls for the deployment of a rich variety of PERSPECTIVES on the same object.

interests. An *interest* is 'anything conducive to the achievement of agents' WANTS, NEEDS and/or purposes' (*SR*: 170). Agents act by definition in terms of their *perceived interests*. These may, however, differ from their *real interests*, which consist in whatever is conducive to their concretely singular flourishing. CP, 'it is in a person's interest simpliciter to come to know what their real interests are' (*D*: 293; see also FREEDOM).

internal realism

The CP clause is necessary because of the ACTIONABILITY requirement and because, from the perspective of the concrete singularity implicit in moral realism, agents – including collectivities of agents in social movements – are emergent totalities, and a different (ethical) principle applies to emergent totalities from that which applies to their constituent parts. This is the *principle of emergent totalities*, which reminds us that morality is only part of the art of living well and suggests that rollback or even backsliding may be justified in specific circumstances, and indeed that a virtuous life sometimes '*requires* the breaking of actualistically formulated or geo-historically specific rules'. Generalising this to all totalities, we have a *principle of foldback* which applies to any movement in which it is recognised that the process concerned is not 'centrist-expressivist-triumphalist-endist', that the principles of actionability and of the *Aristotelian mean* apply. This last calls for only optimal progress (which may be negative, i.e., defensive of hard-won gains with some losses) in the direction specified by the logic of dialectical universalisability. It 'puts a constraint on the constraining of constraints, which may on occasion be a necessary condition for any moral or social progress at all' (*D*: 293; see also DIRECTIONALITY).

The closely related *generic dialectic of interests* and the *dialectic of material interests* are two of the twelve routes, sketched in *Dialectic*, taken by the dialectic of desire to freedom in the transition from form to content. They are forms of the dialectics of agency and discourse (see EMANCIPATORY AXIOLOGY) primarily concerned with structural change, and are likewise dependent upon the truth imposed by the REALITY PRINCIPLE, which engineers a transition from perceived interests and wants to the perception of real interests, and upon the logic of dialectical UNIVERSALISABILITY. Through these dialectics, agents may discover that it is in their purely self-centred interest to be altruistic (cf. the current remarkable international effort to prevent 'bird flu'); in the enlarged sense indicated, they are, indeed, the main motor of the dialectic of freedom (*D*: 269). Their elementary schema is:

> egoistic desire → wants ↔ perceived interests → instrumental reason → EC → knowledge of social structural causes → real interests → desired (wanted) needs → collective agency → EC† ↔ totalising depth-praxis (DIALECTICAL REASON) → universal free flourishing (Adapted from *D*: 289–91)

In this way, what is in one's real *self*-interest coincides with the real self-interest of all. What is above all important about these dialectics is that they effect a transition from merely instrumental to dialectical reason via explanatory critique. This is because the oppressed have a direct material interest in knowledge of power$_2$ relations; they only have to grasp a causal connection between a constraint on their freedom and a power$_2$ relation to get these dialectics going (see also DEPTH [depth-inquiry]).

internal realism. See REALISM.
internal relations. See RELATIONALITY.
international relations. As an academic discipline, international relations (IR) originates in the catastrophe of 1914–18, the so-called Great War. The point of establishing IR was to learn more about the causes of war in order to prevent such a catastrophe from ever occurring again. According to the conventional story, the discipline has since engaged in four major debates. During the twenty years of crises

1918–39 (to paraphrase E. H. Carr [1946]), the Kantian liberals often affiliated with the League of Nations dominated the discussions. In the 1930s political realists began to challenge what they perceived as their 'idealism'. At that time, many German political scientists emigrated to the USA, including Hans Morgenthau, whose *Politics Among Nations: Struggle for Power and Peace* (first edition in 1948, the last and sixth in 1973) constituted a paradigmatic exemplar for the discipline for decades.

However, and increasingly from the 1950s, the positivists began to criticise the metaphysics of classical political realism, in particular their sinister conceptions of human nature. The second debate took place in the 1960s and 1970s when the traditionalists and behaviouralists argued over the correct scientific method (cf. Bull 1969; Kaplan 1966). Soon this debate was over taken by a dispute between the positivist theorists of global interdependence – globalists – and the political realists. Should actors other than states be taken into account? Does economic interdependence matter for peace and war? In the 1980s, various post-positivist meta-theories contested many of the background assumptions of all the previous theories and approaches (e.g., Alker 1981a, b, 1982). A few also brought up points from critical scientific realism, too, particularly on the agency–structure debate (Wendt 1987; Dessler 1989).

However, when the Berlin Wall came down and the Cold War ended, it was postmodernism that divided opinions within the discipline. Some were enthusiasts, many others hated it. The mainstream had taken the existence and identity of sovereign states as unproblematic and focused on discussing and testing theories about systems of domination, balance of power or cooperation between the nation-states. Various forms of critical theory and political economy had already questioned these certainties (Ashley 1981, 1984; Cox 1983). However, it was Richard Ashley's (1987, 1989) turn to radical post-structuralism, quickly followed by the famous edited volume *International/Intertextual Relations: Postmodern Readings of World Politics* (Der Derian and Shapiro 1989), that launched an intense and widely popular debate on identity, sovereignty, the state and geopolitics. Although post-structuralism has lost its drive as the main movement of rebellion, alongside other and often more mainstream post-positivist approaches such as social constructionism (cf. Adler 1997), it has become an established, even if only a minor, part of IR, particularly in Europe.

From a CR point of view, a deeper look at the development of the discipline reveals that IR as a systematic problematic had emerged step by step, element by element, in Europe by the early nineteenth century. Although Kant did not yet use the term 'international', he had systematically articulated the *international problematic* in the 1780s and 1790s. Neither Kant's metaphysical nor his political writings would have been possible without Hume. With his immediate predecessors and contemporaries, from Hobbes and Locke to Descartes and Berkeley, Hume laid down the basis for this problem-field in the mid-eighteenth century. A few decades later, Kant, a central Enlightenment philosopher, together with his immediate predecessor, Rousseau, articulated these conceptions into an acute social problem of war and peace. Kant thought this problem could and should be solved by an arrangement of 'league of nations', free trade, rule of law and republicanism. In this sense, Kant is the beginning of the familiar story. Since the mid-nineteenth century, *realpolitik* thinkers have criticised Kantian 'idealists' or liberalists for advocating dangerous or imperialist actions based on utopianism.

Ontology is the key, although the dividing lines within the discipline are still mostly taken to be primarily epistemological. Heikki Patomäki and Colin Wight (2000) argue that CR ontology highlights the conditions of possibility for a resolution of many of the theoretical, methodological and praxiological cul-de-sacs IR theory currently finds itself in. However, a year earlier, Alexander Wendt (1999) had published his magnum opus, *Social Theory of International Politics*, which is content with merely rearticulating the IR problematic in scientific realist and social constructivist terms. While it became the most widely debated book in IR for years, Patomäki (2002) articulates a systematic CR response to a quest for more emancipatory methodologies, by establishing a way out of the IR problematic. *After International Relations* shows how and why theories based on the international problematic have failed; articulates an alternative, CR research programme; and illustrates how this research programme can be put to work to enable better research and ethico-political practices. Similarly, Wight (2006) provides a systematic CR response to the agency–structure problematic in IR.

In certain ways, there are also parallels between some of the aims of post-structuralism and CR. By showing how the presuppositions of these theories – such as particular collective identities – are historically constructed and political, and by showing the problematic constitutive role the pseudo-Newtonian IR theories themselves play in the practices of diplomacy and global governance, post-structuralism has facilitated a move from IR to world politics proper (cf. Walker 1993: 183). This move is perhaps the most lasting contribution of post-structuralism in the field.

HEIKKI PATOMÄKI

interpretivism. See HERMENEUTICS.
intertextuality. See CRITICAL DISCOURSE ANALYSIS; PERCEPTION.
intra-/inter-. See the entry for the relevant concept; thus, for intra-relationality, see RELATIONALITY.
intrajection. See BEAUTIFUL SOUL, ETC.; DE-AGENTIFICATION; FISSION; MASTER–SLAVE.
intransitive, transitive and **metacritical dimensions**. In philosophy generally, the distinction between transitive and intransitive (f. L. *transitio, transitus*, a going or passing over) corresponds to that between transitive and intransitive verbs in grammar: passing over to or affecting an object, and not doing so, respectively. In Bhaskarian usage, the *transitive dimension* (TD), together with the possibility of distinguishing it from an *intransitive dimension* (ID), was constituted by the first act of REFERENTIAL DETACHMENT in the dawn of human geo-history (PHYLOGENETICALLY) and is continually reconstituted in primary POLYADISATION (ONTOGENETICALLY). In science, the distinction is implied by the transfactuality of causal laws and socialisation into science. The ID is a necessary condition of the intelligibility of scientific change, and the necessity for a TD is an immediate corollary: if the intransitive objects of science exist and act independently of our knowledge of them, the process whereby we come to know them cannot be identical or reducible to them. In phylogenetic terms, referential detachment carried the potential for ALIENATION. Transitivity is thus aligned conceptually with (the possibility of) alienation, DUALISM and split, intransitivity with DUALITY (cf. *EW*: 24, n4).

264 intransitive, transitive and metacritical dimensions

Since there is nothing that cannot in principle be referentially detached, 'whether known, knowable, or not' (*DG*: 399), the ID is synonymous with ONTOLOGY. More specifically, it refers to the objects in the domain of the real (see REALITY) of any human inquiry, theoretical or practical. The TD for its part is synonymous with the epistemological process of any inquiry, but logically must be extended to incorporate everything currently being affected by human praxis. In the physical sciences, the ID is both *existentially* and *causally intransitive*; in the social sciences, existential intransitivity alone mostly applies. This difference is a consequence of the fact that the social sciences are internally related to, or *causally interdependent* with, their subject-matter, in a way that the natural sciences are not. (However, some quasi-social objects possess an intransigence – e.g., the thawing tundra – which makes them 'relatively independent causally', *P*. 51 n.) In other words, existential intransitivity is an a priori condition of any investigation (indeed, of any human act), whereas causal intransitivity is restricted to laws of nature that endure and operate independently of humans. Existential intransitivity thus marks a HIATUS between the TD and ID across the board: 'the concept of existence is univocal: "being" means the same in the human as the natural world, even though the modes of being may radically differ' (*PN3*: 47). It is conferred by existence and the irreversibility of time, not by referential detachment as such, which merely avails itself of it. Once an entity has come into existence at t_1 there is nothing that can happen at time t_2 which can alter the fact and causes of its existence at t_1; it is fully determined and determinate, and can in principle be referentially detached and studied (see also TENSE).

As already indicated, the TD is synonymous with epistemology in its broadest, socially contextualised or materialist (non-idealist) sense (*PN3*: 114), or, better, with the epistemological *process* (see KNOWLEDGE), encompassing everything imbricated with human praxis and currently being affected by it – 'the whole material and cultural infra-/intra-/superstructure of society' (*D*: 218). More specifically, it refers to the ongoing social process of production of knowledge by means of knowledge in any practice or field of inquiry, a process to which the TMSA – as to any social activity – applies; and as such has both INTRINSIC and EXTRINSIC (normative and agentively causal) aspects. In the field of morality, the TD is 'more properly relational, because morality is practical, designating action-guiding relation on or to something', hence is sometimes referred to as the '*transitive–relational dimension*' (*D*: 259).

In sum, '(1) the social production of knowledge (TD continuity) and (2) the intransitive nature of its object (ID/TD difference)' are necessary conditions for scientific (indeed, any intentional) activity, comprising 'the two essential theorems of the TRANSCENDENTAL REALIST account of science' and human praxis generally (*SR*: 92).

The ID/TD difference was probably first conceptualised (in effect) by Marx in distinguishing 'objectivity' as such from the 'objectification' ('objectivation') of human praxis (Bhaskar 2003a: 387; see also REIFICATION). Drawing on the anti-monistic strand in the philosophy of science in the 1970s in relation to the TD and the anti-deductivist strand in relation to the ID (*R*: 180), Bhaskar initially established the distinction via an examination of the logic of discovery in the experimental sciences, where – as already mentioned – the transfactuality of laws and the ongoing socialisation of scientists into the process of science logically entail it. Its conceptualisation, which is central to the critique of the EPISTEMIC FALLACY, entails a further distinction (resembling that made by Althusser between 'thought object' and 'real object') between two kinds of object of knowledge: the relatively changing (and theory-laden) MATERIAL CAUSE of knowledge or *cognitive object* (existing knowledge, theories, models, etc.) which is transformed in the TD, and the relatively unchanging intransitive *real structures and mechanisms* which the inquiry aims to come to know; or the transitive and intransitive objects of knowledge.

It is important to grasp that the two dimensions, whilst really distinct, are not discrete; dialectically speaking, they are 'CONSTELLATIONALLY contrastive terms' (*P*: 151), constituting a constellational IDENTITY. Intent on revindicating ontology in the context of an analysis of experience in science, where causal intransitivity (science-independence) applies, Bhaskar initially overstated the distinction (Bhaskar 2002e: 83–4), focusing on the relations of sheer non-identity or ALTERITY that obtain between language/praxis and what they are about. In subsequent formulations, irreducible non-identity is retained, but explicitly located dialectically within *identity-in-difference*, i.e., epistemology/the TD is seen as constellationally contained within ontology/the ID; or, better, there is 'epistemic/ontic non-identity within ontology' (*D*: 8), where 'ontic' refers to the intransitive objects of specific inquiries and practices (see EPISTEMOLOGY, ETC.). When the matter is viewed in this way, the TD itself may, by valid PERSPECTIVAL SWITCH, be seen as continually passing over into the intransitive, without annulling the distinction. As I type this sentence, I am engaging in a transitive process of production of a dictionary, but at any moment I can referentially detach what I have written, regarding it in its (existential) intransitivity. (This is not affected by the fact that the transitive process has lasted three years – this is the present as indefinitely extendable [see TENSE)].) Moreover, any epistemological process is of course ontologically dependent on the causal powers of a multiplicity of intransitive structures and mechanisms that do not pertain either to the transitive or the intransitive dimensions of the process as such (in the present example, specific powers of reasoning, word-processing, publishing, etc.), yet whose exercise is intrinsic to it. This applies across the board to transitive processes (cf. López 2003a: 85). There is not a transitive dimension 'in here', and an intransitive one 'out there', though of course the causal laws of nature endure and operate independently of us. Everything – including the knowledge-seeker – is within being, of which epistemology/the TD is an emergent stratum.

The *metacritical dimension* (*SR*: 25) was constituted geo-historically when humans developed the capacity to CRITIQUE their theories of knowledge in an ongoing way. In Bhaskar's usage, it refers mainly to the critical, self-reflexive examination of the philosophical and sociological presuppositions of accounts of science and social science (*metacritique*). It is constellationally contained within the TD (within the ID).

intrastructure. See INFRA-/INTRA-/SUPER-STRUCTURE.

intrinsic/extrinsic (f. L. *intrinsecus* and *extrinsecus*, [from] the inside and outside, respectively). Closely related to 'internal/external'. However, the two pairs are not synonymous because what is intrinsic/extrinsic is not confined to a spatial characteristic ('inside' or 'outside' things of a certain type). 'Thus the category "intrinsic" includes some properties of things which lie outside their spatial envelope, e.g., a magnet's field, and others which cannot be identified spatially at all, e.g., a person's charm. And it excludes others which do lie within their spatial envelope, e.g., properties belonging to things of another type' (*RS2*: 77), which are therefore 'extrinsic' to the things concerned. The inclusive aspect of the intrinsic facilitates the later theorisation of the SELF as having an 'outer' dimension that ultimately is none other than the cosmos (*EW*: 6 n; cf. Collier 1991, for whom we have 'bodies-cosmic').

The *intrinsic aspect* (IA) and *extrinsic aspect* (EA) are irreducibly real features (not modes) of the human transitive dimension, constituting a DUALITY or UNITY, the EA constellationally containing the IA. The IA is the normative dimension of any cognitive process, the EA its CAUSAL aspects. Thus 'the category of causality {...} constellationally includes, but is not exhausted by, that of rationality' (*DG*: 400). The two dimensions are presupposed by the intentionality of human AGENCY and its THROWNNESS, respectively. These and some other key characteristics are displayed in Table 27. This does not entail that the relationship between the EA and the IA is asymmetrically inclusive in any cognitive situation; on the contrary, what counts as an adequate causal explanation is adjudicated in the IA, which is just the 'real, if circumscribed, possible and contingently actualised moment in cognition, established by the capacity to reflexively monitor and initiate conduct (and therefore presupposing the causal efficacy of reasons), situating the possibility inter alia of deliberation and/or assessment in the light of norms, standards and more broadly reasoned considerations' (*SR*: 17). CR holds the two in balance, refusing the 'objectivist' reduction of the IA to the EA (characteristic of empiricism) or the 'subjectivist' reduction of the EA to the IA (characteristic of idealism), and emphasising the 'fluidity' (*D*: 279) of their relationship. *Intrinsics* is the theory of the possibility of judgemental rationalism. Corresponding to ontics$_1$ and ontics$_2$ (see EPISTEMOLOGY, ETC.), *intrinsics$_1$* is whatever pertains to the intrinsic features of cognition in general, and *intrinsics$_2$* refers to the same features in some specific inquiry (*SR*: 37). See also ETHICS; HOLY TRINITY; INTRANSITIVE, ETC.; MIND.

intrinsic dialectics, intrinsic extrinsic dialectics. See EMANCIPATORY AXIOLOGY.

introjection. See BEAUTIFUL SOUL, ETC.; DE-AGENTIFICATION; FISSION; MASTER–SLAVE.

intuitional realism. See EMPIRICAL REALISM.

inversion. When illicit, a 2E figure of OPPOSITIONALITY signifying false or illusory 'reversal of hierarchy' (*P*: 242). In ALIENATION and its various forms, including COMMODIFICATION, REIFICATION and FETISHISM, the true relationship between people and the natural order (which includes essential human nature) is inverted – at the level of both theory and social practices and relations, a real ideality comes to be enfolded within material reality that turns the latter upside down. When the UNHAPPY CONSCIOUSNESS of slaves introjectively identifies with the master's IDEOLOGY, it inverts power$_2$ relations in fantasy only. The point is to invert the inversion in practice. See also DEMI-REALITY; TINA SYNDROME.

irrational existent. See ACTUALISM; DEMI-REALITY; SURD.

irrationalism. The view that there are no rational means of preferring (1) one belief or statement to another (*judgemental irrationalism*); (2) one means of knowing to another (*epistemological irrationalism*); or (3) one course of action to another (*irrationalist anti-intellectualism* or *practicism*). That irrationalism is very prevalent today is the almost inevitable consequence of the acceptance of the correct thesis of EPISTEMIC RELATIVITY, often in an exaggerated form, in the absence of ontological (depth-) realism. See also RATIONALISM.

Table 27 Some characteristics of the IA/EA duality

TRANSITIVE DIMENSION	Any cognitive process, more macroscopically agency	Action viewed	Concerned with	Key categories	Not to be confused with	
intrinsic aspect (IA)	intentional aspect (aspiring, aiming) (objectives)	judgementally	validity (assessing, evaluating, criticising)	rationality possibility (geo-historical)	1. present-future 2. self 3. private 4. inner, internal 5. mental	6. praxis 7. freedom 8. philosophy 9. noumenal self
extrinsic aspect (EA)	causal aspect (shaping, conditioning) (objects)	historically	credibility (situating, explaining)	causality actuality (geo-historical)	1. past 2. others 3. public 4. outer, external 5. corporeal	6. structure 7. determination 8. science 9. phenomenal self

irrealism

irrealism encompasses the ensemble of omissive errors committed in Western philosophy, as displayed in Figure 7. In mainstream thought it is basically synonymous with non-REALISM about any class of entities, or CONVENTIONALISM, the view that entities posited in language, science, ethics, etc., are only of conventional, human-made and contingent character, thus denying them sui generis reality (cf. SOCIAL CONSTRUCTIONISM). In CR it embraces conventionalism, but has a more specific meaning: any philosophical position that denegates or excludes some transcendentally or categorially necessary entity, resulting in a series of difficulties and aporiai that are typically dialectically connected in an ideological problematic or TINA COMPROMISE FORMATION. Its short definition is 'non-transcendental realism' (*SR*: 9). It is *not* synonymous with IDEALISM – reductive materialism, e.g., is irrealist about emergence – albeit idealism is typically irrealist about an array of phenomena.

Irrealism denegates or suppresses a series of necessary aspects of reality (e.g., alterity/transfactuality, absence/spatio-temporality, totality/intra-activity, agency/sociality) along the four moments of the concrete universal, and consequently has the effect of destratification, positivisation, detotalisation and de-agentification of reality (*D*: 316 360; see also REALITY PRINCIPLE). As philosophies normally come in the form of theory PROBLEM-field solution sets (*P*: 173; *D*: 313), in terms which affect each other on all these levels, the denial or suppression of one aspect leads to, undergirds and even necessitates further suppressions. Irrealism is endemically aporetic; it produces a mutilation of philosophy or cognition generally by refusing us any purchase on some allegedly ineffable realm of being by an interdiction on the very possibility of saying anything (philosophically) about this world. Although irrealist philosophies have exhibited a vast array of concrete manifestations (*D*: 311 ff.), such that it would be wrong to assume their total congruence both in detail and in principle, they generally have a common denominator in committing some basic mistakes, usually CATEGORIAL, of which the most significant is a *lack* (absence) of various necessary conceptual frames, which entrains aporiai, split, TINA formulations, ad hoc graftings and ideological pliabilities, all as tacit or explicit means of filling this gap.

The fundamental lack is the absence of ONTOLOGY and of a concept of (natural) necessity. Essential to CR's argument against irrealism is an argument for ontology in terms of the necessity of REFERENTIAL DETACHMENT. In order for us to be able to speak, we have to referentially detach the content of speech, and thus ontology is inexorable (*D*: 212). But we can also generalise this argument for referential detachment to the practical order and further formulate a series of transcendental deductions specifying the AXIOLOGICAL NECESSITIES effaced by irrealism. These denials lead to a split ontology and often function to screen out causally efficacious objects – real, actual, simple, constellational, material, ideational, etc. – and can thus serve to underpin various forms of IDEOLOGY.

The absence of a concept of absence (ONTOLOGICAL MONOVALENCE) and of stratification and depth leads to blockism, punctualism, fideism and fixism, and eventually to an ineffable solipsism in the manner of Wittgenstein (*P*: 231). Generally, the irrealist problematic has been formulated in terms that are deeply complex and intertwined, so that it constitutes a self-sustaining 'thicket such that if you enter it anywhere you are embroiled in it everywhere and must collapse – in what I call "reductio ad irrealism" – into a null point from which nothing can be said or done'

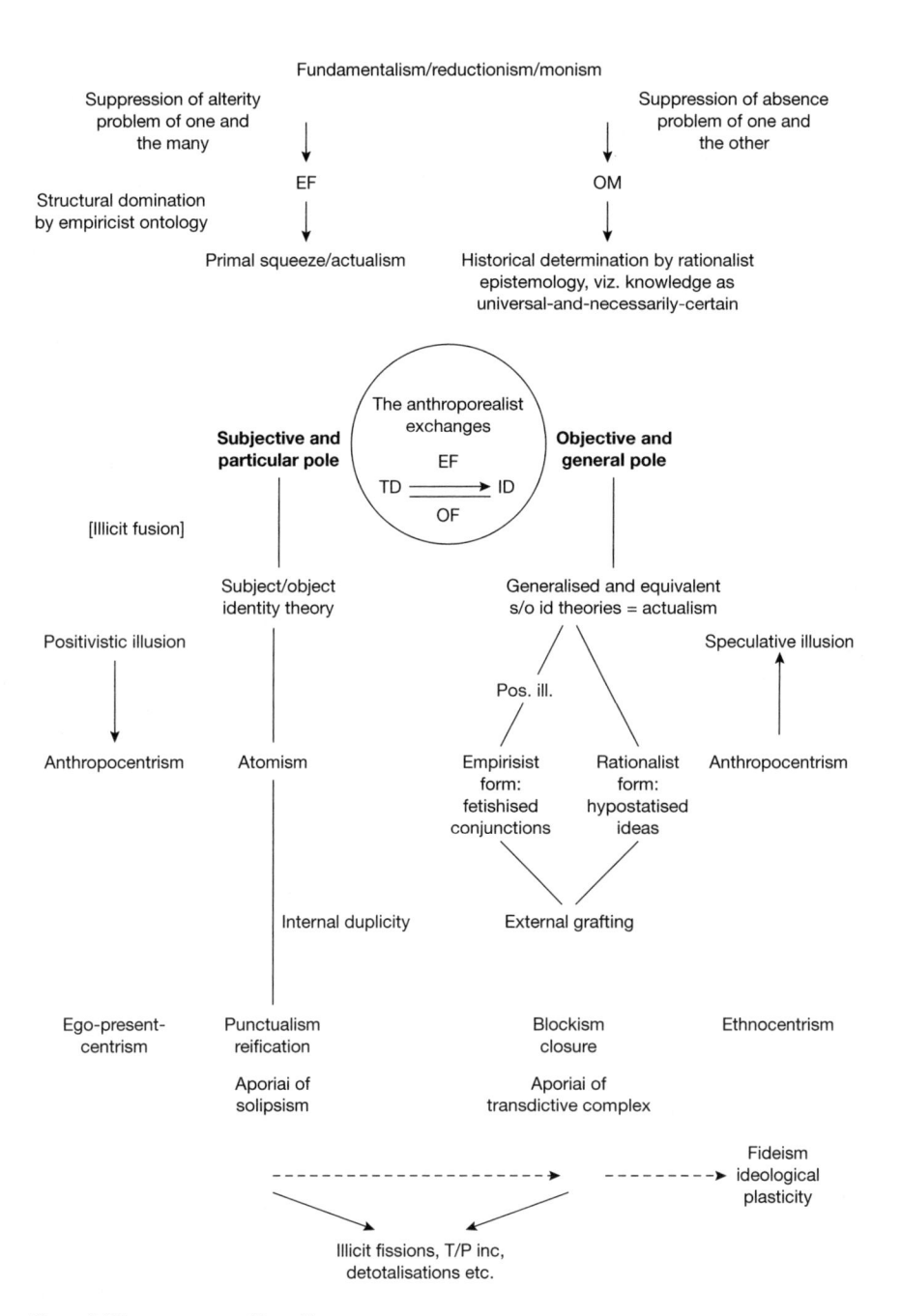

Figure 7 The structure of irrealism
Sources *D*: fig. 2.34, 4.1; *P*: fig. A.13.

(Bhaskar 1997a: 142–3; cf. *P.* 170). The logic of irrealism is thus predicated on the constitutive absence of ontology and on a series of categorical errors entrained by the EPISTEMIC FALLACY and ontological monovalence. It constitutes an emergent false surface categorial structure of reality which overlays and occludes the true deeper one.

Historically, the irrealist problematic has its origins in the primeval quest for an unhypothetical starting point, an *archē*, of knowledge. In the ARCH OF KNOWLEDGE TRADITION, with its quest for certainty and incorrigibility, the upshot is either scepticism or dogmatism, which are linked together as tacit mutually complicit dialectical counterparts (*P.* 193, 225). A third alternative is stoicism (indifference), which characteristically is a recurrent concomitant of the irrealist tradition as the internal inconsistencies become too palpable. This indifference is manifested in both theoretical and practical terms; typically the irrealist is a sceptic in theory and a dogmatist in theoretical practice, but indifferent in ethico-political terms and thus liable to ideological undermining. 'Stoicism purports indifference to reality and power$_2$ relations while Scepticism attempts to deny them' (*D*: 362; see also BEAUTIFUL SOUL, ETC.) Both reactions are disempowering, detotalising and alienating.

The MONISTIC starting point, inaugurated by Parmenides, who denied the possibility both of reference to absences and of change, could not be reconciled with the apparent diversity and transience of the world in which Plato was living. He was thus faced with two main problems: (1) the one and the other (negation and opposition) and (2) the one and the many (problems of universals and induction). While the first problematic leads to the antinomies in the *dialectical* irrealist tradition, the second has been the underlying source of difficulties in the *analytical* irrealist one (*D*: 184 ff., 376).

Parmenidean monism and fear of change was inherited and developed in epistemological terms by Plato and Aristotle. The problem of distinguishing certain knowledge (*epistēmē*) from mere belief (*doxa*) was resolved for Plato by seeing unchanging knowable being as participating in immutable eternal Forms (transcendence) in contradistinction to an impure sensate (empirical) world, and for Aristotle by 'induction supplemented by *nous* or intellectual intuition' (*P.* 5). Plato posited a hierarchy of the Forms, with the idea of the Good as the highest. What is good does not change, and the Forms were deemed superior to the changing world of transient things. The blockism of Parmenides was deconstructed, but at the cost of a split ontology of the permanent set of Forms and the demi-real realm of change, which was devalued and even denied (in the idealist tradition). Aristotle was more materialistic and this-worldly, viewing the epistemically significant as immanent in things themselves, but in the absence of a notion of ontological depth was at a loss to answer the question of how to justify knowledge in view of the flux of transient things and processes of nature. This is PRIMAL SQUEEZE between the domains of metaphysics and experience, exerted on the Platonic–Aristotelian fault-line, ruling out natural necessity and empirically controlled scientific theory. It has had a debilitating legacy for Western philosophy (cf. SCANDAL OF PHILOSOPHY), which has since been dominated by two mutually supportive but incompatible strands: a RATIONALIST epistemology, which strives to formulate knowledge as universal and necessary (cf. the monistic idea of scientific development); and an EMPIRICIST ontology, two currents of which have been elaborated into the anthropocentrism of the POSITIVISTIC ILLUSION, in which philosophy or theory in general is reduced to a naturalistically given simple sense-experience (a fetishism of facts), and the anthropomorphism of the SPECULATIVE ILLUSION, in which the world is by contrast reduced to an autonomised philosophy (a denegation of empirical testing of epistemic accounts), a theme perhaps most consequentially pursued by Hegel. However, both strands belong to the same overall problematic, working themselves out perhaps most significantly in various positivistic accounts, from the radical empiricism of James to the neo-positivism of van Fraassen; in the reductionism inherent in the concept of the empirical world; and in modern epistemological CONSTRUCTIONISM of the postmodern kind. These together constitute the *irrealist ensemble*, the inconsistent combination of subject–object identity theory and dualism (*D*: 383), in various forms of constitutive antagonisms: empiricism vs. rationalism, materialism vs. idealism, etc.

This is the fundamental ensemble inherited from primal squeeze in which the concept of natural necessity and ALETHIC structure is suppressed and we are left without a rationale for knowledge about the real world or for epistemic error. Thus the UNHOLY TRINITY of irrealism could be retroduced to the inability to explicate change and the contingency of knowledge. The absence of a concept of ontological depth and of a real structure of being which could ground induction leads to the TRANSDICTIVE PROBLEM COMPLEX and leaves us with no purchase on judgemental rationality, either. The immediate effect of this is seen in *anthroporealist exchanges*, in which the converse failings of an epistemologisation of being (the epistemic fallacy) and an ontologisation of knowledge (the ontic fallacy) give rise to a TINA compromise form where the ID and TD are seen as unbridgeable. Philosophy has since been stuck on the Platonic–Aristotelian fault-line (*P*: 5). What grounds knowledge? How is the gap between knowledge and being to be bridged? The answer in the irrealist tradition has invariably been some kind of reduction: either to the Forms or concepts (in idealism), to sense experience (in empiricism/positivism) or to intuition. The irrealist ensemble thus comprises the unholy trinity of the epistemic fallacy, ontological monovalence and the primal squeeze (*D*: 199), under the sign of the dominance of monovalence. 'As ontological monovalence, taken literally, entails the exclusion of alterity, otherness, it implies, as primeval monism {Parmenidean}, the identity of thought and being, whether this takes the form of the epistemic fallacy or the ontic fallacy' (*P*: 60). Ultimately this necessitates a FIDEIST response, i.e., the reaction that any beliefs about the organisation of the world cannot be rationally justified, but must be accepted on the basis of faith in God or some other transcendent entity in lieu of alethic truth. And it issues in an ontology of stasis that serves to legitimate existing power$_2$ relations (see DCR).

The problems of philosophy, of which irrealism is at the centre, are for the most part the result of an 'insufficiently non-anthropocentric, differentiated, stratified, dynamic, holistic (concrete) or agentive (practical) view of things' (*D*: 2). Human existence is a radically contingent affair and our knowledge of the world is neither necessary nor certain. What a realist philosophy of science must incorporate is a view of the world as contingent and our knowledge of the world as conditional and subject to CRITIQUE and corrections.

PÄR ENGHOLM

irrealist ensemble. See ANTHROPISM; IRREALISM.
isomorphic, isomorphism. See -MORPH; MORPHO-.

JK

jenseitig, Jenseits. See THIS-WORLDLY/OTHER-WORLDLY.
judgement, judgement form. See EMANCIPATORY AXIOLOGY.
judgemental rationalism/relativism. See CONSTRUCTIONISM; ETHICAL IDEOLOGIES; HERMENEUTICS; HISTORICISM; HOLY TRINITY; NEEDS; UNHOLY TRINITY.
jump point. See TRANSITION.
justification. See ALETHIA; ANALYTICAL PROBLEMATIC; INFERENCE; ONTOLOGY; PHILOSOPHY OF SCIENCE; TRANSCENDENTAL REALISM; TRUTH.
justificationism. See ATOMISM; DETERMINATION; FOUNDATIONALISM; GENETIC FALLACY; PHILOSOPHY OF SCIENCE.
justificationist fallacy. See GENETIC FALLACY.
Kantian level. The stage in the dialectic of EXPLANATION at which models of generative mechanisms are imagined. See also EPISTEMOLOGICAL DIALECTIC.
kenosis, adj. kenotic (Gr., emptying). In theology, Christ's shedding of His divine attributes in order fully to experience human being, whence it can mean 'emptying out' quite generally, e.g., 'the post-Hegelian kenosis of reason' (*D*: 182).
kinetic, n. kinesis (Gr. *kinēsis*, a movement or being moved, a process). Processual or dynamic, e.g., Aristotle's or Hegel's kinetic ACTUALISM.
kinesthesis (Gr. *kinēsis*, movement + *aisthēsis*, perception by the senses). The ability to feel movements of the body. *Kinesthetic rationality*: see REASON.
knowledge, theory of. Epistemology is the philosophical study, analysis and theorisation of knowledge. It asks and answers fundamental questions concerning how we come to know what we know and examines the criteria of judgement for distinguishing 'true' knowledge from mere opinion, belief or prejudice (see DOXA). It also considers the possibility of limits to what can potentially be known.

There are two possible categories of theories of knowledge, the first more general (EPISTEMOLOGY$_1$) with the second (epistemology$_2$) being a more specialised sub-set of the first. This second sort of theory of knowledge is more intimately bound up with science and the philosophy of science. It is developed by a transcendental reflection upon particular knowledge-producing practices and the presuppositions about reality that render such activities intelligible. The knowledge-producing activity most importantly and most commonly reflected upon in this epistemological context is scientific activity. In this CR simply conforms to the usual philosophical practices in the history of thinking on this topic. However, there is also an important distinction to be made here between CR and most other philosophical points of view. That is, there has been a long tradition in philosophy to posit scientific knowledge as the only true knowledge, with the alleged knowledges of all other knowledge-producing practices being consigned to the realm of doxa. CR does not follow in this path and recognises the possibility and importance of other sorts of knowledge-producing activities, artistic practice, for example. It also accords genuine potential knowledge-producing status to the everyday activities carried out by ordinary people. Its deduction of the *INTRANSITIVITY* and *TRANSFACTUALITY* of the objects of knowledge – the defining hallmark of transcendental realist epistemology – follows equally from the nature of human intentional agency as such as from the activity of scientists.

The theories of knowledge of epistemology$_1$ and epistemology$_2$ are both considered to belong to the *transitive* dimension (see below) and accordingly emphasise *FALLIBILISM*. The fallibilism of CR epistemology corresponds with CR ontology. That is, fallibilism not only recognises the omnipresent potential of human error; this fallibility is not best-understood in terms of black and white, truth and error, but rather with a notion of epistemological levels corresponding to ontological complexity and depth: *epistemic relativism*. This gives rise to a notion of truth being conceived not in terms of absolutes but rather in terms of 'true and truer', whereby errors are corrected and greater depth and expanse of explanations are provided in relation to a stratified complex reality.

Epistemology has generally been characterised by a division between two competing schools of thought: *RATIONALISM* and *EMPIRICISM*. That this assertion is only untrue to the extent to which it is a slight exaggeration underlines the distinctiveness and importance of the CR contribution to this branch of philosophy. Both these theories of knowledge depend upon real characteristics of human knowledge-producing potential but exaggerate such characteristics to the exclusion of others and fail properly to contextualise the production of knowledge, in accordance with an impoverished ontology.

CR articulates a realist epistemology with the following features. First, CR's epistemology is explicitly consistent with its ontology. Notwithstanding the emphasis upon ontology in CR, it is a general philosophical requirement that a philosophy's theory of knowledge should be logically consistent with its ontology. In many cases this is only apparently achieved by silence and ambiguity. The relation between CR ontology and epistemology is explicit and relatively detailed. 'Things', structures, relations, powers, mechanisms, the philosophical 'stuff' of existence, exist independently of our cognition; or, to put it another way, reality is independent of our (alleged) knowledges of it. CR articulates this independence through the articulation of the *transitive* and *INTRANSITIVE DIMENSIONS*. Knowledge belongs to the former (with one key exception that will be examined later) and the 'things' of reality to the latter.

A conscious awareness of the preceding leads directly to CR's second feature: *EPISTEMIC RELATIVISM*. Knowledge is inevitably socio-culturally, economically and historically situated. We perceive and know relative to particular PERSPECTIVES. The awareness of this fundamental condition with respect to the production of knowledge has given rise to a frequent and pernicious philosophical error concerning knowledge generally, and scientific knowledge in particular. The fact of this relativism has given rise to a further relativisation of knowledge claims. It is falsely reasoned that, because knowledge is not absolute and timeless, judgement between competing knowledge claims is arbitrary, with scientific assertions being no better or worse than any others. The production of knowledge is a human activity and consequently subject to error, but this does not mean that there is no basis for judgement between competing claims. CR's third epistemological feature is *JUDGEMENTAL RATIONALISM*. Epistemic relativism and judgemental rationalism combine to form CR's distinctive understanding of the relation between the active human component involved in the production of knowledge and an independently existing reality.

The use of the term 'production' signifies this relationship and distinguishes the CR position from the epistemological error historically associated with the use of either 'construction' or 'discovery' in this context. Human beings collectively and actively 'produce' knowledge. They do not simply discover 'facts'. Nor do they 'construct' knowledge in the sense of creation *ex nihilo*. The ordinary English-language sense of the word 'construction' need not have this implication. Rather, it is simply a historical association in the history of the philosophy and sociology of science (most importantly and most mistakenly in the latter). Experience needs active interpretation before it can be turned into knowledge. It is this simple observation that gives the lie to any notion of knowledge being simply a matter of discovery. What makes 'facts' be understood as such is, at least in part, a humanly guided process. The 'facts' of either nature or human social life are not simply perceived; they are perceived through our pre-existing cognitive and theoretical filters.

In this regard, knowledge production as a human activity can be seen as consistent with the more general CR TMSA. CR understands SOCIAL STRUCTURES to be significant features of social life. These structures derive not just from present human activity but from the activity of the past. Human action is both enabled and constrained by the existence of such structures. It also transforms them. This is precisely what occurs with respect to science and other knowledge-producing activities. Knowledge is used to produce knowledge. It is in this one sense that, as mentioned earlier, knowledge can belong to the ID as well as to the TD – given that all social structures are only relatively malleable, as they persist over time. We attempt and sometimes succeed in transforming them, but this is not always an easy task; and this is no less true with respect to discursive structures.

The production of knowledge is economically, historically and sociologically situated. The awareness that knowledge is situated in this manner can lead to an overcompensation, as it were, from the problems of naïve empiricism (the discovery of facts model). Here the active human element and the sociological determinations are stressed so much that a one-sided equation is produced – the nature of reality (natural or social) is entirely left out. 'Reality' thus becomes merely the prognostications of the human mind – the worst form of idealism. For CR, knowledge is knowledge of something – reality.

CR has two schools of thought with respect to the expression of the relationship between knowledge and reality. The older position is called the *correspondence theory of TRUTH*. From this point of view, knowledge consists of propositions that are true and may thus be considered knowledge, if and only if, they correspond with reality. The principal problem with this position consists of the difficulty of specifying exactly what it means for a linguistic or semiological phenomenon (the proposition) to correspond to something of a completely different order (the power or mechanism or 'thing' the alleged knowledge is supposed to be of). A Marxist version of the correspondence theory of truth was articulated by Lenin in *Materialism and Empirio-Criticism* (1909). Here the metaphor of reflection was used to articulate a notion of correspondence. This position is widely believed to have been decisively refuted and thus by extension the 'correspondence theory of truth'. However, many still hold this position and continue to struggle for a better metaphor (and greater conceptual clarity!) for correspondence than reflection.

By way of contrast, the notion of ontological truth, as articulated by Bhaskar and espoused by perhaps the majority of critical realists, has no such problems because truth is understood as itself being a property of reality. Critical realists of the older position would argue that this constitutes a category mistake. Truth is a value and an epistemological property. 'Things' cannot be true or false; they simply are. However, the notion of ALETHIC truth is a manner in which semiological (and other) objections to the correspondence theory of truth can be avoided. Here the alethic truth of any real phenomenon x is the real generative capacity that gives rise to x. It is important to CR epistemology$_2$ that, in spite of human fallibility, science possesses the *potential* to reach such truths. With respect to epistemology$_1$, alethic truth in general would be the totality of real, causal powers that give rise to both actual and empirical events. Another way of putting this (which perhaps adds another dimension to the concept) is to say that truth is the real dialectical ground of a causally efficacious 'thing' or absence. Here we have a dialectic of epistemology and ontology (see EPISTEMOLOGICAL DIALECTIC).

So one set of critical realists would reject the notion of alethic truth as a category error and the other would see the correspondence theory of truth as a failed and outmoded theory of knowledge. CR is thus divided upon some very fundamental questions concerning both epistemology$_1$ and epistemology$_2$. This has in general been a productive rather than damaging split. On one level, it is simply replicating debates found elsewhere (in Marxism for example); and on another it provides an interesting focal point for consideration of a whole different set of debates about whether the notion of alethia can be incorporated within a notion of moral truth. Here the old moral philosophy question of whether one can derive 'ought' from 'is' is re-contextualised and given a new affirmative answer.

GARRY POTTER

L

labour markets. There are currently two broad accounts of labour markets. On the one hand is the *neo-classical* labour market account that emphasises labour supply and demand, and ignores social structures. This account, rooted as it is in the deductive model of explanation, is antithetical to CR. On the other hand is what might be called the *socio-economic* account of labour markets, stemming from heterodox economists such as feminists, institutionalists, Marxists, post-Keynesians, regulationists, social-economists and segmented labour market theorists, as well as non-economists from areas such as organisational theory, sociology of work, labour law, state theory, human resource management, industrial relations and urban geography. Despite diversity, this literature unites around a common theme, namely, that labour markets are EMBEDDED in social structures and the latter cannot be ignored because they make labour markets 'work'. Many of these socio-economists (e.g., Callaghan 1998; Rubery and Grimshaw 1998, 2003) come close to adopting a CR meta-theory, even though, with a few exceptions (e.g., Storper and Walker 1983; Wilson 2005), they do not use CR explicitly. This lack of explicitness leads to two (related) problems. First, the lack of a clear meta-theoretical commitment often leads to an inability to break completely with the deductivism of neo-classical theory and, by extension, with the idea that a labour market is, at least in part, a place where labour supply and demand are functionally related to wages. Second, the nature of the 'embedding' is never clarified, resulting in an ambiguous treatment of the relation between social structures and labour markets. In sum, it remains unclear whether the presence of social structures entirely negates, or merely modifies, the functional relationship between supply, demand and wage rates – and, if the latter, what such a 'modification' might involve. Fleetwood (2006) has recently tried to address these problems by augmenting the socio-economic account with CR. This motivates a complete break with the idea that there are phenomena called 'labour markets' that are embedded in other phenomena called 'social structures'. From this CR-augmented socio-economic account, *labour markets just are, or are exhausted by, the very social structures that constitute them*. The way is then open to use Bhaskar's TMSA or Archer's morphogenetic approach to analyse how agents interact with social structures to make labour market activity possible, and the study of labour markets becomes a branch of social theory (Lawson 2003b).

STEVE FLEETWOOD

lack. See ABSENCE; CONSTRAINT; ILL.

lagged/lapsed. *Lagged efficacy* or *time* refers to delayed causal efficacy, as in starlight or the return of the repressed. It is the temporal analogue of ACTION-AT-A-DISTANCE (*D*: 141). *Lapsed time* occurs when the mail is late or when you are asleep, and, in the opposite direction, is postulated by any 'post-ed' theory such as postmodernism (entailing that the future has in some sense already happened, and the present/modern is past). More generally lapsed refers to anything – e.g., an ANALOGICAL GRAMMAR – that hails from the past but is still (inappropriately) current; lagged to anything – e.g., a decision – that occurs after it was appropriate or rational for it to occur. Thus 'contemporary philosophy is tempora{lly} lapsed, lagged on to antiquated analogies' (*P*: 10). See also TENSE.

Lakatos' moment, Lakatosian moment. The stage in the dialectic of EXPLANATION when a multiplicity of research programmes vie for a deeper level of understanding, leading to the abandonment of degenerative programmes and the further development of progressive ones. See also EPISTEMOLOGICAL DIALECTIC; Lakatos 1970.

lambda (λ) clause or **transform**. See METAPHYSICAL LAMBDA (λ) CLAUSE.

laminated structuratum or **system**. See DEPTH; INTERDISCIPLINARITY; RESEMBLANCE; STRUCTURE.

language. See LINGUISTICS; SIGNS.

language-game. See FORM OF LIFE.

law. See CRITICAL LEGAL STUDIES.

law of bivalence. See VALENCE.

law of (non-)contradiction, **law of excluded middle**. See ANALYTICAL PROBLEMATIC; CONTRADICTION; TABLE OF OPPOSITIONS.

laws, causal or **scientific, laws of nature**. See CAUSAL LAWS.

learning process. Learning processes are powered by (1) *curiosity*, which Bhaskar takes to be a basic instinct or urge; and (2) DESIRE to gain knowledge/absent error (the *DIALECTIC OF COGNITION*) in order to meet wants and needs or absent constraints on flourishing. A *dialectic of learning processes* (or *dialectical learning processes*) constellationally contains the EPISTEMOLOGICAL DIALECTIC. Combined with totalising depth-praxis, it is thus the medium by which 'the dialectic of desire becomes the dialectic of freedom' (*D*: 164) (see EMANCIPATORY AXIOLOGY). In this sense, in the sociosphere it is tantamount to dialectic itself, whose schema is caught in MELD:

> Dialectic may be seen as the (experience of {4D}) the process of (trans)formation and dissolution {2E} of stratified (and differentiated) {1M} totalities {3L}. In the human field it constitutes *a general schema for a LEARNING PROCESS* in which absence (2E), signifying incompleteness, leads to transcendence and a greater totality (3L), in principle reflexively (4D) capable of situating itself and the process whereby it became. (Bhaskar 1997a: 141, e.a.)

Ongoing reflexive learning processes are vital for EUDAIMONIA.

Lebensform. See FORM OF LIFE.

legal studies. Like Marx, Bhaskar has not attempted a theory of law, of academic legal studies, or even of human rights – of which Marx, was, of course, so critical, as in *On the Jewish Question* (Marx 1975: 211–41). Marx's position as to law was that law was 'essentially epiphenomenal, part of the superstructure, a reflection of the viewpoints, the needs and interests' of the ruling class (Kamenka, in Bottomore 1991: 306–7). Pashukanis (1978) argued that 'the legal relationship between subjects is only the other side of the relation between the products of labour which have become commodities'. However, Bhaskar appears to be or have been committed to a version of human rights universalism:

> If moral discourse is, as I have suggested, grounded in historical anthropology {...} then we can allow that it makes sense to ascribe the *existence* of rights (and goods) for all human beings *qua* human beings, in virtue of their possession of a common (although always historically mediated) nature, ultimately grounded in their biological unity as a species {...} even though these rights (and goods) can only come to be formulated as demands, recognised as legitimate and exercised as rights under very definite historical conditions. To collapse a right to the historical conditions of its recognition, realisation or exercise is to commit some ethical form (as distinct from analogy) of the epistemic fallacy, grounded in the actualist collapse of anthropology. (*SR*: 209)

This is Bhaskar's response to MacIntyre's well-known argument which, on Bhaskar's reading, equates the existence of a right with the intelligibility of the making of a claim to it, which is patently always highly specific and local in character. MacIntyre famously compared a belief in universal human rights with a belief in witches or unicorns. Bhaskar's later works essentially elaborate the position indicated in *SR*. Meanwhile Alan Norrie has established himself, in a series of magisterial works, the first published in 1991, as the leading legal scholar employing CR concepts to assist in the critique of aspects of criminal law. Norrie's target is a mode of thinking that he calls

> 'criminal justice thinking', by which I mean a liberal philosophical position which dominates legal thinking about justice, and which has its source in Enlightenment-based, Kantian, ideas. On the philosophical side, the main line of thought is *retributive*, and legal practices and doctrines draw upon this to organise their general principles. (Norrie 1999: 1)

Norrie's critique of 'identity thinking' comes close to that of Adorno, who wrote:

> In large measure, the law is the medium in which evil wins out on account of its objectivity and acquires the appearance of good. {...} Law is the primal phenomenon of irrational rationality. In law the formal principle of equivalence becomes the norm; everyone is treated alike. {...} The total legal realm is one of definitions. (Adorno 1966: 309)

Bhaskar and Norrie explained Norrie's engagement with DCR in their joint introduction to the section on 'Dialectic and dialectical Critical Realism' *ER*: 573–4. Norrie's recent work, they assert,

> argues for a critical realist understanding of social structure and totality as the basis for a dialectical understanding of law's ability to differentiate itself while

retaining its relational connectedness with social, political and economic structures. {...} It is this dialectical positioning of law within *power₂ relations* that engenders a sense of its structuration in the moment that it proclaims its difference. This is another version of the dialectic of structure and agency and it generates an ('external') image of legal practice as structurally shaped but also as possessing its own ('internal') forms (what Norrie calls 'legal architectonics'). {...} In effect Norrie employs DCR to argue once more for law's *entity relationism*.

This highly abstract formulation of law's Janus-faced positioning in society has a more explicit analogue in E. P. Thompson's controversial declaration on the rule of law as an unqualified human good:

> It is true that in history the law can be seen to mediate and to legitimise existent class relations. {...} But this mediation, through the forms of law, is something quite distinct from the exercise of unmediated force. The forms and rhetoric of law acquire a distinct identity which may, on occasion, inhibit power and afford some protection to the powerless. Only to the degree that this is seen to be so can law be of service in its other aspect, as ideology. Moreover, the law in both its aspects, as formal rules and procedures and as ideology, cannot usefully be analysed in the metaphorical terms of a superstructure distinct from an infrastructure. While this comprises a large and self-evident part of the truth, the rules and categories of law penetrate every level of society. {...} As such law has not only been imposed *upon* men from above: it has also been a medium within which other social conflicts have been fought out. (Thompson 1975: 264–7)

On this account, a truly dialectical transformation takes place. The positive corpus of law, announced and imbued with violence as it is, together with its spectral other, the utopian claims of human rights, generally serve the interests of power. But they have attained from time to time, in the course of human struggle, a subversive and thoroughly material, real, content. See also Bowring 2002; MacIntyre 1981; Norrie 1993, 2000a, b.

BILL BOWRING

Leibnizian level (of natural necessity). The final stage in any round of the dialectic of EXPLANATION in science when a NATURAL KIND is defined, securing a real DEFINITION. See also EPISTEMOLOGICAL DIALECTIC.

lemma (Gr., gain, profit). An assumed or demonstrated proposition, something to be taken for granted ('in the bag').

Lessing's principle (G. H. Lessing 1729–81) asserts the unity of history as an unfolding process of reason. It is exemplified in Hegelian preservative SUBLATION (*D*: 105), which the CR concept of sublation sharply critiques.

level. See DEPTH; ONTOLOGY; PRIMACY.

levels of rationality. See RATIONALITY.

liability. See CAUSALITY; TENDENCY.

life-and-death struggle. See MASTER–SLAVE.

limit, limit situation. See DIALECTICAL COMMENT; FIXISM/FLUXISM; TRANSITION.

linear negation. See ABSENCE.

linguistic fallacy. See CULTURE; EPISTEMIC–ONTIC FALLACY; HERMENEUTICS; REFLECTION; UNHOLY TRINITY.

linguistic relativity. See SAPIR-WHORF HYPOTHESIS.

linguistics. The study of all aspects of language structure, function and use, historical development, and acquisition; the science of language (as distinct from the study of languages) announced in Saussure's lectures (1916), where the object of linguistics is seen as *langue*, a self-contained and self-defining structured system of SIGNS – roughly pairings of sound and meaning (see also AUTOPOIESIS; STRUCTURALISM). The most influential contemporary theory in a fragmented field (see, for example, Harris 2003) is that of Chomsky, who introduced GENERATIVE MECHANISMS into an otherwise very traditional framework of linguistic description linked to a rationalist-inspired philosophy of innate ideas (Chomsky 1957, 1966, 2000). Linguistics bristles with key CR themes: the role of ABSTRACTION and idealisation in scientific inquiry, relations between STRUCTURE and PROCESS and social and individual, plus the traditional philosophical problems of meaning and REFERENCE and the relationship between language and thinking. As yet there are few specifically CR contributions to linguistic theory although Chomsky's approach to linguistic theorising has been enthusiastically welcomed as CR in spirit. Keat and Urry (1975, cit. Jones 2003) argue that 'Chomsky's rationalism requires the adoption of a realist view of science'; Collier (1994: 209) argues: 'While CR has no unique or unbreakable relationship with any substantive human-scientific theory, it has, I think, a fairly obvious natural affinity with Chomsky's theories, which is reflected in the terminology both approaches use'. Pateman (1987) is a sustained CR defence of Chomsky, whose conception of 'the heterogeneity of linguistic reality, and the multiple causation of linguistic actuality (speech output and intuitions)' is 'not only correct, but characteristic of a realist approach in ontology'. A recent dissenting voice is Jones (2003), who argues that Chomsky's views are biologically REDUCTIONIST, that CR endorsement of them is 'wholly misconceived' and that Chomsky's linguistics amount to 'pseudo-science' (in the sense of Collier 1994) – 'a perfect example of what can happen if a Humean conception of causality, an unbridled epistemic fallacy, and a dogmatic reductionism based on a "strong ESSENTIALISM" are all rolled up into one and set loose on the world'. The stage is set for a lively debate. CR scholars might find they can usefully contribute to linguistic theory by bringing their non-reductionist view of the social to bear in a critical examination of linguistic concepts and methodology.

PETER JONES

linguistification (of being), vb. linguistify is a consequence of the LINGUISTIC FALLACY.

literary theory. To date, work by critical realists in literary studies is theoretically inclined and not primarily textually based. Most current thinking concerning English and American literature seems divided between those who avow 'high theory' as essential to understanding texts, and those who proselytise a conservative return to descriptive analysis based on the sanctity of the literary text. Both positions respond to the excesses, too numerous to summarise here, of critics inspired by post-structuralist and postmodern theories. Norris (1993) outlines both the hostility of various

literary theory 281

postmodern theorists towards truth-claims and their reductive conflation of the fiduciary and veridical with enlightened universalism. His seminal text describes this as a '"dogmatic relativism" {...} an apt enough description for the kind of reflex response that associates truth-claims of whatever kind with the workings of a brutal and arbitrary will-to-power' (1993: 285). It is in opposition to this context that CR – and subsequently META-REALISM – seeks a robustly different conceptual and theoretical grounding for literary exegesis informed by a principle of an ontological objective, rather than depending on the primordiality of either language or identity.

At the first international conference for CR at Essex University in 1997, after a paper later published as 'Reconsidering literary interpretation' (2001), in responding to questions Philip Tew attempted to synthesise certain concepts from Bhaskar's groundbreaking elaboration of CR with the concept of 'radical realism', detailed by Edward Pols (1992). In so doing, Tew suggested verbally the possibility of a broader concept of 'meta-realism', and subsequently this has been adopted in further developing CR ideas, most notably by Bhaskar (2002c). Tew (2001) 'considers the co-ordinates for a CR methodology for literary interpretation. {...} To do so it is necessary to thematise the literary-critical field dialectically. The concepts attached to textuality, as well as its very existence, rely upon aesthetic or critical interventions *as activities*' (196). Bhaskar outlines two essential points for situating meta-Realism (his preferred spelling) as a critique based on the whole range of actual and possible transformations of being. First, it contrasts with the 'actualism' and 'uniformalism' which derive from 'the remorseless, incessant logic of the capitalist mode of production – with its refrain of the constant repetition of the same' (*RM*: 46). Paradoxically, this logic tends to deny, as do identity politics and postmodernism, most human commonalities. Second, meta-Realism comprehends the object as dependent on an understanding of 'everything which is constitutive of all human beings, something which they all share in common', of mediations that constitute our difference, of the 'geo-historical trajectory' and finally of 'the irreducibility of an essential uniqueness, that every object in the world is actually or potentially unique' (*RM*: 47). Nevertheless there remains a residual universal understanding, a working 'towards sensitised solidarity' (*RM*: 47). Hence any meta-realist interrogation of literature and culture must emphasise, first, that texts simultaneously transcend and yet are grounded in contexts and reality; and, second, that aesthetic and ontological considerations, including the imaginary and the generic, are understood to be governed by certain principles of articulation and reference, including:

- *conventionality* (forms that have precedence and sedimentary influence)
- *an empirical dimension* (a material and objective axis in Bhaskar's sense)
- *actual and possible transformations* (not static or determined simply by form and context, but interrelated to an ongoing ontological transition that constitutes human lives and their experience thereof)
- *universal meanings* (that is, broadly apprehended significations)

This is not a reductive ratiocination, since, as Bachelard says, 'We see the relations between intuition and intellect as more complex than simple opposition. We see them as constantly co-operating when they come into play' (2000: 30–1). Furthermore, important to any such conceptual sense of the literary and the theoretical foundations underlying any exegetical intervention is the avoidance of what Lefebvre describes as 'the confusion between "reality" and "structure"' (2002: 198) together with a positive foregrounding of Pols's exposition (1982: 77) that language cannot supersede language, certainly not by drawing attention to its own reflexive, referential limitations in particular genre manifestations (such as the postmodern novel). Concerning the 'linguistic turn', Norris comments that, 'what remains puzzling is the ease with which this orthodoxy captured the minds of so many otherwise intelligent philosophers and theorists' (1996: 65) and points to the phenomenological aspects underpinning Derrida's deconstructive efforts (52–7, 236–43). This is relevant to the literary-critical field since, as Tew comments, 'literature seeks to transcend one set of relations; but in that very movement away from them, it allows these relations to be inscribed within it, as informing qualities', and 'literature is open, contingent, causal and revisable in its critical relations with the real' (2001: 199, 205). In 'A new sense of reality?' (2003) Tew interrogates certain post-structuralist and postmodernist shibboleths, deploying sources such as Bourdieu, Hegel, Lefebvre and Valdés to supplement a critique drawing on Bhaskar and Pols. He argues that all texts 'presuppose a reality both as constituent of themselves and existentially independent of themselves' and that 'in its very inscription of that which precedes itself, even fiction addresses a set of relations and states, even if it interrogates such conditions through modernist or post-modernist devices (themselves part of an ongoing range of interrelated historical inter-subjective responses' (2003: 46, 48). Tew (2004) further draws upon Bhaskar and Pols to support a meta-realist critique of the contemporary fiction scene.

Philip Tew

Lockean level (of natural necessity). The stage in the dialectic of EXPLANATION at which the facts of a given level are deduced from the intrinsic structure of a deeper level, demonstrating natural necessity. See also EPISTEMOLOGICAL DIALECTIC.

loco-periodisation (alternatively, **local-periodisation** or **locational periodisation**) brings together the distinct concepts of *locality* and *periodicity*. Like GEO-HISTORY, the concept is designed to remind us that the spatial dimension of geo-history is as important as the temporal. It is grounded in the irreducibility of both transformation or change (time) and distanciation or difference (space, spatio-temporal position) (see ALTERITY/CHANGE).

Loco-periodisation need not by any means be entirely subjective, conventional or arbitrary. It possesses a TD/ID bivalence. 'Although the loco-periodisation is a transitive act, the locations and periods are real, grounded in the real differentiations and changes in underlying (explanatorily significant) structures' (*D*: 141), i.e., in the reality of depth-stratification and space–time, tense and process. Loco-periods are thus a species of TOTALITY. They may range from a fleeting episode, such as a conversation, to a geological epoch or the life of a star, and should be thought of as constituting, in any present, processual vertically nested intra-/inter-related hierarchies overlapping horizontally in disjunct fashion. Which we focus on will depend on our cognitive interests.

Loco-periodising in the sociosphere is especially complex. However, the TMSA supplies one non-arbitrary criterion: the transformation as distinct from reproduction of structures, possibly constituting a *TRANSITION* from one kind of social or conceptual system to another and leading to the real DEFINITION of the new structure(s); e.g., 1917 and the 1990s in Russia, or a scientific revolution. Given the reality of social and other stratification, in system-change loco-periodisation will be organised around the rhythmics of deep structures, i.e., those that are causally most significant. See also DIACHRONY/SYNCHRONY; SPACE–TIME.

logic. *Formal logic* (also *standard logic*), the modern form of *classical logic* pioneered by Aristotle, is mainly concerned with the systematisation or axiomatisation of modes of valid reasoning. For its contextualisation and a CR critique, see ANALYTICAL PROBLEMATIC. Its branches include *modal logic*, the study or systematisation of the MODALITIES of necessity and possibility; *deontic logic*, the study of the modalities of obligation and permission; *tense logic* or the logic of temporal concepts, usefully introduced by A. N. Prior (1967) to adapt formal logic for the analysis of change; *relevance logic*, which holds that the premises of a valid inference must be relevant to the conclusion; and *many-valued logic*, which rejects the formal principle of bivalence (see VALENCE), holding that there are more than two truth-values: in addition to true or false, propositions may be indeterminate or undecidable or, indeed, have an infinite degree of truth ranging from completely true to completely false.

The CR critique of formal logic by no means issues in its rejection. On the contrary, it is apt for important kinds of synchronic, though not for depth-or for DIACHRONIC, analysis, and is valued for its rigour and precision. It is a vital moment, though only a moment, in *dialectical logic*, or the dialectic of dialectical and analytical reasoning, which is essential for science and any other rationally transformative praxis, and which necessarily violates the formal principles of identity and non-contradiction at crucial stages (see EPISTEMOLOGICAL DIALECTIC) and demonstrates that formal logical are by no means the only CONTRADICTIONS. The aporiai identified by formal logic mark the limits of its applicability in any such praxis. *Formal logical CONSISTENCY* is overreached by *dialectical* or *developmental consistency*, which is the consistency of a directional PROCESS (e.g., an acorn growing into an oak tree, TR into DCR) rather than of a stable state of affairs. Dialectical logic is thus a 'genuinely multi-dimensional and dynamic logic', made possible by 'a combination of ontological stratification and internally tensed distanciated space–time' (*D*: 63), i.e., by the materialist DIFFRACTION of dialectic. See also EPISTEMIC-ONTIC FALLACY (*logicisation*); EXPRESSIVISM/EXTENSIONALISM (*logical extensionalism/intensionalism*); ONTO-LOGIC; PHILOSOPHY; PHILOSOPHY OF SCIENCE; PROPAEDEUTIC; TRANSCENDENTAL REALISM (modal logic); TRANSITION; UNIVERSALISABILITY.

logic of scientific discovery. See EPISTEMOLOGICAL DIALECTIC.
logical detachment. See DETACHMENT.
logical empiricism or **logical positivism.** See EMPIRICISM; PHILOSOPHY OF RELIGION; PHILOSOPHY OF SCIENCE; PHILOSOPHY OF SOCIAL SCIENCE; TRANSCENDENTAL REALISM.
logical extensionalism. See EXPRESSIVISM (INTENTIONALISM)/EXTENSIONALISM.
logical positivism. See EMPIRICISM.
logicism, logicisation (of being). See EPISTEMIC-ONTIC FALLACY.
logocentric theory of meaning, logocentric theory of reading, logocentrism. See HERMENEUTICS; METAPHYSICS OF PRESENCE.

M

malaise. See DIALECTIC OF MALAISE.

management and organisational studies. The social organisation, and discrete organisations, of work are purposive, and management is concerned with the purpose-related effectiveness of organisation(s). Conceptions of, and research in, management and organisation are co-extensive with the full range of social scientific perspectives and methods; but POSITIVISM, SOCIAL CONSTRUCTIONISM and POSTMODERNISM are the dominant approaches, which CR has begun to challenge. It is an intriguing dialectic that CR, projecting as it does 'universal human emancipation' which is argued to be 'implicit in every act' (*P.* 245), has evidently been gaining ground for over fifteen years within management, the dominant global motivation of which is the servicing of capital to gain wealth for a few through the exploitation of the many. Expectations for democratic, progressive and socially beneficial management and organisation may reasonably be higher amongst the more amenable values of the public or other not-for-profit sectors or in small developing economies 'from Beswick to Brazil' (Wainwright 2003). We do not know much about how influential CR is becoming amongst management practitioners, however, and therefore within the reality of management itself, although there are certainly prominent CR academics now in several British university business schools or departments of management, organisation, NURSING, ECONOMICS and other related disciplines.

CR research in management and organisation draws mainly on CR METHODOLOGY and PHILOSOPHY OF SOCIAL SCIENCE, which rely in turn on 'parent' disciplines including SOCIOLOGY, ANTHROPOLOGY, PSYCHOLOGY, women's studies, ECONOMICS and GEOGRAPHY. Johnson and Duberley's (2000) book on the epistemology of management research advocates a CR approach. Enthusiastic encouragement for the use of CR in evaluation techniques is provided by Pawson and Tilley (1997) and in health-sector management theory by Connelly (2000). Baum commends 'scientific realism' in the Blackwell *Companion to Organisations* (2002), and CR is applauded by Tsoukas and Knudsen in the Oxford *Handbook of Organisation Theory* (2003). Tsoukas (1989) has also commended the validity and generalisability of the results of CR idiographic (case) studies in management, as well as formulating a CR meta-theory of management (Tsoukas 1994), drawing directly on Bhaskar's *RS*. Mingers's (2000) paper on CR as an underpinning philosophy has led to published debate in operational research and information science. Realist and constructivist methodologies have been debated in the *Strategy Management Journal*.

A substantive CR strategic-management study of private-sector domestic appliance and office furniture companies in recession was published by Whittington in 1989. Porter (1993) turned to the structured social ontology of CR in a hospital management study to overcome shortcomings in ethnography, since when a seam of CR research in nursing management has been steadily widening. Ackroyd and Fleetwood's (2000) helpful collection combines six CR management-research methodological papers with six substantive articles from fields including industrial networks and labour-market structuration, which has been thoughtfully reviewed by Junor (2001). Another collection of fifteen useful pieces (five meta-theoretical, five methodological and five substantive) is available as Fleetwood and Ackroyd (2004).

RICHARD SHIELD

management science, often known as **operational research (OR)** focuses on analysing real situations and identifying ways to improve them. Consequently it has a very practical focus. It is often used to look at problems within organisations which necessarily involve many different stakeholders, many different issues. Therefore, most of the problems OR tackles are messy and complex, often entailing considerable uncertainty. These organisational problems may be at the operational or strategic level. In either case, the management scientist works with the clients to find practical solutions. One of the defining features of OR is the use of modelling to enable greater understanding of the problem situation, the underlying assumptions of the different stakeholders and a decision on practical action (Pidd 2002). These models may be mathematical but might equally be diagrammatic. Mingers (2004) points out that the use of models arises from OR's focus on real situations and the concurrent impossibility of experimentation in open systems.

The majority of management scientists have probably not considered their ontological stance. Those outside the discipline tend to assume that because of the strong mathematical orientation of management science those within it are largely positivists, yet few management scientists are satisfied with analyses that simply reflect what is going on at an empirical level. Some – particularly those involved in the area of soft OR (problem structuring methods) (Rosenhead and Mingers 2001) – have aligned themselves with constructionism or critical theory. A few management scientists have used CR to support empirical work, for example in information systems (Dobson 2002; Mutch 2002) and in healthcare management (Kowalczyk 2004).

CR has something to offer OR, for many reasons. OR's focus on real situations often necessitates the integration of objective and subjective information and analyses. An ontology proposing that, while much of the social world is socially constructed, underlying structures exist independently of our knowledge, can support analyses relating to both the transitive and the intransitive dimensions.

Management scientists are increasingly concerned about the usefulness of the analyses they provide, and this concern has broadened their focus from trying to solve an organisation's problem towards increasing understanding of the problem situation. Mingers (2000) suggests that CR is particularly well suited to the approach taken within OR and its practical nature. A retroductive approach is already used implicitly by many management scientists as they dig more deeply into the sources of evidence available to find the underlying causes of particular outcomes. Thus, for instance, statistics are often used as evidence to support an explanation rather than as an explanation in themselves. A more explicit focus on the causal mechanisms, underlying structures and agency may help to clarify thinking further.

One of the biggest challenges for OR is that the problem situation is an open system and yet the model by definition is closed. The need therefore is for abstraction – the ability to retain all the important features of the open system within the closed model. This challenge also arises for critical realists in other fields. The use of a CR ontology ensures that this concern is surfaced and choices are made more explicit.

RUTH KOWALCZYK

Marxism. A diverse and developing tradition revolving around the *oeuvre* of Karl Marx (1818–83) and a growing number of readings and extrapolations of his work. One contested area is the relation between Marxism and CR.

Western Marxism peaked in the 1970s, and the last two decades of the twentieth century were marked by widespread disillusion and disintegration. No sooner had this occurred, however, than new generations, historical circumstances and encounters prompted new Marxist projects and self-reflexive debate. Today, a revitalised Marxism is a major intellectual and political paradigm incorporating a wide variety of projects. Whilst it is true that it has always been characterised by global ambitions, it is important to distinguish between three main levels of abstraction and three leading forms of Marxism: there is a tendency for Marxists to prioritise one level of abstraction – the philosophical, the theoretical or the practical; and conflicting views on the status and limits of Marxism take different forms that can be summarised as classical Marxist, post-Marxist or neo-Marxist (Nielsen 2002).

Marxism is thus no monolith, and attempts to compare it with CRITICAL REALISM face further complexity since CR is itself open and evolving. Unsurprisingly, therefore, there is no straightforward answer to the question about the relationship that commands a consensus. Recent research and discussion has made this clear, but also reflects the fact that it is a crucial and pressing question (Nielsen and Morgan 2006).

Bhaskar's early work on the philosophy of science and social science is strongly influenced by Marx and Marxism, even though he recognises that there is no necessary connection between the two projects. Bhaskar's own view of the relationship between CR and Marxism is that 'Marx's work at its best illustrates critical realism; and critical realism is the absent methodological fulcrum of Marx's work' (*PF*: 143), and he has remained a committed but non-dogmatic and increasingly controversial Marxist throughout his *oeuvre* (*SE*; Bhaskar 2003b). Bhaskar's understanding of a deep relationship between the two traditions that goes well beyond programmatic statements is shared by Collier in his introduction to Bhaskar's early philosophy (1994) and in his own contribution to CR. There is thus a clear but non-partisan Marxist emphasis in Bhaskar's and Collier's work, and in recent years a Marxist strand has gestated within CR. Apart from various individual contributions such as Joseph (1998), Roberts (1999), Marsden (1999) and Creaven (2000), this strand is brought together as an explicit project within CR in Brown et al. (eds) (2002). Joseph and Roberts (eds) (2003) can be seen as representing a similar approach, and recent years have also brought about substantive book-length Marxist contributions by CR scholars such as Jessop (2002b) and Dean (2003).

Other leading critical realists such as Sayer and Lawson do not, however, consider Marxism to be a privileged entry point. Sayer (2000a: 8 n) suggests that '{a}lthough there are many affinities between critical realism and Marxism, the former does not entail the latter', and Lawson (1997, 2003b) associates CR (in economics) with Marx and Marxism but also with a number of non- or arguably anti-Marxist economists such as Keynes and Hayek (cf. Fleetwood 1995). Likewise recent collections that bring together contributions from a number of critical realists, such as Downward (ed.) (2003), Cruickshank (ed.) (2003b), Fleetwood and Ackroyd (eds) (2004), Carter and New (eds) (2004), are by no means exclusively Marxist – if Marxist at all. The dominant view within CR seems to be that it smooths the Marxist/non-Marxist divide and thus is compatible with, and underlabours for, a wide range of critical research programmes in the human sciences.

Marxism and CR are thus both multi-dimensional and have increasingly become so through the last two decades. Marxism spans all major levels of abstraction and takes different forms, and the same is true of CR; but beyond these general similarities and the convergence between CR and Marxism through the grounding of CR in readings and reconstructions of Marx (and others) and a continuing engagement with Marxism there are also tendencies towards systematic divergence. This is clear when CR is (equally or primarily) grounded in non- or anti-Marxist philosophy or substantive theory, but even when this is not the case CR is primarily located at the philosophical level and there are obvious grounds for controversies between critical realists and Marxists prioritising other levels of abstraction. Marxists favouring other philosophical traditions such as Hegelianism, Althusserianism or critical theory will likewise challenge CR on the philosophical level. In sum, CR and Marxism have been, and can be, usefully engaged but they should not be conflated. Simple generalisations about Marxism and/or CR do not further an enlightened discourse. The diverse, and developing and controversial character of both traditions should be taken into account, and thus an open and inclusive perspective on the relation is needed. See also ALIENATION; CAPITALISM; CREATIVITY; CRITIQUE; DCR; DIFFRACTION; DIRECTIONALITY; FETISHISM; GLOBALISATION; HOLY TRINITY; IDEOLOGY; INFRA-STRUCTURE, ETC; INTRANSITIVE, ETC.; POLITICAL ECONOMY.

PETER NIELSEN

master-science. See PHILOSOPHY.

master–slave. The concept of *generalised master–slave-type* (or *POWER₂*) *relations* is a key socio-substantive figure, the hub of which is at 2E. Power$_2$ relations comprise one of the four dimensions of the SOCIAL CUBE and causally implicate the others via their discursive moralisation and ideological legitimation, and manifest themselves at all four planes of the social tetrapolity, as indicated in Table 28. As the leading form of constraint absented by depth-praxis, they thus play a key role in EMANCIPATORY AXIOLOGY. They include, besides class, relations of 'nationality, ethnicity, gender, religious affiliation, sexual orientation, age, health and bodily disabilities generally', which may be internal and internalised (*D*: 333). They may instantiate dialectical CONTRADICTIONS, and entrain hermeneutic and other forms of hegemonic/counter-hegemonic struggle.

Table 28 Modes of master–slave or power$_2$ relations, with some correspondences

Causal-Axiological Chain	1M Non-identity	2E negativity	3L totality	4D transformative agency
Modes of master–slave relations	internalised	dialectical contradiction (e.g., capitalist/worker relation)	mediated power-over persons	personalised power-over (in labour process)
Four-planar social being or human nature (social tetrapolity)	d. (intra-)subjectivity (the stratified person)	c. social relations (concept-dependent, site of social oppositionality)	b. inter-/intra-subjective (personal) relations – transactions with ourselves and others	a. material transactions with nature (making)
Social cube		(1) power$_2$, (2) discursive, (3) normative relations intersecting in (4) ideology	(1) power$_2$, (2) communicative, (3) moral relations	
Alienation (fivefold) from	d. ourselves (5)	c. the nexus of social relations (4)	b. each other (3)	a. the labour process: labour and its product(-in-process) (1) material and means of production (2) more broadly, nature and material objects
Irrealist attitudes to master–slave relations	Stoicism (indifference)	Scepticism (denegation)	Beautiful Soul (alienation)	Unhappy Consciousness (introjection/projection)

master–slave

The master–slave trope has its origins in the Hegelian dialectic of master and slave, which involves a life-and-death struggle for RECOGNITION between two consciousnesses that is given a threefold resolution: (1) immediate – it is the slave who really wins, because he is able to overcome the subject–object distinction (ALIENATION) by recognising himself in the product of his work, while the alienator alienates himself from something essential to his well-being; (2) mediate – mutual forgiveness in an ethical life; and (3) eventual – mutual recognition in a constitutional state.

DCR's dialectic of master and slave (*D*: 334) moves from the dialectics of reconciliation and recognition in Hegel, premised on the persistence of master–slave relations, to dialectics of liberation involving their abolition. A meta-critical dialectic, it extends, generalises and radicalises Hegel's resolution via Marx's (1) transformation of Hegel's system by substituting the alienation of labour for the alienation of spirit and demonstrating that 'the Hegelian state in civil society {. . . is} founded on the alienation and exploitation of labour-power' (*P*: 209), and (2) DIALECTIC OF DE-ALIENATION, which stresses that contradictions must be resolved in practice, and which is itself extended and generalised by Bhaskar to encompass the totality of master–slave-type relations, and radicalised in the DIALECTIC OF DESIRE TO FREEDOM via 'a richer and deeper conception of freedom, democracy, rights, education, health and well-being (in which in particular the concept of FREEDOM is radicalised to include both needs and possibilities for development)' (*D*: 335). It proceeds by immanent and omissive critique of both Hegel and Marx, endorsing Marx's critique of Hegel but critiquing Marx himself for tendencies to practical EXPRESSIVISM and CENTRISM; MONISM (or uni-dimensionality – fixation on class to the neglect of other master–slave-type relations), quasi-ACTUALISM and Prometheanism or TRIUMPHALISM; post-dated ENDISM; and sociological reductionism in ethics (moral irrealism) and a consequent 'failure to complement explanatory critique with concrete utopianism' (*D*: 367; see also HEGEL–MARX CRITIQUE).

Bhaskar gives a twofold rationale for the concept of generalised master–slave-type relations. First, the general attitudes to reality portrayed in the sections of the *Phenomenology* immediately following the dialectic of master and slave apply quite generally to 'contemporary orientations to relations of domination, exploitation, subjugation and control', i.e., Stoicism, Scepticism and the Unhappy Consciousness (see BEAUTIFUL SOUL, ETC.). Second, it 'enables us to pinpoint a characteristic feature of the capitalist mode of production: *the exploitative relation intrinsic to the wage-labour/capital contract is hidden at the level of inter-personal transactions* by FETISHISM and the causally efficacious category mistakes upon which it depends' (*D*: 154). The discourse of masters and slaves, in contrast to that of capitalists and wage-earners, highlights this exploitation and the unfreedom of wage-earners, who have no option but to work for masters. The concept has been criticised by some Marxist critical realists (e.g., Joseph 2002) on the grounds that it treats class on a par with other forms of oppression. However, while it could be deployed in this way, this is not intrinsic to the concept, which in terms of its archaeology is first and foremost a class concept and is fully compatible with the view that class is 'arguably the explanatorily most important {. . .} of master–slave-type relations' (*P*: 209), though not with a view that reduces other forms of oppression explanatorily to class. Moreover, in bringing together in the one concept (1) applicability to all (in Marxist terms) class societies or (in CR terms) *master–slave-type* or *power$_2$-stricken* societies; (2) instantiation of both poles of dialectical CONTRADICTION;

and (3) non-class forms of power$_2$, it fills a lacuna in the Marxist lexicon. It in no way precludes the use of less inclusive concepts, including that of class, but positively invites them.

material causes. See CAUSES.

material interest, dialectic(s) of. See DIALECTIC OF INTERESTS.

material object realism. See REALISM.

materialism in its classical form, resonates strongly with Aristotle's critique of Plato's concept of ideational forms. Aristotle (*Metaphysics*, especially bk VII, 1998) takes the view that the categorisation of types and kinds in reality is explicable in terms of empirical investigation into ESSENCE and substance as characteristics of objects that make them what they are and how they act and change, rather than as essences as forms in contemplation of some non-perceivable ideational template of perfection (see also NATURAL KINDS). Materialism in its most general metaphysical form refers to the view that everything that exists is in some sense material (Lange 1925). Material is usually translated as 'physical', and the term *physicalism* tends to be used interchangeably with materialism in philosophy (Gillett and Loewer 2001). In a contemporary scientific context, the physical is not restricted to its ordinary language use as the aggregated particulate constitution of objects (Smart 1963). It also includes the forces, energies and wave forms that are understood to bind and interact with them. Materialism does not, therefore, reduce to fixed objects but also encompasses processes. Similarly it does not reduce to directly observable entities but also encompasses the understanding that science expands the range of what is perceived. As such, metaphysical materialism at a general level tends to be intertwined with commitments that include: (1) Non-entities are not part of reality because they are immaterial, in the sense that they are not particulate, waves, forces, energies, etc., or some combination of these. (2) Non-entities are not real because they could not, in principle, be brought under an expanded form of perception through scientific method – if they could, they would by definition be material. (3) Non-entities are not real because they are not scientifically testable and explicable – they are in this sense non-empirical. (4) Science thus governs, arbitrates or informs our sense of what is real because science *is* materialist investigation.

From a general realist perspective (1)–(4) are highly plausible but have entailed numerous particular problems in their articulation. Those problems have meant that materialism has accumulated a large amount of unfortunate conceptual baggage that perhaps warrants a terminological shift in order to clarify the grounds of debate. (4) has meant that materialism has been intertwined with PHILOSOPHY OF SCIENCE. Materialism has thus been associated with EMPIRICISM, even though empiricism has idealist forms and tendencies, and with positivism. As such, our understanding of materialism tends to carry particular notions of what it means to identify an entity and its operation in a scientific fashion. Accordingly, dominant strains in (4) and their construction of (3) – especially ideas of observation and testing of event regularity and closed systems – affect (2). For example: Are ideas real? Are conceptual relations as structures real? In what sense are they 'material' in terms of (1)? Marxist materialism focuses on some of these kinds of questions through its critique of Hegel, Left Hegelianism, philosophical method in *The German Ideology*, and scientific methodology in the preface to *Capital* (*P*. 124–32).

In philosophical discourse, different kinds of materialism (*P*. 101) have developed in terms of the tension between issues in (4) and (3). Much of that tension has developed out of a longstanding tension between materialism and IDEALISM (*P.* 102–7). (See also ANALYTICAL PROBLEMATIC and PRIMAL SQUEEZE.) During the seventeenth and eighteenth centuries a strong empiricist concept of materialism emerged based on an ontology that is highly mechanistic and regularised. Such a directly observed and experienced clockwork universe, often illustrated by reference to the work of the mathematician Pierre Simon de Laplace, seemed to leave no room for God (as anything other than first cause), the soul and a free human consciousness. Given the pervasiveness of religious belief, this was a problem for proponents of the new sciences as well as their opponents. A main response was DUALISM. The material universe was constituted by spatially extended substances operating in a clockwork fashion, but there were also non-spatially extended non-substances – spirit, God, soul, the mind or consciousness as something other than body. A materialist–immaterialist or-idealist tension thus emerged whose sociological condition was one of preservation of cherished beliefs. But those beliefs also entailed genuine ontological problems concerning the relationship between materialism, consciousness and will that were themselves indicative of deeper problems concerning ontology of reality based on event regularity in closed systems.

Since the terms of debate in philosophy have been set by materialism in opposition to idealism, subsequent versions of materialism have tended to be constructed with reference to idealism. This is seen at its starkest in philosophy of MIND where the goal of ontological unification has fostered *reductive materialism*, which argues that any seemingly immaterial identified entity x (in this case, consciousness) fully reduces to some other identified material entity y (in this case, physical brain states or neural processes). Reductive materialism, however, is not an empirical statement but rather a philosophical conjecture, subject to problems such as qualia and folk psychology. There is no currently widely accepted neuro-physiological account of consciousness, though there are accounts of brains systems. Reductive materialism tends to be associated with *eliminative materialism*, which is the proposition that false common-sense conceptions of phenomena are displaced by scientifically robust accounts of the same thing (Churchland 1981). In its stronger form (Smart 1963; Rorty 1965), indicative of its close connection to both reductive materialism and Hempel and Oppenheim's symmetry thesis of prediction, eliminative materialism holds that scientific accounts of phenomena will ultimately reduce to more basic accounts in terms of which all material phenomena will be explicable. This position either reaffirms ontological dualism – some things are immaterial – whilst also conjoining this with the proposition that science then becomes inappropriate and the phenomena inexplicable, though perhaps comprehensible, in some other discourse; or affirms a materialist ontological MONISM where accounts in the special sciences will ultimately reduce to those of physics as a science of everything (*PF:* 47–69). As various commentators have noted (Fodor 1974), in ontology, this does not simply imply a more adequate redescription of entities but in many cases defines them out of existence prior to any well-founded science. Moreover, methodologically, it seems to imply that special sciences should shrink as a measure of their success, something for which the sciences themselves provide no evidence.

Many of the problems of materialism, then, are accumulated ontological problems such as closure, event regularity, and accounting for the reality of different types and kinds of entities within the constraints of particular conceptions of science and in conjunction with conflicts with idealism. The principal response in CR has been to restructure ontology of science that resituates materialism in terms of DEPTH-realism, open SYSTEMS, and stratification and EMERGENCE in reality. Much of CR is therefore on a similar path to both non-reductive materialism, which argues that higher-order material entities emerge from but are not reducible to lower-order material entities, and philosophical NATURALISM in the form of scientific realism, which argues that philosophy is continuous with science, and science is materialist in the broad metaphysical sense, but an adequate account of science accounts for its divisions, specialisations and stratifications. However, scientific realism tends to have a more limited commitment to substantive ontological propositions (Kaidesoja 2005). At the same time, though there is greater ontological substance in CR concerning general features of reality that science presupposes by its investigations, there is also a degree of ambiguity in CR, particularly in Bhaskar's work, concerning materialism, precisely in terms of the limits on science and the possibilities of reality that a materialist commitment may entail. In *PN1*: 124–6, for example, when making the case for *synchronic emergent powers materialism*, the form of the constitution of consciousness itself (as opposed to its emergence from a materialist process) is left open. Looked at in a broader sense, this acknowledges a basic problem with materialism in philosophical terms. Metaphysical materialism tends to conflate science, and explicability, testability and perceivability, with materialism, precisely in terms of the way general metaphysical materialism defines itself ([1]–[4] previously). Arguably, this sets an a priori epistemological constraint on both science and reality. To be science is to be materialist, to be real is to be material. This is an ontological assumption but it is not an independently confirmed scientific hypothesis – it is an imposed condition of what science means. In many respects this seems relatively unproblematic, since a materialist notion of science appears to be highly plausible and successful. Moreover, in terms of the combination of (2) and (3) above, the meaning-frame of materialism is highly elastic and would seem to be able to accommodate any new order of being identified – precisely by the capacity to identify it. Thus, if consciousness turns out to entail some radically new emergent entity that is neither particulate nor some currently understood form of energy matrix, force, etc., it need not hold out any particular challenge to a materialist taxonomy. Ideas of spatial extension and substance, conservation of energy and such, might simply adjust as part of the process of bringing the entity into identification (such as the current debate over multi-dimensional string theory). There might still be a case here for moving beyond a materialist conception in terms of shedding accumulated conceptual baggage, as Searle (1999) argues, but this seems more a matter of avoiding the need for endless clarification rather than making some new ontological point.

However, materialism is still left with the problem of (1). Non-entities, if they remain non-entities, cannot simply be identified and thus be material in terms of an expanded definition of what it is to be an entity. This highlights a key aspect of DIRECTIONALITY and that is the need to account for change, PROCESS and TOTALITY without basic conceptual–ontological contradiction (*D*: 4–8). Philosophical materialism is of course sophisticated enough to acknowledge process and change; Marxist dialectical materialism is also a well-developed account of process and change. Bhaskar, however, would argue that neither has a full enough conception of ABSENCE and totality to account for them adequately, despite those developments and acknowledgements. It is on this basis, and perhaps in terms of any plausibility in the philosophy of META-REALITY, that a stronger case for moving beyond a language of materialism might be made. See also (materialist) DIFFRACTION.

JAMIE MORGAN

maya. See DEMI-REALITY; IDEOLOGY; TINA SYNDROME.

meaning, meaningfulness. See CULTURE; DISCOURSE ANALYSIS; ENCHANTMENT; HERMENEUTICS; META-REALITY; PDM; PERCEPTION; REFERENCE; SIGNS.

mechanism, generative. See CAUSAL LAW; CAUSALITY; CONCRETE/ ABSTRACT; COUNTERFACTUAL/TRANSFACTUAL; TENDENCY.

media studies is a multidisciplinary field of research into media from the perspectives of TECHNOLOGY, communication, HISTORY, social institution, CULTURE, language and DISCOURSE. Media can be defined as the technical forms for the production, fixation, transmission, reproduction and commodification of messages and symbols. It includes both mass media and other forms of mediated interaction (Thompson 1995). A common conceptualisation of the field is to separate (and relate) *production* (including its social, political and economic preconditions, the practices of production and coding), *text* (SEMIOTICS, discourses), and *reception* (audiences' interpretation and decoding of messages) (Hall 1980; Marris and Thornham [eds] 1997).

Taking this description of media studies as the point of departure, it is clear that CR approaches are of high relevance. The transformational model of social activity; the model of social stratification, emergence and causal mechanisms; and the CR contributions to cultural analysis could definitely contribute to deepen our understanding of media, culture and society. However, to date there are few concrete examples of applying CR to media studies, and media remains undertheorised from a CR point of view.

Lua (2004) shows that the CR models of open social system and causality can provide a new way of integrating previous research on news production. Different factors influencing the production of news are analysed. Lua questions the dominant CONSTRUCTIONIST view, arguing that news comprises 'overdetermined products', the 'outcome of specific configurations of actualised causal power' (2004: 707). See also MEDIUM THEORY.

CR has been used by media researchers in order to rethink methodological strategies and objects of research (Jensen 2002; Schroder et al. 2003). Berglez (2006) argues that retroduction is a mode of inference required in order to discover structural constitutions of media discourses, and their material implications.

Inspired by a CR conception of generality, Eriksson (2006: 31) argues that audience research 'should aim at producing general knowledge about the constituent properties or transfactual conditions of the process of media consumption'. Audience research has been characterised by a DUALISM between a radical ethnography understanding media consumption and reception as closely related to particular contexts, and the empiricist position defining generality as synonymous with empirical extrapolations and regularities. Eriksson shows that it is possible to transcend this dualism by adopting Bhaskar's standpoint that scientifically significant generality is to be found in the essence of social structures. A stratified model of media consumption is developed in order to understand causal mechanisms at different levels.

MATS EKSTRÖM

mediate, n. mediation (f. L. *medius*, middle). To divide into two equal parts, to be between, to act as an intermediary or medium. Whether in the human world or the natural world simpliciter, 'if A achieves, secures or eventuates in C (either in whole or in part) via or by means of B, then B may be said to mediate their relation' (*D*: 114). Thus a social system, A, is reproduced/changed, C, via the mediation of agency, B. This abstracts, of course, from a complex developmental PROCESS, articulated both vertically (the stratification of agency and social structure, the climatological and atmospheric regime, the material substrate) and horizontally (the presence of the future-in-the-past-in-the-present [see TENSE]). It will involve all manner of mediation, implicating the causally efficacious presence of the past and outside, which may be contradictory as well as supportive, malign as well as benign.

A pivotal category of TOTALITY (3L) and the third moment of the CONCRETE UNIVERSAL ↔ SINGULAR, mediation is thus a form of the NEGATIVITY which unifies the 1M–4D chain of being–becoming (see CAUSALITY), and which may itself be mediated by the other members of the chain. Closely affined to RELATIONALITY, which may take the form either of *inter-action* (involving *contingent mediations*) or *intra-action* (internal relationality) (involving *necessary mediations*), it is the specific contextualised causation which, in a necessarily CONTRADICTORY nexus, orients the rhythmic exercise (absenting processuality) of the causal powers of structures (universality) to concrete singularity. Or it is the *activity* or actualisation of the HOLISTIC CAUSALITY and inter-/intra-action of partial totalities. It may thus be involved in healing – in the overcoming of dichotomy and split – but also in their perpetuation, as in the complex mediations involved in any TINA SYNDROME, in which guise it is tied conceptually, together with ALIENATION and ALTERITY, to HETEROLOGY. The mediating figures par excellence are thus the dialectics of CO-PRESENCE (of absence and presence, as in the presence of the past and outside) and CONSTELLATIONALITY (*P*. 166). Given that diversity is irreducible and essential to creativity and flourishing, in eudaimonian society multiple social mediations between the individual and the realm of universal civic duty would be necessary for the achievement of true unity-in-diversity. Mediation embraces (1) *mediatisation*, which refers either to the process of mediation as such, or, more specifically, to the mediation of information and meanings by the modern media which disembeds space from time (see DISTANCIATION) and may occlude or invert the processes it purports to describe (e.g., the Chechen or Iraqi resistance as terrorism) (see IDEOLOGY); and (2) *VIRTUALISATION* and *hyperrealisation*.

If A achieves C without mediation, A and C are in an *immediate* (*unmediated*) relation. Bhaskar has consistently held that some ACTIONS, including PERCEPTIONS and understandings, must be immediate, otherwise no action or communication would be possible; but that, conversely, such basic actions and understandings are ontologically possible only in virtue of 'an indefinitely extended geo-historical formation of {their} conditions' (cf. COGNITIVE SCIENCE). Hence the immediate and mediate are interdependent moments of a DUALITY, 'characterised by non-arbitrary dialectical *interconnection*' (intra-connection), and 'properly conceived', i.e., furnished with a stratified ontology, 'the programmes of deconstructive SEMIOTICS and reconstructive HERMENEUTICS are not only consistent but interdependent' (*D*: 147 n, e.a.). PMR highlights the immediate, intuitive level to our action and understanding of the world, including of its intrinsic value and deep, fine structure, in the concept of spontaneous right action (e.g., *RM*: 98–100; *MR*: 42–3). This in no way obviates the need for explanation and critique; on the contrary it necessitates them and provides (eristically) an underlying rationale for emancipatory politics.

Demediation is a mode of detotalising abstraction in that it isolates a thing from its context, as in a scientific experiment which instantiates a universal law in a singular (repeatable) instance by preventing or absenting outside interference which would otherwise have occurred. While affording a transcendental deduction of the category of real negation, it is illicit unless accompanied by a reverse, retotalising (*remediating*) movement in thought (see CONCRETE/ABSTRACT).

mediatisation. See MEDIATE.

medical sociology. See SOCIOLOGY OF HEALTH.

medium theory a theory of the causal powers of communications media. Rejecting the standard liberal understanding of communications media as tools with which individual users have a wholly intentional, external and controlling relationship, medium theorists (of whom Jack Goody and Walter Ong are among the most interesting) seek to persuade us that modes of communication – oral, scribal, print, electronic – are extensions of the human senses which bring into play different combinations of sensory practice and which, thereby, have important cognitive, social and organisational effects. They are among the most important means of cultivating, or not, particular dispositions and capacities. The invention of writing, for example, is necessary for the constitution of the strongly individuated, introspective psyche. Frequently misunderstood as a technological determinism, medium theory is viewed more accurately and more fruitfully as an implicit critical realism since, at its best, it grasps the significance of the distinction between media powers and their actualisation in open systems. By clarifying the status of its claims in this way, CR can enhance the critical explanatory power of medium theory. Correlatively, medium theory can enrich CR's understandings of the historicity of actualised human powers. So far, the possibility and desirability of a mutually beneficial relationship between medium theory and CR has not been examined to any significant extent, although communication as subject-constitutive practice is an important theme in Dean (2003) and is explored by Nellhaus (2004) in relation to the emergence of 'deep individualism' in early modern Europe. See also Goody 1977; Ong 1982.

KATHRYN DEAN

MELD. The *ontological–axiological* or *causal–axiological* chain, comprising four *degrees* or *stadia* (pl. Gr. *stadion*, also L. *stadium*, a measure of length): 1M *first moment*, 2E *second edge*, 3L *third level*, and 4D *fourth dimension*. *Moment* signifies something finished, behind us, determinate – a *product*: *transfactual* (structural) *causality*, pertaining to *NON-IDENTITY*; *first is for founding*. *Edge* speaks of the point of transition or becoming, the exercise of causal powers in *rhythmic* (*processual*) *causality*, pertaining to *NEGATIVITY*. *Level* announces an emergent whole with its own specific determinations, capable of reacting back on the materials from which it is formed – *process-in-product*: *holistic causality*, pertaining to

TOTALITY. Dimension singles out a geo-historically recent form of causality – *product-in-process*: human *intentional causality*, transformative AGENCY or praxis.

Thus the MELD moments and corresponding concepts (see Table 29), indicate different modes of ABSENTING or CAUSALITY – of the unfolding of being – constituting the ontological–axiological chain in a *four-sided dialectic* which non-preservatively sublates the Hegelian three-termed dialectic (identity, negativity, totality). Or the theory of MELD builds on the Hegelian CONCRETE UNIVERSAL (universality, particular mediations, singularity [UPS]), crucially adding processuality (such that UPMS, where 'M' stands for 'mediation') and distinguishing *human* agency or concrete singularity (4D) as an emergent level of concrete singularity in general. This is in the spirit of the Marxian critique of Hegel (see HEGEL–MARX CRITIQUE). 2E is the hub of this chain. It not only unifies it in that 'the whole circuit of 1M–4D links and relations' (*DG*: 392) can be derived from absence, but itself reflects the other moments and itself: its concepts constitute a stratified hierarchy (1M), it is 'dynamic in its essence' (2E), 'coheres the system as a whole' (3L), and 'includes transformative praxis, oriented to change' (4D) (*D*: 250). Like all CR concepts, the four moments possess a TD/ID bivalence, speaking both of constructs for thinking being and of real interrelated distinctions within it. As the entry for CRITICAL REALISM makes clear (see esp. Table 15), it is important to view the schema diachronically, not just synchronically; i.e., not just analytically as indicating different moments of being-becoming, but dialectically in terms of (1) the vertical development of the CR system, such that each successor stadion presupposes or preservatively sublates its predecessor: 4D > 3L > 2E > 1M; and (2) the unfolding of being as such (structure is ontologically deeper than its actualisation at 2E, and 3L and 4D speak of emergent levels that are by definition diachronically more recent). MELD four-demensionality mimicks, in short, that of space–time itself. Alphabetical lists of key concepts by stadion and the critique of irrealism (the denegation of 1M–4D) are appended to Table 29.

1M and 3L have a special affinity in so far as they pertain to the possession of causal powers rather than their exercise; conversely, so, too, do 2E and 4D in so far as they pertain to the exercise of causal powers rather than their possession (see CAUSES, Table 9). It is important to note that human capacities, liabilities, powers, structures (e.g., the stratification of the personality) pertain to 1M rather than to 4D, and that their exercise does not always appear at 4D: thus thought is included in 2E to highlight its negativity, and epistemology in 1M to emphasise its constellational inclusion within being and that, without 1M grounds, it is empty. Where concepts clearly bridge all four stadia (as in the case of the unifying concepts of absence/causality/emergence, concrete universal ↔ singular, four-planar social being, the reality principle, relationality, truth and freedom) they appear in all four columns (or lists) in one guise or another. Where they belong equally to several (e.g., mind as emergent result or product [1M], as exercised in thought [2E] or all as emergent process-in-product [3L], [and as engaged intransformative praxis], they are sometimes listed several times. Mostly they appear once only. Some of these, however – e.g., the human capacity for freedom at 1M – can by valid perspectival switch be seen to pertain equally to another stadion, in this case the exercise of the capacity for freedom at 4D. Likewise, in the lists by stadion distinguishing between realist concepts proper and those pertaining to the critique of irrealism, concepts such as virtualisation and mediatisation pertain both to realism

Table 29 Key MELD concepts, with some correspondences

Ontological–Axiological Chain	1M First Moment: Non-Identity	2E Second Edge: Negativity	3L Third Level: Totality	4D Fourth Dimension: Transformative Agency
Formal principle	non-identity	negativity	totality	praxis
Substantive content	infra- intra-/super-structuration	opposition (incl. reversal), transition	holistic causality, reflexivity, universalisability	emancipation; metacritics
Defining concepts	*ontology*: intransitivity, transfactuality (universality), stratification (emergence qua product); SEPM; intrinsic value (enchantment); freedom; alethia	*negativity*: processuality, contradiction, emergence qua process; social relationism, transformationalism; negativity qua absenting constraints/ills; creativity	*totality*: mediation; binding or forming energy, inter-/intra-relationality; maximised by praxis (absents incompleteness); trust	*transformative* (human) *agency*: intentional causality; praxis–poiesis; transitivity; referential detachment; autonomy (self-determination), dialectical reason; right-action
Irrealist errors (forms of scepticism generated by absence of a concept of absence)	*denegation of ontology*: actualism, subject-object identity theory extrusion of objectivity from subjectivity by the epistemic fallacy; anthropism	*denegation of absence*: ontological monovalence dualisms and splits	*denegation of totality*: ontological extensionalism/expressivism (alienations) atomism, punctualism	*denegation of agentive agency*: extrusion of subjectivity from objectivity by reductionism/disembodiment reflexive inconsistency; theory/practice
Modes of illicit abstraction	destratification	deprocessualisation/denegativisation (positivisation)	demediation/detotalisation	desingularisation/deagentification
Irrealist attitudes to master-slave relations	Stoicism (indifference)	Scepticism (denegation)	Beautiful Soul (alienation)	Unhappy Consciousness (introjection/projection)
Forms of split or alienated consciousness	extra-jection (splits the world from thought about it)	retro-jection (splits geo-historical change from the present)	intro-jection (splits the psyche)	pro-jection (splits minds from bodies)
Concrete universal ↔ singular (UPMS)	universality	processuality	mediation	(concrete) *singularity* (not just human)

Polysemy of absence	product	process (incl. subject- or developmental negation)	process-in-product	product-in-process
Causal modes of absence	transfactual causality	rhythmic causality	holistic causality	intentional causality
Modes of radical negation	auto-subversion	self-transformation (self-referentiality, self-emancipation)	self-realisation	self-overcoming
Modes of presence of past/future	existential constitution	co-presence	lagged efficacy	agential perspectivity
Types of tendency	tendency$_a$	tendency$_b$	tendency$_c$	tendency$_d$
Aristotle's causes	material formal	efficient final	formal material	final efficient
Recent turns in thought	realist	red, New Left	Green, feminist	reflexive
Types of politics	emancipatory/ transformative	movement	representative	life
Components of human nature	(1) core universal, grounded in shared genetic constitution, alethic self (DCR) and ground-state (PMR)	(4) changing	(2) historically specific	(3) unique individuality (concrete singularity)
Four-planar social being or human nature (social tetrapolity)	d. (intra-)subjectivity (the stratified person)	c. social relations (concept-dependent, site of social oppositionality)	b. inter-/intra-subjective (personal) relations – transactions with ourselves and others	a. material transactions with nature (making)
Social cube		(1) power$_2$, (2) discursive, (3) normative relations <u>intersecting in (4) ideology</u>	(1) power$_2$, (2) communicative, (3) moral relations	

	a. the labour process: labour and its product(-in-process) (1), material and means of production; (2) more broadly, nature and material objects	b. each other (3)	c. the nexus of social relations (4)	d. ourselves (5)
Alienation (fivefold) from				
Modes of master–slave relations	personalised power-over (in labour process)	mediated power-over persons	dialectical contradiction (e.g., capitalist/worker relation)	internalised
Modes of freedom	agentive; autonomy (self-determination)	totalising; formal legal freedom; well-being, flourishing; eudaimonia (dialectical freedom)	process of absenting constraints/ills	stratified (has degrees and levels); essential autonomy
Modes of alethic truth	contextualised by the dialectic of the singular science concerned	totalising (oriented to maximising explanatory power)	praxis-dependent	ontological stratification, natural necessity, reality principle
Theory of alethic truth (truth tetrapolity) – truth as	(3) expressive-referential (4D and 1M) (TD/ID)	(1) normative-fiduciary (IA)	(2) adequating (warrantedly assertible) or epistemic (2E and 1M) (TD)	(4) ontological, alethic (ID)
Qualities of truth	context-sensitive	totalising	dynamic	grounded
Modes of untruth	(2) *in* an object or being (at that level of reality)	untrustworthiness	(1) *about* an object or being (at any one level of reality)	(3) *of* an object or being *to* its essential nature
Theory of moral alethia – truth as	freedom: free flourishing of each and all	universalisability	dialectical reason	our transcendentally real selves
Judgement form	expressively veracious	fiduciary-imperatival (trustworthy)	descriptive (2E in so far as object is change)	evidential (1M in so far as optimally grounded)
Ethical tetrapolity, logic of	(3) totalising depth-praxis	(1) fiduciariness	(2) explanatory critical theory complex	(4) freedom qua universal human emancipation, the realisation of moral truth (alethia)

	(4) emergence	(2) dialectical contradiction	(1) dialectical connection	(3) negation of the negation
Logic of transition (social)				
Subjectivity/self	transcendentally real self or alethic self	ongoing change (geo-historical constitution)	self-consciousness or reflexive monitoring, incl. internal conversation); internal relations with others	embodied self, site of intentional causality
		illusion of ego (ideological)		
An embodied self (a human being or person)	core universal human *nature*	*rhythmics* of her world-line	particular *mediations*	unique *concretely singular person* (in effect a natural kind)
Qualities of the self	psychic (stratified beliefs, concerns, projects)	mental	emotional	physical
Transformative agency	stratification of mind (consciousness, self-consciousness, pre-conscious, unconscious)	reasons for acting, flow of action/praxis	inter-/intra-action within and among persons	embodied intentional causality (exercised), action/praxis–poiesis
Components of action	will	thought	emotion	making (physical action and objectification; poiesis)
Basic human capacities	freedom	creativity	love	right-action
Presentational dialectics	CR, Hegelian dialectic, and the problems of philosophy – products for transformation (*D*: ch. 1)	Absence, emergence, contradiction, etc.(*D*: ch. 2)	the system of DCR, the totalising dialectic of freedom (*D*: ch. 3)	Metacritique of irrealism (*D*: ch. 4)

Alphabetical list of key concepts by stadion and the critique of irrealism

Note The lists are fairly comprehensive, but not exhaustive. Location of concepts should not be hypostatised.

1M absence (product); abstract, abstraction; alethic truth (aletheia); alterity (difference); analogical grammar; anti-anthropism; atomism, ontological and sociological; autonomy (essential); being; being as intrinsically valuable (enchantment); (human) capacities (e.g., for cognition, language-acquisition, perception, etc.); causal law, causal power (extrinsic aspect); CEP; change (product); constellational identity; core universal human nature; de-onts; depth; depth-realism (vertical realism); disposition; (differentiation of) domains of real > actual > empirical/semiosic; emergence (product); epistemology; essence; explanation; fictive beings; generative mechanism; freedom (human capacity for); four-planar human nature; ground, grounding; intransitivity (ID/TD distinction); intra-structure; hyperreality; level; liability; material and formal causes; method; moral aletheia (truth); moral realism; natural kinds; natural necessity; non-identity; ontological depth; onts; ontological realism; objectivity; ontology; ontological grammar; openness; possibility (*potentia, potentiality*); potential (as for capacity); power; primacy or priority (ontological); product (absence); praxis (qua capacity); psyche; real definition; realism; reality principle (epistemologically mediated alethic truth); referential detachment (the capacity for); relationality (internal relationality); science; social structure; stratification and differentiation; stratification of personality (plane d), four-planar social being; stratum; structuratum; structure; superstructure; synchronic emergent powers materialism (product, powers of mind); PEP; taxonomy; tendency$_a$; throwness; transcendentally real self; transfactuality, transfactual causality (mode of absence), transfactual realism (horizontal realism); truth (ontological/alethic); ultimata; (dialectical) universality (first moment, concrete universal); virtuality; will.

Critique of irrealism and its social context: actualism; anthropism; anthropocentrism; anthropomorphism; anthro-ethno-ego-present-centrism; arch of knowledge; auto-subversion; centrism; conceptual actualism; conceptual realism; constant conjunction; destratification; empirical realism; empiricism; epistemic–ontic fallacy (denegation of ontology, moment of unholy trinity); expressive unity; false beings; fideism; fluxism; fusion; historicism; homology; (subject–object) identity theory; monism (linked with reductionism); positivistic illusion; primal squeeze (moment of unholy trinity); reductionism; regress; scientism; solipsism; speculative illusion; stasis; Stoicism; transcendent realism; transdictive complex; (abstract) universality.

2E absence, absenting (process); aporia; autopoiesis; becoming; bivalence; boundary state; causal chain, causality, causation; change (process); conatus; conjunctural determination; contradiction; constraint; co-presence; creativity; critique; determination; desire; development, developmental consistency; developmental negation, subject-developmental negation; dialectical arguments; directionality; efficient and final causes; dialectical explanation; diffraction (of dialectics); dispositional identity; distanciation; embedding/disembedding; emergence/disemergence (process); entropy/negentropy; epistemic relativity; equivocity; evolution; existential permeation, existential connection; finitude; frontier; gap; geo-historicity; hiatus; hiatus-in-the-duality; intra-structuration (process); loco-periodisation; mentality; mind/thought; needs; negation, negativity; negation of the negation; negative and positive freedom; negative generalisation; nesting; node; δ node, ontological polyvalence; onto-logic; opposition; (ontological) perspectivality; presence of the future; presence of the past and outside; primal scream; primary polyadisation; process (absence, absenting), processuality (second moment, concrete universal); reality principle (epistemologically mediated alethic truth); reciprocity; recursivity; reversal; rhythmic, rhythmic causality (mode of absence); rhythmic identity; self-referentiality or -emancipation; self-transformation; social cube; social relations (plane [c], four-planar social being); spatio-temporality; spread; stretch; stretch-spread; super-structuration; temporality; tendency$_b$; tense; TMSA; threshold, totality; transition; transform; transformation; tri-unity of causality, space and time; truth (adequating); vehicular thrownness; void; wants.

Critique of irrealism and its social context: absolutism; antinomy; anomaly; attachment; blockism; chiasmus; closure; complicity; denegation; de-processualisation; de-negativisation (positivisation); dialectical contradiction; dilemma; duplicity; endism; equivocation; eternisation; fetishism; fixism; foundationalism, fundamentalism; ideological intersect; inversion; master–slave-type (power$_2$) relations; logicism; monism (entails and entailed by ontological monovalence); opposition (ideological); plasticity; problematic of presence; reification; reversal; Scepticism; split; theory–practice inconsistency or contradiction (performative contradiction); Tina syndrome; triumphalism; unseriousness.

3L absence, absenting (process-in-product); alienation; argument; attachment; autonomy; bipolarity; boundary; conceptuality; (general) conceptual scheme; concrete; concrete utopianism; concretion; consciousness; consistency; coherence; connection, connectivity; consequence explanation; constellational identity; constellationality; de-alienation;

dialectical reason (the unity of theory and practice in practice); dialogicality; dual control; (constellational) duality or unity; duality of theory and practice; ecology; emergence (process-in-product); emotion; eudaimonia (highest form of freedom); explanatory critique; flourishing (mode of freedom); formal and material causes; heterocosm; holistic causality (mode of absence, extrinsic aspect); homonymy; hyperrealisation; identity; ideological intersect; indexicality; indivisibility of freedom; inter-relationality, intra-activity (contingent mediations); internal relationality (intra-relationality, intra-activity) (necessary mediations); intra-/inter-subjective relations (plane [b], four-planar social being); love; meaning; mediation (third moment, concrete universal), hence concrete universality as such; mediatisation; rationality, reason; reality principle (epistemologically mediated alethic truth); nexus; permeation; perspectival switch; phronesis; process-in-product (absence, absenting); rationality, reason; reality principle (epistemologically mediated alethic truth); research programme; recognition; reflection, reflexivity (inwardised form of totality); relative autonomy; resonance; self, self-consciousness; self-realisation; social cube; sociosphere; sophrosyne; subjectivity–objectivity; system, systematicity; tendency$_c$; theory; totality, synchronic emergent powers materialism (mind, process-in-product); totalisation (de-totalisation, retotalisation); trust (normative-fiduciary); truth; unity-in-diversity; universalisability; virtualisation; well-being (mode of freedom).

Critique of irrealism and its social context: alienation; analytical problematic; Beautiful Soul; compartmentalisation; (illicit) de-mediation; demi-reality; (illicit) detachment; de-totalisation; dialectical antagonists; duplicity; false consciousness; fission; fragmentation; grafting; hedonism; heterology; heteronomy; ideology; immersion (in meanings), logocentrism; irreflexivity; ontological extensionalism (denegation of totality); punctualism; scepticism; split; supplementarity.

4D absence, absenting (product-in-process); action; actionability; agency (transformative praxis); autonomy (self-determination); axiology; concrete singularity (fourth moment, concrete universal); counter-hegemonic struggle; critique; depth-investigation, depth-praxis, dialectics of freedom; dialectical reason (the unity of theory and practice in practice); emancipation (mode of freedom); emancipatory axiology; embodiment; emergence (product-in-process); final and efficient causes; hermeneutics; hermeneutic circles; hermeneutic counter-hegemonic struggle; humanism; immanence; intentional causality (mode of absence, extrinsic aspect), intentionality; intrinsic aspect; judgemental rationalism; material transactions with nature (making – plane [a] four-planar social being); meta-critics; metaphysics α (underlabouring), metaphysics β (criticism of research conceptual schemas); moral alethia (freedom); poiesis; praxis (qua activity); objectification; prefigurationality; product-in-process (absence, absenting); re-agentification (re-centrification, re-empowerment); reality principle (epistemologically mediated alethic truth); referential detachment (deployment of the capacity for); right-action; self-overcoming; subjectivity; tendency$_d$; this-worldliness; transitivity; truth (expressive-referential); underlabouring.

Critique of irrealism and its social context: anti-humanism; axiological contradictions or inconsistencies; de-agentification (denegation of agentive agency); disembodiment; de-singularisation; eternisation; fact-form; genetic fallacy; hegemony (qua agential project); historicism; hypostatisation; irrationalism (epistemological, judgemental, practical); naturalisation of status quo; other-worldliness; reification; scientism; triumphalism; Unhappy Consciousness.

and to the critique of irrealism, but are listed only once. Social relations belong in 1M in virtue of their being structured and at 3L because of their intra-relationality, but appear at 2E because they are mind- or concept-dependent, continually in process and the site of social oppositionality and contradiction. The moral is that conceptual locations should not be hypostatised.

Plato Etc. added 5C (*fifth component*) to MELD, signifying *sociality*, highlighting that sociality is supervenient on 4D and is derivable 'from the sole premise of intentional agency' (*PG*: 250). This was later dropped, presumably because the point is well made within MELD.

TDCR and the philosophy of META-REALITY add three further stadia to the schema: 5A *fifth aspect* (reflexivity), 6R *sixth realm* ([re-]enchantment), 7Z/A *seventh zone* (non-duality – *a*wakening): MELDARZ or, more euphonically, MELDARA.

See also DIALECTICS; PRESENTATIONAL DIALECTIC.

Meno's paradox. Suggests that knowledge is impossible: 'A man cannot inquire either about that which he knows, or about that which he does not know; for assuming he knows he has no need to inquire; nor can he inquire about that which he does not know, for he does not know about that which he has to inquire' (Socrates/Plato, cit. *PN3*: 153). This is the paradox encapsulated in the concept of the HERMENEUTICAL circle (C1). Plato resolved it by recourse to recollections from a previous life (*anamnēsis*), suggesting the immortality of the soul. For CR it is resolved by the consideration that knowledge-acquisition is '*a pre-existing ongoing social affair*' (*P*: 7) that never starts from scratch (see THROWNNESS) and never arbitrarily stops (see ENDISM [epistemological absolutism]).

metacritical limits. See FIXISM/FLUXISM.

metacritique, metacritique$_1$, metacritique$_2$, metacritical dialectics, metacritical dimension, metacritics. See CRITIQUE; DEPTH; DIALECTICAL ANTAGONISTS; EXPLANATORY CRITIQUE; HEGEL–MARX CRITIQUE; INTRANSITIVE, ETC.; PDM; PERSPECTIVE; PHILOSOPHY.

metaphor. See COGNITIVE SCIENCE; EXPLANATION; GRAMMAR; THEORY, ETC.; TRACE STRUCTURE.

metaphysical idealism. See IDEALISM; IDEALISM VS. MATERIALISM.

metaphysical lambda (λ) clause, or **TINA λ clause,** or **Tina defence mechanism**. 'An escape clause that every axiologically, transcendentally and/or dialectically refutable metaphysical system requires as a safety net, reflecting the posture of weak ACTUALISM' (*DG*: 400).

metaphysical or **ontological materialism**. See CENTRISM; CRITICAL NATURALISM; EMERGENCE; HOLY TRINITY; IDEALISM VS. MATERIALISM; MATERIALISM; MIND; NATURALISM; PDM; PHILOSOPHY OF RELIGION.

metaphysical realism. See IDEALISM VS MATERIALISM; REALISM.

metaphysics, metaphysics α, metaphysics β. See EPISTEMOLOGY, ETC.; PHILOSOPHY.

metaphysics of presence. (1) Heidegger's concept, fundamental to his critique of Western philosophy, for the belief that Being is a fixed essence or property present in things. (2) In Derrida's appropriation, the epistemological belief – determining this understanding of being as presence – in some ultimate sign or presence outside 'the text' (variously God, *Logos*, the Forms, the intrinsic structure of physical reality and/or of all possible thought, the moral law) that underpins all our thought, language and experience, which may achieve unmediated access to it. This ignores that the meaning of words is 'always derived from and deferred through the play of difference among wider networks of terms' (Sayer 2000a: 92). Cf. CENTRISM; FOUNDATIONALISM. The 'metaphysics of presence' is broadly synonymous with *logocentrism*, which refers more precisely to the privileging of speech, as the originator of meaning, over writing on grounds of purity and immediacy to the transcendent sign. (3) Bhaskar broadly endorses Derrida's critique of 'the philosophy of presence' (*P*. 215), which entails PERSPECTIVAL SWITCHES from presence to absence, and himself deploys a more restricted concept of a *REPRESENTATIONALIST–monological problematic of presence*, inaugurated by Descartes and eventually dominated by EMPIRICISM, which was hegemonic in the philosophical discourse of modernity and which he sees as 'curiously repris{ing}' ONTOLOGICAL MONOVALENCE (*P*. 217). *Logocentrism* is employed to refer pejoratively to (1) the view of reading that restricts it to the recovery by the reader of the original meaning of the author; (2) relatedly, the impossible hermeneutic ideal of an identity or correspondence between scientific and lay accounts of a form of life, attainable through immersion in that life. Derrida's critique of logocentrism underplays the stability of discourses and 'the mutuality of hermeneutic interpretation' and overlooks the transcendental necessity of immediate as well as mediate knowledge if there is to be any communication and understanding at all (*P*. 200). The fundamental failing of both Heidegger and Derrida is their lack of a realist, stratified, non-actualist ontology.

meta-principle of totality. See PRINCIPLE OF TOTALITY.

meta-Reality. A new philosophy of identity- or unity-in-difference which intends to complement, rather than to displace, the CR philosophy of DUALITY. It may refer either to the philosophy or to what it is about, but in the present work 'PMR' is normally used to denote the former and 'meta-Reality' the latter. PMR focuses on the 'underlying identity' (non-duality) already argued for in *Dialectic* (301; see also RECOGNITION; SPACE–TIME; ULTIMATA). This is what *meta* (Gr. primary, anterior, first or in the midst of, among, by means of) signifies: the ultimate ingredient sustaining GROUND and ALETHIC truth of reality, which is 'beyond' dualistically understood reality, constellationally encompassing and englobing it. Identity and cognate concepts are now argued to be ontologically, epistemologically and logically prior to non-identity and cognate concepts, there being no way to distinguish non-identicals except in relation to identity. For that reason it is sometimes called a *philosophy of identity* simpliciter. It is to be distinguished sharply from actualist subject–object IDENTITY theory or identity-thinking, for it presupposes the *non-identity* (through *constellational identity*) of the terms. Identity thus in no wise exhausts PMR, for 'it incorporates and retains critical realism as the best account of the realm of duality' or non-identity (*MR*: xv), i.e., it is an essentially preservative SUBLATION of (all the phases of) CR. This means that, in thinking about PMR, most aspects of most entries in *DCR*, e.g., should be presupposed. *Transcendental dialectical critical realism* (TDCR), which was elaborated prior to PMR, now appears as its first moment (5A in the MELDARA schema, see Table 29). While PMR does not fall within the detailed compass of *DCR*, it is important to register how it relates to the overall logic of the system, because it (or something like it) is arguably dialectically necessary for its 'completion' and integrity. This is effected schematically in Table 30 (see also Table 17). What follows are a few comments on some probable current misunderstandings.

(1) PMR is in no sense epistemologically FOUNDATIONALIST, nor is it 'ontological casuistry' (Morgan 2005b). It is largely derived in exactly the same way as CR itself: by a process of duplex immanent critique of the PDM and of its own predecessor phases in CR (including the theory and practice of the broader Left within which they are located, whose Achilles heel is the failure to produce a plausible vision of the good society) – and by transcendental argument from human intentional agency as currently understood. In addition, it combines these two approaches in an argument that TRANSCENDENCE (the CREATIVE move to greater coherence, TOTALITY or EMERGENCE) and transcendental agency are transcendentally necessary conditions for any being at all (*RM*: 268, see a). It is as committed as CR to epistemic relativity.

(2) It is perhaps misleading to say that it concerns 'a level beyond embodiment, beyond direct experience, beyond history and culture', nor is it 'a real Kantian thing in itself' (which has no applicability to the physical world) (Dean et al. 2005: 13, 26). It does concern the level of the absolute, but, as its etymology suggests, we carry meta-Reality around with us – so the argument goes – as mediately ingredient in our bodies and everything we do. *Dialectic* already argued the reality of a transcendental SELF, implicit in our everyday speech and action (see EMANCIPATORY AXIOLOGY). PMR now connects this with the ultimate categorial structure, 'implicate order' (Bohm 1957, cit. *RM*: 184–5, 197) or implicitly conscious field(s) of possibility of the universe (*cosmic envelope*) which is argued both to sustain and to be ingredient in (connect) the whole of being, hence geo-history and our embodied personalities (as their concretely singular *ground-state* on the cosmic envelope, or *dharma*), and to be experienced as a reality on a daily basis in moments of non-ALIENATION and identity consciousness (see also CO-PRESENCE). Apart from the implicitly conscious (supramental) and morally alethic dimension now argued for, this is not so very different from the earlier philosophy in which fundamental levels of being constellationally contain higher ones, just as, say, a table is contained within the field of elementary particles. At the level of ultimata the basic structure of people and the world is identical.

(3) PMR is not religious, although it acknowledges the divine. While Bhaskar's thinking went through a religious phase marked by the appearance of *From East to West*, PMR is a secular THIS-WORLDLY philosophy aspiring to command the assent of people of all faiths and no faith. Bhaskar explicitly states that he is not calling the cosmic envelope God, though it is fine if others do, and that he does not know what lies outside it (e.g., *RM*: 93); however, it is sometimes implied that the cosmic envelope is itself outside space–time (e.g., *MR*: 104). Of course, in contemplating the open structure of possibility – the infinite open totality or absolute – that is the endlessly creative unfolding of being, inaugurated perhaps in Big Bang by 'self-creation out of the void' (*MR*: 109), we all experience it as the sublime. It is now seen as valuable and meaningful in its own right: as our 'inorganic' (Marx) or 'cosmic' BODY (Collier 1991, who aptly also inverts this, conceptualising our own bodies as 'bodies-cosmic'). *Spirituality* is, above all, however, 'a love for being and for life' that yearns to see that possibility unfold, in particular the possibility of universal SELF-realisation in EUDAIMONIA (Bhaskar 2002e). The intrinsic value and meaningfulness of being are entrained by, respectively, the breakdown of the fact/value dichotomy and the collapse of subject–object duality in transcendental identity consciousness (*RM*: 15–16, 270–1).

(4) Nor is PMR anti-Marxist. It retains the practical materialism, epistemological materialism (transcendental realism), and ontological materialism elaborated in the previous works (see IDEALISM VS. MATERIALISM), but qualifies the last by adding that space–time may itself be a derivative form of consciousness (see also Table 30, cosmotheogenesis). This last notion derives in part from Eastern cosmotheogenies, but perhaps above all from the dialectical and spiritual tradition commented on at length in *Dialectic*, and in particular from Hegel. However, Hegel's INFINITE is a closed circle of logical necessity, whereas Bhaskar's is an open spiral of possibility. PMR does run a critique of MATERIALISM, but that is the reductive or mechanical materialism of modernity (irrealist materialism), which Marx himself scathingly criticised. Its conception of realist deep structure of potentiality, overlaid and occluded by irrealist surface categorial structure sustaining and sustained by power$_2$ relations, underpinned by ALIENATION, is in keeping with Marx's emphasis on the creative powers of living labour overlaid and occluded by mystificatory structures of exploitation and control. Like CR, PMR aspires to the abolition of all master–slave-type relations, and its vision of eudaimonia is an elaboration of Marx's 'the free development of each as a condition of the free development of all'. The achievement of this presupposes EMANCIPATORY AXIOLOGY in every detail. PMR is but the culminating moment in a project to effect a paradigm shift at the level of philosophy and *Weltanschauung* – a non-preservative *Aufhebung* or SUBLATION of the philosophical discourses of modernity – as an integral rhythmic in the far broader process of transition to the eudaimonian society. Note that non-preservative sublation is consistent with retention of what is valuable.

Meta-realism is realism that espouses a concept of meta-Reality. See also DESIRE; DETACHMENT/ATTACHMENT; IMMANENCE/TRANSCENDENCE; INFRA-/INTRA-/SUPER-STRUCTURE; LITERARY THEORY; PERCEPTION; PHILOSOPHY OF RELIGION; SELF-REFERENTIALITY; TELEOLOGY/TELEONOMY.

meta-reflexively totalising (self-)situation. See ATTACH/DETACH; DIALECTICAL ARGUMENT; DISTANCIATION; OBJECTIVITY–SUBJECTIVITY; REFLECTION.

method, methodology. The study of research process, particularly of scientific method. Designing research methodology – choosing appropriate practical methods and arranging feasible procedures for securing evidence and producing valid and reliable theory – is problematic. Prevailing empiricist approaches, of which POSITIVISM dominates many present fields of inquiry, employ a non-explanatory concept of cause, deny the social and evaluative character of scientific judgement, sustain a morbid avoidance of ontology and commit the epistemic fallacy by reducing *being* to *knowledge* of (the human experience of) being. PHENOMENOLOGICAL or SOCIAL CONSTRUCTIONIST epistemologies that argue that reality (as well as knowledge of it) is merely culturally or socially produced are similarly at fault. POSTMODERNIST and RELATIVIST alternatives either reduce science to incoherence or commit the GENETIC FALLACY by trying to reduce truth to questions of origin within a paradigm (Cruickshank 2003a: 18).

Accepting a structured realist ontology and a view of science, including social science, as a process of production (of knowledge), CR understands CAUSES as contingently effective, ontologically deep generative powers or mechanisms, and CAUSAL LAWS as the TENDENCIES of things in an open world. Human REASONS are

Table 30 Key TDCR and PMR concepts, with some correspondences

The Ontological-Axiological Chain	1M First Moment: Non-Identity	2E Second Edge: Negativity	3L Third Level: Totality	4D Fourth Dimension: Transformative Agency	5A Fifth Aspect: Reflexivity (TDCR)	6R Sixth Realm: (Re)Enchantment	7A/Z Seventh Zone/Awakening: Non-Duality
Thinking being as	*as such and in general*	*process* + as for 1M	*a whole* + as for 2E	*praxis* + as for 3L	*spiritual* + as for 4D	*enchanted* + as for 5A	*non-dual* + as for 6R
Key PMR concept(s)	demi-reality identity, alethia	co-presence transcendence (creativity)	connectivity reciprocity	spontaneous right-action	spirituality, self-referentiality	re-enchantment intrinsic meaning, value	identity, unity or unity-in-diversity;
Form of reflexivity – immanent critique of	philosophical discourse of modernity (PDM)	PDM + 1M	PDM + 1M, 2E	PDM + 1M, 2E, 3L	PDM + 1M, 2E, 3L, 4D	PDM + 1M, 2E, 3L, 4D, 5A	PDM + 1M, 2E, 3L, 4D, 5A, 6R
Basic human capacities (ground-state qualities)	freedom (enfolded as potential) (negative completion)	creativity	love	spontaneous right-action (without attachment)	fulfilment (self-realisation, enlightenment) (positive completion)		universal fulfilment, peace
Qualities of the self	psychic	mental	emotional	physical	fulfilled intentionality/split intentionality (unfulfilled)	perception (apperception, intuition, and insight)	awakening
Transcendental morphogenesis of action	ground (implicit potential)	emergence (creativity)	identification, unification (love, solidarity)	agency (capacity for right-action)	reflection or fulfilment		
Components of action	will	thought/un-thought	emotion	making, poiesis	fulfilment (or not)		
Creative process	raw materials	transcendence	formation	objectification	reflexivity or reflection of objectification to the maker realisation of will and intentionality of creator		

	transcendental consciousness of real self (loss of ego)	transcendental identification in consciousness (away from subjectivity)	transcendental teamwork or holism	transcendental agency (absorption in activity)	transcendental retreat into self-identity (away from objectivity)
Non-dual components of action (modes of transcendence)					
Principles of spirituality	self-referentiality (primacy of)	simultaneity	complementarity	practical mysticism	radical hermeticism (primacy of self-referentiality entails the liberation and flourishing of all beings)
Conditions for self-realisation	absence of ego (awareness of real self)	clarity of mind unthought	pureness of heart	healthy, energised balanced body, letting go	absence of blind tenacity of belief in the exhaustive nature of physical being
Cosmogeny	polyvalent foundational impulse (from implicit potential)	creativity (transcendental emergence)	love (transcendental identification)	action (transcendental agency)	fulfilled intentionality of the foundational impulse (reflexivity or reflection of objectification back to the creator)
Cosmotheogeny (cycle of cosmic creation)	realm of absolute: self-creation of the creator ex nihilo†	emergence of realm of duality, becoming and time	emergence of realm of demi-reality and of binding nature of love	(commencement of return cycle from alienation) individual self-realisation	individual and universal self-realisation, then god-realisation (theosis)‡; reflexivity

†Corresponding to the descent of consciousness in traditional cosmotheogenies, and to Big Bang in modern scientific theory

‡Corresponding to the ascent of consciousness in traditional cosmotheogenies

Note 7A > 6R > 5A > 4D > 3L > 2E > 1M

themselves often effective causes, and social reality is imbued with language and meaning. Whereas positivists claim that the meaning-laden nature of society and social theory is an obstacle to objective social scientific knowledge, CR argues that this very feature of social ontology makes the science of it possible, although still difficult, because every science must be appropriate to the specificities of the object it studies (Sayer 1984: 4). Critical and interpretive methods must therefore be developed for the social sciences which take into account that its object is concept/discourse/text-dependent (though not concept/discourse/text-determined), value-impregnated and capable of being causally affected by the investigation. Sayer has suggested that a research project be viewed as a triangle of *method*, (real) *object* and *purpose*, in which each corner affects and is affected by both the others. An organisation theorist, Hassard (1993) has advocated multiple-paradigm research, offering a methodologically pluralist example of a case study which operates within four paradigms – functionalism, phenomenology, critical theory and structuralism – and which includes realist ontology.

To date, broadly social science has been the site of most CR methodological activity (more than fifty entries about the humanities and social sciences in this dictionary attest this claim), although it is probably at an early stage. Social scientific methodology attempts to chart the still largely unexplored terrain between PHILOSOPHY OF SOCIAL SCIENCE and the implementation of practical, empirical research projects. This is an ambitious, complex and problematic task, for which CR provides a system of application as well as reasons for optimism. Individual people reproduce and/or transform society, which simultaneously shapes and conditions their own activity. Bhaskar's conception of the stratification of society characterised by EMERGENCE, his transformational model of social activity (TMSA) and RRRE (*r*esolution of a complex event into components, *r*edescription of components, *r*etroduction to possible causes of components and *e*limination of other possible causes) process of EXPLANATION in the open systems of society, initiated the current CR methodological approach. CR social researchers continually ask: What must this or that feature of human life or social reality be like for observed phenomena to be possible and/or changed, and for knowledge of those phenomena and changes to be attainable or improved? They accept the epistemological caution and FALLIBILISM that accompany the transcendental ontological assertion, and they create through retroductive INFERENCE and the construction of MODELS accounts of the causal powers and mechanisms which explain the phenomena in question.

The scope of Bhaskar's TMSA – the limits to NATURALISM – constitutes one continuing problem of CR methodology (see also CRITICAL NATURALISM). A social researcher will want accurately to delineate the abilities and constraints of an agent in a specified context, so the relationship between STRUCTURE and AGENCY must be understood correspondingly well. Bhaskar later modified his model, and the sociologist Archer has produced her more elaborate MORPHOGENESIS/MORPHOSTASIS model, also specifying three types of emergent social properties: structural (material), cultural (belief-systems not reducible to an individual's beliefs) and people (the person, the agent and the actor). For Cruickshank (2003: 113) and Archer social scientific methodology is a matter of formulating '"analytic histories of EMERGENCE"', and social explanation is 'a conceptually mediated and fallible reconstruction of how, over time, social context and individuals {. . .} interacted, {. . .} resulting in change or continuity'. Cruickshank has begun to apply his 'social realist' methodology to the domain of chronic unemployment and the postulate of a deviant social 'underclass' by constructing a domain-specific meta-theory through an immanent CRITIQUE of existing theoretical approaches in that domain.

An alternative attempt to 'conjoin sociological theory and method' has been adopted by Ray Pawson (2000), who has combined Merton's (1968) 'middle-range theory' for theory-driven empirical inquiry with Bhaskar's realism so as to show how to generate hypotheses about social outcomes in terms of underlying generative mechanisms acting in conducive contexts. Pawson gives examples of 'middle-range realism' from the fields of voting practice and criminal recidivism to make the general methodological point that 'theory-building moves up and down a ladder of ABSTRACTION and that the rungs on the ladder are composed of CMO configurations', an explanatory combination of social Mechanisms that produce significant Outcomes in known contingent Contexts, which had been presented in detail by Pawson and Tilley (1997). Other expeditions into social-scientific methodology include the sociologists Danermark et al. (2002), who invoke the pragmatism of Peirce (1932) in search of a path.

Generally recommending realism 'affianced' with social constructionism, the psychologist Harré (1993: 104) has distinguished between *intensive* social research methods, such as case studies and ethnography, and *extensive* techniques, including large-scale surveys and statistical analyses (see also QUANTITATIVE METHODS); this is usefully developed and discussed by Sayer (1984/1992: 241–51). Harré (2001), however, whilst accepting the real being of social structure, would deny it causal power, arguing that social causal capability is the sole preserve of conscious agents.

One of the main points of research and the practice of science, natural and social, is to contribute to the diminution of human illusion, alienation and constraints on flourishing. This inherently emancipatory quality of science is explicated forthrightly by CR (see EMANCIPATORY AXIOLOGY; EXPLANATORY CRITIQUE), which recommends that due awareness of it be acknowledged methodologically, partly by emphasising the explanatory, informative and disseminating roles of research. Reducing the division between researcher and researched tends to challenge the distinction between science and everyday thinking, although master–slave-type social relations sustain the production and abuse of some types of knowledge to reinforce domination and submission. Part of the appeal of overtly CR-informed research might be to challenge the adverse human effects of social divisions (such as class, race and gender). This is an aspect of the practical relational DIALECTIC which obtains between any social investigation and its subject-matter, to which methodology must accommodate itself. Marx thought that social science is necessarily critical and revolutionary, and perhaps the methodological guidance supplied by CR may advance that cause (see MARXISM).

Although CR argues that its philosophy underlabours for *all* science and other forms of investigation, few contemporary *natural* scientists would assert that their outlooks, methods or results are realist, though they may be implicitly. Prominent exceptions include the physicist Penrose (1989, 1994, 2004), who has argued for the transcendental reality of all the universe (see TRANSCENDENTAL REALISM), including particularly of consciousness, and for the intimate but non-reducible emergence of mind from matter.

Another physicist, Bricmont (2001), has commented on the epistemological 'science wars' between sociologists and practitioners of science, rejecting the incoherent reduction of scientific theory to its sociology (ideological power-games and rhetorical exhortation) and encouraging a philosophical harmonisation of theory and practice, with physics and philosophy overlapping productively.

CR is a fallible, reflexive and developing theory or system of philosophy, employing its own methodology, thus critically demonstrating its own problematicity – and its effective, explanatory and CONCRETE UTOPIAN worth – within its original host, an argument-based, largely a priori arts discipline. Crucially, however, CR also shows the liberating way out of the artificial and unnecessary boundaries of traditional a priori philosophical introspection by accepting the interdependence of PHILOSOPHY, society and reality, and by suggesting that scientific knowledge and discourse and philosophical theory and debate – and their respective sociologies and participants – should all actively help and learn from each other.

RICHARD SHIELD

method, quantitative. Quantitative research methods, as traditionally defined, are concerned with the representation and analysis of phenomena as quantities, as opposed to textual interpretations and narratives. These research approaches measure phenomena, and then subsequently present these measurements as variables, which are phenomena that vary by numeric score. The values of these variables are then interpreted as statistics typically distinguished as *non-parametric* or *parametric* data.

Non-parametric data refer to nominal, ordinal or interval data. Nominal data are simply numeric labels from which frequency counts can be established. Ordinal data imply ranks which can be compared but not by degree. Interval data can be compared by degree providing the same scale is used to measure the degree of difference between phenomena, and the scales share the same origin. However, there is no presumption that the origin is independent of measurement. In contrast, parametric data, also referred to as ratio data, presuppose that the degree of difference between variables can be calibrated and that the origin of the scale is common to all measurements. Only ratio data lend themselves to the full range of mathematical operations of adding, subtracting, multiplying and dividing.

Statistical interpretation follows measurement and can proceed according to two main criteria. First, descriptive statistics aim to summarise key features of the numeric magnitudes associated with a *given* set of observations. Descriptions can be univariate, bivariate or multivariate, that is, refer to the patterns of a single variable, or two or more variables, respectively. Non-parametric descriptions calculate or compare the counts or ranks of values of variables. In addition, parametric descriptions can be employed with ratio data to describe measures of central tendency, dispersion or the degree of association between variables.

Second, inferential statistics seek to generalise patterns of variables to a wider set than the sample of observations. This is achieved through a process of hypothesis testing in which, based upon a probability or frequency distribution of the occurrence of the values of a variable or statistic, the researcher attempts to reject a hypothesis about unknown population values from those estimated from the sample of observation. Accepting or rejecting the hypothesis is then subject to error. Non-parametric tests, unlike parametric tests, make no assumptions about the precise nature of the frequency distributions upon which probabilities of errors are calculated. They are therefore more general but have less statistical power.

Despite their variations, the main CR concern over the use of quantitative methods in analysis is made clear in *RS*. Bhaskar's transcendental method of argument challenges the deductive model of EXPLANATION resting on an implicit Humean view in which CAUSAL LAWS are associated with constant conjunctions of atomic events in a closed system – exhibiting both intrinsic and extrinsic closure. Quantitative methods presuppose degrees of closure. Numeric representations assume intrinsic closure. PROBABILITY distributions assume extrinsic closure. This casts doubt upon their relevance to the non-experimental social realm.

This has generated debate about the usefulness of quantitative methods in CR research. For Lawson (1997), a mixture of forms of descriptive statistical analysis coupled with historical and case-study narrative are deemed appropriate because of the excessive closure assumptions implied by inferential statistical work, such as econometrics. This is suggestive of a limited triangulated research strategy. A less restrictive approach is broadly advocated by both Sayer (2000a) and Danermark et al. (2002), who argue that CR is compatible with a wide range of methods, with the key issue being that analysis is matched to the appropriate level of abstraction and the material under investigation. Sayer distinguishes between intensive and extensive research designs. The former is what is typically thought of as social science, i.e., qualitative research, in that it begins with the unit of analysis and explores its contextual relations as opposed to emphasising the formal relations of similarity between them, that is, producing taxonomic descriptions of variables as is the case in the latter, i.e., quantitative design. While it is argued that the approaches have complementary strengths and weaknesses, the causal insights from extensive research will be fewer. Moreover, one is reminded that the validity of the (qualitative) analysis of cases does not rely upon broad quantitative evidence. In contrast, Downward and Mearman (2002, 2003) argue that combining methods is *central* to retroductive activity because specific research methods within intensive and extensive designs differ more in emphasis than in kind through invoking degrees of closure. For example, even 'qualitative' methods, in collating insights and offering stylised interpretations, assume qualitative invariance or intrinsic closure; quantitative methods can also refer to different aspects of the *same* research object and thus are not wedded to particular and different ontological presumptions; and, finally, their combination helps to raise rational belief in a set of mutually supported propositions.

As far as statistical induction is specifically concerned, for example as manifest in econometric analysis, because of the appeal to probability theory and extrinsic closure, Finch and McMaster (2003) in addition to Lawson, recommend non-parametric analysis because of the less stringent measurement assumptions invoked. However, Downward and Mearman argue that non-parametrics still rely upon probability theory and, as such, support Ron's (2002) arguments that traditional econometric methods can help to reveal hidden mechanisms. See also INFERENCE; METHOD; SYSTEM.

PAUL DOWNWARD

methodological circle. See PHILOSOPHY.

methodological individualism and **methodological holism**. See CRITICAL NATURALISM; FISSION/FUSION; METHOD; SOCIAL STRUCTURE; TMSA; VOLUNTARISM.

methodological primacy of the pathological. See ECOLOGICAL ASYMMETRY; PRIMACY.

migration studies. CR has had relatively little impact to date on international debates about migration, at least explicitly. One notable exception is Carter (2000), who explores aspects of transnational migration within a much wider assessment of the prospects for fruitful realist accounts of 'RACE', racism and racialisation. The core element of all theories of regional, national and transnational migration is one simple question: Why do people migrate? To the realist the search is for those generative mechanisms that lie behind the empirical facts of population movements.

The quest for overarching modes of theorisation has taken at least three quite distinct forms. First, there is a resort to classical economic theory in the form of 'push–pull' processes. Push factors may take the form of high levels of un- or under-employment in a sending country. They may alternatively relate to political instability or conflict, resulting in flows of refugees and asylum seekers, or be a function of normative cultural processes. An example of the latter would be social formations based on primogeniture. Pull factors, in contrast, might be economic prosperity and political stability. The receiving country therefore holds out the long-term promise of economic advancement. A temporal conjunction, to varying degrees, of certain of these processes is therefore seen as a potential trigger for population movement.

Second, there have been a variety of Marxist and neo-Marxist models. At the core of these are core-periphery or North–South relations. Actors (potential migrants) are seen as reserve armies of labour and as such represent a key element of an international division of labour. Extreme versions are highly determinist, in terms both of the inevitability of migration under certain conditions and of the structural position assigned to this 'underclass' in the receiving state. (The same form of argument is sometimes used to predict [or explain] a lack of propensity to migrate. Castles [2004], for example, suggests [wrongly, in my view] that the negation of the North–South divide would *necessarily* have this effect.) Neo-Marxist theorists who are concerned that this downplays 'race' and ethnic stratification have flirted with the notion of racialised class fractions. Thus racism and/or racialisation invoke differential outcomes.

Castles (2004: 208) sums up the key generative elements of these first two theoretical approaches as:

- factors arising from the social dynamics of the migratory process
- factors linked to globalisation, transnationalism and North–South relationships
- factors within political systems

In all of the above theorisations the role of social agency is underplayed, individuals characterised as being at the mercy of external forces of which they may, or may not, be aware. They also clearly fail to explain why, say, two individuals in ostensibly similar structural or material positions respond in very different ways to the underlying generative mechanisms. So, third and finally, there are what I have termed elsewhere 'subjectivist' accounts of migration processes (Ratcliffe 2004). These represent the extreme empiricist position in that they accord explanatory primacy to actors' own accounts of decision-making processes. At its simplest this represents a plea for the voices of migrants (or would-be migrants) to be heard, i.e., that we cannot hope to 'explain' migration without paying full attention to the world as it appears to the key social actors.

Although these approaches appear unbridgeable, being grounded in radically differing social ontologies and epistemological positions, there is a possible way forward for realists. This entails a more flexible interpretation of structure and a more nuanced deployment of the empirical data arising from actors' accounts of migration. Expounded more fully in Ratcliffe (ibid.), this relies on the use of the latter (in both spoken and written form, including migrants' letters) to cast light upon, and raise questions about, precisely those historically entrenched inequalities that are seen as constituting generative mechanisms in the migration process.

PETER RATCLIFFE

mind. A central site of philosophical conflict between IDEALISM and MATERIALISM, mind has traditionally been fought in terms of *DUALISM* and *REDUCTIONISM* (Morgan 2004a). Dualists argue that the mind, spirit, or consciousness is separate from the material BRAIN and body. Mind is a private spatially non-extended non-substance. According to Descartes' Cartesian dualism it interacts with substance, according to Leibniz the two are harmonised in advance by God, whilst in Malebranche's occasionalism the two are reconciled by God as they occur. Despite our secular times, dualism still represents a popular position (e.g., Foster 1991) on the mind–body problem, for two main reasons. First, the sense of separation of mind and body has an intuitive appeal. Second, reductionism has its own problems. The influential reductionism of Armstrong, Place and Smart is a form of *central state materialism* (CSM) (*PN1*: 124, 149). CSM argues that mind is neither an exclusively inner environment, nor outwardly expressed behaviour, but rather a real material object whose processes are outwardly causal. Armstrong, Place and Smart's *type and TOKEN IDENTITY theories* of mind argue that specific experiences and intentions are not simply associated with brain states identifiable with particular sub-systems of the brain but can be reduced to them, i.e., there is nothing additional about conscious states. To some degree such physicalist theories are lent credence by neuro-scientific correlations between different brain areas and properties and functions (Nellhaus 2004). Damage to a sub-system can either ligate or affect properties and functions of mind. But physicalism has been unable adequately to account for consciousness itself as a total state of being. In neuroscience this is referred to as the binding problem – how simultaneous processes such as the identification of colour, angles and so on in separate areas of the brain are synchronised into the experience of objects and how the whole is synchronised into a sense of coherent SELF through time and within an environment. Physicalism has also had difficulty in accounting both for intentionality or the difference between a brain state, its location, its content and duration (planning and desires are more than an instance of their action), and also qualia, or the inaccessible properties of someone else's subjective states, such as whether, why and how much pain is felt.

CR's major contribution on this subject is *synchronic emergent powers materialism* (SEPM) (*PN1*: 124–6; *ER*: 601; Collier 1994: 156–60), which shares with Searle's biological naturalism (1998) a refusal of the basic terms of debate set out by dualism and reductionism, on the basis that both are rooted in an ontological idealist–materialist dichotomy. Dualism leaves consciousness as a mysterious non-corrigible intangible, whilst reductionism, in the name of materialism, puts aside the totality of a conscious being which is the very property that gives rise to questioning the nature of mind in the first place. SEPM argues that consciousness is an emergent non-reducible property of the material brain. The term synchronic is used to differentiate the use of the term EMERGENCE from its diachronic application to the evolution of species and their capacities. Thus consciousness is concurrent with a given form of material brain. But this does not in itself indicate a hard claim about the temporal relationships of conscious action and brain materiality. Philosophical and scientific ontologies are different (see EPISTEMOLOGY, ETC.). SEPM is a philosophical solution to the dualist–reductionist conceptual problem, rather than an account of neuro-scientific evidence concerning the structure of consciousness. No further firm claims are made concerning what consciousness is (*PN1*: 124 distances itself from spiritualism, but later TDCR includes numerous controversial claims about consciousness). SEPM is therefore compatible with the possibility that neuro-science will solve the mystery of consciousness, as Churchland and Churchland (2002) have influentially argued; but, unlike Churchlandian physicalism, from a philosophical standpoint critical realists tend to argue that it won't make conceptual sense to talk about it in the language of idealism or materialism. The emergent power of mind that *reasons can be causes* (*PN1*: 102–6) is an important aspect of this. In common-sense terms it is relatively unproblematic to assert that beliefs and desires result in intentional actions and thus consciousness informs individual behaviour, and through the TMSA affects natural and social structures. Denials of this are less about rejecting the idea that consciousness is interactive and more about preserving the integrity of different philosophical positions' understanding of causation and freedom (*PN1*: 106–23). Postmodernism, forms of strong social constructionism, and many political philosophers reject the idea that reasons are causes because they associate causation with Humean constant conjunction. Since consciousness entails choice and the outcome of reasoned action is multi-realised, it is not causal under this definition, and to apply this definition is to open the way for forms of reduction of behaviour to either genetic determinations or behaviouristic stimulus-response to the detriment of cherished notions of FREE WILL, choice and so on. It is in conflict with materialist reduction, therefore, that the idea of causation is rejected, facts and values are split or totally merged (*SR*: 169–211; Collier 1990: ch. 4) and a strong explaining–understanding division is maintained between the natural and human sciences. But the rejection itself accepts a scientistic ontology of scientific practice. The rejection thus creates a disjuncture between reason and the biological mechanisms that give rise to the possibility of reason. Ontological contradiction ensues, in which reason is isolated in an idealist sphere by theorists who would have no wish to describe themselves as idealists. Most merely oppose reductionism, scientific hegemony and social engineering, but lack a coherent alternative to reductive materialism and its eliminative implications. Since CR focuses attention on the generation of the possibility of reason and on the subsequent generative mechanisms of reason, this false choice between causation and freedom, reductive materialism and defensive idealism, is avoided.

JAMIE MORGAN

mind–body dualism. See CRITICAL NATURALISM; DUALISM; IDEALISM VS. MATERIALISM.
mind–body problem. See MIND.
modal realism. See MODALITY; REALISM; TRANSCENDENTAL REALISM.
modality. Modalities (f. L. *modus*, a measure, way) are ways in which entities can exist or be described, the formal analysis of which is the task of various branches of LOGIC. Their main class is that of the *alethic modalities*, which are the necessity, contingency, possibility or impossibility of something's being true. Terms describing them, including 'must', 'may', etc., are called *modal terms*. Propositions asserting what is necessarily or possibly the case are sometimes called *APODICTIC* or *problematic*, respectively. Propositions containing no modal terms are *assertoric* – they merely report matters of 'fact'. (Because the concept of a fact is ideologically saturated, Bhaskar aligns 'assertoric' with 'ontic' [e.g., *SR*: 186].) *Causal modalities* embrace causal necessity and possibility. *Deontic modalities* concern what morally must, should or may be done. *Epistemic modalities* and *doxastic modalities* comprise the modes of *knowing that* and *believing that*, respectively. *Sense modalities* are ways of apprehending the world via each of the five senses.

The *modal fallacy* (a form of LOGICISM) holds that because a proposition is logically necessary it is necessarily true of the world. It is vital to distinguish *logical necessity* and *natural* (or *causal*) *necessity* (the necessary ways of acting possessed by things in virtue of their essential structure; see CAUSALITY; NATURAL KINDS). The *unnecessary* is that which can be changed, which includes essential structures but not the laws that they ground wherever there is a structure of that kind. Something is *transcendentally necessary* if it logically must exist given the existence of some other well-attested phenomenon (e.g., human intentionality). It is *dialectically necessary* if in addition it is necessarily false or limited – 'contradictory, *incoherent* or incomplete in some relevant way, *yet* inexorable or *indispensable*' (*D*: 103; see also CRITIQUE; DIALECTIAL ARGUMENT; IDEOLOGY). Natural necessity > transcendental necessity > dialectical necessity – all of which include *axiological necessity* when epistemologically mediated (*D*: 103; see also AXIOLOGY; REALITY PRINCIPLE). Transcendental and dialectical necessity may exist, like natural necessity, intransitively, without any necessary relation to our subjectivity.

Contingency, and within that *accidentality*, likewise has an epistemic (logical)/ontic bivalence. A proposition is contingent if it is neither impossible nor necessary, i.e., if it can be, but does not have to be, true. An event, etc., or an entity is contingent if its occurrence or existence is not necessary, i.e., if it does not have to occur or exist (things could have been different). External relations are contingent, internal relations (which pertain to intrinsic structures) necessary (see RELATIONALITY). In a different sense, something is contingent on something else if its occurrence or existence depends on it. An *accident* is a property an entity could do without and still be what it essentially is, or an event or state of affairs resulting from a freakish combination of circumstances. It is also used synonymously with contingent in relation to distinguishing a necessary from an accidental sequence of events, and a causal from a non-causal correlation.

Since for CR the necessary ways of acting of things are only contingently manifested in actual events, the domain of the real is also the domain of *possibility*. The actual is thus only a tiny part of the possible, which has ontological, epistemological and logical PRIORITY over it. CR espouses *modal realism*, the view that modal facts exist independently of our knowledge of them, but stops short of endorsing the notion of the actual existence of endless parallel universes (see TRANSCENDENTAL REALISM; Norris 2003:143). See also ANALYTIC/SYNTHETIC; A PRIORI/A POSTERIORI; CAUSAL LAW.

model. See THEORY, ETC.

modernism, modernity, modernisation. See PDM.

moment (f. L. *momentum*, movement, impulse). A 'para-spatial' concept which is the correlative of the 'para-temporal' concept of DETERMINATION (*P*: 83). It signifies (its conventional meanings aside) a part or phase of a whole (TOTALITY), considered either synchronically or in its DIACHRONIC development (see PROCESS; TRANSITION), where parts and whole are both distinct and internally RELATED or essential to each other. Thus universality, processuality, mediation and singularity are moments of the CONCRETE UNIVERSAL ↔ SINGULAR, and transcendental realism, critical naturalism, explanatory critiques and DCR are moments in the development of CR. This usage derives from Hegel. For δ *moment*, see TRANSITION. See also FEYERABEND'S MOMENT; LAKATOS' MOMENT; POPPER'S MOMENT.

monism (f. Gr. *monos*, alone, single in its kind). *Ontological monism* or *holism* (whose complicit dialectical antagonist is *DUALISM*) is the doctrine that there is one, and only one, undifferentiated substance, commonly either spirit (mind) or matter in some guise or other. As such it is an example of TOKEN–token IDENTITY-THINKING, where a singularity of particulars is assumed, issuing in an all-pervasive *token monism*. If there is a plurality of particulars, it issues in a *monadology* or *monadic pluralism*, a form of monism associated historically with Leibniz's theory that the basic constituent elements of the world are an infinitude of ultimately real and indivisible immaterial substances or *monads* (f. Gr. *monas*, a unit), each of which is unique and does not interact with, but none the less somehow harmoniously reflects the state and development of, the others (the whole). After ONTOLOGICAL MONOVALENCE, or the absence of the concept of absence (which is conceptually – although there is mutual entailment – but not historically prior), and together with FOUNDATIONALISM (with which it is profoundly associated) and REDUCTIONISM (which materially implies it), monism is the *ur*-error of the Western tradition, harking back to Parmenides. Monism may be either *reductionist* or *ESSENTIALIST*. Bhaskar's polemic is with both, including Hegel's (spiritual) *constellational monism*, but in PMR he deploys *stratified and differentiated monism* as a descriptor of his own position, which posits an underlying identity at the level of ultimata for the whole of being; ultimately there is only one kind of stuff, with the potential for rich differentiation and emergent strata enfolded within it (*EW*: 86; personal communication). A (materialist) stratified and differentiated monism is also one of the possibilities left open by his theory of MIND (*PN3*: 98). The intent is to avoid any implication of dualism or DOUBLE ASPECT THEORY. (2) More generally, monism refers to the oneness or uniqueness of any approach or subject-matter. Thus the empiricist or positivist theory of scientific development is *monistic* because it sees it occurring solely as a steady linear accretion of knowledge which 'leave{s} meaning and truth-value unchanged' (*RS2*: 130). This is a consequence of DEDUCTIVISM actualistically applied. *Epistemological monism* is the dialectical counterpart of ontological reductionism, as when sociobiology, e.g., explains social phenomena in terms of human genes. The *monistic theory of scientific development* sees knowledge-acquisition as a steady accumulation of positive knowledge; its critique by Kuhn, Lakatos, Sellars and others

318 monistic theory of scientific development

facilitated the emergence of CR in the 1970s. *Anomalous monism* is a theory of MIND premised on an actualist account of causality put forward by Davidson (1970). It says that, while every mental event is identical with some physical event (hence 'monism'), there is no reason to suppose that (intentional) mental concepts and predicates reduce by law to physical ones (hence 'anomalous'). This is because mental concepts are regulated by constitutive principles of rationality, and reasons are not causes – causal laws only connect physical events. The theory beautifully illustrates the dialectical intradependence of monism with dualism, incorporating 'the familiar incoherent matrix of dualistic DISEMBODIMENT in the INTRINSIC aspect and physicalistic reductionism in the extrinsic one' (*P.* 198).

Monism is refuted by the transcendental deduction of ontological bi- and poly-VALENCE and of primary and sempiternal POLYADISATION, whereby it is shown to involve 'a huge illicit FUSION' which – since the monist must detotalise himself – passes over into the 'ultimate fission' of SOLIPSISM (*P.* 217), the FIDEISTIC response to which contributes to the monism vs. dualism problematic or DIALECTICAL ANTAGONISM. See also HEGEL–MARX CRITIQUE; RELATIVE AUTONOMY.

monistic theory of scientific development. See CRITICAL REALISM; EMPIRICISM; INTRANSITIVE, ETC.; MONISM.
monovalence, ontological. See VALENCE.
moral actualism. See ACTUALISM.
moral alethia. See ALETHIA; EMANCIPATORY AXIOLOGY; ETHICS; EUDAIMONIA; FREEDOM.
moral distanciation, moral imagination. See DISTANCIATION.
moral humanism. See HUMANISM.
moral realism. See CRITICAL NATURALISM; CRITICAL REALISM; DCR; EMANCIPATORY AXIOLOGY; ETHICAL IDEOLOGIES; ETHICS; HOLY TRINITY; NEEDS; REALISM.
moral reasoning. See DEA AND DET MODELS; HUMAN NATURE; INTERESTS; NEEDS; PRACTICAL REASON.
moral relativism or **relativity**. See ETHICAL IDEOLOGIES; HOLY TRINITY; NEEDS; PRAGMATISM; UNHOLY TRINITY; WORLD-LINE.
moral truth. See ALETHIA; ETHICS; TRUTH.
morality. See ETHICS.
-morph, morpho- (Gr. *morphē*, form). *Morphology*: the science of form; *morphological*: pertaining to or derived from the history of form. A *homeomorph* (also homoeomorph, homoiomorph) (Gr. *homoios*, like, similar) of x has a structure like that of x, a *homomorph* (Gr. *homos*, same) the same structure, an *isomorph* (Gr. *isos*, equal) an identical structure, a *paramorph* a structure that goes beyond (Gr. *para*, beside or beyond) the structure of x in some sense. In MODEL theory, a *homomorphism* (and the closely related homeomorphism) is 'a structure-preserving mapping from one structure to another' (thus a doll is a hom[e]omorphic model of a baby); 'an *isomorphism* {or FUSION} is a homomorphism that is one-to-one and whose inverse is also a homomorphism' (*CDP.* 392) (as in the EPISTEMIC FALLACY and its inverse ontic fallacy); and a *paramorphism* is a structure-transforming mapping of one or more structures into another (thus the theory of natural selection is a paramorph of the theory of domestic selection and Malthusian population theory, and the desire, in love, to be united with the loved one

may be seen as a paramorph of the desire for de-ALIENATION [*D*: 243]). Hom(e)omorphism, and in particular paramorphism, play an essential role in science. See also MORPHOGENESIS.

morphogenesis/morphostasis. The morphogenetic approach was first advanced by Archer (1979) in the course of explaining the social origins and effects of educational systems. Sociologically, it was a development of Lockwood's (1964) distinction between 'social' and 'system' integration which also implies a stratified social ontology, although this was not explicit in his 1964 article. Although developed independently, it was immediately apparent that this framework constituted the methodological complement of Bhaskar's transformational model of social activity (TMSA). The concept *morphogenesis* indicates that society has no preferred form – contrary to mechanical, organic and cybernetic analogies – but is shaped and re-shaped by the interplay between STRUCTURE and AGENCY. The term was first used by Buckley (1967: 58) to refer to 'those processes which tend to elaborate or change a system's given form, structure or state'. It is contrasted to *morphostasis* which refers to those processes in a complex SYSTEM that tend to preserve the above unchanged.

Its two key features are (1) that structure and agency are analytically separable and (2) are temporally sequenced. This is termed *analytical dualism*, as distinct from *metaphysical DUALISM*, because all social properties are activity-dependent in both their origins and their effects. The morphogenetic premise that structure and agency operate over different time periods is founded upon two simple propositions: that structure necessarily pre-dates the action(s) which transform it; and that structural elaboration necessarily post-dates those actions, as represented in the basic morphogenetic diagram in Figure 8.

Although all three lines are in fact continuous, the analytical element consists in breaking up the flows into intervals determined by the problem in hand. Given any problem and accompanying periodisation, the projection of the lines backwards and forwards would connect up with the anterior and posterior analytical cycles. This is the bedrock of an understanding of systemic properties, of *structuring* over time, which enables explanations of specific forms of structural elaboration to be advanced and thus provides a tool for substantive investigations. See also SOCIAL THEORY.

MARGARET S. ARCHER

multiple control, multiple determination. See DEPTH; DETERMINATION; FREEDOM, ETC.; INFRA-/INTRA-/SUPER-STRUCTURE.
multiple tendentiality. See TENDENCY.

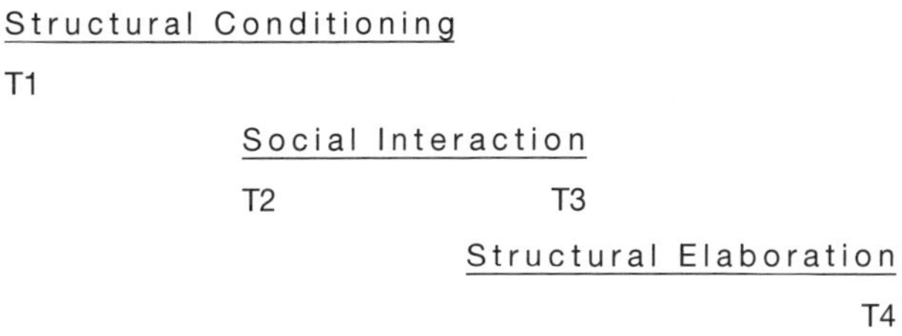

Structural Conditioning
T1
        Social Interaction
        T2   T3
                Structural Elaboration
                      T4

Figure 8 Morphogenesis

N

naïve or **direct realism**. See CRITICAL REALISM; REALISM.

narrative provides 'an account of some process or development in terms of a story, in which a series of events are depicted chronologically' (Sayer 1992: 259). Because of the openness of the world and the directionality of time, there is an inescapable narrative dimension to any account of human affairs. Bhaskar draws a parallel between (progressive) narratives and DIALECTICAL ARGUMENTS in that they 'often take the form of a reflexively monitored episode or life, consisting of a sequence of phases in which each successive moment constitutes a quasi-propositional comment on the alterology or untrue-to-self-or-situation character of the preceding one' (*D*: 105). The TMSA entails that narratives of social structure will have a differential temporality (the *longue durée*) relative to those of individual agents. The need for *grand narratives* – totalising theories of geo-history and social systems – as well as 'local knowledges', if we are to understand much at all, is defended by Sayer (2000a) and is a theme running through much of Bhaskar's work (see, e.g., *PN3*: 44). Other CR discussions of the relative merits of narrative and analysis in social science include Archer (2000), Lawson (1997), Patomäki (2006b) and Porpora (1987). See also HISTORIOGRAPHY; METHOD.

natural kind captures the notion that, contrary to prevailing opinion within the philosophical discourse of modernity, things tend to 'behave the way they do because of what they are' (Groff 2004: 15), that their ways of acting are necessary and immanent in the world rather than contingent and superimposed on it, and dynamic and reactive rather than passive (Ellis 2001). This entails that things have *real* ESSENCES or intrinsic structures (Aristotelian formal CAUSES) which possess causal powers and constitute them as natural kinds, and that the relationship between their intrinsic natures and the way they behave is one of *natural necessity*, i.e., necessity that obtains independently of human beings: 'the activity of a generative mechanism as such or the exercise of a thing's TENDENCIES irrespective of their realisation' (*RS2*: 171). Contrary to the tacit assumption of the Humean account, natural necessity operates independently of *logical necessity*. This presupposes in turn that there is 'an objective world order' (*RS2*: 171) or natural kind structure independent of our identification of it, which we can fallibly come to know (see DEPTH; EPISTEMOLOGICAL DIALECTIC; INTRANSITIVE, ETC.). 'Knowledge of natural necessity is expressed in statements of CAUSAL LAWS; knowledge of natural kinds in real DEFINITIONS. But natural kinds exist, and naturally necessary behaviour occurs, independently of our definitions and statements of causal laws' (*RS2*: 171).

The *nominal essence* of a thing (not to be confused with *nominal DEFINITION*) is its more readily observable properties which are necessary for it to be correctly identified as a thing of a certain type; its real essence is its internal structure or constitution, in virtue of which it 'tends to behave the way it does, *including manifest the properties that constitute its nominal essence*' (*RS2*: 209, e.a.). The distinction derives from Locke, who, however, held that we could not know real essences; it corresponds broadly to the distinction between 'essence' and 'appearance' in Marx, who held that we (fallibly) could. Since they are necessary manifestations of real essences under certain conditions, nominal essences are not, for science, arbitrary or mere matters of convention, any more than real ones are. Thus the durability of gold under certain conditions is internally related to its atomic constitution (see also RESEMBLANCE). Since natural kinds possessing real essences constitute objective *taxa* (pl. Gr. *taxis*, arrangement, division), they provide a basis for non-arbitrary classification or *taxonomic knowledge* which seeks to bring its classifications into line with the real distinctions in the world. There is thus 'a dialectic of explanatory and taxonomic knowledge', of natural necessity and natural kinds, 'within the epistemological dialectic of science' (*D*: 36). Knowledge of natural kinds is fallibly captured in *real definitions*.

Some scientific essentialists (e.g., Ellis 2001) hold that social, and indeed biological, phenomena do not constitute natural kinds. Such a view rests on assumptions that essences are fixed (see FIXISM), and (ultimately) that CAUSALITY in the social world is different from that in the natural world, or perhaps non-existent. This is disputed by CR on several grounds. First, ontological change is transcendentally necessary (see ALTERITY/CHANGE; CONTRADICTION) – other than perhaps some absolute ULTIMATA (if there are such), natures may change, however slowly; if, for example, a proton loses its essential properties, it ceases to exist as a proton, and if something gains the essential properties of a proton it comes to be as a proton. Second, social structures are argued to be material and/or formal CAUSES (Groff 2004; Lewis 2000); thus for Bhaskar, following Marx and Hegel, the capitalist system has an inner dynamic or '*Essence Logic*' (*D*: 333). Popper's 'methodological NOMINALISM' rejects the search for essences on the ground that it involves the *essentialist fallacy* of holding that the only real causes are ultimate ones (Popper 1963: 103). But, given science's 'cosmic incapacity' (*RS2:* 182), this would make '*all* {. . .} scientific claims unverifiable' (*PN3*: 84) and is clearly not a necessary feature of the concept of real essence in an ontologically deep pluriverse, where things possessing essences can be explained in terms of more fundamental things possessing essences without conceivable end.

Since they possess intrinsic structures pertaining to the domain of the real, natural kinds are not to be equated with classes of concretely singular things. They are (within their range), rather, *real UNIVERSALS*: the universal element in the concretely singular. A concrete singularity, for its part, may (come to) possess a unique essence (also known as *haecceity*, f. L. *haecceitas*, 'thisness'), constituted by a universal nature, particular mediations and the rhythmics of its world-line, that is in effect an (emergent) natural kind sui generis, such that the essence of a person, for example, 'consists just in what she is most fundamentally disposed to do (or become)' (*PN3*: 96) (see CONCRETE UNIVERSAL ↔ SINGULAR).

Between universal and de facto singular natural kinds stands the *type*, a distinction which relates to that between causal powers and TENDENCIES: 'a thing possesses

322 natural necessity

powers in virtue of its falling into a natural kind, tendencies in virtue of its being one of a type within that kind. All men (living in certain kinds of societies) possess the power to steal; kleptomaniacs possess the tendency to do so' (*RS2*: 230). Tendency thus conveys the notion of enduring orientation to act in a way that is naturally open to a thing, rather than the bare possibility of transfactual activity; hence 'it leads us to a more precise specification of the natures of particular things (or groups) within kinds' (*RS2*: 235). On such a construal, *ideal types* 'are not subjective classifications of an undifferentiated empirical reality', as they are for Weber, but attempts in the TD to express ('idealisation' in the non-normative sense, or ABSTRACTION) the objective (ID) 'causal structures which account in all their complex and multiple determinations for the concrete phenomena of human history' (Bhaskar 1998a: xvi). Thus for Dean (2003: 43) 'the different modes of capitalism are {...} "real" (not "ideal") types in that {...} each has its own specific tendencies {...} which may or may not be actualised or satisfied in specific social formations but which had to be actualised or satisfied if capitalism were to reproduce itself'. A *token* is a particular concrete instance of a *type*, as a structuratum is of STRUCTURE.

Because of the openness and DEPTH-stratification of the world, CR espouses '*weak*' rather than the '*strong*' *essentialism* that assumes a one-to-one correspondence between essences (possessing causal powers) and the actual and views social phenomena as epiphenomena of a biological substratum (Sayer 1997b: 476). In other words, while things necessarily tend to act in the way they do, 'contingency still lies in the flux of the circumstances in which they act' (*RS2*: 173) and non-reductionism and FREE WILL apply. See also CATEGORY; EMERGENCE; MODALITY.

natural necessity is exemplified by 'a generative mechanism at work' (*SR*: 163). See also ALETHIA; AXIOLOGY; CAUSAL LAW; CAUSALITY; DEFINITION; DETERMINATION; EXPLANATION; MODALITY; NATURAL KIND; REALITY PRINCIPLE; TINA SYNDROME.

naturalisation, or (less strongly) **normalisation** and (more strongly) **eternisation**, makes the social status quo *seem* natural and therefore permanent and ineluctable. It is broadly synonymous with REIFICATION. Naturalisation is generated by the ontic fallacy, which is the implicit social meaning of the epistemic fallacy: the ontologisation, hence naturalisation, of knowledge (*R*: 181). Thus the principle of sufficient practical reason (see DETERMINATION; PRINCIPLE OF INDIFFERENCE) stipulates that there must be grounds for social differences, e.g., inequality, but the *epistemological naturalisation* of the status quo or the reification of the present – bolstered by traditional philosophical analysis of change in terms of difference since Plato, underpinned by fear of change – turns this into 'the requirement that it is *changes, not differences, that must be justified*'. Epistemological naturalisation thus displaces the negativity of change on to ETHICS, which 'becomes the realm of prohibition, of opposition (duty v. desire), dichotomies (fact and value), with action characteristically noumenalised (so that it doesn't conflict with the doctrine of ontological monovalence)' (*P*: 148–9). This is the *negativity* or *negativisation of ethics*. See also EPISTEMIC–ONTIC FALLACY; ETERNISATION; SOCIAL CONSTRUCTIONISM.

naturalism. Over the last century, naturalism 'usually connoted three related ideas: (1) the dependence of social, and more generally human, life upon nature, i.e., materialism; (2) the susceptibility of these to explanation in essentially the same way, i.e. *scientifically*; (3) the cognate character of statements of fact and value, and in particular the absence of an unbridgeable logical gulf between them of the kind maintained by David Hume, Max Weber and G. E. Moore, i.e., ethical naturalism' (Bhaskar 2003a: 425, e.a.). In CR (1) takes the specific form of *ontological* or *emergent powers materialism*, on which 'society is materialised ultimately in virtue of embodied intentional causal agency reacting back on the kinds of materials out of which it was formed' (*D*: 172), and intentional agency is the embodied emergent power of mind. (2) assumes the form of methodological unity-in-diversity among the sciences and vindicates explanatory social science. And (3) is vindicated by EXPLANATORY CRITIQUE, entailing EMANCIPATORY AXIOLOGY, or the view that the good for human being is derivable from our nature as social beings and achievable via totalising depth-praxis. In PHILOSOPHY OF SCIENCE, naturalism is the view that philosophy has little or no relative autonomy from science. *Ethical hyper-naturalism* (an analogue of SCIENTISM) is the view that values are the direct expression of facts and have no role in the development of theory (*SR*: 174). *Anti-naturalism* is the converse of (1)–(3). In 'splitting off the realm of values from the realm of facts (a monstrous detotalisation)', *ethical anti-naturalism* 'denies any possible moral truth grounds' (*P*: 238). See also CRITICAL NATURALISM; ETHICAL IDEOLOGIES; ETHICS; IDEALISM VS. MATERIALISM; NATURALISTIC FALLACY; PHILOSOPHY OF SOCIAL SCIENCE; REDUCTIONISM.

naturalism, critical. See CRITICAL NATURALISM.

naturalism, ethical. See ETHICS; NATURALISTIC FALLACY.

naturalistic *aka* **anti-naturalistic fallacy**. The *naturalistic fallacy* in positivist usage refers to the supposed mistake of deducing conclusions about what ought to be from premises that state only what is the case, or vice versa; *Hume's law* states that such transitions are logically impossible. Called 'naturalistic' by G. E. Moore because he held that value is a non-natural property, but commonly referred to as the *anti-naturalistic fallacy* because it is incompatible with a positivistically conceived naturalism (*hypernaturalism*). Since for CR the transition from is to ought, facts to values (though not vice versa), is licit, its usage is the inverse of positivism's. This is displayed in Table 31.

The dominant ETHICAL IDEOLOGIES in capitalist post/modernity – *emotivism*, *decisionism* and *personalism* – all trade on the naturalistic *aka* anti-naturalistic fallacy, which *de-evaluates reality*, ALIENATING us from it, and which, in leaving ethics ungrounded and immune to explanatory critique, itself implicitly secretes and legitimates a morality: that of the status quo. See also CRITICAL NATURALISM; EMPIRICISM; ETHICS; EXPLANATORY CRITIQUE; FACT–VALUE; NATURALISM.

necessity. See MODALITY.

needs. The concept of needs is one of the most contested in the philosophical lexicon. Advocates of this concept tend to be critical theorists who make a distinction between

Table 31 The naturalistic *aka* anti-naturalistic fallacy

	Positivism	*Critical realism*
Naturalism	F– –/– –>V	F– – – –>V
Anti-naturalism	F– – – –>V	F– –/– –>V
Naturalistic *aka* **anti-naturalistic fallacy**	–(F– –/– –>V)	–(F– – – –>V)

true or real needs and false or artificial needs; the satisfaction of the former advancing, and of the latter impeding, human flourishing. This distinction between true and false needs maps on to that of real and false INTERESTS and is grounded in an argument about the ideological constitution of a false consciousness which is blind to the gap between experienced false needs (WANTS) and real, but unexperienced (unwanted), needs (see REASON). CR participates in this kind of analysis, the locus classicus of which is Marx's work on capitalism (Fraser 1998; Soper 1981). Thus for Bhaskar 'a need is anything (contingently or absolutely) necessary to the survival or well-being of an agent, whether the agent currently "possesses" it or not', by contrast with a WANT, which may be inimical to flourishing; while an INTEREST is 'anything conducive to the achievement of agents wants, needs and/or purposes' (*SR*: 170).

Critics of this broad usage deem its users to be guilty of an unsustainable claim which conceals *either* the necessarily historico-cultural character of needs *or* their status as subjectively chosen preferences. Where needs are deemed to be historico-culturally constructed, a relativist position logically follows. Their glossing as preferences involves a subjectivist position which identifies individual preference-holders as the only qualified judges of what constitutes their own well-being. In either case, there is no objective ground for a critique of actually experienced needs as false. This groundlessness is deemed to be not only true to the facts of the matter but politically desirable in that it counters the dangers of authoritarianism about needs (O'Neill 2004).

While being alert to these dangers, CR, as a philosophy of and for human emancipation, argues that humans have unneeded wants and unwanted needs. Bhaskar's advocacy of social science as EXPLANATORY CRITIQUE (*PNI*) is grounded in this distinction, the purpose of explanatory critique being to dissipate false consciousness about needs by advancing a popular critical attitude to the proliferation of false needs (wants) in capitalist cultures. However, explanatory critique can and should only attempt this where there exist objective possibilities for the subjective experience of true needs (*SR*: 200–11).

CR sets itself the difficult task of sustaining a middle way between relativism and subjectivism; of accepting that human needs are historico-culturally mediated while not forfeiting the possibility of judging actually experienced wants to be antithetical to human flourishing. Different on-going attempts to advance our understanding of this topic include Collier's philosophy of and for moral realism (1999, 2003b). Collier's moral realism embraces non-human nature via an environmental philosophy which owes much to Augustinian ethics. In the field of social theory, Sayer's development of a 'qualified ethical naturalism' (2004) is notable as is the work of Dickens (1992, 2001). Echoing the environmental concerns of Collier, Dickens focuses on the needs of non-human, as well as human, nature. His explanatory critique of capitalism draws on arguments from biology to demonstrate the ways in which class differences are expressed in identifiable incapacities. As does Dickens, Dean (2003) uses a strong Marxist version of CR to contribute to a critical theory of needs. Here the focus is on the contradictions between capitalist and human needs as these have emerged through different modes of capitalism.

KATHRYN DEAN

negation, negativity. See ABSENCE.

negation of the negation. In Marx and Bhaskar, the 'GEO-HISTORICAL transformation of geo-historical forms', intrinsic to any substantial PROCESS. Contrary to Engels's *Dialectics of Nature*, it is not a 'law' of nature, but a synthetic a priori truth of philosophical ontology. In the human world it is exemplified by the TMSA, which entails that geo-historical transformations or TRANSITIONS (transformative negations – absence qua [product-in-]process) are always (1) of some geo-historical form (social structure, conceptual form) that is itself the outcome (absence qua [process-in-]product) of prior transformation, leading to (2) a new form of which the same can be said. Hence the *double negation*. It is thus vital to an understanding of human geo-history; indeed, the negation of the negation correctly specifies what dialectic is: the absenting of absence (*D*: 152). In ideology-CRITIQUE, a *triple negation* is called for (as distinct from being effected): of theoretical ideologies, practical ideologies and the generative structures underpinning them. Geo-historical transformative negations, contrary to Hegel, are rarely wholly, or even essentially, preservative, and may include exogenous sources of change, i.e., are not necessarily radical or linear. Human emancipatory praxis is necessarily geared towards the abolition, not preservation, of structures generative of oppressive social forms.

In Hegelian dialectic, the negation of the negation is the cancellation of CONTRADICTIONS – preservative or additive supersession [negation] of prior conceptual and social forms in higher forms that absents absent, but necessary, concepts [contradictions] in their predecessors, until the system of forms as a whole is perfected – culminating in the reinstatement of the primacy of ANALYTICS and the restoration of positivity in the constellational closure of being as well as geo-history.

Marx's and Bhaskar's understanding of the negation of the negation provides a DIALECTICAL COMMENT (dc′) on, or immanent critique of, Hegel's; i.e., reports theory–practice inconsistencies in it: notably, that the concept of absence which drives his dialectic on is itself lost in 'a pacific sea of positivity' (*D*: 90), and that his closure of geo-history is 'untrue to his theory of truth which would vindicate the unity of theory and practice in practice, that is, in transformation of (socialised) reality to comport it to a rationally grounded notion of it' (*D*: 341). In Marx, the concept also refers specifically to the negation of living labour by capital as objectified labour, requiring a second negation for the free flourishing of labour (Arthur 2003); in terms of the overall geo-historical PROCESS, it is the negation of class society which itself is the negation of an original unity. See also ABSENCE; ALIENATION; NEGATIVE PRESENCE; SUBLATION; TABLE OF OPPOSITION.

negative generalisation extends a concept hitherto understood positively to all its negative instances. (Positive generalisation does the converse.) Thus, in the TMSA as originally formulated, social structures were conceptualised as, among other things, activity-dependent (existing only in virtue of people's activities). Dialecticisation of CR demonstrated, inter alia, the reality of absence and of the presence of the (absent) past and outside. Given this, a PERSPECTIVAL SWITCH from structure to agency indicates, especially in view of the 'depthless atomisation of bourgeois individualism' and the attendant sense of powerlessness in relation to the system, that social structures and forms can survive in the absence of, or even despite, the activity of people, and also in virtue of their inactivity (think of the thawing of the polar ice-caps and of the problem of reversing it). Inactivity is as 'axiologically irreducible as non-being is ontologically':

326 **negative presence/positive absence**

we cannot do everything at once or be aware of all the consequences of acting in one way rather than another (this is the axiological asymmetry [see PRIMACY]). Such considerations effect a *negative generalisation* of the TMSA. This is then *itself* negatively generalised – so sustaining the thesis of activity-dependence – by a spatio-temporal perspectival switch on it, i.e., the negatively generalised TMSA is viewed as a (DISTANCIATED) spatio-temporal stretch-spread or developmental PROCESS, within which structures are maintained by (or without or despite) the activities of the *dead* as well as of the living (see *D*: 158–60). Negative generalisation thus emphasises the hiatus between structure and agency in line with Archer's (1995) refinements of the TMSA, and distances that model further from Giddensian structuration theory (see SOCIAL STRUCTURE).

The negative generalisation of CONSTRAINTS to include remediable absences, ills and needs is fundamental to CR ETHICS and EMANCIPATORY AXIOLOGY.

negative presence/positive absence. In Hegelian dialectic, over a stretch of time *positive absences* (positive contraries or CONTRADICTIONS denoting lack in a conceptual field) are sublated or cancelled but not forgotten: they are held in a cumulative memory store as *negative presences* (negative sub-contraries) and are the key to the Hegelian 'analytical reinstatement' of positivity (see EPISTEMOLOGICAL DIALECTIC). In Marxian and Bhaskarian depth-dialectic, positive absences (transfactually efficacious positive contraries or contradictions at the level of the real) may be simultaneously negative presences (negative sub-contraries at the level of the actual, i.e., not 'realised'); or positive contraries or contradictions may be both really present and actually absent at the same time. Unlike in Hegel, they may themselves be subject to real negation by social praxis (TTTTφ) (this is the NEGATION OF THE NEGATION). Both concepts entail *negative referral*, a concept borrowed from Kosok (1972) to denote REFERENCE to (REFERENTIAL DETACHMENT of) ABSENCE or the absent (the past or the unactualised). For CR, memory is a negative presence, which may of course be causally efficacious, so is the past-in-the-present and the intrinsic outside (see CO-PRESENCE). See also DETACHMENT/ATTACHMENT; ONT/DE-ONT; TABLE OF OPPOSITION.

negative referral. See NEGATIVE PRESENCE.

negativity. See ABSENCE.

negativity of ethics, negativisation of ethics. See NATURALISATION; REIFICATION.

negentropy. See ENTROPY.

neo-colonialism. See COLONIALISM.

neo-liberalism. Although it is often maintained that neo-liberalism is a category used only by its opponents, the members of the Mont Pelerin Society, founded in 1947 around the ideas of Friedrich von Hayek, in fact depicted themselves as 'liberals'. Also Milton Friedman was a member of that society. In the 1940s and 1950s, the nineteenth-century era of classical *laissez-faire* liberalism was definitely over. Keynes and social democracy reigned in the West, Marx and Lenin in the East, and the idea of national liberation and state-led development in the South. Hayek, Friedman and others were simultaneously committed to the ideals of personal freedom and to the free-market principles of neo-classical economics. Key works of this era are Hayek's *Road to Serfdom* (1944) – a work on political philosophy defending individual liberties and freedom of private property against all collectivist ideas – and Friedman's *Essays in Positive Economics*

(1953) – a collection of writings on economic theory defending the idea that markets are efficient and self-regulating, and also advocating new arrangements such as flexible exchange rates and liberalisation of finance.

With the ascendancy of liberal individualist ideas in the 1970s, first in the USA and Chile (Pinochet's military government famously turned to Friedman's Chicago School for economic policy advice), and soon in many other countries all over the world, it was appropriate to characterise these kinds of ideas as 'neo-liberal'. As Hayek (1944: 240) wrote, 'though we neither can wish nor possess the power to go back to the reality of the nineteenth century, we have the opportunity to realise its ideals'. Neo-liberalism advocates the maximisation of individual rights, private property rights, and free-market capitalism. However, radical libertarianism à la Friedman and Hayek is not the only strand of neo-liberalism. It may not even be the mainstream. Only a few neo-liberals would support, say, the legalisation of drugs, gambling, euthanasia or prostitution.

Moreover, the so-called Washington consensus on the right mixture of economic policies in fact includes also an emphasis on efficient taxation (which is often a problem, particularly in the global South) and some social spending (in contrast to, for example, military spending). The label 'Washington consensus' was originally coined by the US economist John Williamson (1990), who argued that it consists of a package of ten general policy principles: fiscal discipline, which normally means fiscal austerity, which in turn translates into cuts in welfare spending; public-expenditure priorities on education, health and infrastructure; tax reform; exchange rates established by markets; competitive exchange rates; trade liberalisation; promotion of foreign direct investment; privatisation; deregulation; and the enforcement of private property rights. In this sense neo-liberalism has been the dominant policy view at the International Monetary Fund (IMF), the World Bank and the US Treasury at the end of the twentieth century and the start of the twenty-first.

In many countries neo-liberal ideas have to compete in multi-party elections; also, in overtly authoritarian contexts, neo-liberalism is dependent at least on passive consent. Therefore various compromises and fusions are inevitable in different fields of propagation and implementation of neo-liberalism. For instance, in development policy the discourse of 'participatory development and good governance' emerged in the programmes of developmental organisations in the 1980s and became ubiquitous in the 1990s. It seems that this discourse is a combination of two separate – and in some respects antagonistic – understandings of development. The first, providing the main framework, is the Western liberalist modernisation theory, which sees the contemporary Western nation-states – and particularly the USA and the UK – as the ultimate aim of development. This theory is in accordance with the Washington consensus. The second, subordinate understanding is the more critical thinking expressed by the Western social movements which in the course of the 1980s started to emphasise the importance of micro-level action-contexts, local resistance and community participation, as well as 'empowerment through participation', both in theory and in practice.

Another similar example is the New Public Management (NPM) discourse. NPM is a broad and complex term used to describe the wave of public-sector reforms apparent throughout the world since the 1980s. Based on neo-classical economics, rational choice theory and managerial schools of thought, new public management seeks to enhance the efficiency of the public sector and the control the government has over it. The basic idea is that markets constitute the most efficient way of allocating resources and providing services to the customers (citizens). This implies, first, a preference for privatisation, for outsourcing public services as well as for public–private partnerships. Second, whenever this is either not possible or not considered desirable, it implies a project of building simulated markets within public organisations. NPM does not claim that government should stop performing certain tasks. The rationale of many social democratic parties for adopting NPM so readily is that NPM is apparently not about whether tasks should be undertaken or not; it is about 'getting things done better' (see, e.g., Osborne and Gaebler 1992).

There is no general CR history, analysis or explanation of neo-liberalism comparable in focus and scope to, e.g., David Harvey's *A Brief History of Neoliberalism* (2005). However, Magnus Ryner's *Capitalist Restructuring, Globalisation and the Third Way: Lessons from the Swedish Model* (2002) comes close. There are also many CR works scrutinising the theoretical underpinnings of neo-liberalism. Bhaskar's analysis of TINA SYNDROME is expressed in highly abstract terms but is clearly targeted also at Margaret Thatcher and her well-known slogan 'There is no alternative' (Tina). Bhaskar's basic idea is that false philosophical theories tend to generate necessitarian interpretations of the world, precluding other possibilities from consideration. However, as the real world consists also of unactualised possibilities, and as new mechanisms, structures and beings may emerge, any necessitarian account is necessarily incomplete and involves various lacks and contradictions. This in turn makes a mixture of defence clauses and Derridean supplements crucial. At a somewhat more concrete level, the main contribution of CR in this context has to do with the critical analysis of the methodology of neo-classical economics, which constitutes a dominant Tina-formation (e.g., Lawson 1997, 2003b). Also CR studies on MANAGEMENT (Ackroyd and Fleetwood 2000; Fleetwood and Ackroyd 2004) are worth mentioning, although they are focused more on methodology than on explaining substantive issues – including the real effects – of neo-liberal governance.

HEIKKI PATOMÄKI

neo-Nietzscheans. Another name for postmodern philosophers and social theorists.
nesting, nestedness. See DISTANCIATION; PROBABILITY; RECURSIVITY; SYSTEM.
neustics. See ONT–DE-ONT.
New Dialectic(s). See SYSTEMATIC DIALECTIC(S).
new realism. See CRITICAL REALISM.
nexus. A combination of aspects of an EVENT whose form co-determines the elements which in turn co-determine each other and so the form. As SYSTEM and TOTALITY are to structures, so nexus is to events. It is normally constituted as a nexus by systemic determination, but may in turn affect the operation of structural mechanisms. See *SR*: 109–11; HOLISTIC CAUSALITY.
Nicod's criterion. Criterion of confirmation, named after Jean Nicod (1889–1924). It is stated in his essay, 'The logical problem of induction', which was first published in French in 1923 and was translated into English in 1930. Nicod claimed that

law-like generalisations are established as probable, if at all, by being confirmed by their favourable instances and are refuted by being invalidated by their unfavourable instances, and argued that induction by enumeration, that is, confirmation by repetition, is the fundamental form of induction. A favourable (positive) instance of a law-like generalisation of the form *All A's are B* is an object which is both A and B, while an unfavourable (negative) instance is an object which is A but not B. As he noted, 'any effect which individual truths or facts can have on universal propositions or laws takes place through these elementary operations, which we shall call *confirmation* and *invalidation*' (1969: 189). It was Carl Hempel, in his 'Studies in the logic of confirmation' (in Hempel 1965), who introduced the expression 'Nicod's criterion' to capture the thesis that universal generalisations are confirmed by their positive instances. Hempel noted that this criterion, combined with two other, intuitively plausible, principles, leads to paradox. It has been called 'Hempel's paradox' or 'the paradox of the ravens' since the example Hempel used to illustrate it was that all ravens are black. The two other principles are the following. *The principle of equivalence*: if a piece of evidence confirms a hypothesis, it also confirms its logically equivalent hypotheses. *The principle of relevant empirical investigation*: hypotheses are confirmed by investigating empirically what they assert. The paradox is generated thus. Take the hypothesis (H): All ravens are black. The hypothesis (H′) *All non-black things are non-ravens* is logically equivalent to (H). A positive instance of H′ is a white piece of chalk. Hence, by Nicod's criterion, the observation of the white piece of chalk confirms H′. Hence, by the *principle of equivalence*, it also confirms H, that is, that all ravens are black. But then the principle of relevant empirical investigation is violated: the hypothesis that all ravens are black is confirmed, not by examining the colour of ravens (or of any other birds), but by examining seemingly irrelevant objects (like pieces of chalk or red roses). So, if the paradox is to be avoided, either Nicod's criterion or at least one of the two other principles should be abandoned. Philosophers differ as to where the blame should be put. But Nicod's criterion is not inviolable. Do positive instances *always* confirm a hypothesis? Consider the following case. Three people give their hat to the cloakroom attendant before the movie. When the first two collect theirs, they leave with each other's hat. These two facts are positive instances of the hypothesis that *all three people will leave with the wrong hat*. But do they confirm it? Clearly not. See also CAUSAL LAW.

STATHIS PSILLOS

Nietzschean forgetting. See REFLECTION.
nihilism. See ETHICAL IDEOLOGIES.
node. See TRANSITION.
nomic. See COUNTERFACTUAL/TRANSFACTUAL.
nominal essence. See NATURAL KIND.
nominalism. See IDEALISM; NATURAL KIND; REALISM; UNIVERSALS/PARTICULARS.
nomological. See COUNTERFACTUAL/TRANSFACTUAL.
nomotheticism. See EXPLANATION.
non-being. See ABSENCE.
non-contradiction, law or principle of. See ANALYTIC PROBLEMATIC; CONTRADICTION.

non-duality, non-dualism. See DEMI-REALITY; DUALITY; INFRA-/INTRA-/SUPER-STRUCTURE; META-REALITY; PDM.
non-existence. See ABSENCE; ONT/DE-ONT.
non-identity. See ALTERITY.
non-locality. See CO-PRESENCE; SPACE–TIME.
non-parity of equivalents. See FISSION/FUSION; PRINCIPLE OF EQUIVALENCE.
non-valence. See AXIOLOGY; VALENCE.
normal science. See EPISTEMOLOGICAL DIALECTIC; PHILOSOPHY OF SCIENCE.
normic. See COUNTERFACTUAL/TRANSFACTUAL.
nursing. A distinct, practice-oriented, knowledge-driven discipline that focuses on improving physical, psychological and social health or the relief of suffering through ethical and appropriate nursing care. Existing work in nursing has thus far used CR to guide empirical research (Porter and Ryan 1996; Tolson 1999) or conveyed the theoretical and philosophical strengths of CR to nursing practice and research (Clark 1998; McEvoy and Richards 2003). The attractiveness of CR to nursing can be attributed to a number of factors related to philosophy and utility.

In terms of ontology, nursing adopts a holistic conception of the human as a core professional value; that is, that humans are dynamic embodied, spiritual, biological, psychological and social beings. Nursing care should, by nature, be holistic and address NEEDS in each of these distinct but inter-linked realms. The ontology of CR is highly congruent with nursing, as it not only allows for the co-existence (and interdependency) of phenomena at different strata but also recognises that phenomena at each strata are equally real and can be changed. This approach can be contrasted to bio-medical approaches, most associated with medicine, which view individuals predominantly or exclusively in reductionist and biological terms. CR therefore offers nursing a holistic, sophisticated and emancipatory ontology.

The opposition of CR to variations of both (judgemental) RELATIVISM and FOUNDATIONALIST OBJECTIVISM is also highly congruent with nursing. Variations of the former are antithetical to a practice-based discipline which purports to be able to influence real and independently existing entities such as pain, pathogens and behavioural patterns. Conversely, the latter movement (most often associated with positivism) offers a naïve, historically questionable, and atomistic approach based on ABSTRACT UNIVERSALISM – it excessively decontextualises and simplifies cognate, culturally embedded beings.

CR gives no inherent preference to either quantitative or qualitative research METHODS. While some proponents demonstrate leanings towards qualitative approaches (Pawson and Tilley 1997; Sayer 2000), there are good examples of CR-driven research that uses both methods (Kazi 2003). Irrespectively, understanding phenomena and complexity takes preference over methodological dogma. This stance is highly congruent with an emerging consensus in health research that previous qualitative–quantitative debates were unconstructive and based on narrow and dated views.

There is great potential for growth in the use of CR in nursing theory and research. The CR projects of explaining and understanding are central to a number of key issues nursing faces, including the promotion of evidence-based practice, understanding patient and professional decision-making, and promoting positive patient experiences.

These areas address complex issues related to the interplay of disparate and large numbers of individual and structural factors operating in open systems and the nature and effects of embodiment, intuitive knowledge, social contexts and organisational settings on behaviour. As a coherent and philosophically sound means to theorise about diverse phenomena and, ultimately, to provide a means to generate knowledge about how such phenomena can be changed, growth in the use of CR will be timely and important to the development of nursing knowledge. See also SELF; SOCIOLOGY OF THE BODY; SOCIOLOGY OF HEALTH AND MEDICINE.

ALEXANDER M. CLARK

O

object relations. See PSYCHOANALYSIS.
objectification, objectivation, objectivication. See INTRANSITIVE, ETC.; REFERENTIAL DETACHMENT; REIFICATION.
objectivity/subjectivity. The *subjectivity/objectivity relation* is one of diremption or ALIENATION in master–slave-type societies, of expressive unity in other non-eudaimonian societies, and of unity-in-diversity in eudaimonia (see DEMI-REALITY). The *PROBLEM of subjectivity and objectivity*, constituted in the demi-real, is resolved by the concepts of REFERENTIAL DETACHMENT, the stratification of the SUBJECT and a meta-reflexively totalising situation, which come together in the concept of a *meta-reflexively totalising (self-)situation* (see REFLECTION). Reversing the conventional order of 'subjectivity/objectivity', such that it reflects the ontological order of EMERGENCE rather than the epistemological order, reminds us that subject and object, language and the world, are CONSTELLATIONALLY contained within subjectivity (level of ontogeny) within an overarching (relativised or relationalised) objectivity (level of ontology, including society and inter-/intra-subjective relations). This also allows a solution to *the* problem of knowledge: we can (fallibly) come to know and express (some of) the world ultimately because we are not split off or alienated from it by any non-illusory DUALISM but an emergent part of it and thus in a position to engage in 'a constant dialectical exchange between theory, observation, and experimental practice' (TRANSCENDENTAL REALISM; cf. *MR*: 89–90). The breakdown of the fact/value dichotomy (see CRITICAL NATURALISM; EXPLANATORY CRITIQUE; EMANCIPATORY AXIOLOGY) also strikes a blow against subjectivity/objectivity dualism, entailing that values inhere in being. Sayer 2000a (58–61, 65 n25) usefully distinguishes three meanings of 'objective' and 'subjective': $objective_1/subjective_1$, value-neutral/value-laden; $objective_2/subjective_2$, (concerning propositions) true/not true; and $objective_3/subjective_3$, pertaining to objects/pertaining to subjects. These loosely correspond to the expressive-referential, adequating or epistemic, and ontological/ALETHIC moments of the truth tetrapolity, respectively. Lacey (2006) has defined $objectivity_1$ as 'the value that only empirical data and appropriate cognitive (epistemic) criteria should be relevant for evaluating the confirmation of scientific theories and hypotheses {. . . which} presupposes that these criteria permit no role for ethical or social values or the interests of the powerful'. See also EPISTEMOLOGICAL DIALECTIC; ONTOLOGY; PERSPECTIVE; Morgan (2005a).
omega (ω) paradox. See PARADOX OF THE HEAP.
omissive critique. See CRITIQUE.

one and the many, problem of. See PROBLEM.

ont/de-ont (Gr. *ont-*, being + *de-*, but [with opposing force]). *Onts* are positive, *de-onts* or ABSENCES negative, existences or presences within the class of *ontics* (objects of specific epistemic inquiries) within ontology (see EPISTEMOLOGY, ETC.); or onts are beings, de-onts non-beings. We thus have the theorems ontology > ontics > onts, h and ontology > ontics > de-onts. Ontics is a more inclusive class than onts or de-onts because it embraces both. De-onts include never anywhere existences (phlogiston), the simply absent (the dead), NEGATIVE PRESENCES (memory, the past and outside), GULFS, SPLITS and VOIDS.

Bhaskar (*D*: 40–1) uses an adaptation of Hare's (1952, 1997) triptych of *tropics*, *neustics* and *phrastics* as a grid to dissect the fine structure of non-being. *Tropics* (Gr. *tropos*, mode) designate a domain of discourse, e.g., the fictional (I) as distinct from the factual (F). *Neustics* (f. Gr. *neuein*, to nod or give a sign of assent) convey attitudes such as acceptance, rejection or indecision (✓, ✗ and /). *Phrastics* (f. Gr. *phrasein*, to declare, propose) denote the ontic content of a proposition, what it is about, which may be positive or negative. The negative is real within all three of these domains in that it can be referred to, and they all may (in different ways) convey negative existential import. This does not, however, abolish the vital differences between them and in particular between the fictional, the factual and the phrastic (cf. *PF*: 121–6; see also REFERENCE). *Ficts* (fictional beings) may be a class of de-ont within discourse (when included within the agent's or society's register of the imagined). Ficts are ontically$_2$ opposed to reality in their assertoric content, but ontologically$_2$ part of reality in their semantic and psycho-social identity (*SR*: 43). *Facts* (L. *factum*, that which has been done or made) are 'established {empirically grounded} results, made, not apprehended' [*D*: 215]). They may be onts or de-onts or, more commonly, both. They are 'potentialities of the conceptual schemes or paradigms governing our enquiries which are actualised in discovery, sustained in (discursive and non-discursive) practice and objectified in sense-experience'. This contrasts with the positivist or empiricist conception of a fact as 'that which is more or less immediately apprehended in sense perception'. Facts in this realist sense are no mere illusions but at bottom spontaneous conceptions, reflecting the way in which the world presents itself to common sense via 'our own prior cognitive reproducts and transforms' (see also COGNITIVE SCIENCE; PERCEPTION). This is the *fact-form*, which is objectively misleading (a real false being irreducible to cognitive mistakes) analogously (in part) to the value-form. The positivist conception of a fact tarts up and reifies our common-sense categories, FETISHISING facts as natural and eternal when they are in reality geo-historically relative social products (*SR*: 280–2; cf. *R*: 9, 60–2; see also REIFICATION).

The term de-ont can be interpreted as calling attention to the double meaning (equivocity) of the Greek words *dein* (vb.) and *deon* (n.) on which *deontology* (the study of moral duty) is formed: (1) to bind, fasten, fetter; and (2) to want, lack, need. These two meanings come together in the pivotal CR concept of absence (de-ont) as constraint (fetter) and as lack or ill-being (need), thereby unifying ethics with the theory of being. Hence Bhaskar sometimes writes *de-ontology* (*P*: 113).

ontic(s). See EPISTEMOLOGY, ETC.; ONT/DE-ONT.

ontic fallacy, ontification, ontologisation. See EPISTEMIC–ONTIC FALLACY; EPISTEMOLOGY, ETC.

334 ontocentrism

ontocentrism. See ANTHROPISM.

ontogenesis/phylogenesis (Gr., birth of [a] being, and birth of [a] kind, respectively). *Ontogenesis* is the formation and development of individual members of a biological species; *ontogeny* the history or science of such formation. They contrast with *phylogenesis* and *phylogeny*, which concern the development of the species. See also POLYAD.

onto-logic. The fundamental orientation or directional tendency of some entity, often a philosophical or other intellectual practice. Thus DIALECTICS is an onto-logic of change, ANALYTICS and IDENTITY THEORY are onto-logics of stasis, and the ontologic of human labour is *POIESIS* (*SR*: 122). The hyphenation attracts attention away from the normal meaning of ONTOLOGY ('the study of being') to a variant meaning of the Greek words comprising it: 'the logic of being'. The usage underlines that philosophies are not just affected by the world, but affect it, and, indeed, while distinct and emergent, are constellationally contained within the world (cf. 'Hegel's {...} *gnoseo-onto-logical* Odyssey' [*D*: 94, e.a.]). Onto-logic (or onto logic) is used in information/communication theory with similar intent. In Hegelian philosophy and theology, it often means 'the logic of Truth'.

ontological–axiological chain. See MELD.

ontological dependence. See DEPTH; EMERGENCE; IDEALISM VS. MATERIALISM; PRIMACY.

ontological depth. See DEPTH.

ontological destratification. See STRATIFICATION.

ontological dialectics. The dialectics of the INTRANSITIVE DIMENSION, which constellationally include *epistemological dialectics* or the dialectics of the transitive dimension.

ontological extensionalism. See EXPRESSIVISM/EXTENSIONALISM.

ontological grammar. See GRAMMAR.

ontological individualism or **social atomism.** See ATOMISM; SOCIAL STRUCTURE.

ontological or **metaphysical materialism.** See CENTRISM; CRITICAL NATURALISM; EMERGENCE; HOLY TRINITY; IDEALISM VS. MATERIALISM; MATERIALISM; MIND; NATURALISM; PDM; PHILOSOPHY OF RELIGION.

ontological monovalence. See VALENCE.

ontological priority. See PRIMACY.

ontological realism (metaphysical realism). See DEPTH; HOLY TRINITY; IDEALISM VS. MATERIALISM; ONTOLOGY; REALISM.

ontological relativism or **relativity.** CR espouses ontological relativity (or *ontological* or *entity relationism*) in that, except for ABSOLUTE constants such as the speed of light, everything is ontologically dependent on its RELATIONS to other things. This is very different from Quinean or Kuhnian ontological relativism, which commits the epistemic fallacy, holding that how the world is depends on our theories. See also HOLY TRINITY; ONTOLOGY; PERSPECTIVE; PHILOSOPHICAL ANTHROPOLOGY; TRANSCENDENTAL REALISM.

ontological stratification. See DEPTH.

ontologisation (of knowledge). See ONTIFICATION.

ontology. CR offers a distinctive, far-ranging and (above all) scientifically informed answer to various ontological issues that have long preoccupied philosophers of the natural and the social/human sciences alike. Those problems arise as soon as one attempts to offer some adequate definition of the term 'ontology'. Traditionally this is the field of philosophic inquiry that has to do with things (Gr. *ta onta*), or with real-world objects and their intrinsic structures, properties, causal powers and so forth. That is to say, it treats of them just in so far as they are taken to exist and to exert such powers quite apart from the scope and limits of human knowledge. This latter is the domain of epistemology, or the branch of philosophy centrally concerned with the question how far and by just what means we are able to acquire such knowledge. Yet there is a difficulty here when it is asked what *justification* we could have for advancing ontological claims if these are thought of as transcending our utmost means of ascertainment. Indeed, it is often argued – by sceptics and anti-realists – that the ontology/epistemology distinction is one that philosophers should let go since it raises the 'problem of knowledge' in so stark a form. For if the truth-value of our various statements is conceived in objectivist terms rather than as falling within the compass of human cognitive grasp, then TRUTH and KNOWLEDGE come completely apart and we are left with no defence against global scepticism. In which case we are better off adopting a sensibly scaled down, epistemic conception that involves no appeal to recognition-transcendent truths. Ontological objectivism is a world well lost when set against the clear and decisive advantage of rebutting the sceptic and bringing knowledge back within reach of human acquisition.

This issue has a long prehistory, starting out with Aristotle's objection to Plato's doctrine of transcendent (suprasensory) essences or forms, and then taken up – by philosophers from Kant to the present – through various attempts to vindicate some realism-compatible theory of knowledge while yielding no hostages to sceptical fortune. However, these attempts have always involved a redefinition of the issue so as to raise ontological questions in epistemological terms and thereby reverse the traditional order of priorities. Such was Kant's answer to Humean scepticism, and such the line of argument typically adopted by later thinkers (like the logical empiricists) who gave up the Kantian claim to somehow reconcile 'transcendental idealism' with 'empirical realism', but whose theories – as Quine was quick to remark – could still be seen to manifest an ill-disguised version of the same DUALISM. Hence the current tendency, much influenced by Quine, to talk of 'ontologies' in the plural and to think of them as 'relative' to various paradigms, frameworks, conceptual schemes, etc. On this account, it is not the way things stand with the world that decides the truth-value of our theories or hypotheses, but rather the range of going theories or hypotheses that decides how things stand with the world. Nowadays the issue is most sharply posed by philosophers of mathematics who argue that we are faced with an ultimate choice between a realist (objectivist) conception of mathematical truths which *ex hypothesi* renders them unknowable and an epistemic conception whereby truth drops out in favour of knowledge within the bounds of formal provability. It is the same choice that is presumed to confront realist philosophers of the natural and social sciences, obliged as they are to explain what could possibly close the gap between truth and knowledge.

For critical realists this is a false dilemma. Rather, we should see that these problems result from a deep-laid dualist mindset that fails to take account of the 'STRATIFIED' nature of reality, that is, the existence of multiple levels and modes of engagement between knower and known. Where naïve (undialectical) realists go wrong is in supposing their case to stand or fall on the argument for a *totally* 'mind-independent' domain of objects, processes and events which cannot be construed as *in any way* subject to change or modification through human agency. This ignores both the extent to which scientific experiments require setting up under highly controlled laboratory conditions so as to exclude disturbing or complicating factors and also the discovery/creation of non-naturally-occurring items – among them transuranic elements or synthetic DNA proteins – which are no less real for their having been produced through such interventionist techniques. Where anti-realists go wrong is in committing the 'EPISTEMIC FALLACY', or the error of supposing that evidence of such involvement – like the effect of observation on the object observed at the microphysical (quantum) level – is sufficient to discredit the claims of scientific realism. Here again there is a failure to grasp some basic lessons from the history of science to date, among them the danger of extrapolating too readily from one to another (e.g., micro- to macrophysical) domain and the need to distinguish *within* those domains between objective REALIA which exist quite apart from the effects of techno-scientific agency and those which come about – increasingly so – through our ability to manipulate nature in various ways. The former are 'INTRANSITIVE' in the sense 'unaffected by our various dealings with them' while the latter are 'transitive' in the sense 'to some degree dependent upon, or brought about by, our techniques of physical intervention'. However, those techniques – and the products thereof – must still be conceived as subject to certain physically determined conditions, i.e., to the jointly enabling and constraining effects of the human interaction with nature's powers and capacities.

Thus CR repudiates that line of sceptical argument which denies that we can reconcile truth and knowledge, or achieve a workable conception of science that finds room for both. On the contrary: what sets this approach apart from various present-day rivals is its ability to explain how science can aspire to objective truth while at the same time making sufficient allowance for those other (among them economic, political, and socio-cultural) factors which bear upon the process of knowledge production. Where positivism went wrong was in setting up an abstract (fetishised) notion of scientific 'fact' and 'method' which completely removed them from the real-world context of material resources, technological powers, and other such crucial (by no means extraneous) factors in scientific research. Moreover, it ignored the often decisive – and equally material – impact of socio-economic and cultural conditions in so far as these may act to promote or to hinder the advancement of scientific knowledge. Where the strong sociologists go wrong, on the other hand, is in swinging right over to the opposite extreme and treating all theories, truth-claims or hypotheses as products of dissimulated class-interest or the epistemic will-to-power. What is missing from both pictures is the CR conception of science as engaging with various aspects, levels and ontological dimensions of reality through a range of likewise specific practices and modes of knowledge-production. Only by paying closer attention to the stratified nature of all such inquiry – that is to say, the complex dialectical relationship between knowledge and the objects of knowledge – can philosophy advance beyond the stage of merely reproducing all those stale dichotomies.

Thus it is important to grasp how science advances through the constant interaction of a great many factors, from physical structures and causal powers to research programmes, experimental techniques, theoretical commitments, and motivating interests of a different (social or ideological) character. In which case the social sciences have a role to play beyond the strictly ancillary role assigned them on the 'two contexts' principle, i.e., one confined to filling out the largely anecdotal 'context of discovery' and steering well clear of any first-order issue concerning the strictly scientific 'context of justification'. What they offer is a better understanding of precisely that relationship between the 'transitive' effects of human agency (of experiment, controlled observation, technological expertise, carefully contrived laboratory set-ups, and so forth) and the 'intransitive' dimension of a physical reality which must be taken to exist and exert its causal powers quite apart from scope and limits of our knowledge. Closely related to this is the issue concerning laws of nature, and the question how far we can ever be justified ('never', as some sceptics would have it) in assuming such laws to exist (see CAUSAL LAW). For there would seem to be something like an inverse relation between the kind of generality or universality claimed on their behalf and their yield in terms of descriptive precision or genuine, case-specific explanatory power. Hence the claim of some anti-realists that such talk is a pointless 'metaphysical' liability since it ignores the whole range of complicating factors (interference effects, outside sources of disturbance, imperfections in our means of observation/measurement) which ensure that no putative 'law of nature' will ever become manifest under even the most rigorously controlled experimental conditions. However, CR maintains, this is just another instance of the false dilemma created by a non-stratified approach that fails to take adequate account of the various ways in which observed phenomena may relate to underlying 'transfactually efficacious' laws or causal regularities.

To be sure, any statement of these latter must always involve a high degree of idealisation, that is, a willingness to discount effects such as those of friction and viscosity (in solid and fluid dynamics), or observer-interference on every scale from the micro- to the macrophysical. Still there is no reason to deny that such laws exist and that their workings are displayed (albeit indirectly or imperfectly so) in the kinds of phenomena available to observation. Hence the CR case against any doctrine that ignores the complexity of our causal interactions with nature and their affordance of a knowledge that goes far beyond such phenomenal or manifest appearances. Where CR breaks new ground is in explaining, not only how the sceptical challenge can be met, but also how the insights of a mature, scientifically informed SOCIOLOGY OF KNOWLEDGE can be harnessed to the realist cause as against those other, more fashionable versions enlisted by postmodernists and social constructivists of various theoretical persuasions. This it does by drawing attention to the complex, dialectical and stratified character of scientific knowledge-production, the transfactual efficacy of laws of nature, and the various aporias that inevitably result from an epistemic (knowledge-based) as opposed to an ALETHIC (objectivist or truth-based) approach.

That it also bids fair to redeem sociology from its 'poor relation' status vis-à-vis the physical sciences – as decreed by the old positivist order of priorities – while not simply turning that order upside-down is a further strong point in its favour. Thus CR answers the objection of those who protest that sociologists of knowledge are advancing strictly preposterous claims when they rest their case on a variety of dubious arguments from disciplines that possess nothing like the weight of tried and tested evidence that typifies the physical sciences. Whilst agreeing on this latter point it grants sociology a vital explanatory role as concerns those various cultural interests, pressures or incentives that have exercised a sometimes decisive impact on the development of scientific knowledge. Here, again, what gives it a clear advantage over other (e.g., positivist or 'strong'-sociological) accounts is the depth of its engagement with crucial issues – such as the role of practical–transformative agency *vis-à-vis* the kinds of constraint exerted by objects, structures or causal forces in the physical domain – which receive nothing like so detailed a treatment elsewhere. It thus puts up the most effective resistance to that currently widespread strain of relativist thinking which tends very often to issue in talk of plural 'ontologies', i.e., linguistic-conceptual schemes, as distinct from the singular (albeit complex and stratified) ontology envisaged by CR. Such talk has many otherwise diverse sources, amongst them post-structuralism, postmodernism, Foucauldian discourse-theory and Kuhnian paradigm-relativism. It is a chief virtue of the CR approach both to diagnose their manifest shortcomings when set against the record of achievement in various fields of scientific inquiry and to offer an alternative, depth-ontological and causal-explanatory account of how such progress has come about. See also COUNTERFACTUAL/TRANSFACTUAL; DETACHMENT; EPISTEMOLOGICAL DIALECTIC; EPISTEMOLOGY, ETC.; REALISM; TRANSCENDENTAL REALISM.

CHRISTOPHER NORRIS

open systems. See CLOSED AND OPEN SYSTEMS.

opposite (n.), opposition (f. L. *ob*, towards, against + *ponere*, to place). An object, state or quality that is the reverse or *contrary* of something else, exemplified in both logical and dialectical CONTRADICTIONS and ANTAGONISMS. Western philosophy is replete with figures denoting different forms of *oppositionality*, including AMBIGUITY, AMBIVALENCE, ANOMALY, ANTINOMY, APORIAI, CHIASMUS, COMPLICITY, COMPROMISE FORMATION, CO-PRESENCE, DENEGATION, DICHOTOMY, DILEMMA, DOMINATION, DUALISM, DUPLICITY, EXISTENTIAL CONSTITUTION, INVERSION, PARADOX, PLASTICITY, SPLIT, SPLIT-OFF, THEORY/PRACTICE INCONSISTENCY, UNSERIOUSNESS (*P*. 242–3); and philosophical PROBLEMS often take the form of an opposition (mind vs. body, etc.), most graphically illustrated in the archetypal *problem of opposites*. This arguably ultimately REFLECTS the contradictory nature of power$_2$-stricken society and its discursive normalisation and moralisation, underpinned immediately by fear of change (cf. Lukács 1919–24; Kant's antinomialism) and ultimately by GENERATIVE SEPARATION. Though the problem of opposites, hence the antinomialism of irrealist thought, can be resolved in theory, it will persist as long as its social grounds do: that is the 'real problem of opposites' (*D*: 246). The figure of theory–practice incompleteness is itself a figure of opposition, betokening a gulf between theory and practice. See also CONSTELLATIONALITY; DIALECTICAL ANTAGONISTS; DUALISM; PROCESS; TABLE OF OPPOSITIONS; TINA SYNDROME.

order, problem of. See HOBBESIAN PROBLEM OF ORDER.

organic composition. In Marx, the (rising) ratio of dead to living labour (fixed to variable capital) in the capitalist process of production – the *organic composition of capital* – which, albeit grounding a TENDENCY for the rate of profit to decline, increases the productivity and potential for creativity of labour. Bhaskar generalises the idea to include ideas themselves and the natural world. The rising *organic composition of nature* (the built environment relative to 'living' nature) 'threatens to tear the world itself apart with ecological contradictions', but a rising *organic composition of ideas* in human history (the stock of ideas deriving from the past relative to living ideas), hence of the potential for intellectual creativity, is helping to make 'a new organisation of the social world' in keeping with universal emancipation possible (*EW*: 69, 97; cf. Bhaskar 1997a: 144). All three are CONTRADICTIONS from the PERSPECTIVE of capital, because the more fixed capital, and infrastructure ideas are created under the aegis of its frenzied totalising dynamic, the higher the ratios rise.

other, the. See ALTERNITY; CO-PRESENCE; PDM; POSTMODERNISM.

other-worldly. See THIS-WORLDLY/OTHER-WORLDLY.

outwith. See CHIASMUS.

overdetermination. See DETERMINATION.

overreaching. See CONSTELLATION; DCR.

P

pan-logicise. See EPISTEMIC–ONTIC FALLACY.
paradigm. See PROBLEM; THEORY, ETC.
paradox (Gr. *para*, beyond, *doxa*, belief). A 2E figure of OPPOSITIONALITY betokening 'incompatibility between established (e.g., epistemological) canons and perceived reality' (*P*: 243). It is 'merely a surface form' of CATEGORIAL error (*P*: x) – typically the EPISTEMIC FALLACY – hence vanishes when the relevant error is abandoned. See also PROBLEM.
paradox, Hempel's (paradox of the ravens). See NICOD'S CRITERION.
paradox, injunctive. See INJUNCTIVE PARADOX.
paradox, Meno's. See MENO'S PARADOX.
paradox, omega. See PARADOX OF THE HEAP.
paradox, Zeno's. See ZENO'S PARADOX.
paradox of the heap or **omega** (ω) or **set-theoretic paradoxes** (e.g., the *paradox of the preface*) 'arise only because we insist in characterising totalities in the same way as we characterise their units, elements or parts. We regard them as aggregates rather than *emergent structures* with sui generis causal powers' (*D*: 319).
paradox of the indeterminacy of explanation. An 'infinite number of descriptions' are compatible with any finite system. But once a higher-order explanans is identified and referentially detached there is no warrant for adding anything superfluous, so the explanation is determinate. When that is itself alethically explained, 'there can be no further *practical* doubt about it' (*P*: 31, e.a.). Hence CR's emphasis on *multi-tiered stratification* (see DEPTH) and ALETHIC truth.
paradox of the preface. See HOLY TRINITY; PARADOX OF THE HEAP.
paradox of the ravens (Hempel's paradox). See NICOD'S CRITERION.
paradox of rule-following or **rule-following paradox**. See PRIVATE LANGUAGE ARGUMENT.
paradoxes of confirmation. See NICOD'S CRITERION for an example.
paradoxes of falsification. If construed actualistically – but not if construed critical realistically – all laws are immediately falsified in open systems.
parallel universes. See PLURIVERSE; TRANSCENDENTAL REALISM.
paramorph. See -MORPH, MORPHO-.
parapanentheism, adj. parapanentheist (f. Gr. *parapan*, on the whole, absolutely, altogether + *en*, in + *theos*, God). Belief in an all-pervasive indwelling or IMMANENT divinity (see *P*: 228); also *panentheism*.

participatory democracy. See EUDAIMONIA; POLITICS.
particular(s). See CONCRETE UNIVERSAL ↔ SINGULAR; MEDIATE; UNIVERSALS/PARTICULARS.
past. See A-SERIES; SPACE–TIME; TENSE.
path-dependency. See CHAOS/COMPLEXITY; SPACE–TIME.
pathological, methodological primacy of the. See ECOLOGICAL ASYMMETRY.
PDM. Acronym for PHILOSOPHICAL DISCOURSE OF MODERNITY.
PEP. Personal emergent power or RELATIONAL property, irreducible to social construction. The concept derives from Archer (2000, 2003b), who uses 'property' and 'power' interchangeably. The key PEPs on Archer's account are *self-CONSCIOUSNESS* or *reflexivity*, *personal identity* and *social identity*, of which the first is of prime importance. See also REFLECTION; SUBJECTIVITY.
perception. (1) The phylogenetically and ontogenetically emergent human capacity to extract and use information about the world through the five senses, together with its exercise in general or in particular instances (the process of perception); also *sense perception*. A vital dimension of *experience*. (2) Understanding (e.g., perception of one's class interests). (3) Intuitive insight. (2) and (3) are not considered further here.

Perceptual REALISM is the view, opposed to (subjective) idealism, that perception gives us access to physical objects that persist in space–time independently of their perception. This is denied by anti-realists, sceptics maintaining that because we never perceive objects directly we can never be sure that they exist; idealists that we can only be aware of ideas or that everything is an expression of ideas. Perceptual realism often goes hand in hand with a *causal theory of perception*, which (where causes are understood positivistically) is the view that the object causes us to perceive it in the sense that we necessarily have a perceptual experience of the relevant sort. Perceptual realism takes three main forms. (a) *Direct realism*, which holds that perception acquaints us directly with objects and FACTS as understood by common sense and positivism (e.g., the cat, and *that* it is on the mat). (b) *Representative* (or *representational*) *realism*, the view that it does so indirectly or mediately via an internal sense-datum underpinning a percept which represents (is like) its external physical cause and is a product of both the object and the conditions in which we view it. R. W. Sellars (1916) called his theory of representative perceptual realism 'CRITICAL REALISM'. Representative realism suffers from the defect of being unverifiable in that, on its own premises, there is no way of directly comparing percept and object (Bhaskar 2003a: 559). (c) *Phenomenalism*, which analyses objects as actual or possible sense-data. This has the disadvantage that there is no way on realist premises of defining sense-data except in terms of material objects, which are not the only kind of object.

It is important to note that Bhaskarian CR, while entailing (without reducing to) perceptual realism, is not a form of (b) (let alone of [a] or [c]), which is framed within the Cartesian–Lockean paradigm of REPRESENTATIONALISM, paradigmatically deploying a model of the individual, contemplative observer divorced from praxis, who can have true knowledge only of the contents of his mind. For CR, by contrast, perception is a dynamic and skilled practical and social activity and accomplishment which may assist in yielding fallible knowledge of the real. While no overall CR theory of perception has yet been elaborated, many of the elements of an adequate theory are arguably present, and Bhaskar has indicated (2003a: 559; *P.* 198 n) that he believes the most promising line of development is suggested by the *ecological perception theory* developed by J. J. Gibson (1950, 1979). Archer (2000: 129, 134) has stressed the role of the body in the ontogenesis of perception. In ascribing reality, CR deploys a perceptual (recognitive) in addition to a causal (demonstrative) criterion. The intelligibility of perception presupposes that the objects (structures, events, behaviours, etc.) that are perceived to exist or occur independently of the perceptual experience, i.e., are existentially if not causally INTRANSITIVE in relation to it. For perception must be of something other than itself, and only thus could it be epistemically significant. Further, for changing perception of the same object (e.g., seeing the sun rise before Kepler and the earth's rim dropping away thereafter) to be possible, objects must be independent of perception (*RS2*: 31). Of course, not all objects and behaviours are perceivable, either intrinsically (e.g., a magnetic field, a social structure, causal powers generally) or because they are beyond the range of our senses; and indeed the continual extension of our powers of perception through sense-extending equipment implies that there is, and probably always will be, a range of entities beyond the reach of those powers. Such objects can thus be *detected* (as distinct from perceived) by material beings such as ourselves only by their effects on material things, and we ascribe reality to them on the basis of a causal criterion alone, via perception of their effects. They can 'only be known, not shown, to exist' (*RS2*: 186). However, because perception is always involved at least indirectly, we can say that in general, for an existential or dispositional (causal powers) claim to be granted, both a demonstrative and a recognitive criterion must be met. This may apply to negative existents (e.g., the erstwhile holes in the New Orleans levies) as well as positive ones (see ONT/DE-ONT). Perception for CR is of course itself caused like anything else, in keeping with ubiquity determinism, but not in the mechanical way presupposed by a Humean understanding of causality, rather conjuncturally in an open-systemic manner by a range of mechanisms pertaining to the natural (the object), practical and socio-cultural orders and the embodied person. Furthermore, it is fallible and may be inadequate or misleading and at the limit false (as in hallucination and illusion), or informed by false and necessary emergent conceptual forms (see TINA SYNDROME), and so subject to CRITIQUE. Finally, Bhaskar agrees with Marx that in power$_2$-stricken society, especially capitalism, the objective phenomenal forms of society itself may be both false and necessary (e.g., the wage- and value-forms) and occlusive of deep structures, such that if we stay at the level of perception, i.e., of the actual/empirical, and do not pursue depth-analysis, we are bound to be misled (*SR*: 304–5; see also DEMI-REALITY; TINA SYNDROME).

Gibson's ecological perception theory was developed independently of Peirce's theory of semiosis (see SIGN), but recognised by Gibson himself to be broadly consonant with it. This in turn is broadly consonant with the semiotic triangle displayed by Bhaskar (Table 32).

Gibson's theory is often classified as a form of direct realism, but this seems misleading. While it does hold that perception is direct in that it is not based on sensation but on direct pick-up of information about invariant properties of the environment, it stresses the dynamic, interactive nature of perception as humans seek out invariants as affordances (e.g., this fruit as edible), while sustaining a distinction

perceptual realism 343

Table 32 The production of perceptual meaning and meaning in general

Gibson	environment source (distal or physical object stimulus)	stimulus invariant (proximal stimulus, the energy impinging on a receptor)	percept (that which is perceived, including affordances)
Peirce	object	representamen	interpretant
Bhaskar	referent	signified	signifier

between the objectivity of the environment and perceptual activity. It thus seems better classed as a *dialectical realism* stressing the developmental interrelation of embodied perception and the environment. This is congruent up to a point with Bhaskar's insistence that some (but by no means all, cf. *RS2*: 240) acts of perception must be basic, in the sense of immediate, for perception to be possible (see MEDIATE), and more so with his PMR view, enshrined in the concept of spontaneous right-action, that there is a non-dual *dimension* to *all* our acts, including perception, implying a general view of perception as a dialectic of immediate and mediate, ground and figure, non-duality and duality in non-arbitrary dialectical interconnection. Immediate perception should not be confused with *epistemic perception* or our *epistemic stance* to the world, whereby we 'read' it as we do a clock, '*as if* it were constituted by facts, i.e., under the descriptions of a theory' (*PF*: 30, 42 n11); such reading constitutes the ideological FACT-FORM. The later Bhaskar stresses that 'we can only understand the world as meaningful in virtue of the *HERMENEUTIC* moment' which is 'paradigmatically the direct reading of the world in a certain way', enabled by unconsciously prestructured 'immediate pre-understandings of the world' (*MR*: 303; cf. COGNITIVE SCIENCE). This is the truth of the idea of *intertextuality* in the linguistic turn, the notion that the world is a text; it is crucial as a starting point for social analysis and critique.

perceptual realism. See PERCEPTION.

performative. A concept introduced into speech act theory by Austin (1962) to distinguish an utterance in which theory and practice are united in that it simultaneously does what it says (e.g., 'I promise' or 'I hereby open this bridge') from a *constative* (f. L. *constat*, it is certain) utterance which merely states something. Constatives, according to Austin, are true or false, performatives neither; but for CR the distinction tacitly endorses a theory–practice, fact–value divorce, for the class of constatives is empty (*SR*: 184–5). All four components of the JUDGEMENT FORM are performative (*P*: 168), and the production of meaning generally or SEMIOSIS is by definition performative (Fairclough et al. 2004). A performative utterance on Austin's later theory is a species of *illocution* or *illocutionary act*. *Performativity* is the quality of being performative. It is made much of within postmodernist subjective idealism, where it often substitutes for a generalised concept of the coherence of theory and practice in practice (see DIALECTICAL REASON): performativity is '*that aspect of discourse that has the capacity to produce what it names*' (Butler 1994, cit. Sayer 2000a: 96). *Performative CONTRADICTION* (also reflexive inconsistency, transcategorial inconsistency) is the basic form of theory–practice inconsistency: the denial in theory or in practice of what is necessary in practice (see also CONSISTENCY; CRITIQUE). The *performative theory of TRUTH* states that to ascribe truth to a statement is to endorse or commend it. Performative can also simply mean 'competent' and 'efficacious', and as such, alongside

344 periodisation

the cognitive, conative, affective and expressive, is one of the five main domains in the Bhaskarian model of the components of ACTION (*D*: 166). See also NATURALISTIC FALLACY.

periodisation. See LOCO-PERIODISATION.
permeation. See RELATIONALITY.
person, personality. See SUBJECTIVITY.
personal identity and **social identity**. See CRITICAL NATURALISM; CULTURE; PEP; SUBJECTIVITY.
personalism. See ETHICAL IDEOLOGIES.
perspectival switch. See PERSPECTIVE .
perspective. Mostly used synonymously in CR with *standpoint* or *vantage point* to refer to (1) any view taken (TD); (2) the framework within some epistemic WORLD-LINE within geo-history from within which any view is taken (TD); and (3) a property of that which is viewed in the ID, not just in the sense that the framework together with its 'contents' can be referentially detached, but also that the world itself is multi-perspectival ('standpointy'), in that it is both multi-faceted, multi-angular or pluriversal and fluid or intra-connected, independently of our subjectivity. (3) is the *constitutive perspectivity* (also *perspectivality*) or intrinsic differentiatedness and inter-/intra-relationality of spatio-temporality (*D*: 174; see also CO-PRESENCE; REFLECTION; STRATIFICATION; TOTALITY); or non-Quinean/non-Kuhnian *ONTOLOGICAL RELATIVITY*. It is important to note that (3) does not follow from (1) and/or (2) – that would be to commit the epistemic fallacy, like Quine (1969) in his concept of ontological relativity and Kuhn (1962) – but from transcendental argumentation establishing the real RELATIONALITY of the world and from relativity theory in modern science. That said, the human epistemological vantage point is itself constellationally contained within the world.

If, therefore, as CR holds, epistemology and method must 'follow the object', changing perspectives – or *perspectival fluidity* and flexibility – is an essential aspect of so doing. A *perspectival switch* accordingly moves, at 3L, (a) from one facet of a totality to another which REFLECTS it, or (b) 'from one *transcendentally* or *dialectically necessary condition* or aspect of a phenomenon, thing or totality to another which is also transcendentally or dialectically necessary for it', as demonstrated by transcendental or DIALECTICAL ARGUMENT (*DG*: 401, e.a.; see also INFERENCE). For example, seeing theory as informing practice, or vice versa, within the DUALITY of theory and practice, effects a *transcendental perspectival switch* (transformative praxis presupposes a theory of transformation and vice versa – see DIALECTICAL REASON), as does seeing alterity as absence or that 'to constrain is to contradict is to absent' (*D*: 281). If you think that social structure and agency are on a par as transcendentally necessary conditions for each other, adopt the perspective of the agent and you will see the massive presence of the past (*D*: 141–2). Seeing the epistemic fallacy as an aspect of anthropocentrism and vice versa effects a *dialectical perspectival switch* (they are, by dialectical argument, false or limited necessities grounded in a common error). Other prominent examples of this kind of switch are the shift from form to content, centre to periphery and figure to ground, in cases where the latter term sustains or grounds the former (e.g., where the centre is sustained by the periphery, or the figure is capital grounded in labour). Perspectival switches may be positive or negative, as the examples show. They may also reverse each other. Thus in the epistemic fallacy a perspectival switch on absence (to presence) eliminates alterity; with the revindication of ontology, a perspectival switch on presence (to absence) reverses this, reinstating alterity. In the latter case the switch is *referential* or realistically *grounded*, in the former not – it is a consequence of 'the subject's epistemic or evaluative interests or her will-to-power (or caprice)' (*D*: 115), valid within its assumptions but ungrounded and irreal. Whether by facilitating immanent critique or emancipatory praxis, perspectival switches, which may be RECURSIVE, play a vital role in articulating the dialectics of freedom. The whole of DCR may be seen, by perspectival switch, 'under the aspect of *Foucauldian strategic REVERSAL*' (*P.* 215, e.a.). *Switch* is sometimes used synonymously with TRANSITION.

Perspectival relativity or *perspectivism* is synonymous with *epistemic relativity* or *relativism*, and consistent with JUDGEMENTAL RATIONALISM. It is established by our THROWNNESS. While our *agentive perspectivality* within the duality of agency and structure encourages switching perspectives from our agency to our thrownness and the apparent immutability of pre-existent structure and context, a further switch to the prefiguration of the future in our transformative praxis can reverse this focus (*D*: 141, 144).

Perspectivism (or *perspectivalism*) is also a variety of the pragmatist theory of TRUTH deriving from Nietzsche and informing post-structuralism, according to which, since there are no facts (including moral facts), only interpretations, truth is 'ultimately an expression of will-to-power, which must be thought, as both necessary and impossible, "under ERASURE"' (*D*: 216). This view, which makes judgemental rationalism theoretically impossible, however much this is contradicted in practice, is arguably self-erasing and a consequence of the absence of a concept of spatio-temporal stratification and distanciation.

Apart from its use as a synonym of perspective, *standpoint* or *AXIOLOGICAL standpoint* refers to the practical interests and commitments deriving from people's experience of class, gender, cultural, occupational and other social relations. *Standpoint theory* is a position within epistemology, well supported in CR, that holds that interested standpoints can provide a more adequate perspective on the social world than ostensibly neutral ones (see also CONTRAST EXPLANATION; FEMINIST THEORY; HERMENEUTICS; KNOWLEDGE).

Wittgenstein (1953: §309) famously opined that the aim of philosophy is 'to shew the fly the way out of the fly-bottle'. Bhaskar reminds us that 'one can only see the fly in the fly-bottle if one's perspective is different from that of the fly' (*RS2*: 8), i.e., the perspective of CRITIQUE and self-reflexivity is indispensable for moving beyond the fly's-eye provincialism of Wittgensteinian FORMS OF LIFE, necessary conditions for which are ontological depth and openness (the bottle is no more appropriate a metaphor than the fly). A metacritique affords a *meta-perspective*. Attempting a God's-eye perspective of the contemplative observer outside the 'bottle' is equally futile, however, if epistemic relativity and the practical materialism entailed by the TMSA are true. Having adopted the perspective of critique, one must, in accordance with the CONSTELLATIONAL containment of epistemology within ontology, *perspectivally retotalise* the scene, i.e., self-reflexively relocate oneself within the 'bottle' as a distinct but essentially related part of what one is viewing, especially in the sociosphere, where 'all investigation has the practical character of a relational DIALECTIC' (*P.* 77).

346 phantasy/fantasy

phantasy/fantasy. See ALIENATION; BEAUTIFUL SOUL, ETC.; CONCRETE UTOPIA; DE-AGENTIFICATION; FISSION/FUSION; INVERSION; VIRTUALITY.

phenomenalism. See PERCEPTION; REALISM.

phenomenology (f. Gr. *phainomenon*, appearance, + *logos*). The study of things as they appear to us – as they are perceived and experienced. In a general sense, *phenomenological* often means 'in terms of (our) experience', or 'descriptive' as opposed to causally explanatory, a contrast deriving from that implicit in the concept of appearances with reality as it is in itself. The *phenomenology of everyday life* refers to our lived experience and consciousness. The *phenomenologicality* of a practice, e.g., science, refers to the lived experience of engaging in it.

(1) In Hegel, the dialectical development of human sentient socialised awareness – as a moment in the ALIENATION and self-recognition of cosmic *Geist* – in its drive to overcome separation of subject and object, towards self-consciousness and freedom as absolute spirit. Each stage of this process unfolds via an Achilles heel CRITIQUE of its predecessor, and the whole constitutes a *dialectical phenomenology*. This is the primary sense in which CR philosophy deploys the concept. 'Following a *phenomenological method*, it starts from actually generated kinds of knowledge and asks what the conditions of their possibility are' (*D*: 232, e.a.). Its task is to make things manifest that were hidden. However, it supplements immanent critique with explanatory critique, repudiating Hegelian notions that dialectical phenomenology is an automatic or parthenogenetic process to which the social world ultimately reduces (see PHILOSOPHY; SPECULATIVE ILLUSION). CR *social phenomenology* is a branch of dialectical phenomenology. Some of Hegel's themes, especially his conception of human subjectivity as situated and embodied, feed into (2).

(2) In the twentieth century, a diverse and complexly developing movement in philosophy (and social theory) inaugurated by Husserl and developed by Heidegger, Merleau-Ponty, Sartre, Levinas and others. It shares important characteristics with HERMENEUTICS, its strengths and some of its weaknesses, and indeed is sometimes characterised as having a hermeneutical wing exemplified by, e.g., Ricoeur and Gadamer, along with transcendental (epistemological), existential (ontological), linguistic and other orientations; and it is an important progenitor of POST-STRUCTURALISM and SOCIAL CONSTRUCTIONISM. Not only are there diverse strands within the movement; there are also significant developments within the thought of each major thinker. Husserl saw himself in the earlier part of his career as the epistemological heir to Descartes, defending the autonomy of the rational subject and philosophy against psychologism and psychology, via a transcendental reduction or bracketing of the reality of objects of everyday acts of consciousnesses in order to investigate the EIDETIC aspects or pure forms of those acts, thus arriving at certain FOUNDATIONS for knowledge. Late in his career he turned to exploration of the *Lebenswelt* (life-world), i.e., of our subjectivity as that of conscious-embodied-beings-in-the-world; and it was this that was mainly developed by his successors, and has had most impact within CR. Heidegger asked how the being of acts of consciousness is related to their objects, thus opening up the terrain of phenomenological ontology.

While contemporary phenomenology is far too proliferous to be caught in a neat definition, it is probably true to say that most phenomenologists share the following insights and/or foci in addition to those phenomenology shares with the older hermeneutical tradition (the pre-understood character of social reality, the non-presuppositionless character of social life and inquiry deriving from our THROWN-NESS, the indexicality of expressions, and the dialogicality of social life and inquiry). (1) The *phenomenological reduction*. While Husserl's particular approach has largely been abandoned, it is important for any thinker to effect a radical change of attitude by turning from Heidegger's ontic to his ontological (see EPISTEMOLOGY, ETC.), i.e., from the realm of things and objectified meanings to the realm of everyday beings and meanings as immediately experienced, in order to achieve an adequate description of our experience as it is in itself. (2) The *aboutness of human intentionality and experience*, our embodied situatedness as natural beings. (3) The question of *the 'meaning of and Being' of beings* (Heidegger). (4) The *privileging of human being* in attempting to answer this question. (5) A certain form of *intuitionism* as 'the principle of all principles': 'whatever presents itself in "intuition" in primordial form (as it were in its bodily reality), is simply to be accepted as it gives itself out to be' (Husserl, cit. *CDP*: 666).

CR shares some of these emphases; Archer in particular has drawn significantly on phenomenological philosophy, especially Merleau-Ponty, in her account of SUBJECTIVITY. However, it criticises phenomenology for tying the explication of Being to human being (i.e., *Dasein*, which is a condition for any being at all) (see ANTHROPISM) and for its proneness to collapse subject and object, epistemology and ontology, language and the world, i.e., for its lack of robust concepts of intransitivity, structure, difference and transfactual efficacy (which are presupposed by its own account of intentionality and the existential), hence of a tradition of EMANCIPATORY AXIOLOGY (see esp. *D*: 229–30). In its tendency to see the life-world as constituted by its occupants, it overlooks that its games and practices are 'always initiated, conditioned and closed outside the life-world itself' (*SR*: 128), i.e., in the past, by social structure and in the future, respectively. In this it remains true to Husserl's uncritical view that philosophy 'does nothing but explicate the sense the world has for us all, prior to any philosophising, and obviously gets solely from our experience – a *sense which philosophy can uncover but never alter*' (Husserl 1929, cit. *PN3*: 155, e.a. by Bhaskar).

philosophical anthropology. Understood broadly as the study of humankind, ANTHROPOLOGY is a constant presence within all philosophical traditions. Yet, while all philosophies will imply *some* kind of anthropology, at *some* levels, some are more explicit about this than others. A characteristic move has been to mark out distinctive features of humanity – whether deriving from the soul, from rational capacities, from the use of language, or elsewhere – which in turn distinguish the human species from others. Yet, especially since Nietzsche, this 'HUMANISTIC' manoeuvre has been put into question. Existentialism resists the reification of any such aspect into a putative human ESSENCE. (Post-)structuralists have presented these features as themselves epiphenomenal: as effects of underlying, perhaps unstable, systems which serve to decentre the subject as the privileged source of agency and meaning. Indeed, thoroughgoing anti-humanisms make a point of avoiding anthropological commitments altogether: for instance Marxisms, like Althusser's, which uphold a certain reading of the later Marx as paradigmatic of the 'subjectless' science of society. Equally, recent philosophers from Habermas to Rorty, and from Rawls to Butler, have invoked a range of reasons why appeal to generalised models of 'HUMAN NATURE' or 'human being' are flawed, and dangerously exclusionary. And yet, like Sartre's existentialism, these positions themselves tend to rest on furtive anthropological commitments: implicit claims about what human beings are like, and how they negotiate their way through the world, are operative even in systems which reject the very idea. As a result, such systems fail to account, for instance, for why it is that beings like us have managed to achieve the patterns of culture which in turn, in some respects, then serve to differentiate us, while other KINDS of being have not.

From a CR point of view, the typical reasons for this blind spot are twofold. The first is a version of the EPISTEMIC FALLACY. Critiques of generalisation about the 'human condition' have tended to take their cue from epistemic claims, and on their basis fallaciously posit various forms of ONTOLOGICAL RELATIVITY. Thus in attempting, rightly, to avoid the projection of chauvinistic, partial models as 'universal' indices of anthropological analysis, many have reverted to a relativism which infers from epistemological differences (arising from tradition, culture or perspective) differences in the structure of reality – human or otherwise – itself (see Quine 1969, for a version of this move, and Davidson 1984, ch. 13, for a critique of the notion of a 'conceptual scheme' on which it depends). The second is the assumption of a 'flat', unstratified conception of human nature. In avoiding crude essentialisms of race, for example, social anthropology has tended, despite itself, to invoke equally crude essentialisms of culture, as if cultures form incommensurable, mutually insulated islets. Viewed thus, the conditions of subjectivity themselves, as well as human NEEDS and capacities in their various aspects, become culturally relative. CR approaches will seek to avoid the reification inherent in the presentation of cultures as closed systems, while at the same time allowing space for substantive anthropological inquiry: particularly, into the ontological aspects of the relation between structure and agency.

As a result, its anthropology owes much to the anthropological commitments, rich but under-theorised, inherent in the work of the early Marx. But it amends them, partly in light of issues raised by clashes between humanist and structuralist interpretations of Marx. Marx's identification of 'species being' in terms of factors with both normative and ontological dimensions – freedom, creative productivity, theoretical activity – provides a backdrop. Marx's early critique of capitalist society hinges on its thwarting of human potential and flourishing, understood in these terms. Crucially, capitalism denies the social aspect of our species being: which itself accounts for the historical variability of human needs and capacities in some respects, but their trans-epochal nature in others. Bhaskar's conception of 'four-planar social being' or human nature (*D*: 152 ff.) might be said to provide a fuller, more nuanced version of the Marxian model. Crucially, it distinguishes $POWER_1$ ('the transformative capacity intrinsic to the concept of agency as such') from $power_2$, the – socially constituted – capacity to 'get one's way' in a social context; this latter being the power inherent in relations of domination. The four planes of human nature comprise (a) material transactions with nature; (b) interpersonal inter- or intra-action; (c) social relations; (d) (stratified) intra-subjectivity. Thus the structure/agency relation is preserved in such a way that moral and social development remain fundamentally open.

Two ambivalences in CR accounts reflect wider debates, in Marxist and ecological theory. First: the extent to which human nature is historicised ('all the way down', or just some of the way down?), or whether there are intrinsic aspects: see Geras 1985, Sayers 1998 for debates within Marxism; Collier 1999, Archer 2000, Norrie 2000a for echoes of this in CR. And, second, the extent to which human attributes make for profound ontological differences between human beings and other species: see Benton 1993, 2001; Soper 1995, 2001. On the 'universalist', humanist side, resisting the overextension of historicism, Archer (2000, 2003b) concentrates on the particularity of human practical orientation towards, and interaction with the world, highlighting core aspects of this which lie prior to the workings of social influence. Archer's notion of the 'internal conversation' is intended as a bulwark against determinism and constructionism, offering as it does a picture in which the human self is the bearer of 'emergent personal powers' which are neither derived exclusively from society (or from biology) nor reducible to it.

GIDEON CALDER

philosophical discourse of modernity (PDM). The revolutionary philosophical discourse, the hallmark of which is the self-defining subject, that accompanied the rise and consolidation of capitalism from its birth in the seventeenth century to its current globalising phase, displacing a philosophical discourse in which people defined themselves in relation to a cosmic order that was intrinsically meaningful, valuable and sacred. Except in the contingent matter of its origins, the PDM is thus in no way a specifically European or Eurocentric phenomenon; it is the philosophical REFLECTION of the root-hogging dynamic of the capitalist mode of production which obtains regardless of which particular CENTRISM holds more superficial sway in any region it invades.

In his later work, Bhaskar explicitly situates the trajectory of CR itself within the wider context of PDM, pointing out that by no means all aspects of the tradition are regressive and that CR itself 'constitutes so many successive critiques' of its phases (*RM*: 166), the rise of each of which was associated with a revolutionary or counter-revolutionary TRANSITION or upheaval. This means (1) that the development of CR is characterised by a process of double immanent CRITIQUE – first of PDM, and second of its own previous phases – isolating and remedying, or suggesting remedies for, incompletenesses in each case; and (2) that CR is not anti-modern in a regressive or undialectical sense.

The phases of PDM (together with their associated socio-political revolution) are: *classical modernism* (CM) (the moment of the bourgeois revolutions of the seventeenth and eighteenth centuries, or 'the advent of world HISTORY', when 'both time and geo-history moved up a gear' [*MR*: 103]); *high modernism* (HM) (the revolutions of 1848 and 1917); *the theory and practice of modernisation* (M) (the defeat of fascism, the revolution of 1949 and the onset of the Cold War and of formal de-colonisation); *postmodernism* (PM) (the revolutionary upheavals of 1968 and the early 1970s in Western Europe, and revolution and counter-revolution in the South, beginning with the Vietnam War); and *bourgeois triumphalism* and its regressive simulacrum, religious and other *fundamentalism* (T/F) (the collapse of the Soviet bloc and the resurgence of imperialism, or the intensification of the second phase of globalisation of capital).

While distinct, all these phases are, like those of CR itself, moments in a developmental process, the later phases of which partially critique and deepen what was implicit in the earlier phases. They are unified above all by a 'tremendous' (*RM*: 123) underlying error: ONTOLOGICAL MONOVALENCE, leading to triumphalism and

350 philosophical discourse of modernity (PDM)

ENDISM (already implicit in the first or base phase), and so constitute a dialectical TOTALITY.

The relationship between these phases and the developing CR non-preservative SUBLATORY critique is indicated schematically in Table 33.

Classical modernism is characterised by (1) *ego- and anthropo-centricity* coupled on to (2) *abstract universality*, both of which are false or illusory; 'indeed, the whole PDM is constituted by the couple of an *ego*, be it an individual or a group, class, gender, nation state (or some complex of these) objectively set against a *manifold*, described in *actualistically universal* terms {presupposing external RELATIONALITY}, which is the object of the ego's action (manipulation and exploitation)' (*RM*: 168, e.a.). In the sociosphere this is manifested in the fact that the discourse both excludes (thereby experiencing difficulty sustaining its own universality) and tacitly includes (thereby losing the category of the modern) a pre- or non-modern other. 'Thus we have the figure of the *intrinsic exterior* or *past*, implicitly secreted in the discourse, below the level of consciousness, as the necessary condition of that discourse, reflecting the exploitation of the excluded in reality, whether the excluded be other individuals, non-bourgeois classes, women, colonised peoples, etc.' (*RM*: 168–9) – a classic example of a TINA FORMATION. CM's totality is therefore (3) detotalised or *incomplete* and *self-refuting* (self-contradictory), and (4) lacking in *REFLEXIVITY*.

These features of CM were the focus of the critique of CM by *high modernism* (HM), broadly endorsed by CR, which reached its apogee philosophically and theoretically in MARXISM and IDEOLOGY CRITIQUE. However, in its proneness to 'SUB-STITUTIONISM' and elitism, deriving from the lack of an organic intelligentsia in the Gramscian sense, HM is itself vulnerable to some aspects of its own critique. HM, like CM, is also – with important exceptions – typically reductively materialist in outlook. The CR critique of this position was inaugurated by the demonstration of the sui generis reality of mind/thought in *PN*.

Modernisation theory and practice was characterised by (5) *unilinearity*, whereby 'developing' countries would inevitably pass through the same stages of economic and political growth as the Western world, and history as a whole is a story of unilinear progress, with Western countries in the vanguard; CR critiques this as a variant of HISTORICISM (in Popper's sense). Intrinsic to unilinearity was (5') a *judgementalism* – completely at odds with modernity's own prevalent view that rational judgements concerning matters of value are impossible – whereby 'developed', 'Western' and 'modern' are superior to anything inconsistent with themselves; and (5'') an accentuated *disenchantment* of the world which had been present in PDM from the outset, whereby the world was increasingly drained of intrinsic meaning and value which were instead sourced to the self-defining modern subject.

Postmodernism, along with the 'new' social movements which accompanied it, is perhaps above all a reaction against the abstractly (actualistically) universalising tendencies of PDM, obliterative of identity and difference, which came to a peak in modernisation theory and practice. 'Unfortunately, in the very accentuation in this phase of a new politics of identity and difference the interconnectedness and unity of humanity and indeed living forms was lost. What was missing here was any conception of a dialectical totality, with the crucial concepts of *dialectical universality* and *concrete singularity* absent' (see CONCRETE UNIVERSAL). That is to say, postmodernism fails to

Table 33 The philosophical discourse of modernity and the CR critique.

	The Philosophical Discourse of Modernity (PDM)		The Critical Realist and PMR Critique		
Moment of PDM	**Defining characteristics**		**CR critique**	**Moment of CR**	**Main stadion and concept(s)**
Classical Modernism (CM)	(1) egocentricity (2) abstract universality		ditto	TR	1M ontology being as structured, differentiated and changing
High Modernism (HM)	(3) incomplete totality (critique of CM) (4) lack of reflexivity (critique of CM)		ditto + substitutionism, elitism, critiques reductive materialism	CN	2E absence, contradiction, process irreducibility of mind/thought
Modernisation theory and practice (M)	(5) unilinearity (5') judgementalism (5'') disenchantment		ditto	EC	3L totality, holistic causality, explanatory critique
Postmodernism (PM)	(6) formalism and (6') functionalism (critique of PDM, stressing identity and difference, and rejecting universality) (7) materialism		ditto + accepts difference but reinstates unity or (dialectical) universality (connection)	DCR	4D transformative praxis and reflexivity, unity-in-diversity
Triumphalism (endism)/ renascent Fundamentalism (T/F)	(8) a purely positive account of reality (ontological monovalence)		ditto + accentuated critique of materialism (implicit consciousness pervades being; direct mind-to-mind causation). False absolute of fundamentalism	TDCR and PMR	5A, 6R, 7A reflexivity, spirituality; enchantment; non-duality (invoking the [never *ending*] infinite or absolute), transcendence, generalised co-presence

Note Columns should be read developmentally, such that (broadly) T/F > PM > M > HM > CM, and PMR > DCR > EC > CN > TR.

distinguish between abstract and dialectical universality, 'jettisoning universality rather than ACTUALISM!' (*RM*: 171, 236). It denegates universality, but is itself a universal modern ideology. The era of postmodernism brought out (6) the *formalism* and (6') *FUNCTIONALISM* of the whole tradition, which assumed the plasticity of nature, treated as an object of instrumental reasoning and practice, together with (7) a *materialism* that was reductionist in theory and mechanical in practice. Philosophically, mechanical materialism downplayed the role of ideas and intentionality in geo-history; substantively, it hinged on a sense of the utter separateness of the ego from the rest of the world, which underwrote manipulative treatment of people as well as world and set up 'a master–slave relationship between the ego or the modern, and the non-ego, other or non- or pre-modern' (*RM*: 171). Postmodernism rejected and displaced this ego, but 'the bearer of the deconstructive discourse remains mysterious, unsusceptible to reflexive situation' (*RM*: 70; see also SUBJECTIVITY). Postmodernism's inability to sustain a coherent totality was mirrored in post-structuralism (also evident in STRUCTURALISM), which detotalises the observer from the field she observes. Neither is therefore able 'to sustain any notion of itself' (*RM*: 172). Crucially, the relationship postmodernism postulates between the postmodern and the modern duplicates the ideological relationship between the modern and the non-modern in PDM as a whole, and its denial of ontology and of the possibility of a judgementally rational assessment of other positions entails that it cannot sustain its own reality, causal efficacy and rationality.

Bourgeois triumphalism is characterised above all by ONTOLOGICAL MONOVALENCE and ENDISM (the entries for which may be consulted for the CR critique), as the ideology of NEO-LIBERALISM proclaims the end of ideology. It may well be, as Bhaskar has suggested, that 'this logic of the "end" has itself a clearly identifiable end. For when everything is commodified, that is, reified and turned into a thing, then the process of commodification must come to a halt; and since there will be nothing more to commodify, there can be no further basis for commodification, that is, the expansion, and thus the very survival, of the regime of commodification, reification and alienation. The only question is whether this regime comes to a halt before the process of commodification is absolute – for when it does, or if it did, there would be no more nature and nothing more to count as a human being' (*RM*: 173). The philosophical critique of regressive *FUNDAMENTALISM* (a cousin of postmodernism, which like it rejects unity and accepts difference, but says 'I'm right and you're wrong' [*RM*: 97]), turns on its provision of a 'false ABSOLUTE' which, summonsed to account by the false bourgeois absolute of power and money (an abstract universal), suppresses creativity, peace, love and freedom (*RM*: 242). See also Taylor 1975, pt I: I.

philosophical ontology. See EPISTEMOLOGY, ETC.; THEORY, ETC.
philosophical psychology. See MIND; PSYCHOLOGY.
philosophy. On the Bhaskarian account, the 'components of an adequate practice of philosophy' (*P*: 214–15) at the present time are:

(1) *Lockean underlabouring* for substantive emancipatory theory and practice, i.e., research and practical programmes oriented to genuine discoveries and/or absenting of constraints on human flourishing and freedom. This involves 'removing a priori objections and clearing conceptual bottlenecks and methodological roadblocks in their way' through formal transcendental investigation of their presuppositions. Occasionally

philosophy might also act as midwife to new sciences and research programmes, especially in the area of the human sciences in their 'perplexed and perilous state', by showing how the cleared or clarified conceptual space could be used 'by choreographing a few sequences for their play' (*PN3*: 167). Underlabouring is the province of *metaphysics* α. This seeks to show the way the world must be for scientific and other practices as conceptualised in experience to be possible, i.e., to elaborate the philosophical ontology or CATEGORIAL structure of the world that they logically presuppose; it concomitantly issues in immanent CRITIQUE of philosophical accounts or scientific and other practices that are incompatible with those presuppositions and therefore inimical to genuine advances in understanding, thereby avoiding the 'bad circularity' of external criteria of knowledge (*SR*: 14). CR philosophy is thus 'critical philosophy *of* and *for*' the sciences (*PN3*: 179) and '*against* the ideologies that threaten them' (*RS2*: 262). (This by no means implies agreement with all that passes for science, nor that science is more worthy than other social practices.) It aspires in particular to contribute to the development of 'a general meta-theory for the *social* sciences', which are lacking in such, 'on the basis of which they will be capable of functioning as agencies of human self-emancipation' (*D*: 2, e.a.).

(2) *Leibnizian metaphysics* β, 'involving the critical assessment of the conceptual and moral hard-cores of research and practical programmes', with emancipatory intent. Whereas metaphysics α 'explicates the presuppositions of practical activities' (and may therefore be thought of as the philosophy of science and other activities), metaphysics β 'elucidates the structure of conceptual fields' in science and elsewhere, i.e., of scientific and other ontologies (and may therefore be thought of as the philosophy of the philosophy of science and other activities). Conceptual schemas cannot be experimentally tested because they provide the very framework within which tests occur, hence epistemic indeterminacy is the order of the day in science (theories are underdetermined by evidence) (see DETERMINATION); neither are they a priori demonstrable. So they can only be assessed indirectly in terms of their developmental consistency, empirical fertility, etc. Philosophy makes its contribution to this via the Leibnizian art of decoding and deciphering conceptual schemes, i.e., deconstructing/reconstructing them, and probing their RESONANCE with their social context. This necessarily involves Aristotelian PROPAEDEUTIC, i.e., both formal logic and its dialectical suspension; and William Morris-style CONCRETE UTOPIANISM in regard to practical programmes. Like metaphysics α, metaphysics β is purely immanent and geo-historically relative. Each interacts with the other. Thus there may be transcendental inquiry into the presuppositions of a scientific conceptual scheme (metaphysics α of β), and transcendental conceptual analysis of a philosophical ontology (metaphysics β of α).

(3) *Kantian transcendental argument*, extended to include Hegelian dialectical PHENOMENOLOGY (i.e., metacritique$_1$, incorporating immanent critique) and Marxian explanatory CRITIQUE (metacritique$_2$) (see also INFERENCE), which 'materially circumscribes' it, repudiating Hegelian assumptions about the final self-sufficiency of philosophy (see SPECULATIVE ILLUSION). The main deficiencies identified in Kant's transcendental procedure are its lack of a conception of a knowable domain of the real (its subjective idealism) – which would ensure that 'what is apodeictically demonstrable is also scientifically comprehensible; that is, what is

synthetic a priori is also (contingently) knowable a posteriori' (*PN3*: 6) – and its lack of a non-anthropocentric and social relational perspective.

(4) *Socratic/Nietzschean gadflying* 'on the neck of the powers that be', incorporating 'both Bachelardian value-commitment and Gramscian optimism of the will and *realism*, informed by concrete utopianism, not pessimism, of the intellect'. Philosophy depends upon the extra-philosophical practical commitments and interests of people (see PERSPECTIVE), which cannot be justified by philosophy alone. Whether mystifying or enlightening, philosophy makes a difference to the world causally.

(5) *Derridean critique of the* METAPHYSICS OF PRESENCE, 'entailing a PERSPECTIVAL SWITCH from every CENTRISM to its periphery, every figure to its dialectical ground'; '*Foucauldian counter-conduct*, entailing resistance to every repression {see REVERSAL}; and *Habermasian* DIALOGICALITY, entailing a willingness to communicate with and listen to all'. DCR as such may indeed be seen as counter-conduct and strategic reversal: of irrealism in philosophy, of master–slave-type relations and constraints on freedom in society.

Metaphysics α and β are (severally and together) radically distinct from traditional metaphysics (what Locke referred to as *master-science*) which proceeds deductively from a small number of ostensibly self-evident and indubitable axioms (cf. FOUNDATIONALISM) to give an account of the necessary truths of reality as such and in general. Such a philosophy sees itself as autonomous, contemplating 'a distinct transcendent realm of its own', 'privileged by some special (high) subject-matter or (superior) mode of truth'. On the CR account, by contrast, 'there can be no philosophy as such or in general, but only the philosophy of particular, historically determinate, social forms' (*SR*: 12): philosophical ontology 'does not have as its subject matter a world apart from that investigated by science; rather, its subject matter just is that world, considered from the point of view of what can be established about it by philosophical argument' (*RS2*: 36). Philosophy is thus *heteronomous* in that it is possible only in relation to praxis other than itself, i.e., must take as its premises existing conceptualisations of other human practices (avoiding commitment to the contents of particular theories), and so is socio-historically conditioned and relative, SYNCATEGOREMATIC, IMMANENT, THIS-WORLDLY and open-ended, establishing 'conditionally necessary {synthetic A PRIORI} truths about our ordinary world as investigated by science' and other practices (*RS2*: 52). This means that it must in the long run be consistent with the results of science, in two respects: in its claims about the world (for, as already intimated, what is synthetically a priori demonstrable is contingently knowable a posteriori) and in its claims about itself (for like any other practice it is susceptible to scientific analysis and explanation). Pertinent here also is the consideration that transcendental arguments are 'merely a species of the wider genus of retroductive argument characteristic {. . .} of scientific activity generally' (*SR*: 11), and indeed of a wide range of human cognitive practices throughout geo-history. Philosophy is not, however, reducible to science or other practices; rather, it enjoys a RELATIVE AUTONOMY which complements, and is arguably indispensable to, them. For, unlike most of the practices on which it depends, it operates a priori, by pure reason (albeit not by pure reason alone); like them, it produces knowledge, 'but it is knowledge of the necessary conditions for the production of knowledge – second-order knowledge' (*PN3*: 8). It may therefore be viewed as a defender of REASON, which leads, via the removal of obstacles to its exercise, in the direction of FREEDOM. Like any cognitive process, it is irreducibly normative (see INTRINSIC/EXTRINSIC).

Both distinct from and interconnected with other practices (indeed internal to the subject-matter of the social sciences, though not of the natural), then, philosophy may be said to be in constellational or dialectical UNITY with them, under the sign of the 'relative primacy' (*SR*: 236 n) of science broadly understood. In the case of method, for example, 'just as there can be no discourse on method in abstraction from the sciences, so there can be no science in abstraction from the possibility of a critical discussion of its method. This is the *methodological circle*, twin-screwing philosophy and science' (*SR*: 19). In delineating the general contours of the categorial structure of the world – to take another example – philosophy illumines the AXIOLOGICAL NECESSITY informing the particular practices of everyday life, thereby promoting self-conscious understanding of what we are doing in our activities. Since its fundamental procedure is immanent critique, CR philosophy has nothing in common with the speculative illusion. This is in no way incompatible with conjecture which, in an effort to break decisively with the philosophy of the status quo, is increasingly resorted to in the later works.

For the CR account and critique of philosophy as it has been practised in the West, whose real definition is 'the Janus-faced {progressive/regressive, rational/rationalising} aporetic and generally unconscious normalisation of the status quo ante' (*P*: 216), see EMPIRICISM, IDEALISM, IDENTITY THEORY, IRREALISM, PROBLEM, RATIONALISM, PDM, VALENCE (ontological monovalence), together with cross-references.

philosophy, problems of. See PROBLEMS OF PHILOSOPHY.
philosophy of history. See FUTURES STUDIES; GEO-HISTORICITY; HISTORIOGRAPHY; and *DCR passim*.
philosophy of identity. See IDENTITY.
philosophy of law. See LEGAL STUDIES.
philosophy of meta-Reality. See META-REALITY.
philosophy of mind. See MIND.
philosophy of religion. The branch of philosophy that addresses questions such as: Does God exist? What can we know about God? What is the relationship between God and nature, and between God and human beings? What in general is the status of religious language, belief and practice? Is religious belief compatible with science?

Materialist worldviews that accompanied the rise of modern science tended not to allow any place for God and to portray religious convictions as simply false and religious practices as relics of the 'unenlightened' past. Some more recent philosophical viewpoints, however, have denied that religious claims can be considered to express either truths or falsehoods. Logical POSITIVISM, for example, drawing on the verification theory of meaning, portrayed religious claims not so much as false, but as unverifiable and therefore meaningless. In contrast, post-Wittgensteinian approaches consider them to be meaningful in virtue of being part of a coherent LANGUAGE-GAME, but one in which assertions of truth (or falsity) do not occur. Prayerful invocations of God, e.g., do not presuppose the truth of God's existence or of any religious tradition's doctrines about how God relates with human beings. Instead they serve social functions, such as calling on a community to centre itself or to remember its own best nature. Ludwig Feuerbach (1804–72) held something like this view as well when he argued that religious claims should not be judged in the same way as scientific ones.

These approaches notwithstanding, much philosophy of religion proceeds in a realist vein, inquiring, e.g., whether or not there are good reasons to accept that God really exists or that the doctrines of any religious traditions can claim the status of truth. Thomistic philosophy is a case in point; it is well known for its defence of the 'five ways', five 'proofs' of the existence of God, proposed by St Thomas Aquinas (1224–74) (Kenny 1981). From a different perspective, connected with its rejection of reductive metaphysical materialism and of positivism, CR also provides justification for proceeding in this vein. The recent spiritual turn in CR, taken by many people associated with Roy Bhaskar (*EW, SE, RM*; see also Hartwig 2001), illustrates that CR can provide a friendly context for entertaining religious views. Long before Bhaskar's spiritual turn, however, the physicist and philosopher of religion Ian Barbour (1966) used the phrase 'critical realism' to describe a philosophical position, distinct from NAÏVE REALISM, which denied that there is immediate access to the world as it is in itself. At the same time, in contrast with idealist thought, Barbour's CR affirmed a world ontologically independent of human observation, which, moreover, humans could know under fallible, historically conditioned descriptions – and, for Barbour, it remains an open question whether or not this world may contain religious realities. Barbour's variety of CR remained little elaborated, and almost unknown among those who have used the term 'critical realism' to label the more elaborated philosophical position developed by Rom Harré and Bhaskar (Porpora 2001a; Shipway 2000, 2002).

For a long time, the latter version of CR had its greatest influence in the social sciences – particularly the line advanced by Bhaskar, which in some ways can be considered an articulation of the philosophy of science implicit in Marxist thought. From this perspective, it may seem ironic that CR has recently made a spiritual turn. It is a turn, however, that has been accompanied by a reassessment of Marx as a spiritual thinker, and one that makes it possible to interpret CR, not in a purely secular way, but in a way that provides insight into the possibility of truth in religion.

There are various, divergent lines of argument associated with the spiritual turn in CR. Bhaskar (*D, EW, SE*, 2002d, e) himself has argued that CR by itself is incomplete, that it needed to be supplemented first by *dialectics* – dialectical critical realism (DCR) – and ultimately by considerations of *transcendence* – transcendental dialectical critical realism (TDCR) and the philosophy of META-REALITY (PMR). In his most recent writings, TDCR and PMR incorporate many principles from Hinduism; PMR aspires to articulate a position 'within the bounds of secularism, consistent with all faiths and no faith' (*RM*: 93). In contrast, Margaret Archer, Andrew Collier and Douglas Porpora together advance a position, more in the spirit of Barbour, that CR is a philosophy of science more hospitable to religious truth claims than either positivism or post-Wittgensteinian philosophy, but that it does not in and of itself offer any way to adjudicate among those claims (Archer, Collier and Porpora 2004; Collier 2001a, 2003a; Porpora 2001c). These critical realists have offered important accounts of the nature of religious experience and its epistemic significance. Additional work on CR and religion has followed. Porpora (2003) and Hugh Lacey (2003a), e.g., have explored affinities between CR and liberation theology that derive from their common interest in EMANCIPATORY PRACTICES, and A. E. McGrath (2001, 2002, 2003) has used Bhaskar's CR as the philosophical underpinning of a major theological treatise (Shipway 2004). See also GEOGRAPHY OF RELIGION; IDEALISM VS. MATERIALISM.

HUGH LACEY, DOUGLAS V. PORPORA

philosophy of science. The branch of philosophy that deals with philosophical and foundational problems that arise within science. It can be divided into two major strands. The one may be called general philosophy of science (or methodology), while the other comprises philosophies of the individual sciences (physics, biology, psychology, economics, etc.). Since the object of general philosophy of science is science in general, its central target is to understand science as cognitive activity. Some of the central questions that have arisen and been thoroughly discussed are: What is the aim and method of science? What makes it a rational activity? What rules govern theory-change in science? What constitutes success in science? How do scientific theories relate to the world? How are concepts formed and how are they related to observation? What is the structure and content of major scientific concepts, such as causation, explanation, confirmation, theory, experiment, model, reduction and so on?

These kinds of questions were originally addressed within a formal logico-mathematical framework, which culminated with the school of logical EMPIRICISM. The central thought of this school was that philosophy of science should be seen as the *logic of science*: the logical-syntactic structure of the basic concepts of science should be laid bare so that their conditions of application should be transparent and intersubjectively valid. Taking a cue from David Hilbert's (1862–1943) formalist approach to mathematics, the logical empiricists aimed at a rational reconstruction of scientific theories as formal axiomatic systems. But, being empiricists, and influenced by the French CONVENTIONALIST philosophers Pierre Duhem (1861–1916) and Henri Poincaré (1854–1912), they took theories to be systems of hypotheses whose ultimate aim was to save the empirical phenomena. The central problem that then arose concerned the status of theoretical terms, that is, of terms which, on the face of it, purport to refer to unobservable entities and events. Logical empiricists favoured the possibility of strict verification by observation as the criterion of meaningfulness, but they soon liberalised it, as it led to the absurd conclusion that most scientific assertions (including empirical generalisations) are meaningless. However, the logical empiricists remained committed to the view that a central task of philosophy of science was to show how a priori knowledge was possible and to separate the analytic (or conceptual) from the synthetic (or factual-empirical) content of scientific theories. Concomitant with this view was the thought that philosophy of science itself is a largely a priori conceptual enterprise aiming to reconstruct the language of science. Following Gottlob Frege (1848–1925), such thinkers rejected psychologism (or naturalism) in favour of justificationism, viz., the view that the target of philosophy of science is to lay down criteria as to how scientists ought to reason, what kind of methods they ought to use and what content the basic scientific concepts ought to have. Consequently, they sharply separated the *context of discovery* from the *context of justification*. This project culminated in Rudolf Carnap's (1891–1970) attempt to devise a formal system of inductive logic and in Carl Hempel's (1905–97) deductive-nomological model of

EXPLANATION. Though Karl Popper's (1902–94) critical rationalism put forward a different conception of scientific method, based as it was on the idea of falsifiability of scientific hypotheses and on the rejection of inductivism, it did share with logical empiricism the hostility to naturalism and the view that philosophy of science is, by and large, a normative and a priori enterprise.

This conception of philosophy of science, dominant until the 1950s, was strongly challenged by three important and influential thinkers. First, Willard van Orman Quine (1908–2000) repudiated the analytic–synthetic distinction, thereby showing that issues about fact and issues about meaning are *not* as sharply separated as it was taken to be by his predecessors. He advanced semantic holism and opened the way for a new conception of how the meaning of theoretical terms is specified: they gain their meaning by being embedded in a theory and by contributing to the theory's empirical content. He also rehabilitated NATURALISM, viz., the view that philosophy is continuous with science and that there is no special philosophical method (by means of a priori conceptual analysis) in virtue of which philosophical knowledge is distinct from, and superior to, the empirical knowledge offered by the sciences. Second, Wilfrid Sellars (1912–89) (1963) attacked INSTRUMENTALISM (the view that scientific theories are mere instruments for the classification and prediction of observable phenomena) and rehabilitated REALISM. He also repudiated FOUNDATIONAL forms of empiricism by revealing and rejecting *the myth of the given*, viz., the view that experiential episodes (the given) directly justify some elite sub-set of one's belief. Finally, Thomas Kuhn (1922–96), in his classic *The Structure of Scientific Revolutions*, showed how any adequate model of science and of scientific knowledge should respect, and be informed by, the actual history of science. This historical turn repudiated the view that philosophy of science was a purely conceptual activity. Kuhn went so far as to challenge the view that theory-change in science is a rule-governed activity, and, taking a cue from Duhem, stressed that understanding science goes hand-in-hand with understanding the role that values play in scientific activity.

In light of these developments, though the central questions that occupy philosophy of science remained intact, the methods followed to address them, as well as the answers that were thought to be legitimate, changed dramatically. What may be called *the naturalist turn* in the philosophy of science allowed the findings of the empirical sciences to have a bearing on, perhaps even to determine, the answers to standard philosophical questions about science. For instance, after a long philosophical ban on CAUSATION, based on empiricist misgiving about the legitimacy and usefulness of the concept, it was taken to be the case that an appeal to causation can cast light on a number of important philosophical issues such as the justification of beliefs, the reference and meaning of theoretical terms and the nature of scientific explanation. As a philosophy of science, CR contributed a lot to the rehabilitation of a non-Humean concept of causation, the defence of realism and naturalism. One particularly interesting strand in the naturalist turn favoured the use of findings in cognitive science in an attempt to understand how theories represent the world, how theories relate to experience and how scientific concepts are formed. Another one was the socialisation of naturalism, given the fact that science is a social activity that takes place within society. Until the 1970s, philosophy of science was still preoccupied with grand theories of how science grows and how theories change. Kuhn himself offered such a well-known theory, based on the thought that long periods of normal science, governed by a dominant paradigm, are punctuated by short but turbulent periods of scientific revolutions which engender new and competing paradigms. Imré Lakatos (1922–74) devised his *methodology of scientific research programmes* in an attempt to combine some of the insights of the Popperian view of science, most notably the thought that theories are abandoned because they are in conflict with experience, with the Kuhnian view that there are no algorithmic rules that govern theory-change.

Though it is noteworthy that these theories were far from the a priori models devised by the logical empiricists and Popper, informed as they were by the actual history of science, they quickly ran out of steam. One issue that became prominent was that the individual sciences might not be similar enough to be lumped together under the mould of a grand unified scheme of how science works. Leaving behind macro-models of science, philosophers of science started to look into the micro-structure of individual sciences. In particular, though physics had been taken to be the paradigm case of science, the diversity among the sciences and the collapse of simple-minded reductive and hierarchical accounts of how science was ordered, led to a mushrooming of interest in the so-called special sciences (biology, psychology, economics, etc.). More special questions started to emerge. Some of them relate to the basic conceptual structure of the sciences (e.g., questions about the concept of evolution, or the nature of psychological explanation, or the status of economic models). Others relate to the commitments that flow from the individual sciences (e.g., Are there laws in the special sciences? What is the status of causal mechanisms? Are there natural kinds in the special sciences?). In a sense, these more specific issues were always integral to the philosophy of science, as for instance can be evinced by the debate among strands within neo-Kantianism about the differences between the natural and the human sciences. But it is also true that the philosophies of the individual sciences have recently acquired a kind of unprecedented maturity and independence. The old positivist dogma of a unified science is still quite attractive. But the more pluralistic view of science that has recently come to the fore suggests that philosophy of science may well abandon the search for master-key answers to general philosophical questions and concentrate on tailor-made answers to specific philosophical questions that arise within the sciences. See also PHILOSOPHY OF SOCIAL SCIENCE; TRANSCENDENTAL REALISM.

STATHIS PSILLOS

philosophy of social science centres upon a number of controversial yet basic questions about the nature of social science. Are the social sciences really scientific? Could they become so if they are not already? Is it *desirable* that the social sciences attempt to be scientific? These questions have given rise to a series of debates virtually as old as the social sciences themselves. Historically these questions have been asked and answered most frequently in relation to natural science; that is: Can we understand social scientific methodologies and explanations as being scientific in something like the same manner as we understand natural science scientificity? Two diametrically opposed answers with respect to the 'possibility of NATURALISM' define the broad contours of debate. Either the social sciences can be scientific in more or less the same fashion as natural science or they cannot (and as an important corollary to this position *should not* attempt to be). While these two positions are directly opposed to one another on one level, they share a common assumption on another: they both accept that the positivist (see below) understanding of natural science is essentially correct. CR strongly argues for the possibility of a scientific understanding of social reality but utterly transforms this debate by thoroughly critiquing the positivist understanding of natural science. Bhaskar's *RS*, as well as *PN*, is thus a seminal text in the philosophy of social science as it is held by critical realists that in order to evaluate the similarities and differences of natural and social science sensibly one must have a correct understanding of natural science to begin with. The 'possibility of naturalism', that is, the possibility for social science to be practised and understood as being scientific in something like the same fashion as natural science, is affirmed by CR. None the less, it grounds this possibility in the particularities of social-scientific subject-matter and our epistemological relationship to it. Social-scientific scientificity is possible precisely because the very characteristics of the human condition that enable social life to be possible also allow its systematic comprehension by human beings. This radically overturns the other side of the debate as well, as the interpretivist understanding of social science denies the possibility of a scientific approach because of such characteristics of social reality.

POSITIVISM was a term first coined by the early sociologist (sometimes referred to as the 'father' of sociology) Auguste Comte, who was a strong believer in both the desirability and the real possibility of sociology being scientific. However, the positivist philosophy of science (in a curious twist of terminology) has actually been most influentially affected by the twentieth-century critics of logical positivism: Karl Popper and Carl Hempel. Their view of science is seen in terms of hypotheses and falsification: the hypothetico-deductive understanding of scientific explanation. The most significant hallmark of scientificity from this perspective is 'risk of prediction'. Scientists produce experimental closure and an attempted falsification of hypotheses. Our scientific explanations and laws are thus our best unfalsified hypotheses concerning generalisations regarding constant conjunctures of events. This perspective accordingly places enormous stress upon a relative symmetry of explanation and prediction. CR demonstrates that this entails an impoverished ontology and a mystification of what actually makes scientific experiment intelligible. Rather than emphasising the empirically observable 'constant conjunctures of events', CR emphasises 'open systems' and a depth-ontology. Scientific explanations go beneath the empirical surface in order to produce explanations of real causal mechanisms. This reflection upon the actual practice of experimentation and explanation go much deeper than the traditional critiques of positivism propounded by the interpretive tradition in the philosophy of social science. These do not assert positivist error with respect to natural science, only the impossibility (and lack of desirability) of its application to social science.

The interpretivist (or HERMENEUTIC) understanding of social science denies the possibility of naturalism for two reasons. First, it notes the centrality of experimentation to natural science and correctly asserts that experimentation is most frequently *not possible* in the social sciences, for both practical and moral reasons. Positivist social science strives for procedures *analogous* to experimentation where such is not possible (as is usually the case). These procedures most commonly involve statistical analysis. The emphasis upon prediction is maintained, as is an emphasis upon measurement and quantification. Interpretivists find this to be on the one hand, dubious, and on the other a misdirecting fetishisation of quantification and prediction, which directs our attention away from what is truly interesting in the social sciences: the understanding of human motivation and agency. A frequent interpretivist critique of positivist social scientific practice is that it rigorously proves only what is already obvious. This perspective preferences *understanding* over purely *causal* explanations and is focused upon the subjective meanings social agents attach to their actions. We can note here that for the critical realist this opposition of understanding and causal explanation does not stand – understanding most significantly involves the comprehension of REASONS; and reasons that are acted on can themselves be causes.

Second, the interpretivist perspective denies the possibility of naturalism because of the unique features of social-scientific subject-matter and crucial differences in the relation of social scientists to their objects of knowledge, i.e., they themselves form a part of it. The reality of human agency means that our social reality is an open system and thus CONSTANT CONJUNCTIONS of events are seldom to be found. Frequent novelty of situation thus renders prediction impossible. Instead, the interpretivist perspective notes that 'rule following' is central to human social behaviour. The possibilities, of either breaking rules or being misinformed concerning them, are only the most obvious manners in which rule following may be distinguished from law like empirical generalisations. However, as can be readily seen from what has gone before, scientific explanation is *not* most significantly a generalisation from empirically observed constant conjunctions of events. The natural world, as well as the social world, is an open system. Novelty is thus not an impediment to scientificity in either realm. In fact, none of the unique features of social reality emphasised by the interpretivist position as presenting barriers to the possibility of social science actually do so when looked at from a CR perspective. Rather, the possibility of social science is grounded in the very nature of the human condition. The same features of social life that render it practically possible and intelligible on a practical daily-life basis are what render it intelligible scientifically.

The CR perspective would thus utterly transform all the principal features of debate between positivists and interpretivists. Let us take prediction as an example. Positivists make it the hallmark of good scientific explanation. The interpretivists assert its impossibility in social science. CR dethrones it from its methodological pride of place on the one hand, while asserting its practical possibility on the other. The imbrication of social scientists with their subject-matter that forms the basis for the interpretivist assertion of the impossibility of social science becomes for the critical realist a condition for its possibility.

The interpretivist tradition in the philosophy of social science has had more recent variations deriving from POST-STRUCTURALISM and POSTMODERNISM. Many of these positions have exaggerated the interpretivist concern regarding the possibly pernicious nature of *even attempting* scientificity in the realm of social explanation. CR thus serves here, as in other areas, as a valuable corrective to postmodernist theoretical excess. See also CRITICAL NATURALISM; TRANSCENDENTAL REALISM; Benton and Craib 2001; Hempel 1965; Popper 1934, 1963; Potter 2000; Winch 1958.

GARRY POTTER

phrastics. See ONT/DE-ONT.

phronesis (Gr. *phronēsis*, good sense or practical wisdom). One of the four cardinal virtues argued for in Plato's *Republic*, alongside justice, courage and balance, 'necessary for DIALECTICAL RATIONALITY, and, as the supreme meta-ethical virtue, for the two-way DIALECTIC OF TRUTH AND FREEDOM. Phronesis almost always requires a degree of *SOPHROSYNE* or balance, thus tying in with the themes of TOTALITY, wholeness and health' (*DG*: 401).

phylogenesis. See ONTOGENESIS/PHYLOGENESIS.

physicalism. See MATERIALISM; MIND.

plasticity. A 2E figure of OPPOSITIONALITY signifying 'susceptibility to a multiplicity of incompatible positions' (*P*: 243). See also DUPLICITY.

Platonic anamnesis. See ALETHIA; MENO'S PARADOX.

Platonic–Aristotelian fault-line. See PRIMAL SQUEEZE; UNHOLY TRINITY.

plenitude. See PRINCIPLE OF PLENITUDE.

pliability ideological plasticity. See DUPLICITY.

pluriverse, dialectical pluriverse. Capture the notion, not so much of (1) (possibly infinite) other universes, which CR is not committed to, though PMR entertains; but of (2) the multiple differentiation or DIFFRACTION of this one (including the '*internal dialectical pluriverse*' [*D*: 276] of the embodied personality), and the irreducibility of difference; and of (3) the real existence, qua possibility, of other and better (social) worlds within this one. See also COSMOGENY; CREATIVITY; DIFFRACTION; TRANSCENDENTAL REALISM.

PMR. Acronym for the philosophy of META-REALITY.

poiesis. See AGENCY.

Poincaré's problem is the problem of transduction. See PROBLEM OF INDUCTION.

polarity. See DUALITY.

political economy. Originally, the study of social activities of production and exchange defined largely in terms of morality, prevailing customs, laws and system of government. Although many thinkers argued – a sentiment shared in part by Marx – that individual self-interested action can lead under certain favourable social conditions, through self-regulating markets, to order and progress, the taken-for-granted assumption of political economy has been that production and exchange are inseparable from society at large. The broad understanding of the nexus of political economy has passed through at least three phases of study. In the second phase it was also challenged by the marginalist revolution that tended to constitute economics as a disconnected sphere, a closed system following its own calculative logic. This is also the origin of the disjunction between economics and political economy that CR is trying to overcome in this field.

The focus of early political economy is the theory of value and its origins. In this first phase in Britain, political economy developed through the works of Locke (1632–1704), Hume (1711–76), Smith (1723–90), Ricardo (1772–1823) and then Marx (1818–83). In France, the so-called physiocrats such as François Quesnay (1694–1774) developed important methodological and theoretical ideas such as the idea of circulation of money and goods, while they also tended to emphasise land as the only or main source of value. The labour theory of value – prevalent in Britain – was more in accordance with the Industrial Revolution. For classical political economists, economic issues were tackled under the rubric of moral philosophy or politics. Most classical political economists were preoccupied with the historical change towards the institutions of capitalist market society. From Smith onwards, analysis of economic conjunctions and developments involved necessarily an analysis of class relations and law. Moreover, the classical school used their ideas about economic developments to explain colonialism, the glorification of the state, international treaties of commerce, and the like; and to make manifold reform proposals that were often controversial in their own time. By and large the classical school never intended to build any separate, compartmentalised discipline of 'economics' although they assumed separate states and societies. However, both the physiocrats and Ricardo paved the way for the methodology of closed-systems modelling of neo-classical economics.

A second phase begins at the time when political economy ceases to be the dominant mode of study. Following arguments from political economists such as Smith for laissez-faire capitalism to replace mercantilism, and for which the iniquitous effects of the Corn Law of 1815 gave a clear impetus, by the late nineteenth century the idea began to emerge that politics and economics now operate according to different principles, though both are in some sense regulated by the nature of the state. The 'discipline' of economics is differentiated from anything political. Based on Robbins's 1932 definition, economics delineates itself more narrowly as the study of the allocation of scarce resources. Economics method is also differentiated, based on the study of individual behaviour in pursuit of given ends, following a 'scientific' or quantifiable modelling form (justified since ends are given and thus the focus could be the 'efficiency' of achievement in terms of given ends amongst alternative means) developed by Stanley Jevons (1871), Léon Walras (1874) and Alfred Marshall (1891). This phase sees political economy become in many ways a subordinate discourse. Its study and influence reduce through the twentieth century as economics expands and political economists begin to expend time and effort in critique of economics' methodological and contextual separations.

Political economy, however, remained vibrant amongst Marxists and some branches of what has become known as heterodox economics, including those working from the legacies of Keynes and Veblen. This vibrancy was given additional impetus in a third phase beginning in the 1970s with the rise of a new focus on the global. The works of André Gunder Frank (1967) and of Immanuel Wallerstein (1979) introduced neo-Marxist world systems theory, and then in the 1980s the work of Robert Cox (1987) introduced a neo-Gramscian approach. Here, an increasing awareness of the significance of international trade, finance, corporate activity, environmental problems, aid and debt and a variety of governance debates within competing ideas of appropriate political ideology, such as neo-liberalism, served to reinvigorate the traditional focus of political economy on politics, policy and power in terms of the role of institutions and processes, but with an additional dynamic to the problem based on contentions concerning the centrality of the nation state itself in an international and global setting. This reinvigoration has not, however reconciled political economy and economics, but has rather occurred outside mainstream economics departments in business studies, sociology, international relations and a variety of other multi-disciplinary forums.

A number of critical realists have made important contributions in this third phase. Heikki Patomäki has done key work on global political economy (see GLOBALISATION), Bob Jessop (2002b) has been a leading English-language exponent

364 political philosophy

of French regulation school political economy, Andrew Sayer (1995, 2005) and John O'Neill (1998) have produced important works on the methodological and conceptual differences between economics and political economy and also moral economy. Various critical realists have contributed to debates concerning Marxist political economy (Fleetwood 2001a) and heterodox economics (Olsen 1996). Neo-Gramscian global political economy shares many CR tenets, while arguing that world-historical processes of production, in particular, determine, in part, forms of state and world orders. At the same time economy is also conceived as deeply political.

HEIKKI PATOMÄKI, JAMIE MORGAN

political philosophy. See POLITICAL THEORY.

political science. CR, let alone DCR, has had little explicit influence in political science. But it informs critiques of behaviouralism, pluralism, and structuralist analyses of power; certain types of Marxist state theory and critical theories of the state and politics; and some critical international relations theory and critical international political economy.

CR distinguishes POWER$_1$ from power$_2$, with the former referring to causal power in general and the latter to power maintained through exploitation, domination, etc. CR also distinguishes the capacities to exercise power from their exercise and empirical evidence for such exercise. Isaac (1987) provides the most extended CR critique of pluralism and its radical critics in political science on the grounds that both sets of theorists are wedded to a Hume–Popper empiricist understanding of causality and scientific explanation. Thus power is causal by virtue of the analysts' identification of regular sequences and this leaves even radical critics unable to show the existence of structural power. Isaac further argues that, rather than power consisting in A getting B to do something B would not otherwise do, social relations of power typically involve both A and B doing what they *ordinarily* do and thereby reproducing relations of domination and exploitation (1987: 86, 96). CR could extend this critique by showing how pluralism NATURALISES political actors rather than investigating the social constitution of identities, defines interests in terms of empirical (revealed) preferences of naturalised actors, and inclines to methodological individualism.

A CR approach to the state treats state power in relational terms based on a DEPTH-ontology rather than in terms of power as an inherent property of sovereign states or of political capacities possessed by state managers. This is already implicit in Marx, Gramsci, Poulantzas and other relational Marxist state theorists. Jessop's strategic-relational state theory develops these arguments further in explicit CR terms (1982, 1990). It focuses on the potential structural powers (or state capacities) inscribed in the state as an institutional ensemble. How and how far such powers (as well as any associated liabilities) are realised depends on the action, reaction and interaction of specific social forces located in and beyond this complex ensemble. It is not the state that acts. Specific sets of politicians and state officials in specific parts of the state system activate specific powers and state capacities inscribed in particular institutions and agencies. These powers or capacities and their actualisation cannot be understood solely through the state but also depend on structural constraints and social forces beyond the state (see also M. Smith 2000). Joseph (2002) offers an explicitly CR re-reading of Gramsci's concept of HEGEMONY as a real structural mechanism that is activated in specific circumstances to enable the ruling class to secure its economic, political and ideological domination. Patomäki (2003) has developed an explicitly CR approach to INTERNATIONAL RELATIONS and international POLITICAL ECONOMY.

(D)CR could be used to critique orthodox realism in international relations theory and practice, rational choice and sociological institutionalism in political science and (international) political economy, and the discourse theory developed by Laclau and Mouffe (1985) and their disciples. It is also relevant, of course, to POLITICAL THEORY.

BOB JESSOP

political theory. Whether CR contributes to political theory is a contentious matter. We are asking whether CR itself can or should make substantial claims about the real of politics, the structures, institutions, organisations and actors of this world and even the policy implications flowing from this. The general tendency is to say that it either can or should not, but this depends upon how CR is understood. If it is conceived of as a philosophy (of social science), then its main role is to focus on ontological, epistemological and methodological issues in the existing spheres of social science. This *UNDERLABOURER* approach would see CR providing a meta-theoretical framework for political theory and analysing its claims rather than offering a political theory of its own. However, it should be obvious that such a task is itself deeply political. In contrast to the natural sciences, the social sciences are deeply integrated into the world that they study so that any critique of political theory has wider implications.

This is clear when we consider Bhaskar's notion of EXPLANATORY CRITIQUE. He argues that a critique of false ideas leads to the question of how and why those ideas were produced and then to the issue of how alternative theories question the organisations, institutions, practices and structures responsible for generating such ideas. This gives political theory a strong normative aspect (about what ought to be), although CR would be critical of most existing normative theory for not developing a depth-ontology (focusing on reasons, obligations, rules, commitments and ethics without sufficient structural grounding). Indeed, against the tendency in normative theory to embrace a subjective or intersubjective ontology of norms and understandings, the CR approach is objective. But, against political theory's dominant positivist notion of objectivity, critical realists like Collier (2003b) distinguish between objectivity and neutrality, showing how Marx, for example, combined objectivity with partisanship in his explanatory critique.

The explanatory critique draws comparisons with Marx's critique of classical political economy which effectively shows how the classical economists' inability to explain things like the production of surplus value, the source of profit or tendencies towards crisis were a product of their own closeness to the system they were attempting to explain. This makes the theoretical intervention inherently political as well as more capable of being objective.

We have perhaps been talking more of the political nature of theory rather than of political theory itself. Parallels have been made between CR and MARXISM and this is the area where CR has been most widely used. Marxists have tended to draw on CR in order to provide a firm ontological grounding or else to develop Marxist method (for example the role of abstraction). But there is no reason why CR cannot be applied to other, non-Marxist, political theories. Going one step further, Marsh et al. (1999), in their

366 politics

study of postwar British politics, have used CR directly. In this work and in a number of other practical applications, CR plays the role of a political science methodology.

As well as contributing to the fields of political methodology, and domestic politics, CR has also contributed to the areas of state theory (POLITICAL SCIENCE), regulation, HEGEMONY (Hay 1996; Jessop 1990, 2002b, 2003a, b; Joseph 2002, 2003a; M. Smith 2000) and, to lesser degree, INTERNATIONAL RELATIONS.

JONATHAN JOSEPH

politics. To have a beneficial social effect, improvement in knowledge presupposes collective effort, i.e., a *politics*. Most generally, politics may be defined as any practice or struggle aimed at transforming the conditions of human ACTION, including resources and social structural sources of determination and possibility (cf. *SR*: 176). *Emancipatory politics* seeks to transform unwanted and unneeded sources of determination, with the aid of realist science, into wanted and needed ones (see EMANCIPATORY AXIOLOGY; EXPLANATORY CRITIQUE; FREEDOM). Given the openness of the world and the causal efficacy of praxis, 'all politics are transitional' (*D*: 269), whether we wish it or not; the point is to orient them towards maximising free flourishing. The politics of TRANSITION towards a EUDAIMONIAN society (itself open and in transition) falls into four main interrelated types corresponding to the moments of the concrete universal (Table 34).

Life politics is conducted at the level of individual world-lines, ideally guided ethically by 'a consequentially derived virtue theory'; *representative politics* expresses 'the needs and interests of different communities', oriented to the preservation of existing freedoms understood as rights; *movement politics* is the politics of social collectivities, oriented to the extension of freedoms understood as rights; and *emancipatory/transformative* politics is oriented to 'fundamental structural change in a rhythmic to eudaimonia'. All would engage in TTTTTTφ 'in a synthetic unity' and aspire to maximise participatory democracy, thereby PREFIGURING universal self-determination (*D*: 120).

CR has a broad orientation to emancipation conceived as the free flourishing of each as a condition of the free flourishing of all. This entails anti-elitism or anti-substitutionism (see SELF-REFERENTIALITY). It does not support any particular political party or movement, but aspires to promote a rich and diversified dialectic of emancipatory movements and emancipatory science capable of democratically effecting significant emancipatory change both globally and locally.

polyad (n., f. Gr. *poly*, many + *-ad*, a group of [people, things, etc.], adj. *polyadic*). A group or set of many. *Primary polyadisation* is a term adapted from post-Vygotskian

Table 34 Types of emancipatory politics, mapped to the concrete universal

Concrete universal	universality	processuality	mediation	concrete singularity
Types of politics	emancipatory/ transformative (dialectically universal)	movement (rhythmic)	representative/ syndicalist (mediated)	life (singular)

Source P: 156–7; *D*: 335.

psychology's concept of *primary dyadisation* to refer to the process, 'necessary for individuation, hence self-identity' (*DG*: 402; *P*: 106), whereby the human infant arrives at a complex, differentiated consciousness of SELF and world. It is possibly intended as a dialectical foil to the non-CR concept, *primary socialisation*. The *primary polyad* comprises the infant and the context of people (the mother in the first instance, constituting a *primary dyad*), material objects and social relations with which she comes into contact. In this context, ABSENCE (lack, want) is experienced as DESIRE, the first intuitive acts of REFERENTIAL DETACHMENT by which the infant separates herself and the act of desiring from the desired object, leading to a recognition of ALTERITY (difference) and to individuation, and inaugurating the DIALECTIC OF DESIRE TO FREEDOM. (See Archer 2000 for a detailed socio-theoretic account of this process.) Difference is therefore ontogenetically PRIOR to IDENTITY, i.e., its priority is transcendentally necessary for self-identity to occur ('whoever AUTOGENETICISED themselves?' [*D*: 230]), and primary polyadisation scientifically grounds 'the phenomenon of the EXISTENTIAL CONSTITUTION or permeation of one entity or category by another' (*D*: 380). This means inter alia that the human self is constituted by its 2E geo-historical processes of formation and the 3L wider context. We are profoundly rhythmic and interdependent creatures from the outset.

Primary polyadisation presupposes that 'mutual TRUST or nurturing care is our basic {*primary*} *existential* or mode of being in the world' and that it is a necessary condition for 'agonistic struggle', which Hegel seems to view as the primary existential (*P*: 3). It is thus a fundamental source of the ethical doctrine that EVIL is unilaterally dependent on the good, which is ontologically prior to it.

In Bhaskar's view, primary polyadisation entails that difference is also *ontologically* prior to identity as such (*D*: 285, 380) within relative being. He sees this, together with an argument demonstrating the irreducibility of difference and change, as transcendentally refuting MONISM, including SOLIPSISM and theories positing an 'original monism' or undifferentiated unity in human history (such as Hegel's) or in the cosmos, both of which must be regarded as open at either end (*D*: 47 n, 355 n). Thus there is *sempiternal* (L. *sempiternus*, f. *semper*-, always + *aeternus*, eternal, everlasting) as well as primary *polyadisation* (*P*: 32). Global intra-dependence as well as the universe itself (a *pluriverse*) are irreducibly plural. For PMR, in particular, ontogeny recapitulates phylogeny, recapitulates cosmogony, and every human creative act is in 'HETEROCOSMIC affinity' with the creation of the universe (*EW*: 55).

polysemy (n., adj. *polysemic*, f. Gr. *poly*, many + *sēma*, a sign). The quality of having many meanings; 'the condensation of many meanings in a single sign – the semiotic counterpart of the conjunctural (or compound) nature, i.e., open systemic character, of events and things in the world' (*RM*: 248).

polyvalence. See VALENCE.

Popper's moment, Popperian moment. The stage in the dialectic of EXPLANATION when all but one of the competing explanatory hypotheses is eliminated as inconsistent with the evidence (see also EPISTEMOLOGICAL DIALECTIC; Popper 1934, 1963).

positive absence. See NEGATIVE PRESENCE.

positive and **negative freedom**. See FREEDOM.

positivise, positivisation (deprocessualisation). See CONCRETE/ABSTRACT.

positivism. See CRITICAL NATURALISM; EMPIRICISM.
positivist(ic) illusion. See SPECULATIVE/POSITIVIST(IC) ILLUSION.
possessive individualism. See HOBBESIAN PROBLEM OF ORDER; IDEOLOGY.
possibility. See CAUSALITY; DISPOSITION; MODALITY; NECESSITY; *POTENTIA*; PRIMACY.
possible futures. See CONCRETE UTOPIA; EMANCIPATORY AXIOLOGY; FUTURES STUDIES; PREFIGURATIONALITY.
possible worlds. See PLURIVERSE; TRANSCENDENTAL REALISM.
postcolonialism. See COLONIALISM.
post-dated endism. See CENTRISM, ETC.
postdisciplinarity. See INTERDISCIPLINARITY, ETC.
post-empiricism, post-positivism. See POST-POSITIVISM.
post-foundationalism. See FOUNDATIONALISM.
posthumanism. See ENTROPY; HUMANISM.
post-Kantian predicament. See REFLECTION.
postmodernism. In *The Condition of Postmodernity*, David Harvey says that postmodernism rejects the idea that there exists 'meta-language, meta-narratives, or meta-theory through which all things can be connected or represented' (Harvey 1990: 45). The world, according to some forms of postmodernism, is fragmented rather than unified and ephemeral rather than permanent. Power relations can be questioned and critiqued only within the boundaries of a particular discursive formation and not by recourse to some sort of normative standpoint based upon a non-discursive 'real' level of reality. Knowing this, it is perhaps unsurprising that critical realists have often criticised postmodernism for its IRREALISM. However, there are at least four reasons why the gap between CR and postmodernism is not so clear-cut as some debates would lead us to believe.

First, postmodernism rejected and rejects positivist ideas about the advancement of science as being based upon the neutral and cumulative acquisition of knowledge by scientists. Postmodernism argues that scientific knowledge is always historically and socially situated, and that it is more often than not guided by accepted methods of scientific discovery and the ideological positions around what counts as 'good science' that accompany this. Even if they do not use the term explicitly, therefore, postmodernists do none the less suggest, like critical realists, that the transitive nature of scientific discovery is crucial and epistemic ALIENATION ineluctable.

Second, many postmodern writers also stress that a real, i.e., INTRANSITIVE, dimension exists and has significant effects upon what might be termed the transitive dimension. Derrida, for example, has argued that 'reality' is always internalised in some form or another within discourse, while Foucault's notion of discursive formation is seen to have realist residues because it seeks to extract the deep structures of discourse. The advantage that critical realists might be said to have over postmodernists in this regard is that the latter rarely give detailed accounts about what this 'real' domain might look like and exactly how it structures and mediates the transitive dimension.

Third, a realist perspective has usefully situated the EMERGENCE and popularity of postmodernism within the causal and historical efficacy of specific structures of capitalism. This is especially noticeable within Marxist approaches. For example, Jameson (1991) argues that postmodernism has arisen under the hegemony of multinational and post-industrial capitalism. This form of capitalism, distinguished as it is by flexible machinery, computerised technology, just-in-time production, niche marketing, local bargaining between capital and labour, a service economy, information and a professional class, is seen as the structural mediator for the wider fragmentation of social life that postmodernism explores and sometimes celebrates. Indeed, it provides the ground for the 'depthlessness', 'desubjectivisation' and 'simulacrum' that characterises a postmodern society that leads, in turn, to a stylisation and pastiche of contemporary societies by, for example, artists. Again, working from a Marxist realist perspective, Woodiwiss (1993) observes that what might be seen as postmodernism is, in reality, the generalisation of the decline of US social modernism to the rest of the world. Woodiwiss describes US social modernism as comprising five structures of 'self-reliance, responsible unionism, opportunity, loyalty and modernity' (Woodiwiss 1993: 3). What Woodiwiss suggests is that the crisis of these five structures during the 1970s in the USA brought about an ideological crisis of 'US modernity' that then came to represent a 'crisis of modernity' *per se* in other parts of the Western world. This then enabled a gap to open up in which a new ideological representation could emerge based upon 'postmodernity'.

Finally, it would nevertheless be erroneous to believe that realism has nothing to gain from postmodernism. Many postmodern thinkers add important and original insights to a CR perspective. For example, critical realists have often neglected a discourse perspective in their analyses. Postmodernism adds to CR in this respect by showing, for example, how the signifying power of language structures the social and natural world in subtle and often powerful ways. Indeed, postmodernists are particularly attuned to those everyday discursive micro-powers that construct subjects so as to create ideological identities in particular contexts. See also ALTERITY; ANTHROPOLOGY; COLONIALISM; CULTURE; DIRECTIONALITY; HERMENEUTICS; HISTORIOGRAPHY; IDEALISM; IRREALISM; LITERARY THEORY; PDM; PHILOSOPHY OF SOCIAL SCIENCE; STRUCTURALISM; UNIVERSALISABILITY; VIRTUALITY; Derrida 1995; Foucault 1977; Joseph and Roberts (eds) 2003; Lash and Urry 1994; López and Potter (eds) 2001; Lyotard 1984; Norris 1997; Woodiwiss 2002.

JOHN M. ROBERTS

post-positivism, post-empiricism. The names given to anti-rationalist currents in the philosophy of science which reject FOUNDATIONALISM, accepting that scientific data are theory- and value-laden or determined. Since this is now granted by the great majority of philosophers of science, positivism is pronounced dead. However, in that many professed post-positivists retain the central Humean notion of a causal law and a correspondingly instrumentalist view of human reason, positivism lives. See also EMPIRICISM.

post-structuralism. See PDM; STRUCTURALISM.

potentia (L. power, ability; efficacy, potency). In Aristotle, *potentiality* (Gr. *dunamis*) is real possibility, a POWER within a thing that strives to become manifest; change is actualisation of potentiality. Bhaskar's usage is similar, but officially rejects Aristotle's TELEOLOGY (whether wholly is eristic) and his ACTUALISM, as is that of Werner Heisenberg, who spoke of 'the unmeasured quantum world as a sea of unrealised potentia, waiting for a trigger to actualise them' (Porpora 2000a: 39). See also CAUSALITY; ULTIMATA.

poverty. Realists have made a number of significant contributions to the poverty debate, but many would not explicitly call themselves realists. They have added to the conceptual framework used to examine poverty, and their approaches contrast vividly with competing ontologies (notably idealist and reductionist approaches), as I will show.

First, early realists focused on human emancipation from oppression (see *SR*). Contemporary impoverishment was seen as historically situated in late capitalism. Details of what level of living would count as 'poor', and what political and social dimensions poverty actually had, were seen as culturally relative and socio-temporally changing. Authors of implicitly realist persuasion have described, for instance, India's social structure (Krishna Bharadwaj, Amit Bhaduri, Uma Kalpagam, Olsen 1996); Asian societies generally (Terry Byres, Henry Bernstein); Latin American social relations (Ruth Pearson, Stephanie Barrientos); and African country histories (Colin Leys, Kwame Ninsin, Jane Harrigan) in realist terms. All these researchers share a common focus on unequal social structures interacting with both human agency and corporate agency.

The dialectics of poverty place it as a malleable, organically changing social relation rather than a mere label for a group, or simply an income cut-off level. Similarly to realists, structuralists in development studies have repeatedly challenged the cultural turn and postmodernism (notably via Tom Brass's contributions). For a review, see Allen and Thomas (2000). Realists see poverty as having primarily a transitive nature rather than being a given, although there is also of course a certain intransitivity to the experience of poverty when one is born into impoverished circumstances. As Bhaskar (*RR*) wrote in 1989, the transitive/intransitive nature of social being is not an either/or polarity (see also Sayer 1992).

People's desire for personal self-respect and corporate honour combines with the real malleability of poverty to make micro empirical research in this area very challenging. Authors tend to link micro and macro evidence, to examine large-scale trajectories, to engage critically with deterministic structuralism, and to challenge naïve notions of government being an anti-poverty agent (e.g., Gruffydd Jones 2001). The paradoxes of how 'poor' people strive to emulate 'elite' behaviours were stressed by Thorstein Veblen, who (like Marx) must be classified as philosophically realist. Veblen's, Marx's and Lenin's (1899) empirical research were CR in their methodology.

The second theme stressed by realists is that poverty is a social relation. Engels (1844), for instance, showed that the manufacturers' employment relations with workers changed over time and thus constructed the workers' poor living conditions in urban Manchester. David Byrne (1999) makes explicit several anti-individualist points, and in doing so follows a well-travelled path similar to that known as the gender and development school (GAD) among development feminists such as Naila Kabeer, Diane Elson, Nancy Folbre and Ruth Pearson. The wage relation, patriarchal social relations, social exclusion and political/social inequality are placed centrally in such research.

Both Marxist and feminist writers challenge the mono-disciplinarity of most economists' approach to poverty. John Brohman (1996) explicitly showed that neo-liberal ideology encourages mono-disciplinarity, and that, by so doing, it tends to mask the real causes and solutions to the poverty problem. Related texts by Hettne (1990) and by Hoogvelt (1997) see globalisation as a *social* trend and poverty explicitly as a social relation embedded in (and accepted by) modern capitalist ideology.

Seeing poverty as a social relation was also associated with the work of Amartya Sen, whose 1965–85 work on famines argued convincingly against a supply-of-food approach and in favour of a social-entitlements approach. In general, realists take a relational approach to poverty, whereas most social constructionists and neo-liberal economists see the poor as a residual category with special characteristics. The Open University taught the 'relational' approach as a theme in a series of courses associated with authors such as Maureen Mackintosh, Gordon White and Alan Thomas (Allen and Thomas 2000). However, it is difficult to place Amartya Sen in the realist school since some of his writings on justice and equality can be seen as individualistic in their underlying ontology. Notable here are his works on capabilities and on economic inequality.

The third orientation of realists in this area is to focus on the transformation of the social world (Subramaniyam 2001), with the label 'poor' placed as potentially an epiphenomenon (i.e., false appearance). This label persistently re-appears in dominant discourses, such as 'redistribution [to the poor] with growth' and 'charity [for the poor] with justice'. These discourses have helped to glorify and spread the modern Western pro-capitalist mentality (cf. Chouliaraki and Fairclough 1999). According to Arturo Escobar and the post-development school, the reform-capitalism discourses simultaneously excuse real suffering and mask the creation of poverty. Realists object to mystification and wish to challenge these particular discourses.

Andrew Sayer's empirical work on capitalism probes two aspects of the hegemonic discourses further. First, he has explored the possibility of challenging capitalism's inequality-generating mechanisms whilst accepting that markets will persist in post-capitalist society (Sayer and Walker 1992). Second, he has argued that the moral aspect of political economy research should be foregrounded and that neo-classical economics is therefore hiding its real normative agenda (Sayer 1995).

Moving away from the discursive aspects of poverty research, a fourth theme among realists is that poverty is not suited to individualistic analysis. Flourishing, as described by Tony Lawson (1997, 2003b), has supra-individualistic aspects and cannot merely 'exist' at the individual level. Other authors who avoid reductionism include Archer (2000), who examines the real effects of one's origins on one's tendencies as an agent; James C. Scott and the Indian subaltern studies school, who see people as socially conditioned and historically located but still capable of political action; and early work by Zygmunt Bauman (1976). Bauman argued that human nature needs to be construed as a 'second nature' to which our socialisation contributes, rather than human nature being merely a genetically or nature-given instinctive nature.

Textbooks on poverty usually construe realists as if they were merely structuralists. It will be better in future to classify the realist school as the 'transformationalists' (following the approach of Byrne 1999). They have much in common with evolutionary economics and old institutionalism. In so far as Marxists and feminists have an empirical political economy orientation, they usually take a realist approach to poverty. Apart from the various academics whose work I have mentioned here, there are millions of activists in SOCIAL MOVEMENTS, non-governmental organisations and the various Social Forums who implicitly take a transformationalist approach to poverty. For such people, the TRANSFORMATIONAL MODEL SOCIAL ACTIVITY is second nature (see *ER* for an introduction).

WENDY OLSEN

power. Used in a variety of contexts in CR. Following earlier work by Harré and Madden (1975; *ER*: 104–19), in *RS1* (172–85, *ER*: 72–7) the term is used for the properties of STRUCTURES or generative mechanisms that account for their tendential behaviour. Such powers exist irrespective of whether they are actualised in events, experienced or perceived in those events, or correctly theorised. This is part of the argument in philosophy of science that science investigates a stratified and open DEPTH-reality and that proposed CAUSAL LAWS are fallible expressions of natural necessity (see CAUSALITY; Collier 1994: 60–3). In conjunction with the concept of EMERGENCE, the term is then applicable to the properties and possibilities of different strata of real entities. Humans have causal powers in a variety of senses. They are living material beings with active powers to do things like jump and run, subject like nearly everything else to the causal powers of other generative mechanisms and strata. They are biologically susceptible to illness, and if they jump off a cliff they may die. However, in virtue of the emergent power of consciousness, humans are capable of rational thought and thus of discovery, invention and planning. Accordingly, they are able to manipulate the causal powers of other generative mechanisms: medicine may cure an illness, and a glider may allow one to take off from a cliff. Medical training, knowledge of aeronautics, access to a competent doctor or the materials to build a glider are historically dependent and socially distributed. As such, particular societies contribute further powers to human activity by socialising the individual to a belief system, by providing systems of knowledge, and by distributing material benefits, rights, duties, opportunities and tasks, usually in an asymmetric way. From a CR perspective, therefore, societies contribute to the constitution of and provide a medium for, embodied positioned practices (*PN1*: 51) that entail either new powers (e.g., marriage as a speech act) and/or distribute the right to exercise powers (e.g., soldiers killing other soldiers as legitimised murder).

It is from this position that CR intersects with traditional arguments about power (control, domination, etc.) and FREEDOM in political philosophy. In *Dialectic* (*P*: 60; *DG*: 402), Bhaskar introduces two concepts of power. $Power_1$ refers to the general causal powers of human agency whose characteristics entail the possibility of human emancipation, such as our capacity to investigate, communicate, plan, construct moral and ethical systems, feel and care for others, and come to agreement based on judgementally rational argument directed at practices that transform our lived circumstances. $Power_2$ refers to negative characteristics such as domination, subjugation, exploitation and control that can be identified in given social structures. Since such $power_2$ characteristics can be constituted and expressed in any social structure in a variety of ways, $power_2$ is compatible with a variety of different ways of theorising the constitution and expression of power (e.g., Lukes [ed.], 1986), such as Lukes's three dimensions (1974). These are that power entails (1) agent A making B do something that she would not otherwise do by the mobilisation of some resource, skill or capacity involving threat, coercion, violence, persuasion, etc.; (2) the construction of compliance by the exclusion of known alternative possibilities, narrowing choice, particularly in policy debate; (3) socialisation such that the 'worldview' of individuals precludes an understanding of real interests. For Bhaskar, the bridge between $power_1$ as possibility and $power_2$ as structurally instantiated forms of domination is provided by the combination of an ontologically naturalistic basis to ETHICS and the concept of negation of real ABSENCE as a transformative solution to social ill. Power$_2$ characteristics in social structures are also aspects of the socialisation of agents who experience them as specific forms of generalisable MASTER–SLAVE-type relations. Manifestations of power$_2$ are felt as lacks or social ills. On the basis of a naturalised ontological underpinning of real ethics, CP, known harm to others when genuinely experienced as such leads to a movement to alleviate the causes of that harm. A real yet often unactualised power of the human is a sense of the good of being in and of itself (*D*: ch. 3; *ER*: ch. 24). It is this human power that means that both master and slave remain unfree whilst social ill persists. EXPLANATORY CRITIQUE explores aspects of the generation of social ill and contributes to arguments for a praxis that transforms the previously reproduced conditions of that ill. In transforming praxis not only are the positive causes of ill eradicated but new understanding and practices are created. Since these new practices flow from the real possibilities of the human for emancipation their constitution is termed the absenting of real or determinate absences directed at the transformation of power$_2$ characteristics based on power$_1$ possibilities. The process itself and the relations of its participants are dialectical because it is the tensions and contradictions of reality, its perpetual incompleteness and, contra ONTOLOGICAL MONOVALENCE, the reality of absence that provide for the possibility of a mutual freedom, where freedom is not the removal of determinations (an impossibility) but the practice of creating wanted determinations.

A variety of critical realists have taken different positions on Bhaskar's argument for power and dialectics. Norrie (e.g., 2004a) has explored how it relates to the problem of freedom and ethics, Joseph (2002: 134–5) has questioned the analytical distinction between power$_1$ and power$_2$ on the basis of a realist theory of hegemony, whilst Roberts (Brown et al. 2002: ch. 12) and Creaven (2002b) have raised issues of its relation to classical Marxism.

Power$_2$-stricken society: see MASTER–SLAVE.

JAMIE MORGAN

power(s), causal. See CAUSALITY.
practical materialism. See AGENCY; ALIENATION; HEGEL-MARX CRITIQUE; HOLY TRINITY; IDEALISM VS. MATERIALISM; SUBJECTIVITY.
practical reason. See ACTIONABILITY; CONSISTENCY; DEA AND DET MODELS; DIALECTICAL REASON; DUALITY OF THEORY AND CRITIQUE; EMANCIPATORY AXIOLOGY; ETHICS.
practical wisdom. See PHRONESIS.
practice. See AGENCY; SUBJECTIVITY.
practicism. See IRRATIONALISM; RATIONALISM.
pragmatism. A predominantly American philosophical tradition initiated by Charles Peirce (see also INFERENCE [abduction]; SIGNS), William James, John Dewey and G. H. Mead, and developed by Quine, Putnam, Davidson, Rorty and others. Best presented as a rich and variegated research programme, held together to a certain degree by a cluster of problematics, but perhaps more in terms of an eponymic lineage as well as a patriotic American breed. In many ways the philosophical counterpart of the American dream, reflecting its positive as well as its negative aspects, incorporating a vision of the omnipotence of humankind and the centrality of practice in the

achievement of this illimitability. Parallel to, exhibiting great similarities with, and subsequently outflanked by LOGICAL POSITIVISM, it initially argued for the eradication of metaphysical pseudo-problems by the elucidation of the tangible consequences of statements, actions, and theoretical and practical stances. Launched by Peirce as a philosophical doctrine, or rather method of philosophising, in which the exact meaning of linguistic forms was identified. By thus making our statements clear, pointless pseudo-problems could be avoided. Just as the early Wittgenstein inaugurated a negative philosophy in which the completed philosophy would entail its own dissolution, the basis of the dissolution of pragmatism's central concepts (practice, meaning, truth) was laid down by Peirce and James. Pragmatism was promoted primarily, not as a method which *solves* problems, but rather which distinguishes real from imaginary problems via the *pragmatic principle*, according to which we should regard anything outside the confines of conceivable practical effects as idle metaphysical quibbles. Ideas as well as practical deeds must always be assessed by reference to their actual practical cash value. This theory of meaning, according to which the meaning of a proposition is exhausted by its conceivable practical consequences (cf. operationalism), was subsequently extended by James into a theory of TRUTH, defining truth as (1) what is good for us in the way of belief (practical truth) and (2) that mental content, ideas, which brings us into accordance with the content of our previous experiences (theoretic truth). This distinction between practical and theoretic truth breaks down, and truth is accorded to any idea which helps us successfully to integrate thinking and acting. Against FOUNDATIONALISM and REPRESENTATIVE REALISM, it is argued that the whole point of inquiry is to enable us to act effectively rather than to represent reality accurately (cf. INSTRUMENTALISM). Thus truth is seen as characterising the relation not between idea and reality, but between parts of an instrumental linguistic apparatus, in which words, notions, concepts are means for the comprehension and successful instrumental linking of different components of experience. Truth is not an antecedent property of anything, but something which is achieved through practical deeds – a process of making true, verifying. This central pragmatist posture towards *temporality* involves a denegation of any object antecedent to the experiential process, and an insistence on *futurity*, epistemologically and ontologically as well as socio-politically.

Still, the two strands in James lead to a split between instrumentalism and coherentism, which is reproduced in a theory/practice inconsistency in which the latter is pursued by philosophers such as Sellars, Quine and Davidson and the former carried through by advocates of pragmatism in other disciplines; and sometimes both are intertwined, as in, e.g., Rorty, where they form an UNHAPPY CONSCIOUSNESS of reductive materialism and radical hermeneutics in which a material world is presupposed as having causal impact upon us, but leaves us no epistemic admission to it, presenting people as simultaneously completely determined (as material objects) *and* perfectly free (as social subjects).

As it did not have a clear principle of demarcation (cf. ANALYTICAL PROBLEMATIC; EMPIRICISM) and lacked intellectual acuity and departmental backing, pragmatism led a shadowy existence from the early 1920s to the 1950s. Under pressure of an increasing internal critique of analytical philosophy, it then gained ground through the work of Sellars, Quine and Davidson, in particular.

In this development of pragmatism into its post-analytical phase, the holistic concept of meaning and the coherence theory of truth take a more solipsistic turn. According to Davidson, the meaning of a proposition is no longer determinable in terms of practical consequences other than those specified inside the specific language-game in which it is conceived, and we could very well construe such a language as infinitely small. Meaning, truth and linguistic form are radically contextualised; idiolects are profuse. Davidson performs a reductio ad absurdum supposedly showing the pointlessness of any concept of truth and ultimately of philosophy as such. Truth is put on a par with meaning, which could only be defined in the linguistic network in which it is conceived. This stance is radicalised by Rorty, who argues that a philosopher could not say much of interest on this issue. Accordingly, Rorty expresses the notion of truth as just the name of a property which all true statements share as a matter of conventional agreement, consigning philosophy to an idle, defeatist position which leaves everything as it is: philosophy becomes intellectual vagrancy, therapy, edification, diversion.

The *pragmatic principle* was at once a battering ram against metaphysical speculation and an obstacle to an appreciation of the complexity of the world, both inanimate and animate, and ultimately a gateway to later ACTUALIST and EXTENTIONALIST tendencies. The predominant current in pragmatism is this anti-representationalism, but there is also a semiotic and realist movement, basing its work on Peirce, which in opposition to Saussurean linguistics stresses the importance of the referent in semantics.

In very broad terms, pragmatism may be said to place knowledge and truth in direct relation to life, repudiating any dualism of theory and practice, fact and value, and judging the worth of ideas and beliefs according to their capacity to meet human needs and interests in a practical way. While it is too complex, protean, and riddled with internal contradiction and ambiguity to be snared in a neat definition, this must be regarded as its core. Other key features include (cf. Wiener 1974) espousal of or emphasis on (1) *EMPIRICISM*, in various guises, ranging from the solipsistically tinged radical empiricism of James to a pluralistic, developmental, problem-solving kind of Dewey; (2) a post-Darwinian understanding of human *experience* as a developmental and geo-historically contingent PROCESS of transaction between organism and environment; (3) a by-and-large *Humean understanding of CAUSALITY*, with an emphasis on probabilistic explanation; (4) epistemic and moral *relativity*; (5) social and political *liberalism*, often blind to the authoritarian aspects of market society; underpinned by (6) *sociological INDIVIDUALISM*. Peirce and Dewey were critical of the Cartesian notion of the self as prior to language and culture, substituting an account of it as a product of social practices. Their work has been an important source of Archer's (2003b) theory of the self (see SUBJECTIVITY) and the *primacy of practice* in the production of the self. *Antinomy of transcendental pragmatism*: see RATIONALISM. *Ultra-pragmatism*: see EMPIRICISM.

PÄR ENGHOLM

praxis–poiesis. See AGENCY.
precautionary principle. See ECOLOGY.
predicative realism. See UNIVERSALS/PARTICULARS.
prediction. See EXPLANATION; HISTORICISM; PHILOSOPHY OF SOCIAL SCIENCE; SOCIAL THEORY.

prefigurationality. In our intentional activities we *prefigure* the future, now, by forming conceptions of the ends we want to achieve. Bhaskar generalises this to the sociosphere, such that, for successful social transformation to occur, it must be prefigured as increasingly shaped possibility of the present-future, mediated by the past (see TENSE). Together with (the principle of) ACTIONABILITY, (the principle of) *prefigurationality* is an important criterion in EMANCIPATORY AXIOLOGY; it is entailed by the requirement for theory–practice CONSISTENCY. It implies 'both means–ends consistency and anticipatory pre-embodiment' in present praxis of 'the values of the end state' (*P*: 66, 143). Contrasts with the principle of (existential) CONSTITUTIONALITY by the past and outside.

prescriptivism. See ETHICAL IDEOLOGIES.

presence. See ABSENCE; METAPHYSICS OF PRESENCE; ONT/DE-ONT.

presence, problematic or **metaphysics of**. See METAPHYSICS OF PRESENCE.

presence of the future. See TENSE.

presence of the past and outside. A 2E figure of oppositionality, 'especially in the form of existential constitution by geo-historical processes of formation' (*P*: 242). See HABITUS; RELATIONALITY; SPACE–TIME; TENSE.

present. See BLOCKISM/PUNCTUALISM; SPACE–TIME; TENSE.

presentational dialectics embraces the mode of exposition and presentation of a system of ideas; to be distinguished from CRITICAL and SYSTEMATIC DIALECTICS. Bhaskar criticises the presentational dialectics of Marx for linearity (*D*: 93). In *Dialectic*, his own broadly corresponds to MELD. Thus ch. 1 deals with CR, Hegelian dialectic and the problems of philosophy – 1M products to be transformed; ch. 2 elaborates dialectic as absence, emergence, contradiction, etc. (2E); ch. 3 sets out the system of DCR and the totalising dialectic of freedom (3L); and ch. 4 critiques (4D) irrealism (cf. *D*: xiii). See also HEGEL–MARX CRITIQUE.

primacy (**priority**). CR has argued from the outset that there is an (unfolding) objective world order comprised of a differentiated and complexly interacting hierarchy of unilaterally dependent domains and strata, or that the world is characterised by *ontological DEPTH-openness*. While higher-order systems or phenomena (entities) can have feedback effects on lower-order (more fundamental) ones, they could not exist without them, whereas the converse does not hold. A lower-order entity thus has *ontological primacy* or *priority* over higher-order ones, and temporal and explanatory (causal) *asymmetry*, far from reflecting a false binary logic which can simply be deconstructed, as post-structuralists claim, is fundamental to being. This commits CR to MODAL REALISM. Principles of *logical* and *epistemological priority* quickly follow. Logically, nothing could be 'manifest in actuality which was not first a possibility' (*MR*: 48), and epistemology is necessarily oriented to the recursive discovery of the generative mechanisms of mechanisms.

Among major asymmetries (dispositional or categorial), many of which are closely bound up with the *irreversibility of time* (entailed by EMERGENCE), are the *temporal asymmetry* of past and present, causes and effects; the *asymmetry of space and time in regard to causality* – causal powers must be possessed and exercised in time but do not necessarily have position in space, e.g., RELATIONS; the entailment in the human sphere of value and practical judgements by factual and theoretical ones but not vice versa (see EXPLANATORY CRITIQUE); and the (at least relative) constellational primacy or priority of the following (it doesn't have one alphabetically according to either the name of the asymmetry or, if they are given, the primary concept – italicised in each case):

- *absenting* (change) over absence (non-being)
- *absence* over presence (negativity over positivity)
- constraints on human action over enablements and the realisation of possibility: there are many more things we cannot or could do than we can. This is *the axiological asymmetry*, offsetable to some extent by emancipatory praxis (made possible by the asymmetry of emancipation).
- *dialectics* over analytics within the dialectic of dialectical and analytical reason
- totality over its aspects, relations over their relata (or the shaping of the part by the whole) (the *ecological asymmetry*)
- our transcendentally real selves over master–slave-type structures of oppression, and action flowing from our essential selves over exploitative and oppressive action (*the asymmetry of emancipation* or of *axiology and emancipation*)
- *entropy* over negentropy
- the *epistemic* over the ontic within the expressive–referential duality of truth, and the *ontological* and *alethic* moment over the other moments in the truth tetrapolity
- *explanation* over prediction in science (entailed by the unfolding openness of the world)
- *internal* over external relations within the dialectics of internal and external relations
- *non-identity* (alterity) over identity within ontogenesis, and *identity* over non-identity within ontology
- *objectivity* over subject–object within subjectivity, or ontology over epistemology
- the *past* over the present
- the *political* over the ethical in the EA of the TD (explanatory primacy), and the *ethical* over the political in the IA of the TD (normative primacy), constituting a fluid constellational unity within the political
- *practice* over discourse and theory in the constitution of subjectivity (Archer 2000; Nellhaus 2004)
- reason (belief) over desire (want, intention) in human action – belief does not have to wait on anything else to issue in action, want must wait on belief (*the psychological asymmetry*)
- *science* over philosophy
- social structure over human agency (*the social asymmetry*)
- masters over slaves in power$_2$-stricken society (the relative *structural asymmetry* of dialectical contradictions, connected with directionality)
- *structures* over events, hence the *possible* (dispositionality) over the actual

The unilateral ontological dependence of social upon natural being is the source of historical materialism's *materialist primacy thesis*. This is a heuristically valuable hypothesis for empirical research which cannot, however, be demonstrated a priori. See also NATURAL KIND; STRATIFICATION; TENDENCY.

primacy of practice. See PRAGMATISM; PRIMACY; SUBJECTIVITY.

primal scream. A concept introduced in *Dialectic* to refer to the first cry of the human infant, signifying inter alia the ABSENCE of the security of the womb and the onset of

378 primal squeeze

DESIRE, inaugurating in any social context, PRIMARY POLYADISATION and the DIALECTIC OF FREEDOM. Though possibly borrowed from the title of Janov (1970), it has no connection with *primal therapy*.

primal squeeze. The third member of the UNHOLY TRINITY. 'The squeeze between the domains of metaphysics (cf. the SPECULATIVE ILLUSION) and the a priori {i.e., RATIONALISM}, and those of experience (cf. the POSITIVISTIC ILLUSION) and the a posteriori {i.e., EMPIRICISM}, ruling out empirically controlled scientific theory and natural necessity alike' (*DG*: 402). It is a consequence of the EPISTEMIC FALLACY mediated by ACTUALISM, such that rationalism and empiricism seem to be the only alternatives. Empirically controlled theory is rendered redundant by a priorism, and natural necessity disappears in the problem of induction. Since it splits off or extra-jects necessity and empirically controlled theory from being (conversely, aspects of being from thought about them), producing an implicit empiricist (flat, destratified) ontology, primal squeeze is a form of ALIENATION and is used synonymously with *ontological destratification*. Historically, it occurs on the *Platonic–Aristotelian fault-line*, which is sometimes a synonym for it (*D*: 182, fig. 2.34). See also ARCH OF KNOWLEDGE; HEGEL–MARX CRITIQUE; IRREALISM.

primary existential. See POLYAD; RECOGNITION; TRUST.
primary polyadisation. See POLYAD.
principle of actionability. See ACTIONABILITY.
principle of the Aristotelian mean. See INTEREST.
principle of bivalence. See LOGIC; REALISM; VALENCE.
principle of causality. See DETERMINATION.
principle of constitutionality. See CONSTITUTIONALITY.
principle of (non-)contradiction. See ANALYTICAL PROBLEMATIC; CONTRADICTION; TABLE OF OPPOSITIONS.
principle of emergent totalities. See INTERESTS; PRINCIPLE OF TOTALITY.
principle of equity. See EQUITY.
principle of equivalence. In general relativity theory, the theorem that being at rest in a uniform gravitational field is equivalent to being in accelerated motion. In formal logic, it treats evidence for a hypothesis as confirmatory of logically equivalent hypotheses (see NICOD'S CRITERION). In CR ethics, it asserts the *parity of equivalents*, that members of the same NATURAL KIND have a right to be treated equally, CP. Thus to say of two creatures that they are human is to grant them the same rights *ab initio* (*D*: 243) – there is nothing in their nature that furnishes any ground for treating them unequally. This provides the basis for an argument for a universal core EQUITY. The principle is violated by Hegel's dialectic of master and slave. More generally, the ethical principle of equivalence states that tokens of a type within the human natural kind have the same rights, e.g., the care of prisoners with mental illnesses should be equivalent to the care of non-prisoners with mental illnesses, or women should get the same wages as men for equal work. As these examples suggest, the principle is systematically breached in capitalist post/modernity, where a principle of *non-parity of equivalents* is often applied. This is a fundamental form of illicit FISSION or detotalisation, ultimately of ALIENATION.

principle of excluded middle. See ANALYTICAL PROBLEMATIC; CONTRADICTION; TABLE OF OPPOSITIONS.

principle of fold-back. See INTERESTS.

principle of identity. See EPISTEMOLOGICAL DIALECTIC; HEGEL–MARX CRITIQUE; IDENTITY.

principle of indifference. A rule of statistical theory which says that, unless we have a reason to expect otherwise, we should be 'indifferent', i.e., each possible event should be assigned an equal probability. Interpreted by some, including Bhaskar, to be a special case of the *principle of insufficient reason*. This states that, if there is no sufficient reason (explanation) for something's being (the case), it is not (the case). It is the negative form of the more familiar *principle of sufficient reason*, otherwise known as the principle of causality or ubiquity determinism (see DETERMINATION; PRINCIPLE OF PLENITUDE): everything has an explanation. The principle of indifference is, therefore, on this interpretation, a logical consequence of the principle of ubiquity determinism. A 1M concept in CR, it speaks of universality and '"metaphysical inertia"' or bound energy, rather than of the processuality of 2E which calls it 'into question and difference' in releasing (unbinding) bound energy, or, in '"metaphysical (neg)ENTROPY"' (*D*: 8). It applies, not actualistically to events and states of affairs, but to relatively enduring and transfactually active mechanisms and structures, which may, however, themselves be transformed (*RS2*: 108, 202, 246–7).

principle of inexorability of ontology states that ontology is inescapable and if not explicitly theorised will be implicitly.

principle of insufficient reason. See DETERMINATION; PRINCIPLE OF INDIFFERENCE.

principle of plenitude (f. L. *plenus*, full). The principle that all possibilities are, or will eventually be, actualised. Commonly assumed in Western philosophy and science before Darwin (see Lovejoy 1936, who coined the term), it is related to (1) the idea of a continuous *great chain* or hierarchy *of being* in which every possible form is, or will be, actualised, an idea – shorn of its actualist and monovalent assumptions – adopted by Collier (1999) in his account of the value of being; and (2) the logic of the *principle of sufficient reason* (see DETERMINATION) (if something exists, there is sufficient reason [an explanation] for it; if not, there is not), applied actualistically. An ONTOLOGICALLY MONOVALENT idea, presupposing the truth of ACTUALISM and ENDISM, it is the 'philosophical social analogue' of the idea of a voidless material-object world, and so 'could perhaps be more aptly labelled "the principle of repletion"' (*D*: 46 n) – a sentiment which exploits the sense in which 'replete' means 'stuffed full' or 'crowded'. While its inapplicability to a world dominated by absences – inequality, scarcity, waste, and ecological constraints – should be obvious, the notion of 'plenitudinous positivity' (*D*: 118) is still common today in such notions as the end of history, the infinitude of difference allegedly accomplished by market society, the plenitudinoos utopia of communism, and the actual existence of all possible worlds. On CR depth-realism, possibility is inexhaustible.

principle of prefigurationality. See PREFIGURATIONALITY.

principle of self-emancipation. See SELF-REFERENTIALITY.

principle of stratification, non-arbitrary. See DCR; DEPTH; EPISTEMOLOGICAL DIALECTIC.

principle of sufficient reason, principle of sufficient practical reason. See DETERMINATION; EMANCIPATORY AXIOLOGY; PRINCIPLE OF INDIFFERENCE.

380 principle of the identity of indescernibles, etc.

principle of the identity of indiscernibles/indiscernibility of identicals. See IDENTITY OF INDISCERNIBLES.

principle of the indivisibility of freedom. See CONDITIONALITY; DCR; FREEDOM, ETC.

principle of totality. The axiom that in a stratified world in open process, in which change and difference are mutually irreducible, 'differentiation is a necessary condition of TOTALITY and diversity of UNITY' (*D*: 270), i.e., for the EMERGENCE of new levels of holistic causality. The *principle of emergent totalities* is the principle of totality applied ethically, viz. totalities require different principles from those that are pertinent to their constituent parts (see INTERESTS). The *meta-principle of totality* is unity-in-diversity – 'free development of the concrete singularity of each {as} a condition for the free development of the concrete singularity of all' (*D*: 298), i.e., moral ALETHIA – 'the ideal of true spirituality', entailing that the truly spiritual person is committed to radical social transformation (*RM*: 116, 224–5).

principle of unity-in-diversity. See UNITY.

priority. See PRIMACY.

private language argument. Deployed by Wittgenstein against Cartesian rationalism and psychologistic theories of meaning, according to which the knowing subject has privileged epistemic access to an inner world of experience, thought and meaning. A private language, logically available to one speaker only, would be required by such a subject to begin discoursing about his private inner experience. But this is not possible, for language is a rule-following activity. A solitary language-speaker would be unable to tell the difference between actually following a rule and merely thinking he was doing so, so language must be social, not private. Correct usage depends upon the public institution of correcting mistakes. The argument is vitiated by ACTUALISM – rule-uniformity is a metonymic displacement of the PROBLEM OF INDUCTION, issuing in the *paradox of rule-following*. It requires supplementing by a depth-realist account of language, intentionality and social context. For the distinction between following a rule, and not doing so – necessary for the correct application of rules – can only be maintained on the assumption that in the former case the rule plays a causal role in generating the action, which the rule-following paradigm denies.

probability. Though not much discussed in CR, probability presents three important problems from its perspective. First, the demonstration of a probabilistic relationship does not permit a distinction between a necessary sequence of events and an accidental one (*PN3*: 10), because the experiences, though repeated, are just singular observational events. Their arrangement is likely a function of manipulation rather than of nature. Second, in speaking of a probabilistic relationship, empiricists conflate the existence of chance, or random elements, with our lack of knowledge of them (Sayer 1992: 191). Third, critical realists claim, the postulation of mechanisms requires a stronger version of causality than the 'causal inference' of empiricism, because it is only probabilistic association; it lacks the categories of 'producing', 'generating' or 'forcing'. Consequently, 'according to the realist theory of causality *qualitative* analysis of objects is required to disclose mechanisms' (Sayer 1992: 179).

Notwithstanding these objections, an abandonment of probability altogether brings its own difficulties, in particular that of how scientists can quantify in open systems, especially populations in the social world. In empiricist approaches, inferences about mechanisms and structures are made on the basis of a probabilistic generalisation from sample to population. If CR is to make headway in population analysis, especially in social science, some solution must be found.

Two main approaches are currently on offer. The first, associated with Ray Pawson, effectively treats frequencies as heuristics and uses sample data as a basis to inform INFERENCE about mechanisms. His C+M=O model treats quantitative data as outcomes which can lead us to EXPLANATORY mechanisms, which will operate dependent on context (Pawson 1989, 2000). In similar vein Dave Byrne (2002) treats variables as 'variate traces' – evidence for, though not actual representations of, phenomena.

Byrne's approach is also conducive to a more thoroughgoing 'ontological' one which, taking a lead from Popper, rejects the empiricist frequency approach to probability, particularly emphasising CR's second objection. In this approach phenomena are seen as probabilistic in character and reality is itself a series of nested probabilities, dependent on other probabilities in a Markov chain in order to be efficacious. Unlike in standard frequency approaches to probabilistic analysis, individual cases are themselves taken to have a probabilistic character which can be known. This approach was theoretically developed by Popper (1959, 1990) but as an empirical research programme is very new. It has been operationalised effectively by Wendy Dyer, through the use of case-based cluster analysis, to explain outcomes in the criminal justice system (Williams and Dyer 2004).

MALCOLM WILLIAMS

problem. Bhaskar claims to have solved most of the textbook *problems of* (Western) *philosophy*. This has been taken for hubris, but clearly, if the problems have been generated within and in virtue of an irrealist *problematic* and, in particular, in virtue of the absence of (adequate) concepts of ontology and alethic truth at 1M, of absence at 2E, of totality at 3L and of agentive agency at 4D (such that the problems of philosophy are unified by the absence of the concept of absence, as set out in Table 35), remedy of these lacks should make their resolution, in theory, possible. The real problem with Bhaskar's claim is perhaps that in professional philosophy, as Popper has noted, 'nothing seems less wanted than a simple solution to an age old philosophical problem' (cit. Morgan 2004b), and the basis for *this* very probably lies in what the paradox-ridden condition of philosophy reflects in the sociosphere (see OPPOSITES). The *problematicity* of philosophy is almost an article of faith, and, since philosophical problems are 'supervenient on life' (*P*: 9), provides an important diagnostic clue to social problems. Problems *matter*, in philosophy because it is causally efficacious, and in everyday life because they place us in *problematic AXIOLOGICAL choice situations* – confronted with them, we do not know how to go on, yet must (see also DILEMMA). Resolving problems of philosophy is intrinsic to immanent CRITIQUE and to the non-preservative SUBLATION of philosophical *problem-fields*. It strengthens the case for CR and emancipation, and against irrealism and the status quo.

Problems discussed in this work are entered or cross-referenced under *problem* and *paradox*. The key to the resolution of most (at 1M) is non-identity or the revindication of (depth-)ontology, facilitating avoidance of HOMOLOGY, i.e., endless regress, by

Table 35 Unification of the problems of philosophy by absence of the concept of absence

CR stadia	1M non-identity absence qua alterity (product)	2E negativity absence qua change (process)	3L totality absence qua holistic causality (process-in-product)	4D transformative agency absence qua intentional causality (product-in-process)
absence of a concept of absence generates	*epistemic fallacy* actualism	*ontological monovalence* monism	*ontological extensionalism*	*reductionist/dualistic disembodiment*
denegation of	ontology	negativity	totality	transformative praxis
modes of denegation of absence (illicit abstraction)	*destratification, fusion* (absence of a necessary distinction) identity theory	*demagicisation (positivisation), deprocessualisation; analysis of change in terms of difference* anthro-ethno-ego-present-centrism	*detotalisation, demediation fission* (absence of a necessary connection)	*de-agentification* desingularisation
ensuing opposition or gulf	flatness/depth	positivity/negativity	atomism/holism	determinism/voluntarism facts/values
key philosophical problems/dilemmas	*the problem of knowledge change (the one and the many), order and chaos, natural kinds and natural necessity)* solipsistic/transdictive problem complex, dilemmas of actualism	(foundationalism) monism *opposites (the one and the other, self-dirempion, negation)* dilemma of de-geo-historicisation	*alienation* the part and the whole	dilemma of *dualism,* dilemma of *reductionism* facts and values
social meaning	there is no depth, transfactuality or alethic truth	there is no history normalisation of the status quo	there are no splits or underlying unities	there is no intentional causality or embodied agency
world-historical problems	*science, social science; order vs. chaos*	*relativity (cognitive, cultural, paradigm); power₂ conflict*	*ideological screening of power₂ conflict and causal connections; alienation*	*agency, emancipation; value, critique*

	1M	2E	3L	4D
some related problems and paradoxes	akrasia; confirmation (ravens or Hempel's, etc.); erasure; falsification; free will and determinism; Goodman's; Grelling's; heap or omega (also 3L); Hobbesian problem of order; indeterminacy of explanation; induction (also 2E); material implication; Meno's; necessary and accidental sequences; other minds; scepticism; self-predicative (Platonic); Poincaré's (transduction); rule-following; subjunctive conditionals; Third Man (regress); universals (taxonomic knowledge)	absence and presence; anti-naturalism and naturalism re social science; fixism vs. fluxism; future contingents; individualism and collectivism re society; intrinsic and extrinsic; language and the world; idealism and materialism; mind and body; reasons and causes; reification and voluntarism re social structure; subjectivity and objectivity; time consciousness (e.g., memory); Zeno's paradoxes of motion	hermeneutical circles; individuation; originality; self-referential [also 1M] (liar, Nietzschean forgetting, preface, set-theoretic); post-festum; predicament of the Beautiful Soul	injunctive, paradoxes of de-agentification and reification
key to resolution in theory	*non-identity* (depth-ontology) avoidance of homology (alethic truth/depth, heterology); natural kinds, natural necessity, transfactuality primary polyadisation; thrownness	*negativity* and *mediation* co-presence; constellational identity (or unity), dispositional and rhythmic identity; SEPM; TMSA; critical or limit situation; meta-reflexively totalising (self-)situation	*retotalisation* (correct understanding of relation of part to whole) distinction structured totalities and aggregates; intra-active frontier; explanatory critique	*re-agentification* emancipatory axiology re-agentification in depth-praxis
key to practical resolution	abolition	of all	power₂	relations

Note The 1M–4D absence of the concept of absence deposits meta-critical voids at each stadion generative of Tina compromise formations which sustain the irrealist problematic. Regardless of their resolution in theory, the problems will persist so long as the power₂ relations that underpin them are not abolished in practice (see OPPOSITES).

Key source P. 244.

384 problem(s) of agency

differentiating structure from its instances. Turning the key involves a non-VALENT strategy of *problematising the question* as a propaedeutic to *de-problematising the situation*: (1) theoretically, by redescribing the alternatives within a transformed *problematic* invoking a deeper level of structure; and (2) practically, by altering the alternatives and their causal grounds. At 2E the key is the reinstatement of negativity. Moreover, the two main classes of problem at 1M and 2E, the *problems of change* and of *opposition*, respectively, are united and resolved by the concept of NEGATIVE REFERRAL (see also TABLE OF OPPOSITION); and, as Table 35 makes plain, the problems of philosophy overall are unified by the concept of absence, with ontological monovalence (2E) – underpinned by *fear of (ontologically oppositional) change* and itself underpinning the EPISTEMIC FALLACY (1M), ALIENATION (3L) and REIFICATION or NATURALISATION of the status quo (4D) – at the hub of their unification, and their *resolution* is unified by the dialectics of CO-PRESENCE of absence and presence.

A *problematic* is the structured field constituted by philosophies, philosophical traditions, theories, etc., within which alone meaningful questions can be asked or problems posed. It will screen out or occlude some questions and problems. It overlaps with PARADIGM, but specifically calls attention to screening. Thus the ANALYTICAL PROBLEMATIC screens out or *de-problematises* existential questions at 2E (ontological monovalence), and essential ones at 1M (actualism and the denial of natural necessity); the critique of actualism *re-problematises* ontology. *Complex* (as in the *transdictive complex*), *problem-field* and *ensemble* (the *irrealist ensemble*) are usually specifically IDEOLOGICAL problematics. *Problem-fields* are typically constituted by DIALECTICAL ANTAGONISMS (e.g., rationalism vs. empiricism). Complex also has a non-pejorative use (e.g., the *explanatory critical theory complex*). A *theory problem-field solution set* (TPF[SS]) (e.g., the ARCH OF KNOWLEDGE TRADITION) comprises the standard theoretical solutions to the problems encountered within a problem-field or complex, together with the means of arriving at them. The concept highlights that philosophies (and philosophy 'as a disciplinary matrix') 'come as packages {...} which are related dialectically to extra-philosophical reality' (*D*: 313; *DG*: 403).

problem(s) of agency. *The* problem of agency is the world-historical PARADOX that people are the agents of geo-history and yet are largely merely carried along by it (see AGENCY; SOCIAL STRUCTURE). More specifically, the *PROBLEM of emancipation*: who are the 'deep totalising conveyors' (*D*: 316) of the DIALECTIC OF FREEDOM to EUDAIMONIA? Here the AXIOLOGICAL commitment entailed by human action and speech is unequivocal: in order to be true to your human nature, it is *you* (no one can substitute for you) who must act *now* in solidarity with others to assist in bringing about a global society – immanent in our practices – in which my free development is as important as your own (see EMANCIPATORY AXIOLOGY; SELF-REFERENTIALITY). The (increasing) interconnectedness of the global village; the unprecedented, if contradiction-fraught, ideological and repressive power of the master-classes; and the capacity of some baneful structures (e.g., the thawing tundra) to persist without and even despite human agency, also entail that this solidarity must embrace the whole species – that today 'the minimal necessary unit for human emancipation is the whole of humanity' (*RM*: 170 n). The threat to the planet posed by the destruction of the environment, the proliferation of weapons of mass destruction and growing global inequities, imbalances and disharmonies lend the task a peculiar urgency and increase the scope for solidarity. It is no longer a question of socialism *or* barbarism: barbarism is here. The future itself is in jeopardy.

The ontological dependence of the DEMI-REAL on the CREATIVITY of people (the *asymmetry of emancipation*: see PRIMACY) entails the possibility of their dispensing with its oppressive and exploitative structures. Among recent developments giving grounds for hope in an escalating dialectic of liberation and disemergence are the rising ORGANIC COMPOSITION of ideas; the development of information technology and post-Fordist changes in the labour process driven by IT, with the requirement for knowledgeable operators; and the 'increasing intra-activity' entailed by globalisation – including of all the sectors of labour, whereby 'the struggles of each sector tend to become the struggles of all' (Hardt and Negri 2005: 125) – and 'the externalisation and globalisation of internal diseconomies (e.g., the ecological effects of pollution)', which render 'transformation tendentially increasingly radically negational in character' (*P*: 152). At the level of philosophy and social theory, CR makes an important contribution by, inter alia, *re-agentifying reality* (i.e., demonstrating the causal efficacy of reasons and the possibility of rational agentive AGENCY, notwithstanding radical decentring), and critiquing reificatory and fetishistic ideologies of de-agentification and disempowerment. Practical advice: in whatever social practices you are deploying your causal agency, do what you can consistently with the claims of a balanced life.

Other than this world-historical problem, major *problems of agency* include the MIND–body problem; whether REASONS can be causes; free will and DETERMINISM; the problem of personal identity (see SUBJECTIVITY); the problems of interpretation (see HERMENEUTICS; MENO'S PARADOX); the problem of SOLIPSISM; and the problem-field of SUBJECTIVITY, and of ERASURE (see also OBJECTIVITY/ SUBJECTIVITY) (*P*: ch. 5.2).

problem of akrasia (f. Gr. *akratos*, without power, lack of [self-]control). In moral philosophy, *akrasia* is weakness of the will, usually discussed in relation to ACTION. The *problem of akrasia* arises in the gap between a concrete axiological judgement prescribing a course of action in a particular situation and the actual performance of the action (*SR*: 186). An akratic action is one that is consciously performed against one's better judgement in the absence of internal malfunction or external compulsion, e.g., having another glass of wine when you judge that you really should not. Moral or normative ACTUALISM makes akratic action – an everyday occurrence – theoretically impossible (*PF*: 152), because the universal application of norms regardless of difference implies that ought = can. This detotalises 'the componental spring{s}' of action by assuming that will is the sole spring (*D*: 321). On depth-open realism akrasia is both possible and resolvable in practice via the concept of CONTRADICTIONS within a MULTIPLY CONTROLLED arena, such that will may be tendentially in opposition to habit, cognition, etc.

problem of alienation. A world-historical PROBLEM, first formulated by Hegel in the dialectic of MASTER AND SLAVE: 'the alienator (by virtue of being alienated from the PRAXIS of the alienated) is alienated from something essential to himself' (*P*: 244). Why does the alienator alienate? Bhaskar's ultimate answer is that he makes a fundamental (category) mistake in believing that it will serve his real interests. Who will de-alienate the alienator? The de-alienation of the alienator as well as the alienated is indispensable for the eudaimonian society. ALIENATION is a mainstay of many other

problem of articulation. See PROBLEM OF INDIVIDUATION.
problem of emancipation. See PROBLEM OF AGENCY.
problem of (Heideggerian) erasure. See ERASURE; REFLECTION.
problem of evil. See DETERMINATION; EVIL.
problem of free will vs. determinism. See DETERMINATION.
problem of identity. See IDENTITY.
problem of individuation and **problem of articulation.** In any kind of research, 'discrete observable items', but not theoretically postulated (unobservable) entities, 'can always be described in ways that are logically independent of each other', i.e., be *individuated* descriptively (*PN3*: 75, n53). How do we individuate real, unobservable items – a magnetic field, social structure, economy – and how specify their articulation internally and in relation to other entities (TOTALITIES)? In general this can only be done relationally: hence the concept of *ENTITY RELATIONISM* and the *relational conception of society*. The BOUNDARIES of particular totalities can only be specified by and within empirical research, and since, on a transcendental realist ontology, it is underlying mechanisms that generate the phenomenal flux of the world, the question of boundaries will in general be subsumed under that of intrinsic causal structure and its articulation with other causal mechanisms. The problems of individuation and articulation are then resolved by the theory with the most causal grip on reality. Where the research is well founded, the boundaries and structures are real (in the ID). However, on the 'intra-active frontier' of 3L HOLISTIC CAUSALITY (*D*: 125), it will sometimes be impossible to individuate entities, e.g., within the fluidity of a physical field. This provides a 3L analogue of 2E genuine occurrences of the PROBLEM OF INDUCTION in the coincidence of identity and change and of identity and difference. See also SYSTEM.
problem of induction. In its elementary Humean form, the problem that no number of instances can confirm an actualistically conceived general law; there can therefore be no warrant that the course of nature will not change tomorrow. It is solved, within limits, when laws are construed as causal powers intrinsic to the structure of things, which, when triggered, act with natural necessity and universality. This solution is limited in that it is restricted to 1M. When 2E is brought into consideration we see that (1) ontologically, the dispositional identity of a thing (e.g., a maturing person) with its *changing causal powers*, and (2) epistemologically, points of TRANSITION (*critical* or *limit situations*) in the EPISTEMOLOGICAL DIALECTIC, entail that there may be genuine occurrences of the problem of induction. The 'course of the deep structure of nature' (*D*: 36) may indeed change, as well as our knowledge of it. This is the co-incidence or *CO-PRESENCE of identity and change*, fundamental to the critique of IDENTITY THEORY. The problem of induction has a DUAL in *Poincaré's problem* or the *problem of transduction*: what justifies the supposition that the laws of nature continue to hold outside closed systems (*SR*: 30)? See also ALETHIA; ARCH OF KNOWLEDGE; HERMENEUTICS; HOBBESIAN PROBLEM OF ORDER; INFERENCE; NICOD'S CRITERION; PRIMAL SQUEEZE; PRIVATE LANGUAGE; SOLIPSISTIC, ETC.; TRANSITION; UNIVERSALS.
problem of knowledge. See IDENTITY; OBJECTIVITY/SUBJECTIVITY; ONTOLOGY; PROBLEM; TRANSCENDENTAL REALISM.
problem of Nietzschean forgetting. See REFLECTION.
problem of opposites. See OPPOSITES; PROBLEM.

problem of order. See HOBBESIAN PROBLEM OF ORDER.
problem of personal identity. See SUBJECTIVITY.
problem of subjectivity and objectivity. See OBJECTIVITY/SUBJECTIVITY; REFLECTION; SUBJECTIVITY.
problem of transduction (Poincaré's problem). See PROBLEM OF INDUCTION.
problem of universals. See UNIVERSALS.
problematic. See PROBLEM.
problematic axiological choice situation. See AXIOLOGY.
problematic of presence. See METAPHYSICS OF PRESENCE.
problem-field. See PROBLEM.
problems of philosophy. See PROBLEM.
process. DIRECTIONAL absenting or change (development), or 'the mode of spatio-temporalising structure' (*D*: 130). It is thus 'where structure meets events {...} the mode of becoming, bestaying and begoing of a structure or thing, i.e., of its genesis in, distanciation over and transformation across space–time. {It} is not an ontological category apart from structure and event; it just is structure (or thing), considered under the aspect of its story – of formation, reformation and transformation – in time', during the course of which it may acquire supervenient causal powers (*SR*: 215; see also ABSENCE; CAUSALITY; DCR; SPACE–TIME). It may be substantial (geo-historical transformation) or non-substantial (action-at-a-distance and intra-activity), and is sometimes deployed to refer to spatial alteriorisation or DISTANCIATION. It is irreducibly TENSED.

A *rhythmic* or *rhythmic process* is a less abstract category designating any specific process consisting in the exercise/impact of the causal powers of a structure or thing, whether micro or macro. There is normally a

> multiplicity of rhythmics at work in a single episode, such as the design of a book jacket, from the spatio-temporality of that process to the narrative of the designer's life, the lagged causal efficacy of her unconscious, her life-cycle as an organism and specifically as a woman, her daily space–time paths, the *longue durée* of differentially structurally sedimented social institutions and that of the social relations upon which they depend, the development of specifically civilised geo-history in the context of human history embedded in the rhythmics of species, genera and kinds, located in the physical development of a solar system, unravelling itself in the entropy of an expanding universe. (*P*: 98–9)

The concept thus possesses a continuous/discontinuous and product/process bivalence associated with the phenomenon of EMERGENCE (and disemergence), and highlights that processes have overlapping spatio-temporalities. Brereton (2004a: 93) has noted its expression in music, considered as 'the universal human capacity to demonstrate consciousness of space and time, by the intentional creation of structuring rhythm'. One might add gymnastics and dance. Lefebvre (e.g., 1995) deploys 'rhythm' in ways similar to Bhaskar's rhythmic to analyse the complex pulsation of the city.

A *dialectical process* (or *rhythmic*) (dp′) is 'any more or less intricate process of conceptual or social (and sometimes even natural) conflict, interconnection and change, in which the generation, interpenetration and clash of oppositions, leading to their transcendence in a fuller or more adequate mode of thought or form of life (or being), plays a key role'. However, such processes are 'not always SUBLATORY (i.e., supersessive), let alone preservative'; nor 'characterised by opposition and antagonism, rather than mere connection, separation or juxtaposition'; nor triadic in form (*D*: 3). Dialectical, as well as non-dialectical, processes constellationally contain 'analytic' or MORPHOSTATIC ones. The *Schillerian dialectic* (Schiller 1795) is a particular kind of dialectical process profoundly embedded in Judaic/Christian/neo-Platonic and Hegelian/Marxist thought whereby world history moves from an original undifferentiated unity or MONISM (Eden, 'primitive communism') via *diremption* or *ALIENATION* (the Fall, class society) to a higher differentiated unity (redemption, communism) and in which the second moment is seen as necessary to the eventual outcome. CR's rejection of monism (see also POLYAD) entails that any such dialectic is local and tensed (as well as open at either end), possessing minimally five, not three, terms; i.e., pre-class society is irreducibly plural, as is class society, and only a world eudaimonian society would possess the quality of genuine unity-in-diversity. In *EW* Bhaskar suggests, in an argument reminiscent of Hegel's interpretation of the Fall, that such a dialectic is a transcendentally necessary condition for humans to become self-consciously aware of their true nature: 'there could be no enlightenment without *avidya* {ignorance} and {. . .} if we are already enlightened, no recognition or realisation of it without a prior forgetting (or fall)' (38, n 23).

Being as such is *processual*. *Processuality* or *rhythmicity* (the second moment of the CONCRETE UNIVERSAL) is the quality of being in process. When a way of thinking denegates process by extruding concepts of absence and change, it *deprocessualises* it.

product, product-in-process, process-in-product, process–product ambiguity. See ABSENCE; PROCESS.

progress. See DIRECTIONALITY.

projection. See ALIENATION; BEAUTIFUL SOUL, ETC.; DE-AGENTIFICATION; FISSION/FUSION; MASTER–SLAVE.

prolepsis, proleptic (Gr. *prolēpsis*, f. *pro-* + *labein*, to take before). Anticipation, anticipatory; taking something future as already existing.

propaedeutic (Gr. *pro-*, before + *paideuein*, to teach) n., adj. (Concerning) preparatory instruction. *Propaedeutics* is the body of principles pertaining to this. The UNDERLABOURING that CR philosophy intends for the various sciences is *propaedeutic* in relation to their own work, and scepticism is 'the *propaedeutic* to immanent critique' (*P*: 10, e.a.). *Aristotelian* or *dialectical propaedeutics* is Bhaskar's term for 'the continual circulation of discourse in and out of the sphere of formal reasoning – in which meanings and truth values remain stable – typical of everyday life, but also necessary for the dialectic of dialectical and analytical reasoning essential to science' (*DG*: 394, e.a.). In the HERMENEUTICS of everyday life and science alike, we necessarily exploit 'cognitive, linguistic and material resources' from the past and outside in our efforts to arrive at a more adequate understanding of the world. Formal reasoning identifies errors and inconsistencies within our existing conceptualisations, and between them and the new material; it can do this only by giving signs a fixed meaning and uniform value. However, in order to move on to a more adequate conceptual determination, we have to *dialectically suspend* formal logic (move out of its sphere) and *bracket* the aporiai it reports, entailing transformative praxis and the unfixing of meaning, 'mediated by inter-subjective dia-logical inconsistencies' (*D*: 136), until we arrive at a new consensus. At this stage, analytical reason may again step in with a formal derivation of the result (in science, the deduction of a causal law, licensing the attribution of natural necessity), but no such derivation could be effected outside the context of the new framework of meaning, i.e., formal logic presupposes a common context of utterance. Analytical is thus 'overreached' by dialectical reason, and the circulation of discourse in and out of the sphere of formal logic is a necessary propaedeutic for any depth-inquiry. See also DIALECTICAL ARGUMENT; EPISTEMOLOGICAL DIALECTIC.

propositionality (quasi-) of theory and practice. See CONSISTENCY; DUALITY; DUALITY OF THEORY AND CRITIQUE; EMANCIPATORY AXIOLOGY; PERSPECTIVE.

psychic ubiquity determinism. See DETERMINISM.

psychoanalysis. A theory and a technique that grew from attempts to understand, and treat by hypnosis, physical symptoms with no organic causes. Freud (1929, 1938) introduced the idea of unconscious motivation to explain these and other symptoms, and argued that we all had experience of primary unconscious processes through dreaming. He also developed the technique of free association to access the contents of the unconscious. He argued that we are instinctively driven to seek pleasure through originally separate biologically given drives and that our experience of the world forces us to adopt a more realistic approach towards pleasure-seeking.

Freud argues that the principal mechanism we use to protect ourselves from the drives towards pleasure is repression – the containment of our socially unacceptable desires within the unconscious. The difficulty of containing these desires and the ways in which these desires can be camouflaged so that they can surface in everyday life gives rise both to sublimations (socially acceptable transformations of the drives, e.g., art, science, sport, etc.,) and symptoms (unproductive distortions of the drives). These processes depend upon symbolisation and take place within what Freud calls the secondary process. Later, when he was more concerned with the dynamics, and the motor of repression itself, Freud introduced a structural theory where inner reality is made up of an Id containing unconscious impulses, an Ego that has grown out of the Id and has accommodated to reality to some degree, and a Superego which develops out of the Id and is based upon the internalisation of, and reactions against, significant others, e.g., mother and father and the social attitudes of the culture and the times. It is the dynamics of these structures that is used to account for the behaviour of the person. Freud argued that the biologically driven pleasure-seeking individual was often asked by society and culture to modify their needs and wants. In Freud's view the moral strictures of society were far too exacting upon the sexual mores of the individual, and it was these unduly stringent requirements that led to neurosis. Development to maturity is seen as inextricably linked to psycho-sexual development with a sequence of oral, anal, phallic and genital phases following each other from infancy to puberty. Arrest at one of these stages and/or regression back to one of them is viewed as a rich source of neurosis. For Freud it was the father who was the most important person in a child's life, and the Oedipus conflict at about the age of five or six, when the child adopts a gendered identity and becomes a fully social being, the significant event in childhood.

Subsequent analytical thinkers and theorists have suggested an earlier origin for many of the processes that Freud saw culminating in the Oedipus conflict, and it is the relationship with the mother or primary caregiver in the first two years of life that has taken over the central role in much post-Freudian theory (e.g., Klein 1984a,1984b; Winnicott 1967; Fairbairn 1952). In Freud's later theory there were two major classes of instinct, Eros and Thanatos, the death instinct. Many analysts have been unwilling or unable to accept the death instinct and the degree to which constitutional aggression is a fundamental human characteristic. One alternative is to see an essentially innocent and constitutionally social infant geared up to making and sustaining relationships whose turn towards pure pleasure-seeking and aggression is based upon failure of early relationships. This viewpoint has been reinforced by the detailed study of babies and their relationship to mother by developmental psychologists.

There is a three-way split within the British Psychoanalytic Society based in 'The Controversial Discussions' (King and Steiner 1992) after Freud's death in 1939, which means that there are three different strands to both training and theory: the Freudian, Kleinian and Independent, the last including figures like Winnicott, Balint and Fairbairn. Bowlby (1988), who was strongly influenced by the Independents, developed ATTACHMENT theory from the intersection of psychoanalysis with animal ethology, developmental PSYCHOLOGY and SYSTEMS theory. Freud is still the most influential psychoanalytical theorist, not least because of the support for his earlier theories by Jacques Lacan. In the UK the Kleinian school is the most coherent alternative group to the Freudians, but a number of Independents are widely influential, in particular Winnicott.

Whether psychoanalysis is a SCIENCE continues to be controversial. However, Collier (1994), Rustin (1991b) and Will (1986) have all argued that it meets CR criteria of scientificity. Kathryn Dean (2003) uses an object relations aspect of Freud's theory to look at and criticise, from a CR perspective, the forms of social parenting of successive capitalist social formations in the post-enlightenment era. The REALITY PRINCIPLE, adapted from Freud, plays a crucial role in CR's EMANCIPATORY AXIOLOGY. See also Clarke 2003; Rustin 1991a.

GRAHAM CLARKE

psychology. A disciplinary domain and professional practice concerning individual cognitive processes and behavioural patterns, often directed to the understanding and amelioration of mental health.

Since it emerged as a separate discipline towards the end of the nineteenth century, psychology has followed a laboratory-experimental paradigm to investigate thinking and behaviour, and clinical work was for many years governed by a model of the psychologist as a 'scientist-practitioner'. This meant that a positivist project to accumulate knowledge about individual psychology, implicitly or explicitly viewed as the manifestation of an underlying universal human nature, was often in tension with the imperative to produce empiricist studies directed to immediate and pragmatic solutions to psychological problems. Critical responses to the mechanistic and reductionist cast of psychological theories and practices, particularly in *radical psychology* of the 1960s, usually appealed to phenomenological philosophy, and humanistic psychology was thus able to constitute itself as a 'third force' in and against the discipline.

The grounding for CR was first developed in social psychology with a *new paradigm* approach which provided a critique of laboratory-experimental approaches without lapsing into simple humanism, an approach captured in the slogan 'for scientific purposes treat people as if they were human beings' (Harré and Secord 1972: 84). Rom Harré was the driving force in this approach, which then spread into other domains of psychology, and inspired the elaboration of explicit connections between new paradigm work and CR (Manicas and Secord 1983). Notwithstanding the claims by advocates of CR to provide a genuinely scientific approach that would not reduce social phenomena to the domain of the individual, however, the main axes of debate over alternative conceptions of psychology have tended to sideline these arguments since the early 1980s. Three reasons can be identified.

First, the political conditions for a revival of a politicised radical psychology do not exist at present. The appeal to CR has been either consigned to the margins of mainstream psychological debate, and arguments about the relationship between CR and revolutionary Marxism in the discipline are not what traditional 'scientific' psychologists want to hear (Willig 1998). There is a correlative exclusion from journals and textbooks of arguments that are deemed to be 'outside' the discipline and which aim to connect socialist feminist perspectives on therapeutic change with CR (New 1996a).

Second, the discipline of psychology is heterogeneous, conceptually and methodologically. It absorbs and neutralises new models of the individual and forms of research with a facility that often surprises those who try to produce alternatives which aim to revolutionise its founding assumptions and professional practice: there are now forms of 'realism' that draw on CR argumentation to warrant laboratory-experimental study (Greenwood 1994). Furthermore, the radical critiques of medical models of mental health and illness which were directed against PSYCHIATRY and psychology also risk being undermined by the deployment of CR as a new vocabulary for quite traditional clinical practice (Littlejohn 2003).

Third, the space for alternative research in psychology has been occupied since the 1980s by *discursive psychology*, an approach which elaborates the 'turn to language' advocated by those involved in the new paradigm in such a way as to reduce investigation to the domain of conversational interaction and to disqualify attempts to ground conversation in any kind of external reality (Edwards and Potter 1992). Those involved in recent 'critical psychology' as an attempt to locate our understanding of psychological processes in political context without sliding into full-blown relativism have unfortunately been compelled to articulate arguments that are compatible with CR on the terrain of 'discourse' (Parker 2002).

IAN PARKER

pulse of freedom. See AGENCY; CONATUS; DCR; DIRECTIONALITY; EMANCIPATORY AXIOLOGY; SUBJECTIVITY.

punctualism. See BLOCKISM/PUNCTUALISM.

PVRS model. A general model of natural selection based on evolutionary change, of which more substantive accounts of natural selection (e.g., those found in biology) may

PVRS model

be considered particular tokens. Developed by Lawson (2003b), the model identifies four components as being essential to any natural-selection-based account: first, a *p*opulation of individuals of some type; second, the existence of *v*ariety within this population; third, a mechanism of *r*eproduction by which later generations of individuals evolve out of earlier generations; and, fourth, a mechanism by which certain members of the population are *s*elected. Lawson employs the model in order to examine the extent to which natural selection is an important mechanism in social theory. By identifying certain social processes as manifestations of the PVRS model he provides qualified support for the application of evolutionary theorising in the social realm. See also FUNCTIONAL EXPLANATION.

PHILIP FAULKNER

QR

quadruplicity, multiple. See CONCRETE UNIVERSAL ↔ SINGULAR.
quale, pl. **qualia** (L. neut. pl. of *qualis*, of what sort having some quality or other). (1) Those properties of CONSCIOUSNESS/MIND which determine 'what it is like' to have sensations, perceptions, etc. (2) Constant conjunctions of atomistic events as apprehended in sense-experience in the EMPIRICIST conception of a causal law (see *RS2*: 211–2).
quantitative methods. See METHOD, QUANTITATIVE.
ρ (rhō) transform. See EPISTEMOLOGICAL DIALECTIC.
race, racism, ethnicity. The term *race* has its origins in the natural sciences of the eighteenth and nineteenth centuries and their efforts to establish objective criteria for the classification of humankind. Race was taken to be a scientific description of naturally occurring divisions amongst people. Much time and effort was expended in producing classifications of races, usually arranged hierarchically and with a range of attributes held to be racially derived. Unfailingly, such classifications reflected wider social relations of power – between the colonisers and the colonised, between master and slave, between ruling classes and working classes, and between men and women – and shared the common feature of anchoring social divisions within nature.

However, the scientific basis for the claims about races proved to be vulnerable to science itself. By the 1930s and 1940s a number of prominent scientists had come to the conclusion that in the understanding of human differences the concept of race was valueless. This has become the commonly held view amongst natural scientists and it undermined the belief – central to *racism* – that races were ontologically real.

Despite its demise as a natural scientific concept, race has persisted in the vocabulary of Anglo-American social science but, deprived of its ontological status, its meaning and relevance for the analysis of the social world have been the subject of intense debate, which concentrates on the referent of the term. Two broad lines of argument have been advanced to defend the concept: those drawing upon interactionist traditions within sociology and those based within postmodernism.

Interactionist sociologies, largely derived from the work of Weber, claim that, though race classifications have no ontological validity, they do continue to function as forms of practical understanding. To study race is therefore to study those forms of popular understanding in which ideas about race and races figure prominently. Postmodernists, on the other hand, argue that racism constitutes a particular form of SUB-JECTIFICATION, one by means of which people come to understand themselves and others as belonging to races. Race and racism, according to this view, are discourses whose meanings are historically variable and socially constituted.

Race concepts have also been challenged and, outside the USA and the UK, largely replaced by the notion of *ethnicity* as a descriptor of group formation. Ethnicity has the great merit of focusing on social processes of group formation and of being unencumbered by associations with nineteenth-century pseudo-science. However, as a social scientific concept, ethnicity remains flawed and for much the same reason as the concept of race, namely that it collapses the subject–object distinction essential to social research. That is to say that both interactionist and postmodernist approaches to race and ethnicity stress their subjective dimension, although for different reasons, whilst none the less continuing to regard them as objective features of the social world. Realism is able to offer a more robust and coherent view of the subject–object relation with regard to race and ethnicity.

The realist has no difficulty with the notion that popular ideas – about races, about astrology, about inequality, about ethnicity – are of relevance to social analysis in so far as people's behaviour is connected in significant ways to the beliefs they hold. What the realist does object to is (1) that such beliefs constitute all there is to be known about the social world; and (2) that such beliefs are capable of empirical demonstration in such a way as to form objective dimensions of group boundaries (such that one can identify 'ethnic groups' on the basis of self-identification and then correlate ethnicity with variables such as social class, voting behaviour and the like).

Realist research into racism and ethnicity has sought to restore the distinction between what people believe, the structural contexts which shape those beliefs and the role of theory in the analysis of social life. This takes a number of forms but key amongst these are: the relationship between ideas about race and ethnicity and discrimination; the implications of the realist view of ethnicity for social research, emphasising particularly the methodological limits of correlational analysis and successionist models of causality in favour of case-study approaches and MORPHOGENETIC models of social development; the critique of notions of race and ethnicity in favour of a sociology of modernity and group formation. See also ANTHROPISM; COLONIALISM; IDENTITY; IDEOLOGY; MASTER–SLAVE-TYPE RELATIONS; Banton 1997; Carter 2000; Fenton 2003; Gilroy 2001; Malik 1996; Porter 1993.

BOB CARTER

radical negation. See ABSENCE.
rational agency, criteria for. See FREEDOM, ETC.; SOLIDARITY.
rational choice theory (RCT). A paradigm with totalising pretensions, RCT is basically the mode of expansion for neo-classical economics, since it transmutes *economic* man [sic] into (abstractly) *universal* man. Economic man came to dominate economics in the late nineteenth century, and has since gained an increasing hold on mainstream economists. Even though this model was never genuinely demarcated and thus always prone to diffuse expansion, it was not until the 1970s that an actual totalising project surfaced in social science. Today RCT is a powerful current in all branches of social study and reform. The essential features of economic and universal man alike are instrumental rationality, measurement and scarcity, and taken together these tenets entail an unlimited drive towards allocating resources/means in ways that produce the best-possible rewards in quantitative terms. Universal man is typically portrayed

formally as a mathematical function that is to be optimised under given conditions that can (in game theory and information-theoretical economics) include other optimising agents with differing endowments. RCT conflicts in just about all possible respects with CR and is a major object of critique and resistance by critical realists. In economics it is a principal aspect of Lawson's (1997, 2003b) critique of the mainstream approach, even though he considers deductivism, and not economic man, to be the core of modern economics. Beyond economics, it is analysed, deconstructed and criticised from various angles in Archer and Tritter, (eds) (2000), where the important relationship between RCT, critical theory and NEO-LIBERALISM is also a guiding interest. See also ATOMISM; INDIVIDUALITY; METHOD; REASON; SOCIAL STRUCTURE.

PETER NIELSEN

rational directionality in geo-history. See DIRECTIONALITY.

rational explanation is the explanation-form whose content is *reason explanations*, i.e. in which reasons are cited as explanatory of human ACTION, whether in a causal (e.g., CR) or non-causal (e.g., HERMENEUTICS) sense. See also EXPLANATION; HERMENEUTICS; TELEOLOGY; WANTS.

rational reconstruction. (1) in analytic philosophy, maths, computer science, etc., 'translation of a discourse of a certain conceptual type into a discourse of another conceptual type with the aim of making it possible to say everything (or everything that is important) that is expressible in the former more clearly (or perspicuously) in the latter' (*CDP*: 774), e.g., empiricist attempts to translate discourse concerning physical objects into discourse concerning immediate objects of sense-experience. (2) Development of a putatively more adequate theory via immanent CRITIQUE, e.g., Bhaskar's reconstruction of the logic of scientific development in *RS*. (3) In Habermas, the translation of intuitive consciousness of linguistic rules and know-how into a logical form to promote communicative competence, a process necessarily involving realism (Bhaskar 2003a: 484). Properly conceived, *deconstruction* in semiotics would be consistent with reconstruction in HERMENEUTICS, constituting with it a DUALITY characterised by non-arbitrary interconnection (*D*: 147). See also PHILOSOPHY OF SCIENCE.

rationalisation. See CRITICAL NATURALISM; TELEOLOGY; WANTS.

rationalism. In philosophy, the view that reason is the main or even the sole source of knowledge. Exponents include Plato; the seventeenth-century 'continental rationalists', Descartes, Leibniz and Spinoza; Hegel after 1806; and, in the twentieth century, Husserl, Popper, Chomsky and most structuralists. *Pre-critical rationalism* is prior to Kant. Historically, rationalism contrasts with EMPIRICISM, the view that experience has primacy in epistemology. Rationalists privilege the A PRIORI and deduction, empiricists the A POSTERIORI and induction. Rationalists typically take the procedures of mathematics or geometry as the paradigm for procedure in all other areas, holding that it is possible to deduce all the laws of nature from a few basic axioms, often equated with 'a fixed innate core of essentially human knowledge' (*SR*: 88). Science would then 'appear as the simple realisation of philosophy or as the automatic product of a practice (or method) authenticated by it' (*PN3*: 7) and knowledge as 'universal-and-necessarily-certain' (*P*: 181), hence incorrigible. This has the corollary 'that science should have no effect on philosophy (cf. Popper, Lakatos)' (*SR*: 236), which

396 rationalism

is contradicted by the history of philosophy and by SOCIOLOGY OF KNOWLEDGE. Rationalism is thus 'classically the victim of the SPECULATIVE ILLUSION' (*D*: 194). A form of ANTHROPOREALISM, in which being is defined in terms of the human attribute of reason, it has tendencies to both epistemological absolutism or ENDISM and FUNDAMENTALISM. (Empiricism is likewise anthroporealist, defining being in terms of human experience, and fundamentalist.) A priori reasoning and deduction do of course play an important role in knowledge production; it is only when they are privileged as the only or main source of knowledge in general to the neglect of other elements that they are problematic. Rationalists find it difficult to explain matter and contingency (since natural is a species of logical necessity), why experiments should be necessary (since laws are deducible a priori without the aid of experience) and how scientific change is possible (since anthroporealism obliterates the TD/ID distinction, tying the world to human being).

Kant tried to reconcile the competing claims of rationalism and empiricism by marrying EMPIRICAL REALISM with transcendental idealism, 'seeing reason as furnishing synthetic a priori principles imposing form on the matter received through the senses' (Bhaskar 2003a: 322). To 'sustain the concepts of the necessity and universality of laws', Kant 'involuted' ontological structure within the human mind (*P*: 204), thus de-ontologising the world and splitting it into a knowable phenomenal and an unknowable noumenal or transcendent realm. Bhaskar's transcendental realism, by contrast, seeks to sublate rationalism and empiricism by showing that knowledge of natural necessity a posteriori can be achieved by scientific depth-investigation.

Rationalism constitutes a dialectical pair with empiricism in that their antagonism over whether laws are deduced a priori by thought alone, or induced a posteriori from sense experience, is grounded in common category mistakes: the EPISTEMIC–ONTIC FALLACY, presupposing ACTUALISM, and ultimately ONTOLOGICAL MONOVALENCE. (From the perspective of actualism, the rationalist and empiricist positions are the only alternatives.) They therefore share many assumptions. Ontological actualism at 1M means that experiment and applied science, ontological stratification and epistemic without ontic change are impossible. Ontological monovalence at 2E makes change both in theory and in the objects of theory impossible. Ontological EXTENSIONALISM at 3L renders impossible HOLISTIC CAUSALITY, EMERGENCE and 'the constellationality necessary to prevent detotalisation of knowledge from its objects, which is an inexorable effect of Cartesianism' (*P*: 195). Dualistic disembodiment (a consequence of rationalism) at 4D 'is tacitly complicit with physicalist reductionism' (a consequence of empiricism). On rationalist assumptions, the illicit FUSION of the epistemic fallacy generates, not EMPIRICAL REALISM, but *objective* and *subjective conceptual realism*, whereby real objects are (1) implicitly seen as constituted, hence deducible by, reason presented as objective but in reality modelled along human lines, and (2) explicitly viewed as constituted by human thought or discourse, respectively. (1) involves *hypostatisation* (Gr. *hypostasis*, that which lies under, substance) of ideas and conceptual forms, i.e., treating them as substances. This illustrates rationalism's tacit complicity with empiricism: ideas are 'brute facts' and facts hypostatised ideas (*D*: 313). (2) is vulnerable to the *antinomy of transcendental pragmatism*: if nature has the status of 'a constituted objectivity, then it cannot yield the historical or empirical ground of the constituting subjectivity or knowledge; conversely, if nature is the historical ground of subjectivity and/or an empirical ground of knowledge, it cannot be regarded solely as a constituted or posited objectivity: it must be essentially *in itself* (and only, so to speak, contingently a possible object of knowledge for us)' (*SR*: 7 n18). It is a species of VOLUNTARISM, (see also SOCIAL CONSTRUCTIONISM). *Theoreticism* or *rationalistic intellectualism*, 'founded on a barely disguised contempt for the agent's point of view', and with 'traditional-authoritarian, utilitarian-technocratic {. . .} and Stalinist variants', tends to see theory as having unmediated efficacy in practice, leading to the *hyper-rationalist* view that practice has no role in the development of theory. (Its opposite is *ultra-positivism* or *ultra-pragmatism*, which sees theory as the direct expression of practice; its converse is *practicism* or *irrationalist anti-intellectualism*, which 'ignores the cognitive bases of action') (*SR*: 170, 174).

Between them, rationalism and empiricism generate PRIMAL SQUEEZE on empirically controlled theory and natural necessity, respectively, resulting in the absence in either of notions of empirically controlled investigation into deeper layers of ontological stratification or empirically ascertainable alethic truth. This has determined the trajectory of Western philosophy since Plato: historical determination by rationalist epistemology, which was 'subjectivised and inwardised by Descartes, to lay the way open for the structural domination of {. . .} empiricist ontology, essentially laid down by Hume, but involuted and modified by Kant' (*D*: 383). Rationalism was the dominant partner historically in that it 'defined the criterion to which empiricism must conform' – knowledge as universal-and-necessarily-certain – thereby fixing actualism and ontological monovalence as 'the ultimately determinant moments in irrealist problematics' (*D*: 194). Empiricism won out because of its resonance with wider currents in modernity: 'bourgeois individualism, Newtonian science and the post-Hegelian kenosis of reason' (*D*: 182).

Today rationalism no longer has credibility 'as a general metaphysic' – 'we can no longer believe that the world is the work of reason or even that reason is at work in the world (that is, that it is there for us to discern, or perhaps for the sake of being discerned by us for it)', and foundationalism has been discredited. It may, however, still have purchase in the phenomenology of science (in moments of CREATIVITY we experience ideas as coming 'from nowhere') and as a regulative ideal for science (captured in Einstein's dictum that 'God is sophisticated but not malicious') (*SR*: 97). Since the nineteenth century, rationalism has been used in a broader sense to indicate a humanistic outlook which denies miraculous causes and sees religion as an expression of human aspirations and morality as ultimately based in conscience rather than in authority, divine or social.

rationality. The various 'levels' of rationality distinguished by Bhaskar are set out in Table 36; for the general concept, see REASON.

reading. See HERMENEUTICS; METAPHYSICS OF PRESENCE.

re-agentification. See AGENTIFICATION.

real, domain of. See REALITY.

real definition. See DEFINITION.

real essence. See ESSENTIALISM; NATURAL KIND.

real negation. See ABSENCE.

realia (f. L. *res*, thing, late L. *realis* [pl. *realia*], real thing). Really existing things or objects.

realism. Historically, realism was a doctrine about the independent and complete existence of UNIVERSALS (properties) (*predicative realism*). It was opposed to nominalism,

398 realism

Table 36 Levels of rationality

Scientific Realism and Human Emancipation *(181 ff.)*		**Dialectic** *(261)*	**Plato Etc.** *(147, 153)*
1. technical	instrumental reason	1. instrumental (including technical)	1. instrumental
2. contextually situated instrumental			
3. practical	practical = criticism, critical reason	2. critical	
4. explanatory-critical (explanatory critique)	explanatory = critique	3. explanatory critical	
5. depth-explanatory critical (ideology critique)	emancipatory reason	4. depth-explanatory critical	2. (a) depth-explanatory
6. depth-rationality (resulting from depth-inquiry)		5. totalising depth-explanatory critical	(b) totalising
		6. dialectical reason	3. totalising depth-explanatory
			4. dialectical or absolute, tendentially approaching:
7. historical rationality	historical reason	7. geo-historical rational directionality	5. geo-historical rational directionality

viz., the view that only particulars exist. Nominalists argued that general terms and predicates are merely names for classifying particulars in terms of their similarities and differences. Realists claimed that universals are real entities referred to by general names and predicates, and argued that they are necessary for grounding the similarities and differences among particulars. Currently, realism has a more general meaning. It affirms the objective reality (existence) of a class of entities, and stresses that these entities are mind-independent. So realism is primarily a metaphysical thesis. But many philosophers think that it has a semantic as well as an epistemic component. The semantic thesis urges that a certain discourse or class of propositions (e.g., about theoretical entities, or numbers, or morals) should be taken at face value (literally), as purporting to refer to real entities. The epistemic thesis suggests that there are reasons to believe that the entities posited exist and that the propositions about them are true. Given this epistemic thesis, realism is opposed to scepticism about a contested class of entities.

Anti-realism can take several forms. One of them is anti-factualism: it takes the contested propositions (e.g., about unobservable entities, or mental states, or numbers) literally but denies that there are facts that make them true. Hence anti-factualism takes the contested propositions to be false and denies that there are entities that these propositions refer to. Mathematical fictionalism and ethical error-theory are species of this form of anti-realism. Another form of anti-realism is non-factualism: the 'propositions' of the contested class are not *really* propositions; they are not apt for truth and falsity; they are not in the business of describing facts. Traditional INSTRUMENTALISM, ethical non-cognitivism and mathematical formalism are cases of this kind of anti-realism. A third popular form of anti-realism comes from Michael Dummett, who argues that the concept of TRUTH is epistemically constrained. Dummettian anti-realism does not deny that the contested propositions (e.g., about numbers) can be (and in fact are) true, but argues that their truth cannot outrun the possibility of verification. This species of anti-realism equates truth with warranted assertibility. Mathematical intuitionism is a case of this form of anti-realism.

Given this Dummettian perspective, realism is the view that every proposition of the contested class is either true or false, independently of anyone's ability to verify or recognise its truth or falsity. Hence realism is taken to subscribe to the logical principle of bivalence. For many realists, realism has nothing to do with truth. But it is arguable that a realist commitment to a non-epistemic conception of truth best captures the realist claim of mind-independence. A non-epistemic conception of truth (the most popular, and controversial, of which is that truth is correspondence with the facts) allows for the possibility of a *divergence* between what there is in the world and what is issued as existing by an epistemically right theory, which is licensed by the (best) evidence or other epistemic criteria. Verificationist views preclude the foregoing possibility of divergence by accepting an epistemic conception of truth. Putnam's *internal realism* claims that questions about the reality of a set of entities or of the truth of a set of propositions are internal to conceptual schemes. It denies that there is one true description of the world and argues that there can be different and equally true such descriptions. This view is consistent with realism if the different descriptions are compatible with each other. But it leads to a relativist anti-realism (via an epistemic conception of truth) if taken to imply that logically incompatible descriptions can all be simultaneously true. *Direct realism* is the view that perception acquaints us directly with physical objects, which exist independently of our perceptions. It is contrasted to *representative realism*, the view that, though there is an independently existing material world, we do not directly perceive external objects; rather, our perception is MEDIATED by sense-data. Realism is opposed to IDEALISM, viz., the view that everything that exists is ideas or made up from ideas. Realism is also opposed to phenomenalism, the view that everything that exists is (actual and possible) sense-data. *Common-sense realism* argues that middle-sized material objects exist (and endure) independently of their being perceived. It also argues that talk about material objects cannot be fully translated, without loss of content, into talk about actual and possible sense-data. The logical positivists declared meaningless the whole issue of realism, arguing that neither realism nor its denial could be verified in experience. But they held on to a concept of EMPIRICAL REALISM, which asserts that *after* we have accepted the framework which posits middle-sized material objects it is an empirical matter what is real and what is not, the chief empirical criterion being that whatever is real is in space and time.

The realist stance can take many specific forms. *Causal realism* is the view that there are genuine and irreducible causal facts that make causal statements true. So it is opposed to the Humean view that causal laws are mere regularities. *MODAL realism* is the view that there are modal facts which make modal statements (e.g., counterfactual conditionals) true. In its most extreme form it accepts the reality of possible worlds other than the actual. *DISPOSITIONAL realism* is the view that objects have real and irreducible dispositions (causal powers). *Categorial realism* is the view that CATEGORIES are constitutive features of the world. *Scientific realism* is the view that most unobservable

400 realism, direct

entities posited by scientific theories exist. *CONSTELLATIONAL realism* is realism about everything. *CRITICAL REALISM* is a thorough realist position which is based on the following claims. 1. Intransitivity: reality consists of independent and enduring objects, structures and generative mechanisms. 2. Transfactuality: causal laws are universal, real and irreducible; they hold with necessity and operate independently of the closed systems (most typically the laboratory) in which they are probed. 3. Stratification: reality is differentiated and layered; the causal mechanisms that operate at each level are self-sustained and not necessarily reducible to lower-level mechanisms. These three claims are the kernel of *TRANSCENDENTAL REALISM. ALETHIC* (from the Greek word for 'truth') *realism* and *moral realism* capture the idea that truths and morality, respectively, are grounded in the world. *Meta-realism, meta-Reality*: see LITERARY THEORY; META-REALITY. *Horizontal* or *transfactual realism* and *vertical* or *theoretical realism*: see DEPTH; DIFFERENTIATION. *Conceptual realism, intuitionist realism, voluntarist realism, transcendent realism*: see EMPIRICAL REALISM. *Eidetic realism*: see EIDOS. See also CATEGORY; ETHICS; IRREALISM; Bhaskar RS3; Devitt 1984; Dummett 1982; Psillos 1999; Putnam 1983; C. Wright 1992.

STATHIS PSILLOS

realism, direct. See REALISM.
realism, internal. See REALISM.
realism, metaphysical or ontological. See DEPTH; HOLY TRINITY; IDEALISM VS. MATERIALISM; ONTOLOGY; REALISM.
realism, moral. See CRITICAL NATURALISM; CRITICAL REALISM; DCR; EMANCIPATORY AXIOLOGY; ETHICAL IDEOLOGIES; ETHICS; HOLY TRINITY; NEEDS; REALISM.
realism, perceptual. See PERCEPTION; REALISM.
reality. A 'potentially infinite TOTALITY of which we know something but not how much' (*D*: 15); while science makes genuine discoveries, it suffers from 'cosmic incapacity' (see also ULTIMATA): it can never know what lies beyond its current frontier, though PHILOSOPHY may establish contingently necessary truths about the general contours of its categorial structure (see also EPISTEMOLOGY, ETC.). The CR concept of reality is a generalised one; given that human being contingently exists, it includes the world of concepts, including false concepts, and all the products of illusion and of imagination, virtualisation and hyperrealisation, including never anywhere existents (fictive beings in their semantic and psycho-social identity) (see ABSENCE; CRITIQUE; IDEOLOGY; ONT/DE-ONT; REFERENCE; TINA SYNDROME) – 'everything is real' (Bhaskar 1997a: 139; cf. *MR*: 44), or, CR is a *CONSTELLATIONAL realism*.

Criteria for ascribing reality: see CAUSALITY; PERCEPTION.

Domains of reality. Transcendental analysis of scientific experiment and human intentional action in general shows that generative mechanisms, the events they generate, and the experiences in which they are apprehended are sui generis real and distinct. They thus – as illustrated in Table 37 – 'constitute three overlapping domains of reality' (*RS2*: 56): the domains of the *real* (d_r) (comprising mechanisms, events and experiences), the *actual* (d_a) (events and experiences) and the *empirical* (d_e) (experiences), such that domain of the real > domain of the actual > domain of the empirical. Bhaskar later extended the domain of the empirical to that of 'the subjective', to embrace

Table 37 Domains of reality

Real	**Actual**	**Empirical**, *more generally* **Subjective** *(the semiosic)*
experiences/concepts and signs events *mechanisms*	experiences/concepts and signs *events*	*experiences/concepts and signs*

concepts as well as the empirical; and Nellhaus has suggested the even wider concept of the (human) 'semiosic' (see SIGN), which seems appropriate *providing* it is borne in mind that the semiosic, like the empirical, also has actual and real dimensions (for a dissenting view, see CRITICAL DISCOURSE ANALYSIS).

It is thus important to note that events and experiences are no less real than mechanisms, and experiences no less actual than events. Experiences (/concepts and signs), however, pertain (in the sense intended) solely to the specifically human subjective. Events and experiences are embraced by the real in so far as they function as mechanisms, e.g., a war may contribute to the formation of a psychic structure generative of post-traumatic stress disorder and human beings themselves, 'like any other empirically given object, are fields of effects' that possess their own causal powers (*PN3*: 111); and concepts and signs may be incorporated in emergent conceptual and semiotic structures and are themselves structurally generated. The distinction between the domains of events and experiences, on the one hand, and mechanisms on the other does not coincide with that between the transitive (epistemological) and INTRANSITIVE (ONTOLOGICAL) dimensions of reality. Mechanisms, events and experiences pertain both to the process of knowledge production and to that which we come to know. This does not gainsay that science takes as its intransitive object proper mechanisms rather than events and experience as such.

In the sphere of morality, a distinction can likewise be made between the domains of the real (dm_r) (the objective good for human kind or moral ALETHIA) and of the actual (dm_a) (actually existing morality).

Absolute reality, relative reality, demi-reality: see DEMI-REALITY; META-REALITY. See also CRITICAL NATURALISM; DCR; REALISM; TRANSCENDENTAL REALISM.

reality principle. (1) in Freud, the means whereby, in the interest of self-preservation, the human mind forms 'a conception of the real circumstances in the external world' (engages in 'reality-testing') and tries to 'make a real alteration to them' via attention, perception, memory, judgement and action (Freud 1911: 302–3). The regulative principle of the secondary process of psychic functioning, it supersedes, both at the level of ontogeny and phylogeny, the primary process, governed by the *pleasure principle*, without, however, abolishing that principle, which in the healthy human being is sublimated rather than repressed. It is thus a distinguishing feature of civilisation, of the mature and responsible human adult by comparison with the child.

(2) Bhaskar incorporates a modified version of this Freudian principle into his DCR system. It is encountered by the first act of REFERENTIAL DETACHMENT, precipitated by experience of desire or lack, during the process of PRIMARY POLYADISATION (at the level of the individual) and during the process of human EMERGENCE geo-historically (at the level of the species). Referential detachment of material objects

402 **reality testing**

brings 'recognition that the referent has causal powers of its own' (*D*: 230) and, when we learn to referentially detach structure, i.e., distinguish intrinsic structure from actual events and come to understand transfactuality (that structures do not always manifest themselves in actuality), leads on to an understanding of TRUTH as ALETHIC. The reality principle is, then, synonymous with (epistemologically mediated) AXIOLOGICAL *necessity* and alethic truth, the causal powers of and real reasons for things in the world (including social things, e.g., power$_2$ relations and countervailing social solidarity) in which we must act; or it is 'alethically attainable truth and practical wisdom' (PHRONESIS) (*D*: 268), and as such is the main motor of the DIALECTIC OF THE EDUCATION OF DESIRE FOR FREEDOM. It thus partakes of all moments of the 1M–4D ontological–axiological chain, and it is particularly important to note that it is dynamic and developing (its 2E aspect). While the causal structures of the natural world are relatively enduring, those of the social world are far more subject to change and development. Further, we deepen our knowledge of the reality principle, including our own powers to be free (or not) via the transitive processes of the dialectics of truth and freedom (themselves axiological needs). Dean (2003) advances the view that, in relation to subjectivation (the process of constitution of subjectivity), 'late' or 'disorganised' capitalism does not have a reality principle as such but a *VIRTUAL reality* principle.

reality testing. See REALITY PRINCIPLE.

***Realrepugnanz,* Kantian**. See CONTRADICTION.

reason. A multi-referential concept that refers to (1) qualities of the intellect associated with coherence, consistency, flexibility and objectivity; (2) the quality of claims about the world; and (3) the quality of actions on and in the world. In the Western philosophical tradition there is a strong tendency to collapse these three aspects in such a way as to naturalise the disengaged, contemplative reason(ing) of the modern Western philosopher and/or scientist. Reason or rationality now becomes the attribute of individuals who engage in interior deliberations about possible courses of ACTION. Here reason is related to individual intentionality, and reasons are viewed as preceding and informing action. In so far as this is the case, individuals and their actions will be deemed rational, rational individuals being individuals who can give an account of the reasons for their actions. In this kind of case, critical realists argue, reasons are causes of action (see MIND). It is this kind of reason-derived and consequential action that CR is interested in promoting since it is in this kind of action that human FREEDOM consists. For CR, the elimination of obstacles to the universal human exercise of reason can be effected in part through the activities of PHILOSOPHY and SCIENCE. This is clear in the work of Bhaskar, for whom philosophy is the underlabourer of science and science is the underlabourer of individuals. Underlabouring, which is necessary because the world does not present itself to us in an immediately intelligible form, involves giving an account of how the world works at the level of the real; a level which lies behind directly experienceable appearances. Underlabouring is necessary because the directly encountered world is not the whole of the world, so experience, unaided by theory, can lead us astray about the true nature of reality. Bhaskar's philosophy of the experimental physical sciences (*RS*) is intended to provide a true account of the rationality of scientific practice. This true account establishes itself against the inadequacies of positivism (inhering in its empiricist ontology) and of idealism (its constructivist ontology). Against positivism, Bhaskar argues that experience alone cannot provide reliable knowledge of the world; against idealism, that the world has a reality independent of our constructions. Rational human action requires recognition of these truths. This theme about the balance of enablement and constraint within which human action is situated informs Bhaskar's philosophy of the social sciences (*PN*), this being a philosophy for a social science whose explanatory adequacy inheres in the explanation of illusory beliefs – 'false consciousness' – about the social world (social science as EXPLANATORY CRITIQUE). Bhaskar's attribution of false consciousness rests not on an acceptance of differential individual endowments of reason but on the necessity of false consciousness for the reproduction of a specific mode of life, namely, that of capitalism. IDEOLOGY as misrepresentation is here a structural given. It has the effect of inhibiting the exercise of an emancipatory rationality.

To provide an explanation of false consciousness is to offer people the means of developing a more truly rational form of REFLEXIVITY than is available by means of their own everyday experiences. It is to give people good reasons for changing those beliefs that misinform them about their own real interests (see NEEDS). What is in question here is the remedying of 'unfulfilled' being through the uncovering of cognitive errors (*SR*: 191). The expert uncovering of cognitive errors is necessary because citizens inhabit a world which is structured so as to misrepresent to them crucial aspects of their everyday reality. So, without philosophical and scientific expertise, the everyday rationality of non-experts remains deficient. It is functional for a way of life which leaves their real interests unsatisfied.

Bhaskar sets out from the universalism of philosophical and scientific understandings of reason, and, in his invocation of psychoanalytic ideas, appears to reject also the distinction between rationality and the emotions. However, he does not engage directly with the latter topic, which receives extended treatment in the work of Margaret Archer and Andrew Collier. Archer's recent work on 'practical knowledge' (2000, 2003b) makes it clear that the distinction is meaningless in relation to activities coming under this description, e.g., dancing, cooking, response to music. Collier (1999, 2003b) explicitly rejects the distinction between rationality and the emotions in favour of an understanding of rationality as a property of the emotions rather than as emotion's 'other'.

Questions about the GEO-HISTORICITY of rationality and about the rationality/emotion relationship are given attention in Dean (2003). Here the conception of KINESTHETIC rationality is used to understand a pre-capitalist, artisanal rationality which is inscribed directly on to the body rather than being mediated through reflexivity. Kinesthetic rationality does not 'know' the split between rationality and the emotions which was brought into being through the EMERGENCE and development of capitalism. The splitting of rationality and the emotions is grounded in the experience of systematic differentiations of relations and activities rendered necessary and possible through capitalist industrialisation. What results is a depersonalised or disengaged mode of 'abstract rationality' – the rationality of philosophy and science on the one hand and, on the other, the atomised, privatised rationality of *homo oeconomicus*. The emancipation to which CR aspires requires the transcendence of these complementary modes of rationality through the re-fusion of rationality and the emotions. Dean proposes the concept of anaclitic rationality (f. Gr. *anaklinein*, to lean one thing back upon the other) to connote the kind of reflexive, public-spirited sociability needed (as a necessary but not sufficient condition) to bring about universal emancipation from capitalist-engendered and -maintained unfulfilled being.

KATHRYN DEAN

reason and emotion. See REASON.
reason explanations. See RATIONAL EXPLANATION.
reasons, real and possible. See CRITICAL NATURALISM.
reasons as causes. See CRITICAL NATURALISM; DETERMINATION; HERMENEUTICS; MIND; REASON; WANTS.
re-centrification, recentrification. See RE-AGENTIFICATION.
recognition. In recent continental European philosophy, recognition tends to be seen as fundamental to sociality and the social bond. This emphasis derives from Hegel Mark II (post-1806), whom Bhaskar criticises for glossing his earlier theme of the desire to be desired (loved) (incorporating desire for de-ALIENATION) as the desire for recognition and reconciliation '(which includes forgiveness about the persistence of MASTER–SLAVE relations)', i.e., of key forms of alienation (*D*: 243); the desire to be desired includes, but is not exhausted by, the desire for recognition. While for CR it is accordingly rather TRUST that is our primary existential, recognition (a 3L figure which analytically involves a doubling of cognition: re-cognition) is none the less seen as an axiological necessity or basic human need/propensity (cf. Sayer 2005, ch. 3), which manifests itself in primary POLYADISATION after REFERENTIAL DETACHMENT. Disputation and argument, for example, presuppose both mutual recognition and some genuine attempt at understanding. Achieving the former, as Hegel emphasised, may involve 'life-and-death' struggle (though this will often be about more than recognition). Hence the Bhaskarian concept of *hermeneutic* HEGEMONIC/counter-hegemonic *struggles*. An agonistic element also plays a prominent role in scientific revolutions as scientists compete for recognition and prestige, and indeed is included by Bhaskar within 'motivation' and 'drive' as a basic component of human action as such (*D*: 164–6). The Hegelian dialectic of reciprocal recognition of rights (which logically entails the right to de-alienation and the abolition of power$_2$ relations [*D*: 367 n]) plays an indispensable role in the CR dialectic of SOLIDARITY, and in EMANCIPATORY AXIOLOGY as such, mediated by the logic of dialectical universalisability: 'I must see an individual or situation as dialectically similar under the appropriate description, which presupposes *recognition* of underlying identities behind manifest but real differences' (*P*: 143, e.a.); and CR itself accordingly moves, at 1M, 'from attention to difference' in TR and CN 'to the recognition of underlying identity' in DCR (*D*: 301, e.a.). But the dialectic of recognition is constellationally contained within the dialectics of care, solidarity and trust, and extended, generalised and deepened (as it logically must be) 'to aspire to the achievement of {...} a true democratic socialist humanism' (*D*: 102) without masters and slaves which will *re-cognise* and realise the latent or partially manifested but really existing tendential rights to universal free flourishing, together with the responsibilities they imply, that are immanent in human practices.
recognition-transcendent truth or **verification-transcendent truth/reality**. See ALETHIA; TRANSCENDENTAL REALISM.
reconstruction. See RATIONAL RECONSTRUCTION.

recursivity. The capacity to achieve a certain outcome within a dialectical process which is then re-applied to that outcome to produce another internally related but distinct outcome, and so on, in principle *ad infinitum* (see AUTOPOIESIS; FINITE/ INFINITE); it is thus a 3L concept. It is a well-known property of the human linguistic and musical faculties and of numerical thinking (e.g., adding one, starting from zero, generates the natural numbers). On the CR account, it is a cardinal property of the human social world as such, for the social mechanisms that generate social phenomena are themselves (emergent) social products (see TMSA) in an endless recursion. Recursivity is also a feature of the non-human world (e.g., the navigational systems of life-forms) and of EMERGENCE as such, which consequently displays *recursive* (or nested) *EMBEDDEDNESS*. But it possesses an added dimension in the human social world in that social structures, which are for the most part non-teleologically or unintentionally produced, may be self-reflexively critiqued and the object of conscious transformation, in a dialectic in principle without end. Thus philosophy and social science themselves have a 'rooted recursivity' (*SR*: 14), as does theory and practice as 'models of' the world are translated into 'models for' practical application in the world (Geertz 1973). This entails a capacity to apply a rule of NEGATION which other animals may not possess, and that both people and institutions are endowed 'recursively with the power to acquire powers' (*D*: 141), or to self-consciously absent constraints on absenting constraints (see DIALECTIC). See also EPISTEMOLOGICAL DIALECTIC; MEDIATE; PERSPECTIVE; STRATIFICATION; SYSTEM; TOTALITY.

redescription. See EXPLANATION.

reductionism. To reduce (f. L. *reducere*, to bring back) X's to Y's is to show that, while X's apparently exist, they are nothing but Y's or parts of Y's; that is, an X and a Y could not occupy the same place and not be identical or one not be part of the other (*RS2*: 181). This is *synchronic explanatory reduction*. CR does not deny its possibility (e.g., the entities of chemistry to physics), but holds that it can only occur after the entities being reduced have been well defined historically and that, if they are emergent, i.e., capable of acting back on the materials from which they are formed, such that the course of nature changes, it leaves their sui generis reality intact; thus if MIND could be shown to be nothing but brain states its status as an emergent power of matter would remain intact. Synchronic explanatory reduction should be distinguished from the *DIACHRONIC explanatory reduction* of an emergent entity, i.e., explanation of the geo-historical processes of its formation out of lower-order entities, which poses no threat to its sui generis causal status. Reductionism often holds that the nature of more complex things – whether causal powers and structures, events and actions, meanings, processes, totalities, concrete singularities, ethics, or concepts, theories and disciplines – can always be explained by simpler or more fundamental things in a manner that does not leave their causal status intact. It implies an IDENTITY of subject-matter between the social and the natural sciences, as distinct from SCIENTISM, which upholds a positivistically conceived UNITY. Reductionism in this sense contrasts with EMERGENTISM and constitutes a DIALECTICAL PAIR with DUALISM, grounded in the more fundamental FISSION/FUSION couple and the duality of the EPISTEMIC–ONTIC FALLACY. It materially implies (idealist or materialist) MONISM. Its many kinds, whether ontological (e.g., the reduction of psychological states to biological states) or epistemological (e.g., the reduction of theory to experience), are illustrated and critiqued throughout this

406 **re-empowerment**

work. *Reificatory reductionism*'s denial of intentional causality at 4D is vulnerable to the DILEMMA that its very statement is 'an intentional causal act with causal effects' (*P*. 244). *Sociological reductionism* moves in the opposite direction, reducing psychological to social states. See also ACTUALISM; DEPTH; INTERDISCIPLINARITY; MIND; SOCIAL CONSRUCTIONISM.

re-empowerment. See RE-AGENTIFICATION.

re-enchantment. See ENCHANTMENT.

reference. A central concept in the theory of meaning. According to Frege's early theory, the reference (semantic value) of an expression is that feature of it which determines whether sentences in which it occurs are true or false. In particular, the reference of a proper name is the object it denotes, the reference (extension) of a predicate is the class of things to which it applies, and the reference (semantic value) of a sentence is its truth-value (truth or falsity). The truth-conditions of a sentence are determined by its syntactic structure and the references (semantic values) of its constitutive expressions. Later, Frege introduced *senses* in his theory of meaning. These are modes of presentation of what is designated. A proper name, for instance, has a reference *and* a sense, i.e., a property (typically a description) which determines what object the name refers to. Frege took the sense of an expression to be what someone who understands the expression grasps, that is, a concept or a thought.

The Fregean orthodoxy was challenged by Kripke (1972), who introduced the causal theory of reference. This theory dispenses with senses (or descriptions) as reference-fixing devices: the reference of a proper name is fixed by a causal–historical chain, which links the current use of the name with an introductory event, i.e., an event (an act of baptism) which associated the name with its bearer. Descriptions associated with the name might be false, and yet the name-users still refer to the named individual, in so far as their use of the name is part of a causal transmission-chain which goes back to the introductory event. The thrust of the causal theory is that the relation between a word and an object is *direct* – a direct causal link – unmediated by a concept. Putnam (1975) extended the causal theory to the reference of natural kind terms. On the causal theory, it is one thing to fix the reference of a term, and it is quite another to find out the exact nature of the *referent* of this term and hence the correct descriptions associated with it.

This last point captures what critical realists call 'referential detachment': the thing referred to (the referent) should be distinguished from the act of referring (the reference) and the beliefs, descriptions or intentions associated with this act (the sense). For CR, the reduction of reference to sense, as in superidealism, issues in judgemental RELATIVISM and makes meaning-change unintelligible; conversely, the reduction of sense to reference tends to epistemic FOUNDATIONALISM and/or ABSOLUTISM and makes meaning-change impossible. More controversially, 'there are no a priori limits on what can be referred to – this is the *generalised concept of reference* and referent' (*DG*: 402–3, e.a.). Thus we can refer to anything imaginable, e.g., talking ducks, and when we do the imaginary or fictional may be inscribed within the real, capable of causally impacting us (see ABSENCE; ONT/DE-ONT). However, while endorsing Frege's ontological realism, CR points out that 'the referent of a sentence may or may not be detachable', i.e., we can get reference wrong, and it is only when we get it right that it has 'the ontological value "true"' (*P*. 53). The limits of our powers of reference comprise a *referential* TOTALITY. See also DETACHMENT; SIGNS; Devitt and Sterelny 1987; Sayer 2000a: ch. 2.

STATHIS PSILLOS

referential detachment. Next to ABSENCE (which, indeed, is implicit in it and powers its geo-historical and ontogenetic EMERGENCE), and alongside ONTOLOGY and differentiation (which it helps to establish), *referential detachment* is *the* base concept of CR. Informally, it is the detachment of the referent (objects, etc., within ontology) from the REFERENCE (the inter-subjective/social referential act within ontogeny phylogeny), 'initially justified by the AXIOLOGICAL need to refer to something other than ourselves' (*D*: 40); conversely, it involves detaching or *absenting* 'oneself or one's discursive act or one's pre-linguistic intuitive acts from what they are about' (*D*: 230). Whether in phylogenetic, ontogenetic or everyday terms, the NEED to referentially detach is constituted by determinate ABSENCE in the form of DESIRE. This means that no philosophy that lacks a concept of determinate absence (like Hegelianism and IRREALISM generally) can think referential detachment. The capacity for referential detachment – entailing a potential for DUALISM, ALIENATION and split – is a phylogenetically emergent human power attendant upon the co-ordinated development of self-CONSCIOUSNESS (REFLEXIVITY), language and INTENTIONALITY, which presuppose a self–world distinction (Brereton 2004b). Referential detachment and everything it entails thus constitutes the transitive dimension and 'establishes {but does not constitute} at one and the same time the EXISTENTIAL INTRANSITIVITY {or ALTERITY} of a being and the possibility of another reference to it, which is a condition of any intelligible discourse at all' (*P*: 52–3). (It is thus 'conceptually close' to 'recognition of the *contingency* of being {...}, and with it critique of ego-present-centricity and ANTHROPISM' [*D*: 230–1, e.a.].) It is 'implicit in all language-use and conceptualised praxis, e.g., playing football' (*DG*: 402–3, e.a.). Its stratified form is ALETHIC TRUTH.

The transcendental argument for referential detachment (hence for existential INTRANSITIVITY and ONTOLOGY) is concise and elegant:

> Realism in the sense that involves existential intransitivity is a presupposition of *discourse* which must be *about* something other than itself, of *praxis* which must be *with* something other than itself or of *desire* which must be *for* something alterior to itself. To someone who doubts whether referential detachment exists just ask them to repeat and/or clarify what they have said, and then ask them what it is that they have repeated or clarified. It must be a referentially detached (social) entity. Any creature capable of differentiation must be capable of referential detachment. (*D*: 212)

Referential detachment is thus also presupposed by PERCEPTION, our everyday activities (e.g., making a cup of coffee), and scientific experience and change. The solipsist or sceptic who doubts that ontology is inexorable and real must logically doubt the existence of doubt, thereby referentially detaching it and committing performative contradiction, hence auto-subversion.

This argument is 'the basis of the argument for the TMSA. But it is equally a critique of every FOUNDATIONALISM {...} SCEPTICISM {...} ABSOLUTISM {...} and

408 reflection

IRRATIONALISM' (*D*: 231). Its concept is implicit in Berger and Pullberg's (1966) concept of *objectivication* (see REIFICATION). Methodological or heuristic advice: 'fully understand why referent must be detached from referential acts and you will be on your way to ontological stratification and alethic truth' (*D*: 375). See also CONSCIOUSNESS; EPISTEMOLOGICAL DIALECTIC; INTRANSITIVE; REALITY PRINCIPLE; REFLECTION; SIGNS.

reflection (f. L. *re-* + *flectere*, to bend back). The process of HOLISTIC CAUSALITY or constitutive intra-relationality in which the whole and parts of TOTALITY mutually 'infect' each other. *Resonance* names the reciprocity or *doubling* of reflection, as when philosophies reflect, and so support, other aspects of the SOCIOSPHERE and vice versa. (*Reflectionism*, on the other hand, or *epistemic reflectionism*, is another name for REPRESENTATIONALISM.) *Reflexivity* is (1) 'the inwardised *form* of totality' at 4D (*DG*: 403, e.a.), i.e., the quality of reflection; and, less commonly, (2) a synonym for totality. In the human embodied person, it is the (prime) personal emergent power (PEP) – transcendentally necessary for genuine accountability, presupposing the causal efficacy of reasons – to self-consciously monitor our activities and deliberate internally upon the always already natural, practical and social context in which they occur. It is thus irreducible to the social or the biological, constituting rather 'the mediatory process between "structure and agency"' that arises from our practical transactions with the objective, above all the natural, world (Archer 2003b: 130). Archer distinguishes three modes of this PEP in late modernity: *communicative*, *autonomous* and *meta-reflexive*, with corresponding *evasive*, *strategic* and *subversive* stances, respectively, towards societal constraints and enablements. The central concern of reflexivity, imposed by the REALITY PRINCIPLE, is theory–practice CONSISTENCY, the most important criterion for philosophy, and human survival and flourishing, hence EMANCIPATORY AXIOLOGY, alike. Indeed, *reflexive consistency* is synonymous with theory–practice consistency, and in turn for DIALECTICAL REASON – the coherence of theory and practice in practice. It is 'systematically connected, in virtue of the intradependence of social being', with TRUST (*D*: 273–4). (*Reflexive inconsistency* is T/P inconsistency.) While all people have the capacity for reflexivity, in geo-historical terms its actualisation as the praxis of self-conscious social change in wanted directions is a recent phenomenon (Dean 2003; see also GEO-HISTORICITY). The coherence of theory and practice in practice at the societal level is notoriously underachieved; we act most reflexively within power$_2$-stricken society when we act from the standpoint of promoting it.

Irreflexivity runs counter to any reflexivising process. *Meta-reflexivity* involves being reflexive about our reflexivity. A *meta-reflexively totalising (self-) situation* is one in which the agent, with the help of a materialist sociology, is able reflexively to DETACH herself and her actions from, and situate them in, the world, monitoring them in full awareness of the TD/ID distinction and of the complexly distanciated character of the world in the modes of ontological depth (in particular the depth-stratification of her own personality) and of spatio-temporality (in particular the presence of the past and outside). This enables the agent to locate herself within the constellational identity of the identity of subject and object within subjectivity (level of ontogeny) within an overarching objectivity (level of ontology) (*D*: 149–50). The concept provides a rational solution, hinging on the stratification of the self, to the *problem of SUBJECTIVITY AND OBJECTIVITY*, and is designed to put paid to irrealist fallacies of LOGICISING, EPISTEMO-

reflexive, reflexivity 409

LOGICISING and LINGUISTIFYING being, resolving 'the *post-Kantian predicament*' (*D*: 79, e.a.) or *self-referential paradox* whereby knowledge of (intransitive) reality is necessary but deemed to be impossible (perhaps most poignantly expressed by Nietzsche, Heidegger and Derrida); knowledge is both necessary and fallibly possible. *Nietzschean forgetting* (that truth is illusory), *Heideggerian ERASURE* (B̶e̶i̶n̶g̶), and *Derridean play* of *différance* all betray a failure to (referentially) detach (absent) the commentator from what is being commented on, manifesting itself in the absence of a concept of intransitivity and ontological depth-openness, in particular of adequate concepts of space–time and the self. The concept also underlines that, notwithstanding we live our lives in the past-in-the-present (and in the intrinsic outside, e.g., the periphery-in-the-centre) – exemplifying 'real material spatio-temporal failure of detachment' (*D*: 139) – we are not fated to remain prisoners of this situation.

reflexive, reflexivity. See REFLECTION.

regularity determinism. See DETERMINISM.

reification (f. L. *res*, thing, Ger. *Verdinglichung*, *Versachlichung*, literally the action or process of turning something into a thing). A concept originating with Marx and developed by Lukács, Weber, Simmel and others. Bhaskar mostly deploys it in the primary meaning accorded it by Marx: as the PROCESS whereby human powers, social relations and products, or human beings themselves, are transformed into (non-social, fixed, NATURALISED) things that appear to be independent of people's control and dominate their lives. It embraces HYPOSTATISATION, which normally refers to the treatment of ideas and conceptual forms and relations as things or substances. Geo-historically, reification is most pronounced in, though not peculiar to, capitalism. Thus the commodification of people's capacity to work (labour-power) turns it into a thing that can be bought and sold. This is closely connected with FETISHISM, which invests the products of labour-power (commodities) with magical or animistic (anthropomorphic) powers, via the mediation of money, as people and their products are accounted 'in terms of abstract units of homogenised commodified time' (*D*: 321). 'Reification, fetishism and hypostatisation' REFLECT or double 'the *dynamic* of commodification' (*D*: 237). Physicalist or socio-biological reductions of the embodied person to neurophysiological drives, genes, etc., are forms of reification, as is the FACT-FORM in empiricist epistemology (which forgets that facts are social products and fetishises their conjunctions) or the treatment of social facts in Durkheimian sociology 'as possessing a life of their own, external to and coercing the individual' (*PN3*: 31). The Humean account of a causal law is reified because it treats constant conjunctions as 'existing quite independently of the human activity necessary for them' (hence fetishises the closed systems within which they occur) (*SR*: 94). Since the whole of being is in 'structured open-systemic flux', its treatment by the ANALYTICAL PROBLEMATIC as static and susceptible of rigid definition is reificatory and has the political effect of reinforcing (reifying) the status quo, for it is change which must be justified from that perspective, not the existing order of things. Drawing on Hannah Arendt, as well as on Marx, Bhaskar and others, Dean (2003: 36) argues that a process of *dereification* ('the dissolution of world into process' at the level of appearances or *deworlding*) goes hand in hand with reification in 'disorganised' capitalism, but this can arguably equally be seen, by valid perspectival switch, as the speeding up and intensification of the process of reification.

All forms of reification are aspects of the wider process of ALIENATION, whereby people are separated from what is essential to their nature, powers, autonomy or flourishing. In thingifying human powers and social things they *DE-AGENTIFY* people.

Reification must be distinguished from (1) *objectivity* or intransitivity as such (the ID); (2) *objectification* or *objectivation*, the imbuing of human powers in a product and its reproduction and/or transformation by human praxis (the TD); and (3) *objectivication*, 'the moment in the process of objectivation in which man establishes distance from his producing and its product, such that he can take cognisance of it and make of it an object of his consciousness' (Berger and Pullberg 1966, cit. *PN3*: 33). Bhaskar later generalised the distinction in (3) to any kind of thing in the concept of REFERENTIAL DETACHMENT. Failure to distinguish (1) and (2) *results* in reification. Following Marx, Bhaskar distinguishes (2) from alienation, 'recognising that alienation is only one possible (geo-)historically determined mode of objectification' (*P*: 108). See also ETERNISATION; VOLUNTARISM.

reinstatement, analytical (Hegel's). See EPISTEMOLOGICAL DIALECTIC; HEGEL–MARX CRITIQUE; NEGATION OF THE NEGATION.

relata. See RELATIONALITY.

relationality (f. L. *relatus*, pl. *relata*, past participle of *referre*, to carry or bring back). *Relations*, other than logical relations between propositions, are properties possessed by things in virtue of their positioning in respect of other things. On the CR account relations are sui generis real and pervasive in both the sociosphere and the rest of nature. '{S}pace-time is a relational property of the meshwork of material beings' (*D*: 53). Relations exist only in virtue of material things, possessing no material substance of their own (i.e., they are 'VIRTUAL' or 'ideal') and so are in principle imperceptible (non-empirical), hence can only be ascribed as real causally, by their effects on material things. Or their material presence consists only in their effects on material things – in the sociosphere, on people and the changes people make to other material things (*PN3*: 30; for a contrary CR view that social relations are 'material', see SOCIAL STRUCTURE). Note that their virtuality entails that relations can never be *wholly* constitutive of things possessing a material dimension. 'An element A is *internally related* to B if B is a necessary condition for the existence of A, whether this relation is reciprocal/symmetrical or not' (*DG*: 399); if it is not a necessary condition, the *relation* is *external* and its determinations in respect of A are contingent. Thus the durability of a lump of gold is internally related under certain conditions to its atomic constitution, but its relation to external forces which might transform it is contingent (see MODALITY). It should be noted that the distinction is not the same as that between INTRINSIC/EXTRINSIC, and that internal relationality does *not* entail explanatory equality among the aspects related, i.e., is consistent with depth-explanation (*PN3*: 43; *SR*: 109–11). Thoroughgoing internality within a SYSTEM or TOTALITY does not eliminate, but depends upon, differentiation and specific causal roles of its elements, and interacting structures within a particular system may be externally as well as internally related. External relationality involves *interaction* among relata. Three kinds of internal relationality or *intra-relationality*, involving *intra-action*, are distinguished: '(1) *existential constitution*, in which one element is essential and intrinsic to another; (2) *{existential} permeation*, in which one element contains another; and (3) *{existential} connection*, in which one element is merely causally efficacious on the other' (*DG*: 399,

relationism, entity relationism or ontological relationism

e.a.). They may produce contradiction and oppositionality, hence are 2E as well as 3L figures.

The *relationality* of the sociosphere ('"society does not consist of individuals [or, we might add, groups], but expresses the sum of the relations within which individuals [and groups] stand'" (*PN3*: 26, adapting Marx) is entailed by the TMSA. Only social practices satisfy the criterion of continuity (always alreadiness) generated by that model (they endure where individuals and groups come and go), and they can only be individuated relationally (*PN3*: 40–1); moreover, social structures (ensembles of internally related practices) are themselves social effects, the identity or meaning (not just identification) of which is dependent on their relational constitution in (changing) systems of difference (see *SR*: 131–2 [where 'rationality' should read 'relationality']; *D*: 155; SIGNS). The sui generis reality of relations in the physical world for its part is entailed by the transcendental deduction of TOTALITY (*D*: 123–5) and the critique of ontological EXTENSIONALISM and more generally of the ANALYTICAL PROBLEMATIC, and confirmed by developments in the modern physical and life sciences, which show that there is 'no special difficulty {...} in ascribing reality to relations on a causal criterion' (*PN3*: 40). In philosophy generally, *relationism* may refer either to epistemological relativism or to the view that relations have a real existence. Within CR it refers to the latter, entailing non-Quinean (non-Kuhnian) *ontological relativism* or relationism (i.e., with the exception of the ABSOLUTE, if there is such, and the speed of light, everything stands in some relation to something else). There may, of course, be relations between relations, as in the ensemble of relations constituting the sociosphere, which the concept of totality is groomed to express. *Entity relationism* is the view that things are to a significant degree 'existentially constituted, and permeated, by their relations with others' (*D*: 125, e.a.) (see also DCR; DIRECTIONALITY; HOLISTIC CAUSALITY; POLYAD; REFLECTION).

CR takes issue with philosophies of pervasive internal relations (everything is internally related to everything else, e.g., the Hegelian tradition) as well as of pervasive external relations (all relations, except between ideas – i.e., all causes – are external, e.g., the Humean tradition). CR is a philosophy of internal *and* external relations, under the sign of the PRIMACY of internal relations. Just as dialecticality constellationally overreaches analyticality (logicality), so internal and necessary overreach external and contingent relations and determinations; indeed, internality corresponds to dialecticality and externality to logicality (*D*: 83 n, EXPRESSIVISM/EXTENSIONALISM). Whether a particular relation is internal or external is of course an in principle open question, for determination by empirical investigation (*PN3*: 42–3). PMR, however, comes close to rejecting the concept of external relationality altogether (*RM*: 234–5).

The central importance of relations for CR is well caught in the slogan epitomising CRITICAL NATURALISM: 'Transformationism. Relationism. Critical naturalism' (*D*: 155). It should be noted that relations depend upon the fundamental category of ABSENCE: 'things are defined by what they exclude (this is not quite true, as intra-activity attests) or, better, by their differences, or, even better, by their position within the system of changing differentiations and differentiating changes' (*D*: 240). 'To be is to be related' (*RM*: 149).

relationism, entity relationism or **ontological relationism**. See DCR; RELATIONALITY; TOTALITY.

relative autonomy. A 3L concept which specifies that the elements of TOTALITY are simultaneously both distinct or non-identical and inter/intra-constituted, permeated or connected, or both *DETACHED* and *ATTACHED*, to varying degrees short of complete autonomy (eclecticism or pluralism), on the one hand, or reduction to an EXPRESSION of the whole (MONISM or holism) on the other (cf. FISSION/FUSION). Both distinctness and interconnection are entailed by the transcendental arguments from intentional agency establishing irreducible ontological stratification (1M) (see DEPTH) and TOTALITY (3L) (see also MEDIATE). The relative autonomy of society from people, and of social structures/practices from the sociosphere, is likewise entailed by the argument from intentional agency establishing the TMSA. The questions of (1) how relative relative autonomy is in any particular case; (2) which relatively autonomous elements comprise a totality; and (3) whether relative autonomy (the repudiation of monism) is compatible with a primacy thesis (the repudiation of eclecticism), i.e., the notion that one social element (e.g., the mode of material production, demography, ideas) ultimately determines the rest of social life, can themselves in the last instance only be established by empirically based research. For the TMSA also entails 'that there can be no general a-historical theory of articulation, that the relations between the elements of the social totality (including the operative forms of causal determination {...}) may change along with the elements themselves' (*SR*: 158–9). There is no a priori reason why the answer to (3) should be negative. See also AUTOLOGY; INFRA-; INTRA-/SUPER-STRUCTURE.

relative being, relative reality. See DEMI-REALITY; META-REALITY.

relativism, relativity. For cultural, epistemic (epistemological), ethical, judgemental, linguistic and ontological relativism, see HOLY TRINITY and the relevant cross-references therein.

religion, philosophy of. See PHILOSOPHY OF RELIGION.

re-moralisation. See CRITICAL NATURALISM.

representationalism. A subjectivist Cartesian–Lockean theory of KNOWLEDGE which tends to give pride of place to a contemplative observer who is disengaged from practice in a PUNCTUALIST 'here-now', having true knowledge only of the contents of his mind, and to sociological individualism. Though related, it is to be distinguished from the *representative REALISM* of the twentieth century, which seeks to avoid the pitfalls of subjectivism and passivity. It is central to what Bhaskar calls the problematic of presence (see METAPHYSICS OF PRESENCE). See also PERCEPTION.

representative realism. See PERCEPTION; REALISM; REPRESENTATIONALISM.

research programme. See THEORY, ETC.

resemblance. At the level of causal powers and STRUCTURES (domain of the real) things may be *identical* to each other. Classification grounded in real identity provides the rationale for the stratification of science. At the level of ordinary things – structurata (particular instantiations of structures) and laminated structurata or systems (concretely singular things comprising, and bonded together by, 'an irreducible plurality of structures' and subject to dual or MULTIPLE CONTROL) – we are in the sphere of UNITY, not identity, and things may *resemble* each other on the basis of structures they possess in common and/or that were common to their genesis and process of formation (e.g., rounded stones on a river bed, the spoken languages of a region). This might be dubbed *strong* or *realist resemblance theory*. In neither case need classification be merely arbitrary, subjective or a matter of convention, though of course we can get it wrong and it will always be more mediated in the latter than in the former case.

Wittgenstein's notion of *family resemblance* is a *weak resemblance theory* reflecting his weak ACTUALISM (*D*: 318; P: 197), which intends to show that the likeness of things does not depend upon realism about deep intrinsic structure, but rather on a network of resemblances at the level of the actual ('build, features, colour of eyes, gait, temperament, etc., etc.' [Wittgenstein 1953: §67]). However, members of a family resemble each other physically in virtue of a common genetic endowment, and, considered as social practices rather than internally, games (Wittgenstein's favourite example) all have it in common that they are designed and played by humans possessing essential powers for the purpose of enjoyable exercise of those powers of one kind or another. While Wittgenstein's examples seem unfortunate, this is not to deny that people do classify on the basis of perceived similarity informed by choice and convention at the level of the actual (e.g., greengroceries), only that it is impossible for classification to be non-arbitrary on realist grounds. Objective resemblance presupposes realism, and where people are mistaken in their classifications their usage can in principle be corrected.

In the philosophy of art, photography, etc., *resemblance theory* argues that artworks *represent* the world in a mediated way (cf. representative REALISM; REPRESENTATIONALISM). Such a claim cannot be grounded in depth-realism in the manner of the objective resemblance of things.

resonance. See REFLECTION.

retotalisation. See CONCRETE/ABSTRACT; EMANCIPATORY AXIOLOGY.

retrodiction, retroduction, retroductive-explanatory arguments. See DIALECTICAL ARGUMENT; EXPLANATION; INFERENCE.

retrojection. See DE-AGENTIFICATION; FISSION/FUSION.

reversal (f. L. *revertere*, to turn back) betokens both *CHANGE* (PROCESS) – *interchange* and annulment – and *OPPOSITION* and/or CONTRADICTION (the reverse, contrary). It is therefore a 2E concept, but also 4D in so far as it involves praxis. Reversals may be *structural* (e.g., abolition of a structure or the reconstitution of powers in re-agentification); *strategic* (e.g., Foucauldian counter-conduct which entails resistance to every oppression); *tactical* (e.g., a perspectival switch from agency to structure in emancipatory struggle); or *conjunctural* (e.g., the suspension of a power$_2$ relation in a crisis). They may be negative as well as positive (e.g., the de-constitution of human powers in de-agentification), but the concept is usually deployed in its positive meaning.

A number of only apparently paradoxical reversals occur within DCR itself: 'from attention to difference to the recognition of underlying identity' at 1M; from anti-anthropism at 1M to humanism at 4D; from dialectical universalisability to particularity and diversity at 3L; from facts to values and from primal scream to the good life in the dialectic of freedom at 4D (*D*: 301, e.a.); and from the 'most (ontologically) negative philosophy' to 'the most (ethically) positive results' (*D*: 284). CR's emphasis on concrete singularity follows Marx in reversing the axis of domination within the Hegelian concrete universal from the universal to the singular.

The *dialectics of reversal* are all those processes in EMANCIPATORY AXIOLOGY whereby power$_2$ relations are rolled back and eventually abolished. They thus embrace counter-hegemonic struggle. They belong to the wider class of *dialectics of opposition*,

414 rhythmic

which, along with the dialectics of TRANSITION, comprise the substantive content of 2E dialectics (*P.* 167).

rhythmic. See PROCESS.

right-action, spontaneous. See ACTION; DIALECTICAL REASON; EVIL; MEDIATE; META-REALITY; PERCEPTION.

rights. See ALIENATION, CRITICAL REALISM, DISPOSITION, EMANCIPATORY AXIOLOGY, EMANCIPATORY MOVEMENTS, EQUITY, EUDAIMONIA, EVOLUTION OF SOCIETY, HUMAN NATURE, LEGAL STUDIES, POLITICS, PRINCIPLE OF EQUIVALENCE, UNIVERSALISABILITY.

RRREI(C) model of applied scientific explanation. See EXPLANATION; METHOD.

S

σ (*sigma*) **transform**. See EPISTEMOLOGICAL DIALECTIC.
saddle points. See TRANSITION.
Sapir–Whorf hypothesis, or **linguistic relativity principle**. The view that the structure of the particular language we speak (and by extension theories we deploy) significantly shapes our understanding of the world. There is considerable disagreement about it within CR, as elsewhere. It is defended by Bhaskar as intrinsic to EPISTEMIC RELATIVITY, which must be accepted both on transcendental and empirical grounds (see *PN3*: 58; cf. *RS2*: 195). On the other hand, Brereton (2004b), for example, drawing on Archer (1988) (who espouses a version of Popper's Third World of Knowledge in the notion that there is only one human Cultural System, comprised of intelligibilia, at any one time), rejects it.

The principle of linguistic relativity is by no means synonymous with the LINGUISTIC FALLACY – the exposition of being in terms of language. It does not entail subjective superidealism, viz., that there is no language- (or theory-)independent world; on the contrary, its intelligibility presupposes the existence of a language-independent world. It is thus perfectly compatible with the TD/ID distinction and the possibility of JUDGEMENTAL RATIONALISM (within linguistic and epistemic relativity). See also CULTURAL ANALYSIS; CULTURE; HOLY TRINITY.

satisfice. To do or choose the good enough rather than the *optimising* (best), or *maximising* (biggest), or *supererogatory* (what is beyond the call of duty). Introduced into economics in the 1950s as part of the discourse of 'modernity's man', and much used subsequently, in philosophy and social theory, in ethics and rational choice theory.

scandal of philosophy. C. D. Broad (1926) famously declared that induction is 'the glory of science but the scandal of philosophy', thereby endorsing Humean SCEPTICISM and Humean and Kantian denegation of ONTOLOGY (Kant himself had declared Humean scepticism a scandal). At about the same time Heidegger averred that, to the contrary, the scandal is that proofs of the existence of an external world should be expected and attempted. This evades the scandal rather than resolving it. The '*real* scandal of philosophy' is IRREALISM and the ANTHROPIC fallacy – 'the exegesis of being in terms of human being' – which underpins scepticism and which Hume, Kant, Broad and Heidegger all commit in various ways (*D*: 205, 214). The real resolution of the scandal lies in the transcendental deduction of the human capacity for REFERENTIAL DETACHMENT.

sceptic, scepticism. See BEAUTIFUL SOUL, ETC.

Schillerian dialectic(s), the Schillerian schema. See PROCESS.
science. See CRITICAL NATURALISM; CRITICAL REALISM; DCR; EPISTEMOLOGICAL DIALECTIC; EXPLANATION; EMANCIPATORY AXIOLOGY; EXPLANATORY CRITIQUE; SCIENTISM; SOCIAL SCIENCE; TRANSCENDENTAL REALISM.
scientific change. See EPISTEMOLOGICAL DIALECTIC; KNOWLEDGE; PROGRESS; THEORY CHOICE.
scientific discovery, logic of. See EPISTEMOLOGICAL DIALECTIC; KNOWLEDGE.
scientific loans. See MODEL; PARAMORPH.
scientific ontology. See EPISTEMOLOGY, ETC.; THEORY, ETC.
scientific realism. See REALISM; TRANSCENDENTAL REALISM.
scientific revolutions. See EPISTEMOLOGICAL DIALECTIC; THEORY, ETC.
scientism. See HISTORICISM.
scotoma, pl. scotomata, adj. scotomatic (Gr. *skotōma*, dizziness associated with dimness of sight, f. Gr. *scotos*, darkness). An isolated area of dimness within the visual field; a blind spot in a philosophy or theory, to which STOICISM is particularly prone.
second edge (2E). See MELD.
self. See SUBJECTIVITY.
self-consciousness. See ALIENATION; CONSCIOUSNESS; DIALECTIC OF CONSCIOUSNESS; IDEALISM; PEP; PHENOMENOLOGY; REFLECTION.
self-determination. See AUTOPOIESIS; FREEDOM.
self-emancipation. See SELF-REFERENTIALITY.
self-esteem, self-love. See ESTEEM.
self-predicative paradox. See PROBLEM.
self-realisation. See ABSENCE; AGENCY; DIALECTICAL REASON; DIRECTIONALITY; EMANCIPATORY AXIOLOGY; FREEDOM; META-REALITY; SUBJECTIVITY.
self-referentiality. A *self-referential* system is AUTOPOIETIC. In the *principle of self-* or *subject-referentiality* or *self-emancipation* the self is conceived as quasi-autopoietic. It states that only we can act, no one can do it for us, and all social change is also self-change; while action implies solidarity in removing constraints on flourishing (see EMANCIPATORY AXIOLOGY), 'solidarity is not substitutionism' (*P*: 143), the educators must educate themselves, as emphasised in the concept of *transformed*, transformative praxis (see TTTTφ): if this is not effected within emancipatory movements, they merely replace one set of master–slave relations with another. Positive freedom 'cannot be imposed from without {...}, it can only come from within' (*RM*: 222). The principle is entailed by the goal of universal autonomy implicit in praxis and speech and by the principle of PREFIGURATIONALITY, which requires means–ends consistency and anticipation of the values of the end-state in our praxis. The philosophy of META-REALITY explicitly argues the PRIMACY of self-referentiality in emancipatory praxis on grounds of the ontological priority of the transcendentally real self over master–slave-type relations of oppression; you change yourself for the better when you act from the ground-state, and in so doing change the condition and outcome of action, which is society. Only when you are at one with yourself, such that your embodied personality expresses your real self, can you know what to do (*RM*: 102–3). The *turn to subjectivity* in this sense is at the heart of the *spiritual turn* (*RM*: 14). This position is arguably already implicitly present in the dialectics of freedom, on which 'self-emancipation' is

indispensable to FREEDOM and emancipatory praxis (*D:* 171, 285; *SR*: 171; cf. *PN3*: 77 n 84). It is not an argument for voluntarism or individualism, since action is always social, 'to be is to be related' and the freedom of each depends on the freedom of all; it is an argument for understanding that our true power is mightier far than that of any army, which cannot do without it: if we change the way we act, it changes (*RM:* 146, 149). *Self-referential incoherence* is radical theory–practice inconsistency or PERFORMATIVE contradiction. See also DIALECTIC OF SOLIDARITY; DIRECTIONALITY; PROBLEM OF AGENCY.

self-reflexivity. See REFLECTION.

semantic holism. A thesis about the nature of representation, espoused notably by Quine, which holds that the meaning of a symbol is relative to the total system of differences in which it is located. See also SCIENTIFIC REALISM.

semiology, semiosis, semiotics. See CRITICAL DISCOURSE ANALYSIS; SIGNS.

sempiternal polyadisation. See POLYAD.

sensationalism. An extreme form of EMPIRICISM whereby all mental states, hence ultimately our knowledge, derive from sensation or sense data, which indeed are often seen as the ultimate and real components of the world.

separation, generative. See GENERATIVE SEPARATION.

SEPM (synchronic emergent powers materialism). See MIND.

seriousness. See OPPOSITES.

set theoretic paradoxes. See PARADOX OF THE HEAP.

sexuality. See SOCIOLOGY OF SEXUALITY.

signified, signifier. See SIGNS.

signs, semiology, semiotics. The terms *semiology* and *semiotics* are often used interchangeably to refer to the philosophical and scientific study of *semiosis*: the relation of signification between sign, mind and object, or, more generally, the production of meaning; but strictly speaking they refer to the theories of Saussure and Peirce, respectively. Often Peirce's ideas are tacked on to Saussure's, but in fact their theories are radically different. For Saussure, the sign consists of two parts: a *signified*, which is a concept, and a *signifier*, which is a mental pattern of a sound or written mark. There is no necessary connection between the two – the same signified may have many signifiers (dog, *chien*, *Hund*, mutt, etc.); this is one way in which language is 'arbitrary', in the sense of socially determined. Signifieds and signifiers each consist of systems of differences, such as between 'cat' and 'kitten' or between /t/ and /d/; this is another way in which language is arbitrary. The abstract structure of a language (*langue*) must be kept distinct from speech (*parole*), and the proper object of semiology is the former.

For CR, Saussurean semiology is wholly unacceptable: the signified and signifier are both wholly cognitive, and provide no form of external reference. Saussure effectively collapses ontology into epistemology, committing the EPISTEMIC FALLACY. The lack of a role for similarity among concepts creates additional problems because it offers no way to explain why, for example, many categories have 'best examples' (this, too, depends on reference to the external world).

Bhaskar has devoted only a paragraph or two to describe the sign (*D*: 222–3; *P*: 52–3), which he defines as tripartite: signifier, signified and referent. He also mentions (without explanation) that each of these elements may have attached to it many other semiotic triangles. Both of these points are strikingly similar to elements of Peirce's theory of semiosis, in which the sign consists of a representamen, object and interpretant (very roughly equivalent to signifier, referent and signified), and each interpretant can, and normally does, serve as the object for another semiotic triangle, leading to unceasing semiosis. The latter entails that perception is always mediated by signs (all objects are known under some description). But the sign's connection to real existence just begins with its object: ultimately, it is possible for the chain of semiosis to re-enter the body in the form of habits. This links Peirce's semiotics to Bourdieu's concept of habitus. Peirce further argues that there are three major types of sign: icons, which are images with some resemblance to the referent; indexes, which point to the referent in the way an effect indicates a cause or a part indicates a whole; and symbols, which are conventionally established, such as words. Saussure's theory only recognises the last. Signs for Peirce are also scalable: for example, a novel is composed of signs, but it also constitutes a sign itself. However, semiosis is not restricted to humans: because a sign depends on the existence of an interpretant (a 'significate effect'), biochemical markers (for example) are signs to the platelet that detects them. Likewise, it is now known that many animals establish symbols, even if rudimentary ones. In other words, semiosis is not inherently anthropic. On the whole, then, Peircean semiotics is compatible with CR.

Because semiosis is integral to human activity, it is an aspect (but only an aspect) of social structures and agency as well as discourse, and any cultural analysis must consider how semiotic products emerge from and act upon structures, agents, and other semiotic products (see CULTURE). Owing to the sign's grounding in and re-entry into embodied practice, semiosis also has important connections to COGNITIVE SCIENCE. Arguably, Bhaskar's third ontological domain – originally the 'empirical', more recently the 'subjective' – is better understood as the semiosic (see REALITY).

Any philosophy of signs and language includes or implies a theory of its social functioning. Bakhtin and (or perhaps as) Voloshinov developed theories of discourse's social character that are vital for CR, particularly regarding language's dynamic character. Language-use involves an interaction with other people, the world, and oneself. Meaning is a 'two-sided act', determined both by the one who uses it and the one for whom it is intended, and words arise and are understood only within specific contexts. Understanding an utterance involves orienting oneself towards it and responding to it. True understanding, then, is dialogic. The immediate social situation and its direct participants govern the utterance's particular form and style, but its deeper structures arise from more durable and basic social dynamics. To theorise the political dimensions of language-use further, one can turn to Gramsci's concept of HEGEMONY, which focuses on the formation of 'common sense' and the naturalisation of power relations.

Owing, on the one hand, to the social, dialogic and consequently polysemic nature of language (and Peircean symbols generally) and, on the other, to the necessary mediation of perception by signs, CR cannot accept a simple correspondence or representational theory of meaning. But it also cannot embrace a view that meaning is secured solely by a discourse's internal coherence. These issues belong to the TRANSITIVE DIMENSION of a theory of meaning. The INTRANSITIVE DIMENSION concerns REFERENCE to an object, which involves the object's causal independence and power of acting upon cognition. The relationship between these dimensions has yet to be fully theorised in CR and is not exhausted by FALLIBILISM. See also CRITICAL

DISCOURSE ANALYSIS; HERMENEUTICS; Bakhtin 1981; Bernard-Donals 1994; Fairclough, Jessop and Sayer 2004; Nellhaus 1998; Peirce 1893–1913; Saussure 1916; Voloshinov 1929.

TOBIN NELLHAUS

singular(ity), concrete. See CONCRETE UNIVERSAL ↔ SINGULAR.
slaughter-bench of history. See CONCRETE UTOPIANISM; GENERATIVE SEPARATION.
social atomism. See ATOMISM; FISSION; SOCIAL STRUCTURE.
social bond. The social bond refers, in Bhaskar, to (1) the *social bondage* effected and maintained by power$_2$ and ideology, and (2) the *SOLIDARITY*, based on the primary existential of TRUST, among subjects$_2$ that has the potential to dispense with (1). It is synonymous with (the social dimensions of) the REALITY PRINCIPLE (*D*: 291).
social change. See EMANCIPATORY MOVEMENTS.
social constructionism (or constructivism). Most generally, the view that social phenomena are as people perceive them to be or as constituted and institutionalised by discourse and social convention (cf. CONVENTIONALISM; VOLUNTARISM). Social constructionism is at one pole of a very pointed debate located within two domains: the structure–agency debate of the social sciences (see AGENCY; SOCIAL STRUCTURE; SUBJECTIVITY) and the determinism–free will debate of philosophy (see DETERMINATION). Although there are different traditions of constructivist ideas, including symbolic interactionism, ethnomethodology, INTERPRETIVISM and PHENOMENOLOGY, all describe in diverse ways and to different extents how the human world is socially and/or discursively constructed. This sometimes involves rejection of the applicability of the category of CAUSALITY to the social world.

In the structure–agency debate there is agreement between most realists that there are aspects of reality which are not manifest to the senses, and that unobservable structures may causally constrain or influence human action to considerable extent. However, social constructionists are united in the standpoint that it is the *interpretation* of social phenomena that shapes outcomes. For example, the early work of Berger and Luckmann (1966) – the founding text of social constructionism in the twentieth century – illustrated that what people think of as the reality of everyday life is created through human action, and is not something that exists independently of human actors. More recently Potter (1995: 98) argues that the world is constituted in one way or another *as people talk it, write it and argue it*.

Critical realists, by contrast, hold that social reality is emergent and stratified; the world is constituted prior to our understanding of it, and prior to our talk or knowledge of it. It is clear that our knowledge about the world (in the TD) and the world which is the object of that knowledge (the ID) are distinct, an understanding from which explanatory critical realism operates. Bhaskar explains that 'the objects from, and by, which knowledge is generated are {. . .} always themselves social products (as is the knowledge generated)' (*RS2*: 185); but, while this (transitive) knowledge is socially produced, it does not produce the objects of knowledge in the ID (of which genuine knowledge is possible).

While there are disagreements regarding structure and agency within CR, from a CR perspective structure and agency are irreducible. Structure is temporally prior to the agency that reproduces and/or changes it, and so has explanatory priority sociologically (Archer 1988/1996). This does not mean, however, that agents are merely passive carriers of structures. Agents are able to learn and strategically adapt practices to alter structures (although there may be unintended consequences). For example, critical realists such as New (2005; FEMINIST THEORY) argue that sex is a real foundation on which gender is socially constructed. The real differences (with causal powers) between the emergent properties of male and female sexed bodies are distinct from the way they are described in gender construction. This chimes with social constructionism's insistence that social phenomena are in some sense human choices and not permanent and 'natural' (see ETERNISATION; NATURALISATION).

The arguments of realists such as New and Archer contest the social constructionist account of the BODY. Where social constructionists would describe the biological as the exception to social construction, Archer (2000: 4) also makes exceptions for the bases of selfhood, reflexivity, thought, memory and emotionality, which are described as 'the resultants of our embodied relations with the world', where 'world' is understood as by no means exclusively social. From a similar perspective, Bhaskar comments: 'As for social constructionism, the embodied personality is a construct, but who does the constructing?' (*RM*: 109). Here is the free-will argument of philosophy, the realm into which Archer shifts the debate. Bhaskar goes on to suggest that people are active agents in the shaping and have the potential to reject or transform it.

It is not just embodiment that is controversial within CR, but also the nature of the social, which has implications for critical realists and how they undertake critical explanatory research. Bhaskar points out that, though sui generis real, 'social structures, unlike natural structures, do not exist independently of the activities they govern' (*PN1*: 48). This calls attention to the distinguishing element of a realist philosophy, which separates ontology from our theories of knowledge (epistemology), and which argues for the existence of a relatively mind-independent reality. Transitive social constructions of knowledge reflect the intransitive ontologically real, but also exist as elements of that real (intransitive) world. While adhering to Bhaskar's formulation, realists such as Sayer posit that some social constructions, for example, certain institutions, often emerge with substantial independence from constructions which external observers have of them. Sayer also warns us that the use of the 'hopelessly misleading metaphor of construction invites idealist slippage', but that 'we have no alternative but to attempt to assess the relative practical adequacy or objectivity of different social constructions'. His version of CR espouses a *weak*, as distinct from *strong*, *social constructionism*, i.e. one that accepts an underlying intransitivity rather than viewing everything as constituted by our current ways of seeing and discoursing (Sayer 2000a: 63, 91–2). See also CRITICAL DISCOURSE ANALYSIS; EPISTEMIC–ONTIC FALLACY; METHOD; ONTOLOGY; PHILOSOPHICAL ANTHROPOLOGY; TECHNOLOGY; TRANSCENDENTAL REALISM; Searle 1995.

DIANE WESTERHUIS

social cube. Constellationally contained – as its 'hub' (*D*: 180) – within the rhythmic cube of four-planar social being (the social tetrapolity) or HUMAN NATURE, the *social cube* is comprised of sub-dimensions of its planes (c) and (b), as set out in Table 38. Like the social tetrapolity itself, it should be thought of as 'a space–time cubic stretch or flow' (*D*: 13), the moments of which are internally related, i.e., as a PROCESS; it is thus open

social evolution 421

Table 38 The social cube, four-planar social being and alienation

Ontological–Axiological Chain	1M Non-Identity	2E Negativity	3L Totality	4D Transformative Agency
Four-planar social being or human nature (social tetrapolity)	d. (intra-) subjectivity (the stratified person)	c. social relations (concept-dependent, site of social oppositionality)	b. inter-/intra-subjective (personal) relations – transactions with ourselves and others	a. material transactions with nature (making)
Social cube		(1) power$_2$, (2) discursive, (3) normative relations	(1) power$_2$, (2) communicative, (3) moral relations	
		intersecting in (4) ideology		
Alienation (fivefold) from	d. ourselves (5)	c. the nexus of social relations (4)	b. each other (3)	a. the labour process: labour and its product (-in-process) (1), material and means of production; (2) more broadly, nature and material objects

at either end, giving it a total of six planes (four positive [(1)–(4)] and two negative [the not-yet and the intransitive determined]).

Putting the sub-dimensions of (c) and (b) together, the social cube comprises (1) power$_2$, (2) discursive/communicative, and (3) normative/moral relations, intersecting in (4) IDEOLOGY, which legitimises and discursively moralises power$_2$ relations. (2) and (3), like the planes of social being, are dimensions of any conceivable society; (1) and (4) (at any rate in the strong sense of categorial error) pertain to DEMI-REALITY only, i.e., would not be present in EUDAIMONIA. They are underpinned by generative separation or ALIENATION. (2) may instantiate dialectical CONTRADICTIONS and includes HEGEMONIC/counter-hegemonic relations. Hegemonic/counter-hegemonic *struggles*, whether hermeneutical or of a more material kind, will typically implicate (2)–(4), but their *site* is (4); or ideology itself – the ideological intersect – has power$_2$, discursive/communicative and normative/moral dimensions or layers (*D*: 349). It should be noted that Bhaskar sometimes uses 'social cube' synonymously with the wider concepts of (i) planes (c) and (b) of the social tetrapolity, and (ii) the social tetrapolity as such. In either usage, but particularly the latter, the totalising nature of the concept makes it particularly useful in critiquing 'one-sided social epistemologies', e.g., non-critical hermeneutics at (b) or discourse theory at (c) (*D*: 163).

social evolution. See EVOLUTION OF SOCIETY.

social form. A key concept in Marxist analysis, systems theory, CR and other kinds of social inquiry denoting a specific mode of organisation of a social content. This concept rests on the separation of form and content in so far as the same 'content' can assume different forms and, within limits, the same form may include different contents. For example, while Marx (1867) asks why wealth in capitalism 'presents itself as an immense accumulation of commodities', others explore the scope and limits of commodification (e.g., Offe 1984).

A *social formation* comprises a coherent set of social forms. It has two main uses in Marxism: (1) the articulation of different modes of production and forms of private and social labour, especially where economic relations are deeply embedded in and hard to disentangle from broader social relations; and (2) an articulated ensemble of social forms that comprise a more or less distinct social order, such as a capitalist social formation shaped by the commodity form, wage form, money form, tax form, state form, nation form, etc. Some theorists contrast the scientific concept of social formation with the ideological notion of a cohesive society (especially when associated with the national state). In this context an important dimension of social change is the conformation or disjunction of forms and the study of transformation.

Form analysis, at least in Marxism, has three aspects. First, it rejects any a priori assumption of the primacy of the logic of capital and/or class struggle in favour of starting with social forms and considering their implications for that logic and/or class formation and class struggle. Second, it suggests that form may well problematise FUNCTION. For example, the fictitious commodification of labour-power poses problems for capital in securing its valorisation and may also disorganise and divide workers (M. Williams 1988). Third, it distinguishes formal and functional adequacy. Thus, while military rule may be functional in repressing popular resistance, the formally adequate (or 'normal') capitalist type of state is a liberal bourgeois democracy (Poulantzas 1970). See also MARXISM; STRATEGIC-RELATIONAL APPROACH.

BOB JESSOP

social formation. See SOCIAL FORM.
social movements. See EMANCIPATORY MOVEMENTS.
social realism. See CRITICAL NATURALISM; SOCIAL STRUCTURE.
social science. Any science of the sociosphere that achieves explanatory knowledge is for CR intrinsically emancipatory because it necessarily issues in CRITIQUE of the status quo; or science in the sociosphere *is* EXPLANATORY CRITIQUE. Knowledge of the causes of their subordination is a necessary condition for rational self-emancipation by slaves, whereas masters have an interest in slaves' continued ignorance of these causes; critical knowledge is thus 'asymmetrically beneficial to the parties involved in relations of domination'. More generally, social science increases knowledge of the possibilities of four-planar social being in terms of self-emancipation. It is thus wrong to regard social science as *equally* a potential instrument of domination and of freedom (*PN3*: 77 n 84; *SR*: 182). See also CRITIQUE; DEPTH; DIALECTIC OF SOCIAL SCIENCE; EMANCIPATORY AXIOLOGY; IDEOLOGY; PHILOSOPHY OF SOCIAL SCIENCE; SOCIAL THEORY; entries for particular disciplines (e.g., SOCIOLOGY).
social structure. Broadly, four different conceptions of social structure can be identified (Porpora 1989). These are structure as (1) patterns of aggregate behaviour that are stable over time; (2) lawlike regularities that govern the behaviour of social facts; (3) rules and resources; and (4) systems of human relations among social positions.

The first conception reflects the social ATOMISM associated with methodological individualism, no longer very prevalent in anthropology or sociology but still dominant in economics (Lawson 1997). The second conception is associated with sociological holism, a position that posits a rigid separation between the realm of individuals and their actions and the realm of collective properties or social facts. Sociological holism tends to be positivistic and to embrace the covering law model of causal EXPLANATION.

CR has always championed a middle course between methodological individualism and sociological holism. On the one hand, the atomism of methodological individualism is inconsistent with CR's emergentist ontology, which considers all reality to be STRATIFIED. Thus, like sociological holism, CR posits levels of structure in human affairs as well. However, in contrast to holism, CR does not view social structural relations as nomothetic regularities and does not regard social structure as a separate realm autonomous from human agency.

CR's break with the covering law model is what allows it to conceptualise social structure in such a way that it remains connected with human agency. The covering law model itself is what has hampered traditional attempts to conceptualise social structure. The covering law model is decidedly DETERMINISTIC so that its application to human affairs precludes any element of what philosophers call 'free will'; the result is a mechanistic model of human behaviour. Thus, as long as CAUSALITY is exclusively understood in terms of the covering law model, there are only two choices: human freedom and responsibility can be upheld by rejecting any causal explanation of human behaviour; alternately, in the name of science and at the expense of free will, deterministic causality in human affairs can be affirmed.

The rejection of all causality in human affairs leads to the position known as VOLUNTARISM, which seems to imply that human behaviour is totally unconstrained socially. Within CR, this position is most associated with Rom Harré (1979; Harré and Varela 1996; Varela 1999). Harré advances a basically post-Wittgensteinian position that holds human action explicable only in terms of reasons and never in terms of causes. From this perspective, social structure is precluded from ever exerting any kind of causal force on human behaviour.

In contrast, as long as the covering law model is embraced, talk of structural causation seems to drag human beings back into a nexus of deterministic laws. It is for this reason that sociological holism has generally been closely tied to positivism, and that even mention of structural causation often tends immediately to be equated with 'structural REDUCTIONISM', the attempt to explain all human behaviour structurally.

Because the CR conception of causality is not deterministic, CR reference to structural causation does not imply any necessary or ineluctable consequences in terms of human behaviour. On the contrary, how humans react to structural causes is always ontologically open in a way that ultimately can only be traced by historical narrative (Porpora 1987). The ineluctable element of narrativity in turn implies a temporal dimension to the structure–agency dynamic that cannot be collapsed (Archer 1995, 1998b).

The temporal dimension is implicit in what might be considered the canonical CR approach to the structure–agency problem, the approach that Bhaskar calls the 'transformational model of social activity' (TMSA). Although there are those within CR who dissent from it or who think its application limited to capitalist systems (e.g., Dean 2000), most critical realists endorse the TMSA in some form.

The TMSA essentially posits a dialectical relation between social structure and individual human agency along the lines of Marx's (1852) famous observation to the effect that people make their history but not under conditions of their own making. Action, according to the TMSA model, is always initiated from a position within a structure, and action always in some way results in some effect on that structure.

The TMSA conceptualises the social as what Archer calls a MORPHOGENETIC process, a process always shaping and shaped by human action. As such, Archer insists, the TMSA is properly understood only if it is unpacked temporally. At each moment in time, some structure predates the action that reproduces or transforms it, and that structured action in turn predates the structural 'elaboration' (reproduction or transformation) it brings about (Archer 1995: 61).

On the surface, the TMSA seems very much in accord with the 'stratification model of social action' associated with Giddens' (1979) 'structuration theory'. For Giddens's stratification model as well, structure is both the 'medium' and 'outcome' of social action. This twofold character Giddens refers to as 'the duality of structure'.

The similarity between the TMSA and the two models did not escape Bhaskar, who cites the stratification model approvingly on at least two occasions. In *PN3*: 35, he suggests that the stratification model and the TMSA are just two sides of the same coin. Other critical realists (Manicas 1987; New 1994; Outhwaite 1987; C. Smith 1990) likewise see a close connection between CR and structuration theory, and both perspectives were strongly represented together in the opening days of the *Journal for the Theory of Social Behaviour*.

Given the apparent affinity between the TMSA and the stratification model, it is perhaps surprising that some critical realists (Archer 1985, 1990, 1995; Porpora 1987, 1989, 1994) have reacted hostilely to the stratification model. Yet, on similar grounds, other critical realists (e.g., Benton 1981) have likewise been dissatisfied with the TMSA model as originally formulated.

At issue has been how social structure should be conceptualised. If Bhaskar alluded to a conception of structure somewhere between conceptions (3) and (4) above, Giddens pressed more unequivocally for conception (3), and indeed popularised it. For Giddens, social structure just is rules and resources.

For Giddens, rules are structures only in their instantiation (Giddens 1979: 64–5). They operate, in other words, only to the extent that they are followed, and thus have what Giddens called only a 'virtual existence'. This idea again seems to align with Bhaskar. According to Bhaskar (*PN3*: 38; *RR*: 95), 'social structures, unlike natural structures do not exist independently of the activities they govern' or 'of the agents' conceptions of what they are doing in their activity'. Elsewhere, sounding very much like Giddens, Bhaskar (*PN3*: 50) even alleges that social structures do not exist apart from their effects. The tilt here in both Giddens and Bhaskar is very much in a hermeneutic direction.

Benton (1981), though sympathetic to the TMSA, raised serious questions about this original formulation. First, according to the CR view, causal powers exist even without being manifested. The same should hold, Benton averred, of social structure. A power relation, for example, still exists even if the power is not actually exercised. Similarly, Benton observed, the activities that sustain a power relation may be other than those the power relation governs. Thus, concluded Benton, social structures can exist independently of the activities they govern or of agents' conceptions of those activities.

In response to these criticisms, Bhaskar conceded some ground to Benton. In an 'Afterword' (*PN3*: 174), Bhaskar specifically admits that structures can exist apart from the activities they govern and without being manifest. Bhaskar still insists, however, that structures are dependent on (some) human activity and, moreover, on agents' conceptions of what they are doing.

The hardest line on these issues has been taken by Archer and Porpora. Explicitly reflecting a Marxist perspective, Porpora is concerned to uphold the fourth conception of social structure, which, he says, allows for 'emergently material' relations. According to Porpora, although all human relations are established and maintained by some human activity, emergently material relations exist – with independent effects of their own – even apart from agents' knowledge of their existence. Examples include relations that – even without agents' knowledge – are exploitative or that otherwise place agents' interests in conflict. Because structuration theory ignores such emergently material relations, Porpora categorises it as an exclusively idealist approach to sociology.

Archer (1995) provides the single most sustained attack on structuration theory – and kindred interpretations of the TMSA. For Archer, the central problem with structuration theory is that it is 'conflationary', 'redefining structure and agency as inseparable'. Archer shows what analysis loses from this conflation. Lost in particular is any kind of dialectic between structure and agency. For us to understand the causal relations between structure and agency, the two must be analytically distinct. Collapsing the distinction thus precludes this possibility.

Lost, then, is our capacity to analyse how structure affects agency, and how agency affects structure. These are processes, Archer reminds us, that unfold over time. The time dimension, too, Archer accuses structuration theory of collapsing. Thus, against Giddens's duality of structure, Archer counterposes the position she calls 'analytic dualism', which holds structure and agency to be dialectically related but ontologically and analytically distinct.

Today, although there still are articles contributing to it (e.g., King 1999; Lewis 2000; Willmott 1997), the debate on structure has largely gone into abeyance. Giddens's once vast influence on social theory has considerably waned, but his understanding of structure remains popular (see Sewell 1992). Although there remains a spectrum of opinion on structure and agency within CR, arguably Archer's analytic dualism has become the dominant position. Bhaskar's NEGATIVE GENERALISATION, of the TMSA in *Dialectic* (154–60), whereby structures can persist without and even despite human agency, takes Archer's central refinements of the TMSA fully on board.

DOUGLAS V. PORPORA

social tetrapolity. Another name for four-planar social being or HUMAN NATURE. Constellationally contains the SOCIAL CUBE.

social theory has a number of distinct, often related, though sometimes conflicting meanings. First, it is used in a general sense to refer to a set of relatively coherent theoretical frameworks or explanations which attempt to account for the social nature of human beings and the character of their social lives, in varying historical and cultural contexts. In this broad sense, social theory relates to issues which potentially span all of the social sciences and humanities. Second, it is sometimes used in a more specific sense to refer to a historically new and distinctive body of thought which has developed

426 social theory

since the modern Enlightenment. This body of thought transcends, but overlaps considerably with, modern sociological theory. It aims to distinguish between and understand different kinds of human societies as whole entities, and to analyse a new kind of society, 'modernity', which emerged in the West over the last few centuries, and which has come increasingly to dominate the rest of the world. Finally, the broad term 'social theory' is sometimes used as a synonym for narrower terms, such as 'political' or 'sociological' theory.

Social theory is important because it sums up what can be said about the social world, its various dimensions, processes of reproduction and transformation, and its links with other aspects of reality. It also defines particular disciplines, fields of study and substantive areas. It is neither a completely unified nor a finished project. It has been characterised not only by various theoretical perspectives but also by a series of disabling DUALISMS which Bhaskar's *PN* and other realist works have sought to overcome.

Building on his *RS*, Bhaskar argued for a CRITICAL NATURALISM as a third way between the relatively unqualified naturalism of positivism, and the anti-naturalism of hermeneutics. His view was that 'the social sciences can be sciences in *the same sense* as the experimental sciences of nature, but *in ways which* are as different from the latter as they are *specific* to the nature of societies' (*D*: 155). He specified the limits to naturalism in terms of key epistemological, relational, critical and ontological differences between the social and natural sciences. The epistemological limit in terms of open social systems means that the development and testing of social theory has to be primarily in terms of its explanatory power rather than predictive ability. The relational limit is that social science, unlike natural science, has a causal interdependency with its subject-matter. The critical limit is that social science is capable of EXPLANATORY CRITIQUE (overcoming the dualism of facts and values), in contrast to natural science. Finally, in terms of ontology, Bhaskar argued that social structures, unlike natural structures, are activity-dependent, concept-dependent, space–time dependent, and social relation-dependent. He also argued against the dualistic views that societies consist fundamentally of individuals or collectivities in favour of a relational conception of society.

Bhaskar's account of social ontology has been important in clarifying the relationship between SOCIAL STRUCTURE and AGENCY. He developed a transformational model of social activity (TMSA) which avoids reducing one to the other or eliding the two as mutually constitutive. His account of human agency similarly challenged the conventional philosophical dualism between reasons and causes with an argument that reasons can be causally explicable and causally efficacious in relation to human behaviour.

Bhaskar's important work on the TMSA was extended and clarified by Archer (1995), who developed an explanatory methodology called MORPHOGENESIS. This approach aimed to bridge the gap between ontology and practical social theory. Based on realist principles of intransitivity, transfactuality and stratification, it involves a non-reductionist, non-conflationist way of studying the interplay between structure and agency across time and space in the reproduction and transformation of social systems. In later work, Archer (2000) also developed an innovative realist account of human agency which goes beyond the dualism of 'undersocialised' and 'oversocialised' models of humankind.

Bhaskar's subsequent dialectical development of CR, and Archer's morphogenetic approach have so far yielded only limited substantive work. However, they avoid the errors and inadequacies of earlier theoretical and philosophical positions, and move beyond them in a way that opens up the possibility of more adequately grounded and emancipatory social scientific theories. See also PDM; PHILOSOPHY; RELATIONALITY; SOCIOLOGY; Callinicos 1999; Danermark et al. 2002; Sayer 2000a; J. Wilson 1983.

Ross Morrow

social work. The organised provision of social services in order to promote social change, problem solving in human relationships and the empowerment and liberation of people from unwanted and unneeded constraints to enhance well-being. Social work is directed to different levels: the individual and family level, the group and organisation level, and the structural level.

CR approaches in social work are not yet very common, but there is a growing interest in the mechanism theme in order to establish causality and arrive at explanations in processes of change. This aspiration, based upon the CR view of a stratified social world with emergent powers and mechanisms, has so far been expressed mainly in terms of theoretical discussions; empirical studies are rare.

The impact of CR thinking in social work research and practice is currently very limited. However, there are signs that CR influence will grow in the future. One journal in the field, *Qualitative Social Work*, explicitly promotes discussion and dialogue on 'emerging examples of critical realism in social work'. Up to the present CR influence is most apparent in the field of evaluation and activity development in social work practice.

The first article about CR in relation to social work (although written in Finnish) is probably Mäntysaari (1991). Independently of Mäntysaari, Anastas and Macdonald (1994) introduced CR in evaluation of social work practice. Since then Pawson and Tilley (1997) have developed CR to an accepted evaluation perspective. Kazi (1998, 2003a, b) and Rostila (2000, 2001) have applied Pawson and Tilley's CR perspective on evaluation in empirical studies. Morén and Blom (2003a, b) have initiated a further development of CR, differing in part from the approach of Kazi and Rostila, as a run-up to evaluation studies. They have elaborated core concepts like 'mechanisms' and 'context' in order to explain how and why client effects emerge in social work practice.

On a more general level, Nygren and Soydan (1997), discussing the link between theory and practice, argue for CR as a highly relevant meta-perspective for social work. Houston (2001a, b), criticising social constructivism in social work practice, proposes CR as a powerful alternative. Blom (2002) argues that CR ontology is compatible with Sartrean existentialism, which is supposed to help explain why clients in social work choose to alter their way of living. Bhaskar (*RM*: 161–2) touches upon a similar theme while discussing co-presence and transcendental identification in work with clients.

Because the research object in social work is both socially constructed and really existing, CR promises to become an important challenger both of resurgent empiricism and of the dominant constructionist approach in social work practice and research. An indicative example is current research on disability (cf. Danermark, 2003). In Sweden, Finland and Britain there are further examples indicating a growing interest in applying

CR to social work evaluation research (Kazi et al. 2002). See also EMANCIPATORY AXIOLOGY; Archer 2000.

Stefan Morén, Björn Blom

sociolinguistics. See LINGUISTICS.

sociological imperialism. A term used by Archer (e.g., 2000), Sayer (e.g., 2005) and other CR social theorists to refer to the widespread modern tendency to see people as exhaustively constituted by the social. See also SOCIAL CONSTRUCTIONISM.

sociology. Broadly, the scientific study of human social life and society in order to describe, explain and understand it, often also bound up with attempts to change society. It developed as an organised discipline in the nineteenth century even though the study of human social life dates back at least to the fourth and fifth centuries BCE. Sociology is scientific in the sense that it tries to develop theories which explain how and why things happen in society, and tests and modifies these theories on the basis of systematic empirical research and logical argument. The discipline is internally diverse, with a variety of meta-theories, research traditions and theoretical frameworks which provide different answers to the questions of what to study in the social world and how best to study it.

The development of realist, and later critical realist, approaches to social science from the 1970s has impacted on sociology in four main areas: meta-theory, theory, methodology and substantive research. In terms of meta-theory, realism and CR offered a critique of, and alternative to, the dominant meta-theories of positivism and idealism, particularly in terms of ontology and epistemology. Bhaskar's critical naturalism offers an important alternative to the relatively unqualified naturalism of positivism and the anti-naturalism of idealist approaches. It also provides the basis for a theory of explanatory critiques whereby the explanations of an objective social science could be shown to imply value judgements. In terms of sociological theory, CR has helped to distinguish SOCIAL STRUCTURES from natural structures more clearly; it has provided a sophisticated account of the relationship between AGENCY and structure which avoids the problem of REDUCTIONISM of one to the other and the conflation of both; it has provided an account of human agency which holds that reasons can be causes, and which avoids the DUALISM of under- and over-socialised views of human beings; and it has developed a MORPHOGENETIC approach which links social ontology to practical social theory in the study of the reproduction and transformation of social systems over time.

In terms of METHODOLOGY, CR is compatible with, and potentially fruitful for such approaches and strategies as interdisciplinary inquiry, methodological pluralism, triangulation, retroduction, and the search for real but unobservable structures, mechanisms and causal powers. Conversely, it is at odds with, and critical of, reductionist approaches such as methodological individualism and methodological holism. In terms of substantive research, CR does not by itself provide a theory of any particular social phenomenon, but its general approach and ideas are becoming evident in a range of sociological areas including studies of CAPITALISM and class, GENDER (see also MASTER–SLAVE), EDUCATION, HEALTH and illness, MANAGEMENT, RACE and racism, EMANCIPATORY MOVEMENTS, the STATE and work. It will be through the development of intellectually exciting social theory, and empirical research, that

CR can be expected to make further challenges to the dominance of postmodernism and the legacy of positivism in sociology. See also SOCIAL THEORY; SOCIOLOGY OF THE BODY, OF HEALTH AND MEDICINE, OF KNOWLEDGE, OF SEXUALITY; Archer, Sharp, Stones and Woodiwiss 1999; Keat and Urry 1982; New 2001b.

Ross Morrow

sociology of the body. Like philosophy, sociology arrived relatively late at the body as a distinct, explicit site of inquiry – although, as with philosophy, assumptions about the nature of embodiment were active in its operations long before it did so. Since the 1970s, most extensively among GENDER theorists (especially those influenced by Foucault), that ground has been made up in new and challenging ways: the body has featured centrally in reassessments both of subjectivity itself, and of the relationship between agency and structure. Whether or not invoking a sex/gender distinction, feminist theorists have been primarily concerned to challenge the essentialist 'reading off' of stereotypical, prejudicial or exclusionary views of women from what de Beauvoir (1953) called 'the data of biology'. Such analysis has been invaluable in revealing the contingency of reified presumptions about the condition of, and appropriate role for, women. It is echoed in analysis of RACE, DISABILITY and HEALTH, all of which are sites where labelling operates to reinforce and to oppress. In the process, ontological and phenomenological questions *tout court* have tended to be subsumed under the heading of 'gender', or equivalent terms of the social operations of discourse and power. There has been a widespread shift from the claim that '*perceptions* of the body are socially constructed' to the claim that 'the *body* is socially constructed' – a move, though prevalent, perhaps best exemplified in the work of Butler (e.g., 1993).

Broadly, CR approaches will be happy with the first claim, but dispute the second. The problem with body-sociology of the full-blown SOCIAL CONSTRUCTIONIST variety is that it operates on an ACTUALIST conception of nature (in one or more of the various senses identified at *D*: 234), and in fear of any form of biological determinism which might derive therefrom. To avoid nature's clutches, it erects a fundamental gap between (inert, blank, meaningless) nature and culture – with the latter as the source of all meaning and social impact. This tends towards the view that the body has no unity, whether phenomenologically or physiologically, being solely the effect of the shifting operations of discourse. In place of this analytical *either/or*, CR works on a dialectical *both/and* conception of the relative significance of nature and culture, avoiding reductive conceptions of either. In dealing with embodied subjectivity, as elsewhere, realists will allow for strata where both positivism and post-structuralism work on depthless models, given significance through verifiable statements or through discourse. The heuristic here is to avoid biological reductionism – and the central state materialism critiqued extensively in *PN* – without denying natural powers; to approach culture as dependent on, but not reducible to, nature. This can be done from different angles. For Archer (2000: 123) the self, as 'an emergent relational property', is realised 'through the necessary relations between embodied practice and the non-discursive environment': she thus locates a pre-discursive aspect to our reflexive embodiment. Assiter (1996) seeks a critical defence of the sex/gender distinction along CR lines. For each – and for all productive CR work on embodiment – the aim is to find a non-reductive reconciliation of elements of first-person experience, social shaping, and ontological preconditions. See also Calder 2005.

GIDEON CALDER

sociology of health and medicine. In the 1950s and 1960s the sociologies of medicine and of health and illness were underpinned theoretically by the American schools of structural-functionalism and symbolic interactionism. It was only in the 1970s that the influence of these schools waned and sociologists both intensified their contributions to less ambitious 'middle-range' theories and to the development and application of a greater diversity of paradigms. Perhaps most pervasive of these paradigms through the 1980s and 1990s have been Foucauldian and other postmodern forms of social constructionism. It is in this context that CR thinking has emerged as a counter to theses that conflate ontology and epistemology, or explicitly subsume the former under the latter.

CR has provided grounds for the restoration of the biological, for re-securing its (ontological) independence from (epistemological) conceptualisations of health and disease. What CR allows is a combination of ontological realism, epistemological relativism, and judgemental rationality (see HOLY TRINITY). Epistemological relativism here asserts that the world can only be known in terms of available descriptions and discourses. This is distinguished from judgemental relativism, which holds that one can neither judge between discourses nor decide that some accounts are better than others. According to CR, it does not follow that, if all knowledge is transitive and fallible, all knowledge is *equally* fallible; no more does it follow that, if all facts are theory-dependent, all are theory-*determined* (Sayer 2000a; Scambler and Scambler 2003).

The sociology of health and medicine discovered CR fairly late, in the 1990s. It has so far been utilised in three substantive areas. In the first of these S. Williams (1999) deploys it to enhance our understanding of chronic illness and disability. The BODY, impaired or otherwise, is a real entity: it cannot be reduced either to the social or to the biological. Disability is the sole product of neither the impaired body nor a socially oppressive society. Rather, it is an *EMERGENT* property involving the interplay of biological impairment, structural enablements/constraints, and socio-cultural interaction leading to structural elaboration/reproduction over time. Structures may be reproduced or critically transformed through social praxis or AGENCY (a vital matter for disability theorists).

A second area is Porter's (1993) ethnographic study of racism and professionalism in a clinical setting. The main participants were (white Irish) nurses and doctors (some of whom were members of racialised minority groups). Racism, Porter avows, is a structural phenomenon, revealing relational power and ontological depth. Its effect on action, however, is not discernible via the constant conjunction of events. Rather, the relationship between structure and action is generative: the former providing the condition for the latter. Racism is a tendency that is realised in some circumstances, but exercised unrealised in others. The (abstractly) universalist-achievement ethos of professionalism, Porter found, typically proved a countervailing mechanism to racism: professional relations manifested in an aura of 'starched propriety' gave the impression nurses 'knew their place' in the occupational hierarchy.

The third area is represented by Scambler's (2002) attempt to move beyond the neo-positivism of much research on health inequalities to interpret well-worked DEMI-REGULARITIES as indicative of class relations as a generative mechanism. More specifically, the 'greedy bastards hypothesis' (GBH) asserts that health inequalities in Britain might be regarded as an unintended consequence of ever-adaptive behaviours of members of its (weakly globalised) power elite informed by members of its (strongly globalised) capitalist-executive. It is the wealthy and powerful who most influence those asset flows most decisive for health and longevity.

These pioneering efforts are promising. With the present decline in commitment to post-structuralist and postmodern sociologies of health and medicine, CR offers a viable alternative grounding for a reconstructed project of modernity. See also Scambler and Higgs 1999; Williams 2003.

GRAHAM SCAMBLER

sociology of knowledge. CR is oriented towards a strong sociological programme of EPISTEMIC RELATIVISM, on the principle that all knowledges can be sociologically explained, with the ontological realist account of the structures underlying the possibilities of both knowing (in the TD) and being known (in the ID). While there has been some substantive work in recent realist-inspired political theory, the development of CR sociology of knowledge has tended to be philosophical.

Classical sociology has long encompassed a concern with the consequences of political and religious knowledges for modern social formation and integration; it also established the principle of the relativity of social knowledges to classes, groups and forms of life. Marx (1867) bequeathed conceptions of both the ideological reproduction of class relations and the subjectification of class subjects (Balibar 1995); Durkheim (1898, 1912) emphasised the function of religion for social cohesion, with liberal individualism securing social cohesion given the modern division of labour; Weber (1905) identified ascetic Protestantism as the primary vehicle of modern rationalisation. The sociology of knowledge emerges as a sub-discipline within sociology to the extent that formalised systems of knowledge are treated as distinct objects with specific structures, an approach most commonly associated with Mannheim's *Ideology and Utopia* (1960) and Scheler's (1980) differentiation of epistemic interests.

These themes have been variously taken up within the Marxist tradition, to which realist philosophy and political theory in this area are most closely related: Gramsci (1971) theorised hegemonic struggles over the organisation and contents of knowledges; Lukács (1919–24) identified the formal, analytical structure of bourgeois class knowledge; Adorno (1966) related identity-thinking to political and economic domination; Althusser (1969) developed his critique of 'theoretical humanism'; Habermas (1972, 1979) took Scheler's notion of epistemic interests into the ideal, speech situation and communication theory; Jameson gave us a theory of the postmodern as the cultural logic of late capitalism. Other streams of sociology have been more narrowly concerned with the knowledges specific to particular institutions, ranging across sociological accounts of the natural sciences to Foucault's (1970, 1977) disciplines.

CR is well placed to make three kinds of contribution to the sociological treatment of knowledge. First, it tempers the insights of relativism with ontological realism, e.g., it grounds accounts of natural scientific change, such as paradigm shifts (Kuhn 1962) and the potential for progressive research programmes (Lakatos 1970), by situating knowledge production within our relations to the existentially autonomous and enduring real objects of the natural sciences.

Second, its meta critical framing of social theory directs sociology towards EXPLANATORY CRITIQUES of cognitive absences. From the perspective provided by epistemic developments in realist philosophy and science, past and current knowledges entail systematic irrealist absences and category errors. Explanations of irrealist forms disclose failures to remove epistemic obstacles, and lead on (CP) to the moral condemnation of social structures dependent on irrealism and whose endurance requires the reproduction of such obstacles. This meta critical perspective also opens up the relation between forms of knowledge and domination (Lukács, Adorno, Althusser) by counterposing the internal relations between irrealism and master–slave relations, on the one hand, with those between realism and human flourishing, on the other.

Third, depth-realist social ontology opens up substantive inquiry into the possibilities afforded by structured social relations for producing specific kinds of knowledge and for liabilities to praxiologies. As against the problem of emergent realism, recent political theory has been more concerned with the epistemic problems of social reproduction in contemporary social formation. (See Jessop [2002b, 2003a] and Joseph [2002, 2003a] on the state and HEGEMONY, and Kathryn Dean [2003] on tendencies to the dissolution of the conditions of theoretical production.)

NICK HOSTETTLER

sociology of sexuality involves the use of different theoretical perspectives and research methods to investigate a range of phenomena brought together under the overarching concept of 'sexuality'. This concept does not have a single, widely accepted definition, and is itself part of the field of study. Its meaning is problematic, shifting and contested, and changes in its meaning lead to different phenomena being defined as sexual or non-sexual in different times and places. To a large extent, the sociology of sexuality is the study of what people define as sexual, and this might conventionally include such things as sexual orientation; sexual identity; sexual desire and fantasies; sexual beliefs, attitudes, values and norms; and sexual behaviour, relationships, communities and institutions. Sociologists generally want to understand these phenomena in relation to society, culture and history, and in relation to other relevant disciplines such as biology and psychology.

The conventional view of the history of the sociology of sexuality is that there was little or no development of it before the Second World War, and that the field first developed sometime during the 1960s or 1970s. However, sociology was part of the discursive explosion about sexuality that characterised Western societies from the eighteenth century. From at least the 1840s onwards, the development and growth of sociology in Western societies has been associated with a more or less continuous but highly uneven and varied approach to the study and understanding of sexuality. By the early twenty-first century, sociologists had studied human sexuality across a wide range of substantive areas, using a variety of different theoretical perspectives and research methods.

Realist and CR approaches and ideas appear to have made relatively little impact on this field so far. Implicit use of CR ideas, however, can be found in certain Marxist and feminist accounts of sexuality. Explicit CR ideas can be found in relation to research methodology, with realism posited as a superior alternative to positivism and hermeneutics, and in the critique and realist reformulation of strong social constructionist accounts of sexuality. Much contemporary sociological theory about sexuality, however, has been influenced by the 'linguistic turn', and has strong roots in the idealist and interpretivist traditions. CR can potentially make a profound contribution here. Its important distinction between the transitive and intransitive dimensions challenges the epistemic and linguistic fallacies of these traditions; its stratified and differentiated ontology of emergent powers promises a better understanding of the relationship between the objects of biology, psychology and sociology with a greater chance of understanding both sexuality and sexual problems; its non-reductionist account of human agency provides an alternative to the oversocialised accounts of human agency prevalent in post-structuralism; and its transformational model of social activity offers scope for a better understanding of the relationship between structure and agency. The development of CR approaches to sexuality may be helped by a growing interest in post-positivist philosophies of science but they will need to contend with the current idealist and discursive orthodoxies in this field. See also SOCIOLOGY; SOCIOLOGY OF THE BODY; Archer 2000; Hull 2006; Morrow 1994, 2003; O'Connell, Davidson and Layder 1994; Vance 1989.

Ross Morrow

sociosphere, adj. sociospherical. An analogue of *biosphere* which calls attention to the organic RELATIONALITY and REFLEXIVITY of the elements of human social TOTALITIES, and their sui generis reality (e.g., *D*: 116).

solidarity. CR ethics are 'a naturalistic ethics of care and solidarity' (*D*: 333; see also RECOGNITION; TRUST). The *dialectic of solidarity* is set out in the entry for EMANCIPATORY AXIOLOGY, in which it plays a vital role. It exemplifies the maxim that 'the slave who knows her slavery must come to know and articulate and achieve her *humanity* before she can become free'. The *dialectic of self and solidarity* (or *selves-in-solidarity*) adds the consideration that she and others 'must satisfy the criteria for rational agency', i.e., must come to know their real self-interest, be disposed to achieve it and be empowered (possess the resources, etc.) to do so (*D*: 294, e.a.; see also FREEDOM). Since not to solidarise in struggling for freedom is to commit theory–practice inconsistency, it articulates 'the unity of immanent critique and dialectical UNIVERSALISABILITY' (*D*: 291). Empowerment includes *amour de soi* or self-ESTEEM, which is a necessary basis for altruism and solidarity, and the liberation of potentialities for flourishing through de-alienation; or solidarity and self-emancipation (autonomy) mutually presuppose each other ('solidarity {. . .} *in* self-emancipation' [*D*: 285, e.a.]). The duplex form of this dialectic reminds us of this, and that solidarity must be sensitive to concretely singular differences, that it implies neither that someone else can act for us (substitutionism) (see SELF-REFERENTIALITY) nor that 'ought' implies 'can', and that solidarising intersubjectivity today is co-present within atomising globalisation whose de-agentifying propensities must be counteracted in a DIALECTIC OF THE EDUCATION OF DESIRE FOR FREEDOM (cf. Dean 2003).

434   solipsistic/transdictive (problem) complex(es)

solipsistic/transdictive (problem) complex(es). *Transdiction* (a term borrowed from Manicas [1987], who got it from Maurice Mandelbaum [*D*: 131 n]) is INFERENCE from the observed to the unobserved, embracing induction and transduction, retroduction and retrodiction, all of which, for CR, are justified by ontological stratification and the possibility of grounding by natural necessity or alethic truth. Together they constitute the *transdictive complex* or PROBLEM-field. In irrealist philosophy the absence of concepts of stratification, natural necessity and alethic truth, manifest in PRIMAL SQUEEZE, generates a cluster of APORIAI in this area of which the PROBLEM OF INDUCTION (in its Humean form, how do we know the course of nature won't change?) is only a special case. This is the *transdictive problem complex*.

Solipsism (f. L. *solus* + *ipse*, only oneself) is the view, closely bound up with egocentrism (or *ego-present-CENTRISM*) and indexicalist PUNCTUALISM, either that nothing exists, or that nothing can be known, outside one's self and the contents of its consciousness. Ontologically, it is thus a species of MONISM, which is entailed by and entails ONTOLOGICAL MONOVALENCE. Epistemologically it leads to SCEPTICISM about everything, including doubt itself, which, if it is serious (unlike the UNSERIOUS scepticism of the modern academy), must eventuate in silence: 'if you cannot say what is not you cannot say anything at all' (*P.* 230). It is refuted by the fact of PRIMARY (and ongoing) POLYADISATION, necessary for personal individuation, and by our THROWNNESS: the dependence of thought, including the solipsistic standpoint itself, on a pre-existing context of social practice which alone renders it meaningful and causally efficacious. It is also practically self-refuting in that the solipsist must contradict his theory (which hence is liable to immanent CRITIQUE) in practice. *Cratylus* (adj. Cratylian), a Sophist contemporary of Socrates, who vainly tried to avoid this contradiction by falling into silence and pointing, is an archetypal solipsistic figure in Western philosophy (see *P.* 53). However, although Parmenidean monism, on the Bhaskarian account, 'has as its only plausible avatar punctualist solipsism and must reduce to it via Cratylian silence' (*P.* 106 n), there is no evidence that any Greek philosopher actually argued for full-blown solipsism, which must be seen rather as profoundly embedded in the *analogical GRAMMARS* of the emergent worldviews that accompanied the rise of capitalism and in particular those informing the philosophy of Descartes.

The transdictive and solipsistic problem complexes are themselves mutually complicit dialectical counterparts within the subject–object IDENTITY theory at the core of IRREALISM, the former pertaining to its objective and general pole (the site of anthropomorphism, cf. RATIONALISM and EMPIRICISM), the latter to its subjective and particular pole (the site of anthropocentrism, cf. empiricism). They have their source in the interlocked primeval philosophical problems of the one and the other, and the one and the many, 'symptomatic of a class-divided and sundered society', on the one hand, and 'metaphoric of diversity and relativity and metonymic of the totality of master–slave relations' (*D*: 356), on the other. Solipsism may, however, take a communal as well as individualist form, as in the solipsism of Wittgensteinian/Winchian FORMS OF LIFE. The absence of concepts of multi-tiered stratification and alethic truth, in the case of the latter, and of primary polyadisation and the pre-existence of social forms as a condition of possibility of subjectivity, in the case of the former, necessitate TINA COMPROMISES with causally impinging reality, which hence can only be accepted on a *FIDEISTIC* basis, i.e., on the basis of faith, authority, conformity to social rules, or voluntarism ('at the limit, caprice or the will-to-power' [*P*: 234]). As belief in that which cannot be rationally justified, fideism thus generates a dualism within monism and subject–object identity theory, splitting it into *anthroporealism* and *transcendent realism*. In their 'systematic illicit intra-dependence' (*P*: 217, e.a.), these constitute the IRREALIST ensemble, which can be sourced to the epistemic fallacy and ultimately to ontological monovalence and the fear of change which underlies the quest for certainty (see FOUNDATIONALISM), and which sets philosophy today 'on the transmutation path through irrationalism to superidealism *or* critical realism' (*D*: 232). See also ANTHROPISM; UNHOLY TRINITY.

sophrosyne (Gr. *sōphrosynē*). In classical Greece, excellence of character and enlightenment in a well-balanced person (a *sōphrōn*, literally one of sound mind), reflexively aware of her self and its social relatedness. No single English word can translate it; *balance* is probably the closest, which is used by Bhaskar interchangeably with sophrosyne to suggest a principle of equipoise for an emergent EUDAIMONIC society as well as for persons. Together with PHRONESIS, it highlights that DCR draws on the pre-modern tradition of dialectical and spiritual enlightenment in its attempt to sublate the paradigm of bourgeois enlightenment and articulate a worldview for a post-power$_2$ order. See also EMANCIPATORY AXIOLOGY; EUDAIMONIA.

space–time. In DCR a complex, multi-layered ontological category. Bhaskar's critical engagement with it marks an important advance compared to CR, which is primarily concerned with 1M issues. He links space and time very closely with CAUSALITY and treats all three as 2E categories of NEGATIVITY (*P*: 67; *D*: 183, 207–8, 243–4, 303, 305, 379–80; *DG*: 392). More precisely, space–time is linked to causality in a tri-unitarian 'CONSTELLATIONAL IDENTITY', with causality being defined as 'TENSED spatialising becoming', 'tensed RHYTHMIC process', or 'tensed spatialised PROCESS' (*D*: 49, 106, 250). Like many concepts in CR (and even more in DCR), time–space is not introduced once and for all but developed at several points, with frequent repetition and confusing phrasal variation, minor or major additions, and, less often, significant new ontological and epistemological arguments. Despite their significance for DCR, however, much more work is required on key space–time categories. In particular, general ontological and epistemological reflections have been over-emphasised at the expense of substantive issues in social theory, where time–space is especially important for a critical, emancipatory politics.

To establish the importance of space–time in DCR, Bhaskar offers a transcendental argument. He deduces the necessity of change from the emergent, entropically expanding nature of the cosmos and concludes that space–time must be an emergent cosmic property (*D*: 124). More simply, and commonsensically perhaps, he claims that CHANGE is real and irreducible (i.e., being is linked to becoming, change must happen) and that this entails discontinuity in space–time. Change in turn is linked with causality, i.e., the capacity to produce (modify, prevent, absent) change. More specifically, 'the causal efficacy of a process constitutes, or rather manifests itself as, its spatio-temporal rhythmic'. This implies 'the *constellational identity of causality*, whether substantial, exercised or not, *and space–time* (in potential process)' (*D*: 77). Developing these points, Bhaskar argues that causality is 'transformative negation in processual (rhythmic) determination' (*D*: 52) and that process is 'spatialised tensing, the mode of becoming [as absenting] or [plain] absenting of effects' (*D*: 53). In short, he links space, time and causality to the theorem of the reality and irreducibility of (always potentially spatialising) tense and the potential and typical spatio-temporality (and hence processuality) of all causal efficacy in the definition of process as 'the mode of absenting which is the becoming and begoing of all effects' (*D*: 54, cf. 299).

At the most abstract-simple level of space–time analysis, Bhaskar presents space–time as an identity *and* as an identity-in-difference. He suggests that, conceptually, time emerges from space (*D*: 258); and, more generally, argues that time is not identical to space, that it can be disembedded from space, and, conversely, that space can be disembedded from time (*D*: 49, 257). He will later add that space can be disembedded from place and time from tense. Four further general points are made about space–time: (a) it is necessarily irreversible because of the impossibility of backwards causation (*D*: 70–1); (b) tensing is real, i.e., past, present, future exist, and also irreducible (*D*: 54, 142); (c) McTaggart's A-SERIES – past, present, and future – and B-series – earlier than, simultaneous with, later than – are equally necessary and also irreducible (*D*: 142, 210); and (d) space–time on any WORLD-LINE is irreducible for the intransitively observed and the transitive observer (*P*: 68). Together these imply additional arguments, such as the 'constitutive dependency of entities (natural as well as social) on their geo-historical processes of formation' (*D*: 68); and the scope for future transformation through action in the present on the basis of inherited conditions into which agential actors are 'thrown' or intentionally place themselves (*D*: 210).

Tying space–time analysis to causality leads to the notion of *RHYTHMIC*. Put simply, this is 'just the spatio-temporal efficacy of the [causal] process' (*D*: 53) or 'a tensed spatialising process' (*P*: 67). Bhaskar presents causality in turn as 'a tensed process in space–time' (*D*: 52). In more complex language: 'the constellational identity of (embodying the distinctions between) structures, mechanisms and spatio-temporal processes manifests itself in the *rhythmic identity of those changing causal powers with their spatio-temporal exercise*' (*D*: 77; cf. *P*: 67). The ontological category of rhythmics entails that things and processes have emergent *spatio-temporalities*. As such, Bhaskar notes, rhythmics 'may have supervenient causal powers of their own' in an 'intra-cosmic constellational unity of spatio-temporality, causality and being' (*P*: 67). Indeed, somewhat obscurely and at the risk of spatio-temporal fetishism, he argues that the rhythmic 'may be a *sui generis* causal power of space–time itself' (*D*: 54) because 'space and time are far from causally inert' (*P*: 67).

Besides such general remarks on the tri-unity of space, time and causality, Bhaskar claims that space–time has (at least) a fivefold character: (1) a referential grid for identifying the presence of existents or non-existents (absences, de-onts); (2) a measure or matrix for locating, dating and individuating entities, events, potentials, fields, etc.; (3) a set of prima facie mutual exclusion relations, turning on difference in extent (space) or change in endurance (time); (4) potentially emergent properties of an existing system of material things and/or relata of a new system of material things, perhaps with causal powers of their own; and (5) as a *rhythmic* that, at the macroscopic level, is generally entropic (*D*: 54; *DG*: 403; cf. *P*: 68). He adds that, since space–time and spatio-temporality are both an identity and identity-in-difference, their components can be disembedded in many ways. For example, space can be disembedded from time (long-distance telephone call, mass mediatisation) and from place (foreign investments); and time can be disembedded from space (a transatlantic flight) and time from tense (abstract labour) (*D*: 70, 258; *P.* 70).

Conversely, given such multiplicities, it is possible for spaces and times (and, a fortiori, spatialities and temporalities) to intersect, overlap or coalesce, often in contradictory or conflictual rather than just divergent ways (*P.* 68; *D*: 54, 79, 142). Obvious examples include world cities and revolutionary conjunctures. Bhaskar also notes four modes in which the past and/or the outside may affect the here and now: (1) EXISTENTIAL CONSTITUTION through containment; (2) the existential pre-existence of the world into which actors are 'thrown'; (3) CO-INCLUSION; and (4) LAGGED, delayed efficacy as in global warming (*P.* 69). The future can also be present in terms of 'the presence-of-the-future-in-the-past-in-the-present' (*D*: 142) or, as Marx put it in the *Eighteenth Brumaire of Louis Bonaparte*, 'men make their own history, but {. . .} not in circumstances of their own choosing'.

Finally, drawing on recent social theory, Bhaskar argues that one should recognise processes such as (1) space–time DISTANCIATION (whether as stretching of social relations in space–time and/or disembedding social relations from space–time) (*D*: 7n) and (2) space–time compression. He also anticipates recent work on the dialectic of path-dependency and path-shaping in claiming that things bear the mark of their GEO-HISTORICAL development but that agential action and transformation, within these limits, is possible (*D*: 210).

Critique

Many of these arguments will be familiar, of course, to natural and social scientists. Bhaskar's use of dialectical as well as analytical reasoning sometimes adds to these disciplines' understanding of space–time; in many cases, his analyses fall short of what they already offer. Indeed, despite its depth, complexity and philosophical sophistication, his arguments on time–space focus mainly on elaborating categories rather than developing substantive analyses. Concrete examples are illustrative and remain undeveloped. Given the richness of the categories, many different empirical outcomes are conceivable with DCR; but we lack explanations as to why particular space–time-causal relations obtain in given circumstances. It is not Bhaskar's aim as an underlabourer to provide substantive analyses, of course; instead it is to critique alternative philosophical positions, such as ANTHROPISM, INDEXICALISM, ATOMISM, PUNCTUALISM, spatio-temporary block universalism (BLOCKISM), closed totalities, etc. (*P.* 68; *D*: 383). In this regard, despite what some regard as unwarranted neologisms and needless obscurity (cf. *P.* xi), there is much brilliance. This said, my critique focuses on the overall rationale of his account and its relevance to the social sciences.

DCR marks a distinct shift in CR with its strong grounding in metaphysical arguments about the cosmos as well as an 'absentialist' concern with the ontological priority of absence over presence. Bhaskar's arguments presuppose the singular formation of the cosmos in the 'Big Bang' – which initiated space–time. Nothing existed and nothing happened before that unique event – unless we identify this as pure and total absence. We cannot otherwise ask about the *before* of the Big Bang or *where* it occurred. Equally there is nothing beyond the bounds of the cosmos – the entropically

438   spatio-temporality

expanding universe constitutes space–time and, outside its expansion, there is no space or time (Yannaras 2004). The origins of space and time in a state of non-locality and non-temporality have major implications for the completeness of Bhaskar's general reflections on time-space and, indeed, on the nature of ABSENCE and presence, ONTS and de-onts, materiality and immateriality, realism and irrealism.

This said, CR should certainly integrate time-space rather than adopt an attitude of 'spatial and temporal indifference' (Sayer 1989: 258). Bhaskar's transcendental argument shows that space–time matters and, indeed, that causality in the known cosmos is inherently a tensed, spatialising process. For DCR, this holds for the natural and social worlds and, a fortiori, the natural and social sciences, enabling critical naturalism about spatio-temporality and the geo-historical earthing of social relations. In emphasising the tri-unity of space, time and causality, Bhaskar avoids spatial and temporal fetishism, because space–time is an emergent, RELATIONAL property of systems of material things. His analysis also facilitates reflexivity on time- and space-bound interactions between the intransitive and transitive domains.

None the less, we need a far richer vocabulary for space and time than Bhaskar offers. To give just five examples: (1) space is a thin category, leading to relative neglect of issues such as scale and multi-scalarity; (2) the account of LOCO-PERIODISATION is oversimplified, ignoring the role of continuities in discontinuity and discontinuities in continuity in favour of identity versus difference or identity-in-difference (*D*: 255); (3) we need more concern with horizons of action and their social variation and their implications for chronotopic governance, strategy and tactics; (4) the use of geographicity and historicity in reflexive learning and the capacity for action; and (5) an emancipatory politics should examine differential capacities to stretch and/or compress time-space and how they are linked to power$_1$ and power$_2$. See also DIACHRONY/SYNCHRONY; ENTROPY/NEGENTROPY; PERSPECTIVE; Crang and Thrift 2000; Giddens 1983; Harvey 1982; Massey 2005; Sayer 2000a; Sheppard and McMaster 2003.

BOB JESSOP

spatio-temporality. See SPACE–TIME; TENSE.
species being. See PHILOSOPHICAL ANTHROPOLOGY.
speculative illusion/positivist(ic) illusion (SI/PI). Inverse, dialectically paired illusions (i.e. lacking an object but real in causal efficacy) – concomitant with the EPISTEMIC–ONTIC FALLACY and the RATIONALIST and EMPIRICIST traditions – that theory, science or (in, e.g., the Hegelian version of the SI) social life generally reduce in the last resort to PHILOSOPHY and sense-experience, respectively. If the SI is ultimately subversive of science, and more generally alterity (presupposing that 'we can somehow check the correctness of our concepts and meaning by comparing them {. . .} with our concepts and meaning' [Norris 2004: 16], the PI is 'the self-abnegation of philosophy in favour of science', thereby apparently effacing the EPISTEMIC–ONTIC FALLACY and producing 'the illusion that there is no illusion' and 'a totally a priori and deductivist account of science as omni-competent' (*SR*: 292–3), or SCIENTISM. While apparent antagonists, the PI and the SI are in reality mutually reinforcing; both presuppose ACTUALISM, and come together in the ARCH OF KNOWLEDGE TRADITION. They entrain the PRIMAL SQUEEZE on realist scientific theory and

speech acts, dialectic of 439

ontology, thereby prohibiting critique and emancipatory social change, and are conducive to TRIUMPHALISM. Bhaskar's account of the SI is fundamentally an endorsement of Marx's critique of Hegel's 'logical mysticism' as the illusion that, by unravelling concepts, we unravel being. His later work stresses that, while illusions are in one sense unreal (in that they have no real object), they are in another sense only too real (in that they are causally efficacious and so part of the world) (see DEMI-REALITY). See also ANTHROPISM; DUAL; IDEOLOGY; HEGEL–MARX CRITIQUE; IRREALISM; REDUCTIONISM.

speech acts, dialectic of. See EMANCIPATORY AXIOLOGY.

spiritual turn, spirituality. See CONCRETE UTOPIANISM; CRITICAL REALISM; ENCHANTMENT; META-REALITY; PHILOSOPHY OF RELIGION; PRINCIPLE OF TOTALITY; SELF-REFERENTIALITY.

split, split intentionality, split-off. See AGENCY; ALIENATION; FISSION/FUSION; GENERATIVE SEPARATION; META-REALITY.

spontaneous right-action. See ACTION; DIALECTICAL REASON; EVIL; MEDIATE; META-REALITY; PERCEPTION.

spread. See DISTANCIATION.

square of oppositions. See TABLE OF OPPOSITIONS.

squeeze, primal. See PRIMAL SQUEEZE.

stadion, pl. stadia. See MELD.

standpoints. See PERSPECTIVE.

state (the), state theory. See HOBBESIAN PROBLEM OF ORDER; POLITICAL SCIENCE; POLITICAL THEORY.

statistical explanation. See EXPLANATION; METHODS, QUANTITATIVE.

statistics, social. See METHODS, QUANTITATIVE; PROBABILITY.

Stoic, Stoicism. See BEAUTIFUL SOUL, ETC.

strata, stratum. See DEPTH; ONTOLOGY; PRIMACY.

strategic-relational approach (SRA). Introduced by Jessop (see his 2005) to overcome the tension in Marxist state theory between emphasis on the logic of capital and the primacy of class struggle. He later extended it to all questions of structure and agency. The SRA has four key features. First, it treats structures as strategically-selective in their form, content and operation; and actions as structurally constrained, more or less context-sensitive, and structuring. A given structure may privilege some actors, some identities, some strategies, some spatial and temporal horizons, and some actions over others. Second, it explores how, if at all, individual and/or collective actors consider this differential privileging through a 'strategic-context' analysis when engaging in action. Third, it examines how strategic reflection and social action reproduce-transform social structures and their emergent properties. Fourth, it notes how reflexively reorganised structural configurations and recursively selected strategies and tactics may interact to produce a relatively enduring structured coherence in social relations. In short, the SRA highlights the dialectic of structurally inscribed strategic selectivities and (differentially reflexive) structurally oriented strategic calculation. The SRA is consistent with other CR approaches, such as Archer's MORPHOGENETIC APPROACH and Bhaskar's TMSA. While these are generic approaches to structure and agency, however, the SRA's Marxist provenance means that it focuses on the strategic selectivity of particular social forms, especially their spatio-temporal specificity, and emphasises the differential material and semiotic capacities of individual and collective actors to engage in reflexive action and social transformation. See also RELATIONALITY; SOCIAL STRUCTURE; TMSA.

BOB JESSOP

strategic reversal, Foucauldian. See PERSPECTIVE, PHILOSOPHY, REVERSAL.
stratification, stratification of science. See DEPTH; DIFFERENTIATION.
stratification of the subject. See AGENCY; DEPTH; SUBJECTIVITY.
stratified monism. See MONISM.
stretch, stretch-spread. See DISTANCIATION; TENSE.
structural elaboration. See MORPHOGENESIS.
structural sin. See DEMI-REALITY; EVIL.
structuralism, post-structuralism. Many of the early works of CR consciously place themselves in the structuralist tradition, and many critical realists would see themselves as some sort of structuralist. However, recent years have also seen an increasing affinity between DCR and some aspects of post-structuralism (Bhaskar 1993a; López and Potter (eds) 2001; Joseph and Roberts (eds) 2003).

Structuralism is a contested tradition, but for critical realists interest is in the nature of the organisation of the social world and how its different parts fit together. With Bhaskar's distinction between the transitive and INTRANSITIVE dimensions, interest is in the nature of the intransitive (social ONTOLOGY) and questions such as structure, STRATIFICATION and TOTALITY. Bhaskar's argument for structure is a transcendental one. Given that knowledge and intentional social activity are possible, this presupposes that the world is structured in a certain way such that its structure is enduring and open to investigation.

Given this, we can say how Bhaskar's transitive/intransitive distinction improves on Althusser's thought object/real object distinction in that the intransitive is a condition of possibility for the transitive to be meaningful (transitive knowledge is of the intransitive object). By contrast, with Althusser's formulation there is a danger that we can only know knowledge (the thought object). The implications of this carry over into post-structuralism with the relation to the intransitive (real object) severed so that there is nothing (meaningful) beyond the text – although this might be said to exist already in the structural linguistics of Saussure where meaning is generated by the relation between words rather than between words and the referent.

There are many similarities between Althusserian structuralism and CR like the emphasis on stratification, overdetermination and, more controversially, theoretical anti-HUMANISM. These have been noted by Collier's comparison of Bhaskar and Althusser (1989, 1990). But critical realists question the way structuralism tends to eradicate the role of agency (agents become no more than bearers of structures and occupants of subject positions). Bhaskar's way of trying to deal with this issue is through the TMSA where structures might be said to be the necessary condition for and reproduced outcome of human agency (*PN2*: 34–5). And, because structures depend on agents for their reproduction, the possibility exists for agents to transform structures. However, problems were raised as to how much this bent the stick the other way (towards the STRUCTURATION theory of Giddens and the idea that structures and agents are mutually constitutive). That Bhaskar argues for an ontological hiatus between STRUCTURE AND AGENCY has allowed for others like Archer (1996) to argue for the PRIMACY of structure.

Recent work has looked at the compatibility of CR and post-structural work from epistemological and ontological angles. The main epistemological agreement is over epistemic relativity, but critical realists argue strongly against this leading to judgemental RELATIVISM (just because we cannot guarantee the correctness of a theory, this does not mean there are no grounds for preferring one theory over another by using some test of explanatory adequacy). Ontological issues include the role of discourse in social reality and the critique of pure presence and IDENTITY.

Deconstruction argues that a simple element cannot be present in and of itself and that each element of a text is interwoven with others creating a chain or trace of differences, syntheses and referrals (Derrida 1981: 26). Critical realists might point to how this extends to non-textual relations as well. Derrida's concept of *différance* can be useful in opening up the uncertainties of meaning and exposing how power relations within discourse privilege certain terms while excluding others. This might be considered consistent with Bhaskar's treatment of struggles over hermeneutic power$_2$ relations (see HEGEMONY).

Norris (1993) has been looking at Derrida from a realist angle for many years. However, critical realists are critical of post-structuralism's ontological stance. Derrida's critique of ontology is an extension of his critique of pure presence and identity. Yet Bhaskar has argued for a much richer conception of ontology based on a critique of what he calls ONTOLOGICAL MONOVALENCE or a purely positive account of reality which analyses real NEGATION in purely positive terms. He opposes that which erases the contingency of existential questions, screens the contingency of being or despatialises and detemporalises accounts of being (*D*: 7, 400–1). This radical presence of an absence at the heart of ontology, the incompleteness of totality (and Bhaskar's reconsideration of Hegel's dialectic) might usefully be compared with some of Žižek's arguments (1999).

Post-structuralists tend to hide behind relativism and contingency in order to avoid ontological questions. Any depth-ontology is rejected in favour of contingent surface relations. This is clear, for example, in Foucault's rejection of his earlier structuralism (discursive formations) in favour of power–knowledge relations, where power is examined in terms of its exercise and effects but not in terms of its underlying causes. (Foucault explicitly rejects a focus on underlying economic relations and centralised institutional [state/sovereignty] relations). Consequently, while critical realists will find post-structuralism well worth engaging with, it is probably fair to say that their stance is more like that of a critical structuralist. See also PDM, POSTMODERNISM.

JONATHAN JOSEPH

structuration. See INFRA-/INTRA-/SUPER-STRUCTURE; SOCIAL STRUCTURE.
structuratum, pl. **structurata**. See STRUCTURE.
structure. 'The structure of a thing is constituted by its causal powers, which, when exercised, manifest themselves as TENDENCIES. A structure will typically be instantiated in a multiplicity of *structurata* {L., structured things}. A concretely singularised individual {including a person} can be understood as a *laminated structuratum*' (*DG*: 404, e.a.). Thus the structure of a building is the configuration of its

elements, its abstract form or type, which it shares with other such entities; a concrete instantiation of it, a building itself, is a structuratum (any structured thing) or more generally a TOKEN of the type. Structures do not exist apart from structurata; they are comparatively enduring, but intrinsically mutable. The distinctions between structure and structuratum, on the one hand, and laminated and other kinds of open systems, on the other, were introduced by Collier (1989: ch. 3, glossary). Structuratum is intended to convey the notion of a 'structured TOTALITY' whose elements have their own distinct reality as contrasted with an 'EXPRESSIVE totality' which con-FUSES them. It may be *simple* (in closed systems) or *compounded* (in open situations). Laminated structurata or SYSTEMS are 'structurata whose elements are necessarily bonded by an irreducible plurality of structures (e.g., human individuals have mental and physical structures; human societies have economic, political and ideological structures)' (ibid.: 194). See also CAUSALITY; CONCRETE/ABSTRACT; SOCIAL STRUCTURE.

structure and **agency**. See AGENCY; SOCIAL CUBE; SOCIAL STRUCTURE; TMSA.

struggle. See CO-PRESENCE; DEPTH; EMANCIPATORY AXIOLOGY; HEGEMONY; IDEOLOGY; RECOGNITION; REVERSAL; SOCIAL CUBE.

studies of education. A heterogeneous collection of disciplinary-based studies of education (history, philosophy, psychology and sociology, as well as 'educational studies'), each defining its object of study and procedures of inquiry differently. Bhaskar argued that, alongside the laboratory, the classroom is one of the most under-analysed 'sites' of science (*RS2*: 247); and CR, by maintaining the possibility of a social science, offers the possibility of a science of education. However, while a large body of work tacitly operates in accordance with CR principles, studies explicitly working from a CR perspective remain limited in number.

Explicit CR has served two main functions in education studies. First, it offers critiques of dominant positivist and 'critical' paradigms. Positivism, underpinning calls for 'evidence-based' policy and practice based on systematic review and randomised controlled trials, reduces education to the empirically measurable. 'Critical' or 'radical' (including Marxist, feminist and multicultural) approaches emphasise the effects of wider social relations of power and, under the influence of postmodernism and standpoint theories, have tended towards idealism and relativism, reducing education to the experiences of knowers. Explicit CR critiques their shared empiricism and advocates instead a non-reductionist analysis of the underlying relations between real learning environments, educational knowledge, social structure and the interior world of the learner. Examples include Corson's pioneering series of CR critiques of such thinkers as Dewey, Popper and Quine (1990, 1991a, 1991b); Scott (2000) on educational approaches generally; and Clegg's critique (2005) of systematic review as a basis for evidence-based policies.

CR's second function has been to offer a sense of mission for educational practitioners and researchers. Though CR emphasises the need to explore underlying structures and generative mechanisms, its purpose in doing so is to effect emancipation, a mission made more explicit with the emergence of DCR and one attractive to many in the education profession. Both these roles have been as an underlabourer. Thus far, CR offers few empirically applicable tools for educational research – though studies are using Archer's morphogenetic theory (1995) to analyse structure–agency relations in educational change (e.g., Willmott 2002), and Archer's analysis of the emergence of educational systems (1979) offers an early example of research based on CR principles – and how its emancipatory potential would be enacted in classrooms remains unclear.

There is, however, a large body of work tacitly based on CR assumptions. In sociology of education, for example, the approaches of Pierre Bourdieu and Basil Bernstein arguably offer theoretical and methodological means for enacting the imperatives of CR's social ontology in empirical research (see Moore 2004). Their theories construct education as an irreducible, emergent and relational social structure and operate with a depth-ontology to explore the underlying structuring principles and generative mechanisms shaping knowledge and pedagogy. In addition, such work shares CR's emancipatory mission and has done much to explore the role of education in social inequality. Thinkers such as Roy Nash (2002, 2004) are exploring the relations of these approaches to CR, though, as yet, relations to DCR and PMR remain unexamined.

Bridging the present gap between explicit CR engagements and CR-compatible research is one key to linking CR's underlabouring role to empirical research and unlocking its emancipatory potential. Tacit CR offers a wealth of conceptual tools for analysing what kinds of curriculum, pedagogy and assessment are emancipatory for which pupils and under what conditions. Conversely, while they remain ontologically and epistemologically agnostic, these approaches can offer little critique of or defence against neo-conservative and postmodernist positions or empiricist applications of their frameworks. The future for CR in education studies lies in making connections between its rich philosophical basis and the wealth of research unaware of its CR compatibility.

KARL MATON, BRAD SHIPWAY

subject. See SUBJECTIVITY.
subject negation, subject-developmental negation. See CONSISTENCY; FIXISM/FLUXISM.
subject–object identity theory. See ACTUALISM; EMPIRICISM; IDENTITY; IRREALISM; KNOWLEDGE.
subject–object duality, subject–predicate duality. See DUALITY.
subject–predicate propositional form. See DCR; FIXISM/FLUXISM.
subject-referentiality. See SELF-REFERENTIALITY.
subjectivation. See REALITY PRINCIPLE; SUBJECTIVITY.
subjective superidealism. See CONVENTIONALISM; EMPIRICAL REALISM; EPISTEMIC–ONTIC FALLACY; HOLY TRINITY; REFERENCE; SAPIR–WHORF; SOLIPSISTIC, ETC., VOLUNTARISM.
subjectivism. See CATEGORY; RATIONALISM; REPRESENTATIONALISM; UNHOLY TRINITY.
subjectivity concerns who and what we are as acting and speaking beings. Who and what is the 'I' that acts and speaks, the *self* who is myself? The classical modernist view that it is the ATOMISTIC ego separated from other egos and the world in a PUNCTUALIST 'here-now' has been critiqued in a range of *DCR* entries (see especially PHILOSOPHICAL DISCOURSE OF MODERNITY), while the very idea of a continuous, unified self has radically been called into question from a range of quarters over the last century, latterly by postmodernism (*P.* 104). But, if we reject such notions of the self, who is it that announces the rejection? The 'I' is notoriously a 'shifter', applying

444 subjectivity

both universally to anyone who utters it and particularistically to one person only, and varying greatly, as a particular, from context to context and time to time (e.g., when someone close to us dies, are we the same person?).

This is the *problem of personal IDENTITY* (*P.* 100, 104–5; *MR*: ch. 2). CR resolves it by placing change and relationality, along with dialectical universality and concrete singularity, at the centre of its conception of the subject/self. Corresponding to the moments of the ontological–axiological chain, as illustrated in Table 39, a human being consists of (1M) a core universal HUMAN NATURE grounded in (but not reducible to) a shared (but also unique) (i) genetic constitution, (ii) transcendentally real or essential or alethic self (DCR) and (iii) ground-state (ultimate implicitly conscious field of potential, thematised as the soul in *EW*) (PMR); (2E) the RHYTHMICS of her WORLD-LINE; (3L) particular MEDIATIONS; constituting her (4D) as a concretely singular person, in effect a unique NATURAL KIND. It is important to bear in mind that 4D > 3L > 2E > 1M; this means that the concretely singular person constellationally contains the other aspects as part of herself. All these elements are (within meta critical limits) changing. As the site of intentional causality or transformative AGENCY, the embodied self is a product-in-*process* at 4D. At 3L it is internally related to other (changing) selves, and the site of complex intra-self RELATIONALITY; in a conception similar to Marx's understanding of nature as our 'inorganic body' (see Collier 1991), both the embodied and the essential self include beings that lie outside their spatial envelope, i.e., they are part of (totalities within) TOTALITY, and REFLEXIVITY itself is the inwardised form of totality. At 2E the whole ensemble is continually being constituted geo-historically, such that it is not only a product-in-process, but *processes*-in-product-in-process – a 'multiply determined totality' (*P.* 61). Its dialectically universal moments at 1M – the stratified mind and psyche; the alethic self; grounded in the deep structure of speech and praxis (see EMANCIPATORY AXIOLOGY); human genetic endowment as a species and the ground-state – are also developing. The self must therefore be construed as *dispositionally identical with its changing causal powers* (an example of the constellational unity of identity and change, which, by perspectival switch, is also the constellational unity of identity and difference in entity relationism; see ALTERITY/CHANGE).

This, then, is who we are on the CR account – changing embodied persons who are also transcendentally real selves, profoundly interconnected with each other and the rest of the cosmos. Our little atomistic ego is therefore rejected as an ILLUSION in the strict sense – a real false being with causal efficacy, an illicit abstraction from 'a much deeper and broader totality' (*EW*: 3). The distinctions between the illusory ego, the embodied self and the transcendentally real self map loosely on to those between the domains of the empirical, the actual and the real, and (in PMR) between demi-real relative reality, relative reality and absolute reality (see META-REALITY; REALITY). In PMR explicitly, and implicitly in CR, the virtuous or enlightened person sheds her ego and brings her embodied self into phase with her essential self, entraining an in principle indefinite expansion of powers and potentials. Although the concept of the transcendentally real self is articulated explicitly only in the works of the spiritual turn, it is strongly implicit in the earlier concept of a (changing and always socially mediated) core human nature, and in the deduction, in *Dialectic* and *Plato Etc.*, of the 1M reality of human needs and potentials, rights and freedoms, from agency and discourse, and the deployment of the logic of dialectical UNIVERSALISABILITY and theory–practice

Table 39 Subjectivity/the self, with some correspondences

Causal–Axiological Chain	1M Non-Identity	2E Negativity	3L Totality	4D Transformative Agency (human concrete singularity)
Concrete universal ↔ singular (UPMS) Polysemy of absence	*universality* product	*processuality* process (incl. subject- or developmental negation)	*mediation* process-in-product	(concrete) *singularity* (not just human) product-in-process
Causal modes of absence	transfactual causality	rhythmic causality	holistic causality	intentional causality
Modes of radical negation	auto-subversion	self-transformation	self-realisation	self-overcoming
Components of human nature	(1) core universal	(4) changing	(2) historically specific	(3) unique individuality
Four-planar social being or human nature (social tetrapolity) Social cube	d. (intra-)subjectivity (the stratified person)	c. social relations (concept-dependent, site of social oppositionality) (1) power₂, (2) discursive, (3) normative relations intersecting in (4)	b. inter-/intra-subjective (personal) relations – transactions with ourselves and others (1) power₂, (2) communicative, (3) moral relations ideology: illusions of the demi-real	a. material transactions with nature (making)
Subjectivity/self	transcendentally real self or alethic self	ongoing change (geo-historical constitution)	self-consciousness or reflexive monitoring, incl. internal conversation; internal relations with others	embodied self, site of intentional causality
An embodied self (a human being or person)	core universal human nature	*illusion of rhythmics* of her world-line	*ego (ideological) particular mediations*	unique *concretely singular person* (in effect natural kind)
Qualities of the self	psychic (stratified beliefs, concerns, projects) stratification of mind (consciousness, self-consciousness, pre-conscious, unconscious)	mental	emotional	physical
Transformative agency	reasons for acting	reasons for acting flow of action/praxis	inter-/intra-action within and among persons	embodied intentional causality (exercised) action/praxis–poiesis

Note See also Figure 2, 'The stratification of agency', which could equally be called 'The stratification of the subject'.

CONSISTENCY to articulate the human moral ALETHIA as the free flourishing of each as a condition of the free flourishing of all. We are true to our essential selves when we act in accordance with this. It is also arguably presupposed by Marx and by emancipatory discourses in general (*RM*: 71, 113).

The *stratification of the self*, or *subject* or *intra-subjectivity*, refers to (1) the triplex stratification of our subjectivity in master–slave-type societies already indicated (illusory ego, embodied self, alethic self); (2) emergent levels of subjectivity manifest in conscious engagement in social practices, reflexive monitoring and accounting for our activities, reflexive monitoring of our lives, and meta-reflexively totalising awareness of our self-situation (see AGENCY; REFLECTION); (3) the depth-stratification of beliefs constituting a psyche; and (4) the stratification of mind, as illustrated in Figure 2 (open biological substratum, unconsciousness, preconscious, CONSCIOUSNESS/self-consciousness). The stratification of mind entails a decentred self, such that, contra Kant, 'transcendental unity of consciousness {...} is an *achievement*, not a presupposition' (*D*: 325). Moreover, the life of subjects in their 'internal dialectical pluriverses' is 'open systemic {and} multiply and conflictually determined', looped through, in the demi-real, by power$_2$ relations and contradictions and subjected to powerful forces of 'ALIENATION, *destratification* or fragmentation' (*P*: 99). *Subjects$_1$* and *subjects$_2$* correspond to POWER$_1$ and power$_2$; while there are arguably no subjects$_1$ in the demi-real (subjects who are not in some way oppressed by power$_2$ relations), it is a necessary category for concrete utopian thinking. The *concrete self* (*D*: 277) is the concretely singular person.

Auto-subversion, self-transformation, self-realisation and *self-overcoming* are moments in the fourfold meaning of radical negation, within the fourfold polysemy of ABSENCE, that it shares with other kinds of negation. Auto-subversion (1M) 'encompasses anything from a counter-CONATIVE TENDENCY to self-destruction, e.g., suicide' (*D*: 106). Self-transformation (2E) is self-change (see also SELF-REFERENTIALITY). Self-realisation (3L) is the fulfilment of one's NEEDS and potentials; in PMR, and at least implicitly in DCR, these are grounded in the transcendentally real self (an analogue of the generative mechanism [*RM*: 196]), such that self-realisation is a coming home to yourself, becoming 'only and fully' who you really are, or enlightened (*RM*: 102–3). Contra Dean et al. (2005: 7), it is not a 'displacement' of freedom but is fundamental to its process. Self-overcoming (4D) is either Nietzschean self-overcoming in the EMERGENCE of a creative 'enhanced humanity', or self-TRANSCENDENCE (auto-EMERGENCE) (*D*: 106).

Archer (2000, 2003b) has developed a complex relational social theoretic account of the ontogenesis of subjectivity that importantly builds on the Bhaskarian position, among others. An attempt is made to display its main contours in Figure 9.

Archer's thoroughgoing account is still in the early stages of reception and debate. Its main thesis, that 'the human being is logically and ontologically prior to the social being, whose subsequent properties and powers need to build upon human ones' (2000: 188), is controversial within CR, which has a strong social constructionist and historicising strand alongside countervailing tendencies. It is both (deliberately) omissive in relation to the Bhaskarian account (the domain of the unconscious); arguably departs from it in some ways (the Archerian subject seems more centred and rational than Bhaskar's 'decentred disunity' whose creative agency is 'the effect of deeper structures' [*PN3*: 113], and far more private, locating our identity primarily in the personal

Figure 9 The ontogenesis of subjectivity as an emergent relational property of human being

THE SOCIAL SELF

Actors [You] (social identity [PEP]; incumbents of social roles, which they play in their own particular way)

↕

Agents (collectives sharing the same life chances within the social system of resources and practices; primary agents [Me] [universal] and corporate or collective agents [interest groups] [**We**]. Necessary mediator in supplying activity with a social purpose and accounting for who acquires which roles.

↕

Persons [I] (personal identity [PEP], social concept of self; supplies activity-potential for Actors); **internal conversation**, concerns re physical well-being (natural order), performativity (practical order), self-worth (social order) – articulation = personal id. Evaluation of (objective) situations in the light of (subjective) concerns and (practical) projects in the light of situations.

↕

Discursive socialisation, social construction, mediated by internal conversation; presupposes the human self

THE HUMAN SELF

The human self/subject [I] (continuous sense of self [universal] – selfhood, powers of reflexivity [incl. internal conversation] or self-consciousness, embodied memory, thought/logic [PEPs] = intentionality; early development is practical and wordless)

↕

Practical differentiation (self and otherness, subject and object, the self and other people, involving practical referential detachment and understanding of transfactual efficacy)

↕

Embodied practical transactions with the natural order (logically and temporally [phylogenetically and ontogenetically] prior to sociality – **primary of practice**)

↕

Shared species potentials and needs at birth, including the capacity to acquire language (which at once is geo-historically emergent from practice, presupposes it ontogenetically and is itself a practical activity)

Source Archer (2000, 2003b).

prioritising of concerns rather than the deep teleology of praxis); and adds to it in others (providing a detailed analysis of the personal emergent powers [PEPs] both of the human and of the social self – stratified as *persons*, *agents* and *actors* – and of the *internal conversation* as the mediator between structure and agency). Overall it is arguably true to the basic trajectory of the Bhaskarian system, espousing a practical materialism whereby subjects are neither biologically pre-given nor the mere bearers of social structures but realised in practice (*the primacy of practice*), together with a realism that 'uncouples the *bases* of human subjectivity from dependency upon any "form of life", and links them instead to the categorial structure of the world' (2000: 312, e.a.), enabling us to grasp the truth of Harold Pinter's saying (2006) that 'civilisation is a betrayal of the dignity, of the essence, of the flame of life'. From within the CR historicising tendency, Dean (2003) has provided an incisive account of the process of constitution of subjectivity (*subjectivation*) in capitalist post/modernity. See also DIALECTIC OF SUBJECTIVITY; OBJECTIVITY/SUBJECTIVITY.

subjectivity/objectivity. See OBJECTIVITY/SUBJECTIVITY.

subjects$_1$ – subjects$_2$ correspond to POWER$_2$ > power$_1$. Subjects$_1$ refers to the essential FREEDOM of all subjects$_2$ – their actually existing potential for freedom. Subjects$_2$ designates their enslavement and subjection to (social) CONSTRAINTS$_2$. Hence subjects$_2$ constellationally contains subjects$_1$.

sublation, dialectical sublation (ds′) (Ger. *Aufhebung*). In Hegel, 'has the threefold meaning of to cancel, preserve and transcend' (*DG*: 404). Thus, in Hegelian dialectic, contradictions are annulled but not forgotten – they are cumulatively preserved as NEGATIVE PRESENCES in an expanded conceptual field which rationally resolves or TRANSCENDS them (see also TABLE OF OPPOSITION). On this conception, 'truth is the whole, the whole is a process and this process is reason': dialectical TOTALITY as dialectical PROCESS as DIALECTICAL REASON (dt′ as dp′ as dr′). Its outcome is 'reconciliation to life in (Hegelian) freedom' (*D*: 21–2). The whole process is one of *preservative determinate sublation*.

The burden of the Bhaskarian critique of this conception is that it is radically linear and immanently TELEOLOGICAL at the level of the actual, reducing being to logical or conceptual necessity, i.e., CENTRIST → TRIUMPHALIST → ENDIST. In real, depth-open geo-history, rational resolutions are only ever partially or essentially preservative; resolutions are not normally rational, let alone invariably so, and are caused by non-reasons as well as by reasons; wholly preservative sublation is the exception rather than the rule – and the practical dialectics of EMANCIPATORY AXIOLOGY imperiously demands *non*-preservative sublation; some processes are purely cumulative rather than transformative-sublatory; and there may often be results of dialectical processes that are anything but sublatory: the status quo ante, mutual ruin, retrogression, disemergence. Some dialectical processes of the sociosphere very probably cannot be sublated – e.g., *Verstehen*, or the pre-interpretation of everyday life, and the fact-form (as such); agency or power$_1$; and a variety of spatio-temporal dimensionalities and processes. They are transcendentally necessary parameters of any conceivable social life. While there is an immanent teleology of human praxis (see EMANCIPATORY AXIOLOGY), it is latent at the level of the real, and so cannot be invoked as uniquely determining the course of actual history.

None the less, Hegelian dialectical sublation of conceptual and socio-cultural forms importantly affords genuine analogues of science's non-arbitrary PRINCIPLE OF

sublime 449

STRATIFICATION and of real material SUPERSTRUCTURATION or EMERGENCE, and is 'a prototype of a genuine EPISTEMOLOGICAL DIALECTIC in science: inconsistency → greater depth and/or totality → restoration of consistency' (*P*: 120). CR treats sublation as a species of *determinate transformative negation* (see ABSENCE) – a *DIALECTICAL RESULT* or outcome of a dialectical PROCESS (which then becomes the starting point in a new process, recursively); or as the process itself. *Preservative* sublation may be *totally* (e.g., 4D > 3L > 2E > 1M), *essentially* (e.g., DCR > EC > CN > TR), or *partially* (e.g., CN > hermeneutics) so. *Non-preservative* sublation (e.g., DCR > Hegelian dialectic, TR > empiricism/rationalism, eudaimonia > demi-reality) may preserve 'what is held to be of value in, even though it is not essential to, the sublated social form' (*D*: 12). The concept of the NEGATION OF THE NEGATION in its Marxist and DCR forms – the geo-historical transformation of geo-historical forms – provides 'a genuine DIALECTICAL COMMENT on Hegelian preservative dialectical sublation' and correctly specifies what dialectic is (the absenting of absence) (*D*: 152). In philosophy and social theory, it is by and large theories and PROBLEMATICs that are sublated: PROBLEMS are *resolved*, PROBLEM-FIELDS *explained*. CR's strategy has been to sublate DUALISMS and splits in social theory and philosophy non-preservatively.

Sublation is very close conceptually to *CONSTELLATIONAL containment*. The former is usually deployed in relation to specific geo-historical dialectical outcomes or processes (usually intra-/inter-theoretical), the latter in addition to more general phenomena – e.g., ontology > epistemology; causes > reasons; power$_1$ > power$_2$. Therefore, constellational containment constellationally contains sublation: constellational containment > sublation. See also HEGEL–MARX CRITIQUE.

sublime. See FINITE/INFINITE; META-REALITY.
substitutionism. See POLITICS; SELF-REFERENTIALITY; SOLIDARITY.
sufficient reason, principle of. See DETERMINATION; EMANCIPATORY AXIOLOGY; PRINCIPLE OF INDIFFERENCE.
summative social universals. See UNIVERSALS/PARTICULARS.
superidealism, super-idealism. See CONVENTIONALISM; EMPIRICAL REALISM; EPISTEMIC–ONTIC FALLACY; HOLY TRINITY; REFERENCE; SAPIR–WHORF; SOLIPSISTIC, ETC.; VOLUNTARISM.
superposition. See CHAOS/COMPLEXITY; TRANSCENDENTAL REALISM.
superstructure, superstructuration. See INFRA-/INTRA-/SUPER-STRUCTURE.
supplement, supplementarity, supplementation. See CONSISTENCY; DCR; DUPLICITY; NEO-LIBERALISM; TINA SYNDROME; VIRTUALITY.
supreme good. See EUDAIMONIA.
surd (f. L. *surdus*, deaf, in the sense of not willing to hear or not understanding, also used to express Gr. *a-logos*, without reason, irrational). In mathematics, an irrational or undecidable number or quantity. Used by Bhaskar for an 'irrational existent', i.e., an entity which cannot be accounted for in a philosophical system's own terms (see ACTUALISM; DEMI-REALITY).
suspension, dialectical suspension. See PROPAEDEUTIC.
sustainability. The concept of sustainability links environment, human needs and the resources situation of FUTURE generations (futurity). Environmentalism raises urgent concerns about the declining state of the local and global ECOLOGIES and

life-support systems, but does not itself supply guidelines for social change. Sustainability can be seen as an ethical and political programme that states that human production, consumption and settlement should respect the real limits of global life support systems and the principles of social justice. It is in this form that sustainability has been taken up by progressive SOCIAL MOVEMENTS. If there can be said to be a sustainability movement, it is contested, diverse and fragmented – as one might expect from what is essentially a developing global research programme for the future of humanity, struggling against considerable odds.

'Sustainable development' has been the dominant discourse used by business and governments. It is regarded by many as meaning 'business as usual' but with increased resource efficiency (Sachs 1999). The UN Earth Summit of 1992 and resulting declaration 'Agenda 21' (UNCED 1992) called for global change, although issues of population and arms spending were kept off the agenda. The UN process continues, though continually hampered by those with a vested interest in destructive forms of development.

Environmentalists argue that understanding sustainability requires an ecological understanding of the historical evolution of life, supplemented by Gaia theory (Lovelock 1979) outlining global life-support feedback processes. Human rights activists argue that effective action for social justice requires understanding of the global systems of society and economy. Global sustainability requires knowledge of POVERTY and underdevelopment; at present developed countries can export environmentally destructive industries and appalling working conditions (see www.rrojasdatabank.net). In this way links are made between sustainability and campaigns for global democratic governance. Some thinkers link dominant forms of production and consumption with ideologies justifying oppression of subordinate groups (Mellor 1997; Braidotti et al. 1994). Sustainability also requires an account of ETHICS that can unite concern for the environment and social justice (Cuomo 1998).

CR work in this area has demonstrated that an epistemology is required that recognises social mediation of knowledge but also the reality of the social and material dimensions of environmental problems (Dickens 1996; Benton 1993). Kate Soper (1995) has contributed an essential discussion of cultural constructions of 'nature'. Critical realists have added to the analysis of capitalism's structural inability to deliver sustainability and the exponential destructive power of economic 'growth' (Gruffydd Jones 2001; Kovel 2002; O'Connor 1996; Næss 2006b). Critical realists have also contributed ethical approaches relevant for sustainability (Collier 1999).

Critical realism can outline schematic relations between different 'levels' of reality (see EMERGENCE) preventing inter-disciplinary studies necessary for sustainability from becoming a chaotic mix of social interpretation and physical facts. The EPISTEMIC FALLACY questions the adequacy of unitary measurement for description of a complex world of open systems. Dialectical critical realism asserts the importance of the unknown and the uncertain, essential for an understanding of environmental precaution (Parker 2001). There is a need for more engagement with SYSTEMS thinking by critical realists, particularly in the context of dialectics and emergence. Critical realists from all disciplines can help social movements to develop sustainability as a more powerful, philosophically informed, holistic vision for the future of humanity.

See also ECOLOGICAL ASYMMETRY; NEEDS; ORGANIC COMPOSITION; PROBLEM OF AGENCY; Clayton and Radcliffe 1998; Martinez-Alier 2003.

JENNETH PARKER

switch, switch situation. See PERSPECTIVE; TRANSITION.
syllogism. See ANALYTICAL PROBLEMATIC.
syncategorematic. See CATEGORY.
synchronic emergent powers materialism (SEPM). See MIND.
synchrony, synchronicity. See DIACHRONY/SYNCHRONY.
synthetic. See ANALYTIC/SYNTHETIC.
system. A system can be defined generically as a collection of entities of any kind (e.g., physical, conceptual, or social) that form a whole, the behaviour of which depends on the relations between the entities more than on the nature of the entities themselves. Systems form nested hierarchies and are characterised by EMERGENT properties, that is, characteristics or behaviours or powers that apply only to the whole rather than to its parts at any particular level. Systems concepts can be applied in any discipline, and this is generally known as the 'systems approach' or 'systems thinking'. It is HOLISTIC and anti-reductionist (Checkland 1981).

Some important systemic concepts are: a *BOUNDARY* that separates the system (S) from its *environment*. The boundary may be actual or conceptual and the environment consists of other systems that affect or are affected by S. S may be described in terms of *STRUCTURE* and *PROCESS*. Structure is the set of components and their relations that are relatively enduring, while process is the dynamic behaviour of the system over time. Interaction within and between systems generally involves *feedback loops*. These may be positive or reinforcing, generating exponential growth or decay; or negative or balancing, generating stability. Interaction may be based on the communication of information rather than on physical causality. Particular patterns of circular causality occur in many different situations and are known as *systems archetypes* (Senge 1990). Systems thinking emphasises the importance of the *observer* in describing and modelling systems, and second-order cybernetics (e.g., AUTOPOIESIS [Maturana and Varela 1980]) and soft systems methodology (SSM) (Checkland and Scholes 1990) adopt an interpretive or constructivist philosophy.

Another key concept relevant to CR is the distinction between *closed* and *open* systems. Historically, natural science tried to deal with closed systems – that is, systems that were insulated from their environment. Laboratory experiments were designed precisely to do this. Systems theory emphasised the importance, outside the laboratory, of open systems that were continually in dynamic interaction with their environments, typically transforming inputs into outputs – a paradigm example being a candle flame. Real-world systems, especially involving humans, were always seen to be open. However, autopoiesis has brought back the idea of closure but in a different form. Autopoietic systems do not have inputs and outputs but transform themselves into themselves, although they do still have interactions with their environments. They are organisationally closed but interactively open. This is also true of any systems that are *self-referential*.

Bhaskar's work tends not to make explicit mention of the systems literature but is fundamentally systemic. In the early CR theory the primary systemic motifs were: the ontological designation of *structures* and MECHANISMS that had powers or tendencies, the operation and interaction of which generated actual EVENTS in an open-systemic world; and the emphasis on *ontological stratification* (including DEPTH-*openness*) and *emergent powers*. In *SR* (109) a system is explicitly defined as a combination of structures whose form causally co-determines the elements which in turn causally co-determine each other and the whole without eliminating differentiation and specific causal roles within it. These concepts can be translated almost directly into the language of systems thinking: systems forming wholes with emergent properties, structure and process, and systemic structure and intra-action generating observed behaviour.

With the move to DCR and the specification of the MELD schema in *Dialectic*, the systemic concepts become even more explicit. The prime moment (1M) includes the concepts mentioned above of structure, generative mechanisms and emergence. The second edge (2E) brings in change and temporality. The key concept of *ABSENCE*, although not a central feature of systems thinking, is certainly compatible with ideas of goal-seeking through error-controlled feedback (removing an absence of equilibrium), and different causal processes cancelling each other out, leading to an absence of change. The third level (3L) is clearly central with its emphasis on totality, HOLISTIC CAUSALITY and recursive embeddings (nested hierarchies in systems terms). The fourth dimension (4D) brings in the realm of agency, embodiment, reflexivity and social transformation, entailing the openness of the future, all of which are concerns especially of second-order, soft systems thinking. Bhaskar also emphasises that in the social world systems are always open and cannot be closed as in a laboratory. Finally, he explicitly draws on the concept of autopoiesis (that is, self-producing systems) as a key feature of the process of knowledge production (*SR*: 54, 83), emergence (*D*: 49) and social reproduction itself (*D*: 156).

In conclusion, we can mention some issues raised by systems thinking which are relevant to CR. First is the question of boundaries. The assumption that there exist objects, mechanisms and structures that interact with each other presupposes the possibility of distinguishing such structures from each other, and thereby presupposes the existence of boundaries (see PROBLEM OF INDIVIDUATION). In systems the question of boundaries is problematic. Whilst it seems obvious for physical objects such as a house, it is less clear for things such as a weather system, a termite colony or an ecosystem, let alone for social systems such as Warwick University, a family, or 'the middle class'.

Second, is whether the structure/process distinction is the same as the real/actual one. A system has a structure (components and relations) that endows it with particular properties (powers or tendencies – i.e., the real). At any time it can be characterised in terms of the states of its components (a state description). Its process, or dynamic behaviour, is then its changes of state. This could be seen as a series of events (i.e., the actual). But the events are not some different kind of thing to the system, as Bhaskar implies, but simply the changes of state of the system relative to a particular time frame.

These issues raise a third – the question of the observer. CR rightly emphasises the independence from humans of the intransitive dimensions of systems and structures in themselves. But systems thinking points out that whenever we humans actually do describe and model, in the transitive dimension, such descriptions will always be relative to the purposes and cognitive characteristics of the observer. We can never actually be independent from the systems we describe – we are inevitably enmeshed in their workings. See also CAUSAL LAW; FUNCTIONAL EXPLANATION; TOTALITY; Ashby 1956; Bateson 1973; Capra 1997; Churchman 1971; Emery 1969; Forrester 1968; Mingers 1995; von Bertalanffy 1971; Weiner 1950.

JOHN MINGERS

systematic dialectic(s). A contemporary Hegelian Marxist school of thought (Arthur 1998, 2002; Reuten and Williams 1989; T. Smith 1993, 1999) which assigns primacy to SYNCHRONY and system over DIACHRONY and process. It is part of a broader movement sometimes called *New Dialectic(s)* which, by contrast, still assigns some importance to Hegel's overall philosophy of history. Bhaskar (1983b: 125) actually appears to have been first to use the concept (Arthur 2006), in relation to Marx's 'quasi-deductive method of exposition' or *PRESENTATIONAL DIALECTICS* – elsewhere referred to as Marx's 'systematic/expository/presentational epistemological dialectics' and distinguished from his '*CRITICAL DIALECTICS*' (*D*: 93). He deploys it similarly in relation to his own transcendentally deductive derivation of the dialectics of freedom from agency and discourse: systematic dialectics concern the overall logic of EMANCIPATORY AXIOLOGY. Presentational dialectics are sometimes distinguished from systematic dialectics as exclusively concerning the mode of exposition. For Bhaskar, both get their full purchase only when inscribed within critical dialectics, constituting a totalising critical project (cf. *D*: 347).

systems, open and closed. See CLOSED AND OPEN SYSTEMS.

T

τ (*tau*) **transform**. See EPISTEMOLOGICAL DIALECTIC.

table of opposition(s). In the traditional Table of Opposition, usually called the Square of Opposition (see e.g., Horn 2006; Parsons 2006), (1) *both e and −e* is ruled out by *contraries* (this is the *law of (non-)contradiction* or the mutual exclusion of opposites); (2) *neither e nor −e* is ruled out by *sub-contraries* (this is the *law of excluded middle*); and (3) both (1) and (2) are ruled out by *contradictories*. These principles are at the heart of the ANALYTICAL PROBLEMATIC. *Dialectical opposites*, being distinct yet inseparable (i.e., sharing a structured causal ground), permit both (1) and (2), over a stretch of time at the level of the actual in Hegel and his EPISTEMOLOGICAL DIALECTIC, and simultaneously at different ontological levels in Marx and Bhaskar or any depth-open realism, thereby completing the traditional Table in that there is no combination of opposites it then does not permit, and demonstrating that change is 'ontologically oppositional' (*P*: 241). Hegel completes the Table by DIALECTICAL ARGUMENT 'in the mode of distanciated time' (cf. NEGATION OF THE NEGATION, the coincidence of identity and change, and the non-VALENT response to [1]), Marx by dialectical argument and explanation 'in the mode of ontological stratification' (cf. DIALECTICAL ANTAGONISTS, and the coincidence of identity and difference in dispositional, rhythmic and constellational IDENTITY) (*P*: 80 n). In Hegel, 'positive contraries (actual and present) become negative sub-contraries (actual and absent)' (*P*: 80), co-existent in the sublating concept (cf., e.g., rationalism and empiricism as sublated within transcendental realism). On depth-open realism, when two or more tendencies counteract each other but nevertheless co-determine a result they are at once real presences (transfactually efficacious) and actual absences (not manifest or 'realised'). In so far as they are tendentially opposed, they are simultaneously (or constitute a unity of) (1) positive contraries at the level of the real (both present): +(*both e and −e*); and (2) negative sub-contraries at the level of the actual (both absent): −(*neither e nor −e*). Both approaches exemplify the DIALECTICS OF CO-PRESENCE or co-inclusion of absence and presence, and are fundamental modes of procedure in CRITICAL REALISM's epistemological dialectic; however, in Hegel the positive contradiction is cancelled to reinstate identity, though not forgotten: it is retained in a cumulative memory store as a *NEGATIVE PRESENCE* '(but not *positive absence* – real negation)' (*D*: 33, e.a.), whereas Marx and Bhaskar hold out the prospect of its real negation in the transformative praxis of the epistemological dialectic and/or, if it is rooted in social contradiction, EMANCIPATORY AXIOLOGY. Combining the two approaches in CR unites the two key PROBLEMS of philosophy at 2E and their resolution, the problem of opposition and the problem of change. 'It is of course the notion of negative subcontraries, turning on the concept of reference to absence {NEGATIVE REFERRAL}, and hence the critique of ONTOLOGICAL MONOVALENCE, which allows the dialectical completion of the table of opposition' (*P.* 80 n).

tacit knowledge. See CONSCIOUSNESS; HERMENEUTICS.

taxa, taxonomy. See EXPLANATION; NATURAL KIND; UNIVERSALS.

TDCR. See META-REALITY.

technical rationality. See RATIONALITY.

technology. The etymological roots of the term technology lie in Gr. *technē* and *logos*, meaning, roughly, the principles of craft or skill. However, in modern times it tends to refer to some part of the process whereby, in the production of useful things, ideas, values and social relations become concretised in material artefacts in such ways as to have important implications for social life. The different definitions which currently abound result in part from this tendency to focus on just one or other aspect of the process, e.g., on the technical artefact, technical activity, technical knowledge or the process of concretisation (Mitcham 1994; Winner 1977). Competing accounts of technology also result from the extent to which *technical* artefacts are distinguished from other kinds of artefact such as art, food, toys, institutions, etc. Contrasting disciplinary approaches to the study of technology can also be seen to focus on one aspect or another of this process. For example, philosophers of technology (e.g., Heidegger, Ellul, Mumford, Borgmann) have tended to focus on *implications* – specifically, on the degree to which technology's growing role in everyday life is responsible for the more dystopian features of modernity. In contrast, the more constructivist sociologists and historians of technology (e.g., Pinch, Bijker, Collins) have been more concerned with the *concretising* aspect of the process. More specifically, they have been concerned with documenting both *how* particular technologies come into being through a process of social negotiation, conflict resolution, etc., and *which* ideas, values and social relations become concretised in particular artefacts.

Little discussion of technology exists in the CR literature. This is surprising not only because technology is such a central category in the study of society and social change, but also because CR would seem to have much to contribute to current debate. For example, the technology literature has found it difficult to avoid the (familiar) errors of determinism and voluntarism. A concern with the revolutionary or disruptive impact of new technology has tended to encourage a deterministic treatment of technical change in which 'the social' is not an essential feature of technology itself – technology becomes social only in and through the uses to which it is put. In response, more constructivist contributions have emphasised the contingent nature of technical change and the irreducibly social nature of not only technology's use, but also its design content and implications, i.e., its very constitution. However, the strong constructivist position, in which technology is treated as a purely social phenomenon, is then unable to explain why technology can have the distinctive and dramatic effects on the social world that motivate technological determinist accounts. Moreover, even where 'dual' conceptions of technology have been advanced, the actual differences between the natural and social domains, upon which such conceptions must ultimately be grounded, have been bracketed (under the influence of the sociology of scientific knowledge 'symmetries'), leaving it difficult to ask any general questions about the nature of technology. The social ontology of CR has much to offer this apparent impasse. An account of technology, couched in terms of the TMSA (which has of course been particularly successful in transcending the determinism/voluntarism debate elsewhere in social science), that theorises the relationship between human agents and material artefacts in a similar manner to that elaborated in the TMSA between human agents and social structure, provides a non-deterministic, irreducibly social and relational account of technology in which the differences between the natural and social components of technology are centre stage (see C. Lawson 2005 for a development of these arguments).

Extending the social ontology of CR to provide an account of technology would seem to offer various opportunities for future research. For example, the conceptualisation (and grounding) of technology as relational would seem to offer the basis for a novel definition of technology itself. All artefacts are relational in some sense. But social relations are more important to the causal powers of some objects than of others (e.g., they are more essential to social objects such as identity cards than technical objects such as hammers). Thus technology could perhaps be distinguished by its reference to that sub-set of (irreducibly relational) artefacts that are not *essentially* relational. In addition, it may prove helpful to recast much of the existing literature on *FUNCTION* in terms of social relationships – indeed, there may be scope for developing the idea of technology fetishism (where the relations constitutive of a material artefact are obscured not by price – as with commodity fetishism – but by function). Furthermore, the concept of ABSENCE appears to play a central role in the dynamic of technological change, a theme yet to be explicitly developed. See also Collier 2003b: ch. 6.

CLIVE LAWSON

teleological explanation, teleological pull. See AGENCY; CAUSES; COLONIALISM; DIALECTICAL INTELLIGIBILITY; DIRECTIONALITY; FUNCTIONAL EXPLANATION; IDEALISM; SUBLATION, TELEOLOGY/TELEONOMY.

teleology/teleonomy (f. Gr. *telos*, an end, purpose or objective, which in philosophical discourse sometimes contrasts with its Latin counterpart, *finis*, as an end that is implicit in the beginning by comparison with an end that has arrived [e.g., of a book, a life – i.e., death] + *logos* and *nomos*, law, respectively; *nomos* should be understood in CR contexts as causal law or mechanism). Taken together, the concepts refer to (1) the philosophical or theoretical doctrine that explains or describes anything in terms of purposes or goal-directedness (TD); and (2) the purposive or goal-directed process itself (ID). The process may range from the movement of the universe, to the evolution of life forms, to the INTENTIONAL activity of human and other animals, to the goal-directedness of a machine. Its driving or attracting force may be external to the cosmos, system, etc. (*TRANSCENDENT teleology/teleonomy*, e.g., Plato, about which CR is agnostic), or internal (*IMMANENT teleology/teleonomy*, e.g., Aristotle, Hegel, Marx, Bhaskar).

For CR, both teleonomic and teleological processes are instances of real negation. In *teleonomy*, directional processes are driven ex ante, 'from behind', by what is implicit in them – by *teleonomic push*; whereas in (immanent) *teleology*), they are pulled 'from in front' by the aspiration to repair some lack or absence (e.g., in Bhaskar, the absence of the moral good for humankind) – by *teleological pull* (*D*: 22, which has 'from in front *and without*' in relation to Hegel's immanent teleology), because teleological pull is '*epistemologically* illegitimate' – it is 'generally only *retrospectively*, ex post, that a stage can

teleology/teleonomy

be seen to be deficient' (*D*: 24, 341 e.a.) – though *ontologically* real; or, it is epistemologically transcendent and ontologically immanent. (Arthur [2006, cf. 2002: 65 f.] arrived at a concept of teleological pull independently of Bhaskar.) As an explanation-form, *teleological/teleonomic explanation* is a species of causal > consequence > FUNCTIONAL EXPLANATION. It may of course be contrasted with (non-teleological) causal explanation in the many contexts where it is not licit.

The tendency in post/modern Western philosophy and science, closely bound up with DISENCHANTMENT, has been to restrict the applicability of *teleological* explanation in particular. Darwin eliminated it from the natural world, and today it is under siege by behaviourism and reductionism in its last mainstream refuge: intentional action. On the CR account, it is mandatory in the sociosphere only in the case of individual human action; however, in virtue of such action, teleology may be immanent at the level of the social real. It should perhaps be underlined that CR construes it, not as an explanation-form in its own right, but as 'a species of ordinary causal explanation' (*PN3*: 84). The causal mechanism in intentional action is a dispositional state of mind, or a 'reason' (beliefs and desires/wants). Teleological explanation in terms of reasons, whether construed as causal or not, is termed *rational EXPLANATION*. 'The door opened because, wanting to go into the kitchen, I pushed it.' Such explanation does not of course explain a great deal, as it must be supplemented by physical and social causes, and the reason I give out to the world or myself may be other than my real reason or motivation, i.e., a rationalisation. All rational explanation can be recast in the form of consequence > FUNCTIONAL EXPLANATION. The mechanism is then the belief that the action will have the intended effect, and so be beneficial or functional for the actor. On the TMSA, social outcomes are, however, largely unintended and so, given chronic contestation and power$_2$ contradictions, teleological functional explanation has only restricted applicability in the sociosphere. There *is* an immanent teleology implicit in human praxis and discourse (see EMANCIPATORY AXIOLOGY) and a tendential rational DIRECTIONALITY which it helps to ground, but it dwells latently at the level of the real – some awareness of which is caught in such popular expressions as 'in the depths of my heart' – and it is highly contingent whether it will be actualised.

In relation to the actual course of geo-history, *teleonomic explanation* is perhaps more promising. Teleonomic or quasi-teleonomic explanation has extensive legitimate application in the life sciences and the study of control and guidance in biological, physical and chemical SYSTEMS displaying persistence towards a goal-state under varying conditions (cybernetics), though many systems can legitimately be described also as teleological – thus a heating system is a 'teleological machine' in that it is oriented to the goal of maintaining the temperature at which the thermostat is set (*PN3*: 117 n), and all organisms have an 'internal teleology' (*D*: 55), i.e., are intrinsically oriented to meeting wants. In geo-history, both teleology and teleonomy, theoretical and practical reason/praxis, will arguably be involved to some extent in any DIRECTIONAL process, whether at the local and regional level or in geo-history considered as a whole. The early Bhaskar describes social processes as 'typically' teleonomic and 'fundamentally non-teleological', this being entailed by the TMSA (*PN3*: 100, 133); dialecticisation, however, arguably marks a shift towards acceptance of significant depth-teleology in the social sphere. Both teleological and teleonomic explanation, when valid, exemplify HOLISTIC CAUSALITY.

458 teleonomic push, teleonomy

PMR marks a return to the idea of teleology in nature by relating human intentionality and its deep ethical content to the ultimate, implicitly conscious, field of pure possibility ingredient in the whole of being – 'the teleology of the ground-state' (*MR*: 265).

teleonomic push, teleonomy. See DIRECTIONALITY; FUNCTIONAL EXPLANATION; TELEOLOGY/TELEONOMY.

tendency. A *causal power* is a potential that may be exercised or not (see CAUSALITY). A *tendency*, in its primary meaning, is a causal power exercised or set in motion – transfactually efficacious (operating in open as well as closed systems) and so normically qualified (it may be offset by the operation of other powers); regardless of the outcome at the level of the actual, something is really going on. (*Causal* or *generative mechanism* may refer to either a power or a tendency, or both. Powers, tendencies and mechanisms are normally instantiated in STRUCTURES.)

Corresponding to the moments of the CONCRETE UNIVERSAL, four distinct but related concepts of tendency can be elaborated (see Table 40). A *tendency$_a$* is the base concept itself, i.e., a transfactually efficacious, normically qualified power; a *tendency$_b$* expresses the *DIRECTIONALITY* of a process; a *tendency$_c$* 'the MEDIATION of a (para-spatial) MOMENT or a (para-temporal) DETERMINATION' (*P*: 83); while a *tendency$_d$* is a generic tendency referring to the role of all tendential effects in the determination of concretely singular outcomes.

Distinctions need to be made between the *enabling, stimulating* and *releasing conditions* of a tendency, which may be either INTRINSIC or EXTRINSIC. Enabling conditions are the positive form of causality possessed by a causal power or powers (as distinct from the negative form of a CONSTRAINT), or the generative mechanisms that give rise to a tendency. Stimulating conditions include all those factors that trigger, facilitate or reinforce the exercise of a tendency, some of which may involve an element of contingency. Releasing conditions refer to the circumstances in which countervailing factors are either absent or weak – there are few or no impediments to the exercise of a tendency.

Drawing on these distinctions, we can distinguish further concepts of tendency corresponding loosely to the moments of the ontological–axiological chain. Here *tendency$_1$* is the same as tendency$_a$; *tendency$_2$* is then a power *ready* to be exercised (its

Table 40 Types of tendency, the concrete universal and the ontological–axiological chain

Concrete universal ↔ singular	universality	processuality	(particular) mediations	concrete singularity
	tendency$_a$ (transfactually efficacy)	*tendency$_b$* (directionality)	*tendency$_c$* (mediation)	(generic) *tendency$_d$* (combined tendential effects in any outcome)
The ontological– axiological chain	1M non-identity *tendency$_1$* (transfactual)	2E negativity *tendency$_2$* (ready)	3L totality *tendency$_3$* (prone)	4D transformative agency *tendency$_4$* (motivated) (e.g., a human disposition)

Note A motivated tendency can obtain outside as well as within 4D.

intrinsic enabling conditions are satisfied); *tendency₃* a power *prone* to be exercised (its extrinsic enabling conditions are satisfied in addition); *tendency₄* a *motivated* power (its intrinsic stimulating and releasing conditions are satisfied in addition); *tendency₅* a power whose exercise is LAPSED or LAGGED (late) (a tendency₃ plus the satisfaction of extrinsic, but not intrinsic, stimulating or releasing conditions). Thus we have concepts of *transfactual tendency, ready tendency, prone tendency, motivated tendency* and *lapsed tendency*. Along this line we can further distinguish *tendency₆*, a power 'realised in all normal circumstances'; *tendency₇*, a power realised in a closed system, normally a laboratory, i.e., a CONSTANT CONJUNCTION of events; and *tendency₈*, a tendency eventually realised in open situations (*P*: 83; *EW*: 37 n).

Jessop's (2003c: 270 n) concept of *double tendentiality* highlights that the structures instantiating tendencies are subject to change, hence must themselves be tendentially reproduced, suggesting a complex dialectical interplay between internal and external RELATIONALITY. *Multiple tendentiality* would seem to fit the case even better.

Turning to the relevance of the concept of a tendency to substantive research, it provides a clear distinction between the CR position and that of EMPIRICAL REALISM (or positivism). The latter presumes that the objects of scientific analysis are CLOSED SYSTEMS. In a closed system, laws may be expressed as regularity statements of the form 'whenever a then b'. By contrast, CR posits that most objects of scientific analysis obtain within open systems, which are subject to interaction and interference between myriad causal mechanisms. Consequently, any given causal mechanism is unlikely to produce empirical effects that can be apprehended in terms of strict regularities, such as, for example, a clear deterministic trend in a data set that continues without variation into the future.

As opposed to a deterministic trend, a tendency is a characteristic manner of acting or transfactual effect of a particular causal mechanism (or set of mechanisms), which our scientific knowledge of the world leads us to believe is in play, whether or not the tendency of interest is revealed in a particular piece of empirical evidence.

Tendencies₁₋₄ seem of most relevance to science. Tendencies₁ are normally qualified powers. Such powers may be possessed unexercised, such that all people have the power to steal, but most refrain from exercising it. Tendency₂ is a stronger notion of tendency, conveyed in the example: 'Kleptomaniacs have a tendency to steal.' The second case implies an enduring predisposition towards a certain type of effect or behaviour. Bhaskar argues that in a closed system there is no difference between these two types of tendency: once their intrinsic conditions are satisfied they are enabled, and they then only require some triggering stimulus to be released or put in motion. The crucial difference between the two types of tendency is revealed in open systems. Here the enduring predisposition is related to the pre-existing satisfaction of intrinsic enabling conditions:

> to say that X has a tendency₂ to do Y is thus to say {...} that most (or the most important) of the intrinsic enabling conditions are satisfied and that when other conditions are satisfied it will do it unless prevented. (*RS2*: 231–2)

Thus in an open system a tendency₁ will need to be stimulated. In other words, it only occurs under a restricted range of conditions. For example, it may be that under normal conditions intrinsic countervailing factors block its operation. In contradistinction, Bhaskar defines a tendency$_2$ as *ready* because intrinsic enabling conditions are satisfied (*D*: 78).

The tendencies Bhaskar defines as *ready*, *prone* and *motivated* (tendencies$_{2-4}$) can be collectively regarded as *robust tendencies*: that is, they all involve an enduring predisposition (Pinkstone 2002). Since tendencies are the 'characteristic ways of acting or effects of mechanisms' (Lawson 1997: 22), we might expect those generative mechanisms with an enduring and pervasive time–space extension to be expressed in tendencies that themselves possess an enduring and pervasive time–space extension (cf. DISTANCIATION). That is, robust tendencies are characterised by intrinsic closure in that intrinsic enabling conditions are irreversibly satisfied whenever the relevant causal mechanism is present. As such, robust tendencies can only be thwarted in their operation by extrinsic, essentially contingent, countervailing factors and consequently would be most likely to give rise to persistent *DEMI-REGULARITIES* in an open system. Notably, endurance implies not only a significant passage of time, but also the *power* to endure in a relatively stable form, despite the effects of other influences and causal processes. The word 'robust' captures this sense of an enduring, pervasive and powerful characteristic or summary effect. At the other end of the spectrum are transient tendencies of the tendency$_1$ variety, which require more context-specific conditions to trigger their action.

In the absence of extrinsic, contingent countervailing factors a robust tendency might give rise to the appearance, at the level of the empirical, of a Humean closed system over some more or less limited time–space terrain. However, unlike the approach of empirical realism, a critical realist decision about whether we are actually dealing with something analogous to a closed system is an a priori ontological question, rather than something that should be accepted on the face value of the appearances associated with a particular set of empirical data (Pinkstone 2003a).

Finally, the significance of particular tendencies may be ranked in an ordinal manner according to the time–space extension of the causal mechanisms that they express. For example, the causal mechanisms that give rise to gravity and the tendency of attraction between physical bodies are a persistent feature of the universe as we know it. Evolutionary tendencies at the biological level must take this into account, so that living things which fly can only do so by means of a much more transient constellation of countervailing forces. Similarly, capitalist society involves a robust tendency continually to expand production and consumption, but faces a stronger, more enduring biologically given tendency for (human) life to flourish only under fairly narrow environmental limits.

BRIAN PINKSTONE, MH

tendential rational directionality. See DIRECTIONALITY.

tense (f. L. *tempus*, time). CR holds that SPACE–TIME is irreducibly *tensed*, i.e., the past, the present and the future are real (see also A-SERIES, ETC.). The *past* is everything that is existentially intransitive, i.e., caused or determined (produced) and determinate (see DETERMINATION; it is not determined in an actualist sense). It is massively present (*in* the present, BINDING it) in any of the four modes, corresponding to 1M–4D, set out in Table 42. It signifies 'a real material spatio-temporal', as distinct from logical, 'failure of DETACHMENT'. The *future* is increasingly shaped, but as yet indeterminate,

Table 42 Modes of presence and absence of the past, future and present

Causal-Axiological Chain	1M Non-Identity	2E Negativity	3L Totality	4D Transformative Agency
Mode of presence	Existential constitution	Co-presence	Lagged efficacy	Agential perspectivity
Presence of the past (transfactual or actual, positive or negative, internally related to other elements or not, agentive or not, the *existentially intransitive determined* (caused, produced) *and determinate*	(a) *product* of a process (mere containment) e.g., past labour in a sculpture, smog (b) *process-in-product* e.g., an essential disposition, a pox mark (c) *product-in-process* e.g., a theory's potential or actual development	of a number of pasts, e.g., of rhythmically differentiated sedimented structures at the opening of Parliament by the Queen – there is a *'simultaneity of non-simultaneous becomings'*	*delayed causal efficacy* e.g., starlight, the return of the repressed, causally efficacious memory	the *world is pre-constituted or always already* – the past is everywhere (as a transcendental perspectival switch from structure to agency makes clear)
Absence of the past		non-preservative negation, material detachment		
Presence of the future mediated by the past (presence-of-the-future-in-the-past-in-the-present) *increasingly shaped possibility* of becoming (not yet determined or determinate)	*possible futures* qua product-in-process, e.g., pregnancy, a genetic code, any impending event (e.g., war), powers and liabilities in general	*the not yet* within the already determined and determinate in any present episode	paradigmatically *intentional agency* – acting in the present, mediated by the past, to bring about a state of affairs that is not yet	*prefigurationality* (the prefiguration of the future in transformative praxis, mediated by the past and producing, not the future, but its content)
Absence of the future		the demise of everything		
Absence of the present	(a) a *distant present* observed (b) *lapsed time* (when asleep, a 'late' event or a 'post-ed' theory)			
Presence of the present the indeterminate *moment of becoming* (indefinitely extendable), demarcating the past from the future	like its absence, and the presence of the future, only philosophically or sociologically significant when mediated by the presence of the past			

possibility of becoming, MEDIATED by the past(-in-the-present). As a notion, the future is 'always either formal or (present-)relative' (hence the concept of the *present-future*), so that we produce, not the future, but its content as we journey through time on our WORLD-LINES. (If time were moving through us, the future would be 'already laid out and fully formed' as in regularity determinism [*SR*: 219].) In due course the future becomes determined and determinate (past). It, too, may be present in any of the four modes in which the past is present. The *present* is the indeterminate (indefinitely extendable) *moment of becoming* between past and future (also *boundary zone* or *DISTANCIATED stretch-spread*), likewise mediated by the past. It 'was becoming when it is in fact in the past, and will be becoming, when it is in the future' (*D*: 256). The duration of any present depends on 'the nature of the episode concerned, e.g., is it a coffee break, a friendship, a life, or a geographical epoch?' (*D*: 256) (see LOCO-PERIODISATION). This entails rejection of any notion of an abstractly universal present, hence past or future, albeit tense is an ontologically real, dialectically universal aspect of any episode; simultaneity is relative, or the world is constituted by 'a multiplicity of differential rhythmic processes', such that we have to talk of 'the simultaneity of non-simultaneous becomings' (*D*: 143, see also PERSPECTIVE).

Tense just is the irreducible and specific (token-reflexive) temporal dimension of the passing over of the indeterminate and not yet (fully) determined to the determined and determinate (hence existentially INTRANSITIVE) in the spatialising PROCESS (or differentiated rhythmics) of becoming, and as such is at the heart of the CR account of space, time and CAUSALITY, which constitute a 'tensed triunity' (*D*: 114; see also TRANSITION). To say that a process is causally efficacious is to say that it is irreducibly tensed. Like time as such, tense is irreversible. A *tensed hiatus* is a quantum leap, registered in lapsed time (see LAGGED/LAPSED), in a process of emergence, e.g., the hiatus between agency and structure (*D*: 277; TMSA). *Tense logic*: see LOGIC.

The (non-anthropic) reality of tense is demonstrated by transcendental deduction from the nature of agency (absenting, causality, change), including intentional agency, and supported by the findings of cosmological theory and an immanent critique of rival views (ACTUALISM; BLOCKISM/PUNCTUALISM; INDEXICALISM) (*D*: 250–8). The essence of the deduction from intentional agency is as follows. In doing something intentionally we bring about a state of affairs that, unless it was freakishly OVERDETERMINED, would not otherwise have occurred. If actualism were true, we could not do this, because the world would be a closed system and there would be a simultaneous conjunction of all times and events (blockism). So there must be a zone of being that is not yet determinate or fully determined, by contrast with what already is determinate and determined. Therefore actualism is false, and open tensed process must be real. Action is necessarily future-oriented (*futurised*). In our plans and intentions we *PREFIGURE* the future. The characteristic structure of action is accordingly:

> present absence → orientation to future → grounding in the presence of the past → praxis (*D*: 255)

theoretical realism. Another name for vertical or DEPTH-realism. See also DIFFERENTIATION.

theoretical reason. See CONSISTENCY; DIALECTICAL REASON; DUALITY OF THEORY AND CRITIQUE.

theoreticism. See RATIONALISM; VOLUNTARISM.

theoretico-practical duality. See CONSISTENCY; DUALITY; EMANCIPATORY AXIOLOGY.

theory choice. It is vital for CR that theories be empirically grounded. In many of the natural sciences this occurs mainly via direct experiment under artificially closed conditions, where theories are subject to a *predictive test*. In the non-experimental, including the social, sciences, direct test situations are unavailable. Problems of *confirmation* and *falsification* of hypotheses and theories are therefore more acute than in the experimental sciences. These can, however, largely be offset. (1) In the comparative method, and in particular the methods of concomitant variation and of difference (see CONTRAST EXPLANATION; DEMI-REGS), combined with statistical techniques, a method of indirect experimental inquiry is available to non-experimental sciences. (2) Contrary to Bhaskar, who holds that 'the social sciences must rely on exclusively explanatory criteria for confirmation and falsification' (1993b: 483), a predictive/postdictive test (though less decisive than in direct experiment) also arguably applies, especially where TENDENCIES are hypothesised to be primary or robust (Benton 1981; Hartwig and Sharp 1999; Pinkstone 2002; Porpora 2001b; cf. Næss 2004).

CR rejects actualist understandings of confirmation and falsification (and the criterion of falsifiability) prevalent in orthodox philosophy of science, whereby laws are construed as empirical regularities confirmed or falsified by their positive instances (e.g., Popper's falsificationism). Hypotheses that do not conform to this format are said by its proponents to be non-falsifiable, hence non-scientific. However, such 'laws' are themselves immediately falsified in open systems along with the theories embodying them, and the history of science is littered with falsified non-existential propositions, such as those relating to phlogiston. When laws are construed as transfactually active and enduring causal powers which sustain a concept of natural necessity in a depth-open world, however, they are confirmed when the real reasons for (generative mechanisms accounting for) any one level of ontological structure have been referentially detached and described. Confirmation is at its strongest when these are then captured in real DEFINITIONS of NATURAL KINDS. In this process extra-evidential considerations such as internal consistency and elegance also play a role (see DETERMINATION [epistemic indeterminacy]). In the HERMENEUTICAL moments of science, meanings are confirmed by dialogue or hermeneutical dialectics.

Within the process of confirmation, there are two vital moments of falsification (absenting): when rival hypotheses are eliminated and when the initial theory is corrected in the light of the new knowledge (see EXPLANATION). If the discoveries are sufficiently far-reaching, they will precipitate a theoretical revolution which falsifies (negates) the initial theory, though 'saving' most of the phenomena it describes and accounting for its ANOMALIES (see also EPISTEMOLOGICAL DIALECTIC).

The primary criteria for theory-choice are thus empirical adequacy and explanatory power. Choose that empirically well-grounded theory which can explain more significant phenomena under its own descriptions than its rivals can explain under theirs. A social theory will in addition explain why inadequate theories continue to be adhered to (see CRITIQUE; EXPLANATORY CRITIQUE).

theory etc. Recorded here are some distinctions necessary for avoiding confusion, starting with the most abstract. All refer to levels of science, considered synchronically

464 theory of action

as a totality, and possess a TD/ID bivalence, i.e., they are worked both *with* (TD) and *on* (ID).

A *general conceptual scheme* (GCS) is the *metaphysical framework* or *scientific ontology* (see EPISTEMOLOGY, ETC.) supplied by some set of ideas and assumptions. It indicates what kinds of fundamental entities there are in the world of interest to science, e.g., atomism and sociological individualism, entity relationism and intersubjectivity.

A GCS generates ('logically, not temporally' [*RS2*: 192]) a *research programme* (RP) incorporating its assumptions and oriented to substantive exploration and sometimes change (e.g., a reductionist programme; historical materialism). A research programme is *progressive* when it is developmentally CONSISTENT and generating theories with greater explanatory power, *degenerative* when riddled with APORIAI and going nowhere.

As already indicated, a RP in turn generates a *theory* or a sequence of theories (T) incorporating *models* (M). Such theories compete with and/or supplant each other, or are devoted to different fields or strata. A model is an imaginative attempt, drawing on metaphor and analogy, to depict visually what some hypothetical entity is 'like' (see Lewis 1996; Martin-Soskice and Harré 1982). A theory may be regarded as basically 'a model {or cluster of models} with existential commitment {...}, i.e., a model in which the entities posited and mechanisms described are conceived as real' (*RS2*: 192). It seeks conceptually to articulate systems of meaning within which REFERENCE and EXPLANATION can occur. *Theoretical$_1$* refers to an entity or behaviour whose existence is in doubt, *theoretical$_2$* to one 'that cannot be directly perceived, either unaided or with the help of sense-extending equipment' (*RS2*: 179). A theory may have problems of *theory-articulation*, *theory-CONFIRMATION* and *theory-application* (theoretical, empirical and METHODOLOGICAL problems, respectively). *Theoreticism*: see RATIONALISM. The stock of models, metaphors, analogies, etc., available to a field of inquiry (including philosophy) or theoretical approach constitute its *analogical GRAMMAR (AG)*. Any RP will in addition work with appropriate METHODS, heuristics and technology and an accepted body of FACTS (MTechF).

All the various levels of science will often not be in harmony with each other. Thus social science has plenty of theories, models, etc., but lacks an adequate GCS, which CR aspires to supply.

Paradigm (f. Gr. *paradeigma*, pattern, literally a showing alongside or beyond), adj. paradigmatic, has at least three overlapping meanings: (1) GCS + RP + T + AG + MTechF, e.g., Copernican astronomy, Newtonian physics; (2) sometimes used synonymously with RP or with the *conceptual scheme* supplied by a RP or some other (e.g., philosophical) body of theory; (3) an archetypal or exemplary instance of its class, useful for assessing other instances, as in: 'suppose philosophers had taken biology or economics as their paradigm of a science rather than physics'; and 'work consists, paradigmatically, in the transformation of given products' (as compared with their reproduction or creation ex nihilo, etc.) (*RS2*: 8, 57). See also PROBLEM (*problematic*, *problem-field*).

theory of action. See EXPLANATORY CRITIQUE.

theory–practice, theory/practice (T/P). See ALIENATION; AXIOLOGY; CONSISTENCY; CONTRADICTION; CRITIQUE; DIALECTICAL REASON; DIRECTIONALITY; DUALITY; EMANCIPATORY AXIOLOGY; EPISTEMOLOGICAL DIALECTIC; FREEDOM; IDEOLOGY; PERFORMATIVE; REFLECTION; TINA SYNDROME; UNIVERSALISABILITY.

theory problem-field solution set (TPF[SS]). See DIALECTICAL ANTAGONISTS; PROBLEM.

third level (3L). See MELD; TOTALITY.

this-world(ly)/other-world(ly). A distinction made in most religions and many philosophical systems whereby the world is divided into two: a sensible or IMMANENT world occupied by the speaker and a co-existent supersensible or TRANSCENDENT world to which the speaker will go after death (the after-life). In German, the corresponding expressions are *das Diesseits* (adv. *diesseits*, adj. *diesseitig*) and *das Jenseits* (adv. *jenseits*, adj. *jenseitig*) (literally, this side and the other side, respectively). While the early Hegel rejected the yearning for, and belief in, *das Jenseits* that was common in his generation, his *Geist*-centred trajectory after 1802 arguably introduced a this-world/other-world split into his system (*D*: 328–9). For CR, philosophy is emphatically this-worldly in that, in keeping with epistemic relativity, there is no philosophy in general, only the philosophy of specific geo-historical social forms, which supply it with premises and a context for its conclusions. It is thus a secular philosophy: 'there is another world, but it is in this one' (Eluard), i.e., the possibility of EUDAIMONIA. This is without prejudice to religious belief. The philosophy of META-REALITY is a this-worldly secular outlook which claims to be compatible with all faiths and no faith. See also BEAUTIFUL SOUL, ETC.; PHILOSOPHY OF RELIGION.

thrownness (Ger. *Geworfenheit*). A Heideggerian term borrowed and stretched by Bhaskar into the 1M–2E concept of *vehicular thrownness* to call attention to the fact, established by the transcendental deduction of the TMSA, that philosophy, science and human life generally are 'pre-existing, ongoing social affair{s}' (*D*: 76, e.a.). The concept highlights (1) our arrival in the (pre-existing, pre-conceptualised) world as materially embodied beings; (2) the spatio-temporality or PROCESSUALITY of our being and understanding (hence 'vehicular': we are thrown into life 'as if into an already moving vehicle' [*D*: 90]); and (3) the ALIENATION of our knowledge and perspectives. While thrownness establishes the explanatory PRIMACY of the political over the ethical, structure over agency, the past over the present (in the EA), Bhaskar sees this as contingently surmountable by the primacy of the ethical over the political (in the IA) within the political in the DIALECTICS OF FREEDOM. Methodological advice: 'see subjects$_2$ as at once "thrown" and as engaged in counter-HEGEMONIC struggles' (*D*: 376). See also AGENCY; CENTRISM, ETC.; EPISTEMOLOGICAL DIALECTIC; FOUNDATIONALISM; HERMENEUTICS; INTRINSIC/EXTRINSIC; SIGNS.

time. See A-SERIES; BLOCKISM/PUNCTUALISM; SPACE–TIME; TENSE.

Tina syndrome. Basically, 'a truth in practice combined or held in tension with a falsity in theory', issuing in emergent error and illusion (*RM*: 84–5). The pivotal figure, underpinned by GENERATIVE SEPARATION, in the Bhaskarian analysis of IRREALISM and more generally power$_2$-stricken social being, or (in PMR) DEMI-REALITY, named ironically for the British prime minister Thatcher's slogan, 'There *is no a*lternative'. The intention is to show that there are, after all, alternatives. The specification of its tendential logic has three basic premises. (1) Impelled by absence in the form of desire, we must act – the AXIOLOGICAL imperative. (2) When we act, we encounter natural necessity (which pertains to the sociosphere and nature alike, embracing inter alia our own fundamental needs, and is existentially and/or causally independent of our beliefs), which efficacious action must take into account. And (3) if we get our account of (2)

466 Tina syndrome

fundamentally wrong and do not correct our mistake, we are consequently forced into a series of endless theoretical and/or practical compromises. The upshot is a cumulative, emergent meshwork of figures and concepts that is incoherent and mystifying yet indispensable to our way of being. The figure is in the line of descent of Freud's account of a *compromise formation* (see also REALITY PRINCIPLE), Marx's theory of IDEOLOGY and ALIENATION, and Hegel's analysis of conceptual and social forms as 'at once both false and necessary ({. . .} a distinguishing feature of DIALECTICAL ARGUMENT)' (*D*: 118). Its metacritical schema is set out in ALIENATION.

The elaboration of a Tina syndrome is set in train when an AXIOLOGICAL necessity (i.e., natural NECESSITY epistemologically mediated in a world in which we must act) is denegated or violated by a theory or (if it is an axiological need) denied overt expression, i.e., when CATEGORIAL error is committed, as in the FUSION of knowledge and being in the epistemic–ontic fallacy. It is entrained, in other words, by theory–practice (axiological) INCONSISTENCY. Thus any anthropocentric subject–object identity theory, for example, if it is to be applicable to the world, 'will have to covertly graft onto or transmute itself into an anthropomorphic correspondence theory, adopting some amalgamation of them or shuttling between the two positions'. This is (1) a *Tina compromise*, involving *anthroporealist exchanges* of thoughts and things (see ANTHROPISM).

If general knowledge is to be possible, such a theory will further require an ACTUALISM postulating, e.g., in its empiricist form, 'the invariant invariance {. . .} of the subjectively defined particulars' (*D*: 117). This is (2) a *Tina connection*, whereby the compromise form is articulated with a *realist* complementary theory or *complement* – an empirical realist one in the example given. In the absence of concepts of primary polyadisation and alethic truth, this inevitably then generates a FIDEISTIC transcendent realism of some kind (e.g., Kant's noumenal realm), producing conceptual alienation and ideological mystification in the form of DUALISM.

Moreover, if the generalisations of the empirical (or conceptual) realist actualism are to apply in an open world, in order to survive it will further have to 'invoke a CP clause inconsistent with itself (for the generalisation cannot be both actual and universal)'. This is (3) a *Tina defence mechanism* or METAPHYSICAL λ *(lambda)* CLAUSE or *safety net*. It resembles Derridean 'supplementarity' which, however, itself requires supplementing with 'the more general concept of a Tina formation {. . .} for the analysis of the effects of the violation of {. . .} axiological necessity' (*D*: 118).

In this way, transcategorial or theory/practice INCONSISTENCY – the denial in theory or practice of what is necessary in practice – entrains 'something like a logic of false consciousness' (an analogue of the logic of discovery in the EPISTEMOLOGICAL DIALECTIC) (*PN3*: 122). (1) + (2) + (3) constitute a *Tina syndrome* or *(compromise) formation* or *complex* – a 2E figure of OPPOSITIONALITY most briefly defined as 'an inconsistent ensemble forced by the REALITY PRINCIPLE' (*P*: 243). Half true and half false, it has a close political cousin in spin or *spinology*.

Because such complexes necessarily presuppose what they (explicitly or implicitly) deny, they are internally contradictory, duplicitous and plastic, involving illicit FUSION and FISSION, FETISHISM and ALIENATION (split). This parallels the fetishism of commodities and fivefold alienation under capitalism, and facilitates their ideological use. Indeed, they *are* IDEOLOGIES in the restricted sense: 'to say that some conceptual

TMSA (transformational model of social activity) 467

or social form is at once both false and necessary {...} incoherent yet indispensable {...} contradictory but dialectically essential is just to say that it is a Tina compromise formation'. Moreover, the unfolding of an irrealist tradition, such as positivism, or of an irrealist dialectic, such as Hegel's, motivated ultimately by fear of change, 'may be regarded as a progressive compounding of Tina compromise upon Tina compromise' (*D*: 118) – 'the multiply mediated compounding of categorial error upon error' (*D*: 363). From such compounding – which loops through all four planes of our social being, and provides the basis for the late Bhaskarian analysis of *avidya* and *maya* (ignorance and illusion) and the DEMI-REAL as ensnaring the authenticity of human being within a meshwork of real false being – only transformed transformative totalising transformist praxis (TTTTφ) can free us. This is possible in virtue of the fundamental asymmetry of axiology and emancipation (see PRIMACY), 'that the emergent, false or oppressive level {of social being, i.e. demi-reality, which might also be called *Tina-reality*}, is unilaterally dependent on the more basic, true, and autonomous order, even though it may dominate and even threaten its existence, just as it typically mystifies, occludes and denies it' (*RM*: 228). See also DCR; DIALECTICAL ANTAGONISTS; DIALECTICAL ARGUMENT; DISCOURSE ANALYSIS; NEO-LIBERALISM.

TMSA (transformational model of social activity). The CR model of social and societal features which presents social action as conditioned, situated and temporalised. It articulates the relationship between SOCIAL STRUCTURE and social AGENCY in a manner that avoids the errors of VOLUNTARISM and REIFICATION, by vindicating the reality and irreducibility of structure and agency as two aspects of the social matrix and by introducing an analytical distinction between them which highlights the processual and TENSED character of the sociosphere. It is further substantialised as an explanatory framework in the SOCIAL CUBE, which specifies the formal structural components of the social sphere, and in the MORPHOGENETIC model which specifically develops the temporal aspect. Conjointly with SEPM, the TMSA situates the human subject as well as social forms in a stratified account of the world in which the objects of the human sciences broadly are *taxonomically* and *causally irreducible* but *interdependent* forms or modes of matter (*SR*: 113, 123 f.). Together they situate the objects of psychology, sociology, etc., as ontologically distinct but constellationally contained forms or aspects of FOUR-PLANAR SOCIAL BEING, embodying all four moments of the concrete universal.

The TMSA can be motivated either (1) by reference to transcendental arguments from intentional and purposeful agency or (2) through an immanent critique of the antinomies of predominant social theory.

(1) The TMSA was formulated by Bhaskar in the course of considering the possibility of NATURALISM in the human sciences (*PN*). Two dominant approaches within social theory are criticised: one naturalist, deeply embedded in a positivist tradition which posits the social sphere as essentially congruent with inanimate nature and biotic structures; one hermeneutic, which presents the social sphere as radically distinct, requiring an entirely different methodological approach. Although they differ in their view of the attainability of a nomothetic social science, both approaches share a presentation of natural science which is essentially positivist and empiricist. What they both fail to see is the inapplicability of this positivist ontology and methodology even to the natural sphere. Further, they do not distinguish naturalism from

468 TMSA (transformational model of social activity)

REDUCTIONISM (which posits an actual identity of subject-matter, thus presaging a non-stratified ontology) and SCIENTISM (which basically shares the methodological approach of reductionism, but is agnostic on ontological matters). Bhaskar formulates a qualified anti-positivist naturalism which presents an account of the sciences that is applicable for the natural as well as the human sciences, but which also acknowledges real distinction between their subject-matters, which at once call for a differentiation of methods as well as providing the basis for their non-reducibility. Bhaskar argues that there are ontological, epistemological as well as relational considerations that indicate these differences, which can be viewed as placing limits on naturalism in the human sciences, but are perhaps better presented as supplying the basis of these sciences and qualifying the format they must take (*RR*: 67, 93). If the first question is: To what extent can society be studied in the same way as nature?, the second and consequent question is: What properties do societies possess that might make them possible objects of knowledge for us? (*RR*: 66, 69).

In this analysis Bhaskar formulates the limits to naturalism in the social sciences, which simultaneously constitute properties of the social realm that make a non-reductive *social* science possible. Only in virtue of their non-natural surplus do social objects become possible objects of an autonomous social science (*PN*: 20 f.). Thus, we see that there are three distinguishing features of the social realm, setting it off from the natural realm, which pose *ontological* limits to naturalism. These concern the *activity-*, *concept-* and *space-time* dependency of social structures. Unlike natural structures, social structures do not exist independently of the activities they govern, nor of the conceptions people have of their activities. Furthermore, social structures are inherently historical and so only relatively enduring, both spatially and temporally (*PN*: 38). An account of the irreducibility of structure and agency quickly follows from the derivation of these limitations. Thus we can establish transcendentally that social forms are a necessary condition for any intentional act. Without a social matrix, intentional agency would have no meaning, but the reasons and motives of the individual agent are quite distinct from the functional imperatives of social structure.

To sum up, social forms are a necessary condition for any intentional act, their *pre-existence* establishes their autonomy as possible objects of scientific investigation and their *causal power* establishes their *reality* (*PN*: 25).

(2) Social theory has customarily been divided between two paradigmatic perspectives, both reductive, regarding the relationship between social agency and social structure. The first is the voluntaristic 'Weberian stereotype', according to which society is nothing but the unintended consequence of intentional and meaningful human behaviour; the second, the reificationist 'Durkheimian stereotype', in which the converse reduction is performed and human subjects are seen as mere artefacts of an alien, external and coercive societal structure. In both schools of thought, the one entity in the society/agency divide is presented as having a mere epiphenomenal status. A third, 'dialectical', model has been formulated in order to correct the apparent lopsidedness of these extremes. Here individuals and society stand in a relation of continuous dialectic, in which individuals produce society, but at the same time are the 'product' of structural constraints. Paradigmatically, society is the objectivation or externalisation of human beings who are the internalisation, the 'embodiment' of society and they are seen as two inseparable aspects of the same entity. This model is

TMSA (transformational model of social activity) 469

mistaken because it actually commits a double reduction in that it advances both a voluntaristic idealism in relation to the understanding of social structure and a mechanistic determinism in relation to the understanding of individuals. What the TMSA achieves is what the third model merely intends, but fails to accomplish. Here people and society are not seen as two moments of the same process, but as two ontologically distinct entities, related in terms of emergence. Society is a necessary and pre-existing condition for human action, and as such both enabling and constraining. On the other hand, society exists only in virtue of human agency. Society is both the ever-present *condition* (material cause) and the continually reproduced (or transformed) *outcome* of human agency. Further, praxis is both work, that is, conscious *production*, and (normally unconscious) *reproduction* of the conditions of production, that is, society. One could refer to the former as the *duality of structure*, and to the latter as the *duality of praxis* (see Figure 10; *PN*: 31 f.; *ER*: 212 f.).

In the formulation of the TMSA, Bhaskar was initially positioned rather close to Giddens's structuration theory, but in collaboration with Archer (cf. MORPHO-GENESIS) an important qualification or, rather, added emphasis is placed on the dimension of temporality, where the distinction between *synchronic* and *diachronic* emergence is crucial. Temporality holds the key to a viable conceptualisation of the interplay of social structure and agency. Although structure is an emergent effect of action, agency is a structurally constrained and enabled mode of action and structures are necessarily prior to social actions. In Archer's parlance, 'structural conditioning' predates the 'socio-cultural interaction' in which the 'structural elaboration' takes place (Archer 1995).

By conceiving social structure as necessarily predating the social action in which it is either reproduced or transformed, the illicit conflations (upward, downward or central in the voluntarist, reificationist and identificationary approaches) are avoided and a formal research model is supplied, for which more substantial explorations could be performed (*ER*: 202). A realist ontology in the social sphere must take into consideration the essential temporal priority, relative autonomy and causal efficacy of social forms, such that the praxis of agents is seen to consist in transformative negation, or essentially a generalised form of production in the Marxian sense, i.e., in work on

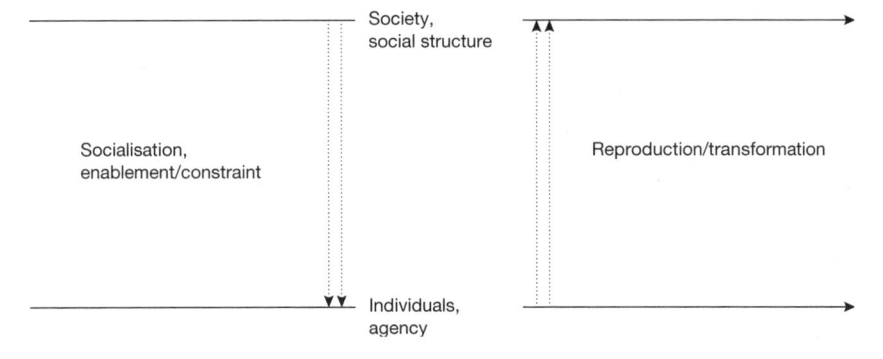

Figure 10 The Transformational Model of Social Activity
Adapted from SR: 126.

pre-given materials, captured by Marx in the saying: 'Men make their own history, but they do not make it just as they please; they do not make it under circumstances chosen by themselves, but under circumstances directly found, given and transmitted from the past' (Marx 1852: 595).

PÄR ENGHOLM

token/type. Terms introduced by C. S. Peirce to designate particulars and kinds or universals (the general), respectively. Thus token physicalism, e.g., holds that every particular is physical; type physicalism that every kind of entity is physical (see NATURAL KIND; STRUCTURE). The distinction is usefully deployed by Bhaskar to dissect the implications of IDENTITY-thinking which constitute the ANALYTICAL PROBLEMATIC (*P*. 138–9), as illustrated in Table 42.

token-reflexive. See APODEICTIC; TRANSCENDENTAL REALISM.

totality. A key, but relatively underelaborated, concept related to epistemic figures of abstract and concrete, analytics and synthetics, as well as to ontic entities such as processes, identities, dispositions and liabilities. In DCR, totality is predominantly a 3L concept in its ontic form, but constellationally it is a MELD concept, spanning structure (1M), emergence (2E) and transformational agency as well as conceptual and meta-critical comprehension in philosophical and scientific discourse at 4D.

Totalities may provisionally be described as SYSTEMS of internal relations, where the relata are linked through, and engaged in, different forms of intra-activity, operating through various forms of HOLISTIC determination. In science as well as in philosophy, we have to distinguish between relatively closed and open, and synchronically and diachronically emergent, totalities, indicative of both stratification and change. Three

Table 42 Implications of identity-thinking

Identity-thinking	Ontological implications/ presuppositions	Epistemological implications/ presuppositions
token-token identity – constancy of particulars	fixism → monism (if single particular); monadology (if plurality of particulars)	fixed subjects, tendentially nominalism
token-type identity – particulars instantiate one type only	actualism, blockism, fetishism (closure) → reductionism, tendentially atomism (if single type)	cognitive triumphalism (subjects are subsumable under known generalisations)
type-type identity – constancy of types	actualism, blockism, fetishism (closure)	reification, naturalisation and eternisation of generalisations
Ensuing theorem	actualism > reductionism > monism/monadology	
Overall effect on analytical problematic	cannot accommodate change	cannot accommodate change
Overall effect	mystification and con-fusion	mystification and con-fusion

forms of intra-activity are distinguished: (1) *existential constitution*, in which one element is essential and INTRINSIC to another; (2) *existential permeation*, in which one element merely contains another; and (3) *causal connection*, in which one element is merely causally efficacious on another (*D*: 123). The notion that things are to a significant extent so constituted, permeated and connected is caught in the concept of *ENTITY RELATIONISM*. Totalities are typically composed of internal relations, whereby the entities constitutive of them are necessary conditions for the existence, not only of the other entities, but also of the totality as a whole.

The universe may be considered an open totality, because it is not only stratified and structured but expanding and changing, such that the tacit BLOCKISM of flat, static ontologies such as that secreted by positivism is not even remotely appropriate. Totalities and their internal relations are at the hub of change qua emergence. Also closely linked to figures of ontic stratification, we may distinguish *sub-totalities*, whose boundaries are more or less permeable, but nevertheless circumscribe RELATIVELY AUTONOMOUS entities, where the internal connections are structurally maintained via FUNCTIONAL or proto-functional processes of NEGENTROPY. Thus biotic systems characteristically manifest various forms of homeostatic process through which steady states are maintained, seemingly in contradiction to basic laws of physics, e.g., the second law of thermodynamics. Sub-totalities are thus separated only relatively, and their boundaries are generally permeable, flexible and changing.

Totalities may be ontological, epistemological and meta-critical, where ontology > epistemology > meta-critics, emergent levels may react back on the levels from which they are formed, and human beings figure as both causal agents and reflexive symbolic producers at 2E and 4D. Thus in the figure of CONSTELLATIONALITY human subjects are seen as products and at the same time transformative agents in a society which is contained within, and an emergent product of, material processes of inanimate matter. However, just as people are both the outcome and the immediate producers of the social structures in which they are embedded, involving a duplex transformation of pre-existent material, society stands to the material and the biotic realm as dependent yet embracing and exerting an influence on the actual operation of mechanisms of the lower orders. Society is overreached by rather than transcends nature, but it is capable of reacting back on the materials out of which it is formed. In this sense the stratification of Being presents a series of hierarchically ordered open totalities in which the lower orders provide the limiting or boundary conditions for the higher ones, and the higher strata determine the range of applicability of the boundary conditions set by the lower ones (see DEPTH; INFRA-/INTRA-SUPER-STRUCTURE). Although laws set limits or impose constraints on the behaviour of the entities enclosed within their range of applicability, they seldom uniquely determine a specific outcome. This is due to the extreme rarity of spontaneously occurring closures in which a specific mechanism operates separately. Mechanisms normally operate jointly with others, and laws specified by scientists point to transfactually efficacious structures, so the outcome is rather to be characterised as TENDENCIES, which are neither contingent nor actual, but necessary, real and constellational. *EVENTS* are thus metaphysically conjunctures, and *things* are correspondingly laminated compounds, both being examples of concrete singulars (see STRUCTURE). These are the objects of applied research, in which the specific constellation of the stimulating factors is weighed. However, the prime interest

472 totality

of 'pure' science is the vertical dimension, in which deep structure, purified of its conjunctural context, is analysed. The object of pure science is CONCRETE UNIVERSALS. Typically the concrete universal manifests or individualises itself via one or more particular differentiations in some concrete singular (*D*: 114).

Any concrete universal has the character of a multiple quadruplicity, with the generative elements of transfactuality (universality), processuality, mediation and singularity, corresponding to the distinctive features of the levels of DCR (*P*: 78; see also MELD), and so processes and events must be presented as constellations of structures bound in open and potentially contradictory totalities. As totalities are emergent structures, and we have to allow for the transfactuality of such structures, the problems of actualism (at 1M) and extensionalism (at 3L) have affinities, as they are the effect of the suppression of natural necessity and lead to monovalence and fixism (at 2E), alienation, detotalisation and split (at 3L) and reification and disembodiment at 4D. Bhaskar speaks of destratification, deprocessualisation, demediation and desingularisation (*D*: 130), and we might add dereflexivisation, as the characteristic modes of 1M–4D *illicit abstraction* and *reduction* effected by the ANALYTICAL PROBLEMATIC, characterised by an undifferentiated (closed) and unstratified (depthless) ontology (see also CONCRETE/ABSTRACT).

The world itself is stratified and open, embracing external and contingent as well as internal and necessary relations, and DETERMINATION is multiple and conjunctural. DCR accordingly distinguishes between *good* and *bad totalities* in terms of whether or not they are open. This is in contradistinction to the Hegelian model of approaching totalities as constellationally closed, such that an open totality is bad because it conjures up the spectre of infinite regress, whereby the epistemic dialectic would issue in more of the same – resulting in a 'bad infinite', leaving knowledge without end, purpose or telos. Hegel thus cannot really envisage a world that is changing and incorporates a correspondingly changing epistemic realm. But this is just what is professed in DCR. Just as worldly process as such (in the ID) is characterised by openness and underdeterminacy, so, too, are the critical and metacritical processes (TD/ID). 'Our totality, unlike Hegel's, is open both synchronically (systematically) and diachronically (to the tensed spatialising causal processes of the future)' (*D*: 273), and the REFLEXIVITY characteristic of human action is the inwardised form of totality. Characteristic of most ideological systems, of which Hegel's is but one, is the urge to effect a constellational closure, to formulate a cosmogony which has a firm epistemological basis, grasps the world as finite and now resolved and fixed epistemically. This phenomenon is characterised by Bhaskar as a FETISHISM of closed systems (*P*: 43); it involves an achieved identity theory and the denegation of alterity, and ultimately leads to ontological extensionalism and blockism. All these irrealist tendencies are implicated by the search for an unhypothetical starting point and a constellational closure. What the analytical problematic cannot grasp is the distinction between analytical and dialectical motifs. What science analytically tries to close and thus resolve at one spatio-temporal stretch-spread must be transformationally open for synchronic and diachronic contextualisation, so that in the long run dialectical and processual rather than analytical CONSISTENCY is attained. Existential forms as well as epistemic configurations are transient and eventually superseded. There is both existential and referential contradiction (*D*: 170; *P*: 138 ff.).

There is a dialectical unity of analytical and dialectical reasoning; the latter overreaches but does not transcend, while depending on, the former, whereas the former without the latter is at a loss and encounters a series of apparently irresolvable aporias. Logical contradictions detected by analytical reason may point to real contradictions which are investigated dialectically to pinpoint their situational characteristics. Scientific as well as philosophical inquiry typically involves dialectical or developmental as well as analytical moments. The same overreaching constellationality is exhibited by the dialectic of nomothetic and taxonomic knowledge. The INDIVIDUATION of things/events/mechanisms is carried through by a specification of their alethic structure (see also EPISTEMOLOGICAL DIALECTIC; TRANSITION). The real is in this sense relational. Identities are paradigmatically dispositional in the sense of being able to do and being susceptible to change.

This epistemic rendering of the ontic stratified structure of being exhibits a characteristic logic in which we start with a chaotic conception of an undifferentiated whole, move through a series of abstractions in which the various codetermining aspects (structures, mechanisms, etc.) are singled out and eventually arrive, in a *retotalising* move, at a more theoretically informed conception of this totality as a concrete totality of distinct yet inseparable factors. The internal development of DCR in this sense reflects this scientific dialectic in which transfactually efficacious or normically qualified powers are distinguished at 1M; their concrete workings in different rhythmics are pursued at 2E; their moulding together to form uniquely laminated compounds, or STRUCTURATA, is studied at 3L; and the intentionality and reflexivity of human subjects are introduced at 4D.

The epistemic determination of concrete totalities can only be effected by substantive research, where the focus will depend on the explanatory objectives, ranging spatio-temporally and horizontally from, say, the emergence of new life forms to the situational logic of a social encounter, and vertically from surface forms to deep structures. The accurate conceptualisation and study of totalities calls for a complex, intertwined, dialectical and recursive process in which at least two moments must be singled out. A moment of *abstraction* is important, because it is the method available to apprehend the deep structure of reality, and a moment of *contextualisation* in which it is considered how these structural features act through, and are acted upon by, a range of causal mechanisms. That is, we must envisage a deep structural *pure* science of 'fundamental' structures and mechanisms and various levels of *applied* investigation into the actual workings of these mechanisms *in situ*. We could thus order the sciences with respect to their degree of abstractness, moving from abstract through intermediate to concrete sciences. Research is intrinsically *totalising* in two reciprocally reinforcing or resonating senses: at the *extensive margin* of inquiry, 'ever wider swathes of reality are seen {...} to be implicated' in a particular object of inquiry; at the *intensive margin*, the object is seen 'to be ever more densely packed' (*SR*: 112; see also *D*: 125–6).

PÄR ENGHOLM

trace structure. A *trace* is basically a negative residue that speaks of change and difference (e.g., pox marks on a human face). Derrida's concept of the *trace structure* of the SIGN is adapted and generalised by Bhaskar to refer to the CO-PRESENCE of absence and presence in the EXISTENTIAL CONSTITUTION of anything by the past

474   transcategorial causality

and outside and in HOLISTIC CAUSALITY (the intra-relationality of systems of difference). The trace structure of the sign is a rich facilitator and source of the conceptual DISTANCIATION and metaphoricity indispensable for the EPISTEMO-LOGICAL DIALECTIC of science.

transcategorial causality. See CAUSALITY.

transcategorial inconsistency. Another name for theory/practice inconsistency (also PERFORMATIVE contradiction, reflexive inconsistency). It is transcategorial because it denies a CATEGORY in theory or practice that is necessary in practice. See also CONSISTENCY; REFLECTION; TINA.

transcendence, transcendent, transcendent realism. See EMPIRICAL REALISM; FIDEISM; HERMENEUTICS; IMMANENCE/TRANSCENDENCE; REALISM; SOLIPSISTIC, ETC.; TINA SYNDROME.

transcendental. See CRITIQUE; IMMANENCE/TRANSCENDENCE; INFERENCE; REALISM; TRANSCENDENTAL REALISM.

transcendental argument. See DIALECTICAL ARGUMENT; INFERENCE.

transcendental deduction, transcendental reasoning. See INFERENCE; PHILOSOPHY; TRANSCENDENTAL REALISM.

transcendental detachment. See ATTACH/DETACH.

transcendental dialectical critical realism (TDCR). See META-REALITY.

transcendental idealism. See IDEALISM.

transcendental identification in consciousness, transcendental identity consciousness. See CO-PRESENCE; CREATIVITY; DEMI-REALITY; IMMANENCE/TRANSCENDENCE; META-REALITY.

transcendental realism (TR) among the most basic tenets of the CRITICAL REALIST project, and one that sets it clearly apart from most present-day approaches to philosophy of the natural and social sciences. It can best be explained by unpacking both terms, seeing how they fit together, and contrasting this specific usage of the phrase with other (on the face of it) similar proposals from theorists of various persuasions. I should emphasise that CR is itself a fairly broad church encompassing some lively differences of view – on this as on other topics – and therefore that my aim in what follows is not so much to present a received (orthodox) account as to pick out the main points in common with primary reference to their formulation in the work of Bhaskar. All the same, the TR thesis is so central to any version of the CR case in whatever domain or field of application that one can reasonably claim to set it out in representative and clear-cut terms.

'REALISM' is easier to define at first approximation. Briefly put, it is the thesis that there exists a real-world physical domain of objects, events, structures, causal powers, and so forth which decide the truth-value of our various statements or theories concerning them and which cannot be treated as in any sense dependent on our current-best or even future-best attainable state of knowledge. It thus comes out very strongly against an epistemic (knowledge-relativised) conception of TRUTH and very strongly in favour of an alethic (truth-based and objectivist) account of KNOWLEDGE. This thesis may be qualified to some extent, e.g., as regards certain artefacts (like transuranic elements, synthetic DNA proteins, or sub-atomic particles produced in cyclotrons) whose existence is indeed a result of applied techno-scientific know-how. However – so the TR argument runs – their production is subject to various physical constraints (such as those of sub-atomic charge, chemical valence, or molecular bonding) which define the range of possible realia and hence determine the scope and limits of effective human intervention. What is crucial here is the complex dialectical relationship between, on the one hand, those 'transfactually efficacious' laws of nature that depend not at all on our various kinds of controlled observation, experimental set-up, manipulative technique, etc., and, on the other, those non-naturally-occurring (but equally law-governed) phenomena that show up under just such specialised, e.g., laboratory conditions. We can make full allowance for the role of human agency in bringing such phenomena about – or in creating the appropriate (physically enabling) environment – while none the less maintaining a realist position with regard to the scientific object-domain along with its intrinsic structural features, dispositional properties, causal powers, and so forth.

Thus one chief sense of the term 'transcendental' in CR parlance is the sense: 'pertaining to an order of objective reality and a range of likewise objective truth-values that may always in principle transcend or surpass the limits of human knowledge'. To this extent TR comes out firmly opposed to any positivist, empiricist, instrumentalist, or other such epistemic approach that would reject the idea of verification-transcendent truths, even if – as in some recent versions of the case – what counts as 'verifiable' is defined in terms of idealised rational acceptability or optimal epistemic warrant 'when all the evidence is in'. However, there is a second, more 'technical' sense which goes back to Kant's distinctive usage of the term in his *Critique of Pure Reason* (1781) and which has to do with the conditions of possibility for knowledge and experience in general. That is to say, it involves a transcendental deduction which accounts for our capacity to acquire such knowledge or to have such experience in terms of certain strictly a priori concepts and intuitions (e.g., those of time, space and causality) that alone make it possible for the mind to impose an intelligible order on the otherwise inchoate flux of sensory impressions. This was Kant's answer to Humean scepticism, or the philosophic 'scandal' (as he saw it) of a radical empiricist outlook that despaired of achieving any adequate solution to the problem of knowledge. Rather, we should see that Hume's quandary need not – could not – arise if philosophy turned its attention to those constitutive powers or faculties of mind which enabled us to bring phenomenal (sensuous) intuitions under concepts of understanding. Where previous thinkers had gone so disastrously wrong – EMPIRICIST and RATIONALIST alike – was in failing to observe this cardinal requirement, and hence falling foul of Kant's cautionary dictum that 'intuitions without concepts are blind' while 'concepts without intuitions are empty'.

Such was the 'Copernican revolution' that Kant claimed to have brought about in epistemology and philosophy of mind. On his account it marked the epochal switch from a mistaken, dead-end, scepticism-inducing concern with what reality is like quite apart from our knowledge of it to a scepticism-allaying concern with how reality must appear to us, given the various a priori forms and modalities of human knowledge. Thus, for Kant, the purpose of transcendental reasoning is to deduce just that range of necessary presuppositions with regard to the structure of phenomenal experience that provide a common framework for our understanding of objects and events in the physical domain and for our self-understanding as conscious, reflective beings whose identity consists in our sense of a continuous spatio-temporal existence. This latter is what Kant refers to as the 'transcendental unity of apperception', i.e., the synthesising power of mind which – however elusive its nature – must be taken as the basic precondition for any such awareness. Moreover, it ensures that our phenomenal intuitions of space, time and causality *cannot but* correspond to the way things stand in reality, since reality *just is* – so far as we can possibly know – the way it appears to us through those given forms of jointly intuitive and conceptual grasp. So knowledge must confine itself strictly to the limits of phenomenal experience if it is to have any hope of defeating the Humean sceptic, or of closing the otherwise unbridgeable gulf between an order of objective or mind-independent (hence unknowable) reality and whatever lies within our epistemic grasp.

To be sure, reason – as distinct from understanding – finds itself impelled to transgress those limits and to posit the existence of a noumenal domain (that of 'things-in-themselves') beyond our phenomenal grasp. Such is, once again, the transcendentally deduced condition of possibility for any notion we can frame – or any scientific theory we propose – with regard to the 'external world'. However, it can serve only as a Kantian 'regulative idea', that is, a source of guidance or orientation for our various cognitive endeavours, rather than playing a constitutive role in the acquisition of knowledge. Hence the title, *Critique of Pure Reason*, epitomising Kant's point that where reason oversteps its appointed bounds – where it presumes to give knowledge of that which can only be thought – then it runs into all manner of dead-end ANTINOMIES or contradictions. That is to say, we can *think* of knowledge as aimed towards an ideal (limit-point) conception of truth at the end of inquiry, but can never actually *achieve* such knowledge, since by very definition it would bring us out on the far side of human attainability. Hence also Kant's claim to have resolved the Humean problem of knowledge by managing to reconcile a full-scale doctrine of 'transcendental idealism' with a somewhat less developed but still (as he argued) indispensable outlook of 'EMPIRICAL REALISM'. In support of the former Kant enlists pretty much the entire conceptual apparatus of the *First Critique*. As concerns the latter his arguments are philosophically unconvincing and do little to support the alethic realist's claim that there exists both a mind-independent (objective) reality and also – in consequence of that – a great range of to us unknown (perhaps unknowable) truths.

Indeed, one can trace all the vexing dilemmas of present-day epistemology to the radical cleavage thus opened up between Kant's 'transcendental idealism' and 'empirical realism', along with his likewise problematical attempt to explain how the manifold of sensuous intuitions can be somehow 'brought under' concepts of understanding. TR breaks the hold of those dilemmas by adopting the alternative that Kant ruled out since he considered it another manifestation of the tendency of reason to overstep its proper (i.e., regulative) limits and lay down the knowledge-constitutive terms and conditions of cognitive inquiry. That is to say, it critiques Kant's critique by maintaining (along with the alethic realist) that truth might always – now as heretofore – transcend or surpass our utmost epistemic powers, while none the less holding *this itself* to be a matter of knowledge borne out by the history of science to date and by our grasp of the complex dialectical process through which science progressively converges on truth under various determinate (e.g., material, techno-scientific and socio-economic) conditions. This makes it 'transcendental' in the Kantian sense of deriving from thought about the very possibility of scientific knowledge and progress in general, but also in the non-Kantian realist sense of allowing us to know – not merely 'think' – how such claims can be warranted or justified.

One source of its explanatory superiority in this regard is the fact that CR allows for an adequately complex or 'stratified' account of the relationship between subject and object, knower and known, or cognitive agency (in a strong sense of that term) and the various physical domains wherein that agency is exercised. Thus it makes a chief point of avoiding the kinds of dead-end DUALIST thinking that have taken such a firm and disabling hold on the discourse of epistemology from Descartes, through Kant, to logical empiricism. This is where CR marks a decisive advance beyond theories – whether in the natural or the human sciences – which endorse some version of the standard dichotomy between, on the one hand, causal explanations and, on the other, approaches that emphasise the rational or normative character of scientific thought. That is to say, it makes due allowance for the constant, many-levelled interaction between physical processes, laws of nature, and the various ways in which these become manifest through experiment, observation, and theory. Where most philosophy of science goes wrong is by ignoring this strictly dialectical relationship between knowledge and the object of knowledge and hence running into a familiar range of dead-end epistemological quandaries. Among them is the positivist fetishisation of 'facts', 'sense-data', 'observation-statements', and so forth, as if these latter could be somehow disjoined from the kinds of theoretically-informed observation or experimental set-up which allow them to emerge under certain specifiable conditions. Thus scepticism is merely the flipside or reactive counterpart of a positivist dogmatism that allows no commerce – no room for this productive two-way exchange – between 'context of discovery' and 'context of justification'. In which case logical positivism/empiricism can best be seen as the latter-day version of an argument which goes right back to Hume's drastic disjunction between 'matters of fact' and 'truths of reason' (analytic truths), and which is sure to insert its sceptical wedge whenever philosophy falls into this way of thinking.

Of course such arguments have long been raised against the logical empiricist programme, not least by those – like W. V. Quine (1969) – who swing right across to the opposite extreme of a radically holistic theory premised on the twin doctrines of the 'theory-laden' character of observation-statements and the 'underdetermination' of scientific theories by the best available evidence. They also play a crucial part in Thomas Kuhn's (1962) paradigm-relativist argument that (in some rather ill-defined sense of the phrase) 'the world changes' for thinkers living before and after a major revolution in the currency of scientific thought. However, such responses merely exacerbate the problem by relativising the 'truth' of any given statement at any given time to the entire body of presently accepted beliefs, conceived as extending all the way – as Quine describes it – from those at the logico-theoretical core to those at the empirical or observational periphery. In which case no belief is 'immune from revision', since even certain axioms of classical deductive logic – like bivalence and excluded middle – might ultimately have to be abandoned under pressure from conflicting empirical evidence, e.g., quantum phenomena such as superposition or wave/particle dualism. From a TR standpoint these post-logical-empiricist developments should be seen as so many symptoms of the deepening crisis in mainstream ANALYTIC philosophy of science rather than as pointing a hopeful way forward from the various problems bequeathed by Kant. Thus they purport to overcome the dilemmas of logical empiricism but only at the cost of embracing a Quinean doctrine of wholesale 'ontological relativism' which finds no room for normative criteria of truth, rationality and progress.

Hence the alternative CR proposal: that we reject the various failed solutions to the problem of knowledge deriving from Kant's likewise failed attempt to square his cardinal theory of transcendental idealism with a fig-leaf version of empirical realism. Rather, we should take the route which Kant was at such pains to close off, i.e., that of transcendental realism or the thesis that truth might always exceed the limits of present-best (or even future-best attainable) knowledge while none the less providing the standard or criterion by which all truth-claims must ultimately be assessed. What makes this possible – so the CR argument runs – is the process of constant dialectical exchange between theory, observation and experimental practice which leaves no room for such merely notional (and scepticism-inducing) dualisms as afflicted the discourse of logical empiricism and which opened the way to their own comeuppance at the hands of Quinean 'ontological relativity' and Kuhnian paradigm-relativism. These latter doctrines can be seen to result from a sceptical over-reaction to the fact that scientific theories always operate at a certain remove from the various complicating factors (e.g., of interference by external forces or disturbing influences of various kinds) which can never be taken fully into account by any theoretical science. Such idealisations – like that of the frictionless solid plane or the inviscid and irrotational medium of fluid mechanics – are the price one pays (necessarily so) when advancing hypotheses beyond the limits of empirical verification. For the sceptic about scientific realism or 'laws of nature' they show that this price is most definitely not worth paying, since every increase in theoretical or causal-explanatory power goes along with an inverse reduction in the way of detailed descriptive or phenomenological yield. That is, there is regular law of diminishing returns whereby any putative advance in the scope and generality of scientific theories must entail a corresponding loss of empirical precision or accountability. In which case – so it seems – we are better off adopting a sensibly scaled down (e.g., 'constructive empiricist') approach that renounces any claim with regard to the existence of real-world objects, properties, or causal powers and instead makes terms with the limits thus imposed on our capacity for forming well-grounded scientific conjectures beyond the strict limits of empirical warrant. Or, again, more drastically, this argument leads to an anti-realist position whereby there is simply no conceiving of objective (verification-transcendent) truths that would somehow surpass the scope of humanly achievable knowledge, proof or ascertainment.

On this view TR entails the downright contradictory pair of propositions (1) that every well-formed (truth-apt) statement has its truth-value fixed quite apart from our best knowledge concerning it, and (2) that veridical knowledge is yet within our cognitive grasp – perhaps at the ideal limit – through various well-tried methods of inquiry. Hence – to repeat – the pyrrhic conclusion embraced by some sceptics and anti-realists, namely that we can *either* have (some notion of) objective truth *or* the idea of 'truth' as epistemically constrained and therefore *ex hypothesi* knowable, albeit at the cost of ruling out any alethic (i.e., objectivist) conception. TR rejects this as a false dilemma and one that has taken hold only in consequence of the widespread EPISTEMIC FALLACY according to which it is strictly inconceivable – a species of logical nonsense – that truth should somehow transcend or elude the compass of optimised human knowledge. Such arguments sometimes go a long way towards granting the force of opposed realist intuitions, e.g., in Crispin Wright's elaborately nuanced proposals for 'superassertibility' and 'cognitive command' as criteria that approximate objectivist truth while keeping it within certain specified epistemic bounds and thus stopping short of making that crucial concession. Indeed, the chief effort of recent work in this 'moderate' anti-realist vein has been to offer formulations of what properly counts as truth in some particular area of discourse which avoid the (supposed) sceptical nemesis of alethic realism and yet make room for some middle-ground approach on terms that the realist might be brought to accept. However – from a CR viewpoint – this amounts to yet another (more elaborately qualified or nuanced) version of the EPISTEMIC FALLACY. That is, it stops crucially short of acknowledging (1) the existence of a real-world (mind- and theory-independent) physical domain along with its sundry objects, structures, laws of nature, causal powers, etc., and (2) the various kinds and levels of human interaction with that physical domain whereby its affordances show up under given (e.g., observational or experimental) conditions.

Thus there is no problem for CR in maintaining *both* an objectivist (alethic or non-epistemic) conception of reality *and* an account of scientific practice that makes full allowance for the role of human agency in revealing certain processes, laws and causal properties whose manifestation (though not their reality) depends on our procedures for finding them out. This exemplifies the close dialectical relationship between TR as a matter of straightforward ontological commitment and the CR 'stratified' conception of reality as that which affords knowledge of the world through various investigative methods and techniques. In so far as these claims appear incompatible or downright contradictory it can only be on account of that deep-laid dualist mindset that has characterised so many episodes of epistemological debate from Kant to logical empiricism. From this point of view, CR looks like a hopeless attempt to square the circle, that is, to explain how we can and do acquire knowledge of objective, hence mind-independent, hence strictly unknowable realia. From a CR standpoint, on the other hand, such objections merely go to show that philosophy has been on the wrong track – and subject to periodic outbreaks of scepticism and anti-realism – since Kant introduced his fateful split between the realms of phenomenal (sensory-cognitive) appearance and noumenal (knowledge-transcendent) reality. If we can only break the hold of this dichotomy – one that has defined the 'problem of knowledge' for philosophers of many, often sharply divergent views – then we shall see that TR involves not so much a squaring of the circle as an adoption of just that approach whose ruling-out by Kant has been the cause of so many subsequent epistemological woes.

This also involves drawing certain MODAL distinctions, as between the orders of contingent ('might-have-been-otherwise') fact, laws of nature which apply (necessarily so) to our own world and all others that physically resemble it, and 'transworld necessary' truths – such as those of logic and mathematics – which cannot be conceived as failing to apply in any possible world whatsoever. Such arguments have been developed chiefly by modal logicians concerned to explicate the logic of necessity and possibility. They have also been deployed by philosophers of science in order to provide an account of causal explanation in counterfactual-dependent terms, that is to say, in terms of what would (or would not) have occurred at some other, physically 'nearby' world in the presence (or absence) of certain antecedent and explanatorily relevant conditions. CR gives added substance to these often rather recondite and speculative claims through its stratified conception of reality and its firm grasp of the various ontological distinctions involved. It also does much to clarify the issue between a hardline modal realist such as David Lewis (1986b), who takes all those possible worlds to co-exist, i.e., to stand ontologically on a par with our own, and 'actualists' who hold that modal talk is a useful heuristic device but who refuse to endorse any such (in their view) wildly extravagant doctrine. For Lewis, 'actual' is a token-reflexive (or deictic) term that functions – in a similar way to words like 'I', 'here', 'now', 'tomorrow', and so forth – always with reference to some individual speaker in some specific time, place, or context of utterance. Thus to call ours the 'actual' world is no more than to locate oneself in relation to just that range of spatio-temporal co-ordinates, causal regularities, and laws of nature that happen to obtain in just those worlds that are physically compossible with ours. What it cannot rule out – unless at the cost of extreme ontological parochialism – is the real (though for us non-actual) existence of all those other possible worlds which differ from our own in certain specifiable respects. So when critics of Lewis charge him with indulging a grossly inflated ontology – when they protest that his distinction between 'actual' and 'real' gives rise to some absurd consequences – he can turn the charge around and ask by what right they deploy modes of counterfactual reasoning if not with reference to something more than a realm of merely abstract, i.e., unrealised possibility. After all – Lewis argues – such reasonings must lack any genuine explanatory force unless they are taken to quantify over various possible (for us non-actual but objectively real) worlds wherein certain physical constants and laws of nature are subject to a process of controlled variation. Otherwise philosophers are getting their arguments on the cheap, that is, exploiting modal-counterfactual talk in a way that – as Russell famously remarked in a different context – has all the advantages of theft over honest toil.

TR allows an approach that enjoys those advantages honestly and gives substantive content to modal claims while avoiding any Lewis-style recourse to the notion of endlessly multiplied divergent counterpart worlds (parallel universes). This it does, to repeat, by drawing a firm and principled distinction between real ('transfactually efficacious') physical constants or laws of nature whose workings are wholly independent of our various investigative methods and, on the other hand, whatever shows up through the deployment of increasingly refined, e.g., technologically enhanced, observational means. Thus it has no need for the kind of far-fetched speculative argument that would assert the reality of worlds which somehow exist in a realm of possibilia spatio-temporally disconnected from ours and hence entirely beyond our epistemic ken. Rather, the TR distinction falls out between objects, structures, properties and powers that pertain to the nature of this-world physical reality, objectively conceived, and the extent to which these are actualised – made manifest – through modes of applied scientific research. That is to say, the 'actual' is not (as in Lewis) a localised sub-set of those various, equally 'real' counterpart worlds that occupy the entire space of logical possibility but a product of certain specific operations – involving (say) electron microscopes, radio telescopes, or particle accelerators – which reveal certain otherwise inscrutable aspects or constituents of physical reality. This also enables the TR approach to give an adequately detailed account of those various hypothetical-conditional or COUNTERFACTUAL-supporting modes of argument that play a large role in causal EXPLANATIONS. Thus, for instance, in explaining why the match ignited when struck one might adduce a whole range of this-world possible conditions under which it would *not* have thus ignited. Among them would count (1) its having been previously doused in water, or (2) its containing, not phosphorus, but some other non-combustible substance, or (3) its having been struck with insufficient force to generate the required frictional heat, or (4) perhaps in an oxygen-depleted atmosphere, or again (5) – at further counterfactual stretch – in one where the oxygen atoms had entered some weird state of quantum superposition.

However, such examples precisely go to emphasise the point that this approach *via* modal or possible-world thought experiments is one that possesses explanatory force only in so far as the worlds in question can be ranked on a scale of relative proximity or distance from the world that we really (not just 'actually') inhabit. That is to say, conditions (1) to (4) of the above series can be thought of as yielding highly plausible realist explanations of why the match might counterfactually not have ignited, and hence of why it did so ignite when those abnormal conditions were absent. Item (5), on the other hand, requires our taking leave of everyday (if not scientifically conceivable) reality, while beyond that lies a whole cornucopia of *logically* possible worlds where so much becomes subject to thought-experimental variation – from the laws of chemical bonding to those of gravitational attraction – that we are simply no longer in the realm of genuine (this-world) causal-explanatory reasoning. These debates in the province of modal logic have mostly been conducted by philosophers of language and science without any overt or acknowledged allegiance to the programme of CR. Still, as I have said, they converge with that programme in a number of crucial respects, not least in their commitment to basic principles of inference to the best causal explanation across all and only those physically possible worlds that most closely approximate to our own. TR has the great virtue, in this context, of sustaining a modal-realist approach with adequate counterfactual-explanatory resources but without having to venture beyond the limits of a plausible physicalist ontology. It also provides a strong alternative to various kinds of SOCIAL-CONSTRUCTIVIST or paradigm-relativist thinking whose appeal derives chiefly from their setting up a typecast, reductively characterised 'realist' opposition which bears no resemblance to the complex, dialectical and stratified CR model. Indeed, its critique of such grossly simplified conceptions – along with their burden of unresolved problems and antinomies – has been among the most striking achievements of work in the CR mode. In so far as TR plays a crucial role in that project (for reasons explained above) it should be seen as a major contribution to present-day, post-empiricist PHILOSOPHY OF SCIENCE. See also DIRECTIONALITY; EPISTEMOLOGICAL DIALECTIC; HOLY TRINITY; INFERENCE.

CHRISTOPHER NORRIS

transcendentally real self. The *transcendentally real*, or *alethic*, *self* is the deep content of praxis, linked in PMR to the ground-state. See ALIENATION; CONCRETE UTOPIANISM; EMANCIPATORY AXIOLOGY; IMMANENCE/TRANSCENDENCE; META-REALITY; SELF-REFERENTIALITY; SUBJECTIVITY.

transdiction. See INFERENCE.

482 transdictive (problem) complex

transdictive (problem) complex. See IRREALISM; SOLIPSISTIC, ETC.
transdisciplinarity. See INTERDISCIPLINARITY, etc.
transduction. See INFERENCE; PROBLEM OF INDUCTION.
transfactual. See COUNTERFACTUAL/TRANSFACTUAL.
transfactual realism. See DEPTH; DIFFERENTIATION.
transform (n.). See ENTROPY; EPISTEMOLOGICAL DIALECTIC; TRANSITION.
transformational model of social activity. See TMSA.
transformative agency. See AGENCY.
transformative negation. See ABSENCE.
transformed, transformative, totalising, transformist praxis. See TTTTφ.
transhumanism. See HUMANISM.
transition/boundary. *Transition* (derivation as for TRANSITIVE), also *transform*, n., is a 2E concept signifying (DIACHRONIC) movement (PROCESS) to and across a (synchronic) *boundary* and issuing in the determinate negation (*transformation*) of the bound entity into something new. Its concept is entailed, in relation to the sociosphere, by the TMSA and, more generally, by EMERGENCE. From the vantage point of transition, the most significant events for science to explain are those transformative of structures, and the most significant structures those that are undergoing transformation (cf. *SR*: 217).

Boundary (or *border*) is that which binds (sets *bounds* or *limits* to) TOTALITY, and must in general be discovered a posteriori. A boundary may be real (as in an entity's spatial envelope, a market boundary [Suzuki 2005] or the boundary between life and death) or virtually real (conceptual) (as in the boundary of a tradition or discipline, a political boundary), and true or false (the alleged logical divide between fact and value). In an open world, boundaries are permeable or intra-active and fluid; in the eudaimonian sociosphere, where they fell would be subordinate to the principle of universal concretely singularised freedom.

A transition may be geo-historical (the transition from one geological epoch to another, from one social system to another, or from knowledge of one stratum of reality to knowledge of another), or dialectically logical (the transition from facts to values, theory to practice, form to content, figure to ground, desire to freedom, CN to DCR). A *transitional concept* is a *bridging concept* which usefully brings together a range of distinct concepts without obliterating them (e.g., 'vertical realism' bridges all the various depth-strata, together with the distinction between philosophical and scientific ontology). *Transitional logic* is dialectical logic, or, less generally, the logic or *dialectics of transitions*; it contrasts with formal logic, the logic of the status quo (cf. DIACHRONIC/SYNCHRONIC). *Transitional rhythmics* or *processes* PREFIGURE or lead to a transition. *Transitional politics* and *transitional praxis* are activities oriented to effecting emancipatory transitions in the sociosphere, which also necessitate *theories of transition*, which philosophy and social theory may play vital roles in articulating.

Any transition (process of determinate negation) will embrace both a *transition state* or *boundary state* or *zone* (a spatio-temporal or temporal moment of indeterminate negation) and, within that, a *point of transition* (a moment of determination, of absolute becoming). The latter is also called a *node* (f. L. *nodus*, knot, knob, joint, tie, bond), or *nodal point* or *limit*, or *turning point*, constituting a *critical* or *limit situation* instantiating a genuine occurrence of the PROBLEM OF INDUCTION. It is the point at which it is no longer possible to say whether A is still A and not B – when A is *switching* to B (see also AXIOLOGY [axiological indeterminacy]). Within this node, in the Hegelianesque EPISTEMOLOGICAL DIALECTIC, there is another node – the δ *node* or δ *moment* – at which 'the hint of the restoration of consistency by an expansion of the pre-existing conceptual field' (*D*: 33) arrives. This is the moment of epistemological TRANSCENDENCE, facilitating a move to a more complete totality. A transition thus exhibits the CO-PRESENCE of presence and ABSENCE in the form of identity and CHANGE and/or identity and DIFFERENCE, and may be indicative of systemic *crisis*. The overall nodal moment or episode constitutes a *nodal* or *limit* or *switch situation*, which will require a PERSPECTIVAL switch to grasp and which may pose genuine difficulties for the determination of boundaries (cf. SYSTEM) and identities (see PROBLEM OF INDIVIDUATION; VALENCE), as well as for formal logic. Its result is a *quantum leap*. There are many kinds of nodal points, including: *junctions* or *connector points* (conjoining two types of process); *branches* (foreclosing some possible developments, opening to others); *jump points* (to another level of possibility); *saddle points* (blocking change); *break points* (ruptures in totalities); and *trigger points* (initiating a transitional rhythmic) (*SR*: 217). The concept exploits the etymological bivalence of 'node' as tying together yet distinguishing, e.g., your hand is both distinguished from and connected to your forearm by your wrist-joint (node). A *nodal question* is usually one concerned with crucial connections between issues. A *nodal line* (of measure relations) is a concept deployed in Hegel's *Logic* in the transition from quantity to quality.

Frontier and (*dialectical*) *threshold* bring the concepts of boundary and transition together. Thus, cosmically speaking 'every moment is a transition, a frontier' (*D*: 189) and, one might add, a threshold. Frontier often carries the additional connotation of a boundary or limit that it is anticipated will be crossed in due course, as in 'the frontiers of knowledge' (*RS2*: 185). A eudaimonian society would be in a permanent state of transition, i.e., its evolutionary frontiers would be open. The *logic of social transition* is (1) dialectical connection (3L); (2) dialectical CONTRADICTION (2E); (3) NEGATION OF THE NEGATION (geo-historical transformation of geo-historical products at 4D); and (4) EMERGENCE (1M). See also CHAOS/COMPLEXITY; CO-PRESENCE; DIALECTICAL LIMIT; LOCO-PERIODISATION; TENSE.

transitive dimension. See INTRANSITIVE, ETC.
transversality. See UNIVERSALS/PARTICULARS.
trigger. See TENDENCY.
trigger points. See TRANSITION.
triple hermeneutic. See HERMENEUTICS.
triumphalism. See CENTRISM → TRIUMPHALISM → ENDISM.
tri-unity of space, time, causality. See SPACE–TIME.
tropics. See ONT/DE-ONT.
trust. A 3L figure closely related to REFLEXIVITY, on the one hand, and SOLIDARITY, care and love, on the other; arguably the human *primary existential*, in the Heideggerian sense of basic 'mode of being in the world' (*P*: 3). It is 'one of the four components of the judgement form, the truth tetrapolity and the ethical tetrapolity' (see EMANCIPATORY AXIOLOGY, Table 19), thereby 'prefiguring a society based on a normative order of trust' (*D*: 263); and it 'underpins the triadic relationship between self- and mutual ESTEEM and existential security' (*DG*: 406) established during the

process of PRIMARY POLYADISATION, on which the components of ACTION are founded (*D*: 166). This suggests that trust is a necessary condition for that other main post/modern candidate for primary existential – 'agonistic struggle' (Hegel), competition and aggression – and indeed for human sociality as such, i.e., it is the basis of the social bond. Its binding power is linked, especially in PMR, with love.

Dialectic distinguishes four kinds of trust, corresponding to 1M–4D: (1) *abstract* trust in expert systems of which one has no knowledge; (2) *mediated* trust in domains, e.g., economics and politics, of which one has some (sceptical) knowledge; (3) *concrete*, 'ideally singularised', trust, exemplified in SOLIDARITY; and (4) *personalised* trust, e.g., friendly, caring or nurturing relationships (*D*: 274). *Trustworthy* and *trustworthiness* have synonyms in *fiduciary* and *fiduciariness* (f. L. *fiducia*, trust), respectively. See also RECOGNITION.

truth. Critical realists disagree with one another about the definition of the concept of truth. That this is so should not be surprising, given that there are deep differences of opinion on the issue amongst philosophers generally. What is notable about the debate amongst critical realists is that it is not limited to the meaning of the concept of truth. There is also disagreement, to a far greater extent than amongst mainstream epistemologists, about the related question of what sort of things can *be* true. Within the mainstream, there is vigorous argument about whether it is sentences, beliefs, propositions or even whole theories that should be understood to be what are called 'truth-bearers'. None the less, participants in the mainstream debate are in agreement that designations of truth-value apply to cognitive-linguistic things – 'things', that is, that are constructed out of language, and used for certain purposes. Within CR, by contrast, there are those who maintain that the concept of truth ultimately applies to physical and social phenomena (see ALETHIA; KNOWLEDGE).

Outside of CR, there are at least six discernible approaches to the concept of truth: correspondence theories, coherence theories, consensus theories, idealised consensus theories, pragmatic theories and deflationary theories. (I will here collect sentences, beliefs, propositions, whole theories and sets of the foregoing under the general category of 'claim', thereby bracketing the debate within analytic philosophy over which kind of cognitive-linguistic object we ought to understand truth-bearers to be.) Before reviewing the contending approaches, however, let me clarify the difference between the concept of truth and the concept of justification. A useful way to grasp the distinction is to say that the terms are responses to different questions. The concept of truth, we may say, is a response to the definitional question: 'What is it that we are saying about a claim when we say that it is or is not true?' The concept of justification, by contrast, is a response to the question: 'What are the reliable *indicators* of a claim's purported truth-value?' In the first case, we are asking what it *means* for a claim to be true. In the second case, we are asking how we shall be able to know if a claim actually *is* true – what the criteria ought to be, in accordance with which we might come to such a judgement. The present discussion concerns competing accounts of the concept of truth, not of justification. This is the case despite the fact that the concepts of truth and justification are clearly related, and intimately so in the special case of the consensus theory of truth, whose proponents argue that for a claim to be true just *is* for it to be judged to be true.

Proponents of the correspondence theory hold that what it means for a claim to be true is that things are, in fact, the way that they are affirmed to be, via the claim. From this perspective, to say that the claim, 'It is snowing outside', is true is to say that it conveys correct information. It really is snowing outside. As Aristotle put it, 'To say of what is, that it is, is true. To say of what is not, that it is, is false.' From a correspondence perspective, what makes a claim be true or false is the relation in which it stands to the state(s) of affairs to which it refers. If the state(s) of affairs to which it refers is as it is said or thought to be, then the claim is true. How one would go about making such a determination – or even if such determinations are in principle possible – is, as noted above, a different matter. None the less, objections to the correspondence theory of truth often amount to a charge that it is incorrect because there is no question-begging way to determine whether or not claims *are* true – that is, to determine whether or not they stand in a relationship of correspondence to that to which they refer.

Proponents of the coherence theory argue that what it means for a claim to be true is that it coheres with our other beliefs. From this perspective, to say that the claim, 'It is snowing outside' is true is to say that it is congruent with the beliefs, say, that it is –5C/22F outside; that water freezes below 0C/32F; and that one is not actually on a movie set. Here what makes a claim be true, if it is, is the relationship in which it stands not to a referent, but to a specified totality of beliefs. And, again, the issue here is what we *mean* when we say that a claim is true, not how we *tell* if a claim is true. One question that may be put to proponents of the coherence theory is: 'Which already accepted beliefs are the normative ones – the ones that figure into the definition of the concept of truth – and why?'

Proponents of consensus theories of truth argue, in one form or another, that what it means for a claim to be true is that it is believed, by some person or group of persons, to be true. Some consensus theorists stipulate that there must be good *grounds* for such belief, i.e., that the 'assent' must be 'warranted'. Consensus theories of this type – what we might call the 'cognitivist' wing of consensus theorising – are sometimes referred to as 'epistemic' theories of truth. From this perspective, what makes 'It is snowing outside' be true is the well-founded-ness of the belief that it is snowing outside. To be clear, the argument is not that belief in the soundness of the claim, 'It is snowing outside', brings on a blanket of white. Well-founded belief is what causes claims to be true; that this is so tells us nothing about what causes snow to fall. In its less cognitivist form, the consensus theory is the view that the concept of truth means, not 'Believed, on good grounds, to be true', but rather simply 'Affirmed (for any one of the many reasons that people affirm things)'. Rorty, for example, holds that 'true' is best understood as meaning something like 'deemed to be praiseworthy'. Here, to say that the claim, 'It is snowing outside', is true is to say that one endorses such a claim, or otherwise approves of it. Note that the relationship between the concept of truth and the concept of justification is unique in the case of what I've called the 'cognitivist' version of the consensus theory. For in this case the definition of the concept of truth is displaced on to the concept of justification: to be true just *is* to be justified. Proponents of the non-cognitivist version, such as Rorty, argue that in fact there are no such things as justified as opposed to unjustified beliefs. Rather, because 'true' simply means 'endorsed', there are simply beliefs that are widely endorsed. A weakness of both varieties of consensus theory is that it is hard to see how they can be made to sustain a serious conception of error, or falsehood.

Proponents of idealised consensus theories argue that what it means to say that a claim is true is not that some actual person or persons do, as a matter of contingent fact, believe it to be true (or otherwise regard it as praiseworthy), but rather that an ideal subject, under specified ideal conditions, *would* believe it to be true. Here, to say that the claim, 'It is snowing outside', is true is to say that an imaginary group of perfectly informed professional meteorologists, for example, operating under no coercion, *would* reach the conclusion that it is snowing outside. Rendering the requisite consensus hypothetical shifts the philosophical weight off the phenomenon of consent per se, and on to the specified idealised conditions. Such a move preserves the concept of truth as a counterfactual norm, but it raises the question of just what the difference is between claims that imagined ideal knowers would affirm and those that they would reject. If the answer is that, being in a perfect epistemic state, they would affirm the claims that are true and reject the ones that are false, then we need to know if the former are simply true by definition – i.e., made so by the very fact that ideal knowers would affirm them – or if they are in some sense 'really' true. If the former, then it is not clear what we have gained by moving from consensus to idealised consensus – for the claims affirmed by non-ideal knowers are also true-by-definition, from the perspective of a non-idealised consensus theory. On the other hand, if they are 'really' true, then an alternate definition of the concept has been tacitly introduced.

Proponents of the pragmatic theory hold that what it means for a claim to be true is that what is claimed is useful. Here, to say that the claim, 'It is snowing outside', is true is to say that the meaning conveyed by it carries some instrumental value. It is worth noting that the pragmatic theory as I've presented it is not the idea that for a claim to be true is for it to be the case that it is, or would be, useful to *believe* that it is true. It is the usefulness of the claim itself that makes it be true, not the usefulness of believing it to be true. The pragmatic theory is vulnerable to the charge that there are all sorts of things that one might want to talk or think about that are not in and of themselves of instrumental value, or towards which one does not wish to adopt an instrumental stance, which at the same time certainly appear to be such that the truth-value of claims about them may be meaningfully assessed. Moreover, there are many claims that are only useful depending upon one's purpose. Finally, it is difficult from the perspective of the pragmatic theory to say why it is that useful claims are in fact useful. Specifically, it would seem to be that they are useful because they are true. But such an opinion is not available to a proponent of the pragmatic theory, who argues instead that claims are true because useful.

Proponents of the deflationary approach argue that the concept of truth is superfluous. To say that the claim, 'It is snowing outside', is true is, from this perspective, simply to reiterate that it is snowing outside. The deflationary approach is similar in a certain way to the correspondence theory, in that proponents of both direct our attention to the weather conditions rather than to the beliefs, preferences and/or objectives of subjects, in their efforts to illustrate through example what is meant by the concept of truth. The crucial difference between the two positions, however, is that the deflationary theorist does not believe that the concept of truth adds any additional meaning to the original assertion. To say that the claim 'It is snowing outside', is true is, from this perspective, equivalent in meaning to saying that it is snowing outside. The deflationary theory raises questions about what claim-making actually is, and whether or not such a practice can be understood, or even carried out, if the concept of truth were to be purged from our meta-theoretical vocabulary.

CR is compatible with a range of theories about the meaning of the concept of truth – though one or another approach may be more consistent with the overall thrust of the position than others. Critical realists are not agreed on the issue. While some critical realists have adopted Bhaskar's approach, others have not; it should not be taken to be definitive for those working within a CR framework.

Bhaskar's treatment of the concept of truth has changed over time. In his first book, *RS*, he argued against the correspondence theory in terms commensurate with the emerging anti-positivist position of the period: the correspondence theory must be rejected because it is impossible to connect individual propositions on to bits of pre-theoretical sense-data to which they could be said to uniquely attach. It is only whole theoretical frameworks that attach to data – and only if one already assumes the soundness of the framework in question. Bhaskar supported the view that the truth or falsity of scientific theories is a function of the judgement of scientists. Thus he affirmed a consensus theory of truth. I have stopped short of calling the early position an ideal consensus theory because there is no suggestion that the scientists in question be perfect, hypothetical knowers, thereby constituting a counterfactual norm. Rather, the claim was that what it means for a theory to be true is that existing scientists say that it is true. (For a dissenting view see *Alethia*)

With *Dialectic* and *Plato Etc.*, Bhaskar introduced the term 'alethic truth' into the vocabulary of DCR. Bhaskar defines alethic truth as 'a species of ontological truth constituting and following on the truth of, or real reason(s) for, or dialectical ground of, *things*, as distinct from *propositions*'. To be the alethic truth of a thing x is to be that feature of the world that causes x to be as it is. That which causes x to be as it is may be either (1) an essential property of x, or (2) an external determinant; these alternatives are not clearly distinguished. The concept of alethic truth thus captures the related ideas that the world has an intrinsic structure and that there is such a thing as natural necessity.

Bhaskar makes two further claims regarding the notion of alethic truth. The first claim is that the concepts of truth and falsehood may be applied to non-ideational material objects and processes. The alethic truth of frozen water, then, would be its molecular structure. The aim of scientific practice, in Bhaskar's view, is to discover such 'alethic truth(s)' – and to represent it/them conceptually, in theories that scientists endorse. The second claim is that alethic truth should be conceived as 'ontological truth', and that such 'ontological truth' is what gives weight to mere 'epistemological truth'. Bhaskar expresses this line of thought through a conceptual model that he calls the 'TRUTH TETRAPOLITY'. The model is an aggregation of different senses of the concept of truth, which Bhaskar asserts are unified through and grounded in the concept/reality of alethic truth. It should be noted that Heidegger used the term earlier in the century to refer to what he, too, thought of as 'ontological' rather than 'epistemic' truth.

Bhaskar's version of the theory of alethic truth is vulnerable to criticism on several scores. From a CR perspective, the charge may be made that 'alethic truth' is simply a misnomer for the concepts of 'causal mechanism' and/or 'real essence', combined with an affirmation that, as correspondence theorists maintain, it is the world to which

488   truth tetrapolity

our claims refer that is the truth-maker of whatever true beliefs we may hold. From a Heideggerian perspective, objections would be raised to the idea that science is a practice through which alethic truth is or could be revealed. See also KNOWLEDGE; REALISM.

RUTH GROFF

truth tetrapolity. Since subjectivity–objectivity within subjectivity is contained within overarching objectivity, such that our situation is a meta-reflexively totalising one, we can say with Hegel that 'truth is the whole', and the theory of the truth tetrapolity is geared to thinking it as such. Its four moments correspond to those of the ontological–axiological chain. The tetrapolity is constituted by (1) normative-fiduciary (3L), (2) adequating (warrantedly assertible or epistemic) (2E), (3) expressive-referential (epistemic-ontic) (4D), (4.1) ontological and (4.2) ALETHIC truth (1M). Any truth-claim that is not exclusively a priori implicates these four moments (cf. the JUDGEMENT FORM). Thus, 'The weather report says that there is a stable high pressure system in place that is unlikely to shift for a week', implies: (1) Trust me, you can act on it – you needn't take your umbrella. (2) Meteorological science has reliable procedures for establishing the existence of weather patterns of this kind. (3) There is indeed a weather report of that kind, which my sentence refers to and expresses. (4.1) There really is a well-established high-pressure system in place. (4.2) It is in the nature of high-pressure systems to preclude the possibility of rain. (4.1) and (4.2) are expressed in, but obtain independently of, language. Corresponding to (2), (3) and (4), falsity as well as truth may be emergent (see CRITIQUE; DEMI-REALITY).

TTTTφ or **d**φ' (where dφ is DIALECTICAL PRAXIS or practically oriented transformative negation). The formula for the theory and practice of EMANCIPATORY AXIOLOGY: '*transformed* (autoplastic), *transformative* (ALLOPLASTIC), *totalising, transformist* (oriented to deep structure global and dialectically universal change) *praxis*' (*D*: 156, e.a.). More fully, *TTTTTT*φ: transformed, transformative, *trustworthy*, totalising, transformist, *transitional* praxis (*D*: 266). TTTTTT theory and praxis underpin the empowering *dialectic of the 7 E's* constituting the *dialectic of the 7E's and 6 T's* (*D*: 121). In PMR the emphasis on self-change in the first 'T' becomes the principle of the primacy of SELF-REFERENTIALITY, which is central to the spiritual turn. However, individual self-change without participation in TTTTTTφ is nugatory, because, by the logic of dialectical UNIVERSALISABILITY, individual depends upon universal self-realisation (*RM*: 238).

type. See NATURAL KIND.

U

υ **(upsilon) transform**. See EPISTEMOLOGICAL DIALECTIC.
ubiquity determinism. See DETERMINATION.
ultimata (L. pl. of *ultimatum*, f. *ultimus*, the most distant, furthest, last) used in two senses by Bhaskar:

(1) *Ultimata for science* or *ultimata-for-us*. These are the most basic ontic levels (kinds of entity) which science must (currently) presuppose (for they may be known only by their material effects) if it is to avoid invoking miracles, viz., the *constellational IDENTITY*, *rhythmic identity* and *dispositional (self-)identity* of a thing's nature with its causal powers, its changing causal powers, and the exercise of its changing causal powers, respectively. They are thus 'characterised by a systematic ontological ambiguity', i.e., of identity/difference and/or identity/change (*P*: 58). In *RS2* (180–1) 'ultimate entities' for science are brought under the umbrella concept of *field(s) of potential*, and it is noted that only their identification, not their existence, 'depends upon the existence of material things in general'. While the concept highlights that being is 'not only just to be able to do, but to be able to become' (*D*: 77, e.a.), it is fundamentally epistemic. Since science suffers from 'cosmic incapacity', i.e., cannot in principle know what lies beyond its farthest frontier and, if it did reach rock-bottom reality, could not be aware of it, it can never know ultimata in any absolute, metaphysical sense ([2], below), and there is 'no reason in principle why there should not be strata of fields {. . .} forever unknown to us' (*RS2*: 181–2, 185).

(2) *Ultimata* in the sense of *ur-stuff* (ABSOLUTE ultimata), the ultimate and essential ingredients of the cosmos. These are referred to as God (the ultim*atum*) or God-stuff (spirit) in *EW* and as the *cosmic envelope/GROUND-STATE* in the philosophy of META-REALITY. Hostettler and Norrie (2003) aver that Bhaskar rejects ultimata in this sense in *Dialectic*, but more plausibly he merely clarifies that his discussion there (76–8) refers to (1) and not to (2). In *RS* he sees no reason for supposing that 'truly ultimate' entities *must* exist, but thinks that, if they do, they must be enduring (the unchanging source of change). In the meta-Reality books, ultimata (2) are conceived as neither 'material' nor 'ideal', in terms of the traditional debate between IDEALISM and MATERIALISM, but, in a conception claimed to transcend this polarity, as *pure* (self-identical, implicitly or 'supra-mentally' conscious) *POTENTIALITY* or dispositionality (fields of possibility) or *ultimate categorial structure* enfolded or involuted in matter, ingredient through and a necessary condition for the whole of being, including human being, where the ground-state of each concretely singular human being is her transcendentally real or

essential/potential self. In cosmotheogenetic terms pure potentiality is the something which was CREATED out of nothing, i.e., its own ultimate ground is the void or complete ABSENCE. In an exponentially expanding and differentiating universe, at least some ultimata (2) must themselves be changing. See also CATEGORY; CAUSALITY; ONTOLOGY; PHILOSOPHY OF RELIGION.

undecidability. See AXIOLOGY; DETERMINATION; LOGIC; VALENCE.

underdeterminacy, underdetermination. See AXIOLOGY; DETERMINATION; FREEDOM; PHILOSOPHY; TOTALITY; TRANSCENDENTAL REALISM.

underlabouring by CR philosophy on behalf of substantive research programmes achieves conceptual clarification and removal of obstacles and objections. It is important to grasp that it accomplishes this, not so much by the methods of analytical philosophy, but by transcendental argument elaborating a philosophical ontology that scientific practices presuppose, which issues in immanent critique of other philosophical accounts of those presuppositions. It is thus a *critical* philosophy of and for science and cognate practices, aspiring in the case of the social sciences to supply the general conceptual schema they currently lack, which is indispensable for the general flourishing of research that would make them effective agents of emancipation (see PHILOSOPHY). See also DCR; DEPTH; EMANCIPATORY AXIOLOGY; INTER-DISCIPLINARITY, ETC.; METHOD; REASON.

Unhappy Consciousness. See BEAUTIFUL SOUL, ETC.

unholy trinity (of IRREALISM). 'The trio constituted by the *EPISTEMIC FALLACY*, *ONTOLOGICAL MONOVALENCE* and *PRIMAL SQUEEZE*. As the epistemic fallacy, mediated by ACTUALISM, determines primal squeeze on empirically controlled theory and natural necessity, the trinity can be seen as a function of a couple. Moreover as ontological monovalence, taken literally, entails the exclusion of ALTERITY (primeval monism) as a PERSPECTIVAL SWITCH on absence, it must be regarded as the primordial failing of Western philosophy' (*DG*: 406). This 'terrible trinity' has structured 'western philosophy up to its contemporary analytical (neo-Humean or neo-Kantian) and postmodernist (post-Nietzschean) forms' (*D*: 199). The pun is intentional – no 'holes–voids–constitutive absences' (*D*: 42 n.) are permitted by these three-in-one dogmas – thus underlining the primacy of ontological monovalence within the trio and its dialectical unity. Elsewhere the metaphor is extended to include 'empirical realism and epistemological subjectivism and irrationalism' as well as the LINGUISTIC FALLACY (the most common form of the epistemic fallacy today) and ontological monovalence in a 'new *unholy tetrapolity*' (*D*: 112) which took shape when FUNDAMENTALISM was finally abandoned and epistemological relativism accepted.

Historically, the unholy trinity is established and occurs on what Bhaskar calls the *Platonic–Aristotelian fault-line*, the consequence of FUNDAMENTALISM or the quest for universal and necessary knowledge in a context where an understanding of the multi-tiered stratification of being was prohibited by ontological monovalence. 'What was required were grounds for the universal *distinct* from the universal concerned and *other* than its instances' (*P*: 5), which neither Platonic Forms nor Aristotelian essences could in principle supply. Philosophy was thus stuck on the fault-line, split off from ontological depth and a concept of natural necessity or ALETHIC TRUTH. Only the HOLY TRINITY of CR – 'the truth and mutual compatibility of ontological realism, epistemological relativism and judgemental rationality' (*MR*: 297) – could heal the split and resolve the myriad PROBLEMS of philosophy to which it gave rise. Bhaskar builds here on Marx's critique of Hegel, the implicit PROLEPTIC rationale of which was in terms of the unholy trinity. See also ARCH OF KNOWLEDGE; HEGEL–MARX CRITIQUE; SOLIPSISTIC, ETC.

unity (f. L. *unus*, one). The quality or condition of forming a whole, as distinct from IDENTITY (f. L. *idem*, the same), the quality or condition of sameness, which is constellationally contained by the concept of unity (*EW*: 5). It expresses the interconnection of the irreducibly distinct elements (e.g., theory and practice, structure and agency) of a DUALITY, CONSTELLATION or TOTALITY in a (normally processual) whole. Any such unity is a *dialectical unity*. Its elements may be antagonistic or opposed (as in the *unity of opposites* in a dialectical CONTRADICTION), or not. In the sociosphere, such a totality 'to be genuine {...} must represent the concrete singularity, and particular differentiations of each individual. The ultimate basis of unity is given by our shared species-being or common humanity and social relationality; and in practice, by our subjugation to the effects of the same or similar sets of oppressive power$_2$ relations' (*DG*: 406). The notion of the unity-in-diversity (or *constellational unity*) of humankind thus brings together all the elements of the concrete UNIVERSAL ↔ SINGULAR. The *dialectic of unity-in-diversity* is the process whereby such unity is achieved. A *constellational unity* is to be distinguished from a *constellational identity*, in that in the latter the elements are contained as 'part of' totality rather than connected.

unity of method. See CRITICAL NATURALISM; NATURALISM.

unity of theory and practice. See DIALECTICAL REASON.

universal, concrete. See CONCRETE UNIVERSAL.

universal whore. Marx's metaphor (*Economic and Philosophical Manuscripts*) for money as the universal medium of exchange of all things, inspired by Shakespeare's reference to it as 'the common whore of mankind' (*Timon of Athens*). Invoked by Bhaskar in relation to HERMENEUTICS to emphasise its equally universal range as medium of communication or intersubjective exchange (*D*: 190).

universalisability pertains, in Kantian ethics, to the *categorical imperative* or the *principle of universality*, which states that you should act only on that principle which you can simultaneously will to be a universal rule of conduct. The universal law invoked here is ACTUALIST or abstract and its principle formal IDENTITY. Universalisability in this sense is false analytical or *abstract universalisability*; it has no regard for the concrete and for specific differences (e.g., in ability), assuming that 'ought' implies, rather than presupposes, 'can' (see also ETHICAL IDEOLOGIES).

For CR 'there are no universal rules' (*RM*: 47), any more than there are universal laws of nature construed actualistically as empirical invariances. (The CR critique of abstractly universalist ethics thus parallels its critique of deductivism [*D*: 113].) There are, rather, universal NEEDS and potentials, rights and FREEDOMS at the level of the real (1M), which, however, are only ever manifest in particular mediated and singularised forms (see HUMAN NATURE), and which actually existing norms may more or less adequately express. Thus CR agrees with the universalists that moral judgements are logically or necessarily universalisable: if you say or imply that someone should φ, then you are yourself committed to φ-ing in materially similar circumstances; this provides both a test for theory–practice CONSISTENCY (hence sincerity) and a

492 universality

criterion for the truth of what you are saying. But CR insists that the universality involved must always be *dialectical*, i.e., transfactual (1M), concrete (3L), ACTIONABLE (4D) and directionally transformative or evolving (2E), siding here more with the particularists or communitarians. The requirement that it be concrete, especially, means oriented to *these* agents in *these* contexts in *these* processes – assertorically, not categorically, imperatival or prescriptive. The CR universal itself, in other words, is a CONCRETE, not an abstract, UNIVERSAL – though distinct, it does not exist separately from the concrete singularities in which it is instantiated and in which it is the moment of unity with members of the same kind. Every concrete singular containing a similar universal element is *dialectically similar*, or its mediated differences are grounded in aspects it shares in common with all others of its kind, constituting it as a unity-in-diversity.

This is the totalising logic of *dialectical universalisability*, which expresses the unity of the CONCRETE UNIVERSAL ↔ SINGULAR:

> One moves from, say, a desire to absent an ill which is functioning as a constraint, to the commitment to absent all dialectically similar ills, and thence to a commitment to absent all ills as such, precisely in virtue of their being {remediable} ills or constraints. This already entails absenting their causes {. . .} and implies a commitment to absenting the absence of the society which will remedy them. (*P*: 147)

This logic – which has 'the causal efficacy of a normative conatus – a tendential drive' (*D*: 293) – and its limits are spelled out in more detail in DIRECTIONALITY and EMANCIPATORY AXIOLOGY.

The *dialectic of universalisability* expresses the SOLIDARITY implied by theory–practice consistent discourse and action. Along with HOLISTIC CAUSALITY and REFLEXIVITY, universalisability is the main substantive principle of 3L dialectics, whose formal principle is TOTALITY. Solidarity implicates, if carried through, the DIALECTIC OF SOCIAL SCIENCE and the DIALECTIC OF MORALITY; i.e., in real geo-historical TRANSITIONS, the logic of universalisability will always be embedded within 'a *dialectic of real geo-historical processes*', i.e. 'a developmental *dialectic of the logic and practice of dialectical universalisability* incorporating *a dialectic of dialectical universalisability and immanent critique*' (*D*: 280, e.a.). Lack of universalisability betokens *theory–practice incompleteness*.

There will be universal rights and freedoms in EUDAIMONIA but they will be historically specific, concretely singularised and open; the norms enshrining them will be the norms of the communities concerned, not those of the abstract universal – expressing (consistently with epistemic relativism and judgemental rationalism) moral ALETHIA as understood by the relevant agents. The great mistake of postmodernism in the ethical field is to fail to distinguish between abstract and dialectical universalisability, which leads it to ditch universality instead of the actualism informing the abstract universality it so justly criticises (see PDM). See also DCR; ETHICS.

universality. See CAUSALITY; CONSTANT CONJUNCTION; COUNTERFACTUAL; DIRECTIONALITY; ETHICS; EXPLANATION; NATURAL KINDS; PDM; TRANSCENDENTAL REALISM; UNIVERSALISABILITY; UNIVERSALS/PARTICULARS.

universals/particulars. Most generally, *particulars* are the ordinary things of the world (as distinct from *singularities*, which are unique particulars), *universals* are properties or abstract entities that can be exemplified by more than one particular at the same time, e.g., the circularity of round objects. Each such object is said to *instantiate* a universal, and the property of circularity is *predicable* of it. *Predicative realism* asserts the existence of universals either independently of particulars (Plato) or as intrinsic to them (Aristotle, CR). In the former case universals are TRANSCENDENT, in the latter IMMANENT. Predicative realism contrasts with *nominalism* (only particulars exist) and *conceptualism* (universals are merely words or concepts).

The universals of most interest to CR and science are the intrinsic STRUCTURES or powers of NATURAL KINDS of things at the level of the real, e.g., the human power of reflexivity, grounded in our common human nature, or the generative mechanisms of a kind of social system. '"Table" and "red" are not real universals in this sense, whereas "gene" and "molecule" are' (*RS2*: 226). In the sociosphere, because the TMSA entails trans-geo-historical continuity (reproduction) as well as social discontinuity (transformation), there may be 'summative social universals' (*SR*: 208), e.g., human NEEDS for sociability, creativity, love and freedom.

Particulars that instantiate real universals are *structurata* (see STRUCTURE). A RESEMBLANCE *theory of universals*, by contrast with realist theory, focuses on likeness at the level of the actual.

Whereas post-structuralism deconstructs the 'binary' oppositionality of universals and particulars, DCR integrates the distinction into the dialectical unity of the CONCRETE UNIVERSAL ↔ SINGULAR as two of its distinct yet interconnected moments (1M universality, 3L mediation).

The problem of universals: do such abstract entities exist, or are they merely words? On the latter assumption, TAXONOMY is arbitrary; on the former, it has a rational basis. The traditional problem of induction may be seen as the problem of the induction of universals conceived actualistically (*P*: 299). See also FIXISM/FLUXISM; REALISM; TOKEN/TYPE.

unthought. See CONSCIOUSNESS; CREATIVITY; META-REALITY.

UPMS, UPS schemas. See CONCRETE UNIVERSAL ↔ SINGULAR.

unseriousness. See DENEGATION.

urban and regional planning (URP) deals with the spatial development of regions (metropolitan as well as rural), cities and neighborhoods, addressing physical, social and economic aspects. The concept of spatial planning is often used as a synonym. Other disciplines, such as urban design and landscape architecture, deal in more detail with a smaller scale of development and with a narrower focus on the physical and visual aspects.

The history of urban planning can be traced at least back to the ancient Sumerian and Babylonian cultures more than 4,500 years ago, where the plans dealt mainly with defence structures, ceremonial edifices and political and cultural buildings contributing to maintain the status quo. Under feudalism, urban planners made the plans for the administrative and market centres for their feudal lords, as well as the residential mansions of the nobility. In the late nineteenth century, improving the health and hygiene conditions in cities became an important part of the rationale for urban planning as a reaction against the risks of epidemics and riots resulting from unhealthy living environments in the industrial cities of laissez-faire capitalism (Kjærsdam 1995).

Contemporary URP represents the efforts of corporate agents (planning agencies and governments at city or regional level) to influence the development of the physical/spatial structures of the territories covered by the plans (notably the built environment and its utilisation). Once established, these structures usually have a considerable durability and have the capability to influence the activities, well-being and economy of those living in or visiting the given areas for a relatively long period. Urban and regional plans envisage the preferred spatial 'answers' in the given contexts to questions such as: For which activities should areas be provided? How large and what kind of areas are needed for the various activities? How should different activities be located in relation to each other? How should the activities be connected? How should the physical environment, in which the activities are established, be designed, in order to attend to both functional, social, aesthetical, economical, and ecological considerations, as well as livability? Which areas should be protected against the construction of buildings or technical infrastructure (Strand 1991)?

In addition to the above substantive issues, URP involves a number of procedural questions, including the organisation of the planning process, the roles of planners vis-à-vis decision-makers, the types of knowledge to be utilised and analyses to be performed, and the instruments for implementation of the plan.

For a plan to be implemented, it needs to be supported by sufficiently powerful agents. The physical/spatial structures of cities and regions are influenced not only by public planning agencies, but also by a variety of other social agents, including private companies, NGOs and individuals. In market societies, the intervention into market processes represented by URP has been justified by the need to promote the collective interests of society; counteract 'externalities' and resolve 'prisoner's dilemma' conditions; and contribute to better information about long-term consequences, and contribute to a better distribution of benefits and burdens than sole market processes can provide (Klosterman 1985). This intervention into market processes, on the one hand, implies a limitation of the freedom of landowners and developers to use areas and construct buildings. On the other hand, such planning is necessary in order to provide the technical and social infrastructure on which further economic development and capital accumulation depend, and living conditions required for the reproduction and attraction of an able and productive labour force. URP thus tends to be resisted by some fractions of capital (notably landowners and the construction industry) while supported by other fractions (notably productive industry and financial capital) (Fogelsong 1996).

Needless to say, URP does not necessarily serve benevolent purposes. Arguably, a main function of planning in capitalist society is to stimulate economic growth, facilitate capital accumulation and contribute to the survival of the prevailing mode of production (Scott and Roweis 1977). The political values on which planning is de facto based often reflect the interests of the power elite rather than underprivileged groups of the population. Planning also often includes top-down strategies limiting the possibilities of local communities to pursue their goals. The content of the plans is a topic of struggle between different population groups with different class interests, cultural backgrounds and ethical values, and the actual outcome of URP reflects the power relationships between such groups.

URP is, however, a potential emancipatory tool which, under social conditions allowing for it, can be used to improve the living conditions of the less privileged and to contribute to an ecologically more sustainable development of cities and regions. Although vested interests and market mechanisms put important limits on the scope for welfare- and environmentally oriented planning in capitalist societies, the political priorities of state, regional and city governments and the professional ideas of planners also matter, and actual urban and regional development differs considerably between cities across and within national contexts (Klosterman 1985).

The fields of knowledge planners draw on when creating urban and regional plans include disciplines within natural and, social science as well as humanities. This calls for a non-reductionist approach, where the perspectives of different disciplines are acknowledged as addressing different levels or aspects of reality instead of being considered as mutually exclusive. Moreover, URP takes place in specific geographical, political and cultural contexts, and the knowledge used in such planning must therefore be adapted to the actual context. Cities are complex systems where a multitude of causal mechanisms at different strata of reality are at work. In such open systems, the possibility of predicting the outcome of a proposed instrument is therefore limited, and positivist conceptions of causality based on universal event regularities are highly inappropriate. At the same time, URP relies heavily on the possibility of predicting the consequences of alternative solutions with at least some degree of precision. Arguably, the spatial structures of cities and urban regions and their relationships with social life and human activities make up 'quasi-closed' systems where the scope for prediction of outcomes of a proposed intervention is clearly lower than in the closed systems of the experiments of the natural sciences, but nevertheless higher than in completely open systems. Anticipation of consequences, which is indispensable in planning, is therefore possible and recommendable, although fallible. The predictions that social research may enable us to make in URP are of the crude 'rule of thumb' type, limited to *qualitative* impact assessments, with modest statements about the *directions* of the influences and maybe their *orders of magnitude* (Næss 2004). The non-exact nature of our knowledge about most of the topics relevant to URP, as well as the continual processes of social change, imply that it is impossible to make exact, quantitative estimates of the various social, environmental and economic impacts of, e.g., a proposed land use or infrastructure development alternative.

PETTER NÆSS

urban studies. Urban studies is a field which concerns the many interrelated social, political and economic processes which generate and mediate a particular spatial setting. It is a field well suited to MULTI-DISCIPLINARY perspectives integrating concepts and methodologies from (traditionally) applied sociology, geography, economics, political economy and cultural studies. Key debates in the field concern the nature of globalisation and its effect on urban development, the changing nature of urban governance, the process of social exclusion, the nature of the welfare state and its relationship to housing and urban policy, citizenship, gender, power and urban democracy, and last, but by no means least, the role of the state in the development of cities.

Given the open, complex and dynamic nature of the urban field, CR is well suited as an ontological approach. It can serve as the ontological springboard for an in-depth, longitudinal case-study strategy which aims to explain the generative causes of urban problems. Such an approach forces researchers thoroughly to conceptualise their object of study and abstract key causal mechanisms underlying urban phenomena over time that influence urban outcomes such as social exclusion, gentrification, housing affordability, urban governance and development.

CR attempts to inspire conceptualisation of urban phenomena through a process of ABDUCTION and RETRODUCTION. Realist abduction involves the contestable postulation of ideas and concepts relevant to the object of study and their use in interpreting and recontextualising phenomena, in order to produce a new description for analysis. First, researchers should aim to discover the relevant structures and mechanisms that explain observable phenomena and regularities. Critical analysis of everyday conceptions of reality should lead to the postulation of these (perhaps unacknowledged) structures and mechanisms (T. Lawson 1997: 196). Examining actual empirical consequences should test these ideas and concepts of causal mechanisms. Finally, the model should be continuously tested and revised until it 'fits' these consequences.

CR has been used to generate new interpretations of urban phenomena, such as differences in international housing policy (J. Lawson 2006), the causes of homelessness (Fitzpatrick 2005), predictability in urban planning (Næss 2004), the spatial form of cities (Terhorst and van de Ven 1997), and differences in housing policy between nations and regions (Dickens et. al., 1985).

One recent example is a study comparing the development of different housing 'solutions': sprawling Australian home ownership and compact Dutch social rental housing have evolved since the nineteenth century (Lawson 2006). It carefully justifies why explanation for different housing solutions lies beyond the description of key events, policies and housing outcomes. It argues the case for a new ontological perspective in housing studies – that of CR, which directs attention towards the *causal tendencies of complex, open and structured housing phenomena*. From this perspective an argument is made for the definition of key social arrangements that promote different housing solutions, to be abstracted from concrete historical case-study research. These dynamic arrangements concern the property rights, circuit of savings and investment as well as labour and welfare relations that develop differently over time and space and are considered to influence alternative pathways in housing history.

In a recent journal article, Fitzpatrick (2005) attempts to re-examine the causes of homelessness, taking ontological inspiration from CR concepts such as stratification, necessity, contingency, counteracting tendencies and causal mechanisms. It is claimed that such an approach promotes careful conceptualisation and will generate a more sensitive explanation and description of homelessness.

Other researchers, such as Næss (2004, 2006a) have used CR ontology as a springboard for critique in urban studies. Næss's constructive, transformative intervention in the field of transport planning focuses on the ontological and epistemological shortcomings of prediction and cost-benefit analysis.

These are merely a few examples of how CR has entered the different realms of urban studies. There are of course many more applications and potential areas for contribution, critique and renewal.

JULIE LAWSON

utopianism. See CONCRETE UTOPIANISM.

V

valence, valency (f. L. *valere*, to be well or strong). A number of important CR concepts orbit around *valence*. As one of 'the myriad figures that oppositionality in philosophy takes' (*P.* 242), *ambivalence* (L. *ambi-*, round, round about) is proneness to both of two or more incompatible positions. Thus 'reflexivity is an ambivalent world-historical phenomenon manifest, on the one hand, as a sense of geo-historicity and, on the other, in the growth of surveillance techniques' (*P.* 79). It is thus distinct from an internal CONTRADICTION in that the one result is not necessarily attained or attainable at the expense of another; and from HOMONYMY (e.g., the process/product homonymy of negation) in that the two or more related things expressed by the conept are incompatible. It is closely related to DUPLICITY.

(*Ontological*) *monovalence* is Bhaskar's term of art for the doctrine that being is purely positive and present. It is well caught by Balzac in *Ursule Mirouët*: 'To modern philosophy, the void does not exist. If there were ten feet of void, the world would cave in! Especially to {mechanical} materialists, the world is full, everything hangs together, is causally connected and functions like a machine.' As Beckett commented in a triple negative a century later with ironic social intent: 'There is no lack of void' (*Waiting for Godot*). The origins of monovalence lie, in the Western tradition, in Parmenidean MONISM and Plato's account of negation and change in terms of difference. A species of MONISM, it is even more fundamental to IRREALISM than the epistemic fallacy (see UNHOLY TRINITY), acting 'ideologically to screen the epistemological and ontological contingency of being and to sequester existential questions generally, as ontological ACTUALISM sequesters "essential" ones. The result is the doubly dogmatically reinforced positivisation of knowledge, and ETERNISATION of the status quo' (*DG*: 401). Monovalence is thus a manifestation of 'ALIENATION, REIFICATION and the fear of change' (*P.* 139). It puts a stop to history on behalf of the master-classes (see ENDISM). An error at 2E blocking change and an adequate conception of space–time via the elimination of absence, it also generates both 1M illicit FUSION (where ALTERITY – a valid perspectival switch on absence – is lost) and 3L illicit FISSION (where TOTALITY is rendered unthinkable, and alienation legitimised), propagating the epistemic fallacy, actualism and PRIMAL SQUEEZE at 1M, ontological extensionalism at 3L and the de-agentification of reality at 4D, unifying the PROBLEMS OF PHILOSOPHY and their metacritique.

Ontological monovalence is refuted by the transcendental deduction of the category of ABSENCE or real negation. This demonstrates that there could be no positive without negative existence, issuing minimally in *ontological bivalence*. However, several further

498 value, values

considerations suggest *ontological polyvalence*. (1) At points of TRANSITION and in regard to some ULTIMATA, we have to speak of the coincidence of IDENTITY and CHANGE and of identity and difference (ALTERITY), i.e., admit a fundamental ambiguity or ambivalence into our ontology in relation to absence and presence. (2) Epistemologically, such existential questions may be *undecidable*, presenting the agent with a *problematic AXIOLOGICAL choice situation*, i.e., where the agent must make a decision (act) but there are no rational grounds for it. *Non-valence*, i.e., questioning whether it is a problematic axiological choice situation after all, then recommends itself as a strategy pending a redescription of the situation in the light of deeper understanding. The (formal logical) *law* or principle *of bivalence* says that there are only two truth values, 'true' and 'false', and that every proposition has only one; cf. the law of excluded middle, with which it should not be confused. See also PRINCIPLE OF PLENITUDE.

value, values. See AXIOLOGY; CRITICAL NATURALISM; CRITICAL REALISM; DCR; EMANCIPATORY AXIOLOGY; EXPLANATORY CRITIQUE; ETHICS; HERMENEUTICS; NATURALISM; NATURALISTIC, ETC.; FALLACY; PRINCIPLE OF PLENITUDE.

value-form. See CRITIQUE; ONT/DE-ONT.

value freedom. See CRITICAL NATURALISM; EXPLANATORY CRITIQUE.

vehicular thrownness. See THROWNNESS.

verification, verifiability, verificationism. See PHILOSOPHY OF RELIGION; PHILOSOPHY OF SCIENCE; REALISM.

verification-transcendent truth(s) or **reality**, or **recognition-transcendent truth**, or **objective truth**. See ALETHIA; TRANSCENDENTAL REALISM.

verificationist fallacy. Confuses scientific descriptions and the processes of arriving at them with what they are about (*PN3*: 78). It is thus a form of the EPISTEMIC FALLACY.

verstehen. See HERMENEUTICS.

vertical realism. See DEPTH.

virtuality, adj. virtual (L. *virtus*, f. *vir*, man: manly excellence, and so generally excellence, capacity, worth, virtue, also valour, courage) has a range of subtly different meanings in everyday use, evidently deriving from understandings that outwardly different things may possess capacities or powers productive of similar effects, thence similar *seeming* effects (as in 'the virtual heat' of wine – Francis Bacon), and centring on the notion of 'not really, but just as if'. This is related to its meaning in optics, where, to put it non-technically, it refers to mirror and other images that apparently have spatial depth but are in reality flat. C. S. Peirce's attempt (1902) at a formal definition – 'a virtual x (where x is a common noun) is something, not an x, which has the efficiency (*virtus*) of an x' (cit. Skagestad 2000) – fits cases such as the virtual memory of a computer, which is not RAM but functions with the same effect, but not, say, that of the virtuality of a film, the causal efficacy of which can hardly be equated with that of the reality it depicts, or which, if it is entirely fictive, cannot be said to have the efficacy of something that is not itself. Most recent discussion of virtuality and *virtual reality* has been clouded by a pall of ACTUALISM.

Notwithstanding the hype, virtual reality has always been with human being – in the NEGATIVE PRESENCES of memory and the positive absences of ideas, mimesis, and semiotic systems, in the teeming products of imagination as well as of fantasy. Indeed, it existed prior to human being in the form of *relations*, which on the Bhaskarian account are virtual or 'ideal' because, while requiring material things as their relata, they have no material substance and yet are sui generis real (causally efficacious) (cf. *RS2*: 237; *PN3*: 50; *SR*: 152; *D*: 164); and probably also in semiotic systems of non-human organisms, if not of the whole physical universe as such, as Peirce famously claimed (Job 2004: 155). Above all a 1M figure of EMERGENCE, as well as a 3L figure of MEDIATION (see also MEDIUM THEORY), in the sociosphere the virtual is best-conceived, not as contrasting with reality, but as constituting a sui generis real emergent stratum of reality with its own irreducible causal powers. (It is a postmodernist myth that within the virtual the distinction between appearance and essence, phenomenal forms and underlying reality, is lost, though virtuality may produce the illusion of depthlessness as well as empty and fragmented selves.) After the dawn of human geo-history, the next quantum leaps in the emergence of virtual reality in the sociosphere came with writing and printing and then with modern electronics. What is new today is not virtual reality as such but above all (1) the augmentation of human powers by audio-visual-kinetic and computer technologies systematically to produce and especially reproduce the virtual and deploy it to new ends or old, whether emancipatory or disemancipatory, with enhanced effect; and (2) the resonance effects of *virtualisation* with increasing COMMODIFICATION, REIFICATION and FETISHISATION, perhaps most directly exemplified in money as the abstract universal measure of value, but also strikingly evident in the disembedding of space from place in the network and the *mediatisation* (see MEDIATE) of social reality, and consequent experience of 'weightlessness', 'de-materialisation', or worldlessness (Dean 2003, for whom [post/modern] virtuality is 'an emergent property of a combination of commodity fetishism and electronics' [115]). Virtualisation in this sense is similar to the concept of 'the spectacle' (Debord 1967) as 'the submission of ever more facets of human sociality to the "deadly solicitations" of the market' (Stallanbrass 2006, citing Retort 2005).

Like philosophy itself, post/modern virtual reality is thus Janus-faced. It may have an IDEOLOGICAL function in relation to other levels of the sociosphere, inverting or otherwise occluding them, or blurring the distinction between virtual and non-virtual reality – in which capacity it is closely affined to the figures of DUPLICITY, supplementarity and graft, with dialectics that are predominantly 1M; or it may present profound insights into them or facilitate the generation of insights via the extension of our powers of communication, hence also of emancipatory praxis.

A virtual reality in which the illusion of non-virtual reality is so complete that it is difficult or impossible for the agent experiencing it to tell the difference between the simulation and the simulated (e.g., by an aeroplane flight simulator) is a *hyperreality*, and *hyperrealisation* is the process of coming to be of hyperrealities. However, the postmodernist notion (Baudrillard 1994) that the hyperreal now encompasses, or ever will, the whole of reality, not just an emergent stratum thereof, i.e., that the very distinction between virtual and non-virtual reality has been effaced from the geo-historical process along with 'the referent', such that we inhabit a giant simulacrum, is ideological in the strict sense: a paradigm of illicit, de-agentifying FUSION, productive of a false, destratified totality, and a characteristic inversion or involution (of reality within virtual reality), just as prefigured in the Hegelian figure of the Unhappy Consciousness whose postmodern descendant fantasises 'a hyperreal world where slaves become masters or media stars' (*D*: 381). To the fantastic kingdom of the hyperreal, as we know – to borrow a phrase from another context which inverts the inversion – 'many are called, but few are chosen'.

virtue theory, virtues. See ETHICS; EUDAIMONIA; PHRONESIS; POLITICS; SELF.

void (n.). An ABSENCE of anything. A *metacritical void* is the absence in some theory of a transcendentally necessary concept, demonstrated by METACRITIQUE, e.g., the absence of the concept of absence. See also VALENCE (ontological monovalence).

volition (will) (f. L. *volere*, to wish, will). Included by Bhaskar in the conative domain of the springs of ACTION (see also DESIRE; WANT), with the proviso that 'it is arguable that the will, comprising the unity of a well-functioning ego, should be accorded a distinct place' (*SR*: 175 n91). Its construal in *Plato Etc.* (66) as (embodied) 'interests, needs, desires', mediated by practical and theoretical reason, accords with this. Whether regarded as constituting a distinct domain or not, the CR account of volition is causal: when 'my arm went up' is subtracted from 'I raised my arm', the remainder is 'I willed or intended it' (see MIND). See also VOLUNTARISM.

voluntarism (f. L. *voluntas*, will) has a range of uses conjugating around the notion of unconstrained will. *Ethical voluntarism* (e.g., emotivism, decisionism; see ETHICAL IDEOLOGIES): standards of right and wrong are chosen by us. *Doxastic voluntarism*: we choose what to believe, our beliefs about the world are free creations of the mind. *Theological voluntarism*: (1) God's will determines what is morally good or bad; or (2) religious belief necessarily involves a substantial amount of choice, i.e., is underdetermined by evidence. *Historical voluntarism*: history happens, or can happen, more or less as we choose it to happen. *Metaphysical voluntarism* or *voluntarist realism*: the fundamental organising principle of the cosmos is a meaningless striving for survival or will to power (Schopenhauer, Nietzsche). Voluntarist realism is an anthroporealist relative of EMPIRICAL REALISM. In moral and legal philosophy, voluntarism is the view that a person is bound by obligations only if they are voluntarily undertaken. In social, legal, political, etc., theory it is the view that legal and other social forms are, or have their origin in, the will of a sovereign legislator or of the people.

In CR, voluntarism normally connotes the idealist views that, contrary to the TMSA, (1) theory has a more or less immediate efficacy in practice, i.e., *theoreticism* (see RATIONALISM), e.g., 'the voluntaristic attempt to build socialism in one country' (*D*: 350); and its converse, (2), *practicism* – the collective decisions of people can ignore or reverse objective geo-hitorical tendencies and obstacles (the Jacobins, Lenin, Mao, Žižek); (3) people make or create the social or their FORMS OF LIFE, which are just an expression of their beliefs and understandings; or (4) the counterpart of (3) within the empiricist tradition, people change the world, including the natural world and its laws, along with their theories (*voluntarist super-idealism* or ontological relativism, e.g., some of Kuhn's formulations). On (3) especially, the actions of people are not to be explicated by causal laws of any kind, but rather HERMENEUTICALLY, by interpretive understanding or by re-thinking their thoughts. It thus entrains either *methodological individualism* or *collectivism*. Because it presupposes a noumenal realm of freedom which has no applicability in the physical world, where it normally assumes regularity DETERMINISM to hold sway, such voluntarism issues in a mind/body DUALISM which disembodies human agency, ceding the body, as it were, to positivism. Thus at 4D its dualistic disembodiment constitutes a de-agentifying dialectical couple with the REIFICATION of physicalist reductionism, which is sublated by the CR concept of *agentive AGENCY*. See also CONVENTIONALISM; HUMANISM; SOCIAL CONSTRUCTIONISM.

voluntarist realism. See EMPIRICAL REALISM; VOLUNTARISM.

W

wage-form. See ALIENATION; CRITIQUE; DIALECTICAL ARGUMENT.
wage-slavery. See MASTER–SLAVE.
wants are rationally assessable, causally efficacious beliefs, dependent on the conative component of ACTION (DESIRE + an AGONISTIC element ['motivation/drive']). To be construed as (exercised) dispositions (TENDENCIES$_2$) to act (*PN3*: 95; *D*: 166, 202). The cognitive and conative bases of action constitute the core components of our *reasons* for action. Their relationship is set out in Table 43.

On this model, INTENTIONALITY is irreducible: people are active by nature, or always already active (see AXIOLOGY). Both desire and the beliefs it transforms into wants or intentions (which may in turn transform it) are generated in a continual stream in the practical conduct of our lives. 'One does what one wants to (or intends) unless prevented. This is a necessary truth. And no further explanation of action *as such* is required' (*PN3*: 96, e.a.). But what one intends, whether one knows it, whether it is in one's power and whether it will actually come to pass, and, if so, what its effects will be, are all contingent, such that EXPLANATION of actions by *reasons* (beliefs and desires/wants) ('rational explanation') will always require supplementation by sociological analysis (see also HERMENEUTICS; PRIMACY), and explanation by stated

Table 44 The relation between the cognitive and conative components of reasons for action

	Cognitive component (belief)	*Conative component (want or intention) (desire + motivation/drive → belief → want)*	*Action*
Modes of tendency	possessed (requires want or desire to be activated)	exercised (automatically manifested in action unless prevented)	realised
Logical asymmetry (depending upon ontological difference)	does not logically presuppose want, desire	logically presupposes belief (e.g., that the action will produce the desired effect)	
Ontological/ axiological difference	once exercised, straightaway manifested in action	must wait on belief	

Source PN3: 95–7

reasons supplementation by depth-psychological analysis to determine whether it is a rationalisation.

Wants constellationally contain INTERESTS and NEEDS on the Bhaskarian account. Needs are (relatively or absolutely) indispensable for survival or flourishing, whereas wants, like perceived as distinct from real interests, may be inimical to it. While it is a necessary truth that people act on their wants, it is not so in regard to their needs and real interests. A *dialectic of wants and needs*, whereby unneeded or unattainable wants are dispensed with, and needs hitherto not cognised or experienced, e.g., for freedom, come to be known and actively wanted, frustrated needs satisfied, and unwanted needs (e.g., for certain forms of labour) diminished, is at the heart of all emancipatory LEARNING PROCESSES, a vital strand in the DIALECTIC OF THE EDUCATION OF DESIRE for freedom and the dialectics of interests.

warranted assertibility. See ALETHIA; INSTRUMENTALISM; REALISM; TRUTH TETRAPOLITY.

weak actualism. See ACTUALISM; RESEMBLANCE.

Whorf hypothesis. See SAPIR–WHORF HYPOTHESIS.

will. See DESIRE; VOLITION; WANTS.

world-historical problem. See PROBLEM; PROBLEM OF AGENCY.

worldlessness. See VIRTUALITY.

world-line(s), world line(s). 'Paths in SPACE–TIME representing the dynamical histories of moving particles' (*CDP*: 867), also termed RHYTHMICS by Bhaskar. Thus the world-line of a power is its rhythmic exercise or GEO-HISTORY (*D*: 252); that of a person is the rhythmic exercise of her powers; her epistemic world-line is the trajectory of her dialectical LEARNING PROCESS, and so on (within her world-line there will be many 'daily space–time paths') (*P*: 98). The world itself as a system (spaceship earth) has a world-line. The concept is intended to marry the principles of non-anthropicity (ontological realism) and epistemic and ethical relativity. Thus spatio-temporal concepts such as 'past', 'present' and 'future', 'real negation' qua absence, and 'here' and 'elsewhere' apply 'for any space–time region for any observer or any world-line from any reference frame', i.e., they are 'not anthropocentric, although naturally any use of [them] will be relativistic'. Because we are located in space–time, EPISTEMIC RELATIVITY 'applies in principle to all our discourse {...} {a}nd the concept of the world-line may be used as a metaphor to illuminate phenomena of epistemic or ethical relativity generally' (*D*: 81 n.). By drawing attention to the fact that we are a part of the reality we describe, etc., it underlines the irreducibility of relativity. Given the RELATIONALITY of being, world-lines must be thought of as criss-crossing and interconnecting, thereby also sharing properties in common. *Divergent* world-lines are EMERGENT TOTALITIES.

XYZ

Zeno's paradoxes of motion purport 'to show that because at any moment of time a thing in motion can only be at one place, motion is impossible. This presupposes a PUNCTUALIST view of time and an aggregative view of TRANSITIONS' (*P*: 11).

zero- or **base-level** (adj.) is a concept necessary for consistent description and measurement of a RELATIONAL and emergent/divergent world (see also HOLY TRINITY; PERSPECTIVE). In the case of divergent WORLD-LINES and RHYTHMICS, or of overlapping/intersecting spatio-temporalities (e.g., the CO-PRESENCE of the past in the present in a building or a rock containing layers of different ages), it is necessary to agree on the use of some *SPACE–TIME* as a yardstick to get an empirical purchase on them. This is not necessarily basic in a physical sense. In Bhaskarian usage, *zero-level action* contains inaction as well as action; *zero-level being*, absence as well as presence (*D*: 45, 47, 140, 163).

zyxa. Last, and first 'optimism of the will and *realism*, informed by concrete utopianism, not pessimism, of the intellect' (*P*: 215), such that freedom is won.

Works by Roy Bhaskar

Note. Although it includes some items not cited in *DCR*, this list is incomplete. The acronyms used to refer to books by Bhaskar appear in square brackets. They are listed alphabetically in 'How to Use This Book', p. 1. *RS* and *PN* without the number of an edition signify any edition of *A Realist Theory of Science* and *The Possibility of Naturalism*, respectively. The full titles of edited collections in which Bhaskar has published may be found in the bibliography of 'Other Works Referred to in the Text', below.

1975a   'Feyerabend and Bachelard: two philosophies of science', *New Left Review* 94.
1975b   'Forms of realism', *Philosophica* 15 (1): 99–127.
1975c   *A Realist Theory of Science*, 1st edn [*RS1*], Leeds: Leeds Books.
1978   *A Realist Theory of Science*, 2nd edn [*RS2*], Brighton: Harvester/Atlantic Highlands, NJ: Humanities.
1979a   *The Possibility of Naturalism: A Philosophical Critique of the Contemporary Human Sciences*, 1st edn [*PN1*], Brighton: Harvester.
1979b   'On the possibility of social scientific knowledge and the limits of naturalism', in J. Mepham and D. Hillel-Ruben (eds), *Issues in Marxist Philosophy*, Vol. III, Oxford: Basil Blackwell; reprinted from *Journal for the Theory of Social Behaviour* 8 (1).
1980   'Scientific explanation and human emancipation', *Radical Philosophy* 26: 16–28; reprinted in *Reclaiming Reality* (1989).
1981a   'The consequences of socio-evolutionary concepts for naturalism in sociology: commentaries on Harré and Toulmin', in U. J. Jensen and R. Harré (eds), *The Philosophy of Evolution*, Brighton: Harvester.
1981b   Entries on Aristotle's theory of causes, determinism, dialectic, empiricism, epistemology, experiment, idealism, Kant's theory of knowledge, laws, materialism, matter and form, metaphysics, models, naturalism, ontology, open texture, Plato's theory of forms, positivism, pragmatism, prediction, rationalism, realism, simplicity, tacit knowledge in B. Bynum and R. Porter (eds), *Dictionary of the History of Science*, London: Macmillan.
1982   'Emergence, explanation, and emancipation', in P. F. Secord (ed.) *Explaining Human Behaviour, Consciousness, Human Action and Social Structure*, Beverly Hills, Calif./London/New Delhi: Sage.
1983a   'Beef, structure and place', *Journal for the Theory of Social Behaviour* 13 (1).

1983b	Entries on contradiction, determinism, dialectics (reprinted with corrections in 1989b), empiricism, idealism, theory of knowledge (reprinted with corrections in 1989b), materialism (reprinted with corrections in 1989b), realism, science, and truth in T. Bottomore (ed.) *A Dictionary of Marxist Thought*, Oxford: Blackwell. (2nd edn 1991.)
1983c	Entries on dialectic and ideology in R. Harré and R. Lamb (eds), *The Encyclopaedic Dictionary of Psychology*, Oxford: Blackwell.
1986	*Scientific Realism and Human Emancipation* [*SR*], London: Verso.
1989a	*The Possibility of Naturalism: A Philosophical Critique of the Contemporary Human Sciences*, 2nd edn, with a postscript [*PN2*], Hemel Hempstead: Harvester Wheatsheaf.
1989b	*Reclaiming Reality* [*R*], London/New York: Verso.
1990	(ed.), *Harré and His Critics: Essays in Honour of Rom Harré with His Commentary on Them*, Oxford: Blackwell.
1991a	*Philosophy and the Idea of Freedom* [*IF*], Oxford/Cambridge, Mass.: Blackwell.
1991b	'Social theory and moral philosophy', appendix 1 in 1991a.
1991c	'Marxist philosophy from Marx to Althusser', appendix 2 in 1991a.
1993a	*Dialectic: The Pulse of Freedom* [*D*], London/New York: Verso.
1993b	Entries on determinism, dialectic, empiricism, theory of knowledge, materialism, model, naturalism, ontology, paradigm, philosophy of science, philosophy of social science, realism, truth in T. Bottomore and W. Outhwaite (eds), *The Blackwell Dictionary of Modern Social Thought*.
1994	*Plato Etc.: The Problems of Philosophy and Their Resolution* [*P*], London/New York: Verso.
1997a	'On the ontological status of ideas', *Journal for the Theory of Social Behaviour* 27 (2/3): 135–47.
1997b	*A Realist Theory of Science* [*RS3*], 3rd edn, London/New York: Verso.
1998a	'General introduction', in M. S. Archer et al. (eds), *Critical Realism: Essential Readings* [*ER*].
1998b	(with Tony Lawson) 'Introduction: basic texts and developments', in M. S. Archer et al. (eds), *Critical Realism: Essential Readings*, Part I [*ER*].
1998c	(with Andrew Collier) 'Introduction: explanatory critiques', in M. S. Archer et al. (eds), *Critical Realism: Essential Readings*, Part III [*ER*].
1998d	(with Alan Norrie) 'Introduction: dialectic and dialectical critical realism', in M. S. Archer et al. (eds), *Critical Realism: Essential Readings*, Part IV [*ER*].
1998e	*The Possibility of Naturalism: A Philosophical Critique of the Contemporary Human Sciences*, 3rd edn [*PN3*], London/New York: Routledge.
1998f	(with Ernesto Laclau) 'Discourse theory vs. critical realism', *Alethia* 1 (2): 9–14; reprinted in his 2002a and in A. Brown et al. (eds) (2002), *Critical Realism and Marxism*.
1999	(with Chris Norris) 'Interview', *The Philosophers' Magazine Online* (http://www.philosophers.co.uk/current/bhaskar.htm), also available in the Bhaskar Archive: http://www.raggedclaws.com/criticalrealism/archive/rbhaskar_rbi.html; excerpt published in *The Philosopher's Magazine* 8, Autumn 1999.
2000a	*From East to West: Odyssey of a Soul* [*EW*], London/New York: Routledge.

506 Works by Roy Bhaskar

2000b 'Introducing transcendental dialectical critical realism', *Alethia* 3 (1): 15–21; reprinted in 2002a.

2001 'How to change reality: story vs. structure – a debate between Rom Harré and Roy Bhaskar', in J. López and G. Potter (eds), *After Postmodernism*; reprinted in 2002a.

2002a *From Science to Emancipation: Alienation and Enlightenment* [*SE*], New Delhi: Thousand Oaks/London: Sage.

2002b *Reflections on Meta-Reality: Transcendence, Emancipation and Everyday Life* [*RM*], New Delhi, Thousand Oaks/London: Sage.

2002c *The Philosophy of Meta-Reality*, Vol. I, *Meta-Reality: Creativity, Love and Freedom* [*MR*], New Delhi, Thousand Oaks/London: Sage.

2002d (with Mervyn Hartwig) 'The philosophy of meta-reality, part 1: identity, spirituality, system' (interview), *Journal of Critical Realism* (incorporating *Alethia*), 5 (1): 21–34; reprinted in 2002a.

2002e (with Mervyn Hartwig) 'The philosophy of meta-reality, part 2: agency, perfectibility, novelty' (interview), *Journal of Critical Realism* (new series) 1 (1): 67–94.

2003a Entries on determinism, dialectic, empiricism, theory of knowledge, materialism, model, naturalism, ontology, paradigm, philosophy of science, philosophy of social science, realism, truth in W. Outhwaite (ed.), *The Blackwell Dictionary of Twentieth Century Social Thought*, 2nd edn.

2003b (with Alex Callinicos) 'Marxism and critical realism: a debate', *Journal of Critical Realism* 1 (2): 89–114.

2006 (with Berth Danermark) 'Metatheory, interdisciplinary and disability research: a critical realist perspective', *Scandinavian Journal of Disability Research*, forthcoming.

Bibliography

Ackroyd, S. and Fleetwood, S. (eds) (2000), *Realist Perspectives on Management and Organisations*, London/New York: Routledge.
Adler, E. (1997), 'Seizing the middle ground: constructivism in world politics', *European Journal of International Relations* 3: 319–63.
Adorno, T. W. (1964/1989), 'Something's missing: a discussion between Ernst Bloch and Theodor W. Adorno on the contradictions of utopian longing', in E. Bloch 1930–1972.
Adorno, T. W. (1966/1990), *Negative Dialectics*, London: Routledge.
Adorno, T. W. (1978), 'Subject-object', in A. Arato and G. Gebhardt (eds), *The Essential Frankfurt School Reader*, Oxford: Blackwell.
Adorno, T. W. and Horkheimer, T. (1979), *Dialectic of Enlightenment*, trans. J. Cumming, London: Verso.
Agar, J. (2005), 'Before critical realism: Kantian empirical metaphysics', *New Formations* 56: 27–39.
Alker, H. R. (1981a), 'Dialectical foundations of global disparities', *International Studies Quarterly* 25 (1): 9–98.
Alker, H. R. (1981b), 'From political cybernetics to global modelling', in R. L. Merritt and B. M. Russett (eds), *From National Development to Global Community*, London: Allen & Unwin.
Alker, H. R. (1982), 'Logic, dialectics, politics: some recent controversies', in H. R. Alker Jr. (ed.), *Dialectical Logic for the Political Sciences*, Amsterdam: Rodopi.
Allen, T. and Thomas, A. (2000), *Poverty and Development into the Twenty-First Century*, Oxford: The Open University/Oxford University Press.
Alston, W. P. (1991), *Perceiving God: The Epistemology of Religious Experience*, Ithaca, NY: Cornell University Press.
Alston, W. P. (1996), *A Realist Conception of Truth*, Ithaca, NY: Cornell University Press.
Althusser, L. (1969), *For Marx*, London: Verso.
Althusser, L. and Balibar, E. (1970), *Reading Capital*, London: New Left Books.
Anastas, J. W. and Macdonald, M. L. (1994), *Research Design for Social Work and the Human Services*, New York: Lexington.
Anderson, J. (1962), *Studies in Empirical Philosophy*, Sydney: Angus & Robertson.
Archer, M. S. (1979), *Social Origins of Educational Systems*, London: Sage.
Archer, M. S. (1982), 'Morphogenesis versus structuration: on combining structure and action', *British Journal of Sociology* 33 (4): 455–83.
Archer, M. S. (1985), 'Structuration versus morphogenesis', in S. N. Eisenstadt and H. J. Helle (eds), *Macro-Sociological Theory*, London/Beverly Hills, Calif.: Sage.
Archer, M. S. (1988/1996), *Culture and Agency: The Place of Culture in Social Theory*, Cambridge: Cambridge University Press.
Archer, M. S. (1990), 'Human agency and social structure: a critique of Giddens', in J. Clark, C. Modgil and S. Modgil (eds), *Anthony Giddens: Consensus and Controversy*, London: Falmer Press, 1990.

Archer, M. S. (1995), *Realist Social Theory: The Morphogenetic Approach*, Cambridge: Cambridge University Press.

Archer, M. S. (1996), 'Social integration and system integration: developing the distinction', *Sociology* 30 (4): 679–99.

Archer, M. S. (1998a), 'The dubious guarantees of social science: a reply to Immanuel Wallerstein', *International Sociology* 13 (1): 5–17.

Archer, M. S. (1998b), 'Introduction: realism in the social sciences', in M. S. Archer et al. (eds), *Critical Realism: Essential Readings*.

Archer, M. S. (2000), *Being Human: The Problem of Agency*, Cambridge: Cambridge University Press.

Archer, M. S. (2002), 'Realism and the problem of agency', *Journal of Critical Realism* (incorporating *Alethia*), 5 (1): 2–10.

Archer, M. S. (2003a), 'The private life of the social agent: what difference does it make?', in J. Cruickshank (ed.), *Critical Realism*.

Archer, M. S. (2003b), *Structure, Agency and the Internal Conversation*, Cambridge: Cambridge University Press.

Archer, M. S., Bhaskar, R., Collier, A., Lawson, T. and Norrie, A. (eds) (1998) *Critical Realism: Essential Readings* [*ER*], London: Routledge.

Archer, M. S., Collier, A. and Porpora, D. (2004), *Transcendence, Critical Realism and God*, London/New York: Routledge.

Archer, M. S. and Outhwaite, W. (eds), (2004), *Defending Objectivity: Essays in Honour of Andrew Collier*, London: Routledge.

Archer, M. S., Sharp, R., Stones, R. and Woodiwiss, T. (1999), 'Critical realism and research methodology', *Alethia*, 2 (1): 12–16.

Archer, M. S. and Tritter, J. (eds) (2000), *Rational Choice Theory: Resisting Colonisation*, London/New York: Routledge.

Aristotle (1998), *The Metaphysics*, London: Penguin.

Armstrong, D. M. (1983), *What Is a Law of Nature?*, Cambridge: Cambridge University Press.

Arthur, C. J. (1998), 'Systematic dialectics', *Science and Society* 62 (3): 447–59.

Arthur, C. J. (2002), *The New Dialectic and Marx's 'Capital'*, Leiden: Brill Academic Press.

Arthur, C. J. (2003), review of *Towards an Unknown Marx: A Commentary on the Manuscripts of 1861–3*, by Enrique Dussel, Yolanda Angulo (trans.), Fred Moseley (ed.), London: Routledge (2001), in *Historical Materialism* 11 (2): 247–63.

Arthur, C. J. (2006), personal communication.

Ashby, W. R. (1956), *An Introduction to Cybernetics*, London: Chapman & Hall.

Ashley, R. K. (1981), 'Political realism and human interests', *International Studies Quarterly* 25 (2): 204–36.

Ashley, R. K. (1984), 'The poverty of neorealism', *International Organization* 38 (2): 225–86.

Ashley, R. K. (1987), 'The geopolitics of geopolitical space: toward a critical social theory of international politics', *Alternatives* 12: 403–34.

Ashley, R. K. (1989), 'Imposing international purpose: notes on a problematic of governance', in E.-O. Czempiel and J. Rosenau (eds), *Global Changes and Theoretical Challenges: Approaches to World Politics for the 1990s*, Lexington, Mass: Lexington Books.

Assiter, A. (1996), *Enlightened Women*, London: Routledge.

Audi, R. (ed.) (2001), *The Cambridge Dictionary of Philosophy*, 2nd edn, Cambridge: Cambridge University Press.

Austin, J. L. (1962), *How to Do Things with Words*, Oxford: Oxford University Press.

Ayer, A. J. (1936), *Language, Truth and Logic*, London: Gollancz.

Bachelard, Gaston (2000), *The Dialectic of Duration*, Manchester: Clinamen Press.

Baker, A. J. (1986), *Australian Realism: The Systematic Philosophy of John Anderson*, Cambridge: Cambridge University Press.

Bakhtin, M. M. (1981), *The Dialogic Imagination: Four Essays*, ed. M. Holquist and Caryl Emerson, trans. M. Holquist, Austin, Tex.: University of Texas Press.

Balibar, Etienne (1995), *The Philosophy of Marx*, London: Verso.

Banton, M. (1997), *Ethnic and Racial Consciousness*, 2nd edn, Harlow: Addison-Wesley Longman.

Barbour, I. (1966), *Issues in Science and Religion*, New York: Harper & Row; Englewood Cliffs, NJ: Prentice-Hall.

Barlow, H. B. (1972), 'Single units and sensation: a neuron doctrine for perceptual psychology?', *Perception* 1: 371–94.

Bateson, G. (1973), *Steps to an Ecology of Mind*, Granada.

Baudrillard, J. (1994), *Simulacra and Simulation*, Ann Arbor, MI: University of Michigan Press.

Baum, J. (ed.) (2002), *The Blackwell Companion to Organisations*, Oxford: Blackwell.

Bauman, Z. (1976), *Towards a Critical Sociology: An Essay on Commonsense and Emancipation*, London/Boston, Mass.: Routledge.

Beaumont, Matthew (2005), *Utopia Ltd.: Ideologies of Social Dreaming in England 1870–1900*, Leiden, Brill.

Bell, W. (2000), *Foundations of Futures Studies*. Vol. 1, *History, Purpose and Knowledge*, New Brunswick, NJ/London: Transaction Publishers.

Benjamin, W. (1940/1973), 'Theses on the philosophy of history', in *Illuminations*, New York: Schocken.

Benton, T. (1977), *Philosophical Foundations of the Three Sociologies*, London: Routledge & Kegan Paul.

Benton, T. (1981), 'Realism in social science: some comments on Roy Bhaskar's *The Possibility of Naturalism*', *Radical Philosophy* 27: 13–21; reprinted in Archer, M. et al. (eds), *Critical Realism: Essential Readings*.

Benton, T. (1993), *Natural Relations: Ecology, Animal Rights and Social Justice*, London: Verso.

Benton, T. (ed.) (1996), *The Greening of Marxism*, New York: Guildford.

Benton, T. (1998), 'Radical politics: neither Left nor Right?', in M. O'Brien, S. Penna and C. Hay (eds), *Theorising Modernity: Reflexivity, Environment and Identity in Giddens' Social Theory*, London/New York: Longman.

Benton, T. (1999), 'Evolutionary psychology and social science: a new paradigm or just the same old reductionism?', *Advances in Human Ecology* 8: 65–98.

Benton, T. (2000), 'An ecological historical materialism', in F. P. Gale and R. M. McGonigle (eds), *Nature, Production, Power: Towards an Ecological Political Economy*, Cheltenham: Edward Elgar.

Benton, T. (2001), 'Environmental philosophy: humanism or naturalism? A reply to Kate Soper', *Journal of Critical Realism* (incorporating *Alethia*) 4 (2).

Benton, T. (2004), 'Realism about the value of nature', in M. S. Archer and W. Outhwaite (eds), *Defending Objectivity*.

Benton, T. and Craib, I. (2001), *The Philosophy of Social Science: The Philosophical Foundations of Social Thought*, London: Palgrave.

Berger, P. and Luckmann, T. (1966), *The Social Construction of Reality*, Garden City, NY: Doubleday.

Berger, P. and Pullberg, S. (1966), 'Reification and the sociological critique of consciousness', *New Left Review* 35.

Berkeley, G. (1710/1878), *The Principles of Human Knowledge*, London: Routledge.

Berglez, P. (2006), 'The materiality of media discourse: on capitalism and journalistic modes of writing', *Örebro Studies in Media and Communication* 4, Örebro University.

Bernard-Donals, M. F. (1994), *Mikhail Bakhtin: Between Phenomenology and Marxism*, Cambridge: Cambridge University Press.

Blau, P. (1977), *Inequality and Heterogeneity: A Primitive Theory of Social Structure*, New York: The Free Press.

Bloch, E. (1930–1972/1989), *The Utopian Function of Art and Literature: Selected Essays*, trs. J. Zipes & F. Mecklenburg, Cambridge, Mass.: MIT Press.

Bloch, E. (1959/1986), *The Principle of Hope*, 3 vols, Oxford: Blackwell.

Blom, B. (2002), 'The social worker–client relationship: a Sartrean approach', *European Journal of Social Work* (5) 3: 23–31.

Bobrow, D. B. (1999), 'Prospecting the future', *International Studies Review* 1 (2): 1–10.

Bohm, D. (1957), *Causality and Chance in Modern Physics*, London: Routledge & Kegan Paul.

Bottomore, T. (ed.) (1991), *A Dictionary of Marxist Thought*, 2nd edn, Oxford: Blackwell.

Bowlby, J. (1988), *A Secure Base*, London: Routledge.

Bowring, B. (2002), review of A. Norrie, *Punishment, Responsibility and Justice*, *Journal of Law and Society* 29: 521–7.

Bourdieu, P. (1984), *Distinction: A Social Critique of the Judgement of Taste*, trans. R. Nice, Cambridge, Mass.: Harvard University Press.

Braidotti, R., Charkiewicz, E., Hausler, S. and Wieringa, S. (1994), *Women, the Environment and Sustainable Development: Towards a Theoretical Synthesis*, London: Zed Books.

Brecht, B. (1964), *Brecht on Theatre: The Development of an Aesthetic*, ed. and trans. J. Willett, New York: Hill & Wang.

Brereton, D. P. (2000), 'Ontic morality', *Alethia* 3 (2).

Brereton, D. P. (2004a), 'Innate virtue, ethical naturalism, and natural law', in S. Glazier (ed.), *The Anthropology of Religion*, Vol. 3, New York: Greenwood Publishers.

Brereton, D. P. (2004b), 'Preface for a critical realist ethnology, Part I: The schism and a realist restorative, Part II: Some principles applied', *Journal of Critical Realism* 3 (1 and 2).

Brereton, D. P. (2005), 'Critical realism', in H. J. Birx (ed.), *Encyclopedia of Anthropology*, 5 vols, Beverly Hills, Calif./London: Sage.

Brewer, J. D. (2000), *Ethnography*, Milton Keynes: Open University Press.

Bricmont, J. (2001), 'Sociology and epistemology', in J. López and G. Potter (eds), *After Postmodernism: An Introduction to Critical Realism*, London: Athlone Press.

Broad, C. D. (1925), *The Mind and Its Place in Nature*, London: Routledge & Kegan Paul.

Broad, C. D. (1926), *The Philosophy of Francis Bacon*, Cambridge: Cambridge University Press.

Brohman, J. (1996), *Popular Development: Rethinking the Theory and Practice of Development*, Oxford/Cambridge, Mass.: Blackwell.

Brown, A., Fleetwood, S. and Roberts, J. M. (eds) (2002), *Critical Realism and Marxism*, London/New York: Routledge.

Browning, G. K. (1996), 'Good and bad infinites in Hegel and Marx', http://www.psa.ac.uk/journals/pdf/5/1/1996/brown.pdf

Buckley, W. (1967), *Sociology and Modern Systems Theory*, Englewood Cliffs, NJ: Prentice-Hall.

Bull, H. (1969), 'International theory: the case for classical approach', in K. Knorr and J. N. Roseanau (eds), *Contending Approaches to International Politics*, Princeton, NJ: Princeton University Press; first published in the April 1966 issue of *World Politics*.

Bull, M. (2000), 'Where is the anti-Nietzsche?', *New Left Review* 3.

Bunge, M. (1959), *Causality: The Place of the Causal Principle in Modern Science*, Cambridge, Mass.: Harvard University Press.

Bunge, M. (1963), *The Myth of Simplicity: Problems of Scientific Philosophy*, Englewood Cliffs, NJ: Prentice-Hall.

Bunge, M. (1977), *Treatise on Basic Philosophy*, Vol. III, *Ontology: The Furniture of the World*, Dordrecht: Reidel.

Bunge, M. (1979), *Treatise on Basic Philosophy*, Vol. IV, *Ontology: A World of Systems*, Dordrecht: Reidel.

Bunge, M. (1980), 'From neuron to behaviour and mentation: an exercise in levelmanship', in H. M. Pinsker and W. D. Willis Jr., *Information Processing in the Nervous System*, New York: Raven Press.

Bunge, M. (1999), 'Ethics and praxiology as technologies', *Techné* 4 (4).
Butler, J. (1990), *Gender Trouble: Feminism and the Subversion of Identity*, London: Routledge.
Butler, J. (1993), *Bodies That Matter*, London: Verso.
Butler, J. (1994), 'Gender as performative: interview', *Radical Philosophy* 66.
Butler, J., Laclau, E. and Žižek, S. (2000), *Contingency, Hegemony, Universality: Contemporary Dialogues on the Left*, London: Verso.
Bynum, B. and Porter, R. (eds) (1981), *Dictionary of the History of Science*, Macmillan: London.
Byrne, D. (1999), *Social Exclusion*, Buckingham: Open University Press.
Byrne, D (2002), *Interpreting Quantitative Data*, London: Sage.
Calder, G. (2005), 'Post-Cartesian anxieties: embodied subjectivity after the linguistic turn', *New Formations*, 56: 83–95.
Callaghan G. (1998), *Flexibility, Mobility and the Labour Market*, Aldershot: Ashgate.
Callinicos, A. (1994), 'Critical realism and beyond: Roy Bhaskar's *Dialectic*', Working Paper 7, Department of Politics, University of York.
Callinicos, A. (1999), *Social Theory: A Historical Introduction*, Cambridge: Polity Press.
Callinicos, A. (2001), entry on Bhaskar and critical realism, in J. Bidet and E. Kouvélakis (eds), *Dictionnaire Marx Contemporain*, Paris: PUF.
Callinicos, A. (2006), *The Resources of Critique*, Cambridge: Polity.
Camic, C. (1986), 'The matter of habit', *American Journal of Sociology* 91 (5): 1039–87.
Capra, F. (1997), *The Web of Life: A New Synthesis of Mind and Matter*, London: Flamingo.
Carr, E. H. (1946/1964), *The Twenty Years' Crisis*, New York: Harper & Row.
Carspecken, F. P. (1995), *Critical Ethnography in Educational Research*, New York, NY: Routledge.
Carter, B. (2000), *Realism and Racism: Concepts of Race in Sociological Research*, London/New York: Routledge.
Carter, B. and New, C. (eds) (2004), *Making Realism Work: Realist Social Theory and Empirical Research*, London/New York: Routledge.
Carter, B. and Sealey, A. (2000), 'Language, structure and agency: what can realist social theory offer sociolinguistics?' *Journal of Sociolinguistics* 4 (1): 3–20.
Castles, S. (2004), 'Why migration policies fail', *Ethnic and Racial Studies* 27 (2): 205–27.
Chalmers, A. F. (1988), 'Is Bhaskar's realism realistic?' *Radical Philosophy* 49: 18–23.
Checkland, P. (1981), *Systems Thinking, Systems Practice*, Chichester: Wiley.
Chidester, D. and Linenthal, E. (1995), 'Introduction', in D. Chidester and E. Linenthal (eds.), *American Sacred Space*, Bloomington, IM: Indiana University Press.
Checkland, P. and Scholes, J. (1990), *Soft Systems Methodology in Action*, Chichester: Wiley.
Chisholm, Roderick M. (1982), *The Foundations of Knowing*, Minneapolis, Minn.: Minneapolis University Press.
Chomsky, N. (1957), *Syntactic Structures*, The Hague: Mouton.
Chomsky, N. (1966), *Cartesian Linguistics*, New York: Harper & Row.
Chomsky, N. (2000), *New Horizons in the Study of Language and Mind*, Cambridge: Cambridge University Press.
Chouliaraki, L. and Fairclough, N. (1999), *Discourse in Late Modernity: Rethinking Critical Discourse Analysis*, Edinburgh: Edinburgh University Press.
Churchland, P. (1981), 'Eliminative materialism and the propositional attitudes', *Journal of Philosophy* 78 (2): 67–90.
Churchland, P. and Churchland, P. (2002), *Brain-wise: Studies in Neurophilosophy*, London: Bradford Books.
Churchman, C. (1971), *The Design of Enquiring Systems*, New York: Basic Books.
Clark A. M. (1998), 'The qualitative–quantitative debate: moving from positivism and confrontation to postpositivism and reconciliation', *Journal of Advanced Nursing* 27: 1242–9.
Clarke, G. (2003), 'Fairbairn and Macmurray: psychoanalytic studies and critical realism', *Journal of Critical Realism* 2 (1).

Clayton, A. M. H. and Radcliffe, N. J. (1998), *Sustainability: A Systems Approach*, London: Earthscan.
Clegg, H. A. (1979), *The Changing System of Industrial Relations in Great Britain*, Oxford: Blackwell.
Clegg, S. (2005), 'Evidence-based practices in educational research: a critical realist critique of systematic review', *British Journal of Sociology of Education*, 26 (3): 415–28.
Coates, P. and Hutto, D. (1996), *Current Issues in Idealism*, Bristol: Thoemmes Press.
Cohen, G. (1978), *Karl Marx's Theory of History: A Defence*, Oxford: Clarendon.
Collier, A. (1989), *Scientific Realism and Socialist Thought*, Hemel Hempstead: Harvester Wheatsheaf.
Collier, A. (1990), *Socialist Reasoning: An Inquiry into the Political Philosophy of Scientific Socialism*, London: Pluto.
Collier, A. (1991/2003), 'The inorganic body and the ambiguity of freedom', *Radical Philosophy* 57: 3–9; reprinted in his *In Defence of Objectivity* (2003).
Collier, A. (1992), 'Marxism and universalism: group interests or a shared world', in R. Attfield and B. Wilkins (eds), *International Justice and the Third World*, London: Routledge.
Collier, A. (1994), *Critical Realism: An Introduction to Roy Bhaskar's Philosophy*, London/New York: Verso.
Collier, A. (1995a), 'The power of negative thinking', *Radical Philosophy* 69: 36–9.
Collier, A. (1995b), 'Realism and formalism in ethics', in M. S. Archer et al. (eds), *Critical Realism: Essential Readings*.
Collier, A. (1998), 'Explanation and emancipation', in M. S. Archer et al. (eds), *Critical Realism: Essential Readings*.
Collier, A. (1999), *Being and Worth*, London/New York: Routledge.
Collier, A. (2001a), *Christianity and Marxism: A Philosophical Contribution to Their Reconciliation*, London/New York: Routledge.
Collier, A. (2001b), 'Real and nominal absences', in J. López and G. Potter (eds), *After Postmodernism*.
Collier, A. (2001c), 'The soul and Roy Bhaskar's thought', *Journal of Critical Realism* (incorporating *Alethia*) 4 (2): 19–23.
Collier, A. (2003a), *On Christian Belief: A Defence of a Cognitive Conception of Religious Belief in a Christian Context*, London/New York: Routledge.
Collier, A. (2003b), *In Defence of Objectivity and Other Essays: On Realism, Existentialism and Politics*, London/New York: Routledge.
Collier, A. (2004), *Marx*, Oxford: Oneworld.
Collier, A. (2005), 'Philosophy and critical realism', in G. Steinmetz (ed.), *The Politics of Method in the Human Sciences*.
Collins, P. H. (1991), *Black Feminist Thought: Knowledge, Consciousness and the Politics of Empowerment*, New York: Routledge.
Commoner, B. (1971), *The Closing Circle: Nature, Man and Technology*, New York: Alfred Knopf.
Comte, A. (1853), *The Positive Philosophy*, Vols 1 and 2, London: Trübner.
Connell, R. W. (1987), *Gender and Power*, Cambridge: Polity.
Connelly J. (2000), 'A realistic theory of health sector management: the case for critical realism', *Journal of Management in Medicine* 14 (5/6): 262–71.
Corson, D. J. (1990), 'Old and new conceptions of discovery in education', *Educational Philosophy and Theory* 22 (2): 26–49.
Corson, D. J. (1991a), 'Bhaskar's critical realism and educational knowledge', *British Journal of Sociology of Education* 12 (2): 223–41.
Corson, D. J. (1991b), 'Educational research and Bhaskar's conception of discovery', *Educational Theory* 41 (2): 189–98.
Cox, R. (1983), 'Gramsci, hegemony and international relations: an essay on method', *Millennium: Journal of International Studies* 12 (2): 205–49.
Cox, R. (1987), *Production, Power and World Order: Social Forces in the Making of History*, New York: Columbia University Press.

Crang, M. and Thrift, N. (eds) (2000), *Thinking Space*, London: Routledge.
Creaven, S. (2000), *Marxism and Realism: Materialistic Application of Realism in the Social Sciences*, London/New York: Routledge.
Creaven, S. (2002a), 'Marxism, realism and dialectic', in A. A. Brown et al. (eds), *Critical Realism and Marxism*.
Creaven, S. (2002b), 'The Pulse of Freedom? Bhaskar's *Dialectic* and Marxism', *Historical Materialism* 10 (2): 77–141.
Creaven, S. (2003), 'Marx and Bhaskar on the dialectics of freedom', *Journal of Critical Realism* 2 (1): 63–94.
Cruickshank, J. (2003a), *Realism and Sociology: Anti-Foundationalism, Ontology and Social Research* London: Routledge.
Cruickshank, J. (ed.) (2003b), *Critical Realism: The Difference It Makes*, London: Routledge.
Cruickshank, J. (2004), 'A tale of two ontologies: an immanent critique of critical realism', *The Sociological Review* 52 (4): 567–85.
Cruickshank, J. (2006), 'Postdisciplinarity and the study of lay normativity: retheorising class in social science' (review article), *Journal of Critical Realism* (5) 1.
Cuomo, Chris (1998), *Feminism and Ecological Communities: An Ethic of Flourishing*, London: Routledge.
Daly, J. (1996), *Marx, Justice and Dialectic*, London: Greenwich Exchange.
Danermark, B. (2002), 'Interdisciplinary research and critical realism: the example of disability research', *Journal of Critical Realism* (incorporating *Alethia*) 5 (1): 56–64.
Danermark, B. (2003), 'Different approaches in assessment of audiological rehabilitation: a meta-theoretical perspective', *International Journal of Audiology* 42: 112–17.
Danermark, B., Ekström, M., Jakobsen, L. and Karlsson, J. C. (2002), *Explaining Society: Critical Realism in the Social Sciences*, London: Routledge.
Davidson, D. (1984), *Inquiries into Truth and Interpretation*, Oxford: Oxford University Press.
Davidson, D. (1970/1980), 'Mental events', in his *Essays on Actions and Events*, Oxford: Clarendon.
Davidson, J. O'Connell and Layder, D. (1994), *Methods: Sex and Madness*, London/New York: Routledge.
Dean, K. (2000), 'Capitalism, psychic immiseration and decentred subjectivity', *Journal for the Psychoanalysis of Culture and Society*, 5(1), pp. 41–56.
Dean, K. (2003), *Capitalism and Citizenship: The Impossible Partnership*, London/New York: Routledge.
Dean, K. (2004), 'Laclau and Mouffe and the discursive turn: the gains and the losses', in J. Joseph and J. Roberts (eds), (2004), *Realism, Discourse and Deconstruction*.
Dean, K. (2005), 'Biology and the new scientific subjectivism: a suitable case for critical realism?', *New Formations* 56: 71–82.
Dean, K., Joseph, J. and Norrie, A. (2005), 'Editorial: new essays in critical realism', *New Formations* 56: 7–26.
de Beauvoir, S. (1953), *The Second Sex*, New York, NY: Knopf.
Debord, G. (1967/1995), *The Society of the Spectacle*, New York: Zone.
Der Derian, J. and Shapiro, M. (1989) (ed.), *International/Intertextual Relations: Postmodern Readings of World Politics*, Toronto: Lexington Books.
Derrida, J. (1981), *Dissemination*, trans. Barbara Johnson, Chicago, Ill.: University of Chicago Press; London: Athlone Press.
Derrida, J. (1995), 'Deconstruction and the other', in R. Kearney (ed.), *States of Mind: Dialogues with Contemporary Thinkers*, New York: New York University Press.
Dessler, D. (1989), 'What's at stake in the agent–structure debate?', *International Organization* 43 (3): 441–73.
Descartes, R. (1637/1965), *A Discourse on Method*, London: Dent.

Descartes, R. (1641/1968), *Meditations*, in R. Descartes, *Discourse on Method and the Meditations*, trans. F. E. Sutcliffe, Harmondsworth: Penguin.
Devitt, M. (1984/1991), *Realism and Truth*, Oxford: Blackwell.
Devitt, M. and Sterelny, K. (1987), *Language and Reality*, Oxford: Blackwell.
Dickens, P. (1992), *Society and Nature: Towards a Green Social Theory*, Philadelphia, Pa.: Temple University Press.
Dickens, P. (1996), *Reconstructing Nature: Alienation, Emancipation and the Division of Labour*, London: Routledge.
Dickens, P. (2000), *Social Darwinism: Linking Evolutionary Thought to Social Theory*, Buckingham: Open University Press.
Dickens, P. (2001), 'Changing nature, changing ourselves', *Journal of Critical Realism* (incorporating *Alethia*) 4 (2): 9–18.
Dickens, P. (2002), *Society and Nature: Ecology, Power and the Transformation of Humanity*, Cambridge: Polity.
Dickens, P., Duncan, S., Goodwin, S. and Grey, F. (1985), *Housing States and Localities*, London/New York: Methuen.
Dirlik, A. (1997), *The Postcolonial Aura: Third World in the Age of Global Capitalism*, Boulder, Colo./Oxford: Westview Press.
Dobson, P. J. (2002), 'Critical realism and information systems research: why bother with philosophy?', *Information Research* 7 (2).
Dobson, P. J. (2003), 'Business process reengineering (BPR) versus outsourcing: critical perspectives', *Systemic Practice and Action Research* 16 (3): 225–33.
Downward, P. (ed.) (2003), *Applied Economics and the Critical Realist Critique*, London: Routledge.
Downward, P., Dow, S. and Fleetwood, S. (2006), 'Transforming economics through critical realism: themes and issues' (review symposium), *Journal of Critical Realism* 5 (2).
Downward, P., Finch, J. H. and Ramsay, J. (2002), 'Critical realism, empirical methods and inference: a critical discussion', *Cambridge Journal of Economics* 26: 481–500.
Downward, P. and Mearman, A. (2002), 'Critical realism and econometrics: constructive dialogue with Post Keynesian economics', *Metroeconomica* 53 (4): 391–415.
Downward, P. and Mearman, A. (2003), 'Critical realism and econometrics: interaction between philosophy and Post Keynesian practice', in Downward (ed.), *Applied Economics*.
Drake, D., Lovejoy, A. O., Rogers, A. K., Santayana, G., Sellars, R. W. and Strong, C. A. (1920), *Essays in Critical Realism: A Co-Operative Study of the Problem of Knowledge*, London: Macmillan.
Dryzek, J. S. (1996), *Democracy in Capitalist Times: Ideals, Limits, and Struggles*, Oxford: Oxford University Press.
D'Souza, R. (2002), 'Sustainable development or self-determination? Asking the hard questions about WSSD', *Social Policy* 34 (4): 23–6.
D'Souza, R. (2004), 'The global commons: but, where is the community?', in R. K. Miller (ed.), *The Informed Argument*, London: Thomson Wadsworth.
D'Souza, R. (2005,6), *Contextualising Interstate Conflicts over Krishna Waters: Law, Science and Imperialism*, Hyderabad: Orient Longmans.
Dummett, M. (1974), *The Justification of Deduction* (British Academy Lecture), Oxford: Oxford University Press.
Dummett, M. (1978), *Truth and Other Enigmas*, London: Duckworth.
Dummett, M. (1982), 'Realism', *Synthese* 52: 55–112.
Dunne, T. (2006), 'Bernard Lonergan (1904–1984)', *The Internet Encyclopedia of Philosophy* (http://www.iep.utm.edu/l/lonergan.htm).
Durkheim, E. (1898/1969), 'Intellectuals and individualism', trans. S. and J. Lukes, *Political Studies* 17 (1): 19–30.
Durkheim, E. (1912/2001), *The Elementary Forms of Religious Life*, Oxford: Oxford University Press.

Eagleton, T. (2000), 'Defending utopia' (review), *New Left Review* 4.
Edgley, R. (1976), 'Reason as dialectic: science, social science and socialist science', *Radical Philosophy* 15: 2–7; reprinted in *ER*.
Edwards, D. and Potter, J. (1992), *Discursive Psychology*, London: Sage.
Edwards, M. A. and Potter, J. (1995), 'Death and furniture: the rhetoric, politics and theology of bottom-line arguments against relativism', *History of the Human Sciences* 8: 25–49.
Edwards, P. (2005), 'The challenging but promising future of industrial relations: developing theory, method and relevance in context-sensitive research', *Industrial Relations Journal* 36 (4): 244–82.
Ekström, M. (1992), 'Causal explanation of social action: the contribution of Max Weber and of critical realism to a generative view of causal explanation in social science', *Acta Sociologica* 35: 107–22.
Elam, D. (1994), *Feminism and Deconstruction: Ms. en Abyme*, London: Routledge.
Eliade, M. (1968), *The Scared and Profane: The nature of religion*, New York, NY: Harvest Books.
Ellis, B. (2001), *Scientific Essentialism*, Cambridge: Cambridge University Press.
Emery, F. (1969), *Systems Thinking*, Harmondsworth: Penguin.
Engels, F. (1844/1892), *The Condition of the Working-Class in England in 1844*, London: S. Sonnenschein.
Engelskirchen, Howard (2004), 'Powers and particulars: Adorno and scientific realism', *Journal of Critical Realism* 3 (1).
Eriksson, G. (2006), 'Rethinking the rethinking: the problem of generality in qualitative media audience research', *Nordicom Review* 27 (1): 31–44.
Fairbairn, W. R. D. (1952), *Psychoanalytic Studies of the Personality*, London: Routledge.
Fairclough, N. and Wodak, R. (1997), 'Critical discourse analysis', in T. van Dijk (ed.) *Discourse as Social Interaction*, London: Sage.
Fairclough, N., Jessop, B. and Sayer, A. (2004), 'Critical realism and semiosis', in J. Joseph and J. M. Roberts (eds), *Realism, Discourse and Deconstruction*, London/New York: Routledge.
Fauconnier, G. and Turner, M. (2002), *The Way We Think: Conceptual Blending and the Mind's Hidden Complexities*, New York: Basic Books.
Fenton, S. (2003), *Ethnicity*, Cambridge: Polity.
Feyerabend, P. K. (1975), *Against Method*, London: New Left Books.
Finch, J. and McMaster, R. (2003), 'A pragmatic alliance between critical realism and simple non-parametric statistical techniques', in Downward (ed.), *Applied Economics*.
Fine, B. (2002), '"Economic imperialism": a view from the periphery', *Review of Radical Political Economics* 34: 187–201.
Fitzpatrick, S. (2005), 'Explaining homelessness: a critical realist perspective', *Housing, Theory and Society* 22 (1): 1–17.
Fleetwood, S. (1995), *Hayek's Political Economy: The Socio-Economics of Order*, London/New York: Routledge.
Fleetwood, S. (ed.) (1999a), *Critical Realism in Economics: Development and Debate*, London/New York: Routledge.
Fleetwood, S. (1999b), 'The inadequacy of mainstream theories of trade unions,' *Labour* 13 (2): 445–80.
Fleetwood, S. (2001a), 'Marx's labour theory of value', *Capital & Class* 73: 41–77.
Fleetwood, S. (2001b), 'Causal laws, functional relations and tendencies', *Review of Political Economy* 13(2): 201–20.
Fleetwood, S. (2006), 'Re-thinking labour markets: a critical realist–socioeconomic perspective', *Capital and Class* 89: 59–89.
Fleetwood, S. and Ackroyd, S. (eds) (2004), *Critical Realist Applications in Organisation and Management Studies*, London: Routledge.
Fodor, J. (1974), 'Special sciences', *Synthese* 28: 77–115.

Fogel, R. W. (1966), '"The new economic history", its findings and methods', *Economic History Review* 19.
Fogelsong, R. E. (1996), 'Planning the capitalist city', in S. Campbell and S. Fainstain, eds, *Readings in Planning Theory*, Cambridge, Mass.: Harvard University Press; Oxford: Oxford University Press.
Foster, J. (1991), *The Immaterial Self: A Defence of the Cartesian Dualist Conception of the Mind*, London: Routledge.
Forrester, J. (1968), *Principles of Systems*, Cambridge, Mass.: MIT Press.
Foucault, M. (1970), *The Order of Things*, New York: Random House.
Foucault, M. (1977), *The Archaeology of Knowledge*, London: Tavistock.
Frank, A.G. (1967), *Capitalism and Underdevelopment in Latin America*, New York: Monthly Review Press.
Fraser, I. (1998), *Hegel and Marx: The Concept of Need*, Edinburgh: Edinburgh University Press.
Freeman, W. J. (1975), *Mass Action in the Nervous System*, New York: Academic Press.
Freud, S. (1911/1995), 'Formulations on two principles of mental functioning', in P. Gay (ed.), *The Freud Reader*, London: Vintage.
Freud, S. (1929/1963), *Civilization and Its Discontents*, London: Hogarth Press.
Freud, S. (1938/1969), *An Outline of Psycho-Analysis*, London: Hogarth Press.
Friedman, M. (1953), *Essays in Positive Economics*, Chicago, Ill.: University of Chicago Press.
Gadamer, H.-G. (1979), *Truth and Method*, London: Sheed and Ward.
Geertz, C. (1973), *Interpretation of Culture*, New York: Basic Books.
Geoghegan, V. (1996), *Ernst Bloch*, London/New York: Routledge.
Geras, N. (1985), *Marx and Human Nature*, London: Verso.
Gewirth, A. (1979), *Reason as Morality*, Chicago, Ill.: University of Chicago Press.
Gibson, J. J. (1950), *The Perception of the Visual World*, Boston, Mass.: Houghton Mifflin.
Gibson, J. J. (1979), *The Ecological Approach to Visual Perception*, Boston, Mass.: Houghton Mifflin.
Giddens, A. (1979), *Central Problems in Social Theory*, London: Macmillan.
Giddens, A. (1983), *The Constitution of Society*, Cambridge: Polity.
Gillett, C. and Loewer, B. (eds) (2001), *Physicalism and Its Discontents*, Cambridge: Cambridge University Press.
Gilroy, P. (2001), *Against Race: Imagining Political Culture beyond the Color Line*, Cambridge, Mass.: Harvard University Press.
Giri, A. K. (2001), 'Social criticism, cultural creativity and the contemporary dialectics of transformations', in A. K. Giri (ed.), *Rethinking Social Transformation: Criticism and Creativity at the Turn of the Millennium*, Jaipur/New Delhi: Rawat Publications.
Godard, J. (1993), 'Theory and method in industrial relations', in R. Adams and N. Meltz, *Industrial Relations Theory: Its Nature, Scope and Pedagogy*, London: The Scarecrow Press.
Goody, J. (1977), *The Domestication of the Savage Mind*, Cambridge: Cambridge University Press.
Gramsci, A. (1971), *Selections from the Prison Notebooks*, ed. and trans. Q. Hoare and G. N. Smith, London: Lawrence & Wishart.
Greenberg, J. R. and Mitchell, S. A. (1983), *Object Relations in Psychoanalytic Theory*, Cambridge, Mass.: Harvard University Press.
Greenwood, J. (1994), *Realism, Identity and Emotion: Reclaiming Social Psychology*, London: Sage.
Groff, R. (2004), *Critical Realism, Post-positivism and the Possibility of Knowledge* London: Routledge.
Habermas, J. (1972), *Knowledge and Human Interests*, London: Heinemann.
Habermas, J. (1979), *Communication and the Evolution of Society*, trans. T. McCarthy, London: Heinemann.
Habermas, J. (1987), *The Philosophical Discourse of Modernity: Twelve Lectures*. Cambridge: Polity.
Hall, S. (1980), 'Encoding/decoding', in S. Hall et al. (eds) *Culture, Media, Language*, London: Hutchinson.

Hammersley, M. (2002), 'Research as emancipatory: the case of Bhaskar's critical realism', *Journal of Critical Realism* (1) 1.
Haraway, D. (1991), '"Gender" for a Marxist dictionary: the sexual politics of a word', in her *Simians, Cyborgs and Women: The Reinvention of Nature*, London: Free Association Books.
Harding, S. (1993), 'Rethinking standpoint epistemology: what is "strong objectivity"?', in L. Alcoff and E. Potter (eds), *Feminist Epistemologies*, New York: Routledge.
Harding, S. (2003), 'Representing reality: the critical realism project', *Feminist Economics* 9 (1): 151–9.
Hardt, M. & Negri, A. (2005), *Multitude: War and Democracy in the Age of Empire*, London: Hamish Hamilton.
Hare, R. M. (1952), *Language and Morals*, Oxford: Clarendon.
Hare, R. M. (1997), *Sorting Out Ethics*, Oxford: Clarendon.
Harman, G. (1995), 'Rationality', in E. E. Smith and D. N. Osherson (eds), *An Invitation to Cognitive Science*, Vol. 3, *Thinking*, Cambridge Mass.: MIT Press.
Harré, R. (1979), *Social Being*, Oxford: Blackwell.
Harré, R. (1986), *Varieties of Realism*, Oxford: Blackwell.
Harre R. (1993), *Social Being*, 2nd edn, Oxford: Blackwell.
Harre R. (2001), 'How to change reality: story v. structure' in López and Potter (eds), *After Postmodernism*.
Harré, R. (2002), 'Social reality and the myth of social structure,' *European Journal of Social Theory* 5: 111–23.
Harré, R. and Madden, E. H. (1975), *Causal Powers: A Theory of Natural Necessity*, Oxford: Blackwell; Totowa, NJ: Rowman & Littlefield.
Harré, R. and Secord, P. (1972), *The Explanation of Social Behaviour*, Oxford: Blackwell.
Harré, R. and Varela, C. R. (1996), 'Conflicting varieties of realism: causal powers and the problems of social structure', *Journal for the Theory of Social Behaviour* 26: 313–25.
Harris, R. (2003), 'On redefining linguistics', in H. G. Davis and T. J. Taylor (eds), *Rethinking Linguistics*, London: Routledge.
Hartmann, H. (1993), 'The unhappy marriage of Marxism and feminism: towards a more progressive union', in A. M. Jaggar and P. S. Rothenberg (eds), *Feminist Frameworks*, New York: McGraw-Hill.
Hartsock, N. (1983), 'The feminist standpoint: developing the ground for a specifically feminist historical materialism', in S. Harding and M. B. Hintikka (eds), *Discovering Reality*, Dordrecht: Reidel.
Hartwig, M. (2000), 'Charging at red flags? Blind spots in Geoff Hodgson's "promised land"', *Alethia* 3 (1) 36–40.
Hartwig, M. (2001), 'New Left, New Age, New Paradigm? Roy Bhaskar's *From East to West*', *Journal for the Theory of Social Behaviour* 31 (2): 139–66.
Hartwig, M. and Sharp, R. (1999), 'The realist third way', *Alethia* 2 (1): 17–23.
Harvey, D. (1982), *The Limits to Capital*, Oxford: Blackwell.
Harvey, D. (1990), *The Condition of Postmodernity*, Oxford: Blackwell.
Harvey, D. (2005), *A Brief History of Neoliberalism*, Oxford: Oxford University Press.
Hassard, J. (1993), *Sociology and Organisation Theory*, Cambridge: Cambridge University Press.
Haug, W. F. (1987), *Commodity Aesthetics, Ideology and Culture*, New York: International General.
Hay, C. (1996), *Re-Stating Social and Political Change*, Buckingham: Open University Press.
Hayek, F. A. (1944), *The Road to Serfdom*, London: Routledge; Chicago, Ill.: University of Chicago Press.
Hegel, G. W. F. (1807/1977), *The Phenomenology of Spirit*, trans. A. V. Miller, Oxford: Oxford University Press.
Heidegger, M. (1927/1967), *Being and Time*, Oxford: Blackwell.

Hempel, C. G. (1965), *Aspects of Scientific Explanation and Other Essays in the Philosophy of Science*, New York/London: Free Press/Collier-Macmillan.

Hettne, B. (1978), *The Political Economy of Indirect Rule: Mysore 1881–1947*, Scandinavian Institute of Asian Studies Monograph Series, No. 32, London: Curzon Press.

Hettne, B. (1990), *Development Theory and the Three Worlds*, Harlow: Longman.

Hicks, G. Dawes (1917), 'The basis of critical realism', *Proceedings of the Aristotelian Society*, substantially reproduced in his 1938.

Hicks, G. Dawes (1938), *Critical Realism: Studies in the Philosophy of Mind and Nature*, London: Macmillan.

Hodgson, G. M. (1999a), 'Marching to the promised land? Some doubts on the policy affinities of critical realism', *Alethia* 2 (2): 2–10.

Hodgson, G. M. (1999b), 'Andrew Collier's promised land', *Alethia* 2 (2): 12–13.

Hodgson, G. M. (2000), 'Motherhood and apple pie', *Alethia* 3 (1): 42–3.

Holloway, J. (2002), *Change the World without Taking Power: The Meaning of Revolution Today*, London: Pluto.

Hoogvelt, A. (1997), *Globalisation and the Postcolonial World: The new political economy of development*, Baltimore, MD: Johns Hopkins University Press.

Horkheimer, M. (1937/1972), 'Traditional and critical theory', in P. Connerton (ed.), *Critical Sociology*, Harmondsworth: Penguin; also in Horkheimer (2002).

Horkheimer, M. (1993), *Critique of Instrumental Reason*, New York: Continuum.

Horkheimer, M. (2002), *Critical Theory: Selected Essays*, New York: Continuum.

Horkheimer, M. and Adorno, T. W. (1972), *Dialectic of Enlightenment*, New York: Herder & Herder.

Horn, L. R. (2006), 'Contradiction', *The Stanford Encyclopedia of Philosophy* (Fall 2006 Edition), E. N. Zalta (ed.), http://plato.stanford.edu/archives/fall2006/entries/contradiction/.

Hostettler, N. and Norrie, A. (2003), 'Are critical realist ethics foundationalist?', in J. Cruickshank (ed.), *Critical Realism*.

Houston, S. (2001a), 'Beyond social constructionism: critical realism in social work', *British Journal of Social Work* 31: 845–61.

Houston, S. (2001b), 'Transcending the fissure in risk theory: critical realism and child welfare', *Child and Family Social Work*, 219–28.

Hull, C. L. (2006), *The Ontology of Sex*, London/New York: Routledge.

Hume, D. (1777/1975), *David Hume's Enquiries Concerning Human Understanding and Concerning the Principles of Morals*, Oxford: Clarendon.

Husserl, E. (1929/1995), *Cartesian Meditations: An Introduction to Phenomenology*, trans. Dorian Cairns, Dordrecht: Kluwer Academic.

Inwood, M. (1992), *A Hegel Dictionary*, Oxford: Blackwell.

Isaac, J. C. (1987), *Power and Marxist Theory: A Realist View*, Ithaca, NY: Cornell University Press.

Isaac, J. C. (1990), 'Realism and reality: some realistic reconsiderations', *Journal for the Theory of Social Behaviour* 20 (1): 1–31.

Jameson, F. (1991), *Postmodernism or the Cultural Logic of Late Capitalism*, London: Verso.

Jameson, F. (2004), 'The politics of utopia', *New Left Review* 25: 35–54.

Janov, A. (1970), *The Primal Scream: Primal Therapy – the Cure for Neurosis*, New York: Dell.

Jarvis, S. (1998), *Adorno: A Critical Introduction*, Cambridge: Polity.

Jensen, K. B. (2002), 'The complementarity of qualitative and quantitative methodologies in media and communication research', in K. B. Jensen (ed.), *A Handbook of Media and Communication Research*, London: Routledge.

Jessop, B. (1982), *The Capitalist State: Marxist Theories and Methods*, Oxford: Martin Robertson.

Jessop, B. (1990), *State Theory: Putting the Capitalist State in Its Place*, Cambridge: Polity Press.

Jessop, B. (2001a), 'The crisis of the national spatio-temporal fix and the ecological dominance of globalising capitalism', *International Journal of Urban and Regional Studies* 24 (2): 323–60.

Jessop, B. (2001b), 'Institutional (re)turns and the strategic-relational approach', *Environment and Planning A* 33 (7): 1213–37.
Jessop, B. (2002a), 'Capitalism, the regulation approach, and critical realism', in A. A. Brown et al. (eds), *Critical Realism and Marxism*.
Jessop, B. (2002b), *The Future of the Capitalist State*, Cambridge: Polity.
Jessop, B. (2003a), 'Critical realism and hegemony: hic Rhodus, hic saltus', *Journal of Critical Realism* 1 (2): 183–94.
Jessop, B. (2003b), 'Putting hegemony in its place', *Journal of Critical Realism* 2 (1) 138–48.
Jessop, B. (2003c), review of *Max Weber's Methodology: The Unification of the Cultural and Social Sciences* by Fritz K. Ringer, Cambridge, Mass.: Harvard University Press, 2000, in *Historical Materialism* 11 (2): 2003.
Jessop, B. (2004), 'Critical semiotic analysis and cultural political economy', *Critical Discourse Studies* 1 (2): 159–74.
Jessop, B. (2005), 'Critical realism and the strategic-relational approach', *New Formations* 56: 40–53.
Jevons, W. S. (1871/1970), *The Theory of Political Economy*, with an introduction by R. Collinson-Black, Harmondsworth: Penguin.
Job, S. (2004), *Beautiful Soul: Analysing the racist nationalist struggle for Russia's psyche*, thesis submitted for the degree of the PhD, University of Sydney.
Johnson, P. and Duberley, J. (2000), *Understanding Management Research: An Introduction to Epistemology* London: Sage.
Jones, B. Gruffydd (2001), 'Explaining global poverty: a realist critique of the orthodox approach', *Journal of Critical Realism* (incorporating *Alethia*) 4 (1): 2–10; reprinted in J. Cruickshank (ed.), *Critical Realism* (2003).
Jones, P. E. (2003), 'Critical realism and scientific method in Chomsky's linguistics', in J. Cruickshank (ed.), *Critical Realism*.
Joseph, J. (1998), 'In defence of critical realism', *Capital and Class* 65: 73–106.
Joseph, J. (2001), 'Hegemony in the fourth dimension', *Journal for the Theory of Social Behaviour* 31 (3): 261–77.
Joseph, J. (2002), *Hegemony: A Realist Analysis*, London/New York: Routledge.
Joseph, J. (2003a), 'Re-stating hegemonic theory', *Journal of Critical Realism* 2 (1): 127–37.
Joseph, J. (2003b), *Social Theory: Conflict, Cohesion and Consent*, Edinburgh: Edinburgh University Press.
Joseph, J. and Roberts, J. M. (eds) (2003), *Realism, Discourse and Deconstruction*, London/New York: Routledge.
Junor, A. (2001), 'Critical realism comes to management', *Journal of Critical Realism* (incorporating *Alethia*) 4 (1): 30–4.
Kaidesoja, T. (2005), 'The trouble with transcendental arguments: towards a naturalisation of Roy Bhaskar's early realist ontology', *Journal of Critical Realism* 4 (1).
Kaidesoja, T. (2006), 'How useful are transcendental arguments for critical realist ontology?', *Journal of Critical Realism* 5 (2).
Kant, I. (1781/1969), *Critique of Pure Reason*, London: Dent.
Kant, I. (1784/1991), 'An answer to the question, What is enlightenment?', in H. Reiss (ed.), *Kant's Political Writings*, Cambridge: Cambridge University Press.
Kant, I. (1983), *Critique of Pure Reason*, trans. A. V. Miller, London: Allen & Unwin.
Kaplan, M. (1966/1969), 'The new great debate: traditionalism vs. science in international relations', in K. Knorr and J. N. Rosenau (eds), *Contending Approaches to International Politics*, Princeton, NJ: Princeton University Press; first published in the October 1966 issue of *World Politics*.
Kazi, M. A. F. (1998), 'Scientific realist evaluation of social work practice', paper presented at the European Society's Conference, International Conference on Evaluation, Rome, Italy.

Kazi, M. A. F. (2003a), 'Realist evaluation for practice', *British Journal of Social Work* 33 (6): 803–8.
Kazi, M. A. F. (2003b), *Realist Evaluation in Practice: Health and Social Work*, London: Sage.
Kazi, M. A. F., Blom, B., Morén, S., Perdal, Anna-Lena and Rostila, Ilmari (2002), 'Realist evaluation for practice in Sweden, Finland and Britain', *Journal of Social Work Research and Evaluation* 3 (2): 171–86.
Keat, R. and Urry, J. (1975/1982), *Social Theory as Science*, London/Boston, Mass.: Routledge & Kegan Paul.
Kenny, A. (1981), *Five Ways: St Thomas Aquinas' Proofs of God's Existence*, Notre Dame, Ind.: University of Notre Dame Press.
Keynes, J. M. (1936), *The General Theory of Employment, Interest and Money*, New York: Harcourt Brace.
King, A. (1999), 'The impossibility of naturalism: the antinomies of Bhaskar's realism', *Journal for the Theory of Social Behaviour* 29 (3): 267–88.
King, P. and Steiner, R. (eds) (1992), *The Freud–Klein Controversies 1941–1945*, London: Routledge.
Kjærsdam, F. (1995), *Urban Planning in History*, Aalborg: Aalborg University Press.
Klein, M. (1984a), *Love, Guilt and Reparation and Other Works 1921–1945*, London: The Free Press.
Klein, M. (1984b), *Envy and Gratitude and Other Works 1946–1963*, London: The Free Press.
Klosterman, R. E. (1985), 'Arguments for and against planning', *Town Planning Review* 56 (1): 5–20.
Kripke, S. (1972), *Naming and Necessity*, Oxford: Blackwell.
Kolakowski, L. (1966/1972), *Positivist Philosophy: From Hume to the Vienna Circle*, Harmondsworth: Penguin.
Kong, L. (1990), 'Geography and religion: trends and perspectives', *Progress in Human Geography*, 14, pp. 355–371.
Kong, L. (2001), Mapping 'new' geographies of religion: politics and poetics in modernity, *Progress in Human Geography*, 25(2), pp. 211–233.
Kosok, M. (1972), 'The formalisation of Hegel's dialectical logic', in A. MacIntyre, ed., *Hegel: A Collection of Critical Essays*, New York: Doubleday.
Kovel, J. (2002), *The Enemy of Nature: The End of Capitalism or the End of the World?*, London: Zed Books/Fernwood Publishing.
Kowalczyk, R. (2004), 'The effectiveness of high-dependency care', in M. Pidd (ed.) *Systems Modelling: Theory and Practice*, Chichester: Wiley.
Kuhn, T. (1962/1970), *The Structure of Scientific Revolutions*, Chicago, Ill.: University of Chicago Press.
Lacey, H. (1997/1998), 'Neutrality in the social sciences: on Bhaskar's argument for an essential emancipatory impulse in social science', *Journal for the Theory of Social Behaviour* 27 (2–3): 213–41; reprinted in M. S. Archer et al. (1998).
Lacey, H. (1999/2004), *Is Science Value Free? Values and Scientific Understanding*, London: Routledge.
Lacey, H. (2000), 'Listening to the evidence: service activity and understanding social phenomena', in C. D. Lisman and I. Harvey (eds), *Beyond the Tower: Concepts and Models for Service Learning in Philosophy*, Washington, DC: American Association for Higher Education.
Lacey, H. (2002a), 'Assessing the value of transgenic crops', *Ethics in Science and Technology* 8: 497–511.
Lacey, H. (2002b), 'Explanatory critiques and emancipatory movements', *Journal of Critical Realism* 1 (1): 7–31.
Lacey, H. (2003a), 'Critical realism and liberation theology', paper presented to the 2003 IACR Annual Conference, Amsterdam.
Lacey, H. (2003b), 'Seeds and their socio-cultural nexus', in S. Harding and R. Figueroa (eds), *Science and Other Cultures: Issues in the Philosophy of Science and Technology*, New York: Routledge.
Lacey, H. (2006), 'The precautionary principle and the autonomy of science', paper presented to the Third Conference of the Parties to the Convention on Biological Diversity Serving as

the Meeting of the Parties to the Cartagena Protocol on Biosafety (COP–MOP/3), Curitiba, Paraná, Brazil.
Lacey, H., Barbosa de Oliveira, M., Berlan, J.-P., and Dantas, M. (2002), 'Alternatives to technoscience and the values of *Forum Social Mundial*', paper presented at a seminar, 'Technoscience, ecology and capitalism' at *Forum Social Mundial*, Porto Alegre, RS, Brazil.
Laclau, E. and Mouffe, C. (1985), *Hegemony and Socialist Strategy*, London: Verso.
Lakatos, I. (1970), 'Falsification and the methodology of scientific research programmes', in I. Lakatos and A. Musgrave (eds), *Criticism and the Growth of Knowledge*, Cambridge: Cambridge University Press.
Lakoff, G. and Johnson, M. (1999), *Philosophy in the Flesh: The Embodied Mind and Its Challenge to Western Thought*, New York: Basic Books.
Lange, F. (1925), *The History of Materialism*, New York: Harcourt.
Laplace, P. S. de (1819/1951), *A Philosophical Essay on Probabilities*, trans. F. W. Truscott and F.L. Emory, New York: Dover.
Larrain, J. (1979), *The Concept of Ideology*, London: Hutchinson.
Lash, S. and Urry, J. (1994), *Economies of Signs and Space*, London: Sage.
Lawson, C. (2005), 'Technology, technological determinism and the transformational model of social activity', in C. Lawson, J. Latsis and N. Martins (eds), *Contributions to Social Ontology*, London: Routledge.
Lawson, J. (2006), *Critical Realism and Housing Studies*, London: Routledge.
Lawson, T. (1997), *Economics and Reality*, London/New York: Routledge.
Lawson, T. (1999), 'Feminism, realism and universalism', *Feminist Economics* 5 (2): 25–59.
Lawson, T. (2003a), 'Ontology and feminist theorising', *Feminist Economics* 9 (1): 119–50.
Lawson, T. (2003b), *Reorienting Economics*, London/New York: Routledge.
Lawson, T. (2005), 'Economics and critical realism', in G. Steinmetz (ed.), *The Politics of Method in the Human Sciences*.
Layder, D. (1994), *Understanding Social Theory*, London: Sage.
Lefebvre, H. (1995), *Writing on Cities*, Oxford: Blackwell.
Lefebvre, H. (2002), *Critique of Everyday Life*, Vol. II: *Foundations for a Sociology of the Everyday*, trans. J. Moore, London/New York: Verso.
Lenin, V. I. (1899/1961), *The Development of Capitalism in Russia*, in *Collected Works*, Vol. 3, Moscow: Progress.
Lenin, V. I. (1909/1972), *Materialism and Empirio-Criticism*, in *Collected Works*, Vol. 17, Moscow: Progress.
Lewis, D. (1986a), 'Causation' in his *Philosophical Papers*, Vol. 2, Oxford: Oxford University Press.
Lewis, D. (1986b), *The Plurality of Worlds*, Oxford: Blackwell.
Lewis, P. (1996), 'Metaphor and critical realism', *Review of Social Economy* 54 (4): 487–506.
Lewis, P. (2000), 'Realism, causality and the problem of social structure', *Journal for the Theory of Social Behaviour* 30 (3): 249–68.
Lewis, P. (ed.) (2004), *Transforming Economics: Perspectives on the Critical Realist Project*, London: Routledge.
Lipton, P. (2004), *Inference to the Best Explanation*, 2nd edn, London: Routledge.
Littlejohn, C. (2003), 'Critical realism and psychiatric nursing: a philosophical inquiry', *Journal of Advanced Nursing* 43 (5): 449–56.
Lloyd, C. (1986a), *Explanation in Social History*, Oxford: Blackwell.
Lloyd, C. (1986b), 'Realism and structuralism in historical theory: a discussion of the thought of Maurice Mandelbaum', *History and Theory* 28: 296–325.
Lloyd, C. (1993), *The Structures of History*, Oxford: Blackwell.
Locke, J. (1690/1975), *An Essay Concerning Human Understanding*, Oxford: Oxford University Press.
Lockwood, D. (1964), 'Social integration and system integration', in G. Zollschan and W. Hirsch (eds), *Explorations in Social Change*, London: Routledge & Kegan Paul.

Lonergan, B. (1957/1992), *Insight: A Study of Human Understanding*, in *The Collected Works of Bernard Lonergan*, Vol. 3, eds F. E. Crowe and R. M. Doran, Toronto: University of Toronto Press.

López, J. (2001), 'Metaphors as principles of "visuality": "seeing" Marx differently', *Journal of Classical Sociology* 1 (1): 69–94.

López, J. (2003a), 'Critical realism: the difference it makes, in theory', in J. Cruickshank (ed.), *Critical Realism*.

López, J. (2003b), *Society and Its Metaphors: Language, Social Theory and Social Structure*, London/New York: Continuum.

López, J. and Potter, G. (eds) (2001), *After Postmodernism: An Introduction to Critical Realism*, London/New York: Athlone.

Lovejoy, A. O. (1936/1962), *The Great Chain of Being*, Cambridge, Mass.: Harvard University Press.

Lovelock, J. (1979), *Gaia: A New Look at Life on Earth*, Oxford: Oxford University Press.

Lua, R. (2004), 'Critical realism and news production', *Media, Culture and Society* 26 (5): 693–711.

Lukács, G. (1919–24/1978), *History and Class Consciousness*, London: Merlin.

Luhmann, N. (1986), 'The autopoiesis of social systems', in F. Geyer and J. van der Zouwen (eds), *Sociocybernetic Paradoxes*, London: Sage.

Lukes, S. (1974), *Power: A Radical View*, London: Macmillan.

Lukes, S. (ed.) (1986), *Power*, Oxford: Blackwell.

Lyotard, J.-F. (1984), *The Postmodern Condition: A Report on Knowledge*, Minneapolis, Minn.: University of Minnesota Press; Manchester: Manchester University Press.

McEvoy P. and Richards, D. (2003), 'Critical realism: a way forward for evaluation research in nursing?', *Journal of Advanced Nursing* 43 (4): 411–20.

McGrath, A. E. (2001), *A Scientific Theology*, Vol. 1, *Nature*, Edinburgh: T. & T. Clark; Grand Rapids, Mich.: Wm B. Eerdmans.

McGrath, A. E. (2002), *A Scientific Theology*, Vol. 2 *Reality*, Edinburgh: T. & T. Clark; Grand Rapids, Mich.: Wm B. Eerdmans..

McGrath, A. E. (2003), *A Scientific Theology*, Vol. 3 *Theory*, Edinburgh: T. & T. Clark; Grand Rapids, Mich.: Wm B. Eerdmans..

MacIntyre, Alisdair (1981), *After Virtue: A Study in Moral Theory*, London: Duckworth.

McLennan, G. (1981), *Marxism and the Methodologies of History*, London: Verso.

Macpherson, C. B. (1962), *The Political Theory of Possessive Individualism: Hobbes to Locke*, Oxford: Oxford University Press.

McTaggart, J. E. M. (1908), 'The unreality of time', *Mind* 17: 457–74; reprinted in S. V. Keeling (ed.), *Philosophical Studies: J. McT. Ellis McTaggart*, Bristol: Thoemmes Press, 1996.

McTaggart, J. E. M. (1927), 'Time', in *The Nature of Existence*, Vol. II, Cambridge: Cambridge University Press.

Maki, U. (2001), 'Economic ontology: what? why? how?', *The Economic World View: Studies in the Ontology of Economics*, Cambridge: Cambridge University Press.

Malik, K. (1996), *The Meaning of Race: Race, History and Culture in Western Society* London: Macmillan.

Mamdani, M. (1996), *Citizen and Subject: Contemporary Africa and the Legacy of Late Colonialism*, Kampala: Fountain; Cape Town: D. Philip; London: J. Currey.

Manicas, P. (1987), *A History and Philosophy of the Social Sciences*, Oxford: Blackwell.

Manicas, P. and Secord, P. (1983), 'Implications for psychology of the new philosophy of science', *American Psychologist* 38: 399–413.

Mandelbaum, M. (1955), 'Societal facts', *British Journal of Sociology* 6: 305–17; reprinted in P. Gardiner (ed.), *Theories of History*, New York: Free Press, 1959.

Mandelbaum, M. (1964), *Philosophy, Science and Sense Perception: Historical and Critical Studies*, Baltimore, Md.: Johns Hopkins University Press.

Mannheim, K. (1960), *Ideology and Utopia*, London: Routledge.

Mäntysaari, M. (1991), Sosiaalibyrokratia asiakkaiden valvojana [Social bureaucracy in controlling clients], *Sosiaalipoliittisen yhdistyksen tutkimuksia* 51, Tampere: Vastapaino.
Marsden, R. (1999), *The Nature of Capital: Marx after Foucault*, London/New York: Routledge.
Marsh, D., Buller, J., Hay, C., Johnson, J., Kerr, P., McAnulla, S. and Watson, M. (1999), *Post-War British Politics in Perspective*, Cambridge: Polity.
Marris, P. and Thornham, S. (eds) (1997), *Media Studies: A Reader*, Edinburgh: Edinburgh University Press.
Marshall, A. (1891/1930), *Principles of Economics*, London: Macmillan.
Martin-Soskice, J. H. and Harré, R. (1982), 'Metaphor in science', in D. S. Miall (ed.), *Metaphor: Problems and Perspectives*, Brighton: Harvester.
Martinez-Alier, J. (2003), *The Environmentalism of the Poor: A Study of Ecological Conflicts and Valuation*, Cheltenham: Edward Elgar.
Marx, K. (1852/1978), *The Eighteenth Brumaire of Louis Bonaparte*, in R. C. Tucker (ed.), *The Marx–Engels Reader*, 2nd edn, New York/London: Norton.
Marx, K. (1867/1976), *Capital*, Vol. I, trans. B. Fowkes, Harmondsworth: Penguin.
Marx, K. (1939/1973), *Grundrisse: Introduction to the Critique of Political Economy*, Harmondsworth: Penguin.
Marx, K. (1971), *Early Texts*, ed. L. Colletti, Oxford: Blackwell.
Marx, K. (1975), *Early Writings*, ed. D. McLellan, Harmondsworth: Penguin.
Marx, K. (1977), *Selected Writings*, ed. D. McLellan, Oxford: Oxford University Press.
Marx, K. and Engels, F. (1970), *The German Ideology*, ed. C. J. Arthur, London: Lawrence and Wishart.
Massey, D. (2005), *For Space*, London: Sage.
Maturana, H. and Varela, F. (1980), *Autopoiesis and Cognition: The Realization of the Living*, Dordrecht: Reidel.
Maturana, H. and Varela, F. (1987), *The Tree of Knowledge*, Boston, Mass.: Shambhala.
Mellor, D. H. (1988), *The Warrant of Induction*, Cambridge: Cambridge University Press.
Mellor, M. (1997), *Feminism and Ecology*, London: Polity.
Menger, C. (1871/1976), *Principles of Economics*, trans. J. Dingwall and B. F. Hoselitz, New York: New York University Press.
Merleau-Ponty, M. (1962/1994), *Phenomenology of Perception*, New York: Routledge.
Merton, R. (1968), *Social Theory and Social Structure*, 3rd edn, New York: Free Press.
Mill, J. S. (1843/1996), *System of Logic* (*Collected Works*, Vol. 7), London: Routledge.
Mingers, J. (1995), *Self-Producing Systems: Implications and Applications of Autopoiesis*, New York: Plenum Press.
Mingers, J. (1999), 'Synthesising constructivism and critical realism: towards critical pluralism', in E. Mathijs, J. Van der Veken and H. Van Belle (eds), *World Views and the Problem of Synthesis*, Amsterdam: Kluwer Academic.
Mingers, J. (2000), 'The contribution of critical realism as an underpinning philosophy for OR/MS and systems', *Journal of the Operational Research Society* 51: 1256–70.
Mingers, J. (2002a), 'Can social systems be autopoietic? Assessing Luhmann's social theory', *Sociological Review* 50 (2): 278–99.
Mingers, J. (2002b), 'Observing organizations: an evaluation of Luhmann's theory of organizations', in T. Hernes and T. Bakken (eds), *Autopoietic Organization Theory: Drawing on Niklas Luhmann's Social Systems Perspective*, Copenhagen: Copenhagen Business School Press.
Mingers, J. (2004), 'Future directions in operations research modelling: critical realism and multimethodology', in S. Fleetwood and S. Ackroyd (eds), *Critical Realist Applications in Organisation and Management Studies*, London: Routledge.
Minnerup, G. (2003), 'Postmodernism and German history', *Debatte: Review of Contemporary German Affairs* 11 (2): 187–212.

Bibliography

Mitcham, C. (1994), *Thinking through Technology: The Path between Engineering and Philosophy*, Chicago, Ill./London: University of Chicago Press.

Mohanty, S. P. (1997), *Literary Theory and the Claims of History: Postmodernism, Objectivity, and Multicultural Politics*, Ithaca, NY: Cornell University Press.

Moll, I. (2004), 'Psychology, biology and social relations', *Journal of Critical Realism* 3 (1).

Moore, G. E. (1939/1993), 'Proof of an external world', in *G. E. Moore: Selected Writings*, London: Routledge.

Moore, R. (2004), *Education and Society: Issues and Explanations in the Sociology of Education*, Cambridge: Polity Press.

Morelli, Mark D. (2003), 'The realist response to idealism in England and Lonergan's critical realism', *Method: Journal of Lonergan Studies* 1 (21): 1–23.

Morén, S. and Blom, B. (2003a), *Insatser och resultat. Om utvärdering i socialt arbete* [Interventions and Results. On Evaluation of Social Work Practice], *Rapport* 48, Institutionen för socialt arbete: Umeå Universitet.

Morén, S. and Blom, B. (2003b), 'Explaining human change: on generative mechanisms in social work practice', *Journal of Critical Realism* 2 (1).

Morgan, J. (2002), 'Philosophical realism in international relations theory: Kratochwil's constructivist challenge to Wendt', *Journal of Critical Realism* 1 (1): 95–118.

Morgan, J. (2003a), '*Empire* inhuman? The social ontology of global theory', *Journal of Critical Realism* 2 (1).

Morgan, J. (2003b), 'What is meta-Reality? Alternative interpretations of the argument', *Journal of Critical Realism* 2 (2): 115–46.

Morgan, J. (2004a), 'Analytical philosophy's contribution to the problem of supervenience (emergence) illustrated using the mind-body problem' (review article), *Journal of Critical Realism* 3 (1).

Morgan, J. (2004b), 'The nature of a transcendental argument: toward a critique of *Dialectic: The Pulse of Freedom*', *Journal of Critical Realism* 3 (2).

Morgan, J. (2005a), 'An alternative argument for transcendental realism based on an immanent critique of Kant' (review essay), *Journal of Critical Realism* 4 (2).

Morgan, J. (2005b), 'Ontological casuistry? Bhaskar's meta-Reality, fine structure, and human disposition', *New Formations*, 56: 133–46.

Morgan, J. (2006), 'The relevance of a transcendental mode of inquiry: a rejoinder to Kaidesoja', *Journal of Critical Realism* 5 (2).

Morgenthau, H. (1948/1973), *Politics among Nations: The Struggle for Power and Peace*, New York: Alfred A. Knopf.

Morris, W. (1891/1912), *News from Nowhere; or, An Epoch of Unrest: Being Some Chapters from a Utopian Romance*, in *The Collected Works of William Morris*, Volume 16, London: Longmans Green.

Morris, W. (1973), *The Political Writings of William Morris*, ed. A. L. Morton, London: Lawrence & Wishart.

Morrow, R. (1994), 'Sexuality as discourse: beyond Foucault's constructionism', *Australian and New Zealand Journal of Sociology* 31 (1): 15–31.

Morrow, R. (2003), *A Sociological Critique of Masters and Johnson's Sex Research and Therapy*, PhD thesis, University of New England, Armidale, NSW.

Mouzelis, N. P. (1995), *Sociological Theory: What Went Wrong?* London/New York: Routledge.

Mutch, A. (2002), 'Actors and networks or agents and structures: towards a realist view of information systems', *Organization* 9 (3): 477–96.

Næss, A. (1989), *Ecology, Community and Lifestyle: Outline of an Ecosophy*, Cambridge: Cambridge University Press.

Næss, P. (2004), 'Prediction, regressions and critical realism', *Journal of Critical Realism* 3 (1): 133–64.

Næss, P. (2006a), 'Cost–benefit analyses of transportation investments: neither critical nor realistic', *Journal of Critical Realism* 5 (1).

Næss, P. (2006b), 'Unsustainable growth, unsustainable capitalism', *Journal of Critical Realism* 5 (2), forthcoming.

Nash, R. (1999), 'What is real and what is realism in sociology?' *Journal for the Theory of Social Behaviour*, 29 (4): 445–66.

Nash, R. (2002), 'A realist framework for the sociology of education: thinking with Bourdieu', *Educational Philosophy and Theory* 34 (3): 273–88.

Nash, R. (2004), 'Can the arbitrary and the necessary be reconciled? Scientific realism and the school curriculum', *Journal of Curriculum Studies* 36 (5): 605–23.

Nellhaus, T. (1998), 'Signs, social ontology, and critical realism', *Journal for the Theory of Social Behaviour* 28 (1): 1–24.

Nellhaus, T. (2000), 'Social ontology and (meta)theatricality: reflexions on performance and communication in history', *Journal of Dramatic Theory and Criticism* 14 (2): 3–39.

Nellhaus, T. (2004), 'From embodiment to agency: cognitive science, critical realism and communicative frameworks', *Journal of Critical Realism* 3 (1): 103–32.

Nellhaus, T. (2005), 'Critical realism and performance strategies', in D. Krasner and D. Saltz (eds), *Staging Philosophy: New Approaches to Theater and Performance*, Ann Arbor, Mich.: University of Michigan Press.

New, C. (1994), 'Structure, agency and social transformation', *Journal for the Theory of Social Behaviour* 24 (3): 187–206.

New, C. (1996a), *Agency, Health and Social Survival: The Ecopolitics of Rival Psychologies*, London: Taylor & Francis.

New, C. (1996b/1997), 'Man bad, woman good? Essentialisms and ecofeminisms', *New Left Review* 216: 79–93, reprinted in L. McDowell and J. Sharp (eds), *Space, Gender, Knowledge*, London: Arnold.

New, C. (1998), 'Realism, deconstruction and the feminist standpoint', *Journal for the Theory of Social Behaviour* 28 (4): 349–72.

New, C. (2001a), 'Oppressed and oppressors? The systematic mistreatment of men', *Sociology* 35 (3): 729–48.

New, C. (2001b), 'Realising the potential: The ESRC Seminar Series on social realism and empirical research', *Journal of Critical Realism* (incorporating *Alethia*) 4 (1): 43–7.

New, C. (2003), 'Feminism and critical realism and the linguistic turn', in J. Cruickshank (ed.), *Critical Realism*.

New, C. (2005), 'Sex and gender: a critical realist approach', *New Formations* 56: 54–70.

Nicod, J. (1923/1969), 'The logical principle of induction', in his *Geometry and Induction*, London: Routledge & Kegan Paul.

Nielsen, P. (2002), 'Reflections on critical realism in political economy', *Cambridge Journal of Economics* 26: 727–38.

Nielsen, P. and Morgan, J. (2006), 'From mainstream economics to the boundaries of Marxism', *Capital & Class* 89.

Niiniluoto, I. (1999), *Critical Scientific Realism*, Oxford: Oxford University Press.

Nolan, P. (2002a), 'A Darwinian historical materialism', in P. Blackledge and G. Kirkpatrick (eds), *Historical Materialism and Social Evolution*, London: Palgrave.

Nolan, P. (2002b), 'What's Darwinian about historical materialism? A critique of Levine and Sober', *Historical Materialism* 10 (2): 143–69.

Nolan, P. (2003), 'Levine and Sober: a rejoinder', *Historical Materialism* 11 (3): 183–200.

Norrie, A. (1993), *Crime, Reason and History*, London: Butterworth.

Norrie, A. (1998), 'The praxiology of legal judgement', in M. S. Archer et al. (eds), *Critical Realism: Essential Readings*.

Norrie, A. (1999), 'Albert Speer, guilt and "the space between"', in M. Matravers (ed.), *Punishment and Political Theory*, Oxford: Hart.

Norrie, A. (2000a), 'Justice and relationality', *Alethia* 3 (1): 2–5.

Norrie, A. (2000b), *Punishment, Responsibility and Justice: A Relational Critique*, Oxford: Oxford University Press.

Norrie, A. (2004a), 'Bhaskar, Adorno and the dialectics of modern freedom', *Journal of Critical Realism* 3 (1).

Norrie, A. (2004b), 'Dialectics, deconstruction and the legal subject', in J. Joseph and J. Roberts (eds), *Realism, Discourse and Deconstruction*.

Norrie, A. (2005), 'Theorising "spectrality": ontology and ethics in Derrida and Bhaskar', *New Formations* 56: 96–108.

Norris, C. (1993), *The Truth about Postmodernism*, Oxford: Blackwell.

Norris, C. (1996), *Reclaiming Truth: Contributions to a Critique of Cultural Relativism*, London: Lawrence & Wishart.

Norris, C. (1997), *New Idols of the Cave: On the Limits of Anti-Realism*, Manchester: Manchester University Press.

Norris, C. (2000), *Quantum Theory and the Flight from Realism: Philosophical Responses to Quantum Mechanics*, London/New York: Routledge.

Norris, C. (2003), 'Response-dependence: what's in it for the realist?', *Journal of Critical Realism* 1 (2): 61–90.

Norris, C. (2004), *Philosophy of Language and the Challenge to Scientific Realism*, London/New York: Routledge.

Norris, C. (2005), *Epistemology: Key Concepts in Philosophy*, London/New York: Continuum.

Norris, C. and Papastephanou, Marianna (2002), 'Deconstruction, anti-realism and philosophy of science: an interview with Christopher Norris', *Journal of Philosophy of Education* 36 (2).

Nygren, L. and Soydan, H. (1997), 'Social work research and its dependence on practice', *Scandinavian Journal of Social Welfare* 6: 217–24.

O'Connor, J. (1996), 'The second contradiction of capitalism', in T. Benton (ed.), *The Greening of Marxism*, New York: Guilford.

Offe, C. (1984), *Contradictions of the Welfare State*, London: Hutchinson.

Oliver, M. (1990), *The Politics of Disablement*, London: Macmillan.

Ollman, B. (1971/1976), *Alienation*, Cambridge: Cambridge University Press.

Ollman, B. (1993), *Dialectical Investigations*, London: Routledge.

Ollman, B. (2003), *Dance of the Dialectic: Steps in Marx's Method*, Urbana, Ill./Chicago, Ill.: University of Illinois Press.

Olroyd, D. R. (1986), *The Arch of Knowledge: An Introductory Study of the History of the Philosophy and Methodology of Science*, London: Routledge.

Olsen W. K. (1996), *Rural Indian Social Relations*, Delhi: Oxford University Press.

Olsen, W. K. (2001), 'Stereotypical and traditional views about the gender division of labour in Indian labour markets', *Journal of Critical Realism* (incorporating *Alethia*) 4 (1): 11–17.

Olsen, W. K. et al. (2002), *The Politics of Money: Towards Sustainability and Economic Democracy*, London: Pluto.

O'Neill, J. (1998), *The Market: Ethics, Knowledge and Politics* London: Routledge.

O'Neill, J. (2004), 'Commerce and the language of value', in Archer and Outhwaite (eds), *Defending Objectivity*.

Ong, W. (1982), *Orality and Literacy: The Technologising of the Word*, London: Methuen.

Osborne, D. and Gaebler, T. (1992), *Reinventing Government*, Reading, Mas.: Addison-Wesley.

Outhwaite, W. (1987), *New Philosophies of Social Science: Realism, Hermeneutics and Critical Theory*, London: Macmillan.

Outhwaite, W. (ed.) (2002), *The Blackwell Dictionary of Modern Social Thought*, 2nd edn, Oxford: Blackwell.

Park, C. (1994), *Sacred Worlds: An introduction to geography and religion*, New York, NY: Routledge.
Parker, I. (ed.) (1998), *Social Constructionism, Discourse and Realism*, London: Sage.
Parker, I. (2002), *Critical Discursive Psychology*, London: Palgrave.
Parker, J. (2001), 'The precautionary principle' in R. Chadwick (ed) *The Concise Encyclopedia of the Ethics of New Technologies*, London: Academic Press.
Parkin, I. (ed.) (1998), *Social Constructionism, Discourse and Realism*, London: Sage.
Parsons, T. (1937), *The Structure of Social Action*, New York: Free Press.
Parsons, Terence (2006), 'The traditional square of opposition', *The Stanford Encyclopedia of Philosophy* (Winter 2006 Edition), E. N. Zalta (ed.), http://plato.standford.edu/archives/win2006/entries/square/.
Pashukanis, E. (1978), *Law and Marxism: A General Theory*, ed. C. Arthur, London: Ink Links.
Passmore, J. (1957/1968), *A Hundred Years of Philosophy*, Harmondsworth: Penguin.
Pateman, T. (1987), *Language in Mind and Language in Society*, Oxford: Clarendon Press.
Patomäki, H. (2002), *After International Relations: Critical Realism and the (Re)Construction of World Politics*, London/New York: Routledge.
Patomäki, H. (2003), 'A critical realist approach to global political economy', in J. Cruickshank (ed.), *Critical Realism*.
Patomäki, H. (2004), *A Possible World: Democratic Transformation of Global Institutions*, London: Zed Books.
Patomäki, H. (2006a), 'Global security: learning from possible futures', in Hans Günter Brauch et al. (eds), *Globalisation and Environmental Challenges: Reconceptualising Security in the Twenty-First Century*, Berlin: Springer-Verlag.
Patomäki, H. (2006b), 'Realist ontology for futures studies', *Journal of Critical Realism* 5 (1).
Patomäki, H. and Wight, C. (2000), 'After post-positivism? The promises of critical realism', *International Studies Quarterly* 44 (2): 213–37.
Pawson, R. (1989), *A Measure for Measures: A Manifesto for Empirical Sociology*, London/New York: Routledge.
Pawson, R. (2000), 'Middle-range realism', *Archives Européennes de Sociologie* 41 (2): 283–325.
Pawson, R. and Tilley, N. (1997), *Realistic Evaluation*, London: Sage.
Peirce, C. S. (1893–1913/1998), *The Essential Peirce: Selected Philosophical Writings*, Vol. 2 (1893–1913), Peirce Edition Project, Bloomington, Ind.: Indiana University Press.
Peirce, C. S. (1932), *Collected Papers*, Vol. 2, ed. C. Hartshorne and P. Weiss, Cambridge, Mass.: Belknap Press.
Penrose, R. (1989), *The Emperor's New Mind: Concerning Computers, Minds and the Laws of Physics*, Oxford: Oxford University Press.
Penrose, R. (1994), *Shadows of the Mind: A Search for the Missing Science of Consciousness*, Oxford: Oxford University Press.
Penrose, R. (2004), *The Road to Reality: A Complete Guide to the Laws of the Universe*, London: Jonathan Cape.
Pidd, M. (2002), *Tools for Thinking: Modelling in Management Science*, 2nd edn, Chichester: Wiley.
Pinkstone, B. (2002), 'Persistent demi-regs and robust tendencies: critical realism and the Singer–Prebisch Thesis', *Cambridge Journal of Economics* 26 (5): 561–83.
Pinkstone, B. (2003a), 'Critical realism and applied work in economic history: some methodological implications', in Downward (ed.), *Applied Economics*.
Pinkstone, B. (2003b), 'Reorienting economics: new horizons', *Journal of Critical Realism* 2 (1).
Pinter, H. (2005), 'Art, truth and politics', Nobel lecture, http://nobelprize.org/literature/laureates/2005/pinter-lecture-e.html.
Pinter, H. (2006), 'Interview', *Newsnight Review*, BBC TV, 23 June.
Pols, E. (1982), *The Acts of Our Being: A Reflection on Agency and Responsibility*, Amherst, Mass.: University of Massachusetts Press.

528 Bibliography

Pols, E. (1992), *Radical Realism: Direct Knowing in Science and Philosophy*, Ithaca, NY: Cornell University Press.
Popper, K. R. (1934/1980/1997), *The Logic of Scientific Discovery*, London/New York: Routledge.
Popper, K. R. (1944/1986), *The Poverty of Historicism*, London/New York: Ark.
Popper, K. R. (1959), 'The propensity interpretation of probability', *British Journal for the Philosophy of Science* 10: 25–42.
Popper, K. R. (1963/1976/1989), *Conjectures and Refutations: The Growth of Scientific Knowledge*, London/New York: Routledge.
Popper, K. R. (1972/1979), *Objective Knowledge: An Evolutionary Approach*, revised edn. Oxford: Clarendon.
Popper, K. R. (1990/1995), *A World of Propensities*, Bristol: Thoemmes.
Porpora, D. V. (1987), *The Concept of Social Structure*, New York: Greenwood Press.
Porpora, D. V. (1989), 'Four concepts of social structure', *Journal for the Theory of Social Behaviour* 19: 195–212.
Porpora, D. V. (1994), 'Cultural rules and material relations', *Sociological Theory* 11: 212–29.
Porpora, D. V. (2000a), 'Quantum reality as unrealised possibility', *Alethia* 3 (2): 34–9.
Porpora, D. V. (2000b), 'The sociology of ultimate concern', *Alethia* 3 (1): 10–15.
Porpora, D. V. (2001a), 'Critical realism, Marxism, and religion: gotta problem with that?', paper presented to the 2001 IACR Annual Conference, Bradford, England.
Porpora, D. V. (2001b), 'Do realists run regressions?', in J. López and G. Potter (eds), *After Postmodernism*.
Porpora, D. V. (2001c), *Landscapes of the Soul: The Loss of Moral Meaning in American Life*, New York: Oxford University Press.
Porpora, D. V. (2003), 'Critical realism and liberation theology: a response to Hugh Lacey', paper presented to the 2003 IACR Annual Conference, Amsterdam.
Porpora, D. V. (2005), 'The spiritual turn and critical realism', *New Formations* 56: 133–46.
Porter, S. (1993), 'Critical realist ethnography: the case of racism and professionalism in a medical setting' *Sociology* 27 (4): 591–609.
Porter, S. (2000), 'Critical realist ethnography', in S. Ackroyd and S. Fleetwood (eds), *Realist Perspectives on Management and Organisations*.
Porter, S. and Ryan, S. (1996), 'Breaking the boundaries between nursing and sociology: a critical realist ethnography of the theory–practice gap', *Journal of Advanced Nursing* 24: 413–20.
Postone, M. (1993), *Time, Labor and Social Domination: A Reinterpretation of Marx's Critical Social Theory*, Cambridge: Cambridge University Press.
Potter, G. (2000), *The Philosophy of Social Science: New Perspectives*, Harlow: Prentice-Hall.
Potter, G. (2006), 'Re-opening the wound: against God and Bhaskar', *Journal of Critical Realism* 5 (1).
Potter, J. (1995), *Representing Reality: Discourse, Rhetoric and Social Construction*, London: Sage.
Poulantzas, N. (1970/1973), *Political Power and Social Classes*, London: New Left Books.
Prior, A. N. (1967), *Past, Present and Future*, Oxford: Oxford University Press.
Psillos, S. (1999), *Scientific Realism: How Science Tracks Truth*, London/New York: Routledge.
Psillos, S. (2002a), *Causation and Explanation*, Chesham: Acumen.
Psillos, S. (2002b), 'Simply the best: a case for abduction', in A. C. Kakas and F. Sadri (eds), *Computational Logic: From Logic Programming into the Future*, Berlin Heidelberg: Springer-Verlag.
Putnam, H. (1975), 'Explanation and reference', in his *Philosophical Papers*, Vol. 2, Cambridge: Cambridge University Press.
Putnam, H. (1981), *Realism, Truth and History*, Cambridge: Cambridge University Press.
Putnam, H. (1983), *Realism and Reason (Philosophical Papers*, Vol. 3), Cambridge: Cambridge University Press.
Putnam, H. (2002), *The Collapse of the Fact/Value Dichotomy and Other Essays*, Cambridge: Harvard University Press.

Quine, W. V. O. (1969/1977), *Ontological Relativity and Other Essays*, New York: Columbia University Press.

Quine, W. V. O. (1970), *Philosophy of Logic*, Englewood Cliffs, NJ: Prentice-Hall.

Ratcliffe, P. (2004), *'Race', Ethnicity and Difference: Imagining the Inclusive Society*, Maidenhead: Open University Press/McGraw-Hill.

Ray, L. and Sayer, A. (eds) (1999), *Culture and Economy after the Cultural Turn*, London: Sage.

Rescher, N. (1987), *Scientific Realism: A Critical Reappraisal*, Dordrecht: Reidel.

Retort (I. Boal, T. J. Clark, J. Matthews, M. Watts) (2005), *Afflicted Powers: Capital and Spectacle in a New Age of War*, London: Verso.

Reuten, G. and Williams, M. (1989), *Value-Form and the State: The Tendencies of Accumulation and the Determination of Economic Policy in Capitalist Society*, London: Routledge.

Riehl, A. (1876–87), *Der philosophische Kriticismus und seine Bedeutung für die positive Wissenschaft*, Vol. I, *Geschichte und Methode der philosophischen Kriticismus*, Vol. II, *Die sinnlichen und logischen Grundlagen der Erkenntniss*, Vol. III, *Zur Wissenschaftstheorie und Metaphysik*, Leipzig: Wilhelm Engelmann.

Roberts, J. M. (1999), 'Marxism and critical realism: the same, similar, or just plain different?', *Capital and Class* 68.

Roberts, J. M. (2002), 'Abstracting emancipation: two dialectics on the trail of freedom', in A. Brown et al. (eds), *Critical Realism and Marxism*.

Robbins, L. (1932), *An Essay on the Nature and Significance of Economic Science*, London: Macmillan.

Rogers, T. (2004), 'The doing of a depth-investigation: implications for the emancipatory aims of critical naturalism', *Journal of Critical Realism* 3(2), PP. 238–69.

Ron, A. (2002), 'Regression analysis and the philosophy of social science: a critical realist view', *Journal of Critical Realism* 1 (1): 119–42.

Rorty, R. (1965), 'Mind-body, identity, privacy and categories', *Review of Metaphysics* 17 (1): 24–54.

Rosen, M. (1982), *Hegel's Dialectic and Its Criticism*, Cambridge: Cambridge University Press.

Rosenhead, J. and Mingers, J. (eds) (2001), *Rational Analysis for a Problematic World Revisited: Problem Structuring Methods for Complexity, Uncertainty and Conflict*, 2nd edn, Chichester: Wiley.

Rostila, I. (2000), Realistinen arviointitutkimus ja onnistumisen pakot [Realist evaluation and the pressures to succeed], in R. Laitinen, *Arvioinnin arkea ja peruskysymyksiä*, Helsinki: Sosiaali- ja terveysturvan Keskusliitto.

Rostila, I. (2001), *Sosiaalisen kuntoutuksen mekanismit. Monet-projektin realistinen arviointi* [Mechanisms of social rehabilitation. Realist evaluation of the Monet project], Helsinki: National Research and Development Centre of Welfare and Health.

Rousseau, J. J. (1754/1958), *A Discourse on the Origin of Inequality*, in *The Social Contract/Discourses*, trans. G. D. H. Cole, London: J. M. Dent.

Ruben, D.-H. (1990), *Explaining Explanation*, London: Routledge.

Rubery, J. and Grimshaw, D. (1998), 'Integrating the internal and external labour markets, *Cambridge Journal of Economics* 22: 199–220.

Rubery, J. and Grimshaw, D. (2003), *The Organization of Employment: An International Perspective*, London: Palgrave Macmillan.

Rubin, G. (1975), 'The traffic in women: notes on the political economy of sex', in R. R. Reiter (ed.), *Towards an Anthropology of Women*, New York: Monthly Review Press.

Rustin, M. (1991a), *The Good Society and the Inner World: Psychoanalysis, Politics and Culture*, London: Verso.

Rustin, M (1991b), 'Psychoanalysis, philosophical realism and the new sociology of science', *The Good Society and the Inner World*, London: Verso.

Ryle, G. (1949), *The Concept of Mind*: London: Hutchinson.

Ryner, M. (2002), *Capitalist Restructuring, Globalisation and the Third Way: Lessons from the Swedish Model*, London: Routledge.

Sachs, W. (1999), *Planet Dialectics*, London: Zed Books.
Saurin, J. (1996), 'Globalisation, poverty, and the promises of modernity', *Millennium: Journal of International Studies* 25 (3): 657–80.
Sayer, A. (1984/1992), *Method in Social Science: A Realist Approach*, London/New York: Routledge.
Sayer, A. (1989), 'The "new" regional geography and problems of narrative', *Environment and Planning D: Society and Space* 7 (3): 253–76.
Sayer, A. (1995), *Radical Political Economy*, Oxford: Blackwell.
Sayer, A. (1997a), 'Critical realism and the limits to critical social science', *Journal for the Theory of Social Behaviour* 27 (4)
Sayer, A. (1997b), 'Essentialism, social constructionism and beyond', *Sociological Review* 45 (3): 453–87.
Sayer, A. (2000a), *Realism and Social Science*, London: Sage.
Sayer, A. (2000b), 'System, lifeworld and gender: associational versus counterfactual thinking', *Sociology* 34: 707–25.
Sayer, A. (2004), 'Restoring the moral dimension in social scientific accounts: a qualified ethical naturalist approach', in Archer and Outhwaite (eds), *Defending Objectivity*.
Sayer, A. (2005), *The Moral Significance of Class*. Cambridge: Cambridge University Press.
Sayer, A. and Walker, R. (1992), *The New Social Economy: Reworking the Division of Labor*, Cambridge, Mass.: Blackwell.
Sayers, S. (1998), *Marxism and Human Nature*, London: Routledge.
Sayers, S. (2003), 'Creative activity and alienation in Hegel and Marx', *Historical Materialism* 11 (1): 107–28.
Saussure, F. de (1916/1983), *Course in General Linguistics*, trans. R. Harris, La Salle, Ill.: Open Court; London: Duckworth.
Scambler, G. (2002), *Health and Social Change: A Critical Perspective*, Buckingham: Open University Press.
Scambler, G. and Higgs, P. (1999), 'Stratification, class and health: class relations and health inequalities in high modernity', *Sociology* 33: 275–96.
Scambler, G. and Scambler, S. (2003), 'Realist agendas on biology, health and medicine: some thoughts and reflections', in S. Williams, L. Birke and G. Bendelow, *Debating Biology: Sociological Reflections on Health, Medicine and Society*, London: Routledge.
Scheler, M. (1980), *Problems of a Sociology of Knowledge*, London: Routledge and Kegan Paul.
Schiller, F. (1795/1967), *Letters on the Aesthetic Education of Mankind*, Oxford: Clarendon Press.
Scott, A. J. and Roweis, S. T. (1977), 'Urban planning in theory and practice: a reappraisal', *Environment and Planning A* 9: 1097–119.
Scott, D. (2000), *Realism and Educational Research: New Perspectives and Possibilities*, London: Routledge.
Scot Henderson, J. (1876), Review of von Hartmann (1875), *Mind* 1 (3): 407–9.
Schroder, K., Drotner, K., Kline, S. and Murray, C. (2003), *Researching Audiences*, London: Arnold.
Schrödinger, E. (1944), *What is Life?* Cambridge: Cambridge University Press.
Searle, J. R. (1995), *The Construction of Social Reality*, New York: Free Press.
Searle, J. R. (1995), 'The problem of consciousness', in C. N. Headley, P. Antonacci and M. Rabinowitz (eds), *Thinking and Literacy: The Mind at Work*, Hillsdale, NJ: Lawrence Erlbaum.
Searle, J. R. (1998), *The Mystery of Consciousness*, London: Granta.
Searle, J. R. (1999), *Mind, Language and Society*, London: Phoenix.
Sellars, R. W. (1916), *Critical Realism: A Study of the Nature and Conditions of Knowledge*, New York: Rand McNally.
Sellars, R. W. (1924), 'Critical realism and its critics', *Philosophical Review* 33: 379–97.
Sellars, R. W. (1929), 'A re-examination of critical realism', *Philosophical Review* 38: 439–55.

Sellars, R. W. (1937), 'Critical realism and the independence of the object', *Journal of Philosophy, Psychology and Scientific Methods* 34: 541–50.
Sellars, R. W. (1939), 'A statement of critical realism', *Revue Internationale de Philosophie* 3: 472–96.
Sellars, W. (1963/1991), *Science, Perception and Reality*, Atascadero, Calif.: Ridgeview.
Sen, A. (1997), 'Human rights and Asian values: what Lee Kuan Yew and Li Peng don't understand about Asia', *The New Republic* 33–40.
Senge, P. (1990), *The Fifth Discipline: The Art and Practice of the Learning Organization*, London: Century Books.
Sewell, W. H. Jr. (1992), 'A theory of structure: duality, agency, and transformation', *American Journal of Sociology* 98: 1–29.
Sheppard, E. and McMaster, R. B. (eds) (2003), *Scale and Geographic Inquiry: Nature, Society, and Method*, Oxford: Blackwell.
Shipway, B. (2000), 'Critical realism and theological critical realism: opportunities for dialogue' *Alethia* 3 (2): 29–33.
Shipway, B. (2002), *Implications of a Critical Realist Perspective in Education*, unpublished PhD thesis, Southern Cross University, Australia.
Shipway, B. (2004), 'The theological application of Bhaskar's stratified reality: the scientific theology of A. E. McGrath', *Journal of Critical Realism* 3 (1).
Shore, B. (1996), *Culture in Mind: Cognition, Culture, and the Problem of Meaning*, Oxford: Oxford University Press, 1996.
Skagestad, P. (2000), 'Peirce, virtuality, and semiotics', http://www.bu.edu/wcp/Papers/Cogn/CognSkag.
Smart, J. J. C. (1963), 'Symposium: materialism,' *Journal of Philosophy* 60 (22): 651–62.
Smith, C. (1990), *Auctions: The Social Construction of Value*, Berkeley, Calif.: University of California Press.
Smith, C. (1996), *Marx at the Millennium*, London: Pluto.
Smith, M. J. (2000), *Rethinking State Theory*, London/New York: Routledge.
Smith, T. (1990), *The Logic of Marx's 'Capital': Replies to Hegelian Criticisms*, Albany, NY: State University of New York Press.
Smith, T. (1993), *Dialectical Social Theory and Its Critics: From Hegel to Analytical Marxism and Postmodernism*, Albany, NY: State University of New York Press.
Smith, T. (1999), 'The relevance of systematic dialectics to Marxian thought: a reply to Rosenthal', *Historical Materialism* 4: 51–65.
Sokal, A. (1996), 'Transgressing the boundaries: toward a transformative hermeneutics of quantum gravity', *Social Text* 14 (1–2): 217–52.
Sokal, A. and Bricmont, J. (1998), *Intellectual Impostures: Postmodern Philosophers' Abuse of Science*, London: Profile.
Soper, K. (1981), *On Human Needs: Open and Closed Theories in a Marxist Perspective*, Brighton: Harvester.
Soper, K. (1990), 'Feminism, humanism and postmodernism', *Radical Philosophy* 55: 11–17.
Soper, K. (1995), *What Is Nature? Culture, Politics and the Non-human*, Oxford: Blackwell.
Soper, K. (2001), 'Realism, humanism and the politics of nature', *Theoria* 98: 55–71.
Sopher, D. (1967), *Geography of Religion*, New York, NY: Prentice-Hall.
Sraffa, P. (1960), *Production of Commodities by Means of Commodities*, Cambridge: Cambridge University Press.
Stallabrass, J. (2006), 'Spectacle and terror', *New Left Review* 37.
Steinmetz, G. (ed.) (2005), *The Politics of Method in the Human Sciences: Positivism and Its Epistemological Others*, Durham, NC/London: Duke University Press.
Storper, M. and Walker, R. (1983), 'The theory of labour and the theory of location', *International Journal of Urban and Regional Research* 7.

532 Bibliography

Strand, A. (1991), *Vedlegg til forslag til forskningsprogram om kommunal planlegging* (Appendix to a proposal for a research program on municipal planning), Oslo: National Committee on Environmental Research/NAVF.

Subramaniyam, V. (2001), 'Critical realism and development programmes in rural south India', *Alethia*, 4(1), pp. 17–23.

Sullivan, D. (2006), 'Hermann Lotze', *The Stanford Encyclopedia of Philosophy* (Spring 2006 Edition), E. N. Zalta (ed.), http://plato.stanford.edu/archives/spr2006/entries/hermann-lotze/

Suzuki, H. (2005), 'Is there something money can't buy? In defence of the ontology of a market boundary', *Journal of Critical Realism*, 4(2), pp. 265–90.

Sweet, W. (2004), 'Jacques Maritain', *The Stanford Encyclopedia of Philosophy* (Spring 2004 Edition), E. N. Zalta (ed.), http://plato.stanford.edu/archives/spr2006/entries/maritain/

Terhorst, P. and van de Ven, J. (1997), 'Fragmented Brussels and consolidated Amsterdam: a comparative study of the spatial organisation of property rights', *Nederlandse Geograifische Studies*, Amsterdam: University of Amsterdam.

Tew, P. (2001), 'Reconsidering literary interpretation' in López and Potter (eds), *After Postmodernism*.

Tew, P. (2003), 'A new sense of reality? A new sense of the text? Exploring meta-realism and the literary-critical field', in K. Stierstorfer (ed.), *Beyond Postmodernism: Re-assessments in Literature, Theory and Culture*, Berlin/New York: Walter de Gruyter.

Tew, P. (2004), *The Contemporary British Novel: From John Fowles to Zadie Smith*, London/New York: Continuum.

Thompson, E. P. (1955/1977), *William Morris: Romantic to Revolutionary*, London: Merlin.

Thompson, E. P. (1975), *Whigs and Hunters: The Origin of the Black Act*, London: Allen Lane.

Thompson, J. B. (1995), *The Media and Modernity: A Social Theory of the Media*, Cambridge: Polity.

Tolson, D. (1999), 'Practice innovation: a methodological maze', *Journal of Advanced Nursing*, 30: 381–90.

Töttö, P. (2004), *Syvällistä ja pinnallista. Teoria, empiria ja kausaalisuus sosiaalitutkimuksessa* [Deep and Superficial. Theory, Empirical Evidence and Causality in Social Research], Tampere: Vastapaino.

Tsoukas, H. (1989), 'The validity of idiographic research explanations', *Academy of Management Review* 14 (4): 551–61.

Tsoukas, H. (1994), 'What is management: an outline of a metatheory', *British Journal of Management* 5 (2): 289–301.

Tsoukas, H. and Knudsen, C. (eds.) (2003), *The Oxford Handbook of Organisation Theory*, Oxford: Oxford University Press.

Uberoi, J. P. S. (1999), *Religion, Civil Society and State: A Study of Sikhism*, Delhi: Oxford University Press.

United Nations Conference on Environment and Development (UNCED) (1992), *Earth Summit*, London: Regent Press.

Van Krieken, R. (1997), 'Beyond the "problem of order": Elias, habit and modern sociology *or* Hobbes was right', http://www.usyd.edu.au/su/social/elias/confpap/order.html#N_1_ n in (Hobbesian) Problem of order.

Vance, C. S. (1989), 'Social construction theory: problems in the history of sexuality', in D. Altman et al. (eds), *Homosexuality, Which Homosexuality? International Conference on Gay and Lesbian Studies*, London: GMP.

Varela, C. R. (1999), 'Determinism and the recovery of human agency: the embodying of persons', *Journal for the Theory of Social Behaviour* 29 (4): 385–402.

Veblen, T. B. (1899), *Theory of the Leisure Class*, New York: Macmillan.

Verstegen, I. (2000), 'Bhaskar and American critical realism', www.raggedclaws.com/criticalrealism/archive/ iverstegen_baacr.html.

Voloshinov, V. N. (1929/1873), *Marxism and the Philosophy of Language*, trans. L. Matejka and I. R. Titunik, Cambridge, Mass.: Harvard University Press; New York: Seminar Press.
von Bertalanffy, L. (1971), *General Systems Theory*, Harmondsworth: Penguin.
von Hartmann, E. (1875), *Kritischer Grundlegungen des Transcendentalen Realismus* [Critical Foundations of Transcendental Realism], Berlin: Duncker.
von Wright, G. H. (1963), *Norm and Action*, London: Routledge.
Wagar, W. W. (2004), *H. G. Wells: Traversing Time*, Middletown, Conn.: Wesleyan University Press.
Wainwright, H. (2003), *Reclaim the State: Experiments in Popular Democracy*, London: Verso.
Walby, Sylvia (1990), *Theorising Patriarchy*, Cambridge: Polity.
Walby, Sylvia (2001), 'Against epistemological chasms: the science question in feminism revisited', in *Signs* 26 (2): 485–509.
Walker, R. B. J. (1993), *Inside/Outside: International Relations as Political Theory*, Cambridge: Cambridge University Press.
Wallace, W. A. (1996), *The Modeling of Nature: Philosophy of Science and Philosophy of Nature in Synthesis*, Washington DC: Catholic University Press of America.
Wallerstein, I. (1979), *The Capitalist World-Economy*, Cambridge: Cambridge University Press.
Walras, L. (1874/1984), *Elements of Pure Economics*, Philadelphia, Pa.: Orion.
Warner, M. M. (1993), 'Objectivity and emancipation in learning disabilities: holism from the perspective of critical realism', *Journal of Learning Disabilities* 26 (5): 311–25.
Weber, M. (1905/1992), *The Protestant Ethic and the Spirit of Capitalism*, London: Routledge.
Weber, M. (1925/1968), *Economy and Society*, ed. G. Roth and C. Wittich, 3 vols, New York: Bedminster.
Weiner, N. (1950), *The Human Use of Human Beings*, New York: Houghton Mifflin.
Wendt, A. (1987), 'The agent–structure problem in international relations theory', *International Organization* (41) 3: 335–70.
Wendt, A. (1999), *Social Theory of International Politics*, Cambridge: Cambridge University Press.
Whittington, R. (1989), *Corporate Strategies in Recession and Recovery*, London: Unwin Hyman.
Wiener, P. P. (1974), 'Pragmatism', in *The Dictionary of the History of Ideas: Studies of Selected Pivotal Ideas*, Vol. 3, ed. P. P. Wiener, New York: Charles Scribner's Sons.
White, H. (1978), 'The historical text as literary artifact', in *Tropics of Discourse: Essays in Cultural Criticism*, Baltimore, Md.: Johns Hopkins University Press.
Wight, C. (1996), 'Incommensurability and cross paradigm communication in international relations theory: what's the frequency Kenneth?', *Millennium* 25 (2): 291–319.
Wight, C. (1999), 'They shoot dead horses, don't they? Locating agency in the agent–structure problematique', *European Journal of International Relations* 5 (1).
Wight, C. (2006), *Agents, Structures and International Relations: Politics as Ontology*, Cambridge: Cambridge University Press.
Wight, C. and Patomäki, H. (2000), 'After postpositivism: the promise of critical realism', *International Studies Quarterly* 44 (2): 213–37.
Will, D. (1986), 'Psychoanalysis and the new philosophy of science', *International Review of Psycho-Analysis* 13: 163–73.
Williams, G. (2001), 'Theorizing disability', in G. Albrecht, K. Seelman and M. Bury (eds), *Handbook of Disability Studies*, London: Sage.
Williams, M. (ed.) (1988), *Value, Social Form and the State*, London: Macmillan.
Williams, M. and Dyer, W. (2004), 'Realism and probability', in B. Carter and C. New (eds), *Making Realism Work*, London: Routledge.
Williams, R. (1977), *Marxism and Literature*, Oxford: Oxford University Press.
Williams, S. J. (1999), 'Is anybody there? Critical realism, chronic illness and the disability debate', *Sociology of Health and Illness* 21(6): 797–819.

Williams, S. J. (2003), 'Beyond meaning, discourse and the empirical world: critical realist reflections on health', *Social Theory & Health* 1: 42–71.

Williamson, J. (1990), 'What Washington means by policy reform', in J. Williamsom (ed.), *Latin American Adjustment: How Much Has Happened?*, Washington, DC: Institute for International Economics.

Willig, C. (1998), 'Social constructionism and revolutionary socialism: a contradiction in terms?', in I. Parker (ed.), *Social Constructionism, Discourse and Realism*, London: Sage.

Willis, P. (1977), *Learning to Labour: How Working Class Kids Get Working Class Jobs*, Farnborough: Saxon House.

Willmott, R. (1997), 'Structure, culture and agency: rejecting the current orthodoxy of organisation theory', *Journal for the Theory of Social Behaviour* 27 (1): 93–123.

Willmott, R. (2002), *Education Policy and Realist Social Theory: Primary Teachers, Child Centred Philosophy and the New Managerialism*, London: Routledge.

Willmott, R. (2003), 'New Labour, school effectiveness and ideological commitment', in J. Cruickshank (ed.), *Critical Realism*.

Wilson, J. (1983), *Social Theory*, Englewood Cliffs, NJ: Prentice-Hall.

Wilson, N. (2005), '"Congratulations it's a labour market!". A causal-explanatory account of the birth of the UK early music musician labour market', paper presented at the International Association for Critical Realism Conference, Cambridge.

Winch, P. (1958), *The Idea of a Social Science and Its Relation to Philosophy*, London: Routledge & Kegan Paul.

Winner, L. (1977), *Autonomous Technology*, Cambridge, Mass.: MIT Press.

Winnicott, D. W. W. (1967), *Playing and Reality*, London: Penguin.

Wittgenstein, L. (1922), *Tractatus Logico-Philosophicus*, London: Routledge.

Wittgenstein, L. (1953), *Philosophical Investigations*, Oxford: Blackwell.

Wodak, R. and Meyer, M. (2001), *Methods in Critical Discourse Analysis*, London: Sage.

Woodiwiss, A. (1993), *Postmodernity USA: The Crisis of Social Modernism in Post-War America*, London: Sage.

Woodiwiss, A. (2002), *The Visual in Social Theory*, London/New York: The Athlone Press.

Woodward, A. (2002), 'Nihilism and the postmodern in Vattimo's Nietzsche', *Minerva* 6.

World Health Organization (1980), *International Classification of Impairments, Disabilities and Handicaps*, Geneva: WHO.

Wright, C. (1992), *Truth and Objectivity*, Cambridge, Mass.: Harvard University Press.

Yannaras, C. (2004), *Postmodern Metaphysics*, Brookline, Mass.: Holy Cross Orthodox Press (in Greek 1993).

Yeung, H. W. (1997), 'Critical realism and realist research in human geography: a method or a philosophy in search of a method?', *Progress in Human Geography*, 21 (1): 51–74.

Žižek, S. (1999), *The Ticklish Subject: The Absent Centre of Political Ontology*, London: Verso.

eBooks – at www.eBookstore.tandf.co.uk

A library at your fingertips!

eBooks are electronic versions of printed books. You can store them on your PC/laptop or browse them online.

They have advantages for anyone needing rapid access to a wide variety of published, copyright information.

eBooks can help your research by enabling you to bookmark chapters, annotate text and use instant searches to find specific words or phrases. Several eBook files would fit on even a small laptop or PDA.

NEW: Save money by eSubscribing: cheap, online access to any eBook for as long as you need it.

Annual subscription packages

We now offer special low-cost bulk subscriptions to packages of eBooks in certain subject areas. These are available to libraries or to individuals.

For more information please contact webmaster.ebooks@tandf.co.uk

We're continually developing the eBook concept, so keep up to date by visiting the website.

www.eBookstore.tandf.co.uk